QUANTUM MECHANICS

QUANTUM MECHANICS

(For BSc. and MSc. Students of Physics and Chemistry)

SECOND EDITION

GURBACHAN S CHADDHA

Formerly
Professor & Head of Physics Department
Punjab Agricultural University
Ludhiana, Punjab

Publishing Globally

NEW AGE INTERNATIONAL (P) LIMITED, PUBLISHERS
LONDON • NEW DELHI • NAIROBI
Bangalore • Chennai • Cochin • Guwahati • Hyderabad • Kolkata • Lucknow • Mumbai
Visit us at **www.newagepublishers.com**

Published by New Age International (P) Ltd., Publishers
First Edition: 2003
Second Edition: 2018
Reprint: 2020

GLOBAL OFFICES

- **New Delhi** — **NEW AGE INTERNATIONAL (P) LIMITED, PUBLISHERS**
  7/30 A, Daryaganj, New Delhi-110002, (INDIA)
  Tel.: (011) 23253771, 23253472, **Telefax:** 23267437, 43551305
  E-mail: contactus@newagepublishers.com • Visit us at www.newagepublishers.com
- **London** — **NEW AGE INTERNATIONAL (UK) LTD.**
  27 Old Gloucester Street, London, WC1N 3AX, UK
  E-mail: info@newacademicscience.co.uk • Visit us at www.newacademicscience.co.uk
- **Nairobi** — **NEW AGE GOLDEN (EAST AFRICA) LTD.**
  Ground Floor, Westlands Arcade, Chiromo Road (Next to Naivas Supermarket)
  Westlands, Nairobi, KENYA, **Tel.:** 00-254-713848772, 00-254-725700286
  E-mail: kenya@newagepublishers.com

BRANCHES

- **Bangalore** — 37/10, 8th Cross (Near Hanuman Temple), Azad Nagar, Chamarajpet, Bangalore- 560 018
  Tel.: (080) 26756823, **Telefax:** 26756820, **E-mail: bangalore@newagepublishers.com**
- **Chennai** — 26, Damodaran Street, T. Nagar, Chennai-600 017, **Tel.:** (044) 24353401, **Telefax:** 24351463
  E-mail: chennai@newagepublishers.com
- **Cochin** — CC-39/1016, Carrier Station Road, Ernakulam South, Cochin-682 016
  Tel.: (0484) 2377303, **Telefax:** 4051304, **E-mail: cochin@newagepublishers.com**
- **Guwahati** — Hemsen Complex, Mohd. Shah Road, Paltan Bazar, Near Starline Hotel, Guwahati-781 008
  Tel.: (0361) 2513881, **Telefax:** 2543669
  E-mail: guwahati@newagepublishers.com
- **Hyderabad** — 105, 1st Floor, Madhiray Kaveri Tower, 3-2-19, Azam Jahi Road, Near Kumar Theater, Nimboliadda Kachiguda, Hyderabad-500 027, **Tel.:** (040) 24652456, **Telefax:** 24652457
  E-mail: hyderabad@newagepublishers.com
- **Kolkata** — RDB Chambers (Formerly Lotus Cinema) 106A, 1st Floor, S N Banerjee Road, Kolkata-700 014
  Tel.: (033) 22273773, **Telefax:** 22275247
  E-mail: kolkata@newagepublishers.com
- **Lucknow** — 16-A, Jopling Road, Lucknow-226 001, **Tel.:** (0522) 2209578, 4045297, **Telefax:** 2204098
  E-mail: lucknow@newagepublishers.com
- **Mumbai** — 142C, Victor House, Ground Floor, N.M. Joshi Marg, Lower Parel, Mumbai-400 013
  Tel.: (022) 24927869, **Telefax:** 24915415, **E-mail: mumbai@newagepublishers.com**
- **New Delhi** — 22, Golden House, Daryaganj, New Delhi-110 002, **Tel.:** (011) 23262368, 23262370
  Telefax: 43551305, **E-mail: sales@newagepublishers.com**

ISBN: 978-81224-3403-3
C-19-05-12079

Printed in India at Print-O-Pack, Delhi.
Typeset at J.R. Enterprises, Delhi.

NEW AGE INTERNATIONAL (P) LIMITED, PUBLISHERS
7/30 A, Daryaganj, New Delhi-110002
Visit us at **www.newagepublishers.com**
(CIN: U74899DL1966PTC004618)

To the Sweet Memory

of

S. Charan Singh

my Taya ji

PREFACE

The author, based on his more than twenty years experience of teaching Quantum Mechanics thought to write a book on the subject. It is an endeavour not the end. The book is suitable to serve as a textbook for B.Sc. and M.Sc. students of physics and chemistry. It actually contains the portion to be covered at senior undergraduate level and comprises a two semester course for M.Sc. students. A thorough knowledge of vectors and special functions is of course assumed. End of almost each chapter contains a problem set. Most of the problems in the set are solved so that a student can have an in depth knowledge of the subject. It is strictly in accordance with the author's conception that no one can learn a subject without solving problems.

The author firmly feels that the reading of this book would make the subject of quantum mechanics which is supposed to be a difficult one, well understood and interesting at the same time. No doubt a large number of books in the subject including books by Indian authors, viz. Ghatak and Lokanathan, Mathews and Venkatesan and Thankappan are available but none of them can be accepted of course in the author's opinion as a single text book for the said level students. With this idea in mind the writing of book was initiated. Original reference wherever necessary are cited so that a student can consult them for further study.

The subject has been introduced in the first two chapters wherein some of the results which could not be explained with the help of classical physics have been quoted. The need to introduce a new mechanics (called quantum mechanics) is emphasized. Underlying principles have been well illustrated. Chapters 3-5 deal with applications of quantum mechanics to one as well as to three dimensional problems. Dirac's ket and bra notation has been introduced in chapter 6. Chapter 7 deals with equations of motion in Schrödinger, Heisenberg and interaction (Dirac) pictures. Chapter 8 through 10 are devoted to angular momenta. In chapter 11, I have talked about invariance, symmetry and conservation laws. Chapter 12 is devoted to approximation methods, viz. WKB approximation (application to bound state and penetration problems), time independent perturbation theory (non-degenerate and degenerate cases), variational method, time dependent perturbation theory and applications. Chapters 13 and 14 cover interaction of radiation with matter (Semi-classical theory) and quantum theory of scattering (methods of partial waves and Born approximation). Chapter 15 deals with identical particles. The problems of hydrogen molecule and helium atom are discussed in particular. Klein-Gordon and Dirac's relativistic equations and their applications are discussed in chapter 16. The last chapter deals with an introduction to the technique of second quantization and in particular it is applied to the non-relativistic Schrödinger field.

The first edition of the book was published in 2003. In this second edition of the book, corrections have been made in large scale in chapter 1 along with some corrections in other chapters. Apart from this six appendices have been added at the end of the book. These would strengthen the contents in the book and at the same time make it self sufficient and self explanatory.

The author is highly thankful to his teachers, Dr. P.K. Sharma while graduate student at Allahabad University and to Dr. B.B. Deb while Ph.D scholar at Indian Institute of Technology, Kanpur. Thanks are also due to all my students who made me to go into the deep understanding of the subject. I would also like to thank Professor Dr. M.B. Singh, Department of Mathematics, Statistics and Physics for his continuous encouragements and help. Finally my special thanks are due to M/s New Age International (P) Ltd., the publisher for its keen interest in making it possible to bring out this edition in its present form.

G.S. Chaddha

CONTENTS

Chapter 1

THE FAILURE OF CLASSICAL PHYSICS (OR THE ORIGIN OF QUANTUM MECHANICS)

1.1 INTRODUCTION

During the period of ten years from 1895-1905 a large number of new phenomena connected with the physics of atomic particles (in general with the microscopic objects, *e.g.* atom-dimension 10^{-10} m, nucleus-dimension 10^{-14} m, elementary particles, *viz.* electron, proton, neutron-dimension 10^{-15} m) are discovered. The atomic dimension is too small to be seen through even a most powerful microscope. However, it is still possible to study the atomic particles if carefully designed experiments are performed and the results thus obtained are analysed. It was found that the Newtonion mechanics (or in general classical physics) is not adequate to understand such experimental results. The attempts to explain these results, namely, black body radiation spectrum, specific heat of solids at low temperature, photoelectric effect, Compton effect, Atomic spectra, *etc.* led to the development of quantum mechanics which gives a very good account of what goes on inside the atoms and nuclei.

1.2 BLACK BODY RADIATION

All solids exhibit nearly the same spectrum of continuous radiation when at the same temperature. The ability of a body to radiate is closely related to its ability to absorb radiation. This is to be accepted since a body at a constant temperature is in thermal equilibrium with its surroundings and must absorb energy from the surroundings at the same rate as it emits energy. For the sake of convenience, one considers an ideal body as the one which absorbs all radiation incident upon it (or a solid with zero reflectivity) regardless of frequency. Such a body is called a black body. (In the laboratory a black body

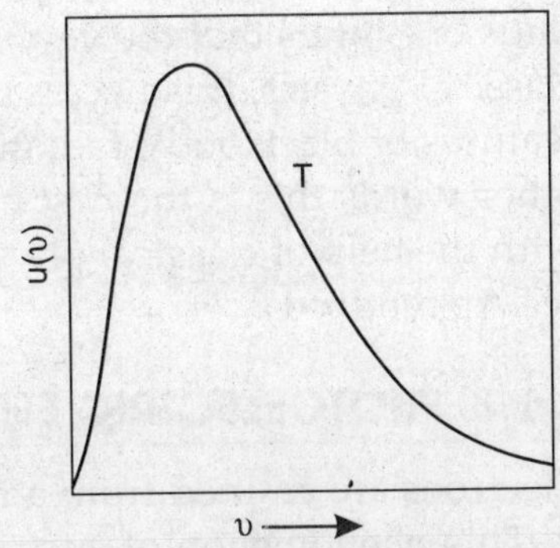

Fig. 1.1 *Spectral energy density as a function of ν for black body at a fixed temperature*

can be approximated by a hollow object (cavity) with a very small hole leading to its interior. The cavity always contains radiation emitted by the walls. One can sample black body radiation simply by inspecting what emerges from the hole). The spectrum of black body radiation is shown in Fig. 1.1 at absolute temperature $T.u(\upsilon)\,d\upsilon$ represents the energy density contributed by radiation in the frequency band υ to $\upsilon + d\upsilon$.

The classical approach treats the radiation inside an enclosure as if it were an assembly of oscillators, each at different frequency. If we use the law of equipartition of energy (or use classical physics), that is, the average energy of oscillator (or energy associated with each degree of vibration) is $k_B\,T$, then the energy per unit volume within the frequency range $d\upsilon$ comes out to be

$$u(\upsilon)\,d\upsilon = 8\,\pi\upsilon^2/c^3\,k_BT\,d\upsilon \quad \text{... (1)}$$

This is the Rayleigh-Jeans formula which predicts that $u(\upsilon)$ should keep on increasing with υ which agrees with the observed distribution shown in Fig. 1.1 at low frequencies. With the further increase of υ, $u(\upsilon)$ of course falls off after reaching a maximum.

In 1901 Planck[1] presented his theory of black body radiation which led to the result

$$u(\upsilon)\,d\upsilon = 8\pi\upsilon^2/c^3\,h\upsilon/(e^{h\upsilon/k_BT} - 1)\,d\upsilon \quad \text{... (2)}$$

and which agrees perfectly with the observed spectrum of black body radiation. The above result was based on the assumption (hypothesis) that the energies of the oscillators are quantized or in other words Planck assumed that oscillators can only have discrete energies $h\upsilon$, $2h\,\upsilon$, ... , *etc.* (The quantum of energy $h\upsilon$ is now called photon) where h is a constant having value 6.626×10^{-34} Joule-second and is called Planck's constant. One may note that in the case of a classical oscillator with say $\upsilon = 1$, $h\upsilon = 6.626 \times 10^{-34}$ Joules $= 4 \times 10^{-15}$ eV which is extremely small and cannot be measured. Hence a classical oscillator assumes continuous energy values. On the other hand for a light wave, $w = 10^{15}$ per second and hence $h\upsilon = 4$eV which is a measurable quantity. Thus quantum physics (or mechanics) is applicable to all microscopic as well as to large objects (macroscopic) whereas classical physics is valid only in the macroscopic limit.

It may be noted that in the limit $h \to 0$, *eqn.*(2) reduces to the Rayleigh-Jeans law. Thus one infers that the very small but nonzero value of h is measure of the failure of classical physics. It is necessary to mention here that the explanation of the observed features of black body radiation led to the introduction of quantum mechanics or in other words this is the first experimental observation which could not be explained with the help of classical physics. (For detailed discussion of black body radiation — see Appendix-1).

1.3 PHOTOELECTRIC EFFECT

Electrons are emitted from a metal surface when radiation of suitable energy falls on it. This phenomenon of such emission of electrons is called photoelectric effect. One observes that

[1]M Planck, Ann. Physik 4,553 (1901)

(i) Electron are emitted as soon as radiation falls on the metal surface. Whole of radiation energy ($h\upsilon$) is transferred to an electron on the surface of metal.

(ii) The kinetic energy of photoelectron emitted is independent of the radiation intensity but depends upon its frequency. A greater intensity of the incident light leads to more photoelectrons but the maximum energy of the emitted electron remains the same.

(iii) The photoelectron energy depends upon the frequency of the incident light. At frequencies below a certain critical (threshold) frequency a characteristic of each particular metal, no electrons are emitted irrespective of its intensity. Above this threshold frequency the photoelectrons have kinetic energy that increases linearly with increasing frequency.

This is contrary to Maxwell's electromagnetic theory (classical physics) according to which if the intensity of light increases, the energy of emitted electrons should increase. This is because of the fact that the force which the incident radiation exerts on the metal electrons and in turn increases their kinetic energy is proportional to the magnitude of its electric field vector $\vec{E}$ which increases when the intensity of the radiation increases as the intensity is proportional to the square of the amplitude of $\vec{E}$.

The above observations were explained by Einstein[2] in 1905 based on the assumption that light is made up of particles called photons of energy $h\upsilon$ and moving with velocity of light. If light of frequency υ falls on the metal and $h\upsilon < e\phi$ which is the characteristic of a particular surface and is called work function ($e\phi = h\upsilon_0$ is the minimum energy needed to dislodge an electron from the metal surface, υ_0 is called threshold or cutoff frequency), electrons are not emitted. If $h\upsilon > e\phi$, the part of the energy of the absorbed photon is used by the electron to become free itself from the metal, the balance energy will appear as the kinetic energy of the outgoing electron. The kinetic energy would be maximum for the least tightly bound electron. Thus

$$T_{max} = h\upsilon - e\phi$$

where T_{max} is the maximum photoelectron energy.

As $$e\phi = h\upsilon_0, \text{ therefore } T_{max} = h(\upsilon - \upsilon_0) \quad \text{... (3)}$$

This is Einstein photoelectric law (or equation), showing a linear relation between frequency and the maximum kinetic energy of photoelectrons and is in agreement with the above observations. It is necessary to mention here that Einstein received the 1921 Nobel prize in Physics for the equation (3) and not for his theory of relativity. (For detailed discussion of photoelectric effect—see Appendix-2).

1.4 SPECIFIC HEAT OF SOLIDS

It is well known that atoms or ions or molecules (we shall refer them as atoms here for sake of simplicity) in solid vibrate about their mean positions due to the thermal energy. Each atom may be thought of as a three dimensional oscillator and its mean

[2] A. Einstein Ann. Physik 17, 132 (1905)

thermal energy should be three times that of one dimensional harmonic oscillator. Classical physics (law of equipartition of energy) predicts later to be $k_B T$ (energy associated with each degree of vibration). Therefore, the thermal energy $E_{\upsilon ib}$ of solid would be $3k_B T$ per atom or $3N_A k_B T = 3RT$ per kmole, where $N_A = 6.023 \times 10^{26}$, called Avogadro number, is number of atoms in 1 kmole of solid and R is the universal gas constant. Therefore, the specific heat per kmole called molar specific heat of solid at constant volume, that is, the rate of increase of thermal energy with temperature per kmole, $C_V = dE_{\upsilon ib}/dT$ then becomes $3R$. At ordinary temperatures this result has been experimentally verified in the case of large number of solids and is referred to as the Dulong-Petit law.

As shown in Fig. 1.2, it has been observed experimentally that when the temperature is lowered sufficiently, the specific heat of solids decrease rapidly instead of remaining constant and indeed goes down to zero as T approaches 0 K. Einstein[3] in 1907 observed that this behaviour can be explained by assuming that the energy of oscillation of any atom in a solid can take only a discrete set of values or is quantized. The quantum of lattic vibrational energy is called phonon. The mean (or average) energy of an oscillator instead of $3k_B T$ would then be $3h\upsilon/(e^{h\upsilon/k_B T} - 1)$. Hence the molar of specific heat of solid at constant volume would be given by

$$\begin{aligned} C_V &= d/dT\,[N_A 3h\upsilon/(e^{h\upsilon/k_B T} - 1)] \\ &= 3R\,[e^{h\upsilon/k_B T}/(e^{h\upsilon/k_B T} - 1)^2]\,(h\upsilon/k_B T)^2 \end{aligned} \quad \text{... (4)}$$

which predicts the rapid decrease of C_V as the temperature is lowered. Of course it has been assumed here that all atoms have the same frequency of oscillation.

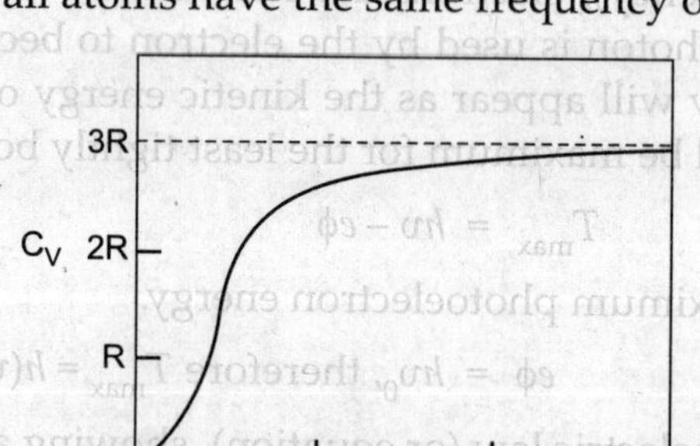

Fig. 1.2 *Molar specific heat of solids at constant volume C_V as a function of temperature*

No doubt the above result is not final and was modified in 1912 by Debye, but it is sufficient for the purpose of displaying the fact that the classical physics was inadequate for explaining the experimental observation and there was need that harmonic oscillator can take only discrete values of energy. (For detailed discussion of the specific heat of solids—see Appendix-4).

---

[3]A. Einstein, Ann.Physik 22, 180 (1907)

1.5 COMPTON EFFECT

Compton in 1923 observed that when X-rays are scattered by the loosely bound electrons in atom, the wavelength of the scattered radiation exceeds that of the incident radiation. The said observation is known as Compton effect[4].

Without going into the details we would just mention that the above result could not be explained with the help of classical physics (electromagnetic theory of radiation).

To explain the above observation, Compton assumed that the X-rays of frequency υ consist of stream of photons of energy $h\upsilon$ and momentum $p = h\upsilon/c$ (called de Broglie relation[5]) which collide against an electron of mass m_0. Following the impact, the energy of photons is reduced to $h\upsilon'$ and its momentum is reduced to $h\upsilon'/c$.

During collision the electron gains energy. Figure 1.3 shows how such a collision is represented. When photon hits electron in the scatterer, it would recoil. Its kinetic energy has to come from the energy of the incident quantum and the later would be left with an energy.

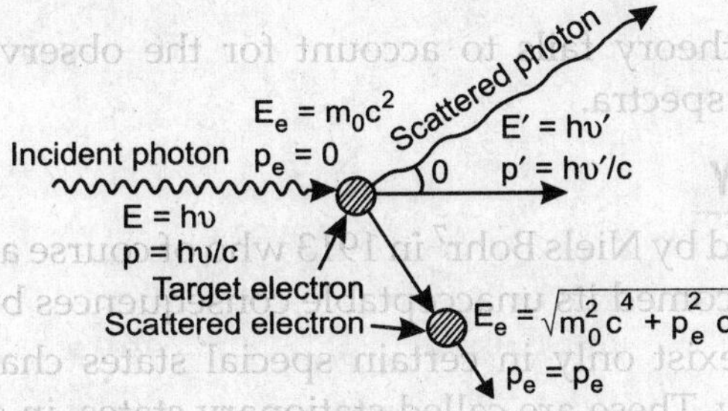

Fig. 1.3 *Compton scattering*

$E' < E$ after the collision in which it gets scattered. The frequency $\upsilon' = E'/h$ of the X-rays so scattered would therefore be less than υ and the corresponding wavelength $\lambda' > \lambda$. By using energy and momentum conservations, the shift in wavelength is found to be

$$\Delta\lambda = \lambda' - \lambda = h/m_0c\,(1 - \cos\theta) \quad \text{... (5)}$$

where θ is the angle of scattering and m_0 is the rest mass of the electron. The above result agrees with the experimental observation. (For detailed discussion of Compton Effect—see Appendix-3).

1.6 THE RUTHERFORD ATOMIC MODEL

Let us try to see how quantum ideas are useful in understanding the atomic structure and atomic spectra.

An atom consists of a small nucleus (size 10^{-14} m) in which its positive charge and nearly all its mass is concentrated. Around this nucleus the electrons revolve in circular

[4]A.H. Compton, Phys. Rev. 21, 483 (1923); 22, 409 (1923)

[5]L.de Broglie, Phil. Mag. 47, 446 (1924); Annals de Physique 3, 22 (1925)

orbits upto distance approximately equal to 10^{-10}m. The said model was proposed by Rutherford[6] in 1911 to explain the scattering of α-particles by thin metallic foils (gold foil of thickness 4×10^{-7}m). Some of the particles are deflected through angles greater than 90°, that is, they emerge in the side of incidence. This large angle scattering was explained based on the Rutherford model.

This picture of atom, however, encountered serious difficulties. In fact such a structure should not be stable at all because the orbit motion of the electrons which are charged particles should emit radiation continuously (as an accelerated charge radiates). The consequent loss of energy should make the paths go spiralling inwards until the electrons collapse into the nucleus within few seconds. But none of these things happen. In fact

(*a*) atoms have tremendous stability

(*b*) also the radiation emitted by atoms is not of continuous character. Instead the most important feature of atomic spectra is the presence of sharp and discrete lines.

Thus the classical theory fails to account for the observed features of atomic stability and the atomic spectra.

1.7 BOHR'S THEORY

The situation was tackled by Niels Bohr[7] in 1913 who of course adopted the Rutherford model of atom but overcomed its unacceptable consequences by postulating that

(*i*) The atom can exist only in certain special states characterized by discrete values of energy. These are called stationary states, in any of which the atom can remain indefinitely without radiating. The energy values associated with stationary states are called energy levels of the atom.

(*ii*) The emission of electromagnetic radiation (or photon) takes place when the atomic electron jumps from one of the stationery states with energy E_i to another with energy $E_f < E_i$. The frequency of radiation emitted is given by

$$E_i - E_f = h\upsilon \quad \text{... (6)}$$

In the inverse process, a photon of appropriate frequency $h\upsilon$ is absorbed by the atom, raising an electron from an orbit of lower energy to one of the higher energy.

By supplementing these postulates by the quantum condition,

$$m\upsilon r = n\hbar \quad (n = 1, 2, 3, \ldots) \quad \text{... (7)}$$

where $\hbar$ is written for the quantify $h/2\pi$. $m\upsilon r$ is the angular momentum of circular orbit, n representing the various stationary states is called quantum number, Bohr was able to calculate the energy levels of the hydrogen atom and thus determined its spectral frequencies.

As indicated above, Bohr's theory (two postulates + quantum condition, $m\upsilon r = n\hbar$), however could only explain the energy levels or discrete lines in the case of hydrogen

[6]E. Rutherford, Phil. Mag. 21, 669 (1911)

[7]N. Bohr, Phil. Mag. 26, 1(1913)

atom but not the presence of fine structure (that is each line consists of a number of lines lying close together, typical spacing being 10^{-3} Å).

1.8 BOHR–SOMMERFELD THEORY (OR OLD QUANTUM THEORY)

One may note that the quantum condition (7) is a special case of the condition

$$\oint p\,dq = nh, \quad (n = 0, 1, 2, 3, \ldots) \qquad \text{... (8)}$$

called Bohr-Sommerfeld quantum rule. Here q is some generalized coordinate and p, the corresponding canonical conjugate momentum. The condition is applicable in the case of periodic motion only and the integration is to be taken over one (full) period of the generalized co-ordinate. Eqn. (7) corresponds to choosing q to be angular position ϕ (angle which radius vector makes with x-axis) varying from 0 to 2π for one period and p its conjugate, the angular momentum in the circular orbit, which is independent of ϕ, being a constant of motion. This can be shown as follows,

$$\oint p\,d\phi = nh$$

or
$$\oint m\upsilon r\,d\phi = nh$$

or
$$\int_0^{2\pi} m\upsilon r\,d\phi = nh$$

or
$$m\upsilon r = nh$$

Bohr's two postulates together with the generalized quantum rule (8) postulated by Sommerfeld[8] in 1916 constitute what is now known as the old quantum theory (or Bohr-Sommerfeld theory). Without going into the details, we would simply mention that the said theory could explain the spectrum of hydrogen atom including its fine structure.

It may be mentioned here that it was actually to explain the observed fine structure in the case of hydrogen atom that Sommerfeld in 1916 extended the Bohr model of hydrogenic atom (H, He^+, Li^{2+}, . . .) to include the elliptical orbits.

The quantum condition (8) has actually two parts.

$$J_\phi = \oint p_\phi\,d\phi = kh, \quad k = 1, 2, 3, \ldots \qquad \text{... (9a)}$$

$$J_r = \oint p_r\,d_r = n_r h, \quad n_r = 0, 1, 2, \ldots \qquad \text{... (9b)}$$

that is, in addition to the angle ϕ which the radius vector makes with the axis, the length of radius vector also varies periodically. $p_r = m\dot{r}$ denotes the momentum associated with the co-ordinate r and is called radial momentum of the electron. $n_r = 0$, corresponds to circular orbit in which case r does not change. The value $k = 0$ was ruled out by Sommerfeld as it corresponds to the motion of electron in a straight line through nucleus.

[8] A. Sommerfeld, Ann. Physik 51, 1 (1916)

One finally gets,

$$E = -2\pi^2 m e^4/(J_r + J_\phi)^2$$

Or the quantized energy values are

$$E = E_n = -me^4/2n^2 \hbar^2 = -e^2/2n^2 a_0, \quad \text{... (10)}$$

$$a_0 = \hbar^2/me^2, \quad n = 1, 2, 3, \ldots$$

where $n = n_r + k$, is the principal quantum number. k and n_r are called azimuthal and radial quantum numbers, respectively.

1.9 LIMITATIONS OF BOHR-SOMMERFELD THEORY

(*i*) It applies only to hydrogenic atoms (H, He^+, Li^{2+}, . . .) but fails when applied to systems such as the helium atom, that is more than one electron systems.

(*ii*) It provides no method for calculating the intensities of the spectral lines. That is, why certain lines are more intense as compared to the others.

(*iii*) It is unable to explain the phenomenon like anomalous Zeeman effect, that is splitting of spectral lines in the weak magnetic field (where the term containing spin of the electron cannot be neglected).

(*iv*) And, most important, it does not permit us to understand, how individual atoms interact with one another to form macroscopic aggregates of matter with the physical and chemical properties we observe.

1.10 DE BROGLIE HYPOTHESIS AND WAVE-PARTICLE DUALITY OF MATTER

The suggestion that matter may have wave like properties was first put forward in 1924 by Louis de Broglie[5]. He argued that if light which consists of waves according to classical physics (light exhibits the well known interference phenomenon which can only be explained by the wave nature of electromagnetic radiation) can behave like particles (The Compton effect gives a clear evidence of the particle nature of radiation). Then it should be possible for matter which consists of particles classically to exhibit wave like behaviour or in other words he proposed that the wave-particle duality may not be a monopoly of light but a universal characteristic of nature.

In the case of photon, energy is given by

$$E = h\upsilon \quad \text{... (11)}$$

where υ is the frequency of radiation. Also energy of photon is expressed as

$$E = (p^2c^2 + m_0^2c^4)^{1/2} = pc$$

as the photon rest mass is zero. Hence

$$pc = h\upsilon$$

Or

$$\lambda = h/p \quad \text{... (12)}$$

which exhibits wave-particle duality in the case of light radiation and is called de Broglie relation.

Accordingly de Broglie proposed the relations, $E = \hbar\,\omega$ $(\omega = 2\pi\upsilon)$ and $p = \hbar\,k\left(k = \frac{2\pi}{\lambda}\right)$ or $\lambda = h/p$ to hold good in the case of a material particle.

In 1927, Davisson and Germer[9] studied the diffraction of electrons (of course of appropriate momentum. For an electron accelerated through 100 volts, which is of the order of X-ray wavelength) from single crystals of Nickel. The diffraction pattern consisted of spots similar to Laue spots obtained in the X-ray diffraction pattern. They showed that the diffraction pattern could be explained if the electrons were assumed to have a wavelength given by the de Broglie relation, $\lambda = h/p$, where p is the momentum of the electron.

The above observations thus demonstrated in a very direct fashion the existence of de Broglie waves associated with electrons (or wave-particle duality even in the case of material particles). Truly speaking electrons are not only the bodies whose wave behaviour can be demonstrated. The diffraction of neutrons and of whole atoms when scattered by suitable crystals has been observed (neutron diffraction is actually a tool for investigating crystal structure). It may once again be pointed out that the finiteness of Planck's constant is the basic point here. Because if h were zero, then no matter what momentum a particle had the associated wave would always correspond to $\lambda = 0$ and would follow the laws of classical mechanics, which can be regarded as the short wavelength limit of wave mechanics in the same way as geometrical optics is the short wavelength limit of wave optics.

1.11 WAVE PACKET AND ITS MOTION

The success of the de Broglie hypothesis that a particle of momentum p is to be associated with a wave of definite wavelength $\lambda = h/p$, leaves one in a very uncomfortable situation. Because a pure harmonic wave (plane wave or travelling wave, $y = e^{i(kx - \omega t)}$) necessarily extends over all space and this fact makes it impossible to get any idea of the position of particle described by such a wave. On the other hand if we consider a wave packet, that is, a wave which is confined to a small region of space, it is then reasonable to suppose that the particle is found within the region of the wave packet (classically whenever we talk about a particle, it simply means a point object having precise position and momentum at every instant of time). The position of particle would then be approximately determined by that of the wave packet, there being an uncertainty in the position which is of the order of the dimension or size of the wave packet.

For the sake of simplicity let us restrict our discussion to one dimension and consider a wavepacket formed by the superposition of plane waves (or harmonic waves) all propogating parallel to ox (or x-axis) and let us represent it by a wave function,

$$\psi(x, t) = (2\pi)^{-1/2} \int_{-\infty}^{+\infty} A(k)\, e^{i(kx - \omega t)}\, dk \qquad \text{... (13)}$$

where $A(k)$ is the Fourier transform of $\psi(x, t)$ and is given by

$$A(k) = (2\pi)^{-1/2} \int_{-\infty}^{+\infty} \psi(x, t)\, e^{i(kx - \omega t)}\, dx \qquad \text{... (14)}$$

[9] C. Davisson and L.H. Germer, Nature 119, 558 (1927); Phys. Rev. 30,705 (1927)

and is called amplitude of the harmonic component with propogation constant k ($= 2\pi/\lambda$) and angular frequency ω in the wave function $\psi(x, t)$.

Let us consider a typical form of the wave packet corresponding to the particle localized within a distance Δx shown in Fig. 1.4, where we have plotted real part of $\psi(x, t)$ (which in general is a complex function) as a function of x at a given instant t. Fig. 1.5 shows the Fourier transform $A(k)$ of ψ plotted as a function of k which has a pronounced peak situated at $k = k_0$ and has a width (defined, for example, at half its maximum value) Δk. It may be mentioned that the wave packet considered here is nothing but Gaussian.

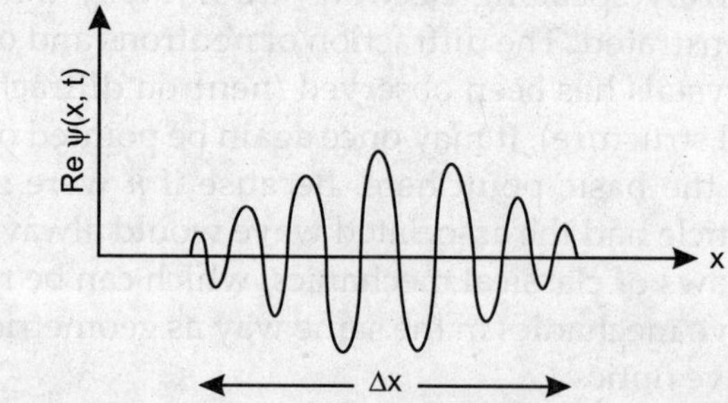

Fig. 1.4 *Form of a typical wave packet*

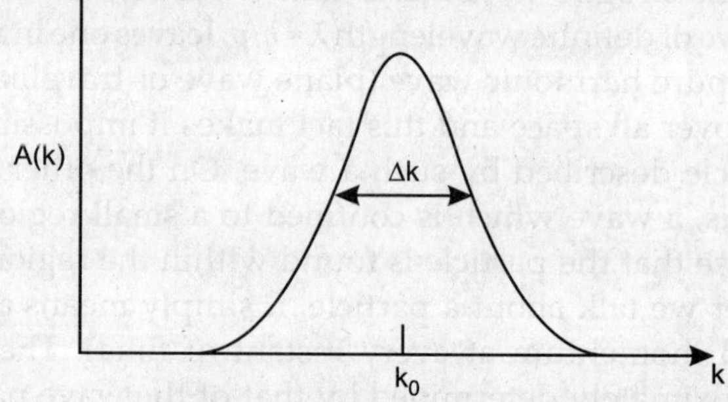

Fig. 1.5 *Shape of the function A(k) centered at $k = k_0$*

We will now show that if the range of k values for which the component waves present in ψ have significant amplitudes, is limited (or if $A(k)$ is negligible except when k lies in a small interval Δk around k_0) then it is possible to assign an average velocity to the wave packet.

The wave packet under the circumstance can then be written as

$$\psi(x, t) = (2\pi)^{-1/2} \int_{k_0 - \Delta k/2}^{k_0 + \Delta k/2} A(k)\, e^{i(kx - \omega t)}\, dk \quad \text{... (15)}$$

where the integral extends only over the significant interval Δk.

Let us assume that $w(k)$ varies slowly (or do not vary significantly) with k in the region of integration so that we may expand $w(k)$ in a Taylor series about $k = k_0$, that is

$$w(k) = w(k_0) + (k - k_0) \left.\frac{dw}{dk}\right|_{k=k_0} + 1/2\,(k - k_0)^2 \left.\frac{d^2w}{dk^2}\right|_{k=k_0} + \ldots \quad \text{... (16)}$$

and can neglect terms involving quadratic and higher powers of $(k - k_0)$

Thus taking, $w(k) = w(k_0) + (k - k_0) \left.\frac{dw}{dk}\right|_{k=k_0}$ and puting $w(k_0) = w_0$ (say), in eqn. (14), one gets

$$\begin{aligned} \psi(x, t) &= (2\pi)^{-1/2} \int_{\Delta k} A(k)\, e^{i\left\{(k-k_0)x + k_0 x - w_0 t - (k-k_0)t \left.\frac{dw}{dk}\right|_{k=k_0}\right\}} dk \\ &= e^{i(k_0 x - w_0 t)} (2\pi)^{-1/2} \int_{\Delta k} A(k)\, e^{i\left(x - \left.\frac{dw}{dk}\right|_{k=k_0} t\right)(k-k_0)} dk \\ &= e^{i(k_0 x - w_0 t)} f(x, t) \qquad \text{... (17)} \end{aligned}$$

where

$$f(x, t) = (2\pi)^{-1/2} \int_{\Delta k} A(k)\, e^{i\left(x - \left.\frac{dw}{dk}\right|_{k=k_0} t\right)(k-k_0)} dk \quad \text{... (18)}$$

Eqn. (17) represents a wave of wavelength $2\pi/k_0$ and frequency w_0 modulated by the integral appearing as a factor $f(x, t)$ which determines the envelope of the wave packet. This factor $f(x, t)$ depends on x and t only through the combination $x - \left.\frac{dw}{dk}\right|_{k=k_0} t$. Hence the wave packet propogates undistorted with velocity.

$$\upsilon_g = \left.\frac{dw}{dk}\right|_{k=k_0} \quad \text{... (19)}$$

called group velocity of the packet since it represents the velocity of motion of a group of waves which make up the wave packet. It may be mentioned here that the packet would propogate undistorted in case the neglect of quadratic and higher order terms in eqn.(16) is justified. However, in case the $w - k$ relationship is linear, case of electromagnetic waves in free space, $w(k) = ck$, c being the speed of light in free space,

$$\begin{aligned} \psi(x, t) &= (2\pi)^{-1/2} \int_{-\infty}^{+\infty} A(k)\, e^{i(kx - wt)}\, dk \\ &= (2\pi)^{-1/2} \int_{-\infty}^{+\infty} A(k)\, e^{ik(x - ct)}\, dk \qquad \text{... (20)} \end{aligned}$$

the packet travels undistorted with speed c. Such a medium for which $w - k$ relationship is linear, is called non-dispersive medium. On the other hand, whenever the $w - k$ relationship is not linear (case of de Broglie waves corresponding to free particle, $w(k) = \frac{\hbar k^2}{2m}$), the medium is said to be dispersive, the wave packet actually spreads with time (or width of the packet increases with time).

Using the relations $E = h\upsilon = \hbar w$ (based on the hypothesis that the relation between the energy E of the particle (material) and the frequency υ of the associated wave is exactly the same as that between the energy of the photon and frequency of light radiation) and momentum p of the particle is associated with the wave of wavelength λ or wave number k, $p = h/\lambda = \hbar k$ (de Broglie relation), we get

$$\upsilon_g = \left.\frac{dw}{dk}\right|_{k=k_0} = \frac{d(E/\hbar)}{d(p/\hbar)} = \left.\frac{dE}{dp}\right|_{p=p_0} \quad \text{... (21)}$$

Now the velocity υ of the particle (whether relativistic or non-relativistic) is given by

$$\upsilon = dE/dp \quad \text{... (22)}$$

For non-relativistic free particle, that is a particle not subjected to any force or a particle whose potential energy is zero or has a constant value at every point in space,

$$E = p^2/2m$$

or

$$dE/dp = p/m = \upsilon \quad \text{... (23)}$$

Similarly in the relativistic case,

$$E^2 = p^2c^2 + m_0^2\, c^4$$

$$\therefore \quad dE/dp = p\,c^2/E$$

Using the relations, $E = m_0c^2/\sqrt{1 - v^2/c^2}$ and $p = m_0\upsilon/\sqrt{1-v^2/c^2}$, the above expression takes the form

$$dE/dp = \upsilon \quad \text{... (24)}$$

Therefore the group velocity of the wave packet is just the velocity which one observes for the associated particle. We have thus come to the conclusion that the velocity of the particle is to be associated with the group velocity of the packet.

1.12 THE UNCERTAINTY PRINCIPLE

The fact that a moving body must be regarded as a de Broglie wave group rather than a localized entity suggests that there is fundamental limit to the accuracy with which we can measure its particle properties. According to Heisenberg uncertainty principle[10], it is impossible to measure simultaneously the position of particle along a particular direction (say x) and also its momentum p_x in this direction with unlimited accuracy. If Δx is the uncertainty in position of a body at some instant and Δp_x is the uncertainty in momentum at the same instant, then

$$\Delta x\, \Delta p_x \geq h \quad \text{... (25a)}$$

Similarly one can write the other two uncertainty relations,

$$\Delta y\, \Delta p_y \geq h \quad \text{... (25b)}$$

[10] W. Heisenberg, Z. Fur Physik 43, 172 (1972)

and

$$\Delta z \, \Delta p_z \geq h \qquad \text{... (25c)}$$

In order to understand the inherent uncertainties associated with the simultaneous measurements of y and p_y, we consider the diffraction of electrons by a long narrow slit of width d as illustrated in Fig. 1.6.

Suppose the electrons with momentum p are coming from a very distant source S, so that before entering the slit the y component of the momentum of the electron is negligible (or in other words a plane wave front is incident on the slit or a parallel beam of electrons is incident on the slit)

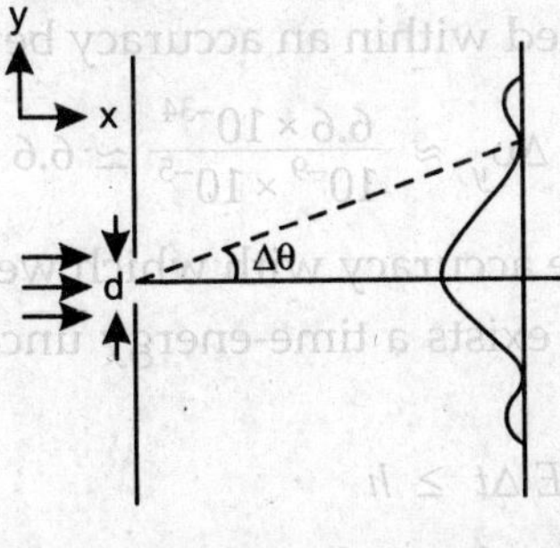

Fig. 1.6 *Diffraction by a single slit*

When the electron passes through the slit or is diffracted by the slit, the uncertainty in its position would be given by

$$\Delta y \approx d \qquad \text{... (26)}$$

Now as soon as the electron enters the slit there would be spreading due to diffraction. The diffraction spreading $\Delta\theta$ (from the wave theory, Ist minimum of the Fraunhofer diffraction pattern is given by $d \sin \Delta\theta = \lambda$, where λ is the de Broglie wavelength of the electron) is approximately given by

$$\Delta\theta \approx \lambda/d \qquad \text{... (27)}$$

As p was the momentum of the electron in the x-direction before entering the slit, it acquires a momentum Δp_y in the y-direction after entering the slit. Hence

$$\Delta p_y/p = \tan \Delta\theta \simeq \Delta\theta \qquad \text{... (28)}$$

From relations (27) and (28)

$$\Delta p_y/p \approx \lambda/d$$

or

$$\Delta p_y \approx p\,\lambda/d \qquad \text{... (29)}$$

Multiplying (26) and (29) and also using the de Broglie relation, one gets

$$\Delta y \Delta p_y \approx h \qquad \text{... (30)}$$

which is called Heisenberg uncertainty relation. The relation (30) of course gives only the order of magnitude. In order to make an exact statement of the uncertainty principle, we must first have precise definition of uncertainty which would be done in the next chapter and would lead to the relation

$$\Delta y \Delta p_y \geq \hbar/2$$

Before we proceed further it may be mentioned here as to why we do not observe the effect of uncertainty principle in our everyday life is because of the smallness of the value of the Planck's constant h. For example, for a small particle of mass 10^{-9} kg if position is determined within an accuracy of 10^{-5} m, then according to the uncertainty principle

$$\Delta y \, m\Delta \upsilon_y \approx h \quad \text{or} \quad \Delta \upsilon_y \approx h/m \, \Delta y,$$

its velocity cannot be determined within an accuracy better than

$$\Delta \upsilon_y \approx \frac{6.6 \times 10^{-34}}{10^{-9} \times 10^{-5}} \simeq 6.6 \times 10^{-20} \text{ m/s}$$

which is much smaller than the accuracy with which we can determine the velocity.

In analogous to (25), there exists a time-energy uncertainty relation according to which

$$\Delta E \, \Delta t \geq h \qquad \text{... (31)}$$

It is necessary to point out here that there is difference in the status of the uncertainty relations (25) and (31). In (25*a*) both x and p_x are dynamical variables (conjugate variables) whereas in (31), energy E is a dynamical variable while t is just a parameter. Accordingly the physical interpretation of eqn. (25) is quite different from that of eqn. (31). The above relation tells us that if ΔE represents the spread in the energy of the particle, then the uncertainty in the instant of passage of particle through a point will be given by

$$\Delta t > h/\Delta E \qquad \text{... (31a)}$$

The relation (31) also implies that an energy determination which has an accuracy ΔE must occupy at least a time interval $\Delta t \sim \frac{h}{\Delta E}$. Hence if a system is in a certain state for a maximum time of Δt, the energy of the system in that state is uncertain by at least $\Delta E \sim h/\Delta t$. For example, the time that an atom retains excess excitation energy before re-emitting it in the form of a photon is ~10^{-8} sec (known as the life time of the excited state). This imposes a limit on the precision with which the excitation energy can be determined. The photon energy would thus be uncertain by an amount

$$\Delta E \sim h/\Delta t \sim \frac{6.6 \times 10^{-34}}{10^{-8}} \sim 6.6 \times 10^{-26} \text{ Joule}$$

Since $\Delta E = h\Delta\upsilon$, the frequency of light emitted would be uncertain by

$$\Delta \upsilon \sim \Delta E/h \sim 1/\Delta t \sim 10^{8} \text{ cycles /sec}$$

This is the irreducible limit to the accuracy with which we can determine the frequency of the radiation emitted by an atom.

PROBLEMS

1.1 Find energy levels of a particle of mass m which is confined to move in the x-direction, using Bohr-Sommerfeld quantum rule. At $x = 0$ and L, the particle undergoes perfectly elastic collisions with the rigid walls. No other forces act on the particle (the problem is equivalent to the motion of a particle in one dimensional box).

1.2 Find out energy levels of one-dimentional harmonic oscillator of angular frequency w using Bohr-Sommerfeld quantum rule.

1.3 Find out energy levels in the case of rigid rotator using Bohr-Sommerfeld quantum rule.

1.4 An electron beam is accelerated through a potential difference Y volts. Find the order of magnitude of the de Broglie wavelength associated with electrons.

1.5 Find de Broglie wavelength of a thermal neutron, that is, a neutron with a speed υ corresponding to the average thermal energy at the absolute temperature T.

1.6 Use uncertainty principle to estimate the ground state energy of a linear harmonic oscillator.

1.7 Estimate the size of hydrogen atom and its ground state energy using uncertainty relation.

1.8 Calculate uncertainty in the momentum of an electron if it is to be confined in a nucleus of radius 10^{-15} m (= 1 fermi). From this find the energy of the electron inside the nucleus.

1.9 (*a*) Consider a pendulum bob of mass 0.1 kg moving at the speed of 3m/sec. Suppose the momentum p_x need not be known more accurately than $\Delta p_x = 10^{-6}\, p_x$, what limitation does the uncertainty principle impose on the simultaneous measurement of x?

(*b*) Compare uncertainties in the velocities of an electron and a proton confined in a 10 Å box.

(*c*) The velocity of a proton in x-direction is measured to an accuracy of 10^{-7} m/s. Determine the limit of accuracy with which the proton can be located simultaneously along the x-axis.

1.10 A 25 g bullet takes 0.5 seconds to reach its target. Regarding the bullet as a mass point, find the order of magnitude of spread of successive shots at the target under the best possible conditions.

Solutions

1.1 The linear momentum of the particle $p_x = p$ would be constant in this case. However, it changes its sign on collision of the particle with the walls.

According to Bohr Sommerfeld quantum rule

$$\oint p_x\, dx = p \oint dx = nh, \quad n = 0, 1, 2, \ldots$$

or

$$p\,2L = nh$$

or

$$p = nh/2L \quad \text{... (1.1)}$$

i.e. only certain values of momentum would be allowed. The corresponding energy values (or energy levels) would be given by

$$E_n = p^2/2m = n^2h^2/8mL^2, \quad n = 0, 1, 2, \ldots \quad \text{... (1.2)}$$

Thus particle in a box can have only discrete energy levels with values E_n given by *eqn.* (1.2) against the continuous energy spectrum in classical physics. The particle can exist in one of the energy states E_n plotted in Fig. 1.7.

For a macroscopic system, say $m = 1$ gm and dimension $L = 1$ cm, from eqn. (1.2), $E_2 - E_1 = 10^{-41}$ eV, which is so small that it cannot be measured or in other words energy spectrum would be continuous. However, if an electron is constraint to move in a box of atomic dimension, the spacing of the energy levels is appreciable and hence

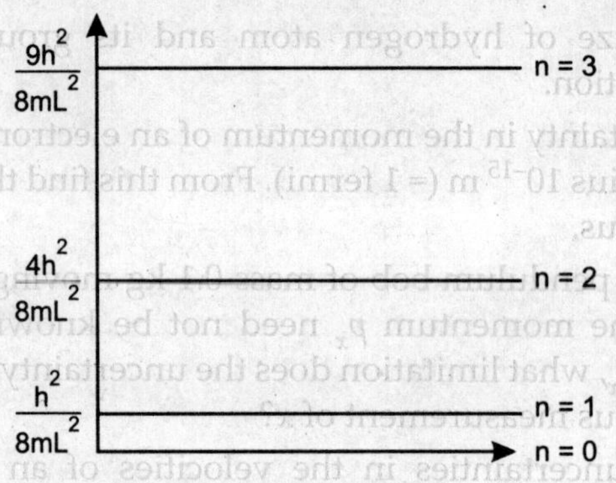

Fig. 1.7 *Energy levels diagram of particle in a one dimensional box*

can be measured. For example, if $L = 10^{-10}$ m and $m = 9.1 \times 10^{-31}$ kg, $E_2 - E_1$ from (1-2) comes out to be 113 eV. This in turn once again shows that the quantum mechanics is applicable in the case of macroscopic as well as in the case of microscopic objects whereas classical physics is applicable only in the cases of macroscopic objects.

1.2 One dimensional harmonic oscillator consists of a particle of mass m bound to the equilibrium position $x = 0$ by a restoring force $F = -kx$. The force constant k and angular frequency $\omega = 2\pi\upsilon$ are connected by the relation $k = m\omega^2$. The particle performs simple harmonic motion with amplitude a_0, its displacement from the origin at time t is $x = a_0 \sin \omega t$.

The total energy of the harmonic oscillator is given by

$$E = p_x^2/2m + 1/2\, kx^2$$

which can be written as

$$x^2/2E/k + p_x^2/2mE = 1 \quad \text{or} \quad x^2/(2E/m\omega^2) + p_x^2/2mE = 1 \qquad \text{... (1.3)}$$

Thus the representative point in phase space, with co-ordinates x and p_x describes an ellipse. The semi-major and semi-minor axes of the ellipse would be given by

$$a = (2E/m\omega^2)^{1/2} \quad \text{and} \quad b = (2mE)^{1/2}, \text{ respectively.}$$

Now applying the Bohr-Sommerfeld quantum rule

$$\oint p_x\, dx = nh, n = 0, 1, 2, ...$$

One gets.

$$\oint p_x\, dx = \pi\, a\, b \text{ (area of ellipse)} = nh \qquad \text{... (1.4)}$$

Substituting values of a and b in above, one gets

$$\pi(2E/m\omega^2)^{1/2}\,(2mE)^{1/2} = nh$$

or

$$E = E_n = nh\upsilon, \quad n = 0, 1, 2, ... \qquad \text{... (1.5)}$$

Thus the energy levels of harmonic oscillator are not continuous but have discrete values and are evenly spaced as shown in Fig. 1.8

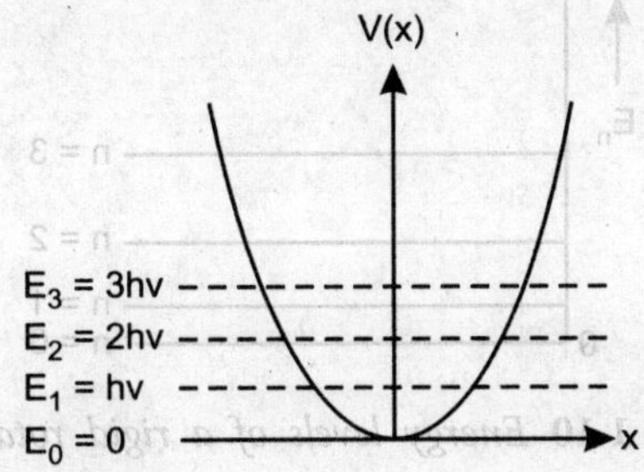

Fig. 1.8 *Energy levels of a harmonic oscillator*

1.3 Two masses m_1 and m_2 are connected by a rigid rod and capable of rotating about an axis passing through the centre of mass and perpendicular to the line joining the two masses, constitute a rigid rotator shown in Fig. 1.9. It is actually a suitable model from which the rotational energy levels of diatomic molecule can be derived.

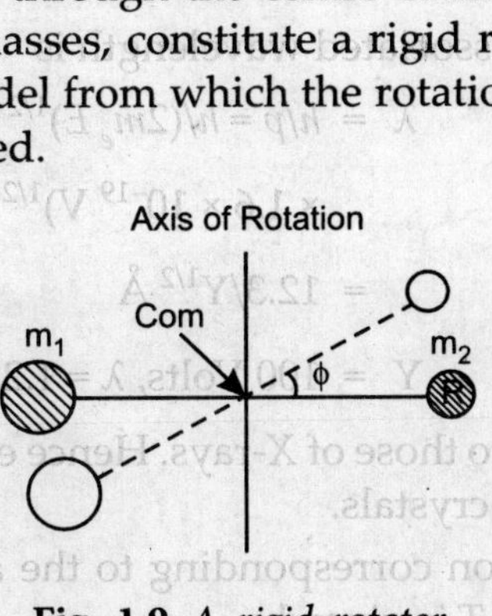

Fig. 1.9 *A rigid rotator*

According to Bohr-Sommerfeld quantum rule

$$\oint p_\phi \, d\phi = n\, h, n = 0, 1, 2, \ldots$$

where ϕ represents the angle of rotation and p_ϕ is the angular momentum, $p_\phi = I\dot{\phi}$, where I is the moment of inertia of the system about axis of rotation and $\dot{\phi}$ is the angular velocity.

There is no potential energy in this case and hence no force. Therefore angular momentum would be constant of motion (torque zero). Hence

$$\oint p_\phi \, d\phi = p_\phi \, 2\pi = n\, h$$

or

$$p_\phi = nh/2\pi, \quad n = 0, 1, 2, \ldots \qquad \ldots (1.6)$$

i.e. the angular momentum of the rotator is quantized.

The energy values (or levels) in this case would be given by

$$E_n = p_\phi^2/2I = n^2h^2/8\pi^2 I, \quad n = 0, 1, 2, \ldots \qquad \ldots (1.7)$$

Thus the energy levels of the rigid rotator are not continuous but discrete as shown in Fig. 1.10.

Fig. 1.10 *Energy levels of a rigid rotator*

1.4 If an electron beam is accelerated through a potential difference V volts, the kinetic energy imparted to the electrons would be given by

$$E = eY = 1.6 \times 10^{-19} \, V \text{ Joules}$$

Since $E = p^2/2m_e$, the associated wavelength is

$$\lambda = h/p = h/(2m_e \, E)^{1/2} = 6.6 \times 10^{-34}/(2 \times 9.1 \times 10^{-31}$$

$$\times 1.6 \times 10^{-19} \text{ V})^{1/2} \text{ m}$$

$$= 12.3/\text{Y}^{1/2} \text{ Å}$$

If $\quad \text{Y} = 100 \text{ Volts}, \lambda = 1.23 \text{ Å}$

which is comparable to those of X-rays. Hence electron diffraction phenomena can be observed with crystals.

1.5 The speed υ of neutron corresponding to the average thermal energy at the absolute temperature T is given by

$$1/2\, m_n v^2 = p^2/2m_n = 3/2\, k_B T$$

where k_B is the Boltzmann constant.

The wavelength which corresponds to such a speed is

$$\lambda = h/p = h/(3m_n k_B T)^{1/2}$$

For

$$T = 300\text{ K}, \lambda \simeq 1.4\text{ Å},$$

a wavelength which is of the order of the distance between atoms in a crystal lattice. A beam of neutrons falling on a crystal therefore gives rise to diffraction pattern analogous to those observed with X-rays.

1.6 The total energy in the case of a linear harmonic oscillator is given by

$$E = p_x^2 / 2m + \frac{1}{2} m \omega^2 x^2 \quad \text{... (1.8)}$$

Let us assume that the uncertainty in the measurement of p_x is p_x itself, that is, $\Delta p_x \sim p_x \sim \hbar/2\,\Delta x$, where we have used the uncertainty relation.

$$\Delta x\, \Delta p_x \sim \hbar/2$$

Now assuming that the particle is confined in the region a and uncertainty in the measurement of x is x itself, that is, $\Delta x \sim x \sim a$.

Thus

$$E = \hbar^2/8\, m\, a^2 + \frac{1}{2} m\, \omega^2 a^2 \quad \text{... (1.9)}$$

The ground state (or the lowest energy state) of the linear harmonic oscillator would correspond to the minimum value of E for which $dE/d\,a = 0$. Hence

$$d\,E/d\,a = -\hbar^2/4\, m\, a^3 + m\, \omega^2\, a = 0$$

which gives,

$$a = (\hbar/2\, m\, \omega)^{1/2} \quad \text{... (1.10)}$$

Substituting the value of a from eqn. (1.10) into eqn. (1.9), one finds that the minimum energy is given by

$$E_{\min} = E_0 = \frac{1}{2}\, \hbar\, \omega \quad \text{... (1.11)}$$

which indeed is the ground state energy of a linear harmonic oscillator. (would be shown in chapter 4). The quantity $1/2\ \hbar w$ is known as the zero point energy and is a consequence of the uncertainty principle.

1.7 The total energy in the case of hydrogen atom is given by

$$E = p^2/2m - e^2/r \quad \text{... (1.12)}$$

Let us assume that the uncertainty in the measurement of p is p itself, that is $\Delta p \sim p \sim \hbar/2\,\Delta x$ where we have used the uncertainty relation

$$\Delta x\, \Delta p \sim \hbar/2 \qquad \text{... (1.13)}$$

Similarly let us assume that the electron is confined to the region $r \sim a$ and let uncertainty in the measurement of r is r itself. This amounts to the fact that $r \sim \Delta x \sim a$. Hence

$$E = \hbar^2/(2m \times 4a^2) - e^2/a \qquad \text{... (1.14)}$$

The ground state (or lowest total energy) in the case of hydrogen atom compatible with the uncertainty relation would correspond to the minimum value of E or would correspond to the condition

$$dE/da = 0$$

which results in

$$a = h^2/4me^2 \;(= a_0/4 = 1/4 \times 1/2 \text{ Å}) \qquad \text{... (1.15)}$$

and tells that the finite size of the atom is due to the uncertainty principle. Thus the lowest energy value is given by

$$E_{\min} = \hbar^2/(8\, m \times \hbar^4/16\, m^2 e^4) - e^2/\,(\hbar^2/4\, me^2)$$

$$= -2\, me^4/\hbar^2 \qquad \text{... (1.16)}$$

It may, however be pointed out that the actual value of the ground state energy of the hydrogen atom is $\frac{-me^4}{2\hbar^2}$. By uncertaintv relation we only get order of the magnitude of the quantity.

1.8 For an electron to be confined with in a sphere of radius $r_0 = 10^{-15}$ m (size of nucleus), the uncertainty in its position may not exceed 10^{-15} m. The corresponding uncertainty in the electron's momentum is

$$\Delta p \approx \hbar/2r_0$$

$$\approx 1.054 \times 10^{-34}/2 \times 10^{-15}$$

$$\approx 0.5 \times 10^{-19} \text{ kg-m/s} \qquad \text{... (1.17)}$$

1.9 If this is the uncertainty in the electron's momentum, the momentum itself must be atleast comparable in magnitude. An electron whose momentum is 0.5×10^{-19} kg-m/s has a kinetic energy T many times greater than its rest energy $m_0 c^2$ and we may therefore use the extreme relativistic formula

$$E = T + m_0 c^2 \approx T \approx p\, c$$

$$\approx \Delta p\, c$$

$$\approx 0.5 \times 10^{-19} \times 3 \times 10^8$$

$$\approx 1.5 \times 10^{-11} \text{ Joules}$$

$$\approx 100\, M\, eY \qquad \text{... (1.18)}$$

Although the electrons do emerge from nuclei in β-decay, they have energies which are only a fraction of this energy. Thus one does not expect the electron to be a basic constituent of a nucleus.

1.10 Let the bullets be fired parallel to the *x*-axis. We would like to find the spread of bullets under the best possible situation, that is minimum spread.

Let us assume that at the time bullet is fired the uncertainty in the *y*-component of its position is Δy and hence the uncertainty in its momentum would be $\hbar/2\Delta y$. After a time *t* the uncertainty in the *y*-component of bullets' position is

$$\Delta y\,(t) = \Delta y + \upsilon t = \Delta y + p_y/m\, t \qquad \text{... (1.19)}$$

Now assuming that the uncertainty in the momentum can be momentum itself,

$$\Delta y\,(t) \approx \Delta y + \hbar/2\,\Delta y\, m\, t$$

At

t = 0.5 sec, the spread would be

$$\Delta y\,(t = 0.5\,) = \Delta y + \frac{\hbar}{2\Delta y m} \times 0.5 \qquad \text{... (1.20)}$$

Minimising this with respect to Δy, one gets

$$1 - \frac{\hbar}{2m\,(\Delta y)^2} \times 0.5 = 0$$

or

$$\Delta y = 1/2 \times (\hbar/m)^{1/2} \qquad \text{... (1.21)}$$

Substituting above in (1.20), one gets,

$$\begin{aligned}\Delta y\,(t = 0.5) &= 1/2\,(\hbar/m)^{1/2} + \hbar/4m \times 2/(\hbar/m)^{1/2}\\ &= (\hbar/m)^{1/2}\\ &= (1.054 \times 10^{-34}/25 \times 10^{-3})^{1/2}\\ &\approx 0.65 \times 10^{-16}\ \text{m}\end{aligned}$$

❑❑❑

Chapter 2

SCHRÖDINGER EQUATION AND PRINCIPLES OF QUANTUM MECHANICS

2.1 SCHRÖDINGER EQUATION

The theoretical exploitation of de Broglie's idea did not await its experimental verification. Immediately after the publication of the hypothesis, Erwin Schrödinger[1] in 1926 proposed that the behaviour of de Broglie waves associated with material particles is governed by a certain differential equation for the wave function ψ, the function which represents the de Broglie wave.

Now we would like to obtain a differential equation for the wave function ψ (actually for the variation of wave function with space and time), called Schrödinger equation which describes the behaviour of particles like electron, proton, neutron etc. (In classical mechanics the motion of a particle is described by Newton's 2nd law or Newton's equation). We cannot really derive Schrödinger equation, just as we cannot derive Newton's equation. The validity of Schrödinger equation or Newton's equation lies in the experimental verification of the results obtained from them. The presentation we are going to give here represents neither a proof nor a derivation of the Schrödinger equation but indicates only a way to get this equation.

Consider motion of a free particle (a particle on which no forces are acting) in one dimension along the x-axis with momentum p. According to de Broglie's relation, a wave of wavelength $\lambda = h/p$ is associated with the particle. Let us assume this wave to be of the form (or the wave is represented by a wave function $\psi(x, t)$).

$$\psi(x, t) = Ae^{i(kx - wt)} \quad \text{... (1)}$$

which is equation of a plane wave or harmonic wave travelling along the positive x-axis with velocity (phase velocity), $v_p = w/k$. Using relations,

and $$p = h/\lambda = \hbar k \quad \text{... (2b)}$$

where $\hbar$ (h cross) $= h/2\pi$, eqn. (1) reduces to

---

[1]E. Schrödinger, Ann. Physik 79, 361, 481 (1926); 80, 437 (1926); 81, 109 (1926).

$$\psi(x, t) = Ae^{i(px - Et)/\hbar} \quad \text{... (3)}$$

Performing partial differentiation w.r.t. x on the above equation, we get

$$\frac{\partial \psi(x, t)}{\partial x} = i/\hbar \quad pAe^{i(px - Et)/\hbar}$$

$$= i/\hbar \quad p\psi(x, t)$$

Multiplying both sides by $-i\hbar$, we get

$$\frac{-i\hbar\, \partial \psi(x, t)}{\partial x} = p\psi(x, t) \quad \text{... (4)}$$

Similarly, differentiating partially eqn. (3) w.r.t. time and multiplying the resulting equation by $i\hbar$ from both the sides, we get

$$\frac{i\hbar\, \partial \psi(x, t)}{\partial t} = E\psi(x, t) \quad \text{... (5)}$$

Eqns. (4) and (5) imply that energy and momentum can be represented by differential operators which act on the wave function $\psi(x, t)$.

$$E \equiv i\hbar\ \partial/\partial t, p \equiv -i\hbar\ \partial/\partial x \quad \text{... (6)}$$

Now the energy of the non-relativistic free particle is given by

$$E = p^2/2m \quad \text{... (7)}$$

By operating on $\psi(x, t)$ gives

$$E\psi(x, t) = (p^2/2m) \quad \psi(x, t)$$

Using the operator equivalence represented by eqn. (6), one gets

$$\frac{i\hbar\, \partial \psi(x, t)}{\partial t} = -\frac{\hbar^2}{2m} \frac{\partial^2 \psi(x, t)}{\partial x^2} \quad \text{... (8)}$$

This is time dependent one dimensional Schrödinger equation for a free particle of mass m.

It is quite important to note that though only harmonic waves of the form (1) were considered in obtaining eqn. (8) but truly speaking solutions of the said equation are not restricted to such waves only.

If ψ_1 is a solution of eqn. (8) and ψ_2 is also the solution, then a linear combination $c_1 \psi_1 + c_2 \psi_2$ also be the solution of the same equation. In general any linear combination or superposition of harmonic waves such as

$$\psi(x, t) = \int A(k)\, e^{i(kx - \omega t)}\, dk \quad \text{... (9)}$$

will also be the solution of eqn. (8). This property (called superposition principle) is a simple consequence of the fact that the wave function, $\psi(x, t)$ and its derivatives appear only linearly in the Schrödinger equation, that is 2nd or higher powers of these quantities do not appear in the equation. One must note that the expression (9) is of the same form as that of the wave packet represented by eqn. (13) of section 1.13. Thus the free particle Schrödinger equation has perfectly good wave packet solutions besides harmonic ones. It also needs mention that it is the group velocity $v_g = dw/dk$ of the wave packet which is equal to the particle velocity and not the phase velocity $v_p = w/k = E/p$, which is the velocity of the individual harmonic wave (1).

However, in the case of a medium (non-dispersive) in which all the component waves (harmonic waves) present in the packet, though having different wavelengths travel with the same velocity, $v_p = v_g$.

Let us now assume that the above procedure remains applicable even if the particle is not free but is in a field characterized by the potential energy function $V(x, t)$. The total energy of the particle then is given by

$$E = p^2/2m + V(x, t) \qquad \text{... (10)}$$

By operating on $\psi(x, t)$, leads to

$$E\psi(x. t) = [p^2/2m + V(x, t)]\, \psi(x, t)$$

and then using eqn. (6). that is the operator equivalence of E and p, one gets

$$\frac{i\hbar\, \partial\psi(x, t)}{\partial t} = -\frac{\hbar^2}{2m}\frac{\partial^2 \psi(x, t)}{\partial x^2} + V(x, t)\, \psi(x, t) \qquad \text{... (11)}$$

This is time dependent Schrödinger equation in one dimension for a particle of mass m moving in a field characterized by a potential energy function $V(x, t)$.

It is necessary to mention here that the manner in which Schrödinger equation was obtained starting from the wave function of a freely moving particle and then the extension of the method to the general case of a particle subjected to arbitrary forces which vary in space and time, that is the case in which $V = V(x, t)$, whether the entire procedure is correct? The answer is, we can solve the resulting equation (the Schrödinger equation) for a variety of physical situations, and compare the results of calculations with the experimental results. If they agree, the postulate embodied in Schrödinger's equation is valid; if they do not agree, the postulate must be disregarded and some other approach would have to be explored. Truly speaking, the Schrödinger equation has turned out to be completely accurate in predicting the results of experiments. One must keep in mind that the eqn. (11) can be used only for the non-relativistic problem and a more elaborate formulation, which we would present at a later stage, is required when particle speeds are comparable with that of light.

2.2 EXTENSION TO THREE DIMENSIONS

In this case eqn. (2*b*) would have the form,

$$\vec{p} = \hbar\vec{k} \qquad \text{... (12)}$$

where $\vec{k}$ is called the wave vector or the propagation vector.

Similarly the harmonic wave associated with the particle must be represented by

$$\psi(\vec{r}, t) = Ae^{i(\vec{k} \cdot \vec{r} - \omega t)} \quad \text{... (13)}$$

which in terms of momentum p and energy E can be written as

$$\psi(\vec{r}, t) = Ae^{i/\hbar(\vec{p} \cdot \vec{r} - Et)} \quad \text{... (14)}$$

Using the procedure same as that used in the one dimensional case, it can be shown that $\vec{p}$ and E can be represented by the differential operators,

$$\vec{p} \equiv -i\hbar\vec{\nabla}, \quad E \equiv i\hbar\, {}^{\partial}/\partial t \quad \text{... (15)}$$

where $\vec{\nabla} = \hat{x}\partial/\partial x + \hat{y}\partial/\partial y + \hat{z}\partial/\partial z$, is called the del operator.

Again following the procedure as that used in the one dimensional case, one leads to the equation.

$$\frac{i\hbar\,\partial\psi(\vec{r}, t)}{\partial t} = -\frac{\hbar^2}{2m}\nabla^2\psi(\vec{r}, t) \quad \text{... (16)}$$

where $\nabla^2 = \frac{\partial^2}{\partial x^2} + \frac{\partial^2}{\partial y^2} + \frac{\partial^2}{\partial z^2}$ is called the Laplacian operator and eqn. (16) the Schrödinger equation for a free particle in three dimensions.

Like in the one dimensional case, any linear combination or superposition of harmonic waves of the form (13), that is

$$\psi(\vec{r}, t) = \int A(\vec{k})e^{i(\vec{k} \cdot \vec{r} - \omega t)}\, d\vec{k} \quad \text{... (17)}$$

would also satisfy the Schrödinger equation.

2.3 SCHRÖDINGER EQUATION IN THE PRESENCE OF POTENTIAL ENERGY FUNCTION $V(\vec{r}, t)$

In this case the particle is subjected to a force $\vec{F}$ expressed as

$$\vec{F} = -\vec{\nabla}V(\vec{r}, t) \quad \text{... (18)}$$

where $V(\vec{r}, t)$ is the potential energy the particle would have if it were at the point $\vec{r}$ at time t. The total energy of the particle in the said case would be given by

$$E = p^2/2m + V(\vec{r}, t) \quad \text{... (19)}$$

On employing the above mentioned procedure and on using the operator equivalence (15), we obtain the equation

$$\frac{i\hbar\,\partial\psi(\vec{r}, t)}{\partial t} = \frac{\hbar^2}{2m}\nabla^2\psi(\vec{r}, t) + V(\vec{r}, t)\,\psi(\vec{r}, t) \quad \text{... (20)}$$

This is three dimensional time dependent Schrödinger equation for a particle moving in force field characterized by the potential energy function $V(\vec{r}, t)$.

2.4 PHYSICAL INTERPRETATION OF ψ

In a light wave the electromagnetic field varies in space and time, in a sound wave pressure varies in space and time; what is it whose variation constitute de Broglie waves. The answer is that the variable quantity ψ called wave function characterizes the de Broglie waves. Let us see little more precisely, what ψ is ? One thing is certain that ψ must in some sense be a measure of the presence of particle. But the function $\psi(x, t)$, in general $\psi(\vec{r}, t)$, itself cannot be direct measure of the probability of finding a particle around position $\vec{r}$ at time t. Because if that is the case, it should be positive and real quantity, but in general ψ is a complex function.

In physical optics though the interference patterns are produced by the superposition of the waves of E and H but the intensity of fringes is measured by E_0^2 (square of the amplitude of E), truly speaking by $(E + E)^* (E + E) = 4|E|^2 = 4E_0^2$. In analogy to this situation, Born assumed that the positive quantity $\psi^* \psi dx\, dy\, dz = |\psi(\vec{r}, t)|^2 d\tau$ measures the probability of finding the particle in the given volume element $d\tau = dx\, dy\, dz$ about the point $\vec{r}$ at time t (called Born interpretation[2], given by him in 1926). $|\psi|^2$ is called position probability density.

In case $|\psi(x, y, z, t)|^2\, dx\, dy\, dz = \frac{1}{2}$, suppose. Then according to the assumption, the probability of finding the particle around the point (x, y, z) in volume element $dx\, dy\, dz$ at time t is $\frac{1}{2}$. This, however, does not mean that the volume has half the particle. The physical meaning of the relation $|\psi(x, y, z, t)|^2\, dx\, dy\, dz = \frac{1}{2}$ is that if the experiment is performed a large number of times, say p, then number of times a particle will be found in the volume element $dx\, dy\, dz$ around (x, y, z) at time t is $p/2$. At a point where $\psi = 0$, the probability of finding the particle is zero, and at the point where $|\psi|^2$ is non-zero, the probability of finding the particle is non-zero.

For a system composed of only one particle, the total probability of finding the particle anywhere in the space at time t of course must be equal to 1 (or the particle is to be found somewhere). Thus we must have

$$\int_{-\infty}^{+\infty}\int_{-\infty}^{+\infty}\int_{-\infty}^{+\infty} |\psi(\vec{r}, t)|^2\, dx\, dy\, dz = 1 \qquad \dots (21)$$

where the integration is over the entire space.

A wave function which satisfies the above condition is called a normalized wave function and the said condition is called normalization condition.

The wave function ψ for which $\int |\psi|^2\, d\tau$ exists is also often called square or quadratically integrable.

In most cases given an unnormalized wave function ψ_u, it is simple matter to normalize it by multiplying it by a normalization factor N. The normalized function is $\psi_N = N\psi_u$, so that on substitution of $\psi = \psi_N$ in eqn. (21), yields

[2]M.Born, Z. Physik 37, 863 (1926); 38, 803 (1926).

$$N^2 \iiint |\psi_u|^2 \, d\tau = 1 \qquad \text{... (22)}$$

N can be found out.

2.5 BOX NORMALIZATION – NON-NORMALIZABLE WAVE FUNCTIONS

There exist wave functions of physical interest which cannot be normalized. For example, if $\psi_u = e^{i(kx-\omega t)}$, then $\int_{-\infty}^{+\infty} |\psi_u|^2 \, dx$ goes to infinity. Thus by applying condition (22), we find that N goes to zero and nothing remains to work with. However, we can use a wave function of this kind if we assume that ψ_u equals zero outside some large but finite region so that N is small but not zero. This way the integral,

$$\int |\psi_u|^2 \, d\tau$$

inside the large but finite extent is finite and when multiplied with N^2 is made to give 1. For example, consider a box to be a cube with edges of length L parallel to x, y, z axes, the wave functions

$$\psi_{\vec{k}}(\vec{r}) = L^{-3/2} e^{i\vec{k}\cdot\vec{r}} \qquad \text{... (23)}$$

are clearly normalized within the box (normalization constant is taken to be $L^{-3/2}$ as

$$\iiint_{-L/2}^{+L/2} |\psi_{\vec{k}}(\vec{r})|^2 \, dx\,dy\,dz = \iiint_{-L/2}^{+L/2} \frac{dx\,dy\,dz}{L^3} = 1) \qquad \text{... (24)}$$

Note that the wave functions, are considered at some instant of time, say $t = 0$. Use the normalized wave function (23) for calculations; after completion of the calculations pass on to the limit of infinite volume for the box, that is, $L \to \infty$ such that there is no L dependence. This procedure is called box normalization.

Let us describe the procedure in little more detail. Instead of the boundary conditions, $\varphi(x) = \varphi(y) = \varphi(z) = 0$ at the walls, one usually takes the periodic boundary conditions with period L. That means if x or y or z is increased by L, the wave function should remain unchanged. Thus

$$\psi(x, y, z) = \psi(x + L, y, z) = \psi(x, y + L, z) = \psi(x, y, z + L) \qquad \text{... (25)}$$

which implies (by substituting (23) into (25)).

$$e^{ik_xL} = e^{ik_yL} = e^{ik_zL} = 1 \qquad \text{... (26)}$$

or

$$k_x = \frac{2\pi}{L} n_x, \; k_y = \frac{2\pi}{L} n_y, \; k_z = \frac{2\pi}{L} n_z \qquad \text{... (27)}$$

where n_x, n_y and n_z are positive or negative integers.

The boundary conditions (25) thus reduce to the requirement that k runs through a discrete set of values (or is quantized), determined by conditions (27). When passing to the limit $L \to \infty$, the distance between two neighbouring values of k tends to zero. Similarly the spacing between the energy eigen values

$$E = \frac{\hbar^2 k^2}{2m} = \frac{\hbar^2}{2mL^2}(n_x^2 + n_y^2 + n_z^2)$$

can be made small by making L large and we return to the free motion of a particle in unbounded space.

2.6 CONSERVATION OF THE TOTAL PROBABILITY

The statement depicted by eqn. (21) amounts to the fact that the particle is sure to be found somewhere in space. This has to be true at all times as long as the particle is stable one which cannot decay or disappear in some other way. Therefore, the total probability must be conserved or should be independent of time, that is

$$\partial/\partial t \int |\psi|^2 \, d\tau = 0 \qquad \text{... (28)}$$

This can be shown in the following manner:

Let us write the Schrödinger equation

$$i\hbar \, \partial\psi/\partial t = -\frac{\hbar^2}{2m} \nabla\psi + V\psi \qquad \text{... (20)}$$

along with its complex conjugate

$$-i\hbar \, \partial\psi^*/\partial t = -\frac{\hbar^2}{2m} \nabla^2\psi^* + V\psi^* \qquad \text{... (29)}$$

where $V(\vec{r}, t)$ is assumed to be real.

Multiplying (20) by ψ^* from the left and (29) by ψ from the right and then subtracting the resulting equations, one gets

$$i\hbar \, (\psi^* \partial\psi/\partial t + \partial\psi^*/\partial t \, \psi) = -\frac{\hbar^2}{2m} \{\psi^* \nabla^2 \psi - (\nabla^2 \psi^*)\psi\}$$

$$= -\frac{\hbar^2}{2m} \vec{\nabla} \cdot \{\psi^* \vec{\nabla}\psi - (\vec{\nabla}\psi^*)\psi\} \qquad \text{... (30)}$$

using vector algebra

The above equation can be written as

$$\partial\rho(\vec{r}, t)/\partial t + \vec{\nabla} \cdot \vec{j}(\vec{r}, t) = 0 \qquad \text{... (31)}$$

where

$$\rho(\vec{r}, t) = \psi^*(\vec{r}, t)\,\psi(\vec{r}, t) \qquad \text{... (32a)}$$

and

$$\vec{j}(\vec{r}, t) = \hbar/2im \, [\psi^* \vec{\nabla}\psi - (\vec{\nabla}\psi^*)\psi] \qquad \text{... (32b)}$$

Eqn. (31) is known as the equation of continuity in fluid dynamics in which ρ represents the density of fluid or number of particles per unit volume and $\vec{j}$, the current density. It is therefore reasonable to interpret ρ given by eqn. (32*a*) as the position probability density in the sense that $\psi^*\psi\, d\tau$ will be the probability of finding the particle in a volume element $d\tau$ and $\vec{j}$ given by eqn. (32*b*) as the probability current density which can be written as

$$\vec{j} = 1/2\,[\psi^*\,\hbar/im\,\vec{\nabla}\psi - \psi\,\hbar/im\,\vec{\nabla}\psi^*]$$

$$= 1/2\,[\psi^*\,\vec{p}_{op}/m\,\psi - \psi\,\vec{p}^{\,*}_{op}/m\,\psi^*]$$

$$= \mathrm{Re}\,[\psi^*\,\vec{p}_{op}/m\,\psi]$$

$$= \mathrm{Re}\,[\psi^*\,\vec{v}_{op}\,\psi] \quad \text{... (33)}$$

where $\vec{v}_{op}$ represents the velocity operator and *Re* stands for the real part of the quantity inside the bracket.

From equn. (32*a*) and (33), one gets

$$\vec{j} = \rho\vec{v}_{op} \quad \text{... (34)}$$

which is same as the expression in the case of fluid dynamics and hence the name current density for $\vec{j}$ is justified.

Integrating eqn. (31) over any arbitrary volume Ω, we get

$$\partial/\partial t\int_\Omega \psi^*\,\psi\,d\tau = -\int_\Omega \vec{\nabla}\cdot\vec{j}\,d\tau = -\int_A \vec{j}\cdot d\vec{s} = -\int_A \vec{j}\,.\,\vec{n}\,ds$$

$$= -\int_A j_n\,ds \quad \text{... (35)}$$

where we have used the Gauss's theorem and replaced volume integral by the surface integral. A is the bounding surface of the region of integration, $\vec{n}$ denotes the unit vector along the outward drawn normal to the surface and j_n the normal component of $\vec{j}$.

In the case of a wave packet for which ψ vanishes at great distances, the surface integral is evidently zero where Ω is the entire space ($\vec{j}$ vanishes as ψ vanishes at infinite bounding surface). Thus from eqn. (35),

$$\partial/\partial t\int |\psi|^2 d\tau = 0 \quad \text{... (28)}$$

which essentially is condition for the conservation of total probability and which also tells us that the wave equation guarantees conservation of normalization: If ψ was normalized at $t = 0$, it would remain normalized at all times.

2.7 EXPECTATION VALUES OF DYNAMICAL VARIABLES

Any physical quantity which can be experimentally measured is referred to as the physical observable or observable (also called dynamical variable). For example, position, momentum, and energy are all observables.

Let us consider a number of particles distributed along the *x*-axis such that there are N_1 particles at x_1, N_2 particles at x_2 and so on, The average position of number of particles (which is nothing but the centre of mass of the distribution) would be given by

$$\bar{x} = (N_1x_1 + N_2x_2 + N_3x_3 + \ldots)/(N_1 + N_2 + N_3 + \ldots)$$

$$= \sum_i N_i x_i / \sum_i N_i \quad \text{... (36)}$$

In quantum mechanics where behaviour of a single particle is involved, let us calculate expectation (average or mean) value denoted by $\langle x \rangle$ of the position of particle confined to the x axis that is described by the wave function $\psi(x, t)$. In this case we must replace the number N_i of particles at x_i by the probability P_i that the particle be found in an interval dx at x_i. This probability is given by

$$P_i = \psi_i^* \psi_i \, dx = |\psi_i|^2 \, dx$$

where ψ_i is the particle wave function evaluated at $x = x_i$. Making this substitution and changing the summations in eqn. (36) into integrations (probability being a continuous variable), we find that the expectation value of the position of the single particle (that is average of the large number of observations made on the position of particle) is

$$\langle x \rangle = \int_{-\infty}^{+\infty} \psi^* x \psi \, dx \Big/ \int_{-\infty}^{+\infty} \psi^* \psi \, dx \quad \text{... (37)}$$

If ψ is a normalized wave function, the denominator of the above equation is equal to the probability that the particle exists somewhere between $x = -\infty$ and $x = +\infty$, and therefore has the value 1. Thus

$$\langle x \rangle = \int_{-\infty}^{+\infty} \psi^* x \psi \, dx \quad \text{... (38)}$$

Similarly we can write for the expectation value of the position vector, or, simply, for the centre of the wave packet,

$$\langle \vec{r} \rangle = \int \psi^* \vec{r} \, \psi \, d\tau \quad \text{... (39)}$$

In general the quantity appearing in place of x will be an operator and why $\langle x \rangle$ is written in the typical form (38) would soon become clear

Similar to eqn. (39), the expectation value of the potential energy $V(\vec{r}, t)$ would be given by

$$\langle V(\vec{r}, t) \rangle = \int \psi^* V(\vec{r}, t) \psi \, d\tau \quad \text{... (40)}$$

Now since, $E = p^2/2m + V$, we can write

$$\langle E \rangle = \langle p^2/2m \rangle + \langle V \rangle$$

which in terms of the differential operators, $\vec{p} \equiv -i\hbar\vec{\nabla}$ and $E \equiv i\hbar \, \partial/\partial t$ reads,

$$\langle i\hbar \, \partial/\partial t \rangle = \langle -\hbar^2/2m \, \nabla^2 \rangle + \langle V \rangle \quad \text{... (41)}$$

Multiplying eqn. (20), that is the Schrödinger equation from the left by ψ^* and integrating over all space, one obtains,

$$\int \psi^* i\hbar \, \partial\psi/\partial t \, d\tau = \int \psi^* (-\hbar^2/2m \, \nabla^2 \psi) \, d\tau + \int \psi^* V \psi \, d\tau \quad \text{... (42)}$$

The last term in this equation is simply $\langle V \rangle$. Thus equn. (40) to (42) would be consistent only if we assume that the method of finding expectation value of a quantity is to replace it by its operator representation, operate it on ψ and multiply from the left by ψ^* and then integrate. Thus we identify the expectation value of energy E and the kinetic energy $T = p^2/2m$ by

$$\langle E \rangle = \int \psi^* (i\hbar\, \partial\psi/\partial t)\, d\tau \qquad \text{... (43)}$$

and

$$\langle p^2/2m \rangle = \int \psi^* (-\hbar^2/2m\, \nabla^2 \psi)\, d\tau \qquad \text{... (44)}$$

2.8 EHRENFEST'S THEOREM[3]

The principle of correspondence between the motion of a wave packet and the motion of a classical particle has been of fundamental significance in the development of quantum mechanics. It will now be shown that the classical equations of motion (or Newton's laws of motion), that is

$$d\vec{r}/dt = \vec{p}/m \qquad \text{... (45a)}$$

$$d\vec{p}/dt = -\vec{\nabla} V \qquad \text{... (45b)}$$

are satisfied exactly by the average motion of a wave packet described by a wave function ψ which is a solution of the Schrödinger equation. Assuming that $\psi(\vec{r}, t)$ is a normalized wave function, the rate of change of the expectation or average value of x for the wave packet (by $\langle \vec{r} \rangle$, we mean a set of three time-dependent numbers, $\langle x \rangle, \langle y \rangle$ and $\langle z \rangle$ and we call the point $\langle \vec{r} \rangle (t)$ the centre of the wave packet (the centre and maximum of a wave packet coincide, if the wave packet has a symmetrical shape) is given by

$$d/dt \langle x \rangle (t) = d/dt \int \psi^* (\vec{r}, t)\, x\, \psi(\vec{r}, t)\, d\tau$$

$$= \int \psi^* x\, \partial\psi/\partial t\, d\tau + \int \partial\psi^*/\partial t\, x\, \psi\, d\tau \qquad \text{... (46)}$$

One may note that the dependence of $\langle x \rangle$ on time comes in through the time dependence of ψ^* and ψ (see r.h.s. of (46), the integration is over space and hence the time dependence is left).

Substituting the values of $\partial\psi/\partial t$ and $\partial\psi^*/\partial t$ from equn. (20) and (29), one gets

$$d/dt \langle x \rangle = -i/\hbar \int \psi^* x (-\hbar^2/2m\, \nabla^2\psi + V\psi)\, d\tau$$

$$+ i/\hbar \int (-\hbar^2/2m\, \nabla^2\psi^* + V\psi^*) x\, \psi\, d\tau$$

$$= i\hbar/2m \int [\psi^* x \nabla^2\psi - (\nabla^2\psi^*) x\psi]\, d\tau \qquad \text{... (47)}$$

Using the vector identity

$$\vec{\nabla} \cdot (\vec{\nabla}\psi^* x\psi) = (\nabla^2\psi^*) x\, \psi + \vec{\nabla}\psi^* \cdot \vec{\nabla}(x\psi),$$

[3] P. Ehrenfest, Z. Physik 45, 455 (1927)

the second part of the r.h.s. of eqn. (47) can be written as

$$\int (\nabla^2 \psi^*)\, x\psi \, d\tau = \int \vec{\nabla} \cdot (\vec{\nabla}\psi^* x\psi)\, d\tau - \int \vec{\nabla}\psi^* \cdot \vec{\nabla}(x\psi)\, d\tau$$

which on using the Gauss's theorem becomes

$$\int (\nabla^2 \psi^*)\, x\psi \, d\tau = \int_A (\vec{\nabla}\psi^* x\psi) \cdot d\vec{s} - \int \vec{\nabla}\psi^* \cdot \vec{\nabla}(x\psi)\, d\tau$$

The surface integral over the infinite bounding surface is zero because a wave packet, $\psi(\vec{r}, t)$ vanishes at large distances. Consequently from above

$$\int (\nabla^2 \psi^*)\, x\psi \, d\tau = -\int \vec{\nabla}\psi^* \cdot \vec{\nabla}(x\psi)\, d\tau \qquad \text{... (48)}$$

Again using the vector identity,

$$\vec{\nabla} \cdot [\psi^* \vec{\nabla}(x\psi)] = \vec{\nabla}\psi^* \cdot \vec{\nabla}(x\psi) + \psi^* \nabla^2 (x\psi^*)$$

the above equation reduces to

$$\int (\nabla^2 \psi^*) x\psi \, d\tau = -\int \vec{\nabla} \cdot [\psi^* \vec{\nabla}(x\psi)]\, d\tau + \int \psi^* \nabla^2 (x\psi)\, d\tau$$

Once again the first integral on the r.h.s. of the above equation can be transformed to a surface integral using Gauss's theorem (with bounding surface at infinity) which would vanish because of the similar reasoning given above. Thus

$$\int (\nabla^2 \psi^*)\, x\psi \, d\tau = \int \psi^* \nabla^2 (x\psi)\, d\tau \qquad \text{... (49)}$$

Substituting above into eqn. (47), we obtain

$$d/dt \langle x \rangle = i\hbar/2m \int [\psi^* \{x\nabla^2 \psi - \nabla^2 (x\psi)\}]\, d\tau \qquad \text{... (50)}$$

Now

$$\begin{aligned} \nabla^2 (x\psi) &= \partial^2/\partial x^2 (x\psi) + \partial^2/\partial y^2 (x\psi) + \partial^2/\partial z^2 (x\psi) \\ &= \partial/\partial x\, \partial/\partial x\, (x\psi) + \partial/\partial y\, \partial/\partial y\, (x\psi) + \partial/\partial z\, \partial/\partial z\, (x\psi) \\ &= x\,\nabla^2 \psi + 2\partial\psi/\partial x \end{aligned}$$

Substituting above into eqn. (50), one gets

$$\begin{aligned} d/dt \langle x \rangle &= -i\hbar/m \int \psi^* \partial\psi/\partial x\, d\tau \\ &= -1/m \langle p_x \rangle \end{aligned}$$

or more generally

$$d/dt \langle \vec{r} \rangle = 1/m \langle \vec{p} \rangle \qquad \text{... (51)}$$

In the similar way

$$\begin{aligned} d/dt \langle p_x \rangle \langle t \rangle &= -i\hbar\, d/dt \int \psi^* \partial/\partial x\, \psi\, d\tau \\ &= -i\hbar \left[\int \psi^* \partial/\partial x\, \partial\psi/\partial t\, d\tau + \int \partial\psi^*/\partial t\, \partial\psi/\partial x\, d\tau\right] \\ &= -\left[\int \psi^* \partial/\partial x\, (-\hbar^2/2m\, \nabla^2\psi + V\psi)\, d\tau \right. \\ &\quad \left. - \int (-\hbar^2/2m\, \nabla^2\psi^* + V\psi^*)\, \partial\psi/\partial x\, d\tau\right] \\ &= \hbar^2/2m \int [\psi^* \nabla^2 (\partial\psi/\partial x) - (\nabla^2\psi^*)\, \partial\psi/\partial x]\, d\tau \\ &\quad + \int [V\psi^* (\partial\psi/\partial x) - \psi^* \partial/\partial x\, (V\psi)]\, d\tau \end{aligned}$$

The first integral on the r.h.s. of the above equation containing Laplacians (∇^2) can be transformed into a surface integral by Green's theorem,

$$\int_{\Omega} (u\,\nabla^2 v - v\nabla^2 u)\, d\tau = \int_{A} (u\,\vec{\nabla}\, v - v\,\vec{\nabla}\, u)\cdot d\vec{s}$$

with bounding surface at infinity and would vanish because of the reasons given above. Thus we are left with

$$\begin{aligned} d/dt\,\langle p_x\rangle &= \int [V\psi^*\,\partial\psi/\partial x - \psi^*\,\partial/\partial x\,(V\psi)]\,d\tau \\ &= \int \psi^*\,\partial V/\partial x\,\psi d\tau \\ &= \langle -\partial V/\partial x\rangle = \langle F_x\rangle \end{aligned}$$

or more generally,

$$d/dt\,\langle \vec{p}\,\rangle = \langle -\vec{\nabla}V(\vec{r},t)\rangle = \langle \vec{F}\,\rangle \qquad \text{... (52)}$$

Equn. (51) and (52) constitute what is known as Ehrenfest's theorem. They provide an example of the correspondence principle since they show that a wave packet moves like a classical particle whenever the expectation values give a good approximation of the classical variables.

2.9 ADMISSIBILITY CONDITIONS ON $\psi(\vec{r}, t)$

Let us try to see as to what are the requirements which ψ must meet such that it can serve as a wave function for a particle in quantum mechanics. We require that

ψ should be finite everywhere ... (53)

Finiteness is needed in order that $|\psi|^2\, d\tau$, for an infinitesimal volume element $d\tau$, should lie between 0 and 1. (If ψ were infinite anywhere, then particle would be localized at that point (that is, there would be no uncertainty in the position of particle), its momentum would then be completely uncertain and the Schrödinger equation would become irrelevant)

ψ should be single valued ... (54)

Single valuedness is required such that the probability density $|\psi|^2$ may be uniquely defined or can have only one value at a particular place and time.

ψ and $\partial\psi/\partial\vec{r}$ (that is, $\partial\psi/\partial y$ and $\partial\psi/\partial z$), should be continuous function of $\vec{r}$ for all $\vec{r}$ ... (55)

Discontinuity in ψ will produce discontinuity in $|\psi|^2\, d\tau$, that is, in the probability of finding the particle in the volume element $d\tau$. But the probability should vary continuously from point to point if no particles are created or destroyed. $\partial\psi/\partial\vec{r}$ (or $\vec{\nabla}\,\psi$) will be related to particle's momentum which must be continuous function of $\vec{r}$ (except at $V = \infty$) so long as V is continuous.

It may be pointed out here that though the above statements are made for $\psi(\vec{r}, t)$ but they apply as well to $\varphi(\vec{r}\,)$. Truly speaking these conditions can be looked upon as statements, like the Schrödinger equation itself, to be tested by experience. These conditions enable us to select the one actual solution to a physical problem from a number of possible solutions to the Schrödinger equation.

2.10 SEPARATION OF VARIABLES; STATIONARY STATES

The time dependent Schrödinger equation in the case in which the potential energy is not a function of time or is represented by $V(\vec{r})$ would be given by

$$i\hbar\, \partial\psi(\vec{r}, t)/\partial t = -\hbar^2/2m\, \nabla^2\psi(\vec{r}, t) + V(\vec{r})\,\psi(\vec{r}, t) \quad \text{... (56)}$$

Let us see if there exist solutions of the above equation of the form,

$$\psi(\vec{r}, t) = \varphi(\vec{r})\, f(t) \quad \text{... (57)}$$

Substituting above in eqn. (56) and dividing both sides of the resulting equation by the product $\varphi(\vec{r})\, f(t)$, we obtain

$$i\hbar/f(t)\, df(t) = 1/\varphi(\vec{r})[-\hbar^2/2m\, \nabla^2\varphi(\vec{r}) + V(\vec{r})\,\varphi(\vec{r})] \quad \text{... (58)}$$

Since the left-hand side of the above equation is a function of time alone while the right-hand side is a function of space alone, this equality is only possible if each of these functions is equal to the same separation constant, say E.

Setting the left-hand side equal to E, we find

$$i\hbar/f(t)\, df(t)/dt = E$$

or

$$df(t)/f(t) = -i/\hbar\, E\, dt$$

Integrating both the sides, one gets

$$ln\, f(t) = -i/\hbar\, Et + ln\, f(0)$$

or

$$f(t) = f(0)\, e^{-i/\hbar^{Et}} \quad \text{... (59)}$$

Let us take $f(0) = 1$ (or $\ln f(0) = 0$). In that case no loss in generality occurs because $f(t)$ will always be used in conjuction with $\varphi(\vec{r})$ to give

$$\psi(\vec{r}, t) = \varphi(\vec{r})\, f(t)$$

and since $\varphi(\vec{r})$ will contain an arbitrary constant say A as multiplying factor. Hence

$$f(t) = e^{-i/\hbar^{Et}} \quad \text{... (60)}$$

In the similar way, $\varphi(\vec{r})$ mast satisfy the equation,

$$-\hbar^2/2m\, \nabla^2\varphi(\vec{r}) + V(\vec{r})\,\varphi(\vec{r}) = E\varphi(\vec{r}) \quad \text{... (61)}$$

which is called time independent Schrödinger equation.

The solution of Schrödinger eqn. (56) can thus be written as

$$\psi(\vec{r}, t) = \varphi(\vec{r})\, e^{-i/\hbar^{Et}} \quad \text{... (62)}$$

with the condition that $\varphi(\vec{r})$ is a solution of (61). The time and space variables are thus said to have been separated. The discussion below would show that the separation constant E introduced earlier is nothing but energy of the particle. If E represents energy of the particle,

$$
\begin{aligned}
E\,\psi(\vec{r},t) &= i\hbar\,\partial\psi(\vec{r},t)/\partial t \\
&= i\hbar\,\partial\{\varphi(\vec{r})\,e^{-i/\hbar^{Et}}\}/\partial t \\
&= E\,\psi(\vec{r},t)
\end{aligned}
$$

The position probability density ρ in the present case using eqn. (62) is given by

$$\rho = |\psi(\vec{r},t)|^2 = |\varphi(\vec{r})|^2 \qquad \text{... (63)}$$

which shows that at every point $\vec{r}$ in space, it is independent of time. Similarly the expectation values of all dynamical variables are also evidently time independent. For this very reason wave functions of the form (62) are said to represent stationary states (the name stationary state is used because none of the particle characteristics change with time) or a system described by a wave function (62) is in an stationary state.

We can also express the time-independent Schrödinger eqn. (61) in an operator form as

$$[-\hbar^2/2m\,\nabla^2 + V(\vec{r})]\,\varphi(\vec{r}) = E\varphi(\vec{r}) \qquad \text{... (61a)}$$

or

$$H\,\varphi(\vec{r}) = E\,\varphi(\vec{r}) \qquad \text{... (64)}$$

where H is the differential operator,

$$H = -\,\hbar^2/2m\,\nabla^2 + V(\vec{r}) \qquad \text{... (65)}$$

and is called Hamiltonian operator.

Eqn. (64) is thus the eigen value equation where E represents the energy eigen value of the operator H and $\varphi(\vec{r})$ is called eigen function belonging to the eigen value E of the operator H.

Before we proceed further let us see, if we assign an arbitrary real value to E in eqn. (64), whether there exist a corresponding eigen function $\varphi(\vec{r})$?. The answer is truly speaking negative. This is because of the fact that as soon as we impose the admissibility conditions of sec 2.8 on $\varphi(\vec{r})$, that is on the solution of eqn. (64), the outcome is that E is restricted to certain specific values only. (This will be seen in the case of a particle in a square well potential). Only these special values are to be considered as eigen values. The set of all such allowed or admissible values of E form what is called the energy spectrum or are called energy levels of the system. Thus the existence of stationary states with discrete energies which were postulated by Bohr in order to account for the nature of atomic spectra, find a rational explanation on the basis of quantum mechanics.

2.11 SUPERPOSITION OF STATIONARY STATES

In order to distinguish between the various possible values of energy E and the corresponding eigen functions $\varphi(\vec{r})$, we label them with the index n. Thus we have,

$$H\varphi_n(\vec{r}) = E_n\varphi_n(\vec{r}) \quad \text{... (66)}$$

and the stationery states of the particle are represented by the wave functions,

$$\psi_n(\vec{r}, t) = \varphi_n(\vec{r})\, e^{-iE_n t/\hbar} \quad \text{... (67)}$$

where $\psi_n(\vec{r}, t)$ is a solution of the Schrödinger eqn. (56). The equation being linear in $\psi(\vec{r}, t)$ as the terms like ψ^2, ψ^3 do not appear, if $\psi_1(\vec{r}, t)$ and $\psi_2(\vec{r}, t)$ are two possible solutions of eqn. (56), then any linear combination.

$$\psi(\vec{r}, t) = a_1\psi_1(\vec{r}, t) + a_2\psi_2(\vec{r}, t) \quad \text{... (68)}$$

where the coefficients a_1 and a_2 are any arbitrary complex constants, is also a possible solution. This expresses the principle of superposition (Indeed this principle of superposition played a basic role in the discussion of wave packet in section 2.1). By a continued superposition, one can build a new $\psi(\vec{r}, t)$, *i.e.*

$$\psi(\vec{r}, t) = a_1\psi_1 + a_2\psi_2 + \ldots + a_n\psi_n \quad \text{... (69)}$$

$$= \sum_n a_n\psi_n(\vec{r}, t)$$

In other words, it simply means that due to the linearity of the Schrödinger equation, it has whole series of other solutions of the form

$$\psi(\vec{r}, t) = \sum_n a_n\psi_n(\vec{r}, t) = \sum_n a_n\varphi_n(\vec{r})\, e^{-iE_n t/\hbar} \quad \text{... (70)}$$

Thus we reach the conclusion that if the system is in some arbitrary state with wave function $\psi(\vec{r}, t)$, then it can be expanded in terms of the eigen functions of Hamiltonian H. A set of wave functions in terms of which such an expansion can be made is called a complete (or closed) set. From eqn. (70) after putting $t = 0$, one gets.

$$\psi(\vec{r}, 0) = \varphi(\vec{r}) = \sum_n a_n\varphi_n(\vec{r}) \quad \text{... (71)}$$

which simply means that any function $\varphi(\vec{r})$ can be expanded or decomposed in terms of the eigen functions $\psi_n(\vec{r})$ of the Hamiltonian.

2.12 DEGENERACY

If the Schrödinger equation has more than one linearly independent solutions (wave functions) for the same energy level E, the energy level is said to be degenerate. If there exists only one eigen function belonging to the given eigen value, the eigen value or state is said to be non-degenerate.

Suppose there are g linearly independent wave functions $\varphi_1(\vec{r}), \varphi_2(\vec{r}), \ldots, \varphi_g(\vec{r})$, belonging to the same energy state, then the energy level is said to be g fold degenerate (or degree of degeneracy of the state is g).

It can be shown that any linear combination of the eigen functions corresponding to the degenerate state, that is

$$\varphi(\vec{r}) = c_1\varphi_1(\vec{r}) + c_2\varphi_2(\vec{r}) + \ldots + c_g\varphi_g(\vec{r}) \qquad \ldots (72)$$

is also an eigen function belonging to the same energy eigen value. We will prove this for $g = 2$. The result can then be generalized.

Suppose $\varphi_1(\vec{r})$, and $\varphi_2(\vec{r})$ are eigen functions belonging to the same eigen value E_1,μ we then have from eqn. (61),

$$-\hbar^2/2m\,\nabla^2\,\varphi_1(\vec{r}) + V(\vec{r})\,\varphi_1(\vec{r}) = E_1\,\varphi_1(\vec{r}) \qquad \ldots (73a)$$

and

$$-\hbar^2/2m\,\nabla^2\,\varphi_2(\vec{r}) + V(\vec{r}) = E_1\,\varphi_2(\vec{r}) \qquad \ldots (73b)$$

Multiplying (73*a*) by c_1, (73*b*) by c_2 and then adding, we get

$$-\hbar^2/2m\,\nabla^2\,(c_1\varphi_1 + c_2\varphi_2) + V(c_1\varphi_1 + c_2\varphi_2) = E_1\,(c_1\varphi_1 + c_2\varphi_2) \qquad \ldots (74)$$

which shows that the linear combination $c_1\varphi_1 + c_2\varphi_2$ is also an eigen function belonging to the same eigen value E_1.

2.13 ORTHONORMALITY OF EIGEN FUNCTIONS

From eqn. (71), the complex conjugate of $\varphi(\vec{r})$, that is $\varphi^*(\vec{r})$, would be given by

$$\varphi^*(\vec{r}) = \sum a_n^*\,\varphi_n^*(\vec{r}) \qquad \ldots (75)$$

Multiplying above by $\varphi(\vec{r})$, and integrating, we obtain

$$\int \varphi^*(\vec{r})\,\varphi(\vec{r})\,d\tau = \sum a_n^* \int \varphi_n^*(\vec{r})\,\varphi(\vec{r})\,d\tau \qquad \ldots (76)$$

Multiplying (71) and (75) and then integrating, we get

$$\int \varphi^*\varphi\,d\tau = \sum a_n^*\,a_n \qquad \ldots (77)$$

where we have supposed that each of the wave functions φ_n is normalized, that is,

$$\int \varphi_n^*\,\varphi_n\,d\tau = 1.$$

Comparing (76) with (77), we have

$$\sum a_n^* a_n = \int \varphi^*\varphi\,d\tau = \sum a_n^* \int \varphi_n^*\,\varphi\,d\tau$$

from which we derive the following formula determining the coefficients a_n in the expansion of the function φ in terms of the eigen functions φ_n.

$$a_n = \int \varphi_n^*\,\varphi\,d\tau \qquad \ldots (78)$$

Substituting φ from (71) into above, we find

$$a_n = \Sigma a_m \int \varphi_n^* \varphi_m \, d\tau$$

from which it is clear that the eigen functions must satisfy the condition

$$\int \varphi_n^* \varphi_m \, d\tau = \delta_{nm} \quad \text{... (79)}$$

where δ_{nm} is the kronecker delta function defined as

$$\delta_{nm} = 1 \text{ for } n = m$$

$$= 0 \text{ for } n \neq m$$

The fact that the integrals of the products $\varphi_n^* \varphi_m$ with $n \neq m$ vanish is called the orthogonality of the functions φ_n. Eqn. (79) is known as the orthonormality condition. Thus the set of eigen-functions φ_n forms a complete set of normalized and orthogonal (or orthonormal) functions.

2.14 CLOSURE RELATION

Any set of functions $\varphi_n(\vec{r})$ which is complete and orthonormal, *i.e.*, for which equn. (71) and (79) are valid has the property of closure (or satisfies the following closure relation),

$$\sum_n \varphi_n(\vec{r})\, \varphi_n^*(\vec{r}') = \delta(\vec{r} - \vec{r}') \quad \text{... (80)}$$

where $\delta(\vec{r} - \vec{r}')$ is the Dirac delta function.

Let us try to prove the above relation. From eqn. (78) the coefficients in eqn. (71) are given by

$$a_n = \int \varphi_n^* \varphi \, d\tau$$

Substituting above back into eqn. (71), we get

$$\varphi(\vec{r}) = \sum_n a_n \varphi_n(\vec{r}) = \sum_n \int [\varphi_n^*(r')\, \varphi(\vec{r}')\, d\tau']\, \varphi_n(\vec{r}) \quad \text{... (81)}$$

Interchanging $\sum_n$ *and* $\int d\tau'$, we obtain

$$\varphi(\vec{r}) = \int d\tau' \;\; \varphi(\vec{r}')\, [\Sigma \varphi_n(\vec{r})\, \varphi_n^*(\vec{r}')] \quad \text{... (82)}$$

The above equation indicates that the value of the integral is just the value, at the particular point $\vec{r} = \vec{r}'$ of the function $\varphi(\vec{r}')$ appearing in the integrand. Clearly, the function which multiplies $\varphi(r')$, must be the Dirac delta function. This is what eqn. (80) states.

Reciprocally, if an orthonormal set $\varphi_n(\vec{r})$ satisfies the closure relation, then it obeys eqn. (71). Writing

$$\varphi(\vec{r}) = \int d\tau' \;\; \varphi(\vec{r}')\, \delta(\vec{r} - \vec{r}') \quad \text{... (83)}$$

and substituting the expression

$$\sum_n \varphi(\vec{r})\,\varphi^*(\vec{r}') = \delta(\vec{r}-\vec{r}')$$

for the delta function into above, we obtain eqn. (82).

To return to (81), all we must do is again interchange summation and integration. This equation then expresses the fact that $\varphi(\vec{r})$ can always be expanded in terms of $\varphi_n(\vec{r})$, with a_n given by eqn. (78).

2.15 HERMITIAN OPERATORS. THEIR EIGEN FUNCTIONS AND EIGEN VALUES

By this time we know that with every physical observable, an operator is associated in quantum mechanics. For example, the operator associated with the momentum of the particle is $-i\hbar\vec{\nabla}$ and similarly, the operator associated with the energy of the particle is $i\,\hbar\partial/\partial t$. All these operators share the property of being linear. An operator is said to be linear if its action on any two functions ψ_1 and ψ_2 is such that

$$A\,(c_1\,\psi_1 + c_2\psi_2) = c_1\,(A\,\psi_1) + c_2\,(A\,\psi_2) \qquad \text{... (84)}$$

where c_1 and c_2 are arbitrary complex numbers.

Let us denote by Q the operator associated with the physical observable q. The eigen value equation, then reads

$$Q\psi_n = q_n\,\psi_n \qquad \text{... (85)}$$

where q_n is the eigen value of the operator Q and ψ_n is the eigen function belonging to the eigen value q_n of the operator Q.

The expectation value of Q in the state ψ_n

$$\begin{aligned}\langle Q\rangle &= \int \psi_n^*\,Q\psi_n\,d\tau \\ &= q_n \int \psi_n^*\,\psi_n\,d\tau \\ &= q_n \qquad \text{... (86)}\end{aligned}$$

where we have used the eigen value eqn. (85) and also the fact that q_n is a number and ψ_n is normalized.

Thus, if ψ_n is an eigen function, expectation and eigen values are identical. Let us now ask what is the requirement which a linear operator must possess such that it can be admitted as a candidate for representing a physical observable. Obviously, it is necessary that the expectation value (or eigen value) of such a linear operator, being the average of the measured values of a physical quantity, must be real. In order to have q real for all wave functions ψ which may represent possible states, we must demand that

$$\int (Q\psi_n)^*\,\psi_n\,d\tau = \int \psi_n^*\,Q\,\psi_n\,d\tau \qquad \text{... (87)}$$

This is because of the fact that

$$\int (Q\psi_n)^*\,\psi_n\,d\tau = \int (q_n\,\psi_n)^*\,\psi_n\,d\tau$$

$$= q_n^* \int \psi_n^* \psi_n \, d\tau = q_n^*$$

and

$$\int \psi_n^* Q \psi_n \, d\tau = q_n \int \psi_n^* \psi_n \, d\tau = q_n$$

and hence $q_n = q_n^*$ which implies that q_n must be real. Operators which satisfy condition (87) are called Hermitian.

2.16 SIMULTANEOUS MEASUREMENTS AND COMMUTING OPERATORS

Let us consider two operators A and B such that

$$A\varphi = \alpha\varphi \qquad \text{... (88a)}$$

$$B\varphi = \beta\varphi \qquad \text{... (88b)}$$

which means, φ is a simultaneous eigen function of A and B belonging to the eigen values α and β, respectively. This amounts to the fact that A and B can be simultaneously measured. As an example, the wave function, $\varphi(x) = (2\pi\hbar)^{-1/2} e^{ipx/\hbar}$ satisfies simultaneously the equations

$$-i\hbar\, \partial\varphi(x)/\partial x = p\varphi(x)$$

and

$$-\hbar^2/2m\, \partial^2\varphi(x)/\partial x^2 = E\,\varphi(x), E = p^2/2m$$

where $-i\hbar\partial/\partial x$ is the operator corresponding to the momentum and $-\hbar^2/2m\, \partial^2/\partial x^2$ is the Hamiltonian operator for a free particle.

Let us multiply (88*a*) from the left by B and (88*b*) from the left by A, we obtain

$$BA\,\varphi = B\,(\alpha\varphi) = \alpha\, B\,\varphi = \alpha\,\beta\,\varphi$$

and

$$AB\,\varphi = A\,(\beta\varphi) = \beta\, A\,\varphi = \beta\,\alpha\,\varphi$$

and by subtraction

$$(AB - BA)\,\varphi = (\beta\,\alpha - \alpha\beta)\varphi = O\,\varphi \qquad \text{... (89)}$$

as α and β are numbers. O is a null operator

If this is to hold for all wave functions belonging to the complete set, $AB - BA$ must be equal to the null operator, or

$$AB - BA = [A, B] = 0 \qquad \text{... (90)}$$

where $[A, B]$ is called commutator of A and B. Two operators satisfying the condition (90) are said to commute. Hence, a necessary condition for two observables to be simultaneously measureable is that they commute.

The converse of this theorem is also true, that is, if A and B are two commuting Hermitian operators, their eigen functions can always be constructed in such a way that they are simultaneous eigen functions. The proof of this would be given at a later stage.

2.17 THE ADJOINT OR SELF ADJOINT OF AN OPERATOR

Given an arbitrary, not necessarily Hermitian operator Q, it is useful to define another operator called the adjoint (or Hermitian) adjoint of Q and denoted by $Q^{\dagger}$ (read as Q dagger), by the requirement.

$$\int f^* Q^{\dagger} g \, d\tau = [\int g^* Qf \, d\tau]^* = \int (Qf)^* g \, d\tau \quad \text{... (91)}$$

where f and g are arbitrary functions.

Comparing eqn. (91) with eqn. (87), we find

$$\int f^* Q^{\dagger} g \, d\tau = [\int (Qf)^* g \, d\tau = \int f^* Q \, g \, d\tau \quad \text{... (92)}$$

that is, Q is Hermitian if it is self adjoint,

$$Q^{\dagger} = Q \quad (93)$$

The converse of this statement is also true, that is, if an operator Q is Hermitian,

$$\int (Qf)^* g \, d\tau = \int f^* Q \, g \, d\tau$$

or

$$[\int g^* Qf \, d\tau]^* = \int f^* Q^{\dagger} g \, d\tau$$

where definition of adjoint has been utilized.

Hence

$$\int f^* Q^{\dagger} g \, d\tau = \int f^* Q \, g \, d\tau$$

or

$$Q^{\dagger} = Q$$

that is, it is self adjoint. Hence Hermitian is synonym for self adjoint.

2.18 SIMPLE THEOREMS ABOUT HERMITIAN OPERATORS

1. Identity operator I, which takes every function into itself, is Hermitian.
2. The sum of two Hermitian operators is Hermitian.
3. If α and β are two Hermitian operators, $\alpha\beta - \beta\alpha$ and $\alpha\beta + \beta\alpha$ would be Hermitian even if α, β do not commute.

2.19 UNCERTAINTY PRINCIPLE-EXACT STATEMENT AND PROOF

In order to make an exact statement of the uncertainty principle, let us first of all give a precise definition of the uncertainty. Let A and B be two operators associated with the physical observables and let $\langle A \rangle$ and $\langle B \rangle$ represent their expectation (or average or mean) values in the state ψ. The uncertainty in A is denoted by ΔA and is defined as the root mean square deviation (in analogy with the standard deviation in statistics), that is

$$\Delta A = (\langle (A - \langle A \rangle)^2 \rangle)^{1/2} \quad \text{... (94)}$$

One must note that by $A - \langle A \rangle$, one actually means $A - \langle A \rangle I$, where I is identity operator. Similarly the uncertainty in B is,

$$\Delta B = (<(B - <B>)^2>)^{1/2} \quad \text{... (95)}$$

Using expression for the mean value, one gets

$$(\Delta A)^2 = \langle (A - \langle A \rangle)^2 \rangle = \int \psi^* (A - \langle A \rangle)^2 \psi \, d\tau \quad \text{... (96)}$$

Since the operators A and B are associated with the physical observables, these are Hermitian (as their eigen values should be real). The operators $A - \langle A \rangle I$ and $B - \langle B \rangle I$ will also be Hermitian because the sum of two Hermitian operators is also Hermitian (I being Hermitian). Hence

$$\begin{aligned} (\Delta A)^2 &= \int \psi^* (A - \langle A \rangle)^2 \psi \, d\tau \\ &= \int \psi^* (A - \langle A \rangle)(A - \langle A \rangle) \psi \, d\tau \\ &= \int (A - \langle A \rangle) \psi]^* (A - \langle A \rangle) \psi \, d\tau \end{aligned}$$

as for a Hermitian operator Q,

$$\int \psi^* Q\psi d\tau = \int (Q\psi)^* \psi d\tau$$

or

$$(\Delta A)^2 = \int |(A - \langle A \rangle) \psi|^2 \, d\tau \quad \text{... (97)}$$

Similarly, one can show that

$$(\Delta B)^2 = \int |(B - \langle B \rangle) \psi|^2 \, d\tau \quad \text{... (98)}$$

From equn. (97) and (98), one gets

$$(\Delta A)^2 (\Delta B)^2 = \int |(A - \langle A \rangle) \psi|^2 \, d\tau \int |(B - \langle B \rangle) \psi|^2 \, d\tau \quad \text{... (99)}$$

Making use of the Schwartz inequality, that is

$$\int |f|^2 \, d\tau \int |g|^2 \, d\tau \geq \left| \int f^* g d\tau \right|^2 \quad \text{... (100)}$$

and eqn. (99) with $f = (A - \langle A \rangle) \psi$ and $g = (B - \langle B \rangle) \psi$, one gets

$$(\Delta A)^2 (\Delta B)^2 \geq \left| \int [(A - \langle A \rangle) \psi]^* [(B - \langle B \rangle) \psi] \, d\tau \right|^2 \quad \text{... (101)}$$

The right hand side of the above equation is equal to

$$\left| \int \psi^* (A - \langle A \rangle)(B - \langle B \rangle) \psi \, d\tau \right|^2$$

where we have made use of the fact that the operator $A - <A> I$ is Hermitian. Taking $A - <A> = \alpha$ and $B - <B> = \beta$, the above inequality can be written as

$$(\Delta A)^2 (\Delta B)^2 \geq \left| \int \psi^* \alpha\beta \, \psi \, d\tau \right|^2$$

which can further be written in the form.

$$(\Delta A)^2 (\Delta B)^2 \geq \left| \int \psi^* \frac{1}{2} (\alpha\beta - \beta\alpha) \psi \, d\tau + \int \psi^* \frac{1}{2} (\alpha\beta + \beta\alpha) \psi \, d\tau \right|^2 \quad \text{... (102)}$$

where 1/2 (αβ – βα) and 1/2 (αβ + βα) are Hermitian operators (The product of two Hermitian operators, that is αβ is Hermitian if and only if they commute or that means if they can be measured simultaneously and accurately).

The right hand side of (102) can be written as

$$\left[\int \psi^* \frac{1}{2} (\alpha\beta - \beta\alpha) \psi \, d\tau + \int \psi^* \frac{1}{2} (\alpha\beta + \beta\alpha) \psi \, d\tau \right]^* \times$$

$$\left[\int \psi^* \frac{1}{2} (\alpha\beta - \beta\alpha) \psi \, d\tau + \int \psi^* \frac{1}{2} (\alpha\beta + \beta\alpha) \psi \, d\tau \right]$$

$$= \frac{1}{4} \left| \int \psi^* (\alpha\beta - \beta\alpha) \psi \, d\tau \right|^2 + \frac{1}{4} \left| \int \psi^* (\alpha\beta + \beta\alpha) \psi \, d\tau \right|^2$$

$$+ \frac{1}{4} \left[\int \psi^* (\alpha\beta - \beta\alpha) \psi \, d\tau \right]^* \times \left[\int \psi^* (\alpha\beta + \beta\alpha) \psi \, d\tau \right]$$

$$+ \frac{1}{4} \left[\int \psi^* (\alpha\beta + \beta\alpha) \psi \, d\tau \right]^* \times \left[\int \psi^* (\alpha\beta - \beta\alpha) \psi \, d\tau \right] \quad \text{... (103)}$$

Now

$$\left[\int \psi^* \alpha\beta \psi \, d\tau \right]^* = [(\alpha\psi)^* \beta\psi \, d\tau]^* \quad \text{as } \alpha \text{ is Hermitian}$$

$$= \int (\beta \psi)^* (\alpha \psi) \, d\tau \quad \text{as} \quad (CD)^* = D^* C^*$$

$$= \int \psi^* \beta\alpha \psi \, d\tau \text{ as } \beta \text{ is Hermitian}$$

Similarly

$$\left[\int \psi^* \beta \alpha \psi \, d\tau \right]^* = \int \psi^* \alpha \beta \psi \, d\tau$$

Hence

$$\left[\int \psi^* (\alpha\beta \pm \beta\alpha) \psi \, d\tau \right]^* = \pm [\psi^* (\alpha\beta \pm \beta\alpha) \psi \, d\tau] \quad \text{... (104)}$$

Substituting above into eqn. (103), makes the last two terms of the said equation vanish. As a result eqn. (102) leads to

$$(\Delta A)^2 (\Delta B)^2 \geq \frac{1}{4} \left| \int \psi^* (\alpha\beta - \beta\alpha) \psi \, d\tau \right|^2 + \frac{1}{4} \left| \int \psi^* (\alpha\beta + \beta\alpha) \psi \, d\tau \right|^2 \quad \text{... (105)}$$

One cannot take αβ – βα = 0 as it would mean that α and β commute and hence they can be measured simultaneously and accurately. However, one may take αβ + βα = 0 without any loss of generality. Truly speaking the equality holds good only if the 2nd term on the right hand side of (105) is non-zero. Hence

$$(\Delta A)^2 (\Delta B)^2 \geq \frac{1}{4} \left| \int \psi^* (\alpha\beta - \beta\alpha) \psi \, d\tau \right|^2 \quad \text{... (106)}$$

Chapter 2

Now

$$\alpha\beta - \beta\alpha = (A - \langle A \rangle I)(B - \langle B \rangle I) - (B - \langle B \rangle I)(A - \langle A \rangle I)$$
$$= AB - BA$$

Hence the inequality (106) reduces to

$$(\Delta A)^2 (\Delta B)^2 \geq \frac{1}{4}\left|\int \psi^* (AB - BA)\, \psi \, d\tau\right|^2 \qquad \text{... (107)}$$

Taking $A = x$ and $B = p_x$, the above inequality results in

$$(\Delta x)^2 (\Delta p_x)^2 \geq \frac{1}{4} |\langle x p_x - p_x x \rangle|^2$$

$$\geq \frac{1}{4} |\langle [x, p_x] \rangle|^2$$

where $[x, p_x] = x\,p_x - p_x\, x$ is commutator of x with p_x and is equal to $i\,\hbar$. Hence

$$(\Delta x)^2 (\Delta p_x) \geq \hbar^2/4$$

or

$$\Delta x\, \Delta p_x \geq h/2 \qquad (108)$$

Taking another case, $A = E$ and $B = t$, one obtains from (107),

$$(\Delta E)^2 (\Delta t)^2 \geq \frac{1}{4} |\langle [E, t] \rangle|^2$$

Now $E \equiv i\,\hbar\, \partial/\partial t$ (the operator equivalence of E) and $[i\,\hbar\,\partial/\partial t, t] = i\,\hbar$. Substitution in above leads to

$$(\Delta E)^2 (\Delta t)^2 \geq \hbar^2/4$$

or

$$\Delta E\, \Delta t \geq \hbar/2 \qquad \text{... (109)}$$

PROBLEMS

2.1 Show that the uncertainty product $\Delta x\, \Delta p_x$ is a minimum $(= \hbar/2)$ for a Gaussian wave packet.

2.2 The ground state of a linear harmonic oscillator of angular frequency ω is, represented by

$$\varphi_0(x) = (m\omega/\pi\hbar)^{1/4}\, e^{-m\omega x^2/2\hbar}$$

What are the expectation values of the position and the momentum in the said state?

2.3 For the ground state of a linear harmonic oscillator, the wave function is given by

$$\varphi_0(x) = (\alpha/\pi)^{1/4} e^{-\alpha x^2/2}, \alpha = m\omega/\hbar$$

Show that the value of the uncertainty product,

$$\Delta x \, \Delta p_x = \hbar/2.$$

2.4 For the first excited state of the harmonic oscillator, the wave function is given by,

$$\varphi_1(x) = (\alpha/4\pi)^{1/4} x \, e^{-\alpha x^2/2}, \alpha = mw/\hbar$$

Show that the value of uncertainty product, $\Delta x \, \Delta p_x = 3/2 \, \hbar$.

2.5 The normalized wave functions in the case of a linear harmonic oscillator are given by

$$\varphi_n(x) = (\alpha/\pi)^{1/4} 2^{-n/2} (n!)^{-1/2} H_n(\alpha^{1/2} x) \, e^{-\alpha x^2/2},$$

$$\alpha = m\omega/\hbar$$

Show that the value of the uncertainty product is

$$\Delta x \, \Delta p_x = (n + 1/2) \, \hbar$$

2.6 Show that for a particle in a one dimensional box of width $L(0 - L)$, that is square well with infinitely high walls, the mean square deviation of the particle is

$$L^2/12 \, (1 - 6/n^2 \pi^2)$$

2.7 Using the operator representation of p_x, show that $(x \, p_x - p_x \, x) \, \psi = i\hbar \, \psi$, where Ψ is an arbitrary function.

2.8 The wave function of a particle is given by

$$\varphi(x) = Ne^{-\alpha x^2}, -\infty < x < +\infty$$

where N and α are constants. Calculate the probability of finding the particle in the region $0 < x < \infty$.

2.9 Show that the momentum operator $\vec{p} \to i\hbar \, \vec{\nabla}$ is Hermitian.

2.10 Show that the product of two Hermitian operators F and G is self adjoint if F and G commute.

2.11 Taking wave function of the 1s electron in hydrogen atom as

$$\varphi_0 = (\pi a_0^3)^{-1/2} e^{-r/a_0},$$

show that the average value of $1/r$ for the said electron would be $1/a_0$.

SOLUTIONS

2.1 Consider, in a one dimensional model, a free particle whose wave function at time $t = 0$ is,

$$\psi(x, 0) = \varphi(x) = (2\pi)^{-1/2} \int_{-\infty}^{+\infty} g(k, 0)\, e^{ikx}\, dk \qquad \text{... (2.1)}$$

where

$$(2\pi)^{-1/2} g(k, 0) = a^{1/2} (2\pi)^{-3/4} e^{-a^2(k-k_0)^2/4} \qquad \text{... (2.2)}$$

which correspond to a Gaussian function centered at $k = k_0$ and multiplied by a numerical coefficient which normalizes the wave function. Accordingly the wave packet (2.1) is called Gaussian.

Let us now calculate $\varphi(x)$. To calculate this, the exponent of (2.1) is written as

$$-a^2/4\,(k-k_0)^2 + ikx = -a^2/4\left(k - k_0 - \frac{2\,ix}{a^2}\right)^2 + ik_0x - x^2/a^2 \qquad \text{... (2.3)}$$

and thus

$$\varphi(x) = a^{1/2}(2\pi)^{-3/4} \int_{-\infty}^{+\infty} e^{-a^2(k-k_0-2ix/a^2)^2/4}\, e^{ik_0x}\, e^{-x^2/a^2}\, dk \qquad \text{... (2.4)}$$

Using the integral,

$$\int_{-\infty}^{+\infty} e^{-\alpha^2(\xi+\eta)}\, d\zeta = \pi^{1/2}/\alpha \qquad \text{... (2.5)}$$

(it can actually be shown that the above integral does not depend on η), one gets

$$\varphi(x) = (2/\pi a^2)^{1/4}\, e^{ik_0x}\, e^{-x^2/a^2} \qquad \text{... (2.6)}$$

Now uncertainty in the measurement of x and p_x is

$$\Delta x = [\langle(x - \langle x\rangle)^2\rangle]^{1/2} = [\langle(x^2 + \langle x\rangle^2 - 2x\langle x\rangle^2)\rangle]^{1/2}$$

$$= [\langle x^2\rangle - \langle x\rangle^2]^{1/2} \qquad \text{... (2.7a)}$$

and

$$\Delta p_x = [\langle p_x^2\rangle - \langle p_x\rangle^2]^{1/2} \qquad \text{... (2.7b)}$$

Now
$$\langle x\rangle = \int_{-\infty}^{+\infty} \varphi^*\, x\, \varphi\, dx = (2/\pi a^2)^{1/2} \int_{-\infty}^{+\infty} x^2 e^{-2x^2/a^2}\, dx$$

$$= 0 \qquad \text{... (2.8)}$$

as integrand is an odd function of x.

$$\langle x^2\rangle = \int_{-\infty}^{+\infty} \varphi^*\, x^2 \varphi\, dx = (2/\pi a^2)^{1/2} \int_{-\infty}^{+\infty} x^2 e^{-2x^2/a^2}\, dx$$

$$= (2/\pi a^2)^{1/2} \times \frac{1}{2}[\pi/2/a^2)^3]^{1/2}$$

$$= a^2/4 \qquad \text{... (2.9)}$$

$$\langle p_x \rangle = \int_{-\infty}^{+\infty} \varphi^* (-i\hbar\, \partial\varphi/\partial x)\, dx$$

$$= (2/\pi a^2)^{1/2} \int_{-\infty}^{+\infty} e^{-ik_0 x}\, e^{-x^2/a^2} (-i\hbar) \times \left[e^{ik_0 x}\, e^{-x^2/a^2}\right.$$

$$\left.(-2x/a^2) + e^{-x^2/a^2} e^{ik_0 x} (ik_0)\right] dx$$

$$= \left[2\, i\hbar/a^2 \int_{-\infty}^{+\infty} e^{-2x^2/a^2}\, x dx + \hbar k_0 \int_{-\infty}^{+\infty} e^{-2x^2/a^2} dx\right] \times (2/\pi a^2)^{1/2}$$

$$= \hbar k_0 \qquad \text{... (2.10)}$$

as the first term vanishes because of the integrand being an odd function of x.

$$\langle p_x^2 \rangle = \int_{-\infty}^{+\infty} \varphi^* (-\hbar^2)\, \partial^2\varphi/\partial x^2\, dx = -(2\hbar^4/\pi a^2)^{1/2}$$

$$\int_{-\infty}^{+\infty} e^{-ik_0 x}\, e^{-x^2/a^2}\, \partial^2/\partial x^2 \left(e^{ik_0 x}\, e^{-x^2/a^2}\right) dx$$

$$= -(2\hbar^4/\pi a^2)^{1/2} \left[-2/a^2 \int_{-\infty}^{+\infty} e^{-2x^2/a^2} dx + 4/a^4 \int_{-\infty}^{+\infty} e^{-2x^2/a^2} x^2 dx\right.$$

$$- 2ik_0/a^2 \int_{-\infty}^{+\infty} x e^{-2x^2/a^2} dx - k_0^2 \int_{-\infty}^{+\infty} e^{-2x^2/a^2} dx$$

$$\left. - 2ik_0/a^2 \int_{-\infty}^{+\infty} x e^{-2x^2/a^2} dx\right]$$

The 3rd and 5th terms in the above equation vanish as the integrands are odd functions of x. Hence

$$\langle p_x^2 \rangle = \hbar^2/a^2 + \hbar^2 k_0^2 \qquad \text{... (2.11)}$$

Using (2.7) to (2.11), one gets

$$\Delta x \Delta p_x = a/2 \times \hbar/a = \hbar/2$$

2.2 $\langle x \rangle = \int_{-\infty}^{+\infty} \varphi_0^* \, x\, \varphi_0\, dx = (m\omega/\pi\hbar)^{1/2} \int_{-\infty}^{+\infty} e^{-m\omega x^2/\hbar}\, x\, dx = 0$ ... (2.12)

as the integrand is an odd function of x.

$$\langle p_x \rangle = \int_{-\infty}^{+\infty} \varphi_0^* (-i\hbar\, \partial\varphi/\partial x)\, dx$$

$$= (-i\hbar) \times (m\omega/\pi\hbar)^{1/2}$$

$$\int_{-\infty}^{+\infty} e^{-m\omega x^2/2\hbar}\, e^{-m\omega x^2/2\hbar} (-mw \times 2\, x/2\hbar)\, dx$$

$$= 0 \qquad \text{... (2.13)}$$

for the same reason.

2.3 From the above problem

$$\langle x \rangle = \langle p_x \rangle = 0 \qquad \text{... (2.14)}$$

Hence

$$\Delta x = [\langle x^2 \rangle - \langle x \rangle^2]^{1/2} = \langle x^2 \rangle^{1/2}$$

and

$$\Delta p_x = [\langle p_x^2 \rangle - \langle p_x \rangle^2]^{1/2} = \langle p_x^2 \rangle^{1/2}$$

$$\langle x^2 \rangle = \int_{-\infty}^{+\infty} \varphi_0^*\, x^2\, \varphi_0\, dx = (m\omega/\pi\hbar)^{1/2} \int_{-\infty}^{+\infty} e^{-m\omega x^2/\hbar}\, x^2\, dx$$

$$= (m\omega/\pi\hbar)^{1/2} \times 1/2\, [\pi/(m\omega/\hbar)^3]^{1/2} = \hbar/2m\omega \qquad \text{... (2.15)}$$

$$\langle p_x^2 \rangle = \int_{-\infty}^{+\infty} \varphi_0^* (-\hbar^2)\, \partial^2 \varphi_0/\partial x^2\, dx$$

$$= (-\hbar)^2 (mw/\pi\hbar)^{1/2} \int_{-\infty}^{+\infty} e^{-m\omega x^2} (-m\omega/\hbar)$$

$$[e^{-m\omega x^2/2\hbar} - mwx^2/\hbar\, e^{-m\omega x^2/\hbar}]\, dx$$

$$= (m\omega\hbar)\,(m\omega/\pi\hbar)^{1/2} \left[\int_{-\infty}^{+\infty} e^{-m\omega x^2/\hbar}\, dx - m\omega/\hbar \int_{-\infty}^{+\infty} x^2\, e^{-m\omega x^2/\hbar}\, dx \right]$$

$$= (mw\,\hbar)\,(mw/\pi\hbar)^{1/2} [\{\pi/(m\omega/\hbar)\}^{1/2} - m\omega/\hbar \times 1/2\, \{\pi/(m\omega/\hbar)^3\}^{1/2}]$$

$$= m\omega\hbar/2 \qquad \text{... (2.16)}$$

Hence

$$\Delta x\, \Delta p_x = (\hbar/2m\omega)^{1/2} \times (m\omega\hbar/2)^{1/2} = \hbar/2$$

2.5 The uncertainty product $\Delta x\, \Delta p_x$ is given by

$$\Delta x\, \Delta p_x = [\langle x^2\rangle - \langle x\rangle^2]^{1/2} \times [\langle p_x^2\rangle - \langle px\rangle^2]^{1/2}$$

Now

$$\langle x\rangle = \int \varphi_n^*(x)\, x\, \varphi_n(x)\, dx$$

Using the recursion relation for the harmonic oscillator wave functions, that is,

$$x\varphi_n(x) = (\alpha)^{-1/2}\left[\{(n+1)/2\}^{1/2}\, \varphi_{n+1}(x) + (n/2)^{1/2}\, \varphi_{n-1}(x)\right]$$

One gets,

$$\langle x\rangle = \int_{-\infty}^{+\infty} \varphi_n^*(x)\alpha^{-1/2}\left[\{(n+1)/2\}^{1/2}\, \varphi_{n+1}(x) + (n/2)^{1/2}\varphi_{n-1}(x)\right] dx \qquad \text{... (2.17)}$$

Use of orthonormality relation for the harmonic oscillator wave functions leads to

$$\langle x\rangle = 0$$

Now

$$\langle x^2\rangle = \int_{-\infty}^{+\infty} \varphi_n^*(x)\, x^2 \varphi_n(x)\, dx = \int_{-\infty}^{+\infty} \varphi_n^*\, x\, x\varphi_n\, dx \qquad \text{... (2.18)}$$

Using recursion relation for the harmonic oscillator wave functions, one gets

$$\langle x^2\rangle = \int_{-\infty}^{+\infty} \varphi_n^*\, x\alpha^{-1/2}\left[\{(n+1)/2\}^{1/2}\, \varphi_{n+1} + (n/2)^{1/2}\varphi_{n-1}\right] dx$$

$$= \{(n+1)/2\alpha\}^{1/2} \int_{-\infty}^{+\infty} \varphi_n^*\, x\varphi_{n+1}\, dx + (n/2\alpha)^{1/2} \int_{-\infty}^{+\infty} \varphi_n^* x\varphi_{n-1}\, dx \qquad \text{... (2.19)}$$

Using once again the recursion relation (2.19), one gets

$$\langle x^2\rangle = \{(n+1)/2\alpha\}^{1/2} \int_{-\infty}^{+\infty} \varphi_n^*\, \alpha^{-1/2}\left[\{(n+2)/2\}^{1/2}\, \varphi_{n+2} + \right.$$

$$\left.\{(n+1)/2\}^{1/2}\, \varphi_n\right] dx + (n/2\alpha)^{1/2} \int_{-\infty}^{+\infty} \varphi_n^*\, \alpha^{-1/2}$$

$$[(n/2)^{1/2}\, \varphi_n + \{(n-1)/2\}^{1/2}\, \varphi_{n-2}]\, dx \qquad \text{... (2.20)}$$

Using orthonormality of harmonic oscillator wave functions 1st and 3rd terms vanish and hence

$$\langle x^2\rangle = (n+1)/2\alpha + n/2\alpha = (n + 1/2)\ \hbar/mw \qquad \text{... (2.21)}$$

Now

$$\langle p_x \rangle = -i\hbar \int_{-\infty}^{+\infty} \varphi_n^* \, \partial\varphi_n/\partial x \, dx$$

Using the recursion relation,

$$\partial\varphi_n/\partial x = -(\alpha)^{1/2} [(n/2)^{1/2} \varphi_{n-1} - \{(n+1)/2\}^{1/2} \varphi_{n+1}]$$

one gets

$$\langle p_x \rangle = i\hbar \int_{-\infty}^{+\infty} \varphi_n^* \, \alpha^{1/2} [(n/2)^{1/2} \varphi_{n-1} - \{(n+1)/2\}^{1/2} \varphi_{n+1}]$$

Using the orthonormality of harmonic oscillator wave functions, one gets

$$\langle p_x \rangle = 0 \qquad \text{... (2.23)}$$

Now

$$\langle p_x^2 \rangle = \int_{-\infty}^{+\infty} \psi_n^2 \, p_x^2 \varphi_n \, dx = -\hbar^2 \int_{-\infty}^{+\infty} \varphi_n^* \, \partial^2\varphi_n/\partial x^2 \, dx$$

$$= -\hbar^2 \int_{-\infty}^{+\infty} \varphi_n^* \, \partial/\partial x \, (\partial\varphi_n/\partial x) \, dx$$

$$= \hbar^2 \alpha^{1/2} \int_{-\infty}^{+\infty} \varphi_n^* \, \partial/\partial x \, [(n/2)^{1/2} \varphi_{n-1} - \{(n+1)/2\}^{1/2} \varphi_{n+1}] \, dx$$

$$= (n\alpha/2)^{1/2} \hbar^2 \int_{-\infty}^{+\infty} \varphi_n^* \, \partial\varphi_{n-1}/\partial x \, dx - \{(n+1)\alpha/2\}^{1/2} \hbar^2 \int_{-\infty}^{+\infty} \varphi_n^* \, \partial\varphi_{n+1}/\partial x \, dx$$

Using the recursion relation (2.22), one gets

$$\langle p_x^2 \rangle = -\alpha (n/2)^{1/2} \hbar^2 \int_{-\infty}^{+\infty} \varphi_n^* \, [\{(n-1)/2\}^{1/2} \varphi_{n-2} - (n/2)^{1/2} \varphi_n] \, dx$$

$$+ \alpha \{(n+1)/2\}^{1/2} \hbar^2 \int_{-\infty}^{+\infty} \varphi_n^* \, [\{(n+1)/2\}^{1/2} \varphi_n - \{(n+2)/2\}^{1/2} \varphi_{n+2}] \, dx$$

Using orthonormality of harmonic oscillator wave functions, one gets

$$\langle p_x^2 \rangle = \alpha \times n/2 \times \hbar^2 + \alpha \times (n+1)/2 \times \hbar^2$$

$$= (n + 1/2) \, m\omega\hbar \qquad \text{... (2.24)}$$

Using equn. (2.17), (2.20), (2.21), (2.23) and (2.24), one gets

$$\Delta x \, \Delta p_x = (\hbar/m\omega)^{1/2} (n + 1/2)^{1/2} (m\omega\hbar)^{1/2} (n + 1/2)^{1/2}$$

$$= (n + 1/2)\, \hbar \qquad \text{... (2.25)}$$

2.6 For the present case, the normalized wave function is given by

$$\varphi_n (x) = (2/L)^{1/2} \sin n\pi x/L \qquad \text{... (2.26)}$$

The mean square deviation of the particle is given by

$$\langle (x - \langle x \rangle)^2 \rangle = \langle x^2 \rangle - \langle x \rangle^2 \qquad \text{... (2.27)}$$

Now

$$\langle x \rangle = \int_0^L x |\varphi_n(x)|^2 \, dx = 2/L \int_0^L x \sin^2 n\pi x/L \; dx$$

$$= 2/L \int_0^L x (1 - \cos 2\, n\pi x/L)/2 \; dx$$

$$= 1/L \; x^2/2 \Big|_0^L - 1/L \, \frac{x \sin 2\, n\pi x/L}{2n\pi/L} \Big|_0^L + 1/L \int_0^L \frac{\sin 2n\pi x/L}{2n\pi/L} \, dx$$

$$= L/2 - 0 - \frac{\cos 2n\pi x/L}{(2n\pi/L)^2} \Big|_0^L = L/2 \qquad \text{... (2.28)}$$

$$\langle x^2 \rangle = \int_0^L x^2 |\varphi_n(x)|^2 \, dx = 2/L \int_0^L x^2 \sin^2 n\pi x/L \, dx$$

$$= 1/L \int_0^L x^2 \; (1 - \cos 2n\pi x/L) \, dx$$

$$= 1/L \;\; x^3/3 \Big|_0^L - x^2/L \, \frac{\sin 2\, n\pi x/L}{2n\pi/L} \Big|_0^L + 1/L = \int_0^L 2\, x \, \frac{\sin 2\, n\pi x/L}{2n\pi/L} \, dx$$

$$= L^2/3 - 0 + 2/L \; x \, \frac{- \cos 2n\pi x/L}{(2n\pi/L)^2} \Big|_0^L - 2/L \int_0^L - \frac{\cos 2\, n\pi x/L}{(2n\pi/L)^2} \, dx$$

$$= L^2/3 + 2/L \; \{-L/(2n\, \pi/L)^2\} + 2/L \, \frac{\sin 2\, n\pi x/L}{(2\, n\pi/L)^2} \Big|_0^L$$

$$= L^2/3 - L^2/2n^2\pi^2 = L^2 \; (1/3 - 1/2n^2\pi^2)$$

Hence

$$\langle x^2 \rangle - \langle x \rangle^2 = L^2(1/3 - 1/2\, n^2\pi^2) - L^2/4 = L^2/12 \; (1 - 6/n^2\pi^2)$$

2.7 $(xp_x - p_x x)\,\psi = -\,i\hbar\,(x\partial/\partial x - \partial/\partial x\; x)\,\psi$

$$= i\hbar\,[x\,\partial\psi/\partial x - \partial/\partial x\,(x\psi)]$$

$$= i\hbar\,[x\,\partial\psi/\partial x - x\,\partial\psi/\partial x - \psi]$$

$$= i\,\hbar\,\psi \qquad \text{... (2.29)}$$

Since ψ is an arbitrary function, the operator $(xp_x - p_x\,x)/i\hbar$ is simply a unit operator. The relation

$$(xp_x - p_x x) \equiv [x, p_x] = i\hbar \qquad \text{... (2.30)}$$

is known as the commutation relation. The quantity

$$[x, p_x] = -\,[p_x, x]$$

is said to be commutator of the two operators x and p_x.

2.8 The probability of finding the particle in the region $0 < x < \infty$ is

$$P = \int_0^\infty \varphi^*\,\varphi\,dx = N^2 \int_0^\infty e^{-2\alpha x^2}\,dx$$

$$= N^2 \times 1/2 \times (\pi/2\,\alpha)^{1/2} \qquad \text{... (2.31)}$$

Imposing the normalization condition

$$\int_{-\infty}^{+\infty} |\varphi|^2\,dx = N^2 \int_{-\infty}^{+\infty} |e^{-2\alpha x^2}\,dx = 1$$

or

$$N^2\,(\pi/2\,\alpha)^{1/2} = 1$$

or

$$N^2 = (2\alpha/\pi)^{1/2} \qquad \text{... (2.32)}$$

Substituting above in (2.31).

$$P = (2\alpha/\pi)^{1/2} \times 1/2 \times (\pi/2\,\alpha)^{1/2} = 1/2$$

2.9 Let us concentrate on the x-component,

$$\iiint (-\,i\hbar\,\partial\Psi/\partial x)^*\,\Psi\,dx\,dy\,dz = i\hbar \iiint \partial\Psi^*/\partial x\,\Psi\,dx\,dy\,dz$$

$$= i\hbar \iint \Psi^*\,\Psi\,dy\,dz\,\Big|_{-\infty}^{+\infty} - i\hbar \iiint \Psi^*\,\partial\Psi/\partial x\,dx\,dy\,dz$$

where we have done integration by parts.

Since the wave function φ vanishes at infinity, the first term on the r.h.s. of the above equation vanishes. Hence

$$\iiint (-\,i\hbar\,\partial\Psi/\partial x)^*\,\Psi\,dx\,dy\,dz = \iiint \Psi^*\,(-\,i\hbar\,\partial\Psi/\partial x)\,dx\,dy\,dz \qquad \text{... (2.33)}$$

which is nothing but condition for an operator to be Hermitian. Hence momentum operator is Hermitian.

2.10 Using eqn. (91),

$$\int \psi^* (FG)^\dagger \psi d\tau = \int (FG\psi)^* \psi d\tau = \int (G\psi)^* F^\dagger \psi \, d\tau$$

$$= \int \psi^* G^\dagger F^\dagger \psi \, d\tau$$

Since ψ is an arbitrary function,

$$(FG)^\dagger = G^\dagger F^\dagger \quad \text{... (2.34)}$$

If F and G are Hermitian, $F^\dagger = F$, $G^\dagger = G$, and hence

$$(FG)^\dagger = GF \quad \text{... (2.35)}$$

Thus the product of two hermitian operators (note the reversal of the order of the factors), is in general not hermitian. However, if F and G commute, that is, if

$$FG = GF \quad \text{... (2.36)}$$

In that case, $(FG)^\dagger = FG$

or FG is self adjoint. That is, the product of two hermitian operators is hermitian only if they commute.

2.11 $\langle 1/r \rangle = \int_0^{\pi} \int_0^{2\pi} \int_0^{\infty} 1/r |\varphi_0|^2 r^2 \sin\theta \, d\theta \, d\phi \, dr$

$$= 1/\pi a_0^3 \int_0^{\infty} r\, e^{-2r/a_0} \, dr \int_0^{\pi} \sin\theta \, d\theta \int_0^{2\pi} d\phi$$

$$= 1/\pi a_0^3 \quad a_0^2/4 \quad 2 \quad 2\pi$$

$$= 1/a_0$$

❑❑❑

Chapter 3

SIMPLE APPLICATIONS OF QUANTUM MECHANICS

3.1 PARTICLE IN A ONE DIMENSIONAL BOX-SQUARE WELL WITH INFINITELY HIGH WALLS

In this chapter, we shall consider some simple systems for which we can find exact solution of the Schrödinger equation determining the stationary states. Such systems are idealizations of systems occurring in nature. The problem of determining the stationary states of motion of a particle of mass m in an external potential field reduces to finding the eigen values of the Hamiltonian operator, that is solving the equation,

$$[-\hbar^2/2m\, \nabla^2 + V(\vec{r})]\, \varphi(\vec{r}) = E\varphi(\vec{r}) \qquad \text{... (1)}$$

or

$$[\nabla^2 + 2m/\hbar^2 \{E - V(\vec{r})]\, \varphi(\vec{r}) = 0 \qquad \text{... (2)}$$

This equation is a linear second-order differential equation. Exact analytical solutions of eqn. (2) can be found only for some forms of the potential energy operator which in the coordinate representation is a function of the coordinates of the particle. The simple systems are those for which the potential energy is constant in whole of the space which actually correspond to the free motion of the particle.

Let us now consider a particle moving (or confined) in a one dimensional box of length L (Fig. 3.1). Mathematically the potential well is described as

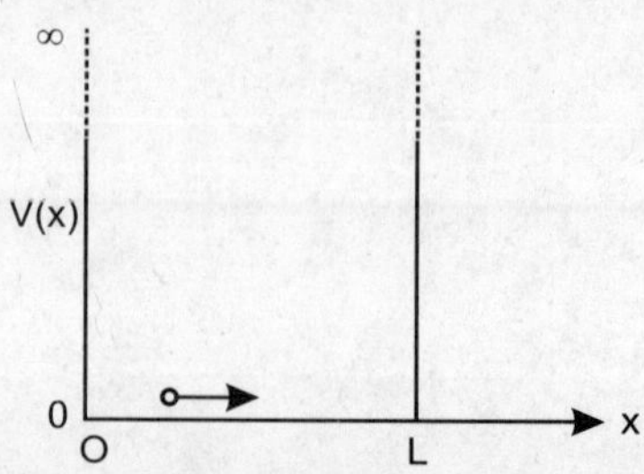

Fig. 3.1 *Particle confined to a box of width L (infinitely deep potential well)*

$$V(x) = 0, 0 < x < L$$
$$= \infty, 0 \le x \text{ and } x \ge L \quad \text{... (3)}$$

and is extremely crude approximation to the problem of electron in atom as only 13.6 eV energy is required to remove the electron from H-atom.

Since $V(x) = 0$ inside the box, the particle is free (no force acts on it). The particle is confined inside the box as $V(x) \to \infty$ at $x = 0$ and at $x = L$ and as the particle cannot have infinite amount of energy. In other words, the boundary conditions on the wave function would be

$$\varphi(x) = 0, 0 \le x \text{ and } x \ge L \quad \text{... (4)}$$

since the probability of finding the particle is zero in these regions.

Our task is to find, what φ is within the box? In this region ($0 < x < L$), eqn. (2) takes the form

$$d^2\varphi(x)/dx^2 + 2m/\hbar^2 E\varphi(x) = 0, \quad 0 < x < L \quad \text{... (5)}$$

as $V(x) = 0$ within the box.

The above equation can be written as

$$d^2\varphi(x)/dx^2 + k^2\varphi(x) = 0, k^2 = 2mE/\hbar^2 \quad \text{... (6)}$$

The form of the above equation is identical with the equation of simple harmonic motion. Thus it has two possible solutions, *viz.* $A \sin kx$ or $B \cos kx$. The general solution of the above equation would then be a linear combination of these two solutions, that is

$$\varphi(x) = A \sin kx + B \cos kx \quad \text{... (7)}$$

where A and B are constants (real) to be determined.

Applying the boundary conditions, $\varphi(x) = 0$ at $x = 0$, we get

$$0 = B \quad \text{... (8)}$$

Similarly for $x = L$, using $\varphi(L) = 0$, we get

$$0 = A \sin kL.$$

This implies either $A = 0$ or

$$kL = (2mE/\hbar^2)^{1/2} L = n\pi, \quad n = 0, 1, 2, \ldots \quad \text{... (9)}$$

If $A = 0$, then $\varphi = 0$, implying there is no particle in the box. Thus, we assume that $A \ne 0$ and from eqn. (9), on squaring both the sides, we get

$$E = E_n = n^2\pi^2\hbar^2/(2m L^2), \quad n = 0, 1, 2, \ldots \quad \text{... (10)}$$

It is clear that n should not be taken equal to zero. This would mean either $L = 0$ from eqn. (9), which is not acceptable or $E = 0$ which we cannot take as it is to be found out (Also for $E = 0$, the solution of eqn. (5) is

$$\varphi(x) = R + Sx$$

Chapter 3

where R and S are constants. The application of boundary conditions (4) implies that both R and S are zero. Hence, the wave function does not exist for $E = 0$. Thus

$$E_n = n^2\pi^2\hbar^2/(2\,mL^2), \quad n = 1, 2, \ldots \qquad \ldots (11)$$

The above expression shows that a particle confined to a box cannot have an arbitrary energy or in other words, it can take only certain energy values. These are known as energy eigen values or energy levels of the system. All this amounts to the fact that the energy of the system is quantized. The integer n in eqn. (11) is referred to as the quantum number. An electron that is described by the wave function with a certain n value is said to be in the quantum state n. The quantum state with the lowest energy ($n = 1$ in the present example) is called the ground state. It should be noted from eqn. (11), that the lowest energy is not equal to zero which is a general result of quantum mechanics applicable to all problems and is completely foreign to the classical mechanics (in which lowest energy is zero).

The wave function corresponding to quantum number n from equn. (7), (8) and (9), is

$$\varphi_n(x) = A \sin kx = A \sin n\pi\, x/L \qquad \ldots (12)$$

The constant A is determined from the normalization condition,

$$\int_{-\infty}^{+\infty} |\varphi_n(x)|^2\, dx = 1$$

Since $\varphi_n(x) = 0$ outside the well, the above condition reduces to ... (13)

$$\int_0^L |\varphi_n(x)|^2\, dx = 1$$

Substituting eqn. (12) into above, we get

$$A^2 \int_0^L \sin^2 n\pi x/L\, dx = 1$$

or $$A^2 \int_0^L (1 - \cos 2\,n\pi x/L)/2)\, dx = 1$$

or $$A^2 \left[x/2 \Big|_0^L - \frac{\sin 2\,n\pi x/L}{4n\pi/L} \Big|_0^L \right] = 1$$

or $$A = (2/L)^{1/2}$$

Thus, the normalized wave functions of an electron in a one-dimentional box are given by

$$\varphi_n(x) = (2/L)^{1/2} \sin n\pi x/L, \quad n = 1, 2, \ldots \qquad \ldots (14)$$

which are solutions of the time independent Schrödinger eqn. (5). These solutions are referred to as the energy eigen functions.

If φ_n and φ_n' are normalized wave functions corresponding to two different states of an electron in a box, for $n \neq n'$.

$$\int_0^L \varphi_n \varphi_{n'}\, dx = 2/L \int_0^L \sin n\pi x/L \quad \sin n'\pi x/L\, dx$$

$$= 2/L \int_0^L [\cos (n-n')\pi x/L - \cos (n+n')\pi x/L]/2\, dx$$

$$= 1/L\,\pi\,[L/(n-n') \quad \sin (n-n')\pi x/L$$
$$- L/(n+n') \quad \sin (n+n')\pi x/L]_0^L$$

$$= 0 \qquad \text{... (15)}$$

Hence, the wave functions for different states of this system are orthogonal. Combining equn. (13) and (15),

$$\int_0^L \varphi_n \varphi_{n'}\, dx = \delta_{nn'} \qquad \text{... (16)}$$

which means that the wave functions of an electron in a box form an orthonormal set. In the illustrative example, we have considered, the wave functions which happen to be real and the Fig. 3.2 shows a plot of energy eigen values and energy eigen functions corresponding to the ground state and the first two excited states.

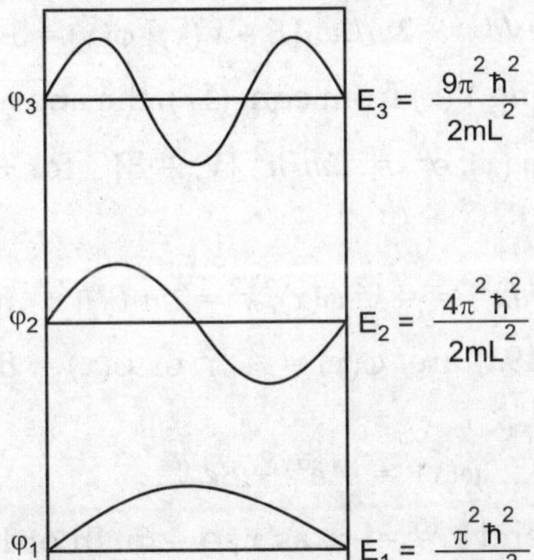

Fig. 3.2 *Energy levels and energy eigen functions for a particle in an infinitely deep potential well*

3.2 PARTICLE IN A SQUARE WELL POTENTIAL WITH FINITE WALLS

We now turn to a more realistic model of an electron in atom though it is still one dimensional. The potential as a function of x is shown in Fig. 3.3 and is mathematically represented as

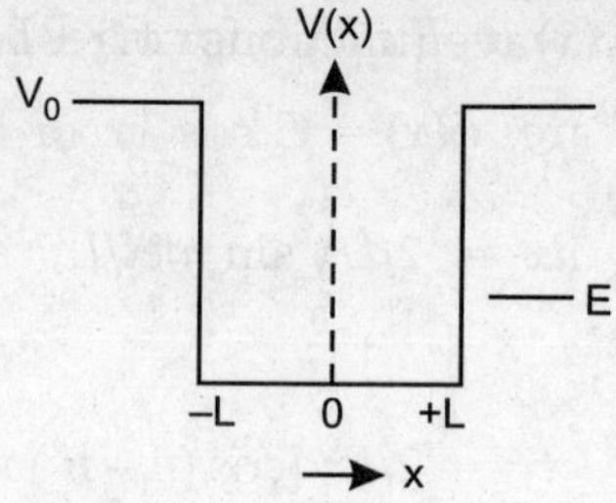

Fig. 3.3 *Potential function for a particle in a box with finite walls*

$$V(x) = 0, \quad \text{for} - L < x < + L \text{ (or } x < |L|) \qquad \text{... (17)}$$

$$= V_0, \quad \text{for} + L < x < - L \text{ (or } x > |L|)$$

where $V(x)$ has been taken to be zero for convenience inside the box though it is constant.

Let us look for solutions corresponding to the bound states, that is, the particle's energy E such that $0 < E < V_0$ (**). Classically E can have any value between 0 and V_0 and the particle is confined (or bound) in the region between $x = -L$ and $x = +L$ (or particle is trapped inside the box) as $V(x) = 0$ in this region and hence K.E. is positive. Outside the well where $V(x) = V_0$, its K.E. $= E - V_0$ becomes negative.

If we want to determine the behaviour of the particle quantum mechanically or want to find $\varphi(x)$, we will have to solve the Schrödinger equation,

$$d^2\varphi(x)/dx^2 + 2m/\hbar^2 \, [E - V(x)] \, \varphi(x) = 0 \qquad \text{... (18)}$$

for all values of x. Substituting $V(x)$ from eqn. (17), the above equation takes the forms,

$$d^2\varphi(x)/dx^2 - \alpha^2 \varphi(x), \; \alpha^2 = 2m/\hbar^2 \, [V_0 - E] \quad \text{for} + L < x < - L \qquad \text{... (19a)}$$

and

$$d^2 \varphi(x)/dx^2 = -k^2 \varphi(x), \; k^2 = 2m \, E/\hbar^2 \quad \text{for} -L < x < +L \qquad \text{... (19b)}$$

The solutions of eqn. (19*a*) are, $\varphi(x) = Ae^{\alpha x}$ or $\varphi(x) = Be^{-\alpha x}$. The general solution would then be given by

$$\varphi(x) = Ae^{\alpha x} + Be^{-\alpha x} \qquad \text{... (20)}$$

Now $e^{\alpha x} \to \infty$ as $x \to \infty$ and $e^{-\alpha x} \to \infty$ as $x \to -\infty$. In order that $\varphi(x)$ is finite for all values of x, we must take $B = 0$ in eqn. (20) and hence the solution would be

$$\varphi(x) = A \, e^{\alpha x}, \quad \text{for } x < -L \qquad \text{... (21)}$$

and similarly, we must take $A = 0$ and the solution would be

**Some people take $V(x) = -V_0$ inside the well, V_0 being positive and $V(x) = 0$ outside the well. Then for bound states, they take $E < 0$, that is negative. Inside the well, E = K.E. + P.E. or E = K.E. + $(-V_0)$ or $V_0 + E$ = K.E. E being negative and $V_0 > E$, K.E. is positive inside the well. Outside the well then K.E. becomes negative.

$$\varphi(x) = B e^{-\alpha x}, \quad \text{for } x > +L \qquad \text{... (22)}$$

The solutions of eqn. (19*b*) are, $\varphi(x) = C \cos kx$ or $\varphi(x) = D \sin kx$. The general solution would then be given by

$$\varphi(x) = C \cos kx + D \sin kx, -L < x < +L \qquad \text{... (23)}$$

For finding the permissible or allowed values of *E*, let us match the expressions for $\varphi(x)$ at the boundaries, that is at $x = -L$ and at $x = +L$. The continuity of $\varphi(x)$ thus gives

$$\text{At } x = -L, A e^{-\alpha L} = C \cos kL - D \sin kL \qquad \text{... (24)}$$

and

$$\text{At } x = +L, B e^{-\alpha L} = C \cos kL + D \sin kL \qquad \text{... (25)}$$

Addition of equn. (24) and (25), yields

$$(A + B) e^{-\alpha L} = 2C \cos kL \qquad \text{... (26)}$$

and similarly subtraction of eqn. (9) from eqn. (8), yields

$$(B - A) e^{-\alpha L} = 2D \sin kL \qquad \text{... (27)}$$

We obtain two more equations from the continuity of $d\varphi(x)/dx$ at boundaries. The differentiation of equn. (21), (22) and (23) thus gives

$$\text{At } x = -L, \alpha A e^{-\alpha L} = Ck \sin k L + Dk \cos kL \qquad \text{... (28)}$$

$$\text{At } x = +L, -\alpha B e^{-\alpha L} = -Ck \sin kL + Dk \cos kL \qquad \text{... (29)}$$

Again addition of equn. (28) and (29), yields

$$\alpha (A - B) e^{-\alpha L} = 2 Dk \cos kL \qquad \text{... (30)}$$

and similarly subtraction of eqn. (29) from eqn. (28), yields

$$\alpha (A + B) e^{-\alpha L} = 2Ck \sin kL \qquad \text{... (31)}$$

Dividing eqn. (30) by eqn. (27), we obtain

$$-\alpha = k \cot kL \qquad \text{... (32)}$$

the result which is valid under the condition $A \neq B$ and $D \neq 0$, otherwise it would give o/o.

Dividing eqn. (31) by eqn. (26), we obtain

$$\alpha = k \tan kL \qquad \text{... (33)}$$

the result which is valid under the condition, $A \neq -B$ and $C \neq 0$, otherwise it would give o/o.

It may be noted that both the equn. (32) and (33) cannot be valid simultaneously as in that case, $k \cot kL = -k \tan kL$ or $\tan^2 kL = -1$ which is impossible as *k* and *L* both are real. Therefore, we have two types of (or classes) of solutions.

Type 1 (or Class 1):

If $D = 0$ then $A = B$ and hence eqn. (32) would not be valid. Therefore with $D = 0$ and $A = B$, we take

$$\alpha = k \tan kL \quad \text{... (34)}$$

and the solution is

$$\varphi(x) = Ae^{\alpha x}, x < -L$$

$$\varphi(x) = Be^{-\alpha x} = Ae^{-\alpha x}, x > +L \quad \text{... (35)}$$

$$\varphi(x) = C \cos kx, -L < x < +L$$

It may be noted that in this case, $\varphi(x) = \varphi(-x)$ for all x or in other words φ is an even function of x (or solutions are symmetric in x).

Type 2 (or Class 2):

If $C = 0$ then $A = -B$ and hence eqn. (33) would not be valid. Therefore with $C = 0$ and $A = -B$, we take

$$-\alpha = k \cot kL \quad \text{... (36)}$$

and the solution is

$$\varphi(x) = Ae^{\alpha x}, x < -L$$

$$\varphi(x) = Be^{-\alpha x} = -Ae^{-\alpha x}, x > +L \quad \text{... (37)}$$

$$\varphi(x) = D \sin kx, -L < x < +L$$

It may be noted that in this case, $\varphi(x) = -\varphi(-x)$ for all x or in other words $\varphi(x)$ is an odd function of x (or solutions are antisymmetric in x).

To find the energy levels corresponding to the symmetric solutions, we have to solve the transcendental eqn. (34) numerically or graphically. And similarly for finding energy levels corresponding to the antisymmetric solutions, we have to solve the transcendental eqn. (36). A simple graphical method for effecting the solutions is described here.

Multiplying eqn. (34) by L and introducing the new variables

$$\lambda = kL, \mu = \alpha L \quad \text{... (38)}$$

one gets

$$\mu = \lambda \tan \lambda \text{ for even eigen states} \quad \text{... (39)}$$

$$\mu = -\lambda \cot \lambda \text{ for odd eigen states} \quad \text{... (40)}$$

with

$$\lambda^2 + \mu^2 = k^2L^2 + \alpha^2L^2$$

$$= 2m/\hbar^2\, EL^2 + 2m/\hbar^2\, (V_0 - E)\, L^2$$

$$= 2m/\hbar^2\, V_0L^2 = \rho^2 \text{ (say)} \quad \text{... (41)}$$

For a given potential well of width $2L$, height V_0 and particle mass m, eqn. (41) describes a circle of radius $\rho = (2mV_0L^2/\hbar^2)^{1/2}$ in λ μ – space. The intersections of this circle in the first quadrant (k, α and L being positive, λ and μ are restricted to positive values only) with the curves (39) yield the energies of even eigen states (or even solutions (35)) and intersections of the same circle with the curves (40) yield the energies of odd states (or odd solutions (37)).

Fig. 3.4 shows plots of $\mu = \lambda \tan \lambda$ and $\mu = -\lambda \cot \lambda$ as a function of λ for three values of $V_0L^2 = \hbar^2/2m$ (that is, $\rho = 1$), $4\,\hbar^2/2m$ (that is, $\rho = 2$) and $12\,\hbar^2/2m$ (that is, $\rho = 2\sqrt{3}$). The shaded circles (intersections) show even solutions and the unshaded circles show odd solutions. Three increasing values of V_0L^2 give all together, one even solution (or symmetric state), two solutions, one corresponding to even (or symmetric) state and one corresponding to odd (or antisymmetric) state and so on. One may note that there is no degeneracy.

Thus it is clear from Fig. 3.4 that for a given particular mass, the energy levels depend on the parameters of the potential well through the combination V_0L^2. For V_0L^2 between zero and $\hbar^2/2m$ (that is for $0 < \rho \leq 1$) there is just one energy level of the first class. For V_0L^2 between $\hbar^2/2m$ and four times this value (that is $1 < \rho \leq 2$) there is one energy level of each class or two altogether. As V_0L^2 increases, energy levels appear successively, first of one class and then of other. Note that, however small V_0 may be, there exists at least one bound state.

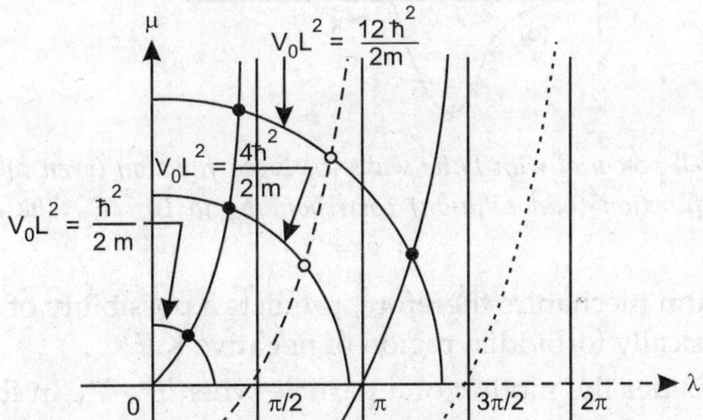

Fig. 3.4 *Graphical determination of the levels in a square well for three values of V_0L^2. The solid curves are corresponding to $\mu = \lambda \tan \lambda$ vs λ and the dashed curves are corresponding to $\mu = -\lambda \cot \lambda$ vs λ*

As an example, let us consider a proton ($m = 1.67 \times 10^{-27}$ kg) inside a square potential well of height $V_0 = 4 \times 10^{-12}$ J and width $2L = 3.65 \times 10^{-15}$ *m*. Hence $\rho = (2m\,V_0L^2/\hbar^2)^{1/2} = 2$, For this value of ρ there would be one symmetric state and one antisymmetric state as depicted by the Fig. 3.4. The corresponding eigen values are approximately 10.8×10^{-19} J and 36×10^{-19} J, respectively.

Fig. 3.5(*b*) shows an even solution (corresponding to $E = E_1$) while Fig. 3.5(*c*) shows an odd solution (corresponding to $E = E_2$). The drawing corresponds to the case when $\rho = 2$, that is, for $\rho < \pi$. The fact that $\varphi(x)$ (and therefore $|\varphi|^2$) is not zero in the regions $x \leq -L$ and $x \geq +L$ is a new result which is not expected on the basis of classical theory.

In these regions the K.E. is negative. Classically the particle should have been turned at the edge of the well and never appear in the negative K.E.

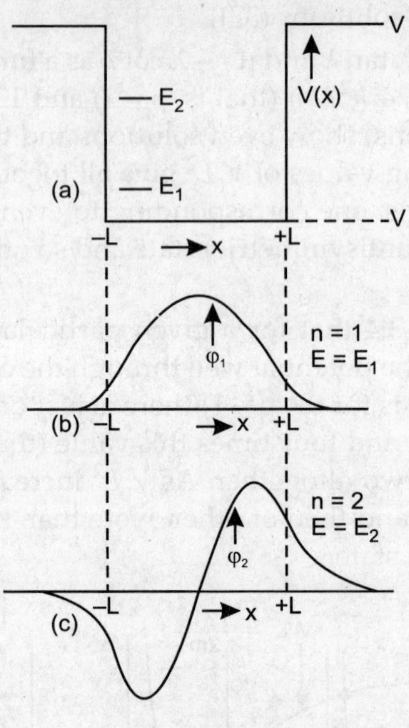

Fig. 3.5 *(a) Square well potential with finite walls (b) Eigen function (even solution) corresponding to $E = E_1$ (c) Eigen function (odd solution) corresponding to $E = E_2$. The drawing corresponds to $\rho = 2$*

regions. The quantum mechanics therefore, predicts a possibility of penetrating some distance into a classically forbidden region of negative K.E.

Let us now consider the motion of a particle when $E > V_0$. In that case, eqn. (18) would read as

$$d^2\varphi(x)/dx^2 + \beta^2\varphi(x) = 0,\ \beta^2 = 2m(E - V_0)/\hbar^2,\ +L < x < -L \qquad \text{... (19a')}$$

Eqn. (18) in the region $+L > x > -L$ would take the form

$$d^2\varphi(x)/dx^2 + k_1^2\varphi(x) = 0,\ k_1^2 = 2mE/\hbar^2,\ -L < x < +L \qquad \text{... (19b')}$$

where $k_1 > k$ (as E in this case is greater than V_0)

The above equn. clearly indicate that the solutions in all the three regions ($x < -L$, $-L < x < +L$ and $x > L$) are now sinusoidal in form. The only difference between the solutions for $x < -L$ and $x > +L$ and the solution for $-L < x < +L$ is that the wavelength of the oscillations is different.

The solutions for different regions must be subjected to the continuity conditions (φ and $d\varphi/dx$ must be continuous) at $x = -L$ and $x = +L$. None of these solutions becomes large without limit (or tend to infinity for $x \to \pm\infty$ as in the case of bound states) and therefore there are no arguments like those in the case of bound states that forced β to have only discrete values (that is, there are no restrictions on the values of

β or E). Hence any energy $E > V_0$ is an eigen value or E has a continuous distribution of allowed values or the energy spectrum is a continuous (not a discrete) set.

To complete the determination of the eigen functions belonging to any $E > V_0$, we have to obtain explicitly the relations among the coefficients appearing in the solutions of (19*a*′) and (19*b*′) which result from the continuity conditions. This of course is a straight forward matter. The details of the similar calculations would be provided in the section 3.3 (*ii*).

Let us content ourselves here simply by pointing out the result that when a particle is incident on the potential well, there is a non-zero probability of its being reflected by the well which is a typical quantum mechanical result (classically a particle with energy $E > V_0$ would keep on going; there is no possibility of reflection).

3.3 POTENTIAL STEP

Consider a potential step (shown in Fig. 3.6) represented by

$$V(x) = 0, \quad x < 0 \qquad \text{... (42)}$$

$$= V_0, \quad x > 0$$

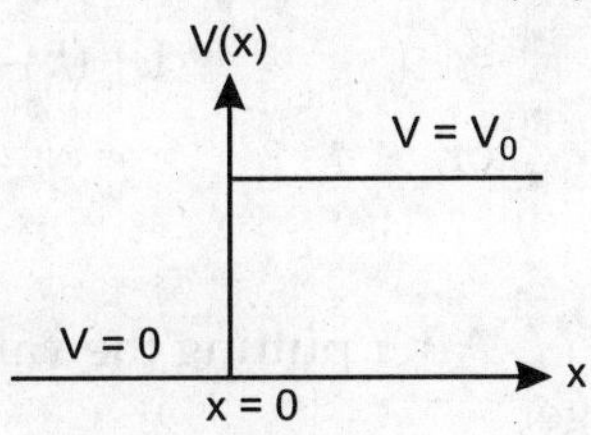

Fig. 3.6 *Potential step*

Let us consider two cases:

(*i*) $0 < E < V_0$. Classically a particle of this energy if it is incident from the left would move freely until it is reflected at the potential step at $x = 0$ because the particle does not have sufficient energy to be in the region $x > 0$ (or it cannot enter the region in which its K.E. is negative). Conservation of energy requires it to turn around, changing the sign of its momentum.

To determine the behaviour of particle quantum mechanically, we will have to solve the Schrödinger equation,

$$d^2\varphi(x)/dx^2 + 2m/\hbar^2 [E - V(x)]\, \varphi(x) = 0$$

which in the present case becomes

$$d^2 \varphi(x)/dx^2 + k^2 \varphi(x) = 0, \; k^2 = 2m\, E/\hbar^2, \; x < 0 \qquad \text{... (43)}$$

and

$$d^2\varphi(x)/dx^2 = \alpha^2 \varphi(x), \; \alpha^2 = 2m\,(V_0 - E)/\hbar^2, \; x > 0 \qquad \text{... (44)}$$

The general solutions of equn. (43) and (44) are

$$\varphi(x) = Ae^{ikx} + Be^{-ikx}, \; x < 0 \qquad \text{... (45a)}$$

$$\varphi(x) = C\, e^{-\alpha x}, \; x > 0 \qquad \text{... (45b)}$$

where A, B and C are constant to be determined.

It may be noted that we have rejected the second linearly independent solution $De^{\alpha x}$ in the region $x > 0$ (that is taken $D = 0$) otherwise φ would have become infinite for large values of x (that is for $x \to \infty$). Since one of the two linearly independent solutions for the same value of E has been excluded, the stationary states for $E < V_0$ are non-degenerate. It may, however be pointed out that the wave functions e^{ikx} and e^{-ikx} multiplied by the time dependent factor $e^{-iEt/\hbar}$ represent the waves propagating in the $+x$ and in the $-x$ directions, respectively.

By matching φ and $d\varphi/dx$ at the discontinuity of the potential, that is at $x = 0$, we have

$$A + B = C \quad \text{... (46a)}$$

$$ik(A - B) = -\alpha C \quad \text{... (46b)}$$

Dividing (46*a*) by (46*b*), we get

$$(A + B)/(A - B) = ik/-\alpha = k/i\alpha$$

or $\{(A + B) - (A - B)\}/\{(A + B) + (A - B)\} = B/A = (k - i\alpha)/(k + i\alpha)$

$$= e^{i\beta} \ (\beta : \text{real}) \quad \text{... (47)}$$

Substituting B/A from eqn. (47) to eqn. (46*a*) gives

$$1 + (k - i\alpha)/(k + i\alpha) = C/A$$

or

$$C/A = 2k/(k + i\alpha) = 1 + e^{i\beta} \quad \text{... (48)}$$

After putting the values of B/A and C/A from equn. (47) and (48) into eqn. (45), we get

$$\varphi(x) = A(e^{ikx} + B/Ae^{-ikx})$$

$$= A(e^{ikx} + e^{i\beta} e^{-ikx})$$

$$= 2Ae^{i\beta/2} [e^{i(kx - \beta/2)} + e^{-i(kx - \beta/2)}]/2$$

$$= 2Ae^{i\beta/2} \cos (kx - \beta/2), \ x < 0 \quad \text{... (49a)}$$

and

$$\varphi(x) = Ce^{-\alpha x} = A \times C/A \ \ e^{-\alpha x} = A\,(1 + e^{i\beta})\, e^{-\alpha x}$$

$$= 2Ae^{i\beta/2} (e^{i\beta/2} + e^{-i\beta/2})/2 \ \ e^{-\alpha x}$$

or

$$\varphi(x) = 2A\, e^{i\beta/2} \cos \beta/2 \ \ e^{-\alpha x}, \ x > 0 \quad \text{... (49b)}$$

Eqn. (49) indicate that the wave function in the case of no degeneracy is real except for an arbitrary constant factor. Hence, a graph may be drawn of such a wave function (Fig. 3.7). The oscillatory solution in the region $V = 0$ (eqn. 49*a*) and the exponentially decaying solution (eqn. 49*b*) in the forbidden region can be joined smoothly at $x = 0$ for all values of E between 0 and V_0, the energy spectrum in this case is, therefore continuous.

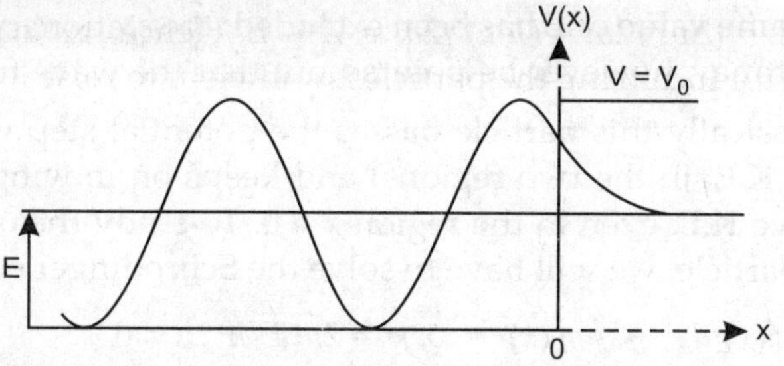

Fig. 3.7 *Eigen function for the potential step corresponding to* $E < V_0$

The solution (45) can be given straight forward interpretation. It represents a plane wave incident from the left with an amplitude A, a reflected wave with an amplitude B propagating toward the left and an exponentially decaying part extending in the region $x > 0$.

The current density is given by

$$j = \hbar/2im\,(\varphi^* \,d\varphi/dx - d\varphi^*/dx\, \varphi) = Re\,(\varphi^*\, \hbar/im\, d\varphi/dx) \qquad ...\ (50)$$

Hence, the current densities associated with the wave functions Ae^{ikx} and Be^{-ikx}, called incident and reflected current densities are given by

$$j_{inc} = \hbar\, k/m\, |A|^2 \qquad ...\ (51)$$

and

$$j_{ref} = \hbar\, k/m\, |B|^2 \qquad ...\ (52)$$

Therefore, the reflection coefficient (Reflectance or Reflectivity)

$$R = j_{ref}/j_{inc} = |B/A|^2 = (B/A)^*\,(B/A) = 1 \qquad ...\ (53)$$

where use has been made of eqn. (47).

Thus the reflection is total, that is, the incident wave is totally reflected and there is no transmission. (Transmission coefficient or transmittance or transmissivity $T = 0$). Also because of the fact that φ is real in the region $x > 0$, j_{train} vanishes, see eqn. (50). In other words, though φ has a finite value in the region to the right of the potential step, there is no permanent penetration or in other words all particles are reflected including those penetrating slightly in the region $x > 0$. Let us try to understand this. For a stationary state, equation of continuity $\partial\rho/\partial t + dj/dx = 0$, reduces to $dj/dx = 0$ which indicates that the current density j is independent of x or has same value at all points x. Now

$$j = \hbar/2im\,(\varphi^*\, d\varphi/dx - d\varphi^*/dx\, \varphi)$$

when calculated with the wave function (49) gives zero. Hence there is no net current anywhere at all. To the left of potential step the relation $|A|^2 = |B|^2$ ensures that the incident and the reflected probability currents cancel one another. If there is no current, there is no momentum in the state (45*a*). In order to observe the particle in the exponential tail (region $x > 0$) it must be localized within the distance of order $\Delta x \simeq \hbar/2\,\Delta p = h/(2 \times 2\pi \times \Delta p) = \lambda/(2 \times 2\pi) = 1/(2 \times 2\pi/\lambda) = 1/2\,\alpha$ where de Broglie relation has been utilized. Hence its momentum must be uncertain by $\Delta p \simeq \hbar/2\Delta x = \hbar\alpha$. Therefore uncertainty in K.E. = $(\Delta p)^2/2m = (\hbar\alpha)^2/2m = V_0 - E$. Hence its final energy E which is E + the added K.E. for localizing the particle is sufficient to raise it into the region $x > 0$.

(*ii*) $E > V_0$ Classically this particle passes the potential step with altered velocity (due to change in K.E. in the two regions) and keeps on moving in the +x direction because of the (+)ve K.E. even in the region $x > 0$. To study the quantum mechanical behaviour of the particle, we will have to solve the Schrödinger equation,

$$d^2 \varphi(x)/dx^2 + k^2 \varphi(x) = 0, \; k^2 = 2m\,E/\hbar^2, \; x < 0 \qquad \text{... (54)}$$

$$d^2 \varphi(x)/dx^2 + k_1^2 \varphi(x) = 0, \; k_1^2 = 2m\,(E - V_0)/\hbar^2, \; x > 0 \qquad \text{... (55)}$$

The general solutions of the above equations are given by

$$\varphi(x) = Ae^{ikx} + Be^{-ikx}, \quad x < 0 \text{ (oscillatory)} \qquad \text{... (56)}$$

$$\varphi(x) = Ce^{ik_1x} + De^{-ik_1x}, \; x > 0 \text{ (oscillatory)} \qquad \text{... (57)}$$

that is, the general solution has oscillatory behaviour over all space.

Two particular solutions are obtained by setting $D = 0$ or $A = 0$. The first one ($D = 0$) represents the wave incident from the left. Reflection occurs at the potential step but there is also transmission to the right. There cannot be a wave propagating in the $-x$ direction as there is no discontinuity in the potential or in other words D should be zero. The second particular solution ($A = 0$) represents a wave incident from right (De^{-ik_1x}), reflected from the potential step (Ce^{ik_1x}) and transmitted to the left (Be^{-ik_1x}). There can not be reflected wave in the region $x < 0$ as there is no discontinuity in the potential or that means A should be zero. Thus φ is doubly degenerate as there are two independent solutions corresponding to the same E. The spectrum of energy eigen values is continuous.

We consider here only the case $D = 0$. Continuity of the wave function and its derivative at $x = 0$ gives

$$A + B = C \qquad \text{... (58)}$$

$$k(A - B) = k_1 C \qquad \text{... (59)}$$

Dividing eqn. (58) by (59), one gets

$$(A + B)/(A - B) = k/k_1$$

or

$$\{(A + B) - (A - B)\}/\{(A + B) + (A - B)\} = B/A = (k - k_1)/(k + k_1) \qquad \text{... (60)}$$

After putting the value of B/A from eqn. (60) into eqn. (58), one gets

$$C/A = 2k/(k + k_1) \qquad \text{... (61)}$$

Hence reflection and transmission coefficient would be given by

$$R = j_{ref}/j_{inc} = \frac{\hbar k/m|B|^2}{\hbar k/m|A|^2} = \frac{(k - k_1)^2}{(k + k_1)^2} = \frac{(1 - k_1/k)^2}{(1 + k_1/k)^2} \qquad \text{... (62)}$$

and

$$T = j_{+rans}/j_{inc} = \frac{\hbar k_1/m|C|^2}{\hbar k/m|A|^2} = \frac{4\,kk_1}{(k+k_1)} = \frac{4k_1/k}{(1+k_1/k)^2} \quad \text{... (63)}$$

It may, however be seen that $R + T = 1$. R and T depend only on the ratio E/V_0. Fig. 3.8 shows the behaviour of R as a function of E/V_0. In terms of the refractive index of the medium

$$\mu = k_1/k\ (= \lambda/\lambda_1) = (1 - V_0/E)^{1/2}, \quad \text{... (64)}$$

R and T can be written as

$$R = \{(1-\mu)/(1+\mu)\}^2 \quad \text{... (65)}$$

$$T = 4\mu/(1+\mu)^2 \quad \text{... (66)}$$

It should be noted that the possibility of reflection of particle ($R \neq 0$) at the step when the energy of the particle is sufficient to surmount the step is a purely quantum mechanical result.

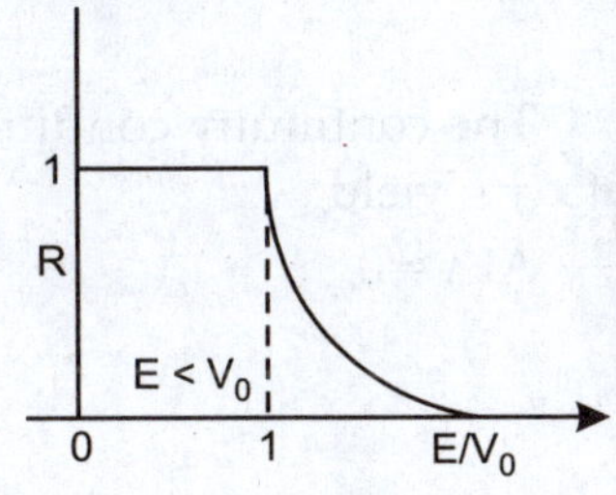

Fig. 3.8 *Reflection coefficient R for the Potential step as a function of E/V*$_0$

3.4 RECTANGULAR (OR SQUARE) POTENTIAL BARRIER

The potential energy variation in the case of a rectangular potential barrier shown in Fig. 3.9 is given by

$$V(x) = 0,\ x < 0 \text{ and } x > L \quad \text{... (67)}$$

$$= V_0,\ 0 < x < L$$

Let us consider two cases

(*i*) $0 < E < V_0$. Classically a particle of energy E if it is incident from the left would be reflected at the barrier as it cannot enter the region ($0 < x < L$) in which its K.E. is negative. To describe the behaviour of particle quantum mechanically, we will have to solve the Schrödinger equation,

$$d^2\varphi(x)/dx^2 + 2m/\hbar^2\,[E - V(x)]\,\varphi(x) = 0$$

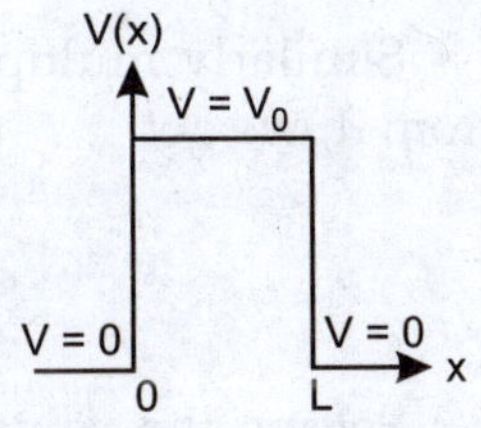

Fig. 3.9 *A rectangular potential barrier of width L and height V*$_0$

which takes the forms,

$$d^2\varphi(x)/dx^2 + k^2\varphi(x) = 0,\ k^2 = 2mE/\hbar^2,\ x < 0 \text{ and } x > L \quad \text{... (68)}$$

and

$$d^2\varphi(x)/dx^2 = \alpha^2\varphi(x),\ \alpha^2 = 2m(V_0 - E)/\hbar^2,\ 0 < x < L \quad \text{... (69)}$$

The general solutions of these equations are given by

$$\varphi(x) = Ae^{ikx} + Be^{-ikx},\ x < 0 \quad \text{... (70)}$$

$$\varphi(x) = Ce^{\alpha x} + De^{-\alpha x},\ 0 < x < L \quad \text{... (71)}$$

$$\varphi(x) = Fe^{ikx} + Ge^{-ikx},\ x > L \quad \text{... (72)}$$

Notice that we allow for waves travelling in both the directions for $x < 0$ to represent the incident and the reflected waves. We must also allow for both $e^{\alpha x}$ and $e^{-\alpha x}$ terms in the region $0 < x < L$ because x is finite and there is no danger of φ becoming infinite. We have only a wave travelling from left to right for $x > L$ as there cannot be any wave travelling from right to left (or reflected wave) since there is no discontinuity in the potential. Hence we must set $G = 0$. The solution, therefore would be

$$\varphi(x) = Fe^{ikx}, x > L \quad \text{... (73)}$$

The continuity conditions (that is, φ and $d\varphi/dx$ should be continuous) at $x = 0$ and at $x = L$ yield,

At $x = 0$,

$$A + B = C + D \quad \text{... (74)}$$

$$ik(A - B) = \alpha(C - D) \quad \text{... (75)}$$

At $x = L$,

$$Ce^{\alpha L} + De^{-\alpha L} = Fe^{ikL} \quad \text{... (76)}$$

$$\alpha(Ce^{\alpha L} - De^{-\alpha L}) = ikFe^{ikL} \quad \text{... (77)}$$

There are number of ways of solving these equations. Let us multiply eqn. (74) by α and then add it to (75), and subtract (75) from it, we get

$$C = [(\alpha + ik)A + (\alpha - ik)B]/2\alpha \quad \text{... (78)}$$

$$D = [(\alpha - ik)A + (\alpha + ik)B]/2\alpha \quad \text{... (79)}$$

Similarly multiplying (76) by α, and then adding it to (77) and subtracting (77) from it, we get

$$C = [(\alpha + ik)\, e^{-(\alpha - ik)L}\, F]/2\alpha \quad \text{... (80)}$$

$$D = [(\alpha - ik)\, e^{(\alpha + ik)L}\, F]/2\alpha \quad \text{... (81)}$$

Setting the values of C and D equal to each other from equn. (78) and (80) and similarly from equn. (79) and (81), we get

$$(\alpha + ik)A + (\alpha - ik)B = (\alpha + ik)\, e^{-(\alpha - ik)L} F \quad \text{... (82)}$$

$$(\alpha - ik)A + (\alpha + ik)B = (\alpha - ik)\, e^{(\alpha + ik)L} F \quad \text{... (83)}$$

From eqn. (83),

$$B/A = (\alpha - ik)/(\alpha + ik)\, [e^{(\alpha + ik)L} F/A - 1] \quad \text{... (84)}$$

Putting above value of B/A into (82) yields

$$(\alpha + ik) + (\alpha - ik)^2/(\alpha + ik)\, [e^{(\alpha + ik)L} F/A - 1] = (\alpha + ik)\, e^{-(\alpha - ik)L} F/A$$

or

$$(\alpha + ik)^2 + (\alpha - ik)^2\, [e^{(\alpha + ik)L} F/A - 1] = (a + ik)^2\, e^{-(\alpha - ik)L} F/A$$

or

$$(\alpha + ik)^2 - (\alpha - ik)^2 = F/A\, [(\alpha + ik)^2\, e^{-(\alpha - ik)L} - (\alpha - ik)^2\, e^{(\alpha + ik)L}]$$

or

$$F/A = \frac{4ik\alpha}{[(\alpha + ik)^2 e^{-(\alpha - ik)L} - (\alpha - ik)^2 e^{(\alpha + ik)L}]}$$

After multiplying the numerator and denominator by $e^{(\alpha - ik)L}$,

$$F/A = \frac{4\, ik\alpha e^{(\alpha - ik)L}}{[(\alpha + ik)^2 - (\alpha - ik)^2 e^{2\alpha L}]}$$

$$= \frac{4\, ik\alpha e^{(\alpha - ik)L}}{[(\alpha^2 - k^2)(1 - e^{2\alpha L}) + 2\, ik\alpha\,(1 + e^{2\alpha L})]} \quad \text{... (84')}$$

Putting the value of F/A from above into eqn. (84), we get

$$B/A = \frac{(\alpha^2 + k^2)(e^{2\alpha L} - 1)}{[(\alpha^2 - k^2)(1 - e^{2\alpha L}) + 2ik\alpha\,(1 + e^{2\alpha L})]} \quad \text{... (85)}$$

It may be mentioned here that in case one is interested in finding C/A and D/A, this can be achieved by substituting the value of F/A from (84) into equn. (80) and (81).

From (85), the reflection coefficient (or the probability of reflection) is given by

$$R = j_{ref}/j_{inc} = \frac{\hbar k/m|B|^2}{\hbar k/m|A|^2} = |B/A|^2 = (B/A)^*(B/A)$$

$$= \frac{(\alpha^2 + k^2)^2 (1 - e^{2\alpha L})^2}{[(\alpha^2 - k^2)^2 (1 - e^{2\alpha L})^2 + 4k^2\alpha^2 (1 + e^{2\alpha L})^2]}$$

After dividing the numerator and denominator by $(1 - e^{2\alpha L})^2$, one gets

$$R = \frac{(\alpha^2 + k^2)^2}{(\alpha^2 - k^2)^2 + 4k^2\alpha^2 \{(1 + e^{2\alpha L})/(1 - e^{2\alpha L})\}^2}$$

$$= \frac{(\alpha^2 + k^2)^2}{(\alpha^2 - k^2)^2 + 4k^2\alpha^2 \left(\frac{1 + e^{4\alpha L} + 2e^{2\alpha L}}{1 + e^{4\alpha L} - 2e^{2\alpha L}} - 1\right) + 4k^2\alpha^2}$$

$$= \frac{(\alpha^2 + k^2)^2}{(\alpha^2 + k^2)^2 + 4\, k^2\alpha^2 \{4(e^{2\alpha L} + e^{-2\alpha L} - 2)\}}$$

$$= \frac{(\alpha^2 + k^2)^2}{(\alpha^2 + k^2)^2 + 4k^2\alpha^2 \;\; 1/\left(\frac{e^{\alpha L} - e^{-\alpha L}}{2}\right)^2}$$

$$= \frac{(\alpha^2 + k^2)^2}{[(\alpha^2 + k^2)^2 + 4k^2\alpha^2/\sin h^2 \alpha L]} \quad \text{... (86)}$$

After substituting the values of α^2 and k^2, one gets

$$R = V_0^2/[V_0^2 + 4E(V_0 - E)/\sin h^2 \alpha L]$$

$$= \left[1 + \frac{4E(V_0 - E)}{V_0^2 \sin h^2 \alpha L}\right]^{-1} \quad \ldots (87)$$

The probability of finding the particle in the region $x > 0$, is given the name transmission coefficient and using eqn. (84′) is given by

$$T = j_{trans}/j_{inc} = \frac{\hbar k/m|F|^2}{\hbar k/m|A|^2} = |F/A|^2 = (F/A)^*(F/A)$$

$$= \frac{16k^2\alpha^2 e^{2\alpha L}}{[(\alpha^2 - k^2)^2(1 - e^{2\alpha L})^2 + 4k^2\alpha^2(1 + e^{2\alpha L})^2]}$$

$$= \frac{16k^2\alpha^2}{[(\alpha^2 - k^2)^2(e^{-2\alpha L} + e^{2\alpha L} - 2) + 4k^2\alpha^2(e^{-2\alpha L} + e^{2\alpha L} + 2)]}$$

Adding and subtracting $4k^2\alpha^2(e^{-2\alpha L} + e^{2\alpha L} - 2)$ from the denominator, one gets

$$T = \frac{16k^2\alpha^2}{[(\alpha^2 + k^2)^2(e^{2\alpha L} + e^{-2\alpha L} - 2) + 16k^2\alpha^2]}$$

$$= \frac{4k^2\alpha^2}{\left[(\alpha^2 + k^2)^2\left(\frac{e^{\alpha L} - e^{-\alpha L}}{2}\right)^2 + 4k^2\alpha^2\right]}$$

$$= \frac{4k^2\alpha^2}{[(\alpha^2 + k^2)^2 \sin h^2 \alpha L + 4k^2\alpha^2]} \quad \ldots (88)$$

Putting the values of α^2 and k^2, one gets

$$\mathrm{T} = \left[1 + \frac{V_0^2 \sin h^2 \alpha L}{4E(V_0 - E)}\right]^{-1} \quad \ldots (89)$$

One may, however check that $R + T = 1$. There are two interesting situations in which equn. (86) to (89) become simpler. Consider the purely formal limit in which $\hbar \to 0$. The quantity $\hbar$ is a physical constant, but we can consider it as a mathematical variable in order to examine the classical limit of our formulas. As $\hbar \to 0$, k and α approach infinity and hence $T \to 0$, $R \to 1$, which is, of course, the proper behaviour of a classical particle with $E < V_0$. The other interesting limit occurs for high and wide barrier, that is, when $\alpha L >> 1$. In that case $\sin h\alpha\ L \approx \frac{1}{2} e^{\alpha L}$, hence from (89) after neglecting 1 in comparison to the other term which is very large, one gets

$$T \simeq \frac{4\,E\,(V_0 - E)}{V_0^2 \left[\frac{1}{2}\, e^{\{2\,m\,(V_0 - E)/\hbar^2\}^{1/2} L}\right]^2}$$

$$\simeq 16\,E/V_0\,(1 - EV_0)\, e^{-2\{2m\,(V_0 - E)/\hbar^2\}^{1/2} L} \quad \text{... (90)}$$

The variation of the wave function $\varphi\ (x)$ (the real part of φ) vs x is shown in Fig. 3.10. The wave function has an exponential form in the forbidden region inside the barrier. But there is also an oscillatory wave to the right of

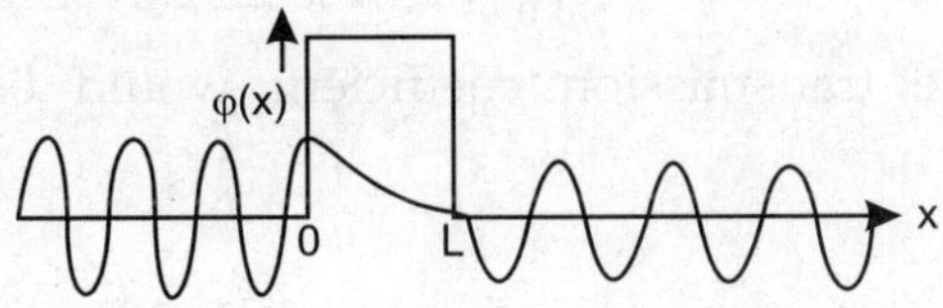

Fig. 3.10 *The real part of $\varphi(x)$ as a function of x for $E < V_0$ showing exponential decay in the barrier region and decrease in amplitude of transmitted wave*

barrier. Thus we conclude that there is finite probability of particle penetrating the barrier and appearing on the other side. The ability (or phenomenon) of particle to penetrate the barrier when $E < V_0$ is a quantum mechanical result and is termed as tunnel effect.

The tunnel diodes used in electronic circuits are based on this phenomenon. The tunnel effect forms the basis for understanding the emission of α-particles (that is, nucleus of Helium atom) from heavy nuclei. The concept also explains the tunneling of electrons through potential barrier between two superconductors.

(ii) $E > V_0$. Classically a particle with this energy will pass through the barrier. If we want to study the behaviour of the particle quantum mechanically, we will have to solve the Schrödinger equation in the relevant region. Our equations are basically unchanged but α becomes imaginary in the region $0 < x < L$, that is in the barrier region. Hence we may write $\alpha = i\beta$ where $\beta = [2m\,(E - V_0)/\hbar^2]^{1/2}$, is a real number. Substituting $\alpha = i\beta$ into equn. (86) and (88), one gets

$$R = [1 + \{4E(E - V_0)\}/\{V_0^2 \sin^2 \beta L\}]^{-1} \quad \text{... (91)}$$

and

$$T = [1 + \{V_0^2 \sin^2 \beta L\}/\{4\mathrm{E}\,(E - V_0)\}^{-1} \quad \text{... (92)}$$

One may see that once again $R + T = 1$. For $E \to V_0$, $\sin \beta L \approx \beta L$, and hence from eqn. (92).

$$T = [1 + (mV_0 L^2/2\hbar^2)]^{-1} \quad \text{... (93)}$$

For increasing E ($E > V_0$), the transmission coefficient oscillates as is clear from eqn. (92). It may be noticed that the transmission coefficient becomes unity (or the barrier becomes transparent) whenever in eqn. (92),

$$\beta L = n\pi, \quad n = 1, 2, 3, \ldots$$

or

$$[2m\,(E - V_0)/\hbar^2]^{1/2} L \;=\; n\pi$$

(or $2\pi/\lambda \quad L = n\pi$ or $L = n\lambda/2$, that is, the width of the barrier is integral multiple of half of the deBroglie wavelength in the barrier region).

Squaring both the sides of the above equation, one gets

$$E = V_0 + (n^2\pi^2\,\hbar^2/2m\,L^2)$$
$$= V_0\,[1 + (n^2\pi^2\hbar^2/2mV_0L^2)] \qquad \text{... (94)}$$

From eqn. (92), the transmission coefficient would be minimum when $\beta L = (2n + 1)\pi/2$, $n = 0, 1, 2, \ldots$

or

$$E = V_0\,[1 + (2n + 1)^2\,\pi^2\hbar^2/8mV_0L^2] \qquad \text{... (95)}$$

The minimum value of T would be given by

$$T = T_{min} = [1 + V_0^2/4E(E - V_0)]^{-1} = \left[1 + \frac{1}{4\,E/V_0\,(E/V_0 - 1)}\right]^{-1} \qquad \text{... (96)}$$

where $\sin^2 \beta L$ has been put equal to 1 in (92).

As E decreases below V_0, T decreases monotonically as from eqn. (89),

$$T = [1 + \{V_0^2 \sin h^2\,\alpha L/4E\,(V_0 - E)\}]^{-1}$$

and when $\alpha L >> 1$, $\sin h\alpha L = (e^{\alpha L} - e^{-\alpha L})/2 \simeq \frac{1}{2}\,e^{\alpha L}$. Hence the second term in the denominator of T dominates and therefore 1 can be neglected. One ends with the result,

$$T \simeq \frac{4\,E\,(V_0 - E)}{V_0^2\,(1/2\,e^{\alpha L})^2} = 16\,E/V_0\;\;(1 - E/V_0)\,e^{-2\alpha L} \qquad \text{... (97)}$$

The behaviour of T vs E/V_0 for a given value of $2mV_0L^2/\hbar^2$ is shown in the Fig. 3.11. It may however be noted that even in the case of $E > V_0$ there is finite probability of reflection which is a quantum mechanical result.

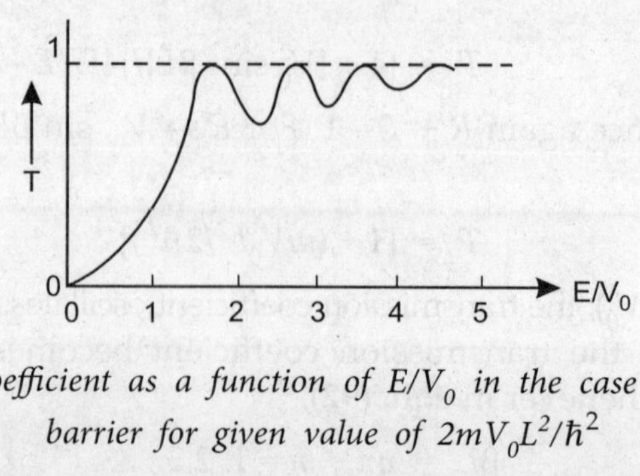

Fig. 3.11 *Transmission coefficient as a function of E/V_0 in the case of a rectangular potential barrier for given value of $2mV_0L^2/\hbar^2$*

3.5 APPLICATION OF BARRIER PENETRATION—ALPHA DECAY

Nuclei which contain 210 or more nucleons are so large that the short range nuclear forces that hold them together (or which are of attractive nature and are approximately proportional to their mass number A, that is, number of nucleons in them) are hardly able to counter balance the mutual repulsion (which is roughly proportional to Z^2, that is, square of the atomic number) of their protons. Such unstable nuclei (to be designated as parent nuclei) emit alpha particles ($^4_2$He nuclei) and get converted into new nuclei (called daughter nuclei) with atomic number less by two and mass number less by four and in turn increasing their stability by reducing their size. The alpha emission or decay is represented as

$$^{226}_{88}\text{Ra (Radium)} \rightarrow {}^{222}_{86}\text{Rn (Radon)}$$
$$+ {}^4_2\text{He (alpha particle with energy} \simeq 5 \text{ MeV)} \quad \ldots (98)$$

While a heavy nucleus can, in principle, spontaneously reduce its size by alpha decay, there remains the problem of how an alpha particle can actually escape from the nucleus. Fig. 3.12 shows a plot of the potential energy $V(x)$ of an alpha particle as a function of its distance x from the centre of daughter nucleus, Rn (Radon) in this case. Also indicated on the graph is the distance r_0 at which the strong interaction (*) with the Rn nucleus takes over ($r_0 = 1.1\ [(222)^{1/3} + (4)^{1/3}] \approx 8.41\ F$) and then remains constant throughout the nucleus. It is around this distance in the surface region of the parent Ra nucleus that the alpha particle can be considered to be formed. The graph immediately tells why the alpha particle finds it difficult to escape even if it is formed.

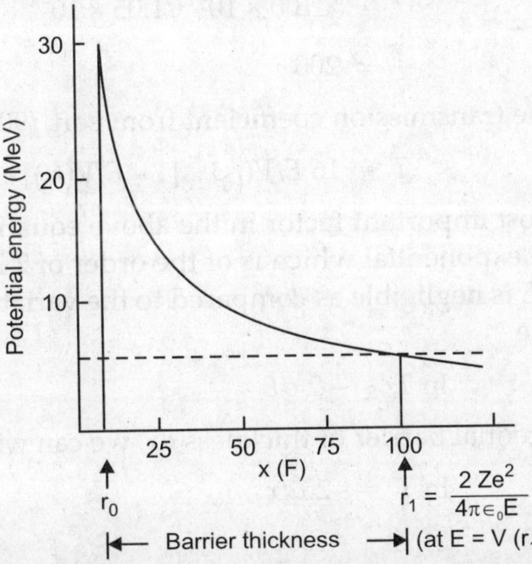

Fig. 3.12 *Potential energy of an α particle as a function of its distance from the centre of nucleus of charge Ze (that is, Ze is the nuclear charge (Z + 2) e, the charge of parent nucleus minus the α-particle charge of 2e)*

*This strong interaction normally represented by V_0 is approximately equal to – 50 MeV in this case but has not been shown in the figure as none of the results to be derived depend on the magnitude of V_0.

At $x = r_0$, the height of the potential barrier (Coulomb potential) is approximately equal to 30 MeV (= $2Ze^2/4\pi\epsilon_0 r_0$) which is equal to the work that must be done against the repulsive electrostatic force to bring an alpha particle from infinity ($x \to \infty$, $V(x) \to 0$) to a position adjacent to the Rn nucleus but just outside the range of its attractive nuclear forces. This height is very much greater than the energy of the emitted alpha particle which is about 5 MeV. We may, therefore regard an alpha particle in such a nucleus as being inside the box whose walls require an energy of 30 MeV to be surmounted. The energy of the alpha particle emitted by Ra nucleus being only about 5 MeV as mentioned earlier is 25 MeV short of the energy needed for escape, that is, classically it cannot decay or cross the barrier. Also classically, it cannot enter the barrier and is free to move only at distances greater than r_1 where r_1 is given by

$$E = 2Ze^2/4\pi\epsilon_0 r_1, \ (Z = Z_{\text{Radon}}) \qquad \text{... (99)}$$

r_1 is the classical distance of closest approach to the nucleus of an alpha particle of energy E (which is totally K.E. at large x) coming from the outside. One may calculate the value of r_1 from the above relation which comes out to be approximately 100 F for Rn. Quantum mechanics, however, allows it to tunnel through and we now estimate this important tunneling probability.

In the present case,

$$\begin{aligned} \alpha L &= [2m\,\{V(x = r_0) - E\}/\hbar^2]^{1/2}\, L \\ &= (2 \times 4 \times 1.6 \times 10^{-27} \times 25 \times 10^6 \times 1.6 \times 10^{-19})^{1/2} \\ &\quad \times 100 \times 10^{-15}/1.05 \times 10^{-34} \\ &\simeq 200 \end{aligned}$$

that is, $\alpha L \gg 1$. Hence the transmission coefficient from eqn. (90) would be given by

$$T = 16\, E/V(r_0) \quad [1 - E/V(r_0)]\, e^{-2\alpha L} \qquad \text{... (100)}$$

Since $\alpha L \gg 1$, the most important factor in the above equation is the exponential. The factor in front of the exponential which is of the order of 2 is not significant since its variation with V and E is negligible as compared to the variation in the exponential itself. Hence we can write

$$\ln T \simeq -2\,\alpha L \qquad \text{... (101)}$$

For a rectangular potential barrier of thickness dx, we can write

$$\ln T = -2\alpha dx \qquad \text{... (102)}$$

where

$$\begin{aligned} \alpha^2 &= 2m/\hbar^2 \quad [V(x) - E] \\ &= 2m/\hbar^2 \quad [(2Ze^2/4\pi\epsilon_0 x) - E] \end{aligned}$$

making α, a function of x.

This is expression for the transmission coefficient (or tunneling probability) of a rectangular barrier. The actual barrier encountered by alpha particle has an exponential tail (or is variable in height) as shown in Fig. 3.13. We can approximate it as consisting of many rectangular barriers of decreasing

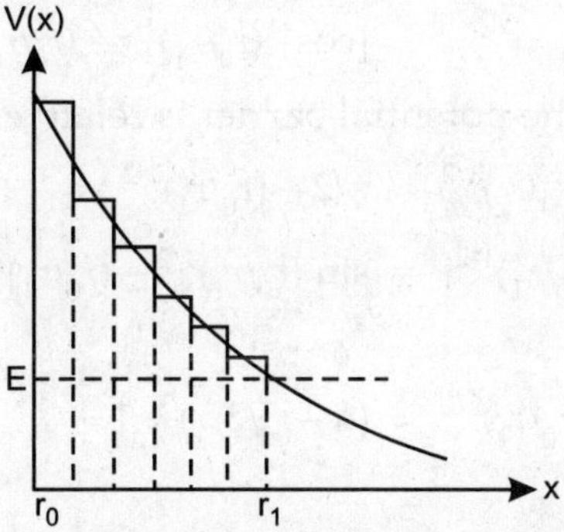

Fig. 3.13 *Barrier approximated by series of rectangular barriers*

height and obtain the total probability by summing the tunneling probability of each barrier over the region between r_0 and r_1. In this entire region, of course $E < V$. Hence taking the summation over all the rectangular potential barriers, one gets

$$\ln T = -2\int_{r_0}^{r_1} \alpha(x)\,dx$$

$$= -2\int_{r_0}^{r_1} (2m/\hbar^2)^{\frac{1}{2}} (2Ze^2/4\pi \in_0 x - E)^{1/2}\,dx$$

Making use of eqn. (99), one leads to

$$\ln T = -2\,(2m\,E/\hbar^2)^{1/2} \int_{r_0}^{r_1} (r_1/x - 1)^{1/2}\,dx$$

Putting $x = r_1 \cos^2 \theta$, $dx = r_1\, 2\cos \theta\,(-\sin\theta\, d\theta)$ and also changing the limits to θ (at $x = r_0$, $\theta_0 = \cos^{-1}(r_0/r_1)^{1/2}$, and at $x = r_1$, $\theta = 0$), one gets

$$\ln T = -2(2m\,E/\hbar^2)^{1/2} r_1 \int_0^{\theta_0} 2\,(1/\cos^2\theta - 1)^{1/2} \sin\theta \cos\theta\, d\theta$$

Since

$$(1/\cos^2\theta - 1)^{1/2} = (1 - \cos^2\theta)^{1/2}/\cos\theta = \sin\theta/\cos\theta,$$

$$\ln T = -2(2m\,E/\hbar^2)^{1/2}\, r_1 \int_0^{\theta_0} 2\sin^2\theta\, d\theta$$

$$= -2\,(2m\,E/\hbar^2)^{1/2}\, r_1 \int_0^{\theta_0} (1 - \cos 2\theta)\, d\theta$$

$$= -2\,(2m\,E/\hbar^2)^{1/2}\, r_1\,(\theta_0 - \sin\theta_0 \cos\theta_0)$$

After putting the value of E from eqn. (99),

$$\ln T = -2\,[2m/\hbar^2\; 2Ze^2/4\pi\epsilon_0 r_1]^{1/2}$$

$$[\cos^{-1}(r_0/r_1)^{1/2} - (r_0/r_1)^{1/2}(1 - r_0/r_1)^{1/2}] \quad \text{... (103)}$$

Because of the fact that the potential barrier is relatively wide, $r_1 >> r_0$,

$$\cos^{-1}(r_0/r_1)^{1/2} \approx \pi/2 - (r_0/r_1)^{1/2}$$

as $\quad \cos\{\pi/2 - (r_0/r_1)^{1/2}\} = \sin(r_0/r_1)^{1/2} \simeq (r_0/r_1)^{1/2}$

Also

$$(1 - r_0/r_1)^{1/2} \approx (1 - r_0/2r_1) \approx 1$$

Hence from eqn. (103)

$$\ln T = -2\,[2m/\hbar^2 \quad 2Ze^2/4\pi\epsilon_0 r_1]^{1/2}\,[\pi/2 - 2(r_0/r_1)^{1/2}]$$

Replacing r_1 by $r_1 = 2Ze^2/4\pi\epsilon_0 E$ and simplifying,

$$\ln T = 4e/\hbar\,(m/\pi\epsilon_0)^{1/2} Z^{1/2} r_0^{\;1/2} - e^2/\hbar\epsilon_0\,(m/2)^{1/2} ZE^{-1/2}$$

After substituting the values of all the constants, one gets

$$\ln T = 2.97\, Z^{1/2} r_0^{\;1/2} - 3.95\, ZE^{-1/2} \quad \text{... (104)}$$

where E is in MeV and r_0 is in fermis.

The decay constant λ (the decay probability per unit time) is given by

$$\lambda = \nu T$$

where ν is the number of times per second an alpha particle within a nucleus strikes the potential barrier around it and T is nothing but the probability that the particle would be transmitted through the barrier. If we suppose that at any moment only one alpha particle exists as such in a nucleus and it moves back and forth in the nuclear diameter.

$$\nu = v/2\,r_0$$

where v is the alpha particle velocity when it finally leaves the nucleus. Substituting the typical values for $v = 2 \times 10^7$ m/sec and $r_0 \simeq 10$ F, respectively, we get

$$\nu = 10^{21}\,\text{sec}^{-1}$$

Hence

$$\lambda = 10^{21}\,T$$

Using the value of $\ln T$ from eqn. (104), one gets

$$\ln\lambda = \ln 10^{21} + 2.97\, Z^{1/2} r_0^{\;1/2} - 3.95\, ZE^{-1/2} \quad \text{... (105)}$$

The form of the above relation resembles with the empirical Geiger—Nuttal law and is in good agreement with the experimental observations. However, it may be noted that the above discussion (or quantum mechanical theory developed by Gamow[1] in 1928) is limited to one dimensional case only whereas in actual sense the motion of alpha particle is three dimensional.

PROBLEMS

3.1 Find the energy levels and wave functions of a particle moving in a one dimensional box described by the potential energy function,

$$V(x) = 0, -L < x < +L \text{ (or } |x| < L)$$
$$= \infty, x \geq L, x \leq -L \text{ (or } |x| > L)$$

3.2 A particle is trapped in a one dimensional box whose length is L. Its energy is $\pi^2\hbar^2/(2mL^2)$. Where in the box, is the particle most likely to be found? (or probability of finding the particle is maximum).

3.3 Consider a particle of mass 9.1×10^{-31} kg and energy $E = 342$ eV confined in a one dimentional box of length 10^{-10} meter. Calculate

(*a*) the quantum number n of this state of the system and (*b*) the energy required to take the system to its next excited state.

3.4 Find out the ground state energy (or also called zero point energy) of a particle confined in one dimensional box of length L using uncertainty relation.

3.5 Consider a particle confined inside a 3-dimensional box represented by the potential energy function.

$$V(x, y, z) = 0, \text{ for } 0 < x < L, 0 < y < L, 0 < z < L$$
$$= \infty, \text{ elsewhere}$$

Obtain for the particle, (*a*) the energy eigen values, (*b*) the wave functions corresponding to the energy eigen values and (*c*) the degeneracy of the first and second excited states.

3.6 Calculate the eigen values and eigen functions for a particle confined in a two dimensional square box of length L, assuming that the particle is free to move inside the box, that is

$$V(x, y) = 0, 0 < x < L, 0 < y < L$$
$$= \infty, L \leq x \leq 0, L \leq y \leq 0$$

3.7 In the case of a square potential well with finite walls, consider the limiting case $V_0 \to \infty$ and obtain energy levels and energy eigen functions.

3.8 Consider a beam of electrons of energies (*i*) 0.03 eV and (*ii*) 0.045 eV incident on a one dimensional potential barrier of height 0.035 eV. Calculate the percentage of electrons transmitted in each case.

---

[1] G. Gamow, Z. Physik 51, 204 (1928).

3.9 A beam of electrons of energy 2.5 eV falls on one-dimensional reactangular potential barrier of heights 5 eV and width 1 Å. Calculate the percentage of electrons transmitted.

3.10 A one kg ball rolls slowly on a level track, then encounters a hump of sinusoidal cross-section, 1 m long and 0.1 m high. The kinetic energy of the ball is neglected. What is the probability that it can overcome the hump?

3.11 Consider a square well potential given by

$$V(x) = -V_0, 0 < x < L$$

$$= 0, x < 0 \text{ and } x > L$$

Calculate the reflection and transmission coefficients when particle is incident from the left with energy $E > 0$. At what energies of the electron will there be no reflection?

3.12 A particle is restricted to move along a circular track on which the potential is constant. Find the allowed energy values and the energy eigen functions. Show that the energy levels except the ground state are doubly degenerate.

SOLUTIONS

3.1 Within the box ($-L < x < +L$), the one-dimensional Schrödinger equation takes the form

$$d^2 \varphi(x) dx^2 + k^2 \varphi(x) = 0, k^2 = 2m\, E/\hbar^2 \quad \text{... (3.1)}$$

where V has been taken to be zero in this region.

The above equation has two possible solutions, *viz.* $A \sin kx$ or $B \cos kx$. Hence the general solution of the above equation would be given by

$$\varphi(x) = A \sin kx + B \cos kx \quad \text{... (3.2)}$$

Applying boundary conditions ($\varphi = 0$ at $x = \pm L$), we get

$$0 = A \sin kL + B \cos kL$$

and

$$0 = -A \sin kL + B \cos kL$$

from which by subtracting and adding, we obtain

$$A \sin kL = 0, B \cos kL = 0 \quad \text{... (3.3)}$$

Now we do not want both A and B to be zero. Since this would give the physically uninteresting solution $\varphi = 0$ everywhere (as in this particular problem, particle is to be found some where in the box). Also, we cannot make both $\sin kL$ and $\cos kL$ equal to zero for a given value of k(or E) as both sine and cosine functions for the same argument cannot be zero. There are thus two possible classes of solutions†.

For the first class,

$$A = 0 \text{ and } \cos kL = 0 \quad \text{... (3.4a)}$$

and for the second class,

$$B = 0 \text{ and } \sin kL = 0 \quad \text{... (3.4b)}$$

Thus

$$kL = n\pi/2 \quad (3.5)$$

where n is an odd integer for the first class and an even integer for the second class, The two classes of solutions and their energy eigen values are then

$$\varphi_n(x) = B \cos n\pi x/2L,\ n = 1, 3, 5, \ldots \quad (3.6a)$$

and

$$\varphi_n(x) = A \sin n\pi x/2L,\ n = 2, 4, 6, \ldots \quad (3.6b)$$

(The constants A and B can easily be found out in each case so that the eigen functions $\varphi_n(x)$ are normalized. A and B come out to be $(1/L)^{1/2}$) and

$$(kL)^2 = (2mE/\hbar^2)\, L^2 = (n\pi/2)^2$$

or

$$E = E_n = n^2\pi^2\hbar^2/8mL^2 \quad \text{... (3.7)}$$

in both the cases, that is $n = 1, 2, 3, \ldots$

3.2 The energy of the particle being $\pi^2\hbar^2/(2mL^2)$ implies that the particle is in the state $n = 1$. Probability density in the present case would be given by

$$|\varphi(x)|^2 = 2/L \quad \sin^2 \pi x/L \quad \text{... (3.8)}$$

For finding the value of x at which the probability is maximum, one has to differentiate (3.8) w.r.t. x and equate the result to zero. This would give the value of x corresponding to either maximum or minimum.

$$d/dx\,(2/L \sin^2 \pi x/L) = 0$$

or

$$2/L \times 2 \sin \pi x/L \cos \pi x/L \times \pi/L = 0$$

or $\quad \sin 2\pi x/L = 0 \Rightarrow 2\pi x/L = m\pi,\ m = 0, 1, 2, \ldots$

Hence

$$x = Lm/2 = 0 \text{ for } m = 0$$

$$= L/2 \text{ for } m = 1 \quad \text{... (3.9)}$$

$$= L \text{ for } m = 2$$

†Whenever, the potential energy function V(x) = V(–x) that is symmetric with respect to x, the solutions divide into two classes. See problem 11.1 for the discussion.

Differentiating $\sin 2\pi x/L$ once again w.r.t. x, one gets

$$d/dx \sin 2\pi x/L = 2\pi/L \cos 2\pi x/L$$

$$= +2\pi/L \text{ for } x = 0 \qquad \text{... (3.10)}$$

$$= -2\pi/L \text{ for } x = L/2$$

Hence the probability of finding the particle is maximum at the middle of the well.

3.3 Energy of the nth energy state in the present case is given by

$$E_n = n^2\pi^2\hbar^2/(2mL^2),\ n = 1, 2, 3, \ldots \qquad \text{... (3.11)}$$

Putting the given data,

$$342 \times 1.6 \times 10^{-19} = \frac{(3.14)^2 \times (1.054 \times 10^{-34})^2\, n^2}{2 \times 9.1 \times 10^{-31} \times (10^{-10})^2}$$

one gets, $n = 3$

Energy required to take the system to $n = 4$, would be given by

$$E_4 - E_3 = \pi^2\,\hbar^2/2mL^2 \quad (4^2 - 3^2)$$

$$= (608 - 342) \text{ eV}$$

$$= 266 \text{ eV} \qquad \text{... (3.12)}$$

3.4

$$\Delta x = L,$$

hence

$$\Delta p_x = \hbar/2L \qquad \text{... (3.13)}$$

K.E. or lowest energy or zero point energy of the particle would be given by

$$E = (\Delta p_x)^2/2m = \hbar^2/8mL^2 \qquad \text{... (3.14)}$$

3.5 The Schrödinger equation in the present case would be given by

$$(\partial^2/\partial x^2 + \partial^2/\partial y^2 + \partial^2/\partial z^2)\,\varphi\,(x, y, z) + 2m/\hbar^2 E\,\varphi\,(x, y, z) = 0,$$

$$0 < x < L, \quad 0 < y < L, 0 < z < L \qquad \text{... (3.15)}$$

Using the method of separation of variables, we write

$$\varphi(x, y, z) = X(x)\,Y(y)\,Z(z)$$

Substituting above in eqn. (3.15) and dividing the resulting equation by $X(x)\,Y(y)\,Z(z)$, we obtain

$$\frac{1}{X}\frac{d^2 X}{dx^2} + \frac{1}{Y}\frac{d^2 Y}{dy^2} + \frac{1}{Z}\frac{d^2 Z}{dz^2} = -2mE/\hbar^2 \qquad \text{... (3.16)}$$

In this equation, each term in the sum depends only on one variable and therefore each term must be equal to a constant. Let

$$\frac{1}{X}\frac{d^2 X}{dx^2} = -k_x^2, \quad \frac{1}{Y}\frac{d^2 Y}{dy^2} = -k_y^2, \quad \frac{1}{Z}\frac{d^2 Z}{dz^2} = -k_z^2 \qquad \text{... (3.17)}$$

with

$$k_x^2 + k_y^2 + k_z^2 = 2mE/\hbar^2 \qquad \text{... (3.18)}$$

The solution of the X part of eqn. (3.17) is given by

$$X(x) = A \sin k_x x + B \cos k_x x \qquad \text{... (3.19)}$$

Applying the boundary conditions, $X(x = 0) = X(x = L) = 0$, one gets $B = 0$ and $\sin k_x L = 0$ or

$$k_x = n_x \pi/L, \quad n_x = 1, 2, 3, \ldots \qquad \text{... (3.20)}$$

The results for Y and Z are similar. Hence

$$k_y = n_y \pi/L, \quad n_y = 1, 2, 3, \ldots \qquad \text{... (3.21)}$$

$$k_z = n_z \pi/L, \quad n_z = 1, 2, 3, \ldots \qquad \text{... (3.22)}$$

Using (3.20) to (3.22) and (3.18), one gets

$$E_{n_x n_y n_z} = \pi^2 \hbar^2/2mL^2 \left(n_x^2 + n_y^2 + n_z^2\right); \; n_x, n_y, n_z = 1, 2, 3, \ldots \qquad \text{... (3.23)}$$

The corresponding normalized wave functions are,

$$\varphi(x, y, z) = X(x)\, Y(y)\, Z(z) = (8/L^3)^{1/2} \sin n_x \pi x/L \; \sin n_y \pi y/L \sin n_z \pi z/L \qquad \text{... (3.24)}$$

The first excited state is triply degenerate as

for $n_x = 1, n_y = 1, n_z = 2$

or

for $n_x = 1, n_y = 2, n_z = 1$

or

for $n_x = 2, n_y = 1, n_z = 1$,

the energy eigen value is given by

$$E = \frac{\pi^2 \hbar^2}{2mL^2}(1^2 + 1^2 + 2^2) = \frac{3\pi^2 \hbar^2}{mL^2} \qquad \text{... (3.25)}$$

The second excited state is also triply degenerate as for

for $n_x = 1, n_y = 2, n_z = 2$

or

for $n_x = 2, n_y = 2, n_z = 1$

or

for $n_x = 2, n_y = 1, n_z = 2$,

the energy eigen value is given by

$$E = \frac{\pi^2 \hbar^2}{2mL^2}(1^2 + 2^2 + 2^2) = \frac{9\pi^2 \hbar^2}{2mL^2} \qquad \text{... (3.26)}$$

3.6 The time-independent Schrödinger equation for a particle inside the box is

$$\partial^2 \varphi(x, y)/\partial x^2 + \partial^2 \varphi(x, y)\, \partial y^2 + 2m/\hbar^2 E\varphi(x, y) = 0 \qquad \text{... (3.27)}$$

where $V(x, y)$ has been taken to be zero in the case of a free particle. Using the method of separation of variables, we write

$$\varphi(x, y) = X(x)\, Y(y) \qquad \text{... (3.28)}$$

and substituting above in eqn. (3.27) and then dividing the resulting equation by $\varphi(x, y)$, we get

$$\frac{1}{X(x)}\frac{d^2 X(x)}{dx^2} + \frac{1}{Y(y)}\frac{d^2 Y(y)}{dy^2} = \frac{-2mE}{\hbar^2} \qquad \text{... (3.29)}$$

In the above equation each term in the sum depends only on one variable and therefore each term must be equal to a constant. Let

$$\frac{1}{X(x)}\frac{d^2 X(x)}{dx^2} = -k_x^2, \; \frac{1}{Y(y)}\frac{d^2 Y(y)}{dy^2} = -k_y^2 \qquad \text{... (3.30)}$$

with

$$k_x^2 + k_y^2 = 2mE/\hbar^2 \qquad \text{... (3.31)}$$

The solution of the X-part of eqn. (3.30) is given by

$$X(x) = A \sin k_x L + B \cos k_x L \qquad \text{... (3.32)}$$

Using the boundary conditions $\varphi(x = 0) = \varphi(x = L) = 0$, one gets $B = 0$ and

$$\sin k_x L = 0$$

or

$$k_x = n_x \pi/L, \quad n_x = 1, 2, \ldots \qquad \text{... (3.33)}$$

Similarly, we can obtain

$$k_y = n_y \pi/L, \quad n_y = 1, 2, \ldots \qquad \text{... (3.34)}$$

Using (3.33), (3.34) and (3.31), the energy eigen values in the present case are found to be

$$E_{n_x n_y} = \pi^2\hbar^2 (n_x^2 + n_y^2)/2mL^2, \; n_x, n_y = 1, 2, 3, \ldots \qquad \text{... (3.35)}$$

The corresponding normalized wave functions are

$$\varphi(x, y) = X(x)\, Y(y) = (2/L) \sin n_x \pi x/L \sin n_y \pi y/L \qquad \text{... (3.36)}$$

3.7 For $V_0 \to \infty$, $\alpha \to \infty$. Hence $\varphi(x) \to 0$ for $x \leq -L$ and similarly for $x \geq +L$ which it should as the particle cannot have infinite amount of energy or cannot exist in the region in which energy is infinite.

The two classes of solutions would be,

If
$$D = 0, \varphi(x) = C \cos kx \quad \text{... (3.37)}$$

and after applying the boundary condition

$$\varphi(L) = C \cos kL = 0$$

or

$$kL = \pi/2, 3\pi/2, 5\pi/2, \ldots$$
$$= 1 \times \pi/2, 3 \times \pi/2, 5 \times \pi/2, \ldots$$

or

$$k = 1 \times \pi/2L, 3 \times \pi/2L, 5 \times \pi/2L, \ldots \quad \text{... (3.38)}$$

If
$$C = 0, \varphi(x) = D \sin kx \quad \text{... (3.39)}$$

and after applying the boundary condition,

$$\varphi(L) = D \sin kL = 0$$

or

$$kL = \pi, 2\pi, 3\pi, \ldots$$
$$= 2 \times \pi/2, 4 \times \pi/2, 6 \times \pi/2$$

or

$$k = 2 \times \pi/2L, 4 \times \pi/2L, 6 \times \pi/2L, \ldots \quad \text{... (3.40)}$$

The even and odd solutions still alternate in energy, the energy eigen values are

$$E = E_n = \hbar^2 k^2/2m = \hbar^2 (n\pi/2L)^2/2m$$
$$= n^2\hbar^2 \pi^2/8\, mL^2, \quad n = 1, 2, 3, \ldots \quad \text{... (3.41)}$$

Putting the values of k from (3.38) into (3.37) and from (3.40) into (3.39).

$$\varphi(x) = C \cos n\pi x/2L, \quad n = 1, 3, 5, \ldots \quad \text{... (3.42)}$$

and

$$\varphi(x) = D \sin n\pi x/2L, \quad n = 2, 4, 6, \ldots \quad \text{... (3.43)}$$

The constants C and D can be found out from the normalization condition,

$$\int_{-L}^{+L} |\varphi(x)|^2\, dx = 1$$

and they are found to be $(1/L)^{1/2}$.

3.8 (*i*) This corresponds to $E < V_0$ case. Hence the reflection coefficient $R = 1$ and the transmission coefficient $T = 0$ or in other words all the electrons incident would be totally reflected (and nothing would be transmitted).

(*ii*) This corresponds to $E > V_0$ case. Hence the reflection coefficient.

$$R = \frac{(1-k_1/k)^2}{(1+k_1/k)^2} = \frac{[1-\{(E-V_0)/E\}^{1/2}]^2}{[1+\{(E-V_0)/E\}^{1/2}]^2}$$

$$= \frac{[1-\{(0.045-0.035)/0.045\}^{1/2}]^2}{[1+\{(0.045-0.035)/0.045\}^{1/2}]^2}$$

$$\simeq 0.13$$

Therefore, $T = 1 - R = 0.87$

That is, 13% of the incident electrons would be reflected and the rest 87% would be transmitted.

3.9 The situation corresponds to $E < V_0$ case. Hence the transmission coefficient is given by

$$T = [1 + V_0^2 \sin h^2 \alpha L/\{4E\,(V_0 - E)\}\,]^{-1}$$

$$= [1 + \sin h^2 \alpha L/\ \{4E/V_0)\,(1 - E/V_0)^{-1}$$

$$\alpha L = [2m\,(V_0 - E)/\hbar^2]^{1/2}\,L$$

$$= (2 \times 9.1 \times 10^{-31} \times 2.5 \times 1.6 \times 10^{-19})^{1/2} \times 10^{-10}/$$

$$(1.054 \times 10^{-34})$$

$$\simeq 0.8$$

Hence, $\sin h\,\alpha L = \sin h\,0.8 = 0.888$

Therefore,

$$T = [1 + (0.888)^2/\{4 \times 0.5 \times (1 - 0.5)]^{-1}$$

$$\simeq 0.56$$

That is, 56% of the electrons incident would be transmitted.

3.10 The potential energy function is sinusoidal as shown in the Fig. 3.14 and is mathematically represented by

$$V(x) = m\,g\,h \sin \pi x/L \qquad \text{... (3.44)}$$

one may verify the $V(x)$ values at $x = 0$, $L/2$ and at L. The case corresponds to $E < V_0$. The transmission coefficient or the tunneling probability in the case of rectangular potential barrier for

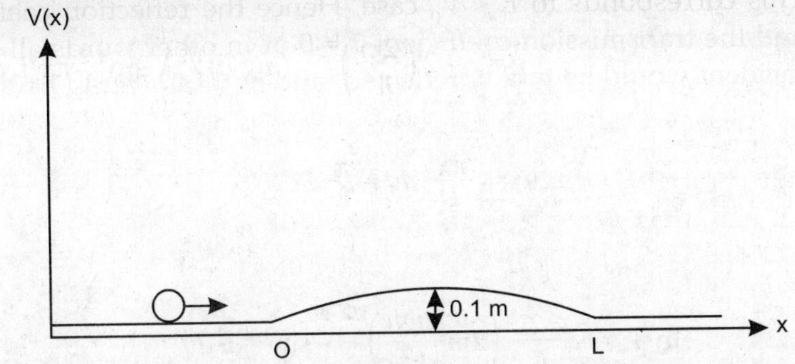

Fig. 3.14 *Sinusoidal potential barrier*

$$\alpha L = [2m(V_0 - E)/\hbar^2]^{1/2} L$$

$$= [2 \times 1 \times (1 \times 9.8 \times 0.1)]^{1/2} \times 1/(1.054 \times 10^{-34})$$

or $$\alpha L \gg 1$$

where E has been taken equal to zero as K.E. of the ball is to be neglected and V is already zero on the level track, is given by

$$T = 16\,(E/V_0)\,(1 - E/V_0)e^{-2\alpha L}$$

or $$\ln T = -2\alpha L$$

where the first term has been neglected compared to the second one and the result is actually for the rectangular potential barrier. In the present case the barrier can be approximated by a series of barriers of thickness dx (see Fig. 3.15) each and the total probability can be obtained by summing the tunneling probability of each barrier over the region between 0 and L. Hence

$$\ln T = -2\int_0^L \left[\frac{2m}{\hbar^2} V(x)\right]^{1/2} dx \qquad \dots (3.45)$$

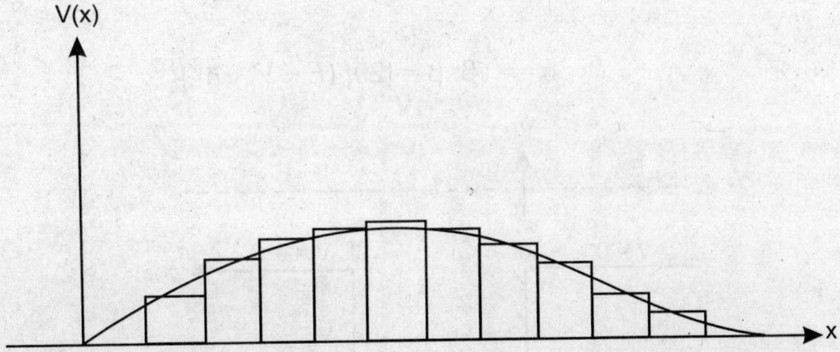

Fig. 3.15 *Barrier represented by series of rectangular barriers*

$$= -2\left(\frac{2m}{\hbar^2}\right)^{1/2} (mg\,h)^{1/2} \int_0^L \sin^{1/2} \pi x/L\,dx$$

Putting, $$y = \frac{\pi x}{L},\ dy = \frac{\pi}{L}\,dx$$

Hence

$$\ln T = \frac{-2L}{\pi}\left(2m\frac{mgh}{\hbar^2}\right)^{1/2} \int_0^{\pi} \sin^{1/2} y\,dy$$

$$= \frac{-2L}{\pi}\left(2m^2\frac{gh}{\hbar^2}\right)^{1/2} \sqrt{2}\,\pi \text{ as} \int_0^{\pi} \sin^{1/2} y dy = \gamma\left(\frac{3}{4}\right)\gamma\left(\frac{1}{4}\right)$$

$$= \frac{-4\,Lm}{\hbar}(gh)^{1/2}$$

$$= \frac{-4\times 1\times 1\times(9.8\times 0.1)^{1/2}}{1.05\times 10^{-34}}$$

$$\simeq -4\times 10^{34}$$

or

$$T = e^{-4\times 10^{34}}$$

3.11 The given square well is shown in Fig. 3.16. Truly speaking the square barrier becomes square well if in eqn. (67), V_0 is replaced by $-V_0$ in the region $0 < x < L$, V_0 being positive. Hence in the present case $k^2 = 2mE/\hbar^2$ and

$$\alpha^2 = 2m\,(-V_0 - E)/\,\hbar^2,\ 0 < x < L$$

or

$$\alpha = i\beta,\ \beta = \{2m\,(E + V_0)/\hbar^2\}^{1/2} \quad \text{... (3.46)}$$

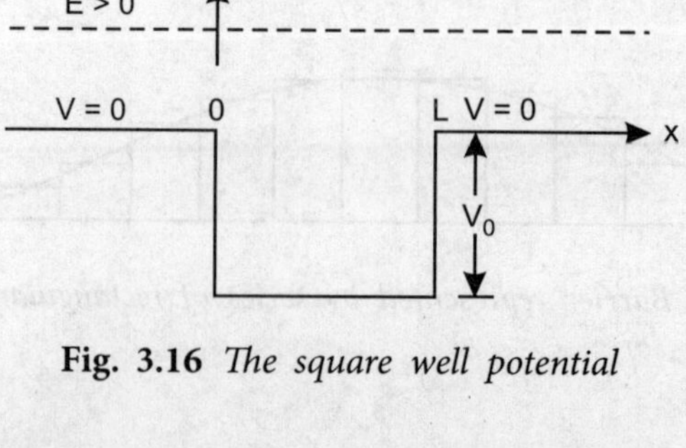

Fig. 3.16 *The square well potential*

Accordingly substituting $\alpha = i\beta$ in (86), one gets

$$R = 1 \Big/ \left[1 + 4k^2 \alpha^2 / (\alpha^2 + k^2)^2 \times 1 \Big/ \left(\frac{e^{\alpha L} - e^{-\alpha L}}{2}\right)^2\right]$$

$$= 1 \Big/ \left[1 + 4k^2 (-\beta^2) / (k^2 - \beta^2)^2 \times 1 \Big/ \left(\frac{e^{i\beta L} - e^{-i\beta L}}{2}\right)^2\right]$$

$$= 1 \Big/ \left[1 + 4k^2 \beta^2 / (k^2 - \beta^2)^2 \times 1 \Big/ \left(\frac{e^{i\beta L} - e^{-i\beta L}}{2i}\right)^2\right]$$

$$= 1 \Big/ \left[1 + \frac{4E(E+V_0)}{V_0^2 \sin^2 \beta L}\right]$$

$$= \left[1 + \frac{4E(E+V_0)}{V_0^2 \sin^2 \beta L}\right]^{-1} \qquad \ldots (3.47)$$

Similarly

$$T = 1 \Big/ \left[1 + \frac{(\alpha^2 + k^2)^2}{4k^2 \alpha^2} \times \left(\frac{e^{\alpha L} - e^{-\alpha L}}{2}\right)^2\right]$$

$$= 1 \Big/ \left[1 + \frac{V_0^2}{4E(E+V_0)} \times \sin^2 \beta L\right]$$

$$= \left[1 + \frac{V_0^2 \sin^2 \beta L}{4E(E+V_0)}\right]^{-1} \qquad \ldots (3.48)$$

From eqn. (3.47), the reflection coefficient R becomes zero (or the transmission coefficient T becomes unity) when

$$\sin \beta L = 0 \quad \text{or} \quad \beta L = n\pi, \quad n = 1, 2, 3, \ldots$$

or

$$\beta^2 L^2 = n^2 \pi^2$$

Putting value of β^2 from (3.46)

$$E_n = n^2 \pi^2 \hbar^2 / 2mL^2 - V_0 \qquad \ldots (3.49)$$

3.12 Let 0 be chosen as the origin and the coordinate x varies around the circular track (see Fig. 3.17). Let 'a' be the circumference of the ring. The Schrödinger equation in the present case would be given by

$$\frac{d^2 \varphi}{dx^2} + \frac{2mE}{\hbar^2} \varphi = 0$$

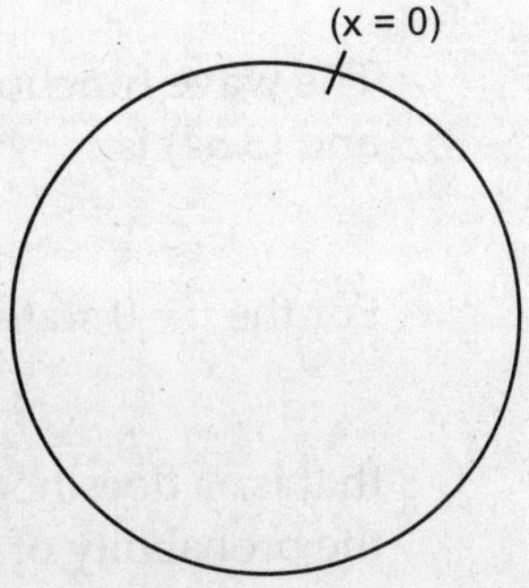

Fig. 3.17 *Particle moving in a ring of constant potential energy*

where V has been taken to be zero on the ring. Therefore,

$$\frac{d^2 \varphi}{dx^2} + k^2 \varphi = 0, \; k^2 = \frac{2mE}{\hbar^2} \quad \text{... (3.50)}$$

The general solution of the above equation would be given by

$$\varphi(x) = A \sin kx + B \cos kx \quad \text{... (3.51)}$$

where A and B are constants to be determined. Applying the boundary condition (or the continuity condition)

$$\varphi(x = 0) = \varphi(x = a),$$

one gets

$$B = A \sin ka + B \cos ka \quad \text{... (3.52)}$$

Application of the boundary condition that $d\varphi/dx$ should be continuous, that is,

$$\left.\frac{d\varphi}{dx}\right|_{x=0} = \left.\frac{d\varphi}{dx}\right|_{x=a} \text{ leads to}$$

$$Ak = Ak \cos ka - Bk \sin ka \quad \text{... (3.53)}$$

Multiplying eqn. (3.52) by Bk and eqn. (3.53) by A and then adding the resulting equations, one gets

$$\cos ka = 1$$

Therefore,

$$ka = (2mE/\hbar^2)^{1/2} a$$

$$= 2n\pi, \; n = 0, \pm 1, \pm 2, \ldots \quad \text{... (3.54)}$$

as $\cos(-\theta) = \cos(\theta)$. On squaring both the sides, the energy levels would be given by

$$E = E_n = n^2 \hbar^2/2ma^2, \quad n = 0, \pm 1, \pm 2, \ldots \quad \text{... (3.55)}$$

The wave function corresponding to the quantum number n from equs (3.51) and (3.54) is

$$\varphi_n(x) = A \sin 2n\pi x/a + B \cos 2n\pi x/a \quad \text{... (3.56)}$$

For the $n = 0$ state (or ground state) for which $E = 0$, the wave function is

$$\varphi_0 = B = \text{constant} \quad \text{... (3.57)}$$

that is, φ does not change round the ring in the lowest state or in other words the probability of finding the particle anywhere on the ring is equal. Thus $n = 0$ is a possible state. It may also be noted that except $n = 0$ state, all other states are doubly degenerate because of n having (+) ve as well as (–) ve integral values (For example for $n = 1$, $\varphi_1 = A \sin 2\pi x/a + B \cos 2\pi x/a$ and for $n = -1$, $\varphi_{-1} = -A \sin 2\pi x/a + B \cos 2\pi x/a$).

A and B can be determined from the normalization condition,

$$\int_0^a (A \sin 2n\pi x/a + B \cos 2n\pi x/a)^2 \, dx = 1$$

or

$$(A^2 + B^2)\, a/2 = 1$$

or

$$A^2 + B^2 = 2/a \qquad \text{... (3.58)}$$

The simplest way of representing A and B to satisfy the above equation is to take

$$A = (2/a)^{1/2} \cos \alpha, \; B = (2/a)^{1/2} \sin \alpha$$

where $\tan \alpha = B/A$. Therefore the normalized wave functions are given by

$$\varphi_n(x) = (2/a)^{1/2} \cos \alpha \sin 2n\pi x/a + (2/a)^{1/2} \sin \alpha \cos 2x\pi n/a$$

$$= (2/a)^{1/2} \sin (2n\pi x/a + \alpha), \quad n = 0, \pm 1, \pm 2, \ldots \qquad \text{... (3.59)}$$

❑❑❑

Chapter 4

THE LINEAR HARMONIC OSCILLATOR

4.1 SCHRÖDINGER EQUATION AND ENERGY EIGEN VALUES

Harmonic motion occurs whenever a system of some kind vibrates about an equilibrium position. The system could be a point mass supported by a spring, a diatomic molecule (atoms of molecule vibrate about their equilibrium position), atom or ion in a crystal lattice—there are truly speaking countless examples both in the macroscopic or in the microscopic realms. The potential energy in the case of a linear harmonic oscillator is given by

$$V(x) = \frac{1}{2} kx^2 \qquad \text{... (1)}$$

and is plotted in Fig. 4.1. k, the force per unit displacement is called spring constant and $(k/m)^{1/2} = w$ represents the classical frequency of oscillations, m being the mass of the oscillator.

Schrödinger equation for the harmonic oscillator would, therefore be given by

$$\frac{d^2 \varphi}{dx^2} + \frac{2m}{\hbar^2} \left(E - \frac{1}{2} kx^2\right)\varphi = 0$$

where E is the energy of the oscillator. The above equation can be written as

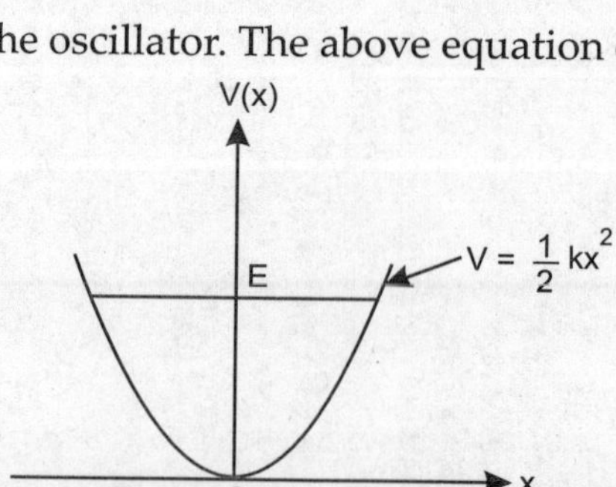

Fig. 4.1 *Potential energy of harmonic oscillator as a function of x, the displacement from the equilibrium position (taken as origin)-parabolic potential*

$$\frac{d^2\varphi}{dx^2} + (\beta - \alpha^2 x^2)\,\varphi = 0, \beta = 2m\,E/\hbar^2, \alpha = \frac{(mk)^{1/2}}{\hbar} = \frac{mw}{\hbar} \quad \text{... (2)}$$

It is not an easy task to solve the above equation. One thus simplifies it by introducing dimensionless variable ξ such that

$$\xi = \sqrt{\alpha}\, x$$

as the dimension of α are $MT^{-1}/(MLT^{-2} \times LT) = 1/L^2 = m^{-2}$

Now $$\frac{d}{dx} = \frac{d}{d\xi} \times \frac{d\xi}{dx} = \sqrt{\alpha}\,\frac{d}{d\xi}$$

Hence $$\frac{d^2}{dx^2} = \frac{d}{dx} \times \frac{d}{dx} = \frac{\alpha d^2}{d\xi^2}$$

Substituting above in eqn. (2) and dividing the resulting equation by α, one gets

$$\frac{d^2\varphi}{d\xi^2} + (\lambda - \xi^2)\,\varphi = 0 \quad \text{... (3)}$$

where $$\lambda = \beta/\alpha = 2E/\hbar w \quad \text{... (4)}$$

We begin the solution of eqn. (3) by finding the asymptotic form that ϕ must have as $\xi \to \pm\infty$. Let us write eqn. (3) as follows,

$$\frac{d^2\,\varphi/d\xi^2}{(\xi^2 - \lambda)\varphi} = 1$$

As $\xi \to \pm\infty$, $\xi^2 > \lambda$, the above differential equation reduces to

$$\frac{d^2\,\varphi}{d\xi^2} = \xi^2\varphi \quad \text{... (5)}$$

which has solutions of the form

$$\varphi_\infty = e^{\pm\xi}/2 \quad \text{... (6)}$$

Since

$$\frac{d^2\varphi_\infty}{d\xi^2} = (\xi^2 \pm 1)\,\varphi_\infty \xrightarrow[\xi \to \pm\infty]{} \xi^2\,\varphi_\infty$$

As $\xi \to \pm\infty$, $\varphi_\infty = e^{+\xi^2}/2$ becomes arbitrarily large (infinity) which violates the boundary condition. Hence we retain

$$\varphi_\infty = e^{-\xi^2}/2 \quad \text{... (7)}$$

as the asymptotic form of φ. But we have to find φ such that it is valid for all values of ξ from $-\infty$ to $+\infty$ (classically an oscillator with energy E vibrates back and forth between

$x = -A$ and $x = +A$, where E and A are related by $E = \frac{1}{2} kA^2$ or A is nothing but the amplitude of the motion)

Hence let us try a solution of the type

$$\begin{aligned} \varphi &= f(\xi)\, \varphi_\infty \\ &= f(\xi)\, e^{-\xi^2}/2 \end{aligned} \qquad \text{... (8)}$$

where $f(\xi)$ is some function of ξ which is yet to be found out. By inserting the φ of eqn. (8) into eqn. (3) and dividing the outcome by $e^{-\xi^2}/2$, we obtain

$$f''(\xi) - 2\xi f'(\xi) + (\lambda - 1) f(\xi) = 0 \qquad \text{... (9)}$$

where f' and f'' denote the first and second derivatives of f with respect to ξ, respectively.

The standard procedure for solving the above differential equation is to assume a series solution for $f(\xi)$, namely

$$\begin{aligned} f(\xi) &= A_0 + A_1\xi + A_2\xi^2 + A_3\xi^3 + \ldots \\ &= \sum_{n=0}^{\infty} A_n \xi^n \end{aligned} \qquad \text{... (10)}$$

and then to determine the values of the coefficients A_n. Differentiating f with respect to ξ, yields

$$\begin{aligned} f'(\xi) &= 1A_1 + 2A_2\xi + 3A_3\xi^2 + \ldots \\ &= \sum_{n=1}^{\infty} n A_n \xi^{n-1} \end{aligned}$$

By multiplying this equation by ξ, we obtain

$$\begin{aligned} \xi f'(\xi) &= A_1\xi + 2A_2\xi^2 + 3A_3\xi^3 + \ldots \\ &= \sum_{n=0}^{\infty} n A_n \xi^n \end{aligned} \qquad \text{... (11)}$$

as $n = 0$ does not add anything to the series.

The second derivative of f with respect to ξ is

$$\begin{aligned} f''(\xi) &= 1.2\, A_2 + 2.3\, A_3\xi + 3.4\, A_4\xi^2 + \ldots \\ &= \sum_{n=2}^{\infty} n(n-1) A_n \xi^{n-2} \end{aligned}$$

which can be written as

$$f''(\xi) = \sum_{n=0}^{\infty} (n+1)(n+2) A_{n+2} \xi^n \qquad \text{... (12)}$$

Substituting eqns. (10), (11) and (12) into eqn. (9), we obtain

$$\sum_{n=0}^{\infty}[(n+1)(n+2)A_{n+2}-(2n+1-\lambda)A_n]\xi^n = 0 \qquad \text{... (13)}$$

This equation would hold good for all values of ξ, if the quantity inside the brackets is zero for all values of n, that is if the coefficient of each ξ^n is zero. Thus, we obtain the condition,

$$(n+1)(n+2)A_{n+2} = 2(n+1-\lambda)A_n$$

or

$$A_{n+2}/A_n = (2n+1-\lambda)/\{(n+1)(n+2)\} \qquad \text{... (14)}$$

This is a recursion formula which enable us to find the coefficients A_2, A_4, A6, . . . in terms of A_0 and the coefficients A_3, A_5, A_7, . . . in terms of A_1. Since the recursion formula connects only terms that differ in index by 2 one leads to power series for $f(\xi)$ consisting either entirely of even powers of ξ or entirely of odd powers of ξ (that is, n is either even or odd) and thus correspond to either even or odd solution. Thus two solutions are obtained.

One for $\qquad A_1 = 0$ and A_0 finite,

$$f_0(\xi) = A_0 + A_2\xi^2 + A_4\xi^4 + \ldots \qquad \text{... (15)}$$

called even solution (n – even or f_0 is even function of ξ) and another for $A_0 = 0$ and A_1 finite,

$$f_1(\xi) = A_1\xi + A_3\xi^3 + A_5\xi^5 + \ldots$$

$$= \xi(A_1 + A_3\xi^2 + A_5\xi^4 + \ldots) \qquad \text{... (16)}$$

called odd solution (n – odd or f_1 odd function of ξ) and the two solutions are, therefore linearly independent. The φ for even n would be given by

$$\varphi = f_0(\xi)\, e^{-\xi^2}/2 \qquad \text{... (17)}$$

and for odd n,

$$\varphi = f_1(\xi)\, e^{-\xi^2}/2 \qquad \text{... (18)}$$

and the general solution of eqn. (3), therefore would be given by the linear combination of the above two solutions, that is,

$$\varphi = e^{-\xi^2}/2\,[Af_0(\xi) + B f_1(\xi)] \qquad \text{... (19)}$$

In order to understand if such a solution can describe a physical state or in order to see that φ does not become infinite for ξ (or n) $\to \infty$ because if ξ – is large the dominant terms in series (for f_0 or f_1) are those for which n is large, let us consider the asymptotic behaviour of f_0 and f_1. From eqn. (14)

$$A_{n+2}/A_n \xrightarrow[n\to\infty]{} 2/n \qquad \text{... (20)}$$

which of course holds good for f_0 and as well as for f_1.

Let us consider a series,

$$e^{\xi^2} = 1+\frac{\xi^2}{1!}+\frac{\xi^4}{2!}+\ldots+\frac{\xi^n}{(n/2)!}+\frac{\xi^{n+2}}{(n+2/2)!}+\ldots \quad \text{... (21)}$$

The ratio of the coefficients of ξ^{n+2} and ξ^n is given by

$$\frac{a_{n+2}}{a_n} = \frac{1/(n+2/2)!}{1/(n/2)!} = \frac{(n/2)!}{(n+2/2)!}$$

$$= \frac{(n/2)!}{(n/2+1)\,(n/2)!}$$

$$= 1/(n/2+1) \xrightarrow[n\to\infty]{} 2/n \quad \text{... (22)}$$

which is precisely the same as the ratio A_{n+2}/A_n in (20). That means the infinite series (15) for even n behaves like e^{ξ^2} and the infinite series (16) for odd n behaves like $\xi\, e^{\xi^2}$. It may be noted that the infinite series (15) behaves like e^{ξ^2} only for large n otherwise it is convergent. Same is the case for (16). This simply amounts to the fact that if series (15) and (16), that is f_0 and f_1 are allowed to run to an unlimited number of terms the wave function given by eqn. (17), that is

$$\varphi = f_0(\xi)\, e^{-\xi^2}/2 = e^{\xi^2}\, e^{-\xi^2}/2 = e^{+\xi^2}/2$$

and similarly the wave function given by eqn. (18), that is

$$\varphi = f_1(\xi)\, e^{-\xi^2}/2 = \xi e^{+\xi^2}/2$$

becomes infinite for ξ large and violates the boundary condition. Consequently, there is no conceivable choice of the coefficients A and B which would prevent φ (ξ) in (19) from diverging. This behaviour can be avoided if beyond a certain coefficient, one of the two series (15) or (16) terminates and becomes a polynomial (not an infinite series) of degree n (or the series terminates at power n of ξ). By setting the coefficient of the other one equal to zero in (19), we obtain physically meaningful solutions. We can do it by putting

$$A_{n+2} = 0,\; A_n \neq 0 \quad \text{... (23)}$$

and in that case both (17) and (18) are now finite everywhere. Then from (14), $\lambda = 2n + 1$ for any value of n. Hence the energy levels are given by

$$2\,E/\hbar w = 2n+1$$

or

$$E = E_n = (n+1/2)\,\hbar\, w, \qquad n = 0, 1, 2, \ldots \quad \text{... (24)}$$

where n is called vibrational quantum number.

E can therefore assume only the discrete values (called energy eigen values) $1/2\ \hbar w$, $3/2\ \hbar w$, $5/2\ \hbar w$, . . . The energies are thus indeed quantized. The energy levels here are evenly spaced (Fig. 4.2), unlike the energy levels of a particle in one dimensional box whose spacing diverges. The constant spacing between successive levels, $\hbar w$, is exactly what had been postulated by Planck. One may note that when $n = 0$,

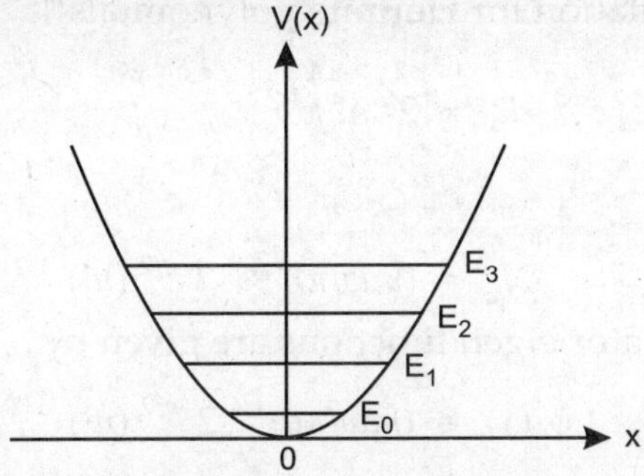

Fig. 4.2 *Energy levels E_n in the case of one dimensional harmonic oscillator for n = 0, 1, 2 and 3*

$$E_0 = \frac{1}{2}\hbar w \qquad \text{... (25)}$$

which is lowest value of the energy (ground state energy) of the oscillator. This value is called the zero point energy because a harmonic oscillator in equilibrium with its surroundings would approach an energy $E = E_0$ and not $E = 0$ as the temperature approaches 0 K and is actually the consequence of the uncertainty principle.

4.2 HARMONIC OSCILLATOR WAVE FUNCTIONS

With $\lambda = 2n + 1$, eqn. (9) takes the form,

$$H''_n(\xi) - 2\xi - 2\xi H'_n(\xi) + 2n H_n(\xi) = 0 \qquad \text{... (26)}$$

where we have written $H_n(\xi)$, the polynomial in place of $f(\xi)$, the infinite series. The above equation is called Hermite equation and $H_n(\xi)$ the Hermite polynomial of degree n which is given by (if one wants one may write $H_n(\xi)$ for n even and n odd separately).

$$H_n(\xi) = (-1)^n e^{\xi^2} \frac{d^n}{d\xi^n} e^{-\xi^2}, \quad n = 0, 1, 2, \ldots \qquad \text{... (27)}$$

The first few Hermite polynomials from above are given by

$$H_0 = 1, H_1 = 2\xi, H_2 = 4\xi^2 - 2 \text{ and } H_3 = 8\xi^3 - 12\xi$$

The complete harmonic oscillator eigen functions are, therefore of the form.

$$\varphi_n(x) = N_n H_n(\sqrt{\alpha}\, x) e^{-\alpha x^2/2} \qquad \text{... (28)}$$

where N_n is an as yet undetermined normalization constant. One can determine N_n by the normalization condition,

$$\int_{-\infty}^{+\infty} |\varphi_n(x)|^2 dx = 1 \qquad \text{... (29)}$$

Substituting the value of $\varphi_n(x)$ from eqn. (28) into above, one gets

$$\int_{-\infty}^{+\infty} N_n^2/\sqrt{\alpha} \quad H_n^2(\xi) e^{-\xi^2} \; d\xi = 1$$

Using the following relation for Hermite polynomials††

$$\int_{-\infty}^{+\infty} H_n^2(\xi)\, e^{-\xi^2}\, d\xi = 2^n\, n!\, \pi^{1/2}$$

one gets,

$$N_n = (mw/\pi\hbar)^{1/4}\, 2^{-n/2}\, (n!)^{-1/2} \qquad \text{... (30)}$$

Hence harmonic oscillator eigen functions are given by

$$\varphi_n(x) = (mw/\pi\hbar)^{1/4}\, 2^{-n/2}\, (n!)^{-1/2}\, H_n(\sqrt{\alpha}\, x)\, e^{-\alpha x^2/2},$$

$$\alpha = mw/\hbar, \quad n = 0, 1, 2, \ldots \qquad \text{... (31)}$$

Note that for each eigen value (given n) there exists only one eigen function. Therefore the eigen values of the linear harmonic oscillator are non-degenerate. The harmonic oscillator eigen functions for $n = 0, 1, 2$ and 3 are plotted in Fig. 4.3. In each case the range to which a particle oscillating classically with the same energy E_n would be confined; is indicated; evidently the particle is able to penetrate into the classically forbidden region.

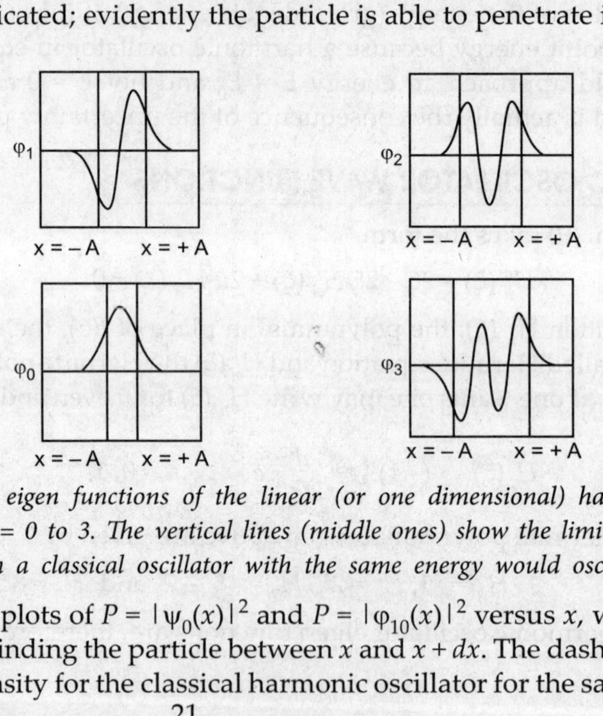

Fig. 4.3 *Normalized eigen functions of the linear (or one dimensional) harmonic oscillator for quantum numbers n = 0 to 3. The vertical lines (middle ones) show the limits –A to +A between which a classical oscillator with the same energy would oscillate*

Fig. 4.4 shows plots of $P = |\psi_0(x)|^2$ and $P = |\varphi_{10}(x)|^2$ versus x, where $|\varphi(x)|^2\, dx$ is the probability of finding the particle between x and $x + dx$. The dashed curves indicate the probability density for the classical harmonic oscillator for the same energy values, that is $\frac{1}{2}\hbar w$ in the first case and $\frac{21}{2}\hbar w$ in the later case. Classically, the particle is most

††Orthonormality relation for Hermite polynomials is

$$\int_{-\infty}^{+\infty} H_n(\xi)\, H_m(\xi)\, e^{-\xi^2}\, d\xi = 0, \quad n \neq m$$

likely to be found at the ends of its motion (P is proportional to $1/v$, v is minimum there,

$$v = (k/m)^{1/2} (A^2 - x^2)^{1/2},$$

which can be obtained from $d^2x/dt^2 = -k/m\, x$ by integration). Exactly the opposite behaviour is observed by the quantum mechanical oscillator in its lowest energy state. As shown in the Fig. 4.4(*a*) (*i*), $P = |\varphi_0(x)|^2$ has its maximum value at $x = 0$ and also (*ii*) there is a finite probability of finding the particle outside the classically permitted region. However, this disagreement between classical and quantum mechanical probabilities becomes less and less marked for large n. Fig. 4.4 (*b*) shows that $P = |\varphi_{10}(x)|^2$ (for $n = 10$) exhibits an oscillatory behaviour and the average probability density resembles the one obtained classically. Thus classical and quantum pictures begin to resemble for large n in agreement with the correspondence principle.[1]

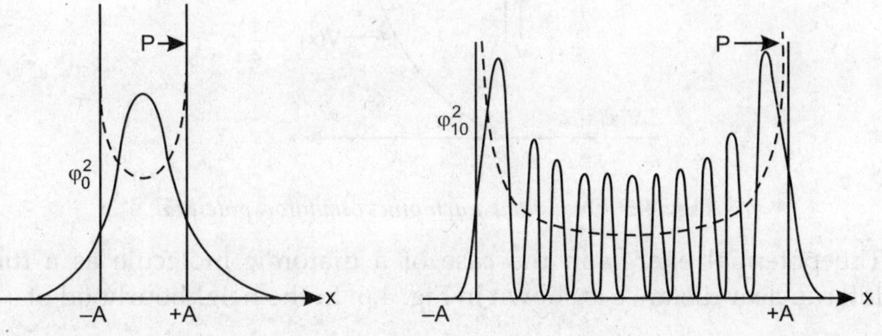

Fig. 4.4 *Probability density as a function of x (a) in the n = 0 state and (b) in the n = 10 state. Dashed curves correspond to the classical harmonic oscillator for the same energy values*

PROBLEMS

4.1 For a linear harmonic oscillator, show that the expectation value of the potential energy in the nth state is given by $\frac{1}{2}(n + 1/2)\hbar w = E_n/2$.

4.2 Show that for a linear harmonic oscillator in the ground state the probability of finding the particle outside the classical limits is approximately 0.16 (or 16%).

4.3 Consider a particle of mass m in two dimensional potential, $V = \frac{1}{2}k(x^2 + y^2)$.

Find the allowed energy eigen values and the energy eigen functions of the system. Also show that the first excited state is two fold degenerate.

4.4 Find the energy eigen values of a particle moving in a potential $V(x, y, z) = \frac{1}{2}k(x^2 + y^2 + z^2)$ or $V(r) = \frac{1}{2}kr^2$, that is in a spherically symmetric potential. It is case of a 3-*d* harmonic oscillator or of an isotropic oscillator. Also find the degeneracies of the first and the second excited states of the oscillator.

[1] N. Bohr, Nature 121, 580 (1923)

4.5 Calculate the zero point energy of a system consisting of a mass of 10^{-3} kg connected to fixed point by a spring which is stretched 10^{-2}m by a force of 0.1 Newton, the particle being constraint to move only along x-axis.

4.6 Find energy levels and energy eigen functions of a particle of mass m moving in a potential

$$V(x) = \frac{1}{2}kx^2, x > 0$$

$$= \infty, x < 0$$

shown in the Fig. 4.5

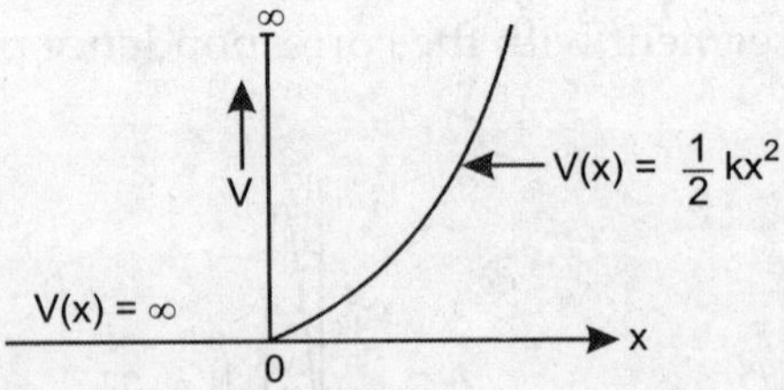

Fig. 4.5 *One sided harmonic oscillator potential*

4.7 The potential energy in the case of a diatomic molecule as a function of internuclear distance is shown in Fig. 4.6. In the neighbourhood of

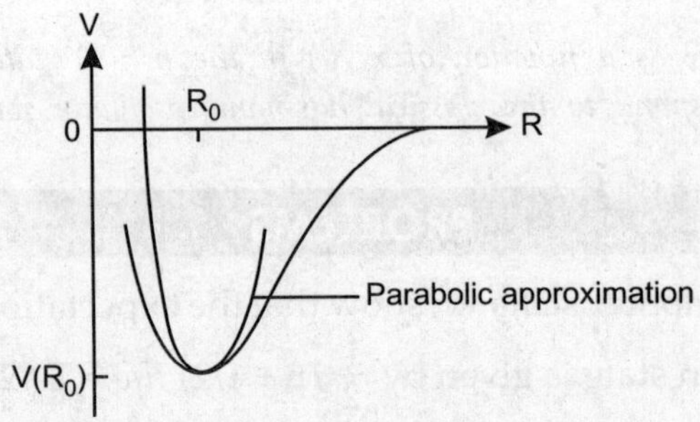

Fig. 4.6 *The potential energy of a diatomic molecule as a function of inter-nuclear distance*

the minimum of this curve which corresponds to the equilibrium configuration, the shape of the curve can be approximated as parabola. In this region,

$$V = V(R_0) + \frac{1}{2}k(R - R_0)^2$$

where R_0 is the equilibrium separation of the two atoms. Classically $V(R_0)$ is energy that must be given to the two atoms in order to separate them and is called dissociation energy.

One can think of the vibration in the diatomic molecule to be caused by a spring of force constant k between the two atoms of masses m_1 and m_2 as shown in the

Fig. 4.7 and the two atoms are constrained to move back and forth relative to their centre of mass. Find the possible eigen values of the system. What is the ground state wave function?

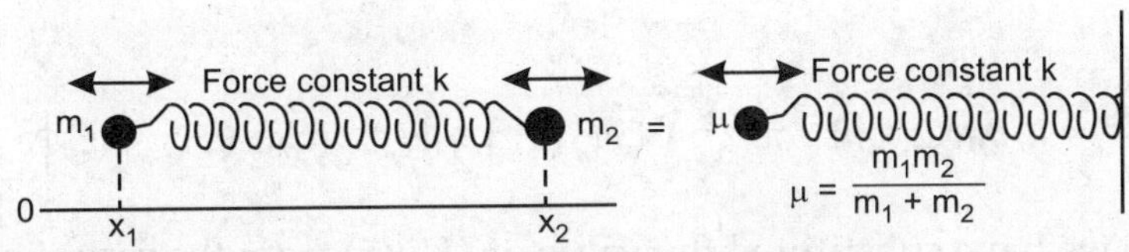

Fig. 4.7 *A two body oscillator which is model for vibration of a diatomic molecule. The system behaves like an ordinary harmonic oscillator with the same spring constant but with the reduced mass* μ

4.8 What are the energy levels and energy eigen functions of a particle moving in a potential

$$V(x) = \frac{1}{2}kx^2 + Cx$$

where C and k are constants.

4.9 Obtain recursion relation for the harmonic oscillator wave functions starting from the recursion relation for the Hermite polynomials.

4.10 Find energy levels and the energy eigen functions of a particle moving in a potential

$$V(x) = \frac{1}{2}kx^2 + C$$

where C and k are constants.

4.11 Calculate the vibrational energy levels of CO molecule assuming the force constant k of the bond in CO to be 187 N/m.

4.12 Assuming that the H_2 molecule behaves exactly like a harmonic oscillator with a force constant 573 N/m, find the vibrational quantum number corresponding to its dissociation energy 4.5 eV.

SOLUTIONS

4.1 Expectation value of the potential energy in the case of a linear harmonic oscillator is given by

$$\langle V \rangle_n = \frac{1}{2}k\langle x^2 \rangle_n = \frac{1}{2}k \int_{-\infty}^{+\infty} \varphi_n^*(x)\, x^2\, \varphi_n(x)\, dx \qquad \text{... (4.1)}$$

where

$$\varphi_n(x) = (mw/\pi\hbar)^{1/4}\, 2^{-n/2}\, n!^{-1/2}\, H_n(\xi)\, e^{-\xi^2}/2,$$

$$\xi = (mw/\hbar)^{1/2}\, x,\; k = mw^2 \qquad \text{... (4.2)}$$

Therefore

$$\langle x^2\rangle_n = (mw/\pi\hbar)^{1/2}\, 2^{-n}\, n!^{-1} \int_{-\infty}^{+\infty} e^{-\xi^2} H_n(\xi)\, H_n(\xi)\, \xi^2\, \hbar/mw\, (\hbar/mw)^{1/2}\, d\xi$$

$$= \hbar/mw\, 2^{-n}\, n!^{-1}\, \pi^{-1/2} \int_{-\infty}^{+\infty} \xi H_n(\xi) \left[\frac{1}{2} H_{n+1}(\xi) + nH_{n-1}(\xi)\right] e^{-\xi^2}\, d\xi \quad \text{... (4.3)}$$

where we have substituted the value of $\xi H_n(\xi)$ from the recursion relation for the Hermite polynomials, that is

$$H_{n+1}(\xi) - 2\xi\, H_n(\xi) + 2n\, H_{n-1}(\xi) = 0 \quad \text{... (4.4)}$$

Again substituting the values of $\xi\, H_{n+1}(\xi)$ and $\xi\, H_{n-1}(\xi)$ from the above into (4.3), one gets

$$\langle x^2\rangle_n = \hbar/mw \quad 2^{-n}\, n!^{-1/2}\, \pi^{-1/2} \int_{-\infty}^{+\infty} H_n(\xi)$$

$$\left[\frac{1}{2}\left\{\frac{1}{2} H_{n+2}(\xi) + (n+1) H_n(\xi)\right\}\right.$$

$$\left. + n\left\{\frac{1}{2} H_n(\xi) + (n-1) H_{n-2}(\xi)\right\}\right] 3^{-\xi^2}\, d\xi$$

$$= \hbar/mw\;\; 2^{\,n}\, n!^{\;-1/2}\, \pi^{\;-1/2} \int_{-\infty}^{+\infty} H_n(\xi) \left[\frac{1}{4} H_{n+2}(\xi)\right.$$

$$\left. + \left(n + \frac{1}{2}\right) H_n(\xi) + n(n-1)\, H_{n-2}(\xi)\right] e^{-\xi^2}\, d\xi$$

Using the orthonormality relation for Hermite polynomials,

$$\int_{-\infty}^{+\infty} H_n(\xi)\, H_m(\xi)\, e^{-\xi^2}\, d\xi = \pi^{1/2}\, 2^n\, n!, \text{ if } n = m$$

$$= 0, \qquad \text{if } n \neq m$$

one gets,

$$\langle x^2\rangle_n = \hbar/mw\left(n + \frac{1}{2}\right) \quad \text{... (4.5)}$$

and hence

$$\langle V \rangle_n = \frac{1}{2} k\hbar/mw \left(n + \frac{1}{2}\right) = \frac{1}{2} mw^2 \hbar/mw \left(n + \frac{1}{2}\right)$$

$$= \frac{1}{2}\left(n + \frac{1}{2}\right)\hbar w = E_n/2 \qquad \text{... (4.6)}$$

4.2 The ground state has the energy $E_0 = \frac{1}{2}\hbar w$. Classically the energy of the oscillator is $\frac{1}{2}mw^2A^2$, where A is the amplitude of motion. Accordingly A corresponding to the ground state energy is given by

$$\frac{1}{2}\hbar w = \frac{1}{2} mw^2 A^2$$

or

$$A = \pm(\hbar/mw)^{1/2} \qquad \text{... (4.7)}$$

The probability of finding the particle in the classical limits would be given by

$$P_C = \int_{-A}^{+A} |\varphi_0(x)|^2 \, dx$$

where the ground state wave function

$$\varphi_o(x) = (mw/\pi\hbar)^{1/4} e^{-\xi^2/2}, \; \xi = (mw/\hbar)^{1/2} x \qquad \text{... (4.8)}$$

Hence

$$P_c = (mw/\pi\hbar)^{1/2} \int_{-1}^{+1} e^{-\xi^2} \, (\hbar/mw)^{1/2} \, d\xi$$

where we have changed the limits from

$$x = \pm A \text{ to } \xi = (mw/\hbar)^{1/2} \times \pm(\hbar/mw)^{1/2} = \pm 1$$

Accordingly,

$$P_C = \pi^{-1/2} \int_{-1}^{+1} e^{-\xi^2} \, d\xi$$

$$= 2\pi^{-1/2} \int_0^1 e^{-\xi^2} \, d\xi$$

$$= 2\pi^{-1/2} \int_0^1 (1 - \xi^2 + \xi^4/2! - \xi^6/3! + \xi^8/4! - \ldots) \, d\xi$$

$$= 2\pi^{-1/2}[\xi - \xi^3/3 + \xi^5/10 - \xi^7/42 + \xi^9/216 - ...]_0^1$$

$$= 2\pi^{-1/2}[1 - 0.333 + 0.1 - 0.024 + 0.004]$$

$$= 0.84$$

That is, the probability of finding the particle in the classical region is about 84%. Therefore, the probability of finding the particle outside the classical limits would be about 16%.

4.3 The Schrödinger equation for two dimensional harmonic oscillator would be given by

$$\frac{\partial^2 \varphi(x, y)}{\partial x^2} + \frac{\partial^2 \varphi(x, y)}{\partial y^2} + \frac{2m}{\hbar^2}\left[E - \frac{1}{2}k(x^2 + y^2)\right]\varphi(x, y) = 0 \qquad ... (4.9)$$

Substituting

$$2E/\hbar w = \lambda, \quad \rho = \left(\frac{mw}{\hbar}\right)^{1/2} x, \ \eta = \left(\frac{mw}{\hbar}\right)^{1/2} y, \ k = mw^2 \qquad ... (4.10)$$

one gets,

$$\left(\frac{\partial^2}{\partial \rho^2} + \frac{\partial^2}{\partial \eta^2}\right)\varphi(\rho, \eta) + [\lambda - (\rho^2 + \eta^2)]\,\varphi(\rho, \eta) = 0 \qquad ... (4.11)$$

Using the method of separation of variables, that is, putting

$$\varphi(\rho, \eta) = X(\rho)\, Y(\eta)$$

into above equation, one gets

$$\left(\frac{Yd^2 X(\rho)}{d\rho^2} + \frac{Xd^2 Y(\eta)}{d\eta^2}\right) + [\lambda - (\rho^2 + \eta^2)]\, X(\rho)\, Y(\eta) = 0$$

On dividing the above equation by XY, we obtain

$$\left(\frac{1}{X}\frac{d^2 X}{d\rho^2} - \rho^2\right) + \left(\frac{1}{Y}\frac{d^2 Y}{d\eta^2} - \eta^2\right) = -\lambda \qquad ... (4.12)$$

Note that the first factor is function only of x and the second factor is function only of y and the sum is a constant. This would be possible only when each factor is equal to a constant. This gives us the following two equations,

$$\frac{d^2 X}{d\rho^2} + (\lambda_x - \rho^2)\, X = 0 \qquad ... (4.13)$$

$$\frac{d^2 Y}{d\rho^2} + (\lambda_y - \eta^2)\, Y = 0 \qquad ... (4.14)$$

where the constants λ_x and λ_y are related by the relation,

$$\lambda_x + \lambda_y = \lambda \qquad ... (4.15)$$

Now each of the eqns. (4.13) and (4.14) is a differential equation for a linear harmonic oscillator for which the normalized wave functions are known. Thus finite solutions are possible only if

$$\lambda_x = 2n_x + 1, \lambda_y = 2n_y + 1, n_x, n_y = 0, 1, 2, \ldots \quad \text{... (4.16)}$$

Substituting (4.16) into (4.15) and the outcome into eqn. (4.10), the energy levels in the present case would be given by

$$E = E_n = \left(n_x + \frac{1}{2}\right)\hbar w + \left(n_y + \frac{1}{2}\right)\hbar w$$

$$= \left(n + \frac{1}{2}\right)\hbar w, n = n_x + n_y = 0, 1, 2, 3, \ldots \quad \text{... (4.17)}$$

Similarly,

$$X(x) = N_{n_x} e^{-\alpha x^2/2} H_{n_x}(\sqrt{\alpha}\, x), n_x = 0, 1, 2, 3, \ldots$$

$$Y(y) = N_{n_y} e^{-\alpha y^2/2} H_{n_y}(\sqrt{\alpha}\, y), n_y = 0, 1, 2, 3, \ldots$$

where $\alpha = mw/\hbar$.

Therefore,

$$\varphi_{n_x n_y} = N_{n_x} N_{n_y} e^{-\alpha(x^2+y^2)/2} H_{n_x}(\sqrt{\alpha}\, x) H_{n_y}(\sqrt{\alpha}\, y)$$

$$= \left(\frac{mw}{\pi\hbar}\right)^{1/2} 2^{-(n_x+n_y)/2} n_x!^{-1/2} n_y!^{-1/2}$$

$$H_{n_x}(\sqrt{\alpha}\, x) H_{n_y}(\sqrt{\alpha}\, y) e^{-\alpha(x^2+y^2)/2} \quad \text{... (4.18)}$$

The first excited state would correspond to $n = 1$. That is, to $n_x = 0, n_y = 1$ and $n_x = 1. n_y = 0$. Hence is two fold degenerate.

4.4 The Schrödinger equation in the present case is given by

$$\nabla^2 \varphi(x, y, z) + 2m/\hbar^2 \left[E - \frac{1}{2}k(x^2 + y^2 + z^2)\right] \varphi(x, y, z)$$

$$= 0, k = mw^2 \quad \text{... (4.19)}$$

on substituting,

$$\frac{2E}{\hbar w} = \lambda, \rho = \left(\frac{mw}{\hbar}\right)^{1/2} x,$$

$$\eta = \left(\frac{mw}{\hbar}\right)^{1/2} y \text{ and } \sigma = \left(\frac{mw}{\hbar}\right)^{1/2} z \quad \text{... (4.20)}$$

into above, we get

$$\left(\frac{\partial^2}{\partial\rho^2} + \frac{\partial^2}{\partial\eta^2} + \frac{\partial^2}{\partial\sigma^2}\right) \varphi(\rho, \eta, \sigma)$$

$$+ [\lambda - (\rho^2 + \eta^2 + \sigma^2)] \varphi(\rho, \eta, \sigma) = 0 \quad \text{... (4.21)}$$

Using the method of separation of variables, that is, by substituting,

$$\varphi(\rho, \eta, \sigma) = X(\rho)\, Y(\eta)\, Z(\sigma) \qquad \text{... (4.22)}$$

into above and by dividing the resulting equation by $X(\rho)\, Y(\eta)\, Z(\sigma)$ and rearranging the terms, one obtains

$$\left(\frac{1}{X}\frac{d^2X}{d\rho^2} - \rho^2\right) + \left(\frac{1}{Y}\frac{d^2Y}{d\eta^2} - \eta^2\right) + \left(\frac{1}{Z}\frac{d^2Z}{d\sigma^2} - \sigma^2\right) = -\lambda \qquad \text{... (4.23)}$$

The first factor is a function of x only, the second factor is a function of y only and third of z only and the sum is equal to a constant. This would be possible only when each factor becomes equal to a constant. This gives us three equations,

$$\frac{d^2 X}{d\rho^2} + (\lambda_x - \rho^2)\, X = 0 \qquad \text{... (4.24)}$$

$$\frac{d^2 Y}{d\eta^2} + (\lambda_y - \eta^2)\, Y = 0 \qquad \text{... (4.25)}$$

$$\frac{d^2 Z}{d\sigma^2} + (\lambda_z - \sigma^2)\, Z = 0 \qquad \text{... (4.26)}$$

with
$$\lambda_x + \lambda_y + \lambda_z = \lambda \qquad \text{... (4.27)}$$

Each of the eqns. (4.24) to (4.26) is a differential equation for a one dimensional harmonic oscillator for which finite solutions are possible only if

$$\lambda_x = 2n_x + 1,\ \lambda_y = 2n_y + 1,\ \lambda_z = 2n_z + 1 \qquad \text{... (4.28)}$$

where n_x, n_y and $n_z = 0, 1, 2, 3, \ldots$

Substituting (4.28) into (4.27) and the result into eqn. (4.20), the energy eigen values would be given by

$$E = E_n = (n + 3/2)\, \hbar w,\ n = n_x + n_y + n_z = 0, 1, 2, 3, \ldots \qquad \text{... (4.29)}$$

The energy eigen functions in the present case would be given by

$$\begin{aligned} \varphi_{n_x n_y x_z}(x, y, z) &= N_{n_x} N_{n_y} N_{n_z}\, e^{-\alpha(x^2 + y^2 + z^2)/2} \\ &\quad \times H_{n_x}(\sqrt{\alpha}\, x)\, H_{n_y}(\sqrt{\alpha}\, y)\, H_{n_z}(\sqrt{\alpha}\, z) \\ &= \left(\frac{mw}{\pi\hbar}\right)^{3/4} 2^{-(n_x + n_y + n_z)/2}\, n_x!^{-1/2}\, n_y!^{-1/2}\, n_z!^{-1/2} \\ &\quad \times H_{n_x}(\sqrt{\alpha}\, x)\, H_{n_y}(\sqrt{\alpha}\, y)\, H_{n_z}(\sqrt{\alpha}\, z)\, e^{-\alpha(x^2 + y^2 + z^2)/2} \end{aligned} \qquad \text{... (4.30)}$$

The degeneracies of the first few energy levels are listed in Table 4.1.

Table 4.1 Degeneracies in the case of an isotropic harmonic oscillator for n = 0, 1, and 2.

n	n_x	n_y	n_z	*Degeneracy*
0	0	0	0	1
1 (first excited state)	1	0	0	3
	0	1	0	
	0	0	1	
2 (second excited state)	2	0	0	6
	0	2	0	
	0	0	2	
	1	1	0	
	0	1	1	
	1	0	1	

Degree of degeneracy of a level with given n.

For fixed n_z and n, the number of pairs (n_x, n_y) for which $n_x + n_y = n - n_z$, are

$$(n_x = 0, n_y = n - n_z); (n_x = 1, n_y = n - n_z - 1);$$

$$(n_x = 2, n_y = n - n_z - 2); \ldots; (n_x = n - n_z, n_y = 0)$$

Thus the total number of required pairs would be $n - n_z + 1$. Now each n_z can vary from 0 to n. Hence degree of degeneracy

$$= \sum_{n_z=0}^{n} n + 1 - n_z$$

$$= n \sum_{n_z=0}^{n} 1 + \sum_{n_z=0}^{n} 1 - \sum_{n_z=0}^{n} n_z$$

$$= n(n+1) + (n+1) - n(n+1)/2$$

$$= (n+1)\,[(n+1) - n/2]$$

$$= (n+1)\,(n+2)/2$$

4.5 Spring constant,

$$k = F/x$$

$$= 0.1/0.01$$

$$= 10 \text{ N/m},$$

hence

$$w = (k/m)^{1/2} = 10^2 \text{ sec}^{-1}$$

Zero point energy,

$$E_0 = \frac{1}{2}\hbar w = \frac{1}{2} \times 1.05 \times 10^{-34} \times 10^2 \text{J}$$

$$= 0.5 \times 10^{-32} \text{ J}$$

4.6 The boundary conditions in this case are going to be different than those in the case of a usual harmonic oscillator. The solution in this case would be given by

$$\varphi = f_1(\xi)\, e^{-\xi^2/2} \qquad \text{... (4.31)}$$

which must vanish at $x = 0$ (only odd solution has been retained as the even solution f_0 (ξ) which does not satisfy the boundary conditions that is, which does not vanish at the origin has been discarded). In order that the solution does not become infinite for large ξ (or n) in the region $x > 0$, the series f_1 (ξ) is to be terminated at ξ^n (or f_1 becomes a polynomial say H_n (ξ) instead of an infinite series). This amounts to

$$\lambda = 2n + 1$$

or

$$E_n = \left(n + \frac{1}{2}\right)\hbar w, \quad n = 1, 3, 5, \ldots$$

The eigen functions in the present case are given by

$$\varphi_n(x) = (\alpha/\pi)^{1/4}\, 2^{-n/2}\, n!^{-1/2}\, H_n(\sqrt{\alpha}\, x)\, e^{-\alpha x^2/2},$$

$$\alpha = mw/\hbar, \quad n = 1, 3, 5, \ldots \qquad \text{... (4.32)}$$

4.7 From Fig. 4.7 considering one-dimentional motion, we have

$$x_2 - x_1 = R_0 + x \text{ (or } R - R_0 = x) \qquad \text{... (4.33)}$$

$$\frac{m_1\, d^2 x_1}{dt^2} = kx \qquad \text{... (4.34)}$$

$$\frac{m_2\, d^2 x_2}{dt^2} = -kx \qquad \text{... (4.35)}$$

Multiplying eqn. (4.34) by m_2 and (4.35) m_1, respectively and subtracting the results so obtained, one gets

$$m_1 m_2 \left(\frac{d^2 x_2}{dt^2} - \frac{d^2 x_1}{dt^2}\right) = -(m_1 + m_2)\, kx \qquad \text{... (4.36)}$$

Differentiating eqn. (4.33) twice w.r.t. time, one has

$$\frac{d^2 x_2}{dt^2} - \frac{d^2 x_1}{dt^2} = \frac{d^2 x}{dt^2} \quad \text{... (4.37)}$$

Combining eqns. (4.36) and (4.37), one gets

$$\frac{\mu dx^2}{dt^2} = -kx \quad \text{... (4.38)}$$

where $\mu = m_1 m_2/(m_1 + m_2)$ is called the reduced mass of the molecule. Thus the frequency of oscillation of such a two body oscillator is

$$w = (k/\mu)^{1/2} \quad \text{... (4.39)}$$

and the potential energy function can be written as

$$V(x) = V(R_0) + \frac{1}{2} mw^2 x^2$$

$$= C + \frac{1}{2} mw^2 x^2 \quad \text{... (4.40)}$$

where C is a constant.

The Schrödinger equation in the present case would be given by

$$\frac{d^2 \varphi}{dx^2} + \frac{2\mu}{\hbar^2}\left(E - \frac{1}{2} mw^2 x^2 - C\right)\varphi = 0$$

Proceeding in the same manner as in the case of a linear harmonic oscillator and substituting

$$\lambda = 2(E - C)/\hbar w, \ \alpha = \mu w/\hbar,$$

one finally gets

$$E - C = E_n - C = \left(n + \frac{1}{2}\right)\hbar w$$

or

$$E_n = \left(n + \frac{1}{2}\right)\hbar w + C, \quad n = 0, 1, 2, 3, \ldots \quad \text{... (4.41)}$$

Note that all the vibrational states have energies greater than C, that is, $V(R_0)$.
Similarly the wave functions in the present case would be given by

$$\varphi_n(x) = (\alpha/\pi)^{1/4}\, 2^{-n/2}\, n!^{-1/2}\, H_n(\sqrt{\alpha}\, x)\, e^{-\alpha x^2/2},$$

$$n = 0, 1, 2, 3, \ldots \quad \text{... (4.42)}$$

Hence

$$\varphi_0(x) = (\alpha/\pi)^{1/4}\, e^{-\alpha x^2/2}$$

4.8 The potential energy in this case is

$$V(x) = \frac{1}{2}kx^2 + Cx$$

$$= \frac{1}{2}k\,[x^2 + (C/2k)\,x]$$

$$= \frac{1}{2}k\,(x + C/k)^2 - C^2/2k \qquad \text{... (4.43)}$$

The Schrödinger equation, therefore would be given by

$$\frac{d^2\varphi\,(x)}{dx^2} + \frac{2m}{\hbar^2}\left[E - \frac{1}{2}k\left(x + \frac{C}{k}\right)^2 + \frac{C^2}{2k}\right]\varphi(x) = 0$$

Putting $x_1 = x + C/k$, the above equation reads,

$$\frac{d^2\,\varphi}{dx_1^2} + \frac{2m}{\hbar^2}\left[E + \frac{C^2}{2k} - \frac{1}{2}kx_1^2\right]\varphi = 0 \qquad \text{... (4.44)}$$

Proceeding in the same manner as that in the case of a linear harmonic oscillator, one gets

$$E_n = \left(n + \frac{1}{2}\right)\hbar w - C^2/2k,\ w = (k/m)^{1/2},\ \ n = 0, 1, 2, 3, \ldots \qquad \text{... (4.45)}$$

and similarly

$$\varphi_n(x_1) = (\alpha/\pi)^{1/4}\,2^{-n/2}\,n!^{-1/2}\,H_n(\sqrt{\alpha}\,x_1)\,e^{-\alpha x_1^2/2},\ \alpha = mw/\hbar,$$

$$n = 0, 1, 2, 3, \ldots \qquad \text{... (4.46)}$$

4.9 The recursion relation for Hermite polynomials is given by

$$H_{n+1}(\xi) - 2\,\xi\,H_n(\xi) + 2n\,H_{n-1}(\xi) = 0 \qquad \text{... (4.47)}$$

The harmonic oscillator wave functions are given by

$$\varphi_n(\xi) = (mw/\pi\hbar)^{1/4}\,2^{-n/2}\,n!^{-1/2}\,H_n(\xi)\,e^{-\xi^2/2} \qquad \text{... (4.48)}$$

Therefore for $n = n + 1$,

$$\varphi_{n+1}(\xi) = (mw/\pi\hbar)^{1/4}\,2^{-(n+1)/2}\,(n+1)!^{-1/2}\,H_{n+1}(\xi)\,e^{-\xi^2/2} \qquad \text{... (4.49)}$$

and similarly for $n = n - 1$,

$$\varphi_{n-1}(\xi) = (mw/\pi\hbar)^{1/4}\,2^{-(n-1)/2}\,(n-1)!^{-1/2}\,H_{n-1}(\xi)\,e^{-\xi^{2/2}} \qquad \text{... (4.50)}$$

Putting values of H_{n+1}, H_n and H_{n-1} from eqns. (4.48) to (4.50) into eqn. (4.47), we get

$$2^{(n+1)/2}\,(n+1)!^{1/2}\,\varphi_{n+1}(\xi) - 2\xi\,2^{n/2}\,n!^{1/2}\,\varphi_n(\xi)$$
$$+\,2n\,2^{(n-1)/2}\,(n-1)!^{1/2}\,\varphi_{n-1}(\xi) = 0$$

Dividing above equation throughout by $2^{(n+1)/2}\,n!^{1/2}$, we get the required recursion relation,

$$(n+1)^{1/2}\,\varphi_{n+1}(\xi) - 2^{1/2}\,\xi\,\varphi_n(\xi) + n^{1/2}\,\varphi_{n-1}(\xi) = 0 \qquad \text{... (4.51)}$$

4.11 The vibrational energy levels of a diatomic molecule are given by

$$E_n = \left(n + \frac{1}{2}\right)\hbar w, \qquad n = 0, 1, 2, 3, \ldots$$

where the angular frequency of vibration

$$w = (k/\mu)^{1/2}$$

in which k is the spring constant of the bond in CO and $\mu = m_1 m_2/(m_1 + m_2)$ is the reduced mass of the CO molecule with m_1 and m_2 being the masses of ^{12}C and ^{16}O atoms. Therefore

$$\mu = (19.9 \times 10^{-27} \times 26.6 \times 10^{-27})/(19.9 \times 10^{-27} + 26.6 \times 10^{-27})$$
$$= 1.14 \times 10^{-26}\ \text{kg}$$

and

$$w = \{187/(1.4 \times 10^{-26})\}^{1/2}$$
$$= 12.81 \times 10^{13}/\text{sec}$$

The vibrational energy levels of CO would, therefore be given by

$$E_n = \left(n + \frac{1}{2}\right) \times 1.05 \times 10^{-34} \times 12.81 \times 10^{13}$$

$$= \left(n + \frac{1}{2}\right) \times 13.45 \times 10^{-21}\ \text{J}$$

$$= \left(n + \frac{1}{2}\right) \times 8.4 \times 10^{-2}\ \text{eV},\ n = 0, 1, 2, 3, \ldots$$

❑❑❑

Chapter 5

CENTRAL POTENTIAL PROBLEMS

5.1 THE RIGID ROTATOR

Consider a system consisting of two particles of masses m_1 and m_2 joined by a weightless link of length r. Let us assume that this system rotates about an axis passing through its centre of mass and normal to the plane containing the two masses. Such a system is called rigid rotator and is illustrated in Fig. 5.1. The term rigid is used due to the fixed distance between the masses. It describes a rotating diatomic molecule where

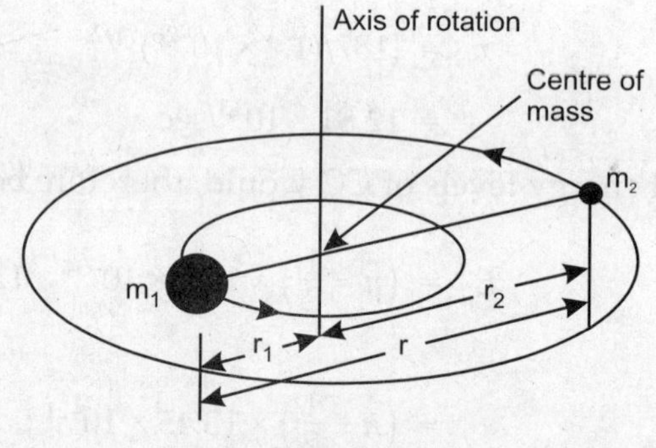

Fig. 5.1 *A rigid rotator-rotation of a diatomic molecule*

(*i*) The bond length is assumed to remain unchanged during the rotation, that is, $r = r_e$, where r_e represents the distance between two atoms in the equilibrium position of the molecule. This in turn amounts to neglecting their vibrations.

(*ii*) The axis of rotation is not fixed (or is free to take up any position), that is, the atoms are not constrained to remain in one plane only.

Let us assume that the centre of mass of the system is fixed at the origin of our coordinate system or in other words, we are considering centre of mass coordinate system. Let the distance of m_1 from the centre of mass be r_1 and that of m_2 be r_2, then $m_1r_1 = m_2r_2$ and $r_e = r_1 + r_2$. Hence

$$r_1 = m_2 r_e/(m_1 + m_2) \qquad \text{... (1)}$$

$$r_2 = m_1 r_e/(m_1 + m_2)$$

The kinetic energy of the rotating molecule is given by

$$T = \frac{1}{2} m_1 v_1^2 + \frac{1}{2} m_2 v_2^2$$

where v_1 and v_2 are the linear velocities of masses m_1 and m_2, respectively. If ω is the angular velocity and since r_1 and r_2 remain unchanged during the rotation,

$$T = \frac{1}{2} m_1 \omega^2 r_1^2 + \frac{1}{2} m_2 \omega^2 r_2^2$$

$$= \frac{1}{2} I\omega^2 \qquad \text{... (2)}$$

where $I = m_1 r_1^{\,2} + m_2 r_2^{\,2}$ is the moment of inertia of the system about an axis passing through the centre of mass and perpendicular to the plane containing the masses. On substituting the values of r_1 and r_2, moment of inertia takes the form $I = \mu r_e^{\,2}$, $\mu = m_1 m_2/(m_1 + m_2)$ is called reduced mass of the molecule. The case is thus equivalent to a one body problem of mass μ rotating at a distance r_e from the axis (or origin). Fig. 5.2 shows the state of affair in spherical polar coordinates.

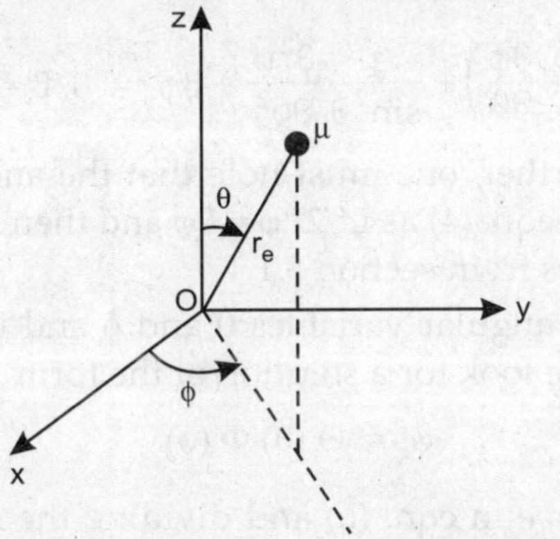

Fig. 5.2 *Spherical coordinates for an equivalent one body problem*

Eqn. (2) can also be written as

$$T = L^2/2I \qquad \text{... (3)}$$

where $L = I\omega$ is the angular momentum of rotation.

If no external forces act on the rotator (masses may be assumed to be neutral), we may put $V = V(r_e) = 0$ (This may be regarded as a special case of spherically symmetric field as in this case V depends upon distance r_e, the radial coordinate). Therefore the Schrödinger equation,

$$H\varphi = E\varphi \quad \text{... (4)}$$

leads to

$$-\frac{\hbar^2}{2\mu}\nabla^2\varphi = E\varphi \quad \text{... (5)}$$

where the kinetic energy has been written as $T = \frac{1}{2}\mu v^2 \left(= \frac{1}{2}\mu\omega^2 r^2\right)$

Substituting the value of ∇^2 in spherical polar coordinates in eqn. (5), one gets

$$\frac{1}{r^2}\frac{\partial}{\partial r}\left(r^2\frac{\partial\varphi}{\partial r}\right) + \frac{1}{r^2}\frac{1}{\sin\theta}\frac{\partial}{\partial\theta}\left(\sin\theta\frac{\partial\varphi}{\partial\theta}\right)$$

$$+ \frac{1}{r^2\sin^2\theta}\frac{\partial^2\varphi}{\partial\phi^2} + \frac{2\mu}{\hbar^2}E\varphi = 0$$

and using the fact that $r = r_e$ is fixed, the radial part of the above equation vanishes and hence

$$\frac{1}{\sin\theta}\frac{\partial}{\partial\theta}\left(\sin\theta\frac{\partial\varphi}{\partial\theta}\right) + \frac{1}{\sin^2\theta}\frac{\partial^2\varphi}{\partial\phi^2} + \frac{2\mu r^2}{\hbar^2}E\varphi = 0$$

or
$$\frac{1}{\sin\theta}\frac{\partial}{\partial\theta}\left(\sin\theta\frac{\partial\varphi}{\partial\theta}\right) + \frac{1}{\sin^2\theta}\frac{\partial^2\varphi}{\partial\phi^2} + \beta\varphi = 0, \beta = 2I\frac{E}{\hbar^2} \quad \text{... (6)}$$

Before we proceed further, one must note that the another way of obtaining the above equation is to write eqn. (4) as $L^2/2I\ \varphi = E\varphi$ and then substitute the value of L^2 in spherical polar coordinates from section 8.1.

Eqn. (6) contains two angular variables θ and ϕ and is solved by the method of separation of variables. We look for a solution of the form

$$\varphi = \Theta(\theta)\,\Phi(\phi) \quad \text{... (7)}$$

After substituting above in eqn. (6) and dividing the resulting equation by $\Theta\,\Phi$, we get

$$\left[\frac{1}{\sin\theta}\frac{d}{d\theta}\left(\sin\theta\frac{d\Theta}{d\theta}\right) + \beta\Theta\right]\frac{\sin^2\theta}{\Theta} = -\frac{1}{\Phi}\frac{d^2\Phi}{d\phi^2} \quad \text{... (8)}$$

The left side of the above equation is a function of θ alone and the right side of ϕ alone, but both sides are equal. This is only possible when both sides are put equal to the same separation constant, say m_l^2. We thus obtain a pair of differential equations, each in one variable as

$$\frac{d^2\phi}{d\phi^2} + m_l^2\,\phi = 0 \quad \text{... (9)}$$

and

$$\frac{1}{\sin\theta}\frac{d}{d\theta}\left(\sin\theta\,\frac{d\Theta}{d\theta}\right)+\left(\beta-\frac{m_l^2}{\sin^2\theta}\right)\Theta=0 \qquad \text{... (10)}$$

Eqn. (9) has $e^{m_l i\phi}$ and $e^{-m_l i\phi}$ as solution. Hence,

$$\Phi = Ae^{\pm im_l\phi} \qquad \text{... (11)}$$

Now for Φ to satisfy single valuedness condition, it must have the same value for $\phi = 0$ and $\phi = 2\pi$ as both these values correspond to the same physical point. That is

$$\Phi(\phi) = \Phi(\phi + 2\pi)$$

or $$Ae^{\pm im_l\phi} = Ae^{\pm im_l(\phi+2\pi)}$$

or $$e^{\pm 2im_l\pi} = 1$$

or $$\cos 2\pi m_l + i\sin 2\pi\, m_l = 1 \qquad \text{... (12)}$$

as $\cos(-\theta) = \cos\theta$. Eqn. (12) will be true if and only if

$$m_l = 0, \pm 1, \pm 2, \pm 3, \ldots \qquad \text{... (13)}$$

called the magnetic quantum number (see section 8.3)

Since Φ depends on m_l, it is better to write

$$\Phi_{m_l} = Ae^{im_l\phi} \qquad \text{... (14)}$$

as solution of eqn. (9) with m_l given by eqn. (13). Note that since both of m_l are included, it is not necessary to mention $e^{-iml\phi}$ separately.

We now turn to eqn. (10). The change of variables.

$$\xi = \cos\theta,\ F(\xi) = \Theta(\theta) \qquad \text{... (15)}$$

transforms it into

$$\frac{d}{d\xi}\left[(1-\xi^2)\frac{dF}{d\xi}\right]+\left[\beta-\frac{m_l^2}{(1-\xi^2)}\right]F = 0 \qquad \text{... (16)}$$

which has been obtained by writing first part of eqn. (10) as

$$\frac{d}{d(\cos\theta)}\left[\sin^2\theta\frac{d}{d(\cos\theta)}\right]$$

Eqn. (16) is of the same form as associated Legendre equation. Since θ can have values from $0 - \pi$, the said equation will have significance for ξ from -1 to $+1$.

The equation has singular points at $\xi = \pm 1$ as $1 - \xi^2 = 0$ and hence $m_l^2/(1-\xi^2) \to \infty$. For $m_l = 0$, of course there is no such difficulty. Multiplying eqn. (16) by $(1 - \xi^2)$ and making the limit $\xi^2 \to 1$ (or $\xi \to \pm 1$), we conclude that $m_l^2\, F(\xi) \to 0$. The vanishing of $F(\xi)$ when $\xi^2 \to 1$ for $m_l \neq 0$ suggest that $F(\xi)$ contains a factor $(1-\xi^2)^a$ where $a > 0$. To solve eqn. (16), we therefore look for solutions of the form,

$$F(\xi) = (1-\xi^2)^a\, G(\xi),\ a > 0 \qquad \text{...(17)}$$

with $G(\xi)$ which is some function of ξ remaining finite at $\xi \to \pm 1$. Substituting above into eqn. (16), we obtain

$$(1-\xi^2)\frac{d^2G(\xi)}{d\xi^2} - 2(2a+1)\xi\frac{dG(\xi)}{d\xi} +$$

$$\left[\beta - 4a^2 - 2a + \frac{(4a^2\, m_l^2)}{(1-\xi^2)}\right]G(\xi) = 0 \quad \text{... (18)}$$

The choice

$$4a^2 = m_l^2\left(or\ |\ m_l\ | = 2a\ or\ a = \frac{|\ m_l\ |}{2}\right) \quad \text{... (19)}$$

helps to eliminate the inconvenient last term in eqn. (18) and we are left with

$$(1-\xi^2)\frac{d^2G(\xi)}{d\xi^2} - 2\xi(|\ m_l\ |+1)\frac{dG(\xi)}{d\xi} +$$

$$[\beta - |m_l|\ (|m_l|+1)]\ G(\xi) = 0 \quad \text{... (20)}$$

Let us solve the above equation by Sommerfeld expansion method, that is by assuming power series solution,

$$G(\xi) = \sum_{n=0}^{\infty} A_n \xi^n \quad \text{... (21)}$$

Substituting (21) into (20), we obtain

$$(1-\xi^2)\ \Sigma n(n-1)\ A_n\ \xi^{n-2} - 2\xi(|m_l|+1)\ \Sigma n\ A_n\ \xi^{n-1} +$$

$$[\beta - |m_l|(|m_l|+1)]\Sigma A_n\ \xi^n = 0 \quad \text{... (22)}$$

For the above equation to hold good for all values of ξ the coefficient of the every power of ξ should vanish separately. Equating the coefficient of to zero, we get

$$(n+1)\ (n+2)\ A_{n+2} - n(n-1)\ A_n -$$

$$2n(|m_l|+1)\ A_n + [\beta - |m_l|\ (|m_l|+1)]A_n = 0$$

or $$(n+1)\ (n+2)\ A_{n+2} + [\beta - (|m_l|+n)^2 - (|m_l|+n)]\ A_n = 0$$

or $$(n+1)\ (n+2)\ A_{n+2} + [\beta - (|m_l|+n)\ (|m_l|+n+1)]\ A_n = 0$$

or $$A_{n+2}/A_n = \frac{[(|\ m_l\ |+n)(|\ m_l\ |+n+1)-\beta]}{(n+1)(n+2)} \quad \text{... (23)}$$

This is called recursion relation. From which, we have

$$A_{n+2}/A_n\ \overrightarrow{n\to\infty\ 1} \quad \text{... (24)}$$

that is, the coefficients tend to equality ($A_{n+2} = A_n$) for large n and hence the series (21) diverges for $\xi = 1$ for large n. This behaviour which finally leads to an unacceptable solution can be avoided if the infinite series $G(\xi)$ is terminated beyond a certain coefficient or if it becomes a polynomial say of degree n (or if the series terminates at power n of ξ) instead of an infinite series. We can do it by putting

$$A_{n+2} = 0,\ A_n \neq 0$$

in eqn. (23). This amounts to

$$\beta = (|m_l| + n)(|m_l| + n + 1) \quad \text{... (25)}$$

It has already been seen through eqn. (13) that $|m_l| = 0, 1, 2, 3,...$ and n (see eqn. 21) of course is again 0, 1, 2, 3,.... Thus the sum $|m_l| + n = l$ (say) would also have values 0, 1, 2, 3.... Therefore we can write

$$\beta = l(l+1), \quad l = 0, 1, 2, 3, ... \quad \text{... (26)}$$

Now using the value of β from eqn. (6), one gets

$$\frac{2IE}{\hbar^2} = l(l+1)$$

or

$$E = E_l = l(l+1)\,\hbar^2/2I, \; l = 0, 1, 2, ... \quad \text{... (27)}$$

in which l is called rotational or orbital quantum number (see section 8.2). The E_l values are the energy eigen values of the rigid rotator. It is customary to set

$$B = \hbar/4\pi I \quad \text{... (28)}$$

where B is called the rotational constant and has the dimensions of frequency. Thus

$$E_l = Bh\,l(l+1), l = 0, 1, 2, 3 ... \quad \text{... (29)}$$

Fig. 5.3 shows the first few energy levels of the rigid rotator. The separation of the two adjacent levels, l and $l-1$ is equal to,

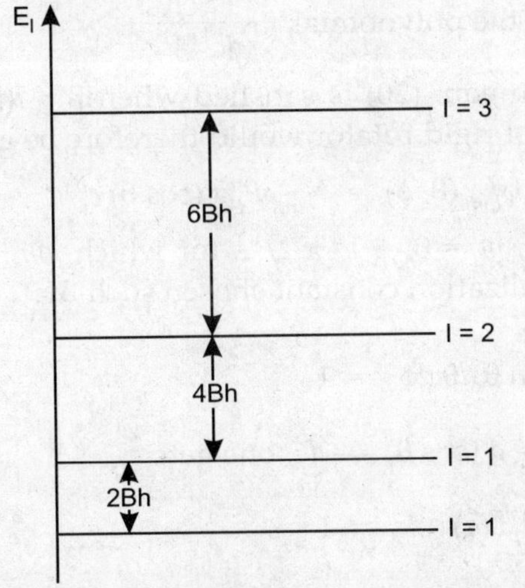

Fig. 5.3 *Energy levels of a rigid rotator for l = 0, 1, 2 and 3. Each level is separated from the next lower level by 2Bhl*

$$E_l - E_{l-1} = Bh[l(l+1) - l(l-1)]$$

$$= 2Bh\,l \quad \text{... (30)}$$

Note that the energy level spacing in this case shows an increase proportional to l in contrast with the constant spacing in the case of a harmonic oscillator. It must also be noted that the lowest state for which $l = 0$ (and obviously $m_l = 0$) has zero energy and this is permissible as $Y_l^{m_l}(\theta, \phi)$, the eigen function in the case of rigid rotator (to be shown in the next section) becomes Y_0^0 which does not depend upon θ and ϕ or there is no variation in the wave function over the surface of sphere (of radius r_e) for different values of θ and ϕ.

5.2 RIGID ROTATOR WAVE FUNCTIONS AND DEGENERACY

With $\beta = l(l + 1)$, eqn. (16) takes the form

$$\frac{d}{d\xi}\left[(1-\xi^2)\frac{dP_l^{|m_l|}}{d\xi}\right]+\left[l(l+1)-\frac{m_l^2}{(1-\xi^2)}\right]P_l^{|m_l|} = 0$$

or
$$(1-\xi^2)\frac{d^2P_l^{|m_l|}(\xi)}{d\xi^2}-2\xi\frac{dP_l^{|m_l|}(\xi)}{d\xi}+\left[l(l+1)-\frac{m_l^2}{(1-\xi^2)}\right]P_l^{|m_l|}(\xi) = 0 \quad \text{... (31)}$$

which is associated Legendre's equation. Its solution would be associated Legendre's polynomial $P_l^{|m_l|}(\xi) = (1-\xi^2)^{|m_l|/2}\frac{d^{|m_l|}P_l(\xi)}{d\xi^{|m_l|}}$, where $P_l(\xi)$ is the Legendre's polynomial) of degree l. Note that the above equation has been obtained by replacing the infinite series $G(\xi)$ by the polynomial, $\frac{d^{|m_l|}P_l(\xi)}{d\xi^{|m_l|}}$.

With this substitution eqn. (20) is satisfied when $\beta = l(l + 1)$ is put in that. The complete eigen function of rigid rotator would therefore be given by

$$\varphi_{l,m_l}(\theta, \phi) = N_{lm_l}P_l^{m_l}(\cos\theta)\,e^{im_l\phi} \quad \text{... (32)}$$

where $l = 0, 1, 2, \ldots, m_l = 0, \pm 1, \pm 2, \ldots$ for which $|m_l| \le l$ as $|m_l| = l - n$. The coefficient N_{lm_l} is a normalization constant chosen such that

$$\int_0^{\pi}\int_0^{2\pi}|\varphi_{l,m_l}(\theta,\phi)|^2\sin\theta\, d\theta\, d\phi = 1$$

Replacing $\sin\theta d\theta$ by $-d(\cos\theta) = -d\xi$, one gets

$$2\pi|N_{lm_l}|^2\int_{-1}^{+1}|P_l^{m_l}(\xi)|^2 d\xi = 1$$

or

$$2\pi|N_{lm_l}|^2 2/(2l+1)\frac{(l+m_l)!}{(l-m_l)!} = 1 \quad \text{... (33)}$$

Hence the complete eigen function of the rigid rotator would be given by

$$\varphi_{l,m_l}(\theta,\phi) = \frac{(-1)^{m_l}(2l+1)}{4\pi}\left\{\frac{(l-m_l)!}{(l+m_l)!}\right\}^{1/2}P_l^{m_l}(\cos\theta)\,e^{im_l\phi}$$

$$= Y_l^{m_l}(\theta, \phi), l = 0, 1, 2, ...$$

and $$m_l = 0, \pm 1, \pm 2, \ldots \quad ... (34)$$

with $P_l^{m_l}(\cos\theta) = 0$ if $|m_l| > l$. Also note that

$$Y_l^{-m_l}(\theta, \phi) = (-1)^{m_l}[Y_l^{m_l}]^* \quad ... (35)$$

The functions $Y_l^{ml}(\theta, \phi)$ are called spherical harmonics. The phase factor $(-1)^{m_l}$ has been chosen to agree with that most commonly used in literature.

Note that for a given value of l, $|m_l| = l - n$, n being zero or an integer, can take values from $0 - l$. That is, $m_l = -l, -l+1, 0, ... l-1, +l$. Thus there are $(2l + 1)$ possible values of m_l and hence same number of eigen functions for the given value of l. Hence for every value of l there are $(2l + 1)$ states of the rigid rotator having same energy as it depends only on l. The degeneracy is therefore, $(2l + 1)$ fold. Accordingly the first four spherical harmonics or eigen functions are given by

$$l = 0: \quad 2l + 1 = 1; m_l = 0$$

$$Y_0^0 = (4\pi)^{-1/2}$$

$$l = 1: 2l + 1 = 3; m_l = -1, 0, +1$$

$$Y_1^{-1} = \left(\frac{3}{8\pi}\right)^{1/2} \sin\theta\, e^{-i\phi}$$

$$Y_1^0 = \left(\frac{3}{4\pi}\right)^{1/2} \cos\theta$$

$$Y_1^1 = -\left(\frac{3}{8\pi}\right)^{1/2} \sin\theta\, e^{i\phi}$$

5.3 HYDROGEN ATOM (OR HYDROGEN LIKE ATOMS, VIZ. He^+, Li^{++}, Be^{+++}, - - - OR ONE ELECTRON ATOMS)

As another illustration of the central potential (or spherically symmetric potential or potential which depends upon the separation of two particles only) problem, let us consider the problem of hydrogen atom. It is essentially a two body problem consisting of a positively charged nucleus (in fact proton) and a negatively charged electron, both of them moving about their centre of mass under the action of their mutual attraction characterized by the potential energy function.

$$V(\vec{r}_1, \vec{r}_2) = \frac{-Ze^2}{|\vec{r}_1 - \vec{r}_2|}$$

or $$= V(|\vec{r}_1 - \vec{r}_2|)$$

$$V(r) = -Ze^2/r$$

where $\vec{r}_1$ and $\vec{r}_2$ are the position vectors of the electron. Ze is the charge of the nucleus. This way we can deal with the H-atom and with the He^+, Li^{++}, Be^{+++} ions in one step by simply choosing the atomic number $Z = 1, 2, 3, 4, \ldots$ $\vec{r} = \vec{r}_1 - \vec{r}_2$ is called the relative coordinate and r being the separation between the two particles.

The Schödinger equation, therefore for the present two body problem reads.

$$\left[\frac{-\hbar^2}{2m_1}\nabla_1^2 - \frac{\hbar^2}{2m_2}\nabla_2^2 + V(r)\right]\varphi(\vec{r}_1, \vec{r}_2) = E_T\varphi(\vec{r}_1, \vec{r}_2) \quad \ldots (36)$$

where first and the second terms represent kinetic energies corresponding to the electron and the nucleus, respectively. E_T represents the energy of the total system.

Let us also introduce the centre of mass coordinate R, defined by

$$\vec{R} = (m_1\vec{r}_1 + m_2\vec{r}_2)/M \quad \ldots (37)$$

where $M = m_1 + m_2$ is the total mass of the system.

Now

$$\vec{\nabla}_1 = \frac{\partial}{\partial \vec{r}_1} = \frac{\partial}{\partial \vec{r}}\frac{\partial \vec{r}}{\partial \vec{r}_1} + \frac{\partial}{\partial \vec{R}}\frac{\partial \vec{R}}{\partial \vec{r}_1} = \frac{\partial}{\partial \vec{r}} + \frac{m_1}{M}\frac{\partial}{\partial \vec{R}}$$

$$= \vec{\nabla}_r + \frac{m_1}{M}\vec{\nabla}_R$$

as $\frac{\partial \vec{r}}{\partial \vec{r}_1} = 1$, $\vec{r}_2$ being independent of $\vec{r}_1$

and

$$\vec{\nabla}_2 = \frac{\partial}{\partial \vec{r}_2} = \frac{\partial}{\partial \vec{r}}\frac{\partial \vec{r}}{\partial \vec{r}_2} + \frac{\partial}{\partial \vec{R}}\frac{\partial \vec{R}}{\partial \vec{r}_2} = \frac{-\partial}{\partial \vec{r}} + \frac{m_2}{M}\frac{\partial}{\partial \vec{R}}$$

$$= -\vec{\nabla}_r + \frac{m_2}{M}\vec{\nabla}_R$$

as $\frac{\partial \vec{r}}{\partial \vec{r}_2} = -1$. Therefore

$$-\frac{\hbar^2}{2m_1}\nabla_1^2 - \frac{\hbar^2}{2m_2}\nabla_2^2$$

$$= -\frac{\hbar^2}{2m_1}\left(\vec{\nabla}_r + \frac{m_1}{M}\vec{\nabla}_R\right)\cdot\left(\vec{\nabla}_r + \frac{m_1}{M}\vec{\nabla}_R\right)$$

$$-\frac{\hbar^2}{2m_2}\left(-\vec{\nabla}_r + \frac{m_2}{M}\vec{\nabla}_R\right)\cdot\left(-\vec{\nabla}_r + \frac{m_2}{M}\vec{\nabla}_R\right)$$

$$= -\frac{\hbar^2}{2M}\nabla_R^2 - \frac{\hbar^2}{2\mu}\nabla_r^2 \quad \ldots (38)$$

where $\mu = m_1m_2/(m_1 + m_2)$ is called reduced mass of the system. Note that

$$\nabla_R^2 = \frac{\partial}{\partial X^2} + \frac{\partial}{\partial Y^2} + \frac{\partial}{\partial Z^2} \text{ and } \nabla_r^2 = \frac{\partial^2}{\partial x^2} + \frac{\partial^2}{\partial y^2} + \frac{\partial^2}{\partial z^2}$$

in which (X, Y,Z) are coordinates of the centre of mass and (x, y, z) are the relative coordinates.

Substituting eqn. (38) into eqn. (36), it can be written in terms of $\vec{R}$ and $\vec{r}$ as

$$\left[-\frac{\hbar^2}{2M}\nabla_R^2-\frac{\hbar^2}{2\mu}\nabla_r^2+V(r)\right]\varphi(\vec{r},\vec{R})= E_T\varphi(\vec{r},\vec{R}) \quad \text{... (39)}$$

Let us now use the method of separation of variables (one thinks that it may be possible to separate the variables in eqn. (36) by assuming

$$\varphi(\vec{r}_1,\vec{r}_2) = \varphi_1(\vec{r}_1)\varphi_2(\vec{r}_2)$$

but the answer is negative. With this in mind, the centre of mass and relative coordinates have been introduced) and write.

$$\varphi(\vec{r},\vec{R}) = \varphi(\vec{r}_1)\chi(\vec{R}) \quad \text{... (40)}$$

Substituting above into eqn. (39) and dividing the resulting equation by $\varphi(r)\chi(\vec{R})$, one gets

$$-\frac{\hbar^2}{2M}\frac{\nabla_R^2\chi(\vec{R})}{\chi(\vec{R})}+\left[-\frac{\hbar^2}{2\mu}\frac{\nabla_r^2}{\varphi(\vec{r})}+V(r)\right]= E_T \quad \text{... (41)}$$

Note that the first term in the above equation depends only upon $\vec{R}$ and the second (in square bracket) only on ~r and the sum is equal to E_T. This would be possible only when

$$-\frac{\hbar^2}{2M}\nabla_R^2\chi(\vec{R}) = E_R\chi(\vec{R}) \quad \text{... (42)}$$

and

$$\left[-\frac{\hbar^2}{2\mu}\nabla_r^2+V(r)\right]\varphi(\vec{r}) = E_r\varphi(\vec{r}) \quad \text{... (43)}$$

where E_R and E_r are constants and are connected by the relation

$$E_R+E_r = E_T \quad \text{... (44)}$$

Eqn. (42) tells us that the centre of mass of the two particles moves like a free particle of mass M or it describes the motion of H-atom as a whole with E_R referred to as the energy of translation. Eqn. (43) describes the relative motion of two particles and is same as the equation for the motion of a single particle of mass μ moving in a field characterized by the potential $V(r)$. Thus the two body central force problem has effectively been reduced to a one body problem. E_r called the relative energy is the difference between the total energy E_T and E_R. Since in the H-atom problem, we are interested in the energy levels and energy eigen functions associated with the motion of electron relative to nucleus (the proton), we can take the centre of mass to be at rest. Also from now onwards, we would remove the subscript r on E.

As V is function only of r, it is convenient to solve the problem in spherical polar coordinates (r, θ, ϕ). Using the relation,

$$\nabla_r^2 = \frac{1}{r^2}\frac{\partial}{\partial r}\left(r^2\frac{\partial}{\partial r}\right)+\frac{1}{r^2 \sin\theta}\frac{\partial}{\partial\theta}\left(\sin\theta\frac{\partial}{\partial\theta}\right)+\frac{1}{r^2\sin^2\theta}\frac{\partial^2}{\partial\phi^2}$$

eqn. (43) takes the form

$$\left[\frac{1}{r^2}\frac{\partial}{\partial r}\left(r^2\frac{\partial}{\partial r}\right)+\frac{1}{r^2 \sin^2\theta}\frac{\partial}{\partial\theta}\left(\sin\theta\frac{\partial}{\partial\theta}\right)+\frac{1}{r^2\sin^2\theta}\frac{\partial^2}{\partial\phi^2}\right]$$

$$\varphi(r, \theta, \phi) + \frac{2\mu}{\hbar^2}[E - V(r)]\varphi(r, \theta, \phi) = 0 \qquad \text{... (45)}$$

Once again using the method of separation of variables, we look for the solution,

$$\varphi(r, \theta, \phi) = R(r)\, Y(\theta, \phi) \qquad \text{... (46)}$$

Substituting above into eqn. (45) and multiplying the resulting equation by r^2/φ, we get

$$\frac{1}{R(r)}\frac{d}{dr}\left(r^2\frac{dR}{dr}\right)+\frac{2\mu r^2}{\hbar^2}[E - V(r)] = \frac{-1}{Y(\theta,\phi)}\left[\frac{1}{\sin\theta}\frac{\partial}{\partial\theta}\left(\sin\theta\frac{\partial}{\partial\theta}\right)+\frac{1}{\sin^2\theta}\frac{\partial^2}{\partial\phi^2}\right]Y(\theta, \phi) \qquad \text{... (47)}$$

The left side of the above equation is a function of r only and the right side is function of angular coordinates. This is only possible if both of them are equal to the same separation constant, say β. Referring to the eqns. (6) and (26) of section 5.1 and eqn. (34) of section 5.2, one gets $\beta = l(l+1)$ and $Y(\theta, \phi) = Y_l^{m_l}(\theta, \phi)$.

Hence eqn. (47) to be called radial equation takes the form,

$$\frac{1}{r^2}\frac{d}{dr}\left(r^2\frac{dR}{dr}\right)+\frac{2\mu}{\hbar^2}\left[E - V(r) - l\,(l+1)\frac{\hbar^2}{2\mu r^2}\right]R(r) = 0 \qquad \text{... (48)}$$

Note that so far our treatment is general for any two body central field problem. Now for the H-atom case, we put $V(r) = -Ze^2/r$ in the above equation, Ze being the nuclear charge. This way the treatment would hold good for hydrogen like systems, He^+, Li^{++}, Be^{+++} ... as well. Therefore

$$\frac{1}{r^2}\frac{d}{dr}\left(r^2\frac{dR}{dr}\right)+\frac{2\mu}{\hbar^2}\left[E + \frac{Ze^2}{r} - l\,(l+1)\frac{\hbar^2}{2\mu r^2}\right]R(r) = 0$$

or

$$\frac{1}{r^2}\frac{d}{dr}\left(r^2\frac{dR}{dr}\right)+\left[\frac{2\mu E}{\hbar^2}+\frac{2\mu Ze^2}{r\hbar^2}-\frac{l\,(l+1)}{r^2}\right]R(r) = 0 \qquad \text{... (49)}$$

In order to solve the above equation, let us introduce a dimensionless variable $\rho = \alpha r$ (α to be defined little later). Dividing eqn. (49) by α^2 and by changing the variables from r to ρ, one gets

$$\frac{1}{\rho^2}\frac{d}{d\rho}\left(\rho^2\frac{dR}{d\rho}\right)+\left[\frac{2\mu E}{\hbar^2\alpha^2}+\frac{2\mu Ze^2}{\rho\alpha\hbar^2}-\frac{l(l+1)}{\rho^2}\right]R(\rho) = 0$$

or

$$\frac{1}{\rho^2}\frac{d}{d\rho}\left(\rho^2\frac{dR}{d\rho}\right)+\left[-\frac{1}{4}+\frac{\lambda}{\rho}-\frac{l(l+1)}{\rho^2}\right]R(\rho) = 0 \quad \text{... (50)}$$

where

$$\alpha = \left(\frac{-8\mu E}{\hbar^2}\right)^{1/2} \text{ and has dimensions cm}^{-1}$$

and

$$\lambda = \frac{2\mu Ze^2}{\alpha\hbar^2} = \frac{Ze^2}{\hbar}\left(\frac{\mu}{-2E}\right)^{1/2} \quad \text{... (51)}$$

after putting the value of α.

Note that two sets of solutions are possible in the case of present problem. The first set applies when $E < 0$ and corresponds to the bound states and the second set applies when $E > 0$ and corresponds to the unbound states. Here we are going to deal in detail with the bound states. Thus α is real. Also note that the choice of the number 1/4 in eqn. (50) is purely arbitrary. However, it helps in making the calculations simple.

Let us first examine the behaviour of $R(\rho)$ in the asymptotic region, that is, for $\rho \to \infty$. In this case, eqn. (50) reduces[+] to

$$\frac{d^2R(\rho)}{d\rho^2}-\frac{1}{4}R(\rho) = 0 \quad \text{... (52)}$$

Apart from a constant, the solution of the above equation is $R(\rho) = e^{\pm\rho/2}$. We should take only the negative sign in the exponent and reject $e^{\pm\rho/2}$ otherwise $R(\rho)$ would become infinite for $\rho \to \infty$ and would violate the condition that the solution should be finite everywhere. Thus the solution for the large values of ρ would be $R(\rho) = e^{-\rho/2}$.

To study the behaviour of radial function $R(\rho)$ near the origin ($\rho \to 0$), let us put $R(\rho) = u(\rho)/\rho$ in eqn. (50). The radial eqn. then reduces to

$$\frac{d^2u(\rho)}{d\rho^2}+\left[-\frac{1}{4}+\frac{\lambda}{\rho}-\frac{l(l+1)}{\rho^2}\right]u(\rho) = 0$$

For $l \geq 1$ (or $\neq 0$), the term $l(l+1)/\rho^2$ is large (or diverges) for small ρ and so dominates over the other terms. Thus in the neighbourhood of the origin we have, to a good approximation

$$\frac{d^2u(\rho)}{d\rho^2}-\frac{l(l+1)}{\rho^2}u(\rho) = 0 \quad \text{... (53)}$$

[+] $$\frac{1}{\rho^2}\frac{d}{d\rho}\left(\frac{\rho^2 dR}{d\rho}\right)=\frac{1}{\rho^2}\left(\frac{\rho^2 d^2R}{d\rho^2}+2\rho\frac{dR}{d\rho}\right)\xrightarrow[\rho\to\infty]{}\frac{d^2R}{d\rho^2}$$

which has the solution except for a constant,

$$u(\rho) = \rho^s \quad \text{... (54)}$$

where S is yet to be determined.

Substituting above into eqn. (53), we obtain

$$S(S-1) - l(l+1) = 0$$

or

$$S^2 - l^2 - S - l = 0$$

or

$$(S+l)(S-l-1) = 0$$

Thus

$$S = -l \quad \text{or } l+1$$

and hence

$$R(\rho) = \frac{u(\rho)}{\rho} = \rho^{-\overline{l+1}} \text{ or } \rho^l$$

The former is not acceptable as it makes $R(\rho)$ diverge as $\rho \to 0$. Thus we are left with the solution

$$R(\rho) = \rho^l \quad \text{as } \rho \to 0 \quad \text{... (55)}$$

Since the solution $R(\rho)$ has to be valid for all values of ρ from $0 - \infty$ ($-\infty$ is not be taken as r is positive in spherical polar coordinates), it is therefore taken as

$$R(\rho) = e^{-\rho/2} \rho^l G(\rho) \quad \text{... (56)}$$

where $G(\rho)$ is some function of p and is to be determined.

Substituting eqn. (56) into eqn. (50) and multiplying the resulting equation by ρ, we get

$$\rho G''(\rho) + (2l + 2 - \rho) G'(\rho) + (\lambda - \overline{l+1}) G(\rho) = 0 \quad \text{... (57)}$$

For solving the above equation, let us assume the series solution,

$$G(\rho) = \sum_{k=0}^{\infty} a_k \rho^k \quad \text{... (58)}$$

On substituting above into eqn. (57), we get

$$\rho \Sigma k(k-1) \rho^{k-2} a_k + (2l + 2 - \rho) \Sigma k\, a_k \rho^{k-1} + (\lambda - \overline{l+1}) \Sigma a_k \rho^k = 0$$

This is only possible when the coefficient of every power of p vanishes separately. Equating the coefficient of p^x to zero, we get

$$k(k+1)\, a_{k+1} (2l+2)(k+1)\, a_{k+1} - k\, a_k + (\lambda - \overline{l+1})\, a_k = 0$$

from which, we obtain the recursion relation,

$$\frac{a_{k+1}}{a_k} = \frac{\{(k+l+1) - \lambda\}}{\{(k+1)(k+2l+2)\}} \quad \text{... (59)}$$

For large values of k (that is, the asymptotic behaviour of series (26)), the above equation reduces to

$$a_{k+1}/a_k \xrightarrow[k \to \infty]{} 1/k \qquad \text{... (60)}$$

Let us consider the infinite series,

$$e^{\rho} = 1 + \rho + \frac{\rho^2}{2!} + \ldots + \frac{\rho^k}{k!} + \frac{\rho^{k+1}}{(k+1)!} + \ldots$$

The ratio of the coefficients of successive terms in the above series,

$$\frac{b_{k+1}}{b_k} = \frac{\frac{1}{(k+1)!}}{\frac{1}{k!}} = \frac{1}{(k+1)} \xrightarrow[k \to \infty]{} \frac{1}{k} \qquad \text{... (61)}$$

which is same as eqn. (60). Thus the series $G(\rho)$ behaves like e^{ρ} for large k. Therefore

$$R(\rho) = \rho^l e^{-\rho/2} e^{\rho} = \rho^l e^{\rho/2} \qquad \text{... (62)}$$

which means, for ρ large $R(\rho) \to \infty$ or diverges. This violates the boundary condition that the solution should be finite everywhere. To avoid this, we have to make the series $G(\rho)$ terminate (or it should be a polynomial instead of an infinite series). Let us assume that the series terminates at the power k of ρ, that is

$$a_{k+1} = 0, \; a_k \neq 0 \qquad \text{... (63)}$$

which amounts to (using eqn. (59)),

$$\lambda = n = k + l + 1 \qquad \text{... (64)}$$

Since k and l are zero and positive integers, n will have the values 1, 2, 3, . . . and that l cannot exceed $n - 1$ (or $l = 0, 1, 2, \ldots, n - 1$). The number n is called total or principal quantum number.

Putting the value of λ from eqn. (64) into eqn. (51), we get

$$\frac{Ze^2}{\hbar}\left(\frac{\mu}{-2E}\right)^{1/2} = n$$

Therefore, the energy eigen values or energy levels in the present case are given by

$$E = E_n = -\frac{\mu Z^2 e^4}{2\hbar^2 n^2} = \frac{-Z^2 e^2}{2a_0 n^2}, \; n = 1, 2, 3, \ldots \qquad \text{... (65)}$$

where

$$a_0 = \frac{\hbar^2}{\mu e^2}.$$

These E_n values for $Z = 1$ are just the energy levels of the hydrogen atom and are displayed in Fig. 5.4. a_0 being equal to the radius of the first orbit for hydrogen in the Bohr model, is called Bohr radius of the atom. The ground state (or the lowest state) of the hydrogen atom corresponds to $n = 1$ with energy – 13.6 eV. That is, the total energy needed (or work to be done) to separate the electron and the nucleus at infinite distance apart and also at rest is 13.6 eV. This energy is known as the ionization energy of the hydrogen atom in the ground state and is also the binding energy of the hydrogen atom. Note that as n increases, $|E_n|$ decreases as $1/n^2$ and correspondingly En = $-|E_n|$ increases converging to the limit $E_n = 0$ as $n \to \infty$.

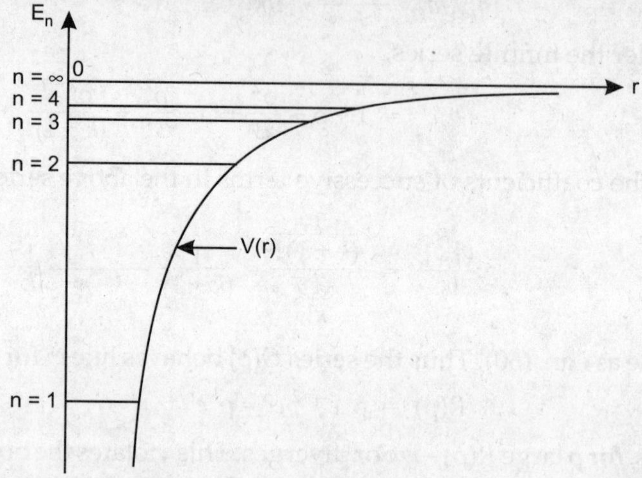

Fig. 5.4 *Energy level diagram for H-atom in relation to the potential*

5.4 EIGEN FUNCTIONS OF HYDROGEN LIKE ATOMS AND DEGENERACY

With $\lambda = n$, eqn. (57) takes the form

$$\rho L''(\rho) + (2l + 2 - \rho) L'(\rho) + (n - l - 1) L(\rho) = 0 \qquad \text{... (66)}$$

where the infinite series $G(\rho)$ has been replaced by the polynomial $L(\rho)$. The equation is same as the associated Laguerre equation,

$$\rho L_i^{j''}(\rho) + (j + 1 - \rho) L_i^{j'}(\rho) + (i - j) L_i^j(\rho) = 0 \qquad \text{... (67)}$$

Comparing above with eqn. (66), we get

$$2l + 2 = j + 1 \qquad \text{... (68a)}$$

$$n - l - 1 = i - j \qquad \text{... (68b)}$$

From the above equations, we get

$$i = n + l, \quad j = 2l + 1$$

and hence

$$L_i^j(\rho) = L_{n+l}^{2l+1}(\rho)$$

and the degree of polynomial is $n + l - (2l + 1) = n - l - 1$. Thus the solution of eqn. (66) would be associated Laguerre polynomial, $L_{n+l}^{2l+1}(\rho)$. Correspondingly $R(\rho)$, the radial function for given ρ and l from eqn. (56) would be given by

$$R(\rho) = R_{nl}(\rho) = N_{nl}\rho^l e^{-\rho/2} L_{n+l}^{2l+1}(\rho) \qquad \text{... (69)}$$

and the complete hydrogen like atom wave functions are then given by

$$\varphi_{nlm_l}(r, \theta, \phi) = R_{nl}(\rho) Y_l^{m_l}(\theta, \phi)$$
$$= N_{nl}\, \rho^l e^{-\rho/2} L_{n+l}^{2l+1}(\rho)\, Y_l^{m_l}(\theta, \phi) \qquad \text{... (70)}$$

where

$$\rho = \alpha r = \left(\frac{-8\mu E}{\hbar^2}\right)^{1/2} r$$

Putting $E = E_n$ from eqn. (65) and also using the relation $\hbar^2/\mu e^2 = a_0$, we get

$$\rho = \frac{2Z}{na_0} r \qquad \text{... (71a)}$$

or

$$r = \frac{na_0}{2Z} \rho \qquad \text{... (71b)}$$

The normalization constant N_{nl} is such that

$$\int_0^{\pi}\int_0^{2\pi}\int_0^{\infty} \varphi^*_{nlm_l}(r, \theta, \phi)\, \varphi_{nlm_l}(r, \theta, \phi)\, r^2 \sin\theta\, d\theta\, d\phi\, dr = 1 \qquad \text{... (72)}$$

Substituting from eqn. (70) into above, we get

$$\int_0^{\infty} R_{nl}^2(r) r^2 dr \int_0^{\pi}\int_0^{2\pi} |Y_l^{m_l}(\theta, \phi)|^2 \sin\theta\, d\theta\, d\phi = 1$$

Since the spherical harmonics is already normalized,

$$\int_0^{\pi}\int_0^{2\pi} |Y_l^{m_l}(\theta, \phi)|^2 \sin\theta\, d\theta\, d\phi = 1$$

Hence

$$\int_0^{\infty} R_{nl}^2(r) r^2 dr = 1$$

or

$$|N_{nl}|^2 \int_0^{\infty} \rho^{2l} e^{-\rho} |L_{n+l}^{2l+1}(\rho)|^2 r^2 dr = 1$$

Using eqn. (71b) and using the relation,

$$\int_0^{\infty} \rho^{2l+2} e^{-\rho} |L_{n+l}^{2l+1}(\rho)|^2 d\rho = 2n \frac{\{(n+l)!\}^3}{(n-l-1)!} \quad \text{... (73)}$$

for the associated Laguerre polynomials, we get

$$N_{nl} = \pm\left(\frac{2Z}{na_0}\right)^{3/2}\left[\frac{(n-l-1)!}{2n\{(n+l)!\}^3}\right]^{1/2} \quad \text{... (74)}$$

Thus

$$R_{nl}(r) = -\left(\frac{2Z}{na_0}\right)^{3/2}\left[\frac{(n-l-1)!}{2n\{(n+l)!\}^3}\right]^{1/2} e^{-\rho/2}\rho^l L_{n+l}^{2l+1}(\rho) \quad \text{... (75)}$$

where the negative sign is chosen so as to make $R_{10}(r)$ (which contains $L_1^1 = -1$) positive. Thus the total wave function for hydrogen like atom would be given by

$$\varphi_{nlm_l}(r, \theta, \phi) = R_{nl}(r)\, Y_l^{m_1}(\theta, \phi) \quad \text{... (76)}$$

where $R_{nl}(r)$ and $Y_l^{m_l}(\theta, \phi)$ are given by eqns. (34) and (75), respectively. The quantum numbers, $n = 1, 2, 3, ..., +\infty$; $l = 0, 1, ..., n-1$; and $m_l = -l, -l+1, \ldots, 0, 1, \ldots l-1, +l$ are called total (or principal) quantum number, orbital quantum number and the magnetic quantum number, respectively. Let us mention that it is customary to denote $l = 0, 1, 2, 3, \ldots$ values by the letters *s, p, d, f*. . . , respectively in spectroscopic notation.

Since the energy eigen values (65) depend only on the principal quantum number n, they are degenerate with respect to l and m_l. For a given value of n, l can vary from 0 to $n-1$ and for each of these l values, m_l can vary from $-l$ to $+l$ including zero, that is it can have $(2l + 1)$ values. Hence the total degeneracy (or number of independent wave functions) of the energy level E_n will be given by

$$\sum_{l=0}^{n-1}(2l+1) = 2\sum_{l=0}^{n-1} l + n$$

$$= 2n(n-1)/2 + n = n^2 \quad \text{... (77)}$$

Thus the number of allowed states for a given value of n is n^2. We shall subsequently discover that when one takes into account the spin of the electron into account the degeneracy of the state with principal quantum number n becomes $2n^2$.

The ground state wave function in the case of H-atom is obtained by putting $Z = 1$, $n = 1$ (and hence $l = 0$ and $m_l = 0$) in eqn. (76) and by also putting the value of ρ from (71*b*), that is

$$\varphi_{100}(r, \theta, \phi) = (1/\pi a_0^{\,3})^{1/2} e^{-r/a_0} \quad \text{... (78)}$$

which depends only on r and is independent of θ and ϕ. It is also called 1s (as $l = 0$) orbital in spectroscopic notation and is non degenerate.

For the first excited state, $n = 2$ and therefore the allowed values of l are 0 and 1. For $l = 0$, $m_l = 0$ and for $l = 1$, m_l can have three possible values, namely, – 1, 0, + 1. Thus the four wave functions corresponding to the first excited state are given by

$$\varphi_{200}(r, \theta, \phi) = \varphi_{2s} = \left(\frac{1}{32\pi a_0^3}\right)^{1/3}\left(2 - \frac{r}{a_0}\right)e^{-r/2a_0} \quad \text{... (79)}$$

$$\varphi_{210}(r, \theta, \phi) = \left(\frac{1}{32\pi a_0^5}\right)^{1/2} r \cos\theta \, e^{-r/2a_0} \quad \text{... (80)}$$

$$\varphi_{211}(r, \theta, \phi) = \left(\frac{1}{64\pi a_0^5}\right)^{1/2} r \sin\theta \, e^{i\phi} \, e^{-r/2a_0} \quad \text{... (81)}$$

$$\varphi_{21-1}(r, \theta, \phi) = \left(\frac{1}{64\pi a_0^5}\right)^{1/2} r \sin\theta \, e^{-i\phi} \, e^{-r/2a_0} \quad \text{... (82)}$$

and thus the first excited state is four fold degenerate.

The functions (81) and (82) being complex, have the disadvantage that we cannot draw pictures of them in real space. However, we can obtain functions which are equally good solutions of eqn. (45) with $V(r) = -Ze^2/r$ by taking the following linear combinations of spherical harmonics having the same l values. Therefore

$$\frac{1}{2^{1/2}}(Y_1^1 + Y_1^{-1}) = \left(\frac{3}{4\pi}\right)^{1/2} \sin\theta \cos\phi$$

$$-\frac{i}{2^{1/2}}(Y_1^1 - Y_1^{-1}) = \left(\frac{3}{4\pi}\right)^{1/2} \sin\theta \sin\phi \quad \text{...(83)}$$

Since $\sin\theta \cos\phi$ and $\sin\theta \sin\phi$ represent the angular dependence of x and y components of r, respectively, the above two combinations are labelled as p_x and p_y states, respectively. Similarly (80) containing r cas θ is designated as p_z. Thus (80) to (83) can be written as

$$\varphi_{210} = \varphi_{2p_z} \quad \text{... (84)}$$

$$\varphi_{211} = \varphi_{2p_x} = \left(\frac{1}{32\pi a_0^5}\right)^{1/2} r \sin\theta \cos\phi \, e^{-r/2a_0} \quad \text{... (85)}$$

and

$$\varphi_{21-1} = \varphi_{2p_y} = \left(\frac{1}{32\pi a_0^5}\right)^{1/2} r \sin\theta \sin\phi \, e^{-r/2a_0} \quad \text{... (86)}$$

It is, however necessary to mention here that our treatment of the H-atom was somewhat approximate as it did not take account for the fine structure of the hydrogen spectrum and also for the spin. Both of these would be explained by the relativistic quantum mechanics (Dirac theory).

5.5 PROBABILITY DENSITY IN THE GROUND STATE (OR 1s ORBITAL) OF HYDROGEN ATOM

The ground state wave function of the hydrogen atom is given by

$$\varphi_{nlm_l} = \varphi_{100} \text{ (or } \varphi_{1s}) = (1/\pi a_0^3)^{1/2} e^{-r/a_0} \quad \text{... (87)}$$

The probability density for an electron in this orbital (or state) would then be given by

$$|\varphi_{1s}|^2 = 1/\pi a_0^3 \, e^{-2r/a_0} \quad \text{... (88)}$$

It should be noted that $|\varphi_{1s}|^2$ is independent of θ and ϕ and therefore the electron distribution in 1s orbital is spherically symmetric. Thus, if $P(r)\, dr$ represents the probability of finding the electron in a spherical shell between r and $r + dr$, then

$$P(r)\, dr = 4\pi r^2\, dr\, |\varphi_{1s}|^2 = 4r^2/a_0^3 e^{-2r/a_0} dr \quad \text{... (89)}$$

The maximum value of $P(r)$, called radial probability density occurs at a value of r which is found from the equation,

$$\frac{dP(r)}{dr} = [r^2(-2/a_0) + 2r]e^{-2r/a_0} = 0$$

or

$$(r - r^2/a_0) = 0$$

or

$$r = a_0 \quad \text{... (90)}$$

That is, the maximum radial density occurs at $r = a_0$ or in other words, the most probable distance of the electron from the nucleus is precisely what has been predicted from the Bohr's theory. Fig. 5.5 shows the plot of $P(r)$ as a function of r/a_0.

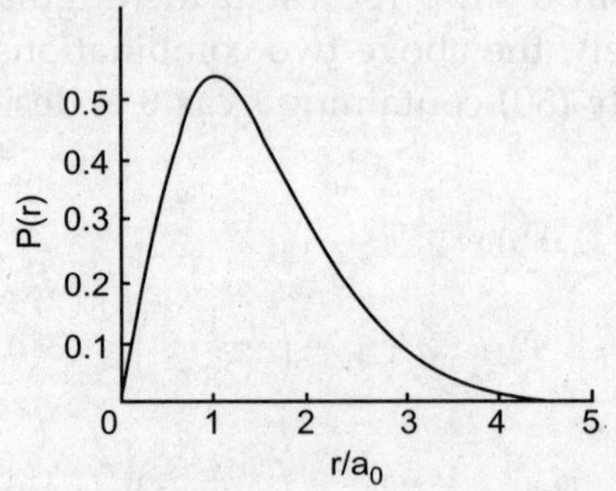

Fig. 5.5 *Radial Probability density P (r) for n = 1 state of hydrogen atom*

5.6 ISOTOPES OF HYDROGEN

The physical systems which are very close to hydrogen atom are its two isotopes, deuterium and tritium. In deuterium, the proton is replaced by a nucleus having the same charge but possessing one neutron in addition to the proton and in tritium the proton is replaced by a nucleus having the same charge but possessing two neutrons in addition to the proton. The mass of deuterium nucleus is roughly $2m_p$, and that of tritium nucleus, $3m_p$.

The reduced masses μ_d and μ_t, therefore become

$$\mu_d = m_e \times 2m_p/(m_e + 2m_p) = m_e(1 + m_e/2m_p)^{-1}$$

$$\simeq m_e(1 - m_e/2m_p) \quad \text{... (91)}$$

$$\mu_t \simeq m_e(1 - m_e/3m_p) \quad \text{... (92)}$$

Since $m_e/m_p << 1$, it is clear that μ_d and μ_t are approximately equal to m_e (or equal to the reduced mass of the hydrogen atom, $\mu = m_e\, m_p/(m_e + m_p) \simeq m_e$). Accordingly μ_d and μ_t can be replaced by μ without introducing much error. Thus the energy levels given by eqn. (65) are practically the same in all the three cases that is, hydrogen, deuterium and tritium. Nevertheless, there are slight differences, of the order of thousandth in relative value and can be detected experimentally by the optical spectrograph of sufficient resolution.

5.7 THREE DIMENSIONAL SQUARE WELL POTENTIAL

The potential energy variation in the case of a three dimensional spherically symmetric square well potential displayed in Fig. 5.6 is given by

$$V(r) = -V_0, \; r < r_0 \quad \text{... (93)}$$

$$= 0, \quad r > r_0$$

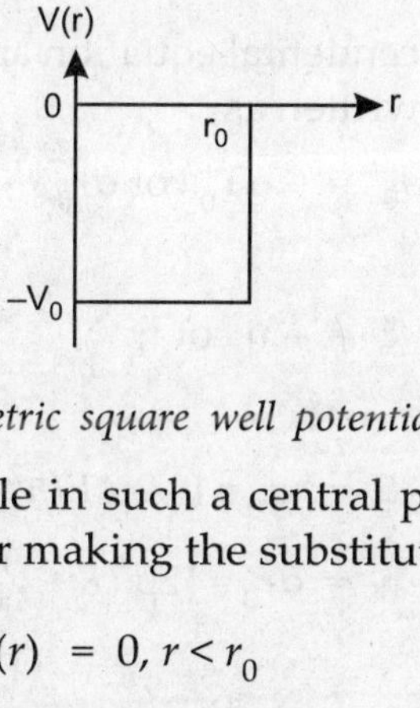

Fig. 5.6 *A spherically symmetric square well potential of depth V_0 and radius r_0*

For bound states of a particle in such a central potential, $-V_0 < E < 0$. Then the radial Schrödinger eqn. (48) after making the substitution $R(r) = u(r)/r$, takes the form

$$\frac{d^2u(r)}{dr^2} + \left[\alpha^2 - \frac{l(l+1)}{r^2}\right]u(r) = 0, \; r < r_0 \quad \text{... (94a)}$$

$$\frac{d^2u(r)}{dr^2} - \left[\kappa^2 + \frac{l(l+1)}{r^2}\right]u(r) = 0, \; r > r_0 \quad \text{... (94b)}$$

where

$$\alpha^2 = \frac{2\mu(E + V_0)}{\hbar^2} \text{ and } \kappa^2 = \frac{-2\mu E}{\hbar^2}$$

Let us consider two cases:

(i) $l = 0$ case, that is, the s-state

The general solution of eqns. (94) would be given by

$$u(r) = A \sin \alpha r + B \cos \alpha r, \; r < r_0 \quad \text{... (95a)}$$

$$u(r) = Ce^{-\kappa r} + De^{+\kappa r}, \quad r > r_0 \quad \text{... (95b)}$$

Since $R(r) = u(r)/r$ is not to become infinite for $r = 0$, B must be zero.

Also $R(r)$ has to vanish (or has to be finite) as $r \to \infty$, D must be zero.

Therefore

$$u(r) = A \sin \alpha r, \; r < r_0 \quad \text{... (96a)}$$

$$u(r) = Ce^{-\kappa r} \quad r > r_0 \quad \text{... (96b)}$$

The constants A and C have to be chosen in such a fashion that both $R(r)$ (or $u(r)$) and dR/dr (or $du(r)/dr$) are continuous (or match) at $r = r_0$. Applying the continuity conditions, one gets

$$A \sin \alpha r_0 = Ce^{-\kappa r_0} \quad \text{... (97a)}$$

$$\alpha A \cos \alpha r_0 = \kappa Ce^{-\kappa r_0} \quad \text{... (97b)}$$

Dividing the above two equations by each other, one gets the following equation (or condition) for the determination of allowed energy eigen values,

$$\kappa = -\alpha \cot \alpha r_0 \quad \text{... (98)}$$

The above equation is a transcendental equation and can be solved by graphical or by numerical methods. It can be written as

$$\kappa r_0 = -\alpha \Upsilon_0 \cot \alpha r_0$$

or

$$\xi = -\eta \cot \eta \quad \text{... (99)}$$

where

$$\xi = \kappa r_0 = (-2\mu E/\hbar^2)^{1/2} r_0,$$

$$\eta = \alpha r_0 = [2\mu (E + V_0)/\hbar^2]^{1/2} r_0$$

and hence

$$\xi^2 + \eta^2 = 2\mu V_0 r_0^{\,2}/\hbar^2 \quad \text{... (100)}$$

Fig. 5.7 shows the curves corresponding to eqn. (99) in a coordinate system $\xi = \kappa r_0$ $\eta = \alpha r_0$. The solution of the eigen value problem is obtained by determining the points of intersection of these curves with the arc plotted according to eqn. (100). Only points of intersection falling into the first quadrant of the plane are of interest, as $\kappa > 0$ and $\alpha > 0$. The circle corresponds to a well depth and radius satisfying the inequalities

$$\frac{\pi^2}{4} \le \frac{2\mu V_0 r_0^2}{\hbar^2} < \frac{9\pi^2}{4}$$

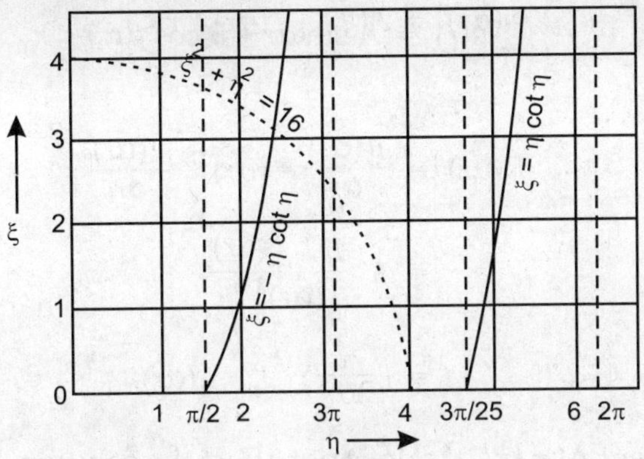

Fig. 5.7 *A graphical solution of eqn. (98) or (99) with* $\xi^2 + \eta^2 = 16$

In that case, there is one intersection, that is only one solution (odd solution) or bound state (state with negative energy) since η lies between π/2 and 3π/2. This energy can be determined from the value of ξ corresponding to the intersection of the curves through the formula

$$E = \hbar^2\xi^2/2\mu r_0^2 \quad \text{... (101)}$$

It may be noted that the problem for $l = 0$ is the same as the one dimensional square well with finite walls for odd solutions.

(*ii*) $l \neq 0$ case

To solve eqn. (94*a*) for an arbitrary non-zero value of *l*, we make the substitution.

$$u(r) = r^{1/2}\, v(r) \quad \text{... (102)}$$

which leads to

$$r^{1/2}\frac{d^2v(r)}{dr^2} + r^{-1/2}\frac{dv(r)}{dr} + \left[a^2 - \frac{l(l+1)}{r^2}\right]r^{1/2}v(r) - \frac{1}{4}r^{-3/2}v(r) = 0$$

or

$$(\alpha r)^2\frac{d^2v(\alpha r)}{dr^2} + (\alpha r)\frac{dv(\alpha r)}{d(\alpha r)}\left[(\alpha r)^2 - \left(l+\frac{l}{2}\right)^2\right]v(\alpha r) = 0,\ r < r_0 \quad \text{...(103)}$$

and similarly

$$(i\kappa r)^2\frac{d^2(i\kappa r)}{d(i\kappa r)^2} + (i\kappa r)\frac{dv(i\kappa r)}{d(i\kappa r)} + \left[(i\kappa r)^2 - \left(l+\frac{l}{2}\right)^2\right]v(i\kappa r) = 0,\ r > r_0 \quad \text{... (104)}$$

Eqn. (103) is Bessel's equation of order ± (*l* + 1/2) having argument α*r*, its solution, therefore would be given by

$$v(\alpha r) = A'\, J_{l+1/2}(\alpha r) + B'\, J_{-l-1/2}(\alpha r),\ r < r_0$$

where *A*′ and *B*′ are arbitrary constants.

Since $J_{-l-1/2}(\alpha r) \to \infty$ for $r \to 0$, *B*′ must be zero. Hence

$$v(\alpha r) = A' J_{l+1/2}(\alpha r), \quad r < r_0$$

and thus

$$R(\alpha r) = \frac{u(\alpha r)}{\alpha r} = (\alpha r)^{1/2} \frac{v(\alpha r)}{\alpha r}$$

$$= A' \frac{J_{l+1/2}(\alpha r)}{(\alpha r)^{1/2}}$$

$$= \left(\frac{\pi}{2\alpha r}\right)^{1/2} J_{l+1/2}(\alpha r)$$

where the coefficient $A' = (\pi/2)^{1/2}$ does not affect the solution. Therefore

$$R(\alpha r) = A\, j_l(\alpha r), \quad r < r_0 \qquad \text{... (105)}$$

where $j_l(\alpha r) = \left(\frac{\pi}{2\alpha r}\right)^{1/2} J_{l+1/2}(\alpha r)$ is termed as the spherical Bessel function of order l.

Similarly, the solution of eqn. (104) which is also Bessel's equation of order $\pm (l + 1/2)$ but with argument $i\kappa r$, would be given by

$$v(i\kappa r) = A'' J_{l+1/2}(i\kappa r) + B'' J_{-l-1/2}(i\kappa r), \; r > r_0$$

Note that both $J_{l+1/2}$ and $J_{-l-1/2}$ are to be taken into account as in this case the point $r = 0$ is not included. A'' and B'' are orbitrary constants. Since the spherical Neumann function is defined as

$$\eta_l(x) = (-1)^{l+1}\left(\frac{\pi}{2x}\right)^{1/2} J_{-l-1/2}$$

$$R(i\kappa r) = B\, j_l(i\kappa r) + C\eta_l(i\kappa r), \; r > r_0 \qquad \text{... (106)}$$

that is, the linear combination of $j_l(x)$ and $\eta_l(x)$ is the solution in this case which can be represented by spherical Hankel functions of first and second kinds.

Since the boundary condition in this case demands that the wave function must approach zero (or is finite) in the limit $r \to \infty$, only Hankel functions of first kind denoted by $h_l^{(1)}x = j_l(x) + i\eta_l(x)$ which decreases exponentially with r can be taken as solutions. The Hankel functions of second kind which increase exponentially with r are not acceptable. Therefore

$$R(i\kappa r) = D\, h_l^{(1)}(i\kappa r), \; r > r_0 \qquad \text{... (107)}$$

The functions (105) and (107) and their derivatives (or which amounts to logarithmic derivatives) have to join continuously at $r = r_0$. Thus

$$\left.\frac{1}{j_l(\alpha r)} \frac{dj_l(\alpha r)}{dr}\right|_{r=r_0} = \left.\frac{1}{h_l^{(1)}(i\kappa r)} \frac{dh_l^{(1)}(i\kappa r)}{dr}\right|_{r=r_0}$$

or

$$\frac{\alpha}{j_l(\alpha r)} \frac{dj_l(\alpha r)}{d(\alpha r)}\bigg|_{r=r_0} = \frac{i\kappa}{h_l^{(1)}(i\kappa r)} \frac{dh_l^{(1)}(i\kappa r)}{d(i\kappa r)}\bigg|_{r=r_0}$$

or

$$\frac{\alpha}{j_l(\alpha r_0)}\left[j_{l-1}(\alpha r_0) - \frac{(l+1)}{\alpha r_0} j_l(\alpha r_0)\right] = \frac{i\kappa}{h_l^{(1)}(i\kappa r_0)}\left[h_{l-1}^{(1)}(i\kappa r_0) - \frac{l+1}{i\kappa r_0} h_l^{(1)}(i\kappa r_0)\right]$$

Cancelling second term on both sides, one gets

$$\frac{\alpha j_{l-1}(\alpha r_0)}{j_l(\alpha r_0)} = i\kappa \frac{h_{l-1}^{(1)}(i\kappa r_0)}{h_l^{(1)}(i\kappa r_0)} \quad \text{... (108)}$$

Combining above with eqn. (100), a set of simultaneous transcendental equations is reached which can be solved graphically or numerically by a method similar to the case $l = 0$ and the energy eigen values E can be determined. Let us see this for the case $l = 1$.

In the said case, eqn. (108) takes the form,

$$\frac{\alpha\, j_0\,(\alpha r_0)}{j_1\,(\alpha r_0)} = i\kappa \frac{h_0^{(1)}\,(i\kappa r_0)}{h_1^{(1)}\,(i\kappa r_0)}$$

Substituting for $j_0, j_1, h_0^{(1)}$ and $h_1^{(1)}$ into above, one gets

$$\left\{\frac{\alpha\,(\sin(\alpha r_0)}{\alpha r_0}\right\} \Bigg/ \left[\frac{\sin \alpha r_0}{\alpha^2 r_0^2} - \frac{\cos \alpha r_0}{\alpha r_0}\right] = i\kappa\left(-\frac{e^{-\kappa r_0}}{\kappa r_0}\right) \Bigg/ i\left(\frac{1}{\kappa r_0} + \frac{1}{\kappa^2 r_0^2}\right) e^{-\kappa r_0}$$

After simplification and after dividing L.H.S. of the above equation by $\dfrac{\sin \alpha\, r_0}{\alpha r_0}$, one gets

$$\frac{\alpha}{(1/\alpha r_0 - \cot \alpha r_0)} = \frac{\kappa^2 r_0}{(1 + \kappa r_0)}$$

$$\text{or} \quad \frac{\cot \alpha r_0}{\alpha r_0} - \frac{1}{\alpha^2 r_0^2} = \frac{1}{\kappa^2 r_0^2} + \frac{1}{\kappa r_0} \quad \text{... (109)}$$

Let us define

$$\begin{aligned}(\xi^2 + \eta^2)^{1/2} &= (2\mu V_0/\hbar^2)^{1/2}\, r_0 \\ &= k_0 r_0 \\ &= x_0 \end{aligned} \quad \text{... (110)}$$

$$\text{and} \quad \alpha/k_0 = (1 + E/V_0)^{1/2} = \delta \quad \text{... (111)}$$

Therefore,

$$\kappa r_0 = \kappa k_0 r_0 / k_0 = x_0\, \kappa / k_0$$

$$= x_0 (-E/V_0)^{1/2}$$

$$= x_0 (1 - \delta^2)^{1/2} \quad \text{... (112)}$$

Eqn. (109) with the help of eqns. (110 – 112) can be written as

$$\frac{1}{x_0 \delta \tan x_0 \delta} - \frac{1}{\delta^2 x_0^2} = \frac{1}{x_0 (1-\delta^2)^{1/2}} + \frac{1}{x_0^2 (1-\delta^2)}$$

or
$$\frac{1}{\tan x_0 \delta} = \frac{1}{x_0 \delta} + \frac{\delta}{(1-\delta^2)^{1/2}} + \frac{\delta}{x_0 (1-\delta^2)}$$

or
$$\tan x_0 \delta = \frac{x_0 \delta}{\left[1 + \frac{\delta^2}{(1-\delta^2)}\{1 + x_0 (1-\delta^2)^{1/2}\}\right]} = f_1 \quad \text{... (113)}$$

Taking $x_0 = 10$, draw the curves, $y = \tan x_0\ \delta$ and $y = f_1$ in the region $0 < \delta < 1$ as shown in Fig. 5.8. The points of intersections give the required solutions.

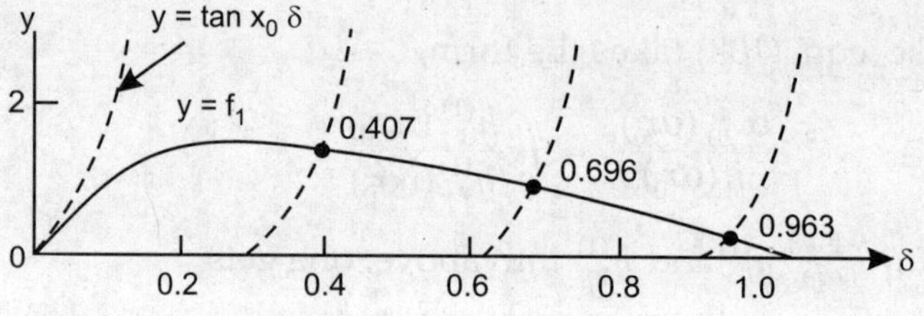

Fig. 5.8 *Graphical solution of eqn. (113) for $x_0 = 10$*

5.8 APPLICATION TO DEUTERON PROBLEM

This is the nucleus of heavy hydrogen consisting of two nucleons, a proton and a neutron, each of nearly the same mass m_N, giving $\mu = \frac{1}{2} m_N$. (or $2\mu = m_N = M$, say). The deuteron binding energy is 2.23 MeV, a figure that can be obtained either from the discrepancy in mass between m_{deuteron} and $m_p + m_n$ $(E = (m_p + m_n)c^2 - m_d c^2)$ or from photodisintegration experiments which show that only gamma rays with energy $h\upsilon \geq 2.23$ MeV can break deuteron into its constituents, that is proton and neutron. The deuteron is the smallest nucleus from which one learns many interesting nuclear properties such as the nature of nuclear forces (about which we have a limited knowledge even today as compared to the Coulomb force between proton and electron in the case of hydrogen atom), binding energies etc. just as the hydrogen atom is the smallest atom from which one derives valuable information of the atomic structure and atomic spectra.

The potential energy of the deuteron (or the potential energy of either nucleon with respect to the other) depends upon the distance r between neutron and proton roughly as shown by the solid line in Fig. 5.9. The positive value of the potential simply reflects the inability of the nucleons to come closer than about half fermi. The negative potential decreasing with r and becoming zero at a short distance indicates that the force is attractive and short range. Let us assume that the force is central (this assumption is not quite correct since central force cannot account for the quadrupole moment of the deuteron. As the quadrupole moment is small, the assumption is approximately correct) and can be represented by a spherical square well as shown by dashed line in Fig. 5.9 and represented mathematically by

$$V(r) = -V_0, \quad r \leq r_0$$
$$= 0, \quad r > r_0$$

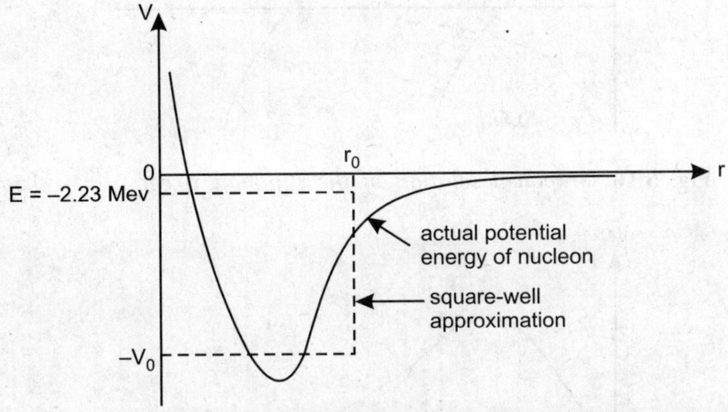

Fig. 5.9 *The actual potential energy of either proton or neutron in deuteron and the square well approximation to this potential energy as function of the distance between proton and neutron*

where V_0 and r_0 represent the strength and range, respectively of the interaction holding the deuteron together.

There is considerable evidence to suggest that the ground state of the deuteron is $l = 0$ or *s*-state (the total angular momentum of the deuteron $l = 1$ is due effectively to the spins of the two nucleons which compose it and thus the nucleons in the ground state of the deuteron have zero orbital angular momentum, *i.e.* the ground state is an s-state which is spherical symmetric or depends on *r* only).

Proceeding in the manner same as that in sec. 5.7 for $l = 0$, one gets the following relation for the depth and the range of the potential.

$$\cot \alpha r_0 = -\kappa/\alpha \qquad \text{... (114)}$$

where $\alpha^2 = M(E + V_0)/\hbar^2 = M(V_0 - B)/\hbar^2$ and $\kappa^2 = MB/\hbar^2$, $B = 2.23$ MeV being the binding energy of the deuteron.

Eqn. (114) can be written as

$$\cot \eta = -\kappa r_0/\eta \qquad \text{... (115)}$$

where $$\eta = \alpha r_0 = [M(V_0 - B)/\hbar^2]^{1/2}\, r_0$$

For $B = 2.23$ MeV and $r_0 \leq 2$ fermi, $\kappa = 0.232$ (fermi)$^{-1}$ and hence $\alpha r_0 \lesssim 0.5$. If we then draw the curves $y = \cot \eta$ and $y = -0.5/\eta$, the intersections give the roots of eqn. (115) and from Fig. 5.10 it is clear that the roots are slightly larger than $(n + 1/2)\pi$, where

$n = 0\,1, 2, \ldots$ If we choose $V_0 >> B$ (or $|E|$), then κ/α is small and cot αr_0(or cot η) is small and negative; αr_0 (or η) is then slightly greater than $\pi/2$. The situation is illustrated in Fig. 5.11 where $u(r)$ is plotted as a function of r. The inside (sine function in eqn. (96*a*) function has reached a maximum and just started to decrease

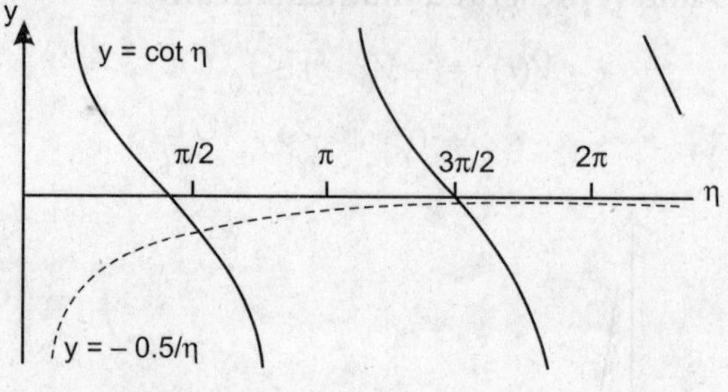

Fig. 5.10 *Graphical solution of the equation η cot η = – 0.5*

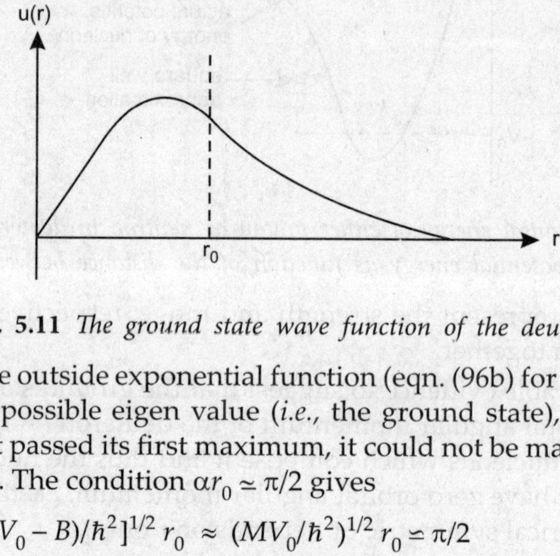

Fig. 5.11 *The ground state wave function of the deuteron*

where it joins the outside exponential function (eqn. (96b) for $r > r_0$). This condition represents the first possible eigen value (*i.e.*, the ground state), since, if the interior function had not yet passed its first maximum, it could not be made to fit a decreasing exponential outside. The condition $\alpha r_0 \simeq \pi/2$ gives

$$[M(V_0 - B)/\hbar^2]^{1/2}\, r_0 \approx (MV_0/\hbar^2)^{1/2}\, r_0 \simeq \pi/2$$

since we are assuming that B (or $|E|$) $<< V_0$. Hence

$$V_0 \simeq \pi^2\hbar^2/4Mr_0^{\,2} \qquad \ldots (116)$$

Using $r_0 \simeq 2$ fermi, we obtain $V_0 \simeq 25$ MeV for the nuclear potential well.

Taking $\alpha r_0 = (n + 1/2)\,\pi = 3\pi/2, 5\pi/2, \ldots$ would yield excited states of the system as for $\alpha r_0 > \pi$ or $> 2\pi$, wave function (eqn. (96a) for $r < r_0$) would have nodes at π and 2π ($\sin \alpha r_0 - \sin \pi = 0$ and also $\sin \alpha r_0 = \sin 2\pi = 0$) and thus would not be wave function of the ground state (corresponding to the fundamental mode in classical mechanics). We therefore have a condition.

$$\pi/2 \le \alpha r_0 < \pi \qquad \ldots (117)$$

To obtain a better approximation for αr_0, we take $\alpha r_0 = \pi/2 + \epsilon$ and substitute in eqn. (115),

$$(\pi/2+\in)\cot(\pi/2+\in) = -\kappa r_0$$

If $\in$ is small, $\cot(\pi/2+\in) \simeq -\in$, hence from above $\in \simeq 2\kappa r_0/\pi$. Therefore

$$\alpha r_0 = \pi/2 + 2\kappa r_0/\pi$$

or

$$V_0 \simeq \frac{\pi^2\hbar^2}{4Mr_0^2} + \frac{2\hbar}{r_0}\left(\frac{B}{M}\right)^{1/2} \quad \text{... (118)}$$

which is approximately equal to 35 MeV.

PROBLEMS

5.1 The $l=0 \to l=1$ rotational absorption line occurs at 1.153×10^{11} Hertz in $^{12}C^{16}O$ molecule and at 1.102×10^{11} Hertz in $^{?}C\ ^{16}O$ molecule. Find the mass number of the unknown carbon isotope.

5.2 A $^{200}Hg\ ^{35}Cl$ molecule emits a 4.4 cm photon when it undergoes a rotational transition from $l = 1$ to $l = 0$. Find the interatomic distance in this molecule. Given the masses of ^{200}Hg and ^{35}Cl are 3.32×10^{-25} kg and 5.81×10^{-26} kg, respectively.

5.3 Find energy eigen values and energy eigen functions in the case of a plane rotator.

5.4 The ground state wave function in the case of hydrogen atom is given by

$$\varphi_{100}(\text{or } \varphi_{1s}) = (1/\pi a_0^3)^{1/2}\, e^{-r/a_0}$$

where $a_0 = \hbar^2/\mu e^2$ is the Bohr's radius. Find the probability of the 1s electron in a sphere of radius $r = a_0$.

5.5 Assuming that the interaction between n and p that makes up a deuteron can be represented by the potential energy function

$$V(r) = -V_0\, e^{-r/r_0}$$

find depth of the attractive potential well taking $l = 0$ (that is corresponding to the s-state).

5.6 A particle is locked up in a sphere of radius r_0 where no force acts on the particle (or the particle is free inside the spherical box of radius r_0). The potential energy variation is thus given by

$$V(r) = 0, \quad 0 < r < r_0$$

$$= \infty, \quad r > r_0,$$

find energy eigen values and the energy eigen functions of the particle in such a potential.

5.7 The interaction between the two atoms of a diatomic molecule can be described to a good approximation by a potential function

$$V(r) = -C/r + D/r^2$$

where C and D are positive constants. The said potential is called Kratzer potential and is displayed in Fig. 5.12. Taking μ as the reduced mass of the molecule, show that the energy levels would be given by

$$E = \frac{-\mu C^2}{2\hbar^2} \times \frac{1}{\left[n_r + \frac{1}{2} + \left\{\gamma^2 + \left(l + \frac{1}{2}\right)^2\right\}^{1/2}\right]^2}$$

where

$$\gamma^2 = 2\mu D/\hbar^2$$

and n_r is a non-negative integer.

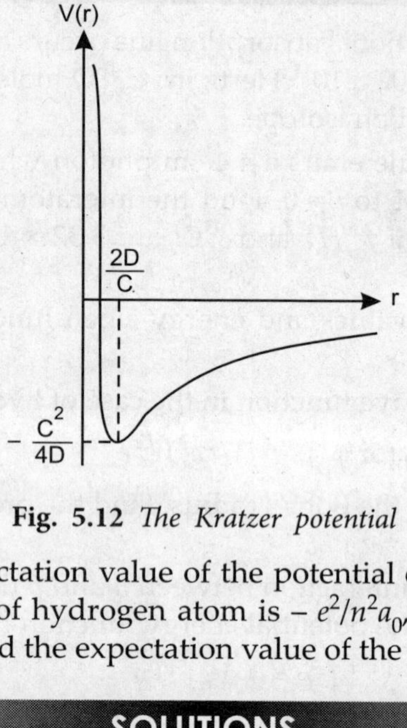

Fig. 5.12 *The Kratzer potential*

5.8 Show that the expectation value of the potential energy of an electron in the nth quantum state of hydrogen atom is $-e^2/n^2a_0$, a_0 being the Bohr's radius. From this result, find the expectation value of the kinetic energy.

SOLUTIONS

5.1 $E_1 - E_0 = \hbar^2/I_1 = h\upsilon_1 = \hbar^2/\mu_1 r_e^2$

and

$$\hbar^2/I_2 = h\upsilon_2 = \hbar^2/\mu_2 r_e^2$$

Dividing one by the other,

$$\mu_2/\mu_1 = 1.153/1.102$$

Assuming n to be the mass number of the unknown carbon isotope, $[n \times 16/(n + 16)] / [12 \times 16/(12 + 16)] = 1.153/1.102$

which on solving gives $n = 13$.

5.2 $$h\upsilon = E_1 - E_0 = \hbar^2/I_{HgCl}$$

or $$I_{HgCl} = \hbar/2\pi\upsilon = \hbar\lambda/2\pi c$$

or $$\mu r_e^2 = \hbar\lambda/2\pi c$$

Hence

$$r_e^2 = \hbar\lambda/2\pi c\mu$$

Putting, $\mu = 33.2 \times 5.81 \times 10^{-26}/(33.2 + 5.81)$ kg, $\lambda = 0.044$ m and values of the other constants, the interatomic distance comes out to be

$$r_e = 2.23 \text{ Å}$$

5.3 Plane rotator is the one in which the two rotating masses are always confined to a plane, say *xy*. That is, the axis of rotation is fixed. θ in this case would be 90°. It is example of a rigid diatomic molecule rotating in the θ = π/2 plane. The Schrödinger equation in this case would, therefore be given by

$$\frac{d^2\varphi(\phi)}{d\phi^2} + \frac{2IE}{\hbar^2}\varphi(\phi) = 0 \qquad \text{... (5.1)}$$

In this case $\varphi(\phi) = \Phi(\phi)$ and hence

$$\frac{1}{\Phi(\phi)}\frac{d^2\Phi(\phi)}{d\phi^2} = \frac{-2IE}{\hbar^2} = -m^2 \text{ (say)} \qquad \text{... (5.2)}$$

The solution of the above equation is

$$\Phi(\phi) = A\,e^{im\phi} \qquad \text{... (5.3)}$$

where *A* is an arbitrary constant that can be determined from the normalization condition.

The eigen values of the problem are determined from the requirement that Φ should be a single valued function of position. That is, the value of Φ should be the same at $\phi = 0$ and at $\phi = 2\pi$. From this it follows that

$$(2IE/\hbar^2)^{1/2} = m,$$

that is,

$$E = E_m = m^2\hbar^2/2I, \qquad m = 0, \pm 1, \pm 2, \ldots \qquad \text{... (5.4)}$$

The plane rotator can take only these values of energy. Note that the expression for energy levels is same as that obtained classically. The eigen functions of the plane rotator are

$$\Phi_m(\phi) = Ae^{im\phi}, \qquad m = 0, \pm 1, \pm 2, \ldots \qquad \text{... (5.5)}$$

which means, two eigen functions (one corresponding to the positive value of *m* and the other corresponding to the negative value of *m*) linearly independent from each other belong to every E_m energy value except $E_m = 0$, that is, the ground state. Thus in the present case every state except the ground state is doubly degenerate. The result follows from the fact that for every energy level $E_{m'}$ the rotator can rotate both in the positive and in the negative direction.

5.4 The required probability in the sphere of radius a_0 (using eqn. 89) would be given by

$$P = \int_0^{a_0} p(r)\,dr = 4/a_0^3 \int_0^{a_0} e^{-2r/a_0} r^2\,dr \qquad \text{... (5.6)}$$

On integrating by parts, we get

$$P = \frac{4}{a_0^3}\left[\frac{r^2 e^{-2r/a_0}}{-2/a_0}\Big|_0^{a_0} - \int_0^{a_0} 2r\frac{e^{-2r/a_0}}{-2/a_0}\,dr\right]$$

$$= \frac{4}{a_0^3}\left[-\frac{a_0}{2}r^2\frac{e^{-2r/a_0}}{a_0}\Big|_0^{a_0} + a_0\left\{r\frac{e^{-2r/a_0}}{-2/a_0}\Big|_0^{a_0} - \int_0^{a_0}\frac{e^{-2r/a_0}}{-2/a_0}\,dr\right\}\right]$$

$$= \frac{4}{a_0^3}\left[-\frac{a_0^3 e^{-2}}{2} - \frac{a_0^3 e^{-2}}{2} + \frac{a_0^2}{2}\frac{e^{-2r/a_0}}{-2/a_0}\Big|_0^{a_0}\right]$$

$$= \frac{4}{a_0^3}\left[-a_0^3 e^{-2} - \frac{a_0^3}{4}(e^{-2}-1)\right]$$

$$= (-5e^{-2}+1)$$

$$= 0.323$$

5.5 The radial Schrödinger eqn. (eqn. (48)) in the present case ($l = 0$) reads,

$$\frac{d^2R(r)}{dr^2} + \frac{2}{r}\frac{dR(r)}{dr} + \frac{2\mu}{\hbar^2}(E + V_0 e^{-r/r_0})R(r) = 0 \qquad \text{... (5.7)}$$

The form of above equation becomes much simpler if instead of $R(r)$ and r the following new variables are introduced,

$$u(r) = r\,R(r) \quad \text{and} \quad x = e^{-2r/r_0} \qquad \text{... (5.8)}$$

The function $u(x)$ then satisfies the following differential equation,

$$\frac{x^2 d^2u(x)}{dx^2} + \frac{x\,du(x)}{dx} + (\gamma^2x^2 - \kappa^2)\,u(x) = 0 \qquad \text{... (5.9)}$$

where

$$\gamma^2 = 8\mu V_0 r_0^2/\hbar^2 \text{ and } \kappa^2 = -8\,\mu E r_0^2/\hbar^2 \qquad \text{... (5.10)}$$

Eqn (5.9) is nothing but Bessel's equation of order $\pm k$ having argument γx. Its general solution, would, therefore be

$$u(x) = AJ_\kappa(\gamma x) + BJ_{-\kappa}(\gamma x)$$

where A and B are arbitrary constants

For $r \to \infty$ that is for $x \to 0$, as only J_κ vanishes but $J_{-\kappa}$ diverges, B is to be taken zero such that $u(0)$ vanishes (or is finite). Further for $r \to 0$ or $x \to 1$, $u(x = 1) = 0$. This is possible when

$$J_\kappa(\gamma) = 0 \qquad \text{... (5.11)}$$

which determines the energy eigen values.

The quantity γ is determined by the potential V_0 and the parameter r_0, the κ belonging to value γ can be calculated from (5.11). Then the eigen value E can be easily determined from the defining expression of κ.

Let us apply the results to the deuteron problem. In this case, the reduced mass

$$\mu = m_n m_p/(m_n + m_p) = 0.837 \times 10^{-27} \text{ kg}$$

and the range of nuclear forces

$$r_0 = 2 \times 10^{-15} \text{ m}$$

The experimental value of the binding energy E of the deuteron is

$$E = -2.23 \text{ MeV.}$$

Substituting the above data into eqn. (5.10), one obtains

$$\kappa = 0.998 \text{ and } V_0 = 2.24\,\gamma^2 \text{ MeV} \quad \text{... (5.12)}$$

The solution of the transcendental eqn. (5.11) for this value of κ is $\gamma = 3.82$, which on substitution into (5.11) yields

$$V_0 = 32.7 \text{ MeV}$$

5.6 The radial Schrödinger equation is given by

$$\frac{d^2R(r)}{dr^2} + \frac{2}{r}\frac{dR(r)}{dr} + \frac{2\mu}{\hbar^2}\left[E - V(r) - \frac{l(l+1)\hbar^2}{2\mu r^2}\right]R(r) = 0$$

After dividing above by $\alpha^2 = 2\mu E/\hbar^2$ and putting $p = \alpha r$, one gets

$$\frac{d^2R(\rho)}{d\rho^2} + \frac{2}{\rho}\frac{dR(\rho)}{d\rho} + \left[1 - \frac{l(l+1)}{\rho^2}\right]R(\rho) = 0,\ 0 < r < r_0 \quad \text{... (5.13)}$$

which is spherical Bessel's equation. The general solution is given by

$$R(\rho) = A\,j_l(\rho) + B\,n_l(\rho) \quad \text{... (5.14)}$$

where $j_l(\rho)$ and $\eta_l(\rho)$ are spherical Bessel and spherical Neumann functions, respectively and are defined by the following relations,

$$j_l(\rho) = \left(\frac{\pi}{2p}\right)^{1/2} J_{l+1/2}(\rho) \quad \text{... (5.15)}$$

$$n_l(\rho) = (-1)^{l+1}\left(\frac{\pi}{2\rho}\right)^{1/2} J_{-(l+1/2)}(\rho)$$

where J's are Bessel functions of order $\pm(l + 1/2)$.

Further for $\rho \to 0$,

$$j_l(\rho) = \frac{\rho^l}{(2l+1)!!} \quad \text{... (5.16)}$$

$$n_l(\rho) = (2l-1)!!\,(1/\rho)^{l+1}$$

where $(2l + 1)!! = 1.3.5 \ldots (2l - 1)(2l + 1)$

From eqn. (5.16), it is clear that B should be taken equal to zero in eqn. (5.14), otherwise $R(\rho)$ would diverge at $\rho = 0$. Hence

$$R(\rho) = Aj_i(\rho) \qquad \ldots (5.17)$$

As the particle cannot escape from the sphere (or cannot have infinite amount of energy), the wave function $R(\rho)$ has to be zero, both on the surface of the sphere and everywhere outside of it. From the boundary condition $R(\alpha r_0) = 0$, the following condition is reached from eqn. (5.17),

$$j_l(\alpha r_0) = 0 \qquad \ldots (5.18)$$

From this transcendental equation, α can be determined and then the energy eigen values E from the relation, $\alpha^2 = 2\mu E/\hbar^2$ can be obtained.

For a given angular momentum quantum number l, eqn. (5.18) has infinite number of solutions. For numbering them a radial quantum number n_r can be introduced.

For $\qquad l = 0$, eqn. (5.18) gives

$$j_0(\alpha r_0) = \sin \alpha r_0/(\alpha r_0) = 0$$

which amounts to

$$\alpha r_0 = n_r \pi, \quad n_r = 1, 2, 3, \ldots$$

Using the relation, $\alpha^2 = 2\mu E/\hbar^2$ leads to energy eigen values

$$E = E_{nr} = n_r^2 \pi^2 \hbar^2/2\mu r_0^2, \quad n_r = 1, 2, 3, \ldots \qquad \ldots (5.19)$$

For $l = 1$, eqn. (5.18) gives

$$j_1(\alpha r_0) = 0$$

which amounts to

$$\left(\frac{\pi}{2\alpha r_0}\right)^{1/2} J_{l+1/2}(\alpha r_0) = 0$$

or

$$\left(\frac{\pi}{2\alpha r_0}\right)^{1/2} \left(\frac{2}{\pi \alpha r_0}\right)^{1/2} \left(\frac{\sin \alpha r_0}{\alpha r_0} - \cos \alpha r_0\right) = 0$$

or

$$\tan \alpha r_0 = \alpha r_0 \qquad \ldots (5.20)$$

The solutions of the above transcendental equation are given by the intersections of the curves $y = \tan \alpha r_0$ and $y = \alpha r_0$ displayed in Fig. 5.13. These are

$$\alpha r_0 \simeq (n_r + 1/2)\pi, \quad n_r = 1, 2, 3, \ldots \qquad \ldots (5.21)$$

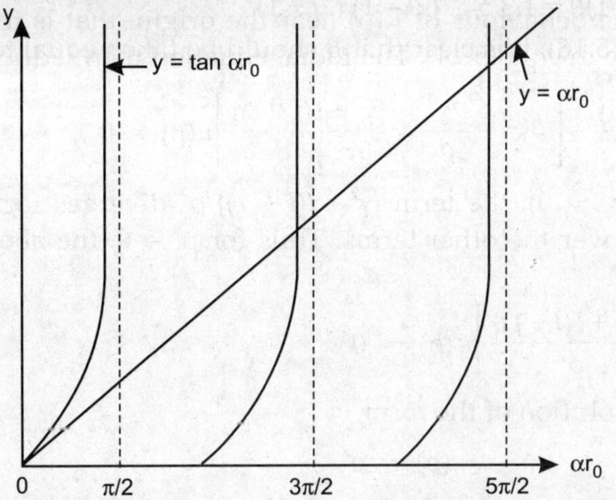

Fig. 5.13 *Graphical solution of eqn. (5-20)*

from which energy eigen values corresponding to $l = 1$ can be obtained. Similarly one can obtain the transcendental equations for higher l values and determine the corresponding energy eigen values. The energy eigen functions corresponding to different l values are given by $A\, j_l(\rho)\, Y_l^{m_l}\,(\theta, \phi)$.

5.7 By putting the value of $V(r)$ in eqn. (48), the radial Schrödinger equation takes the form

$$\frac{1}{r^2}\frac{d}{dr}\left(r^2\frac{dR}{dr}\right)+\left[\frac{2\mu E}{\hbar^2}+\frac{2\mu C}{\hbar^2 r}-\frac{2\mu D}{\hbar^2 r^2}-\frac{l(l+1)}{r^2}\right]R(r) = 0$$

Introduce the variable $\rho = \alpha r$, $\alpha = C/2D$. Dividing the above expression by α^2, one gets

$$\frac{1}{\rho^2}\frac{d}{d\rho}\left(\rho^2\frac{dR}{d\rho}\right)+\left[\frac{2\mu E}{\hbar^2\alpha^2}+\frac{2\mu C}{\hbar^2\rho\alpha}-\frac{2\mu D}{\hbar^2\rho^2}-\frac{l(l+1)}{\rho^2}\right]R(\rho) = 0 \quad \text{... (5.22)}$$

Making the following subsitution

$$\gamma^2 = 2\mu\, D/\hbar^2, \quad \kappa^2 = -\,2\mu E/\hbar^2\alpha^2 \quad \text{... (5.23)}$$

eqn. (5.22) takes the form,

$$\frac{1}{\rho^2}\frac{d}{d\rho}\left(r^2\frac{dR}{d\rho}\right)+\left[-\kappa^2+\frac{2\gamma^2}{\rho}-\frac{\gamma^2+l(l+1)}{\rho^2}\right]R(\rho) = 0 \quad \text{... (5.24)}$$

Let us first examine the behaviour of $R(\rho)$ in the asymptotic region, that is, for $\rho \to \infty$. In this case eqn. (5.24) reduces to (see the case of hydrogen atom),

$$\frac{d^2R(\rho)}{d\rho^2} - \kappa^2 R(\rho) = 0 \quad \text{... (5.25)}$$

The solution of the above equation is obviously, $R(\rho) = e^{-\kappa\rho}$ ($e^{+\kappa\rho}$ has been rejected as it violates the condition that the solution should be finite everywhere).

To study the behaviour of $R(\rho)$ near the origin, that is for $\rho \to 0$, let us put $R(\rho) = u(\rho)/\rho$ in eqn. (5.24). The radial equation then reduces to

$$\frac{d^2 u(\rho)}{d\rho^2} + \left[-\kappa^2 + \frac{2\gamma^2}{\rho} - \frac{\{\gamma^2 + l(l+1)\}}{\rho^2} \right] u(\rho) = 0$$

For $l \geq 1$(or $l \neq 0$), the term $\{\gamma^2 + l(l + 1)\}/\rho^2$ diverges for small ρ and hence dominates over the other terms. Thus for $\rho \to 0$, the above equation can be written as

$$\frac{d^2 u(\rho)}{d\rho^2} - \left[\frac{\{\gamma^2 + l(l+1)\}}{\rho^2} \right] u(\rho) = 0 \qquad \text{... (5.26)}$$

which has solution of the form

$$u(\rho) = \rho^{\lambda} \qquad \text{... (5.27)}$$

Substituting above into eqn. (5.26) gives

$$\lambda(\lambda - 1) = \gamma^2 + l(l+1)$$

This is a quadratic equation for λ, which has the (+)ve root (see details in hydrogen atom problem)

$$\lambda = 1/2 + \{\gamma^2 + (l + 1/2)\}^{1/2}$$

Hence

$$R(\rho) = \rho^{\lambda-1} \text{ for } \rho \to 0$$

Solution of (5.24) can thus be taken as

$$R(\rho) = e^{-\kappa\rho} \rho^{\lambda-1} v(\rho) \qquad \text{... (5.28)}$$

where $v(\rho)$ is some function of ρ and is to be determined.

Proceeding in the fashion similar to that in the case of hydrogen atom, one gets

$$\kappa = \gamma^2/(n_r + \lambda) \qquad \text{... (5.29)}$$

where n_r is a non-negative integer.

From the above the energy eigen value E is given by

$$E = -\hbar^2 \alpha^2 \kappa^2/2\mu = -2\mu C^2 \kappa^2/4\gamma^2 \hbar^2$$

$$= \frac{-\mu C^2}{2\hbar^2} \times \frac{1}{[n_r + 1/2 + \{y^2 + (l + 1/2)^2\}^{1/2}]^2},$$

$$n_r = 1, 2, \ldots \qquad \text{... (5.30)}$$

Let us investigate the energy levels for $\gamma << 1$, that is, for $D \to 0$.

$$E = -\frac{\mu C^2}{2\hbar^2} \times \left[n_r + \frac{1}{2} + \left(l + \frac{1}{2}\right)\left\{1 + \frac{\gamma^2}{(l + 1/2)^2}\right\}^{1/2} \right]^{-2}$$

$$= -\frac{\mu C^2}{2\hbar^2} \times \left[n_r + \frac{1}{2} + \left(l + \frac{1}{2}\right) + \frac{\gamma^2}{(2l+1)} \right]^2$$

$$= -\frac{\mu C^2}{2\hbar^2} \times \frac{1}{n^2}\left[1 + \frac{\gamma^2}{n(2l+1)}\right]^{-2}$$

$$= -\frac{\mu C^2}{2\hbar^2} \times \frac{1}{n^2}\left[1 - \frac{2\gamma^2}{n(2l+1)}\right] \qquad \text{... (5.31)}$$

where $n = n_r + l + 1$.

For the limiting case $\gamma = 0$ (or $D = 0$), the energy levels correspond to those of the hydrogen atom. This is actually expected as the Kratzer potential changes into Coulomb potential by the substitution $D = 0$ and $C = Ze^2$.

5.8 The normalized eigen function of the hydrogen atom in the nth state is given by

$$\varphi_{nlm_l}(r, \theta, \phi) = -\left[\left(\frac{2}{na_0}\right)^3 \frac{(n-l-1)!}{2n\{(n+l)!\}^3} \frac{(2l+1)}{4\pi} \frac{(l-m_l)!}{(l+m_l)!}\right]^{1/2} \times$$

$$e^{-\rho/2} \rho^l L_{n+l}^{2l+1}(\rho) P_l^{m_l}(\cos\theta) e^{im_l\phi}$$

where $\rho = 2r/na_0$.

The expectation value of the potential energy ($V = -e^2/r$), in the nth quantum state would, therefore be given by

$$\langle V \rangle = \int \varphi^*_{nlm_l}(-e^2/r)\varphi_{nlm_l}\, d\tau$$

$$= e^2\left[\left(\frac{2}{na_0}\right)^3 \frac{(n-l-1)!}{2n\{(n+l)!\}^3} \frac{(2l+1)}{4\pi} \frac{(l-m_l)!}{(l+m_l)!}\right] \times$$

$$\int_0^\infty \int_0^\pi \int_0^{2\pi} \frac{1}{r} e^{-\rho} \rho^{2l} |L_{n+l}^{2l+1}(\rho)|^2 |P_l^{m_l}(\cos\theta)|^2 r^2 \sin\theta\, d\theta\, d\phi\, dr$$

Using the relations,

$$\int_{-1}^{+1} |P_l^{m_l}(\cos\theta)|^2 d(\cos\theta) = \frac{2}{(2l+1)} \frac{(l+m_l)!}{(l-m_l)!}$$

and $\int_0^{2\pi} d\phi = 2\pi$, one gets

$$\langle V \rangle = -e^2 \frac{2}{na_0} \frac{(n-l-1)!}{2n\{(n+l)!\}^3} \int_0^\infty e^{-\rho} \rho^{2l} |L_{n+l}^{2l+1}(\rho)|^2 \rho d\rho$$

$$= -e^2 \frac{2}{na_0} \frac{(n-l-1)!}{2n\{(n+l)!\}^3} \frac{\{(n+l)!\}^3}{(n-l-1)!}$$

$$= -e^2/n^2 a_0$$

Hence

$$\langle T \rangle = \langle H \rangle - \langle V \rangle = -e^2/(2n^2 a_0) + e^2/(n^2 a_0) = e^2/(2n^2 a_0)$$

Chapter 6

DIRAC'S KET AND BRA NOTATION

6.1 SUPERPOSITION PRINCIPLE

The principle states that if φ_1 and φ_2 are wave functions describing any two states of a given system, then corresponding to any linear combination of the two wave functions there exists a state of the system. By a linear combination of $\varphi_1, \varphi_2, \ldots, \varphi_r$, we mean any quantity of the form $c_1\varphi_1 + c_2\varphi_2 + \ldots + c_r\,\varphi_r$ where $c_1, c_2, \ldots, c_r$ are arbitrary complex numbers. The r functions $\varphi_1, \varphi_2 \ldots, \varphi_r$ are said to be linearly independent (or form a linearly independent set) if none of them can be expressed as a linear combination of the others, or equivalently, if the only way to make $c_1\varphi_1 + c_r\,\varphi_2 + \ldots + c_r\,\varphi_r$ vanish is by taking all the constant coefficients to be zero.

The superposition principle is a fundamental principle of quantum mechanics to which there is no correspondence in classical mechanics. For example, in classical physics, superposition of a vibrating string on itself gives, a mode with twice the amplitude and have four times the energy of the initial state of vibration ($(A + A)^2 = 4A^2$), whereas in quantum mechanics, the superposition of a state on itself gives the same state φ_1 and $c\varphi_1$, where c is an arbitrary non-zero complex number, correspond to the same state.

The superposition principle also has deep mathematical significance. This inference comes from the fact that the principle is a fundamental characteristic of vectors representing points in space. If $\vec{V}_1$ and $\vec{V}_2$ are the position vectors of any two points, then $c_1\vec{V}_1 + c_2\vec{V}_2$ is again a vector which represents some point in space. Therefore, it is possible to think of wave functions as vectors of some kind. The ordinary vectors with which we are familiar, need only three real numbers for the complete specification of each one, *viz.* the components v_1, v_2, v_3 specify a vector $\vec{V}$ completely. But each wave function—vector requires an infinite set of complex numbers for its specification, namely the values of φ at all the points $\vec{r}$. Therefore, we say that the wave function vectors are infinite dimensional. The set of all wave functions of a given system, therefore forms an infinite dimensional complex linear vector space. The term linear over here is used in the sense that any linear combination of wave function vectors is also a wave function vector in the same space.

In this geometrical picture of wave functions, the quantities $\varphi(\vec{r})$ for various $\vec{r}$ are thought of as different components of the wave function vector φ. Here the continuous variable $\vec{r}$ serves as a label distinguishing the different components of φ and is counterpart of the index i labelling the components v_i of an ordinary vector $\vec{V}$. With this identification, the integral $\int |\varphi(\vec{r})|^2 d\vec{r}$ ($d\vec{r}$ represents the volume element), called norm becomes the analogue of $\Sigma_i v_i^2$ and it may therefore be viewed as square of the length of the wave function vector. Because in ordinary space the scalar product of the vector with itself is the square of the magnitude or length of the vector and because $\varphi(\vec{r})$ is complex, the absolute square ($|\varphi(\vec{r})|^2$) has to be taken to get a real length and of course instead of sum over i, integral over $\vec{r}$ is to be taken as $\vec{r}$ is a continuous variable. More generally $\int \phi^*(\vec{r})\varphi(\vec{r})d\vec{r}$ is defined as the scalar (or inner) product of ϕ and φ (where the asterisk denotes complex conjugation) analogous to $\Sigma_i u_i v_i$ (or $u_x v_x + u_y v_y + u_z v_z$) for ordinary vectors. In view of this, the notation (ϕ, φ) as the counterpart of $\vec{U}\,.\,\vec{V}$ is often used for such integrals.

From the definition, the scalar product of ϕ with φ is

$$(\phi, \varphi) = \int \phi^*(\vec{r})\varphi(\vec{r})d\vec{r} \qquad \text{... (1)}$$

From the above definition, using the relation $(FD)^* = D^*F^*$, it follows that

$$(\phi, \varphi) = (\varphi, \phi)^* \qquad \text{... (2)}$$

Also from eqn. (1)

$$(\phi, c\varphi) = c(\phi, \varphi) \; (c : \text{arbitrary complex number}) \qquad \text{... (3)}$$

that is, the scalar product is linear with respect to the post factor (In (x, y), x is called the prefactor and y the post factor).

From eqns. (2) and (3), it can be shown that

$$(c\,\phi, \varphi) = (\varphi, c\,\phi)^* = [c(\varphi, \phi)]^* = c^*(\varphi, \phi)^* = c^*(\phi, \varphi)$$

or

$$(c\,\phi, \varphi) = c^*(\phi, \varphi) \qquad \text{... (4)}$$

that is, the scalar product is not linear (or is antilinear) with respect to the prefactor. Also from eqn. (2),

$$(\varphi, \varphi) = (\varphi, \varphi)^* \qquad \text{... (5)}$$

implying that the scalar product (φ, φ) is a real non-negative number. That is, the norm of φ,

$$(\varphi, \varphi) \geq 0 \qquad \text{... (6)}$$

with equality sign holding good if and only if ($\varphi \equiv 0$ or if it is a null vector (The null vector denotes the origin of the vector space).

6.2 DIRAC NOTATION

It has been observed in the last section that the values of the Schrödinger wave function $\varphi(\vec{r})$ at various points $\vec{r}$ are equivalent to the components of a vector. While the components are numbers which depend on the choice of the coordinate system (called basis in the vector space), the vector itself is an abstract geometrical object having significance independent of coordinate systems. In the quantum mechanical context, the abstract mathematical object which is associated with the quantum state, is a state vector. The different kinds of wave functions for a given state, are simply the sets of components of the state vector as referred to different coordinate systems or bases in the state vector space.

Following Dirac, the state φ of the dynamical system is represented by the symbol $|\varphi\rangle$ and is read as 'ket φ' which is counterpart of the notation $\vec{V}$ for a vector in three dimensions. The state vectors constitute a complex vector space, i.e. any linear combination of ket vectors is also a ket vector in the same space. For example, if we have two ket vectors $|\varphi\rangle$ and $|\chi\rangle$ then the linear combination $c_1\,|\varphi\rangle + c_2\,|\chi\rangle$ is also a ket vector in the same space, c_1 and c_2 being arbitrary complex numbers.

Corresponding to every vector $|\varphi\rangle$ there corresponds a conjugate vector $\overline{|\varphi\rangle}$ (read as ket φ bar), for which Dirac used the notation $\langle\varphi\,|$, called 'bra φ'. The conjugate of bra vector is a ket vector and vice versa.

$$\overline{\langle\varphi\,|} \equiv |\varphi\rangle, \overline{|\varphi\rangle} \equiv \langle\varphi| \qquad \text{... (7a)}$$

Also conjugate of $c|\varphi\rangle$ is not $c\langle\varphi\,|$ but

$$\overline{c|\varphi\rangle} = c^*\overline{|\varphi\rangle} = c^*\langle\varphi| \qquad \text{... (7b)}$$

that is, the correspondence between $|\varphi\rangle$ and $\langle\varphi|$ is not linear one, where c is a complex number.

The relation between any vector and its conjugate vector may be understood easily if we consider examples of their representations.

In the Schrödinger representation,

$$|\varphi\rangle \to \varphi(\vec{r}) \text{ and } \langle\varphi| \to \varphi^*(\vec{r}) \qquad \text{... (8)}$$

In the matrix representation, $|\varphi\rangle$ is represented by a column vector (infinite component column), that is

$$|\varphi\rangle \to \begin{bmatrix} \varphi_1 \\ \varphi_2 \\ \cdots \\ \cdots \\ \cdots \end{bmatrix} \qquad \text{... (9a)}$$

and $\langle\varphi|$ then goes over into the row vector (Transpose complex conjugate),

$$\langle\varphi| = [\varphi_1{}^*, \varphi_2{}^* \ldots] \qquad \text{... (9b)}$$

The definition ensures the scalar product, $\varphi_1{}^*\,\varphi_1 + \varphi_2{}^*\,\varphi_2 + \ldots$

6.3 HILBERT SPACE

The scalar product in the Dirac notation is written as $|\phi|\varphi\rangle$ which is equivalent to $\int \phi^* (\vec{r})\varphi(\vec{r})d\vec{r}$ in the Schrödinger notation. From which it follows that $\langle\phi|\varphi\rangle \neq \langle\varphi|\phi\rangle$ but

$$\langle\phi|\varphi\rangle = \langle\varphi|\phi\rangle^* \quad \text{... (10a)}$$

and hence $\langle\psi|\varphi\rangle$ is a real non-negative number, that is $\langle\varphi|\psi\rangle \geq 0$, the equality holds (or is equal to zero) if $|\varphi\rangle \equiv 0$ ... (10b)

From the definition (of the scalar product), it also follows that the scalar product of two vectors is linear in the second vector or post factor,

$$\langle\phi\,(c|\varphi\rangle + c'|\chi\rangle) = c\langle\phi|\varphi\rangle + c'\langle\phi|\chi\rangle \quad \text{... (10c)}$$

and antilinear in prefactor (see eqn. (4)), that is

$$\langle c\phi_1 + c'\phi_2|\varphi\rangle = c^*\langle\phi_1|\varphi\rangle + c'^*\langle\phi_2|\varphi\rangle \quad \text{... (10d)}$$

Any infinite dimensional vector space such that between any two vectors of the space a scalar product with properties (10*a*) – (10*d*) is defined, is called a Hilbert space. The space of ket vectors (or bra vectors) representing quantum mechanical states is thus a Hilbert space.

6.4 EIGEN VALUES AND EIGEN VECTORS

As an illustration of the way one works with the Dirac notation, we present here the proof of the orthogonality theorem.

A ket, $|\varphi_{A'}\rangle$ (also expressed as $|A'\rangle$) is called an eigen vector or eigen ket, of the operator A if

$$A|\varphi_{A'}\rangle = A'|\varphi_{A'}\rangle \quad \text{... (11)}$$

The above equation is called the eigen value equation and the number A' which characterizes the eigen vector is called an eigen value. Let us restrict ourselves to the eigen value problem for the Hermitian operators only. At the same time, let us assume that there is only one linearly independent ket for each eigen value (this is the non-degenerate case). Denote the eigen values of A by A', A'', . . . and the corresponding eigen vectors by $|\varphi_{A'}\rangle, |\varphi_{A''}\rangle$ and so on.

Let the operator A in eqn. (11) is Hermitian (or self adjoint), then

$$\langle\varphi_{A''}|A|\varphi_{A'}\rangle = \langle\varphi_{A'}|A|\varphi_{A''}\rangle^* \quad \text{... (12)}$$

Using (11), eqn. (12) becomes

$$A'\langle\varphi_{A''}\,|\,\varphi_{A'}\rangle = \left(A''\langle\varphi_{A'}|\varphi_{A''}\rangle\right)^*$$

Using (10*a*), the above equation takes the form

$$A'\langle\varphi_{A''}|\varphi_{A'}\rangle = A''^*\langle\varphi_{A''}|\varphi_{A'}\rangle$$

or

$$(A' - A''^*)\langle\psi_A''|\psi_A'\rangle = 0 \quad \text{... (13)}$$

Substituting $A' = A''$ and since $\langle\psi_A'|\psi_A'\rangle > 0$ (equality sign is not to be taken otherwise $|\psi_A'\rangle$ will be a null vector and eqn. (11) then would not have any meaning), from the above equation, it follows that

$$A' = A'^* = \text{real}.$$

i.e. all eigen values of a Hermitian operator are real.

In the light of the above conclusion, eqn. (13) may be written as

$$(A' - A'')\langle\psi_A''|\psi_A'\rangle = 0$$

which shows that the eigen vectors belonging to two different eigen values ($A' - A'' \neq 0$ or $A' \neq A''$) are orthogonal or

$$\langle\psi_A''|\psi_A'\rangle = 0 \quad \text{... (14)}$$

called the orthogonality condition.

Since Schrödinger equation is linear, if φ is a solution, then $c\varphi$ is also a solution where c is an arbitrary complex number. One can always choose the constant c such that

$$\int_{-\infty}^{+\infty} |c\varphi|^2 dx = \int_{-\infty}^{+\infty} |\varphi_n|^2 dx = 1,$$

which is the normalization condition. Similarly due to the linearity of A, we can normalize the eigen vectors arbitrarily. In general, we write

$$\langle\varphi_{A''}|\varphi_{A'}\rangle = \delta_{A'A''} \quad \text{... (15)}$$

called the orthonormality relation. $\delta_{A'A''}$ being Kronecker delta function defined through the condition

$$\delta_{A'A''} = 0,\ A' \neq A''$$

$$= 1,\ A' = A''$$

6.5 BASIS IN HILBERT SPACE (COMPLETENESS OR CLOSURE PROPERTY)

Let $|\phi_1\rangle$, $|\phi_2\rangle$, . . . be a linearly independent set of states in the Hilbert space, i.e. no linear combination of these vectors vanish except when all the coefficients vanish. Let the set be also a complete one, in the sense that any arbitrary vector $|\varphi\rangle$ of the space can be expressed in the form,

$$|\varphi\rangle = c_1|\phi_1\rangle + c_2|\phi_1\rangle + ... = \sum_i c_i|\phi_i\rangle \quad \text{... (16)}$$

or that means, the set spans the space.

Then the set $|\phi_1\rangle, |\phi_2\rangle, ...$ can be chosen as a basis or coordinate system for the Hilbert space just as one can use set of three linearly independent vectors $\hat{x}, \hat{y}, \hat{z}$ as basis for the ordinary three dimensional space.

Any vector $\vec{V}$ in the 3*d* physical space can be written as,

$$\vec{V} = v_x\hat{x} + v_y\hat{y} + v_z\hat{z}$$

$$= (\hat{x}\,.\,\vec{V})\,\hat{x} + (\hat{y}\,.\,\vec{V})\,\hat{y} + (\hat{z}\,.\,\vec{V})\hat{z}$$

where $\hat{x}, \hat{y}, \hat{z}$, are unit vectors along x, y, z directions. This also shows that the set $\hat{x}, \hat{y}, \hat{z}$ is complete. Note that the vectors $\hat{x}, \hat{y}, \hat{z}$, being orthogonal are automatically linearly independent and thus a complete orthonormal set forms a basis.

The set of numbers c_1 (analogy $(\hat{x}\,.\,\vec{V}) = v_x$), c_2, . . . in eqn. (16) are components of $|\varphi\rangle$ with respect to the $|\phi_i\rangle$ basis. Note that since eigen vectors of the Hermitian operators are orthogonal and since we can normalize them, the eigen vectors form a basis for the Hilbert space.

Now taking the scalar multiplication of the both sides of eqn. (16) with $\langle\phi_j|$, one gets

$$\langle\phi_j|\varphi\rangle = \Sigma_i\, c_i\langle\phi_j|\phi_i\rangle$$

$$= \Sigma_i c_i\, \delta_{ij}$$

where it has been assumed that the basis is orthonormal. Therefore expression for the components c_j becomes,

$$\langle\phi_j|\varphi\rangle = c_j \qquad \text{... (17)}$$

Using above, eqn. (16) can be written as

$$|\varphi\rangle = \Sigma_i\langle\phi_i|\varphi\rangle|\phi_i\rangle$$

or

$$|\varphi\rangle = \Sigma_i\,|\phi_i\rangle\,\langle\phi_i|\,\varphi\rangle,$$

as we can place the number $\langle\phi_i|\varphi\rangle$ after the ket, $|\phi_i\rangle$

From above,

$$|\varphi\rangle = (\Sigma_i\,|\phi_i\rangle\,\langle\phi_i|)|\varphi\rangle \qquad \text{... (18)}$$

Thus the operator $\Sigma_i\,|\phi_i\rangle\,\langle\phi_i|$ acts on every ket $|\varphi\rangle$ to give same ket $|\varphi\rangle$. Since $|\varphi\rangle$ is arbitrary, it follows that

$$\Sigma_i\,|\phi_i\rangle\,\langle\phi_i| = I \qquad \text{... (19)}$$

where I denotes the identity operator. The above relation is called, the closure relation. It may be mentioned here that the closure relation.

$$\Sigma_i\,\phi_i\,(\vec{r})\,\phi_i{}^*\,(\vec{r}\,') = \delta(\vec{r} - \vec{r}\,')$$

(where $\delta(\vec{r} - \vec{r}\,')$ is the Dirac delta function) is the Schrödinger representation of eqn. (19).

6.6 SIMULTANEOUS EIGEN KETS OF COMMUTING OPERATORS

As an another illustration of the way one works with the Dirac notation, we present here the proof of the theorem that if two Hermitian operators commute, they have simultaneous eigen kets.

Let A and B represent Hermitian operators corresponding to two physical observables. It is given that the two operators commute, *i.e.*

$$AB = BA \qquad \text{... (20)}$$

If $|\varphi_\alpha\rangle$ is an eigen ket of A belonging to the eigen value α, we have

$$A\,|\varphi_\alpha\rangle = \alpha|\varphi_\alpha\rangle \qquad \text{... (21a)}$$

Applying B to both sides of the above equation, we get

$$BA\,|\varphi_\alpha\rangle = B\alpha\,|\varphi_\alpha\rangle \qquad \text{... (21b)}$$

Using eqn. (20), that is, replacing BA on left side of the above equation by AB.

$$A(B\,|\varphi_\alpha\rangle) = B\alpha\,|\varphi_\alpha\rangle = \alpha(B|\varphi_\alpha\rangle) \qquad \text{... (22)}$$

as α is number or B is linear.

The above equation expresses the fact that $B\,|\varphi_\alpha\rangle$ is also an eigen ket of A belonging to the same eigen value α. If suppose $|\varphi_\alpha\rangle$ corresponds to a non-degenerate state (or if α is a non-degenerate eigen value) then $B|\varphi_\alpha\rangle$ must be multiple of $|\varphi_\alpha\rangle$, *i.e.*

$$B|\varphi_\alpha\rangle = \beta|\varphi_\alpha\rangle \qquad \text{... (23)}$$

where β is a number. The above equation implies that $|\varphi_\alpha\rangle$ must be eigen ket of B. Hence all eigen kets of A which correspond to the non-degenerate states, must also be eigen kets of B. That is, both the operators will have simultaneous eigen kets.

6.7 CHANGE OF BASIS; UNITARY TRANSFORMATION

Let $|\varphi_i\rangle$ represet a set of orthonormal kets, *i.e.*

$$\langle\varphi_i|\varphi_j\rangle = \delta_{ij} \qquad \text{... (24)}$$

Consider a transformation,

$$|\phi_i\rangle = \sum_j U_{ij}\,|\varphi_j\rangle \qquad \text{... (25)}$$

which transforms the basis $|\varphi_i\rangle$ into the basis $|\phi_i\rangle$.

If the kets $|\phi_i\rangle$ also form an orthonormal set, *i.e.*, if

$$\delta_{ij} = \langle\phi_i|\phi_j\rangle$$

$$= \sum_m\sum_n U^\dagger_{im}U_{jn}\langle\varphi_m\mid\varphi_n\rangle$$

where use has been made of eqn. (25) and where $U^\dagger$ represents the transpose complex conjugate of U or Hermitian adjoint of U. Using eqn. (24), the above equation becomes

$$\delta_{ij} = \sum_m\sum_n U^\dagger_{im}U_{jn}\delta_{mn} = \sum_n U^\dagger_{in}U_{jn}$$

which amounts to

$$U^{\dagger} U = 1$$

or

$$U^{\dagger} = U^{-1} \quad \text{... (26)}$$

which is definition of unitary operator (an operator U is unitary if its inverse U^{-1} is equal to its adjoint $U^{\dagger}$). The operator U is therefore unitary. The eqn. (25) thus represents a unitary transformation and U_{ij} is called unitary matrix. Therefore the unitary matrix transforms the basis $|\varphi_i\rangle$ into the basis $|\phi_i\rangle$.

6.8 LINEAR HARMONIC OSCILLATOR

The problem has already been discussed starting from the Schrödinger equation in its full details. However, we would discuss it once again to make the reader more familiar with the Dirac ket and bra notation. The Hamiltonian in the present case is given by

$$H = \frac{p^2}{2m} + \frac{1}{2} m w^2 x^2 \quad \text{... (27)}$$

where x and p satisfy the commutation relation,

$$[x, p] = i\hbar \quad \text{... (28)}$$

Our aim here is to solve the eigen value equation

$$H|E\rangle = E|E\rangle \quad \text{... (29)}$$

where $|E\rangle$ is an eigen ket of operator H belonging to the eigen value E.

To solve the above equation (or to diagonalize the Hamiltanian H), let us introduce the operator a and its adjoint (transpose complex conjugate) $a^{\dagger}$ (read as a dagger) through the equations,

$$a = 1/(2m\hbar w)^{1/2}(mwx + ip) \quad \text{... (30a)}$$

and

$$a^{\dagger} = 1/(2m\hbar w)^{1/2}(mwx - ip) \quad \text{... (30b)}$$

From the above equations,

$$x = (\hbar/2mw)^{1/2}(a + a^{\dagger}) \quad \text{... (31a)}$$

and

$$p = -i(m\hbar w/2)^{1/2}(a - a^{\dagger}) \quad \text{... (31b)}$$

and hence x and p are hermitian as $x^{\dagger} = x$, $p^{\dagger} = p$.

Multiplying (30a) with (30b) and then using eqns. (27) and (28), we get

$$\hbar w\, a a^{\dagger} = \frac{1}{2m}[m^2 w^2 x^2 + p^2 - imw(xp - px)]$$

$$= H + \frac{1}{2}\hbar w \quad \text{... (32)}$$

Similarly, one can show that

$$\hbar w a^{\dagger} a = H - \frac{1}{2}\hbar w \quad \text{... (33)}$$

Addition and subtraction of eqns. (32) and (33) leads to

$$H = \frac{1}{2}\hbar w (a^{\dagger} a + a a^{\dagger}) \quad \text{... (34)}$$

and

$$a a^{\dagger} - a^{\dagger} a = [a, a^{\dagger}] = 1 \quad \text{... (35)}$$

Multiplying eqn. (32) from the right and eqn. (33) from the left by a and then subtracting the resulting equations, one gets,

$$a H - H a = [a, H] = \hbar w a \quad \text{... (36)}$$

Similarly, one can show that

$$a^{\dagger} H - H a^{\dagger} = [a^{\dagger}, H] = -\hbar w a^{\dagger} \quad \text{... (37)}$$

Defining $|P\rangle = a|E\rangle$ and then using eqns. (29) and (33), we get

$$\begin{aligned} \hbar w \langle P|P\rangle &= \hbar w \langle E|a^{\dagger} a|E\rangle \\ &= \langle E|H - 1/2\, \hbar w|E\rangle \\ &= (E - 1/2\, \hbar w) \langle E|E\rangle \end{aligned}$$

Now $\langle E|E\rangle$ and $\langle P|P\rangle$ being norms of the vector are positive numbers.

Hence

$$E - 1/2\, \hbar w \geq 0$$

or

$$E \geq 1/2\, \hbar w, \quad \text{... (38)}$$

the equality sign holding good (or $E = 1/2\, \hbar w$) if and only if $|P\rangle = a|E\rangle = 0$ or if $|P\rangle$ is a null ket. That all eigen values of H should be positive, also follows from eqn. (27) since the expectation values of x^2 and p^2 should be positive.

Further substituting the value of $H a$ from eqn. (36) and then using eqn. (29), we get

$$\begin{aligned} H a|E\rangle &= (aH - \hbar w\, a)|E\rangle \\ &= (aE - \hbar w\, a)|E\rangle \\ &= (E - \hbar w)\, a|E\rangle \end{aligned} \quad \text{... (39)}$$

The above equation tells us that if $|E\rangle$ is an eigen ket of H, then $a|E\rangle$ is also an eigen ket of H belonging to the eigen value $(E - \hbar w)$ provided $a|E\rangle$ is not a null ket which will occur only if $E = 1/2\, \hbar w$.

Thus $E - \hbar w$ is an eigen value of H provided $a|E\rangle \neq 0$. Similarly one can show that $E - 2\hbar w$, $E - 3\hbar w$, ... are also eigen values of H provided $aa|E\rangle$, $aaa|E\rangle$,... are not null kets.

Continuing in this way, we obtain a series of eigen values E, $E - \hbar w$, $E - 2\hbar w$, $E - 3\hbar w$, ... which cannot extend to infinity as it would then contain eigen values contradicting eqn. (38). Further they can terminate only at $E = 1/2\ \hbar w$ (or finally to be denoted by $E_0 = 1/2\ \hbar w$).

In the similar manner, using eqns. (29) and (37), one can show that

$$H\,a^\dagger|E\rangle = (a^\dagger H + \hbar w\, a^\dagger)\,|E\rangle$$

$$= (a^\dagger E + \hbar w\, a^\dagger)\,|E\rangle$$

$$= (E + \hbar w)\, a^\dagger\,|E\rangle \qquad \text{... (40)}$$

which implies that if $|E\rangle$ is an eigen ket of H belonging to the eigen value E, then $a^\dagger|E\rangle$ will also be an eigen ket of H belonging to the eigen value $(E + \hbar w)$ provided $a^\dagger\,|E\rangle$ is not a null ket. If it is a null ket, then use of eqns. (29) and (32) gives

$$0 = \hbar w\, a\, a^\dagger\,|E\rangle$$

$$= (H + 1/2\ \hbar w)\,|E\rangle$$

$$= (E + 1/2\ \hbar w)\,|E\rangle$$

which implies, $E = -1/2\ \hbar w$, a result which contradicts eqn. (38) as $|E\rangle$ by definition is not a null ket. Thus if E is an eigen value, then $E + \hbar w$ is always another eigen value of H (and so are $E + 2\ \hbar w$, $E + 3\ \hbar w$ and so on). Since eigen values must terminate at 1/2 $\hbar w$, we have

$$E = E_n = (n + 1/2)\,\hbar w, \quad n = 0, 1, 2, ... \qquad \text{... (41)}$$

a result which has already been obtained earlier starting from the Schrödinger equation.

Let us now denote the eigen ket belonging to the eigen value $(n + 1/2)\,\hbar w$ by $|n\rangle$.The eigen value equation, therefore becomes

$$H\,|n\rangle = (n + 1/2)\,\hbar w\,|n\rangle, \quad n = 0, 1, 2... \qquad \text{... (42)}$$

We assume that the eigen kets $|n>$ are normalized so that

$$\langle m|n\rangle = \delta_{mn} \qquad \text{... (43)}$$

Now since $a\,|E\rangle$ is an eigen ket of H belonging to the eigen value $E - \hbar w$ ($= (n + 1/2)\,\hbar w - \hbar w = (n - 1/2)\,\hbar w$) it implies that $a\,|n\rangle$ would be an eigen ket of H belonging to the eigen value $(n - 1 + 1/2)\,\hbar w = (n - 1/2)\,\hbar w$. Thus $a\,|n\rangle$ must be a multiple of $|n - 1\rangle$:

$$a\,|n\rangle = \alpha_n\,|n - 1\rangle$$

In order to calculate α_n, we calculate square of the length of $a\,|n\rangle$, that is,

$$\langle n|a^\dagger\, a|n\rangle = |\alpha_n|^2\langle n - 1\,|\,n - 1\rangle$$

$$= |\alpha_n|^2 \qquad \text{... (44)}$$

where use has been made of the orthonormality condition (43).

Further using eqns. (33), (42) and (43), we get

$$\hbar w\langle n|a^\dagger a|n\rangle = \langle n|H - 1/2\,\hbar w \mid n\rangle$$

$$= \langle n|(n+1/2)\hbar w - 1/2\,\hbar w \mid n\rangle$$

$$= n\,\hbar w\,\langle n|n\rangle$$

$$= n\,\hbar w$$

Therefore,

$$|\alpha_n|^2 = n$$

and as a result eqn. (44) reads as

$$a|n\rangle = n^{1/2}|n-1\rangle \qquad \text{... (45)}$$

where we have taken $\alpha_n = (n)^{1/2}\, e^{i \times \text{phase factor}}$ the absolute value of which is $n^{1/2}$. In the similar fashion, it can be shown that

$$a^+|n\rangle = (n+1)^{1/2}\;|n+1\rangle \qquad \text{... (46a)}$$

or

$$|n+1\rangle = a^\dagger\,|n\rangle/(n+1)^{1/2} \qquad \text{... (46b)}$$

Thus from the above, if $|0\rangle$ denotes the ground state eigen ket, then

$$|1\rangle = \frac{a^\dagger}{1^{1/2}}|0\rangle, \quad |2\rangle = \frac{a^\dagger}{2^{1/2}} \quad |1\rangle = \frac{a^{\dagger 2}}{1^{1/2}2^{1/2}} \quad |0\rangle = \frac{a^{\dagger 2}}{2!^{1/2}}|0\rangle$$

and hence

$$|3\rangle = \frac{a^{\dagger 3}}{n!^{1/2}}|0\rangle, \ldots$$

In general, the nth quantum state of the oscillator would be given by

$$|n\rangle = \frac{a^{\dagger n}}{n!^{1/2}}|0\rangle \qquad \text{... (47)}$$

Based on the result, $E_n - E_0 = n\,\hbar w$, we can think of an excited state $|n\rangle$ of the oscillator as containing n quanta of energy $\hbar w$, in addition to the zero point energy 1/2 $\hbar w$ (or the oscillator in the state $|n\rangle$ is an assembly of n non-interacting particles, each of energy $\hbar w$). The different states of the oscillator correspond to different number of particles. The operator $a^\dagger$ according to eqn. (46a) creates a quantum of energy (or a particle) and is, therefore called the creation operator. Similarly, the operator a according to eqn. (45) annihilates a quantum of energy (or a particle) and therefore is called annihilation or destruction operator (also see chapter 17). The relations, (45) and (46), that is, $a\,|0\rangle = 0$ and $a^\dagger\,|0\rangle = |1\rangle$, are consistent with this interpretation since a particle cannot be destroyed when no particle is present as in $|0\rangle$ but a particle can be created even when no particle is present initially. The interpretation of $N = a^\dagger\, a$ as the number operator is, then suggested by eqn. (55) of chapter 17.

It is in view of the above interpretation that the representation defined by the basis vectors $|n\rangle$, $n = 0, 1, 2, \ldots$ is named as the occupation number representation. Note that the particle number n is actually the quantum number characterizing the state of excitation of the oscillator and it is truly speaking the matter of convenience that it is called the particle number.

Using eqns. (43) and (45), we get

$$\langle m|a|n\rangle = (n)^{1/2}\langle m|n-1\rangle$$
$$= (n)^{1/2}\,\delta_{m,\,n-1} \quad \ldots (48)$$

Similarly from eqns. (43) and (46), it follows that

$$\langle m|a^{\dagger}|n\rangle = (n+1)^{1/2}\langle m|n+1\rangle$$
$$= (n+1)^{1/2}\,\delta_{m,\,n+1} \quad \ldots (49)$$

Therefore, the matrices a and $a^{\dagger}$ would be given by

$$a = \begin{pmatrix} a_{00} & a_{01} & a_{02} & a_{03} & a_{04} & \cdots \\ a_{10} & a_{11} & a_{12} & a_{13} & a_{14} & \cdots \\ a_{20} & a_{21} & a_{22} & a_{23} & a_{24} & \cdots \\ a_{30} & a_{31} & a_{32} & a_{33} & a_{34} & \cdots \\ \cdots & \cdots & \cdots & \cdots & \cdots & \cdots \end{pmatrix}$$

$$= \begin{pmatrix} 0 & \sqrt{1} & 0 & 0 & 0 & \cdots \\ 0 & 0 & \sqrt{2} & 0 & 0 & \cdots \\ 0 & 0 & 0 & \sqrt{3} & 0 & \cdots \\ 0 & 0 & 0 & 0 & \sqrt{4} & \cdots \\ \cdots & \cdots & \cdots & \cdots & \cdots & \cdots \end{pmatrix} \quad \ldots (50)$$

where the first row corresponds to $m = 0$ and the first column to $n = 0$. Also eqn. (48) has been used.

Similarly using eqn. (49), one gets

$$a^{+} = \begin{pmatrix} 0 & 0 & 0 & 0 & 0 & \cdots \\ \sqrt{1} & 0 & 0 & 0 & 0 & \cdots \\ 0 & \sqrt{2} & 0 & 0 & 0 & \cdots \\ 0 & 0 & \sqrt{3} & 0 & 0 & \cdots \\ 0 & 0 & 0 & \sqrt{4} & 0 & \cdots \\ \cdots & \cdots & \cdots & \cdots & \cdots & \cdots \end{pmatrix} \quad \ldots (51)$$

Thus,

$$N = a^{\dagger}a = \begin{pmatrix} 0 & 0 & 0 & 0 & 0 & \cdots \\ 0 & 1 & 0 & 0 & 0 & \cdots \\ 0 & 0 & 2 & 0 & 0 & \cdots \\ 0 & 0 & 0 & 3 & 0 & \cdots \\ 0 & 0 & 0 & 0 & 4 & \cdots \\ \cdots & \cdots & \cdots & \cdots & \cdots & \cdots \end{pmatrix} \quad \ldots (52)$$

that is, it is diagonal with eigen values 0, 1,2, ...

Substituting from eqns. (50) and (51) into eqns. (31), one can obtain the following matrices for x and p,

$$x = (\hbar/2mw)^{1/2}\begin{pmatrix} 0 & \sqrt{1} & 0 & 0 & 0 & \ldots \\ \sqrt{1} & 0 & \sqrt{2} & 0 & 0 & \ldots \\ 0 & \sqrt{2} & 0 & \sqrt{3} & 0 & \ldots \\ 0 & 0 & \sqrt{3} & 0 & \sqrt{4} & \ldots \\ \ldots & \ldots & \ldots & \ldots & \ldots & \ldots \end{pmatrix} \quad \ldots (53)$$

and

$$p = -i(m\hbar w/2)^{1/2}\begin{pmatrix} 0 & \sqrt{1} & 0 & 0 & 0 & \ldots \\ -\sqrt{1} & 0 & \sqrt{2} & 0 & 0 & \ldots \\ 0 & -\sqrt{2} & 0 & \sqrt{3} & 0 & \ldots \\ 0 & 0 & -\sqrt{3} & 0 & \sqrt{4} & \ldots \\ \ldots & \ldots & \ldots & \ldots & \ldots & \ldots \end{pmatrix} \quad \ldots (54)$$

❑❑❑

Chapter 7

QUANTUM DYNAMICS

7.1 THE EQUATIONS OF MOTION

In classical dynamics, we first of all determine the variables such as position coordinates arid velocities that specify the state of a physical system and then describe the development of the system in time. The equations of motion which describe the development of the physical system in time are, therefore, differential equations in these variables. The state of a physical system in quantum mechanics is represented by a wave function. The equation of motion, therefore could be a differential equation for the wave function. However, the observable quantities are not the wave functions but the expectation values of a set of Hermitian operators corresponding to the dynamical variables (the observable quantity is E which is expectation value of the operator $\hat{H}$). Therefore, the equation of motion should be concerned with the development in time of these expectation values.

Now the expectation value of an operator $\hat{A}$ at time t in the state represented by the normalized wave function $\psi(t)$ actually $\psi(\vec{r}, t)$, is given by

$$\langle \hat{A} \rangle (t) = \int \psi^* \hat{A} \psi d\tau \qquad \text{... (1)}$$

where the notation $\langle \hat{A} \rangle$ (t) implies that the expectation value $\langle \hat{A} \rangle$of the operator $\hat{A}$ is a number which depends on t. Therefore, the variation with time of $\langle \hat{A} \rangle$can be viewed as arising in one of the following ways:

(*i*) The wave function ψ changes with time but $\hat{A}$ remains unchanged.

(*ii*) $\hat{A}$ changes with time but ψ remains unchanged.

(*iii*) Both $\hat{A}$ and ψ change with time.

Correspondingly, we have the Schrödinger, the Heisenberg and the Interaction (or also called Dirac) pictures of time development. Of course, the variation with time of $\langle \hat{A} \rangle$ calculated in any of these pictures must agree with the observed rate of variation of $\langle \hat{A} \rangle$.

7.2 THE SCHRÖDINGER PICTURE

In this picture the wave function ψ describing the quantum mechanical state of the system is a function of time while the operator $\hat{A}$ is not. The time development of ψ is obtained by solving the time dependent Schrödinger equation,

$$i\hbar\, \partial\psi_s(t)/\partial t \;=\; \hat{H}(t)\, \psi_s(t) \qquad \text{... (2)}$$

where the subscript S corresponds to the wave functions in the Schrödinger picture. $\hat{H}(t)$ being Hamiltonian of the system.

Often, $\hat{H}$ does not depend explicity on time. In that case from the above equation, one obtains

$$\partial\psi_s(t)/\psi_s(t) \;=\; -\, i/\hbar\, \hat{H}_s \partial t$$

where $\hat{H}_s$ denotes the time independent Hamiltonian in the Schrödinger picture. Integration of the above equation leads to

$$\psi_s\,(t) \;=\; e^{-i/\hbar \hat{H}_s t}\, \psi_s\,(0) \qquad \text{... (3)}$$

where ψ_s (0) describes the state of the system at time $t = 0$ and where the exponential of an operator is defined through the power series,

$$i^{\hat{\theta}} \;=\; 1 + \hat{\theta} + \frac{\hat{\theta}\,.\,\hat{\theta}}{2!} + \frac{\hat{\theta}\,.\,\hat{\theta}\,.\,\hat{\theta}}{3!} + \ldots . \qquad \text{... (4)}$$

If the eigen functions of $\hat{H}_s$ are denoted by $\varphi_n(\vec{r})$, then

$$\hat{H}_s\; \varphi_n(\vec{r}) \;=\; E_n \varphi_n(\vec{r}) \qquad \text{... (5)}$$

and therefore from eqns. (3), (4) and (5), one gets the relation,

$$\psi_s(t) \;=\; e^{-it\hbar E_n t}\, \psi_s(0) \qquad \text{... (6)}$$

which can be used to study the time development of states.

Now the expectation value of an operator $\hat{A}_s$s in the state ψ_s (t) (actually $\psi(\vec{r}, t)$) is given by

$$\langle \hat{A}_s \rangle(t) \;=\; \int \psi_s^*(t) \hat{A}_s \psi_s(t)\, d\tau \qquad \text{... (7)}$$

Hence the time rate of change of the expectation value of $\hat{A}_s$s will be

$$\frac{d}{dt}\langle \hat{A}_s \rangle \;=\; \frac{d}{dt}\int \psi_s^*(t) \hat{A}_s \psi_s(t)\, d\tau\,.$$

$$=\; \int \frac{\partial\psi_s^*(t)}{\partial t} \hat{A}_s \psi_s(t)\, d\tau + \int \psi_s^*(t) \hat{A}_s \frac{\partial\psi_s(t)}{\partial t}\, d\tau \qquad \text{... (8)}$$

where the term containing the variation of $\hat{A}_s$ with time (that is, the term,

$$\int \psi_s^* \partial\hat{A}_s/\partial t\; \psi_s d\tau$$

has been taken to be zero as the operator in the present picture does not depend explicity on time.

Substituting the values of $\partial\psi_s/\partial t$ and $\partial\psi_s^*/\partial t$ from the Schrödinger equation,

$$i\hbar\partial\psi_s(t)/\partial t = \hat{H}_s\psi_s(t) \qquad \ldots (9)$$

and from its complex conjugate into eqn. (8), one gets

$$\frac{d}{dt}\langle\hat{A}_s\rangle = \frac{1}{i\hbar}\int[-(\hat{H}_s\psi_s)^*\hat{A}_s\psi_s + \psi_s^*\hat{A}_s\hat{H}_s\psi_s]d\tau$$

As $\hat{H}_s$ is Hermitian, the above equation leads to

$$\begin{aligned}\frac{d}{dt}\langle\hat{A}_s\rangle &= \frac{1}{i\hbar}\int[-\psi_s^*\hat{H}_s\hat{A}_s\psi_s + \psi_s^*\hat{A}_s\hat{H}_s\psi_s]d\tau \\ &= \frac{1}{i\hbar}\int\psi_s^*(\hat{A}_s\hat{H}_s - \hat{H}_s\hat{A}_s)\psi_s d\tau \\ &= \left\langle\frac{1}{i\hbar}[\hat{A}_s, \hat{H}_s]\right\rangle \qquad \ldots (10)\end{aligned}$$

where $[\hat{A}_s, \hat{H}_s]$ is called the commutator of $\hat{A}_s$ and $\hat{H}_s$.

From eqn. (10), it follows that

$$d\langle\hat{A}_s\rangle/dt = 0, \text{ if } [\hat{A}_s, \hat{H}_s] = 0,$$

that is, an observable $\langle\hat{A}_s\rangle$ is independent of time or is a constant of motion if the corresponding operator commutes with the Hamiltonian. If $\hat{A}_s \equiv \hat{H}_s$, then $[\hat{H}_s, \hat{H}_s] = 0$ and hence

$$d\langle\hat{A}_s\rangle/dt = d\langle\hat{H}_s\rangle/dt = 0$$

or in other words, if $\hat{H}_s$ is not an explicit function of time, $\langle\hat{H}_s\rangle$ = energy, is a constant of motion or conserved.

By comparing eqn. (10) with the equation of motion for the dynamical variable A_s in classical mechanics, we would see that the expectation value of the operators obey the same equation of motion in quantum mechanics as obyed by dynamical variables in classical mechanics provided we identify the commutator bracket divided by $i\hbar$ with the classical Poisson bracket. The underlined statement refers to Ehrenfest's theorem.

The equation of motion of a dynamical variable as written in terms of the Poisson bracket is given by

$$dA_s/dt = \{A_s, H_s\} \qquad \ldots (11)$$

where

$$\{A_s, H_s\} = \sum_i\left(\frac{\partial A_s}{\partial q_i}\frac{\partial H_s}{\partial p_i} - \frac{\partial A_s}{\partial p_i}\frac{\partial H_s}{\partial q_i}\right)$$

denotes the Poisson bracket of classical dynamical variables $A_s = A_s(q_i, p_i)$ and $H_s = H_s(q_i, p_i)$.

The comparison of eqn. (10) with eqn. (11) leads to the conclusion that the classical equation of motion goes over into the quantum mechanical equation of motion if the replacement

$$\{A_s, H_s\} = \frac{1}{i\hbar}\langle[\hat{A}_s, \hat{H}_s]\rangle \qquad \ldots (12)$$

is made.

7.3 THE HEISENBERG PICTURE

In order to distinguish the wave functions and the operators in the Heisenberg picture from those in the Schrödinger picture, we will use here the subscript '*H*'. Thus ψ_H and $\hat{A}_H$, respectively denote a wave function and an operator in the Heisenberg picture. In this case the wave function ψ_H is time independent whereas the operator $\hat{A}_H$ depends on time.

Substituting $\psi_S(t)$ from eqn. (3) into eqn. (7), we get

$$\langle \hat{A}_S \rangle = \int \psi_s^*(0)\, e^{i\hat{H}_s t/\hbar}\, \hat{A}_s e^{-i\hat{H}_s t/\hbar}\, \psi_s(0)\, d\tau$$

Defining the operator $\hat{A}_H(t)$ by the following relation (that is, the operators in the Heisenberg and Schrödinger pictures are related by the equation).

$$\hat{A}_H(t) = e^{i\hat{H}_s t/\hbar}\, \hat{A}_s e^{-i\hat{H}_s t/\hbar} \qquad \text{... (13)}$$

We obtain from the above

$$\langle \hat{A}_s \rangle = \int \psi_H^* \hat{A}_H(t) \psi_H d\tau \qquad \text{... (14)}$$

which is nothing but the expectation value of $\hat{A}_H(t)$, that is, $\langle \hat{A}_H(t) \rangle$. From eqn. (14) we conclude that the expectation values remain the same if one considers that the wave function depends upon time and the operator is independent of time (that is, if one considers Schrödinger picture) or if the operator depends upon time but assumes that the wave function is independent of time (that is, if one considers Heisenberg picture), as expected.

From eqn. (13), the total time derivative of $\hat{A}_H(t)$ is given by

$$\frac{d}{dt}\hat{A}_H(t) = \frac{i}{\hbar} e^{i\hat{H}_s t/\hbar} (\hat{H}_s \hat{A}_s - \hat{A}_s \hat{H}_s)\, e^{-i\hat{H}_s t/\hbar}$$

Note that the term $\partial \hat{A}_H/\partial t$ has not appeared as $\hat{A}_s$ is independent of time. Multiplying both sides of the above equation by $i\hbar$, one gets

$$\begin{aligned} i\hbar \frac{d}{dt}\hat{A}_H(t) &= e^{i\hat{H}_s t/\hbar} (\hat{A}_s \hat{H}_s - \hat{H}_s \hat{A}_s)\, e^{-i\hat{H}_s t/\hbar} \\ &= (e^{i\hat{H}_s t/\hbar}\, \hat{A}_s e^{-i\hat{H}_s t/\hbar})\, \hat{H}_s - \hat{H}_s\, (e^{i\hat{H}_s t/\hbar}\, \hat{A}_s\, e^{-i\hat{H}_s t/\hbar}) \\ &= \hat{A}_H(t)\, \hat{H}_s - \hat{H}_s \hat{A}_H(t) \\ &= [\hat{A}_H(t), \hat{H}_s] \end{aligned} \qquad \text{... (15)}$$

Let us show that if the Hamiltonian is assumed to be independent of time in Schrödinger representation, then it is also independent of time in Heisenberg representation.

$$\begin{aligned} \hat{H}_H(t) &= e^{iH_s t/\hbar}\, \hat{H}_s e^{-i\hat{H}_s t/\hbar} \\ &= e^{iH_s t/\hbar}\, e^{-i\hat{H}_s t/\hbar}\, \hat{H}_s \\ &= \hat{H}_s \end{aligned}$$

Thus eqn. (15) reads

$$i\hbar\, \partial \hat{A}_H(t)/\partial t = [\hat{A}_H(t), \hat{H}_H] \quad \text{... (16)}$$

which gives the time dependence of an operator in the Heisenberg picture and is called Heisenberg equation of motion.

Note that the above equation of motion closely resembles the classical Hamilton's canonical equation of motion written in terms of Poisson bracket notation, that is

$$dA_H/dt = \{A_H, H_H\} \quad \text{... (17)}$$

where $\{A_H, H_H\}$ denotes the Poisson bracket of classical dynamical variables $A_H = A_H(q_i, p_i)$ and $H_H = H_H(q_i, p_i)$. In case $A_H = A_H(q_i, p_i, t)$ one more term $\partial A_H/\partial t$ would appear in eqn. (17). However q_i and p_i are functions of time. By comparing eqn. (16) with eqn. (17), we see that the classical equation of motion goes over into the quantum mechanical equation of motion if dynamical variables are made operators and the replacement

$$\{A_H, H_H\} \rightarrow \frac{1}{i\hbar}[\hat{A}_H, \hat{H}_H] \quad \text{... (18)}$$

is made.

The equation of motion for the expectation value of $\hat{A}_H(t)$ can be found by finding the total time derivative of $\langle \hat{A}_H(t) \rangle$ from eqn. (14). Hence

$$\frac{d}{dt}\langle \hat{A}_s \rangle = \frac{d}{dt}\langle \hat{A}_H(t) \rangle = \frac{d}{dt}\int \psi_H^* \hat{A}_H(t)\psi_H d\tau$$

$$= \int \psi_H^* \frac{d}{dt}\hat{A}_H(t)\psi_H d\tau$$

Using eqn. (16), the above equation leads to

$$\frac{d}{dt}\langle \hat{A}_s \rangle = \frac{d}{dt}\langle \hat{A}_H(t) \rangle = \int \psi_H^* \frac{1}{i\hbar}[\hat{A}_H, \hat{H}_H]\psi_H d\tau$$

$$= \langle 1/i\hbar\, [\hat{A}_H, \hat{H}_H] \rangle \quad \text{... (19)}$$

Thus the equations of motion for the expectation values are the same both in the Heisenberg picture and as well as in the Schrödinger picture (see eqn. (10)), as expected. It may, however be pointed out that the basic equation of motion in the Schrödinger picture is

$$i\hbar\partial\psi_s(t)/\partial t = \hat{H}_s\psi_s(t) \quad \text{... (20)}$$

where it is assumed that the Hamiltonian of the system does not depend explicity on time or $\partial \hat{H}_s/\partial t = 0$ and in the Heisenberg picture it is eqn. (16).

7.4 THE INTERACTION PICTURE (OR DIRAC'S PICTURE)

In this picture both the operators and the wave functions change with time. The picture is useful when the Hamiltonian can be split up into two parts. This could be the case, for example, when the system is put under an external field. The time independent part represents the Hamiltonian of the system in the absence of the external field and the time dependent part arises from the presence of the external field. We will discuss such cases in the time dependent perturbation theory.

Let the total Hamiltonian,

$$\hat{H}(t) = \hat{H}_0 + \hat{H}'(t) \qquad \text{... (21)}$$

where $\hat{H}_0$ is independent of time and $\hat{H}'(t)$ denotes the Hamiltonian characterizing the interaction of the system with an external time dependent field.

Starting with the Schrödinger eqn. (2), we have

$$i\hbar\, \partial\, \psi_s(t)/\partial t = \{\hat{H}_0 + \hat{H}'(t)\}\, \psi_s(t) \qquad \text{... (22)}$$

where the subscript S of course refers to the wave function in the Schrödinger picture.

The wave function $\psi_{\text{int}}(t)$ in the interaction picture is defined from the wave function $\psi_s(t)$ in the Schrödinger picture by

$$\psi_{\text{int}}(t) = e^{i\hat{H}_0 t/\hbar}\, \psi_s(t) \qquad \text{... (23)}$$

or

$$\psi_s(t) = e^{-i\hat{H}_0 t/\hbar}\, \psi_{\text{int}}(t) \qquad \text{... (24)}$$

Substituting above into eqn. (19), we obtain

$$i\hbar e^{-i\hat{H}_0 t/\hbar} \frac{\partial}{\partial t} \psi_{\text{int}}(t) + i\hbar \times -i\hat{H}_0/\hbar\, e^{-i\hat{H}_0 t/\hbar}\, \psi_{\text{int}}$$

$$= \{\hat{H}_0 + \hat{H}'(t)\}\, e^{-i\hat{H}_0 t/\hbar}\, \psi_{\text{int}}(t)$$

Cancelling the term containing $\hat{H}_0$ on both the sides, we get

$$i\hbar\, e^{-i\hat{H}_0 t/\hbar} \frac{\partial}{\partial t} \psi_{\text{int}}(t) = \hat{H}'(t)\, e^{-i\hat{H}_0 t/\hbar}\, \psi_{\text{int}}(t)$$

Multiplying the above equation by $e^{i\hat{H}_0 t/\hbar}$ from the left and writing

$$\hat{H}'_{\text{int}}(t) = e^{i\hat{H}_0 t/\hbar}\, \hat{H}'(t)\, e^{-i\hat{H}_0 t/\hbar} \qquad \text{... (25)}$$

The equation of motion in the interaction picture is obtained as

$$i\hbar \frac{\partial}{\partial t} \psi_{\text{int}}(t) = \hat{H}'_{\text{int}}(t)\, \psi_{\text{int}}(t) \qquad \text{... (26)}$$

The above equation shows that $\psi_{\text{int}}(t)$ changes in time only if $\hat{H}'$ (that is, the perturbation) is present. Following the rule (25), an arbitrary operator $A'(t)$ changes into

$$\hat{A}_{\text{int}}(t) = e^{i\hat{H}_0 t/\hbar}\,\hat{A}(t)\,e^{-i\hat{H}_0 t/\hbar} \quad \text{... (27)}$$

Note that the $\hat{H}_0$ and $\hat{H}_s$ are identical. From the above equation, the time rate of change of $\hat{A}_{\text{int}}(t)$ would be given by

$$\frac{d}{dt}\hat{A}_{\text{int}}(t) = \frac{1}{i\hbar}(\hat{A}_{\text{int}}\,\hat{H}_0 - \hat{H}_0\,A_{\text{int}}) + \left\{\frac{\partial \hat{A}(t)}{\partial t}\right\}_{\text{int}}$$

$$= \frac{1}{i\hbar}[\hat{A}_{\text{int}}(t), \hat{H}_0] + \left\{\frac{\partial \hat{A}(t)}{\partial t}\right\}_{\text{int}} \quad \text{... (28)}$$

It may be noted that the term $\{\partial\hat{A}(t)/\partial t\}_{\text{int}}$ would not appear if $\hat{A}(t) = \hat{A}_s$. Eqns. (26) and (28) exhibit the fact that in the interaction picture both wave functions and the operators move but their time developments are governed by the different portions of the total Hamiltonian. The change in time of the wave function is dictated by $\hat{H}'_{\text{int}}(t)$ and that of operators by $\hat{H}_0$. Thus the interaction picture is intermediate between the Schrödinger picture in which only the wave functions change in time and the Heisenberg picture in which operators are subjected to the time development.

The equation of motion for the expectation value of $\hat{A}_{\text{int}}(t)$ can be obtained as follows:

$$\frac{d}{dt}\langle\hat{A}_{\text{int}}\rangle = \frac{d}{dt}\int \psi^*_{\text{int}}(t)\,\hat{A}_{\text{int}}(t)\,\psi_{\text{int}}(t)\,d\tau$$

$$= \int \frac{\partial\psi^*_{\text{int}}}{\partial t}\hat{A}_{\text{int}}\psi_{\text{int}}\,d\tau + \int \psi^*_{\text{int}}\hat{A}_{\text{int}}\frac{\partial\psi_{\text{int}}}{\partial t}\,d\tau$$

$$+ \int \psi^*_{\text{int}}\frac{\partial\hat{A}_{\text{int}}}{\partial t}\psi_{\text{int}}\,d\tau$$

Substituting the value of $\partial\psi_{\text{int}}/\partial t$ from eqn. (26) and the value of $\partial\psi^*_{\text{int}}/\partial t$ from its complex conjugate into the above equation, we get

$$\frac{d}{dt}\langle\hat{A}_{\text{int}}\rangle = \left\langle\frac{1}{i\hbar}[\hat{A}_{\text{int}}, \hat{H}_{\text{int}}]\right\rangle + \left\langle\frac{\partial A_{\text{int}}}{\partial t}\right\rangle_{\text{int}} \quad \text{... (29)}$$

❑❑❑

Chapter 8

ORBITAL ANGULAR MOMENTUM AND SPIN

8.1 ORBITAL ANGULAR MOMENTUM

The orbital angular momentum of a particle in quantum mechanics is defined just as it is in classical mechanics, namely

$$\vec{L} = \vec{r} \times \vec{p} \qquad \text{... (1)}$$

where $\vec{r}$ and $\vec{p}$ represent the position and linear momentum of the particle, respectively. The components of the classical orbital angular momentum are given by

$$\begin{aligned} L_x &= yp_z - zp_y, \\ L_y &= zp_x - xp_z, \\ L_z &= xp_y - yp_x, \end{aligned} \qquad \text{... (2)}$$

Since in quantum mechanics $\vec{p}$ is to be replaced by the differential operator, $\vec{p} \to i\hbar\,\vec{\nabla}$, the components of $\vec{L}$ will take the form.

$$\begin{aligned} L_x &= -i\hbar\left(\frac{y\partial}{\partial z} - \frac{z\partial}{\partial y}\right), \\ L_y &= -i\hbar\left(\frac{z\partial}{\partial x} - \frac{x\partial}{\partial z}\right), \\ L_z &= -i\hbar\left(\frac{x\partial}{\partial y} - \frac{y\partial}{\partial x}\right) \end{aligned} \qquad \text{... (3)}$$

Let us transform the above equations to a system of spherical polar coordinates with the same origin using the relations (chain rule)

$$\frac{\partial}{\partial x} = \frac{\partial}{\partial r}\frac{\partial r}{\partial x} + \frac{\partial}{\partial \theta}\frac{\partial \theta}{\partial x} + \frac{\partial}{\partial \phi}\frac{\partial \phi}{\partial x},$$

$$\frac{\partial}{\partial y} = \frac{\partial}{\partial r}\frac{\partial r}{\partial y} + \frac{\partial}{\partial \theta}\frac{\partial \theta}{\partial y} + \frac{\partial}{\partial \phi}\frac{\partial \phi}{\partial y} \qquad \text{... (4)}$$

$$\frac{\partial}{\partial z} = \frac{\partial}{\partial r}\frac{\partial r}{\partial z} + \frac{\partial}{\partial \theta}\frac{\partial \theta}{\partial z} + \frac{\partial}{\partial \phi}\frac{\partial \phi}{\partial z}$$

From the Fig. 8.1, the spherical polar coordinates of the point P are related to the cartesian coordinates by

$$\begin{aligned} x &= r \sin\theta \cos\phi, \\ y &= r \sin\theta \sin\phi, \\ z &= r\cos\theta \end{aligned} \qquad \text{... (5)}$$

Therefore,

$$r = (x^2 + y^2 + z^2)^{1/2} \qquad \text{... (6)}$$

$$\cos\theta = z/r = z/(x^2 + y^2 + z^2)^{1/2} \qquad \text{... (7)}$$

$$\tan\phi = y/x \qquad \text{... (8)}$$

On differentiating (6) and then using eqn. (5), we get

$$\frac{\partial r}{\partial x} = \frac{1}{2}(x^2 + y^2 + z^2)^{-1/2}\, 2x$$

$$= x/r = \sin\theta\cos\phi \qquad \text{... (9a)}$$

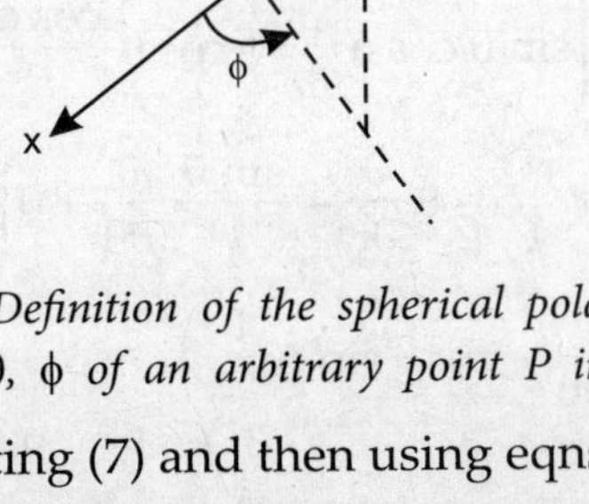

Fig. 8.1 *Definition of the spherical polar coordinates r, θ, ϕ of an arbitrary point P in space*

Similarly on differentiating (7) and then using eqns. (6) and (5),

$$-\sin\theta \frac{\partial \theta}{\partial x} = z\left(-\frac{1}{2}\right)(x^2 + y^2 + z^2)^{-3/2}\, 2x$$

or

$$\frac{\partial \theta}{\partial x} = z/(x^2 + y^2 + z^2)^{1/2}\;\; x/(x^2 + y^2 + z^2)\sin\theta$$

$$= z/r \frac{r \sin\theta \cos\phi}{r^2 \sin\theta}$$

$$= \cos\theta \cos\phi / r \qquad \text{... (9b)}$$

Similarly differentiating (8) and then using (5), we get

$$\sec^2\phi \frac{\partial\phi}{\partial x} = -y/x^2$$

or

$$\frac{\partial\phi}{\partial x} = \frac{-r\sin\theta\sin\phi}{r^2\sin^2\theta\cos^2\phi\sec^2\phi} = -\frac{\sin\phi}{r\sin\theta} \qquad \text{... (9c)}$$

Proceeding in the similar fashion, one can show that

$$\frac{\partial r}{\partial y} = \sin\theta\sin\phi, \frac{\partial\theta}{\partial y} = \cos\theta\frac{\sin\phi}{r}, \frac{\partial\phi}{\partial y} = \frac{\cos\phi}{r\sin\theta} \qquad \text{... (10)}$$

$$\frac{\partial r}{\partial z} = \cos\theta, \frac{\partial\theta}{\partial z} = -\frac{\sin\theta}{r}, \frac{\partial\phi}{\partial z} = 0 \qquad \text{... (11)}$$

Substituting (5) and (4) into eqn. (3) and then substituting the above relations into the resulting equations, one gets

$$L_x = -i\hbar\left[r\sin\theta\sin\phi\left\{\cos\theta\frac{\partial}{\partial r} - \frac{\sin\theta}{r}\frac{\partial}{\partial\theta} + 0\right\}\right.$$

$$\left. - r\cos\theta\left\{\sin\theta\sin\phi\frac{\partial}{\partial r} + \cos\theta\frac{\sin\phi}{r}\frac{\partial}{\partial\theta} + \frac{\cos\phi}{r\sin\theta}\frac{\partial}{\partial\phi}\right\}\right]$$

$$= i\hbar\left[\sin^2\theta\sin\phi\frac{\partial}{\partial\theta} + \cos^2\theta\sin\phi\frac{\partial}{\partial\theta} + \cot\theta\cos\phi\frac{\partial}{\partial\phi}\right]$$

$$= i\hbar\left[\sin\phi\frac{\partial}{\partial\theta} + \cot\theta\cos\phi\frac{\partial}{\partial\phi}\right] \qquad \text{... (12)}$$

$$L_y = -i\hbar\left[r\cos\theta\left\{\sin\theta\cos\phi\frac{\partial}{\partial r} + \cos\theta\frac{\cos\phi}{r}\frac{\partial}{\partial\theta} - \frac{\sin\phi}{r\sin\theta}\frac{\partial}{\partial\phi}\right\}\right.$$

$$\left. - r\sin\theta\cos\phi\left\{\cos\theta\frac{\partial}{\partial r} - \frac{\sin\theta}{r}\frac{\partial}{\partial\theta} + 0\right\}\right]$$

$$= -i\hbar\left(\cos\phi\frac{\partial}{\partial\theta} - \cot\theta\sin\phi\frac{\partial}{\partial\phi}\right) \qquad \text{... (13)}$$

$$L_z = -i\hbar\frac{\partial}{\partial\phi} \qquad \text{... (14)}$$

Defining square of the magnitude of L by the equation,

$$L^2 = L_x^2 + L_y^2 + L_z^2 = L_xL_x + L_yL_y + L_zL_z$$

and substituting for L_x, L_y and L_z from eqns. (12 – 14) into above,

$$L^2 = -\hbar^2\left[\left(\sin\phi\frac{\partial}{\partial\theta}+\cot\theta\cos\phi\frac{\partial}{\partial\phi}\right)\left(\sin\phi\frac{\partial}{\partial\theta}+\cot\theta\cos\phi\frac{\partial}{\partial\phi}\right)\right.$$

$$\left.+\left(\cos\phi\frac{\partial}{\partial\phi}-\cot\theta\sin\phi\frac{\partial}{\partial\phi}\right)\left(\cos\phi\frac{\partial}{\partial\theta}-\cot\theta\sin\phi\frac{\partial}{\partial\phi}\right)+\frac{\partial}{\partial\phi}\frac{\partial}{\partial\phi}\right]$$

$$= -\hbar^2\left[\sin^2\phi\frac{\partial^2}{\partial\theta^2}+\cot^2\theta\cos^2\phi\frac{\partial^2}{\partial\phi^2}+\sin\phi\,(-\operatorname{cosec}^2\phi)\cos\phi\frac{\partial}{\partial\phi}\right.$$

$$+\cot\theta\cos^2\phi\frac{\partial}{\partial\theta}-\cos\phi\sin\phi\,(-\operatorname{cosec}^2\phi)\frac{\partial}{\partial\phi}$$

$$\left.+\cot\theta\sin^2\phi\frac{\partial}{\partial\theta}+\cos^2\phi\frac{\partial^2}{\partial\theta^2}+\cot^2\theta\sin^2\phi\frac{\partial^2}{\partial\phi^2}+\frac{\partial^2}{\partial\phi^2}\right]$$

$$= -\hbar^2\left[\frac{\partial^2}{\partial\theta^2}+\cot\theta\frac{\partial}{\partial\theta}+\frac{\partial^2}{\partial\phi^2}+\cot^2\theta\frac{\partial^2}{\partial\phi^2}\right]$$

$$= -\hbar^2\left[\frac{1}{\sin\theta}\frac{\partial}{\partial\theta}\left(\sin\theta\frac{\partial}{\partial\theta}\right)+\frac{1}{\sin^2\theta}\frac{\partial^2}{\partial\phi^2}\right] \quad \text{... (15)}$$

The above results would be basis for the investigation of the eigen value problems for the orbital angular momentum.

8.2 EIGEN VALUES AND EIGEN FUNCTIONS OF L^2

Operation of L^2 on $\varphi(r, \theta, \phi)$ in view of eqn. (15), gives

$$L^2\varphi(r,\theta,\phi) = -\hbar^2\left[\frac{1}{\sin\theta}\frac{\partial}{\partial\theta}\left(\sin\theta\frac{\partial}{\partial\theta}\right)+\frac{1}{\sin^2\theta}\frac{\partial^2}{\partial\phi^2}\right]\varphi(r,\theta,\phi)$$

$$= -\hbar^2\left[\frac{1}{\sin\theta}\frac{\partial}{\partial\theta}\left(\sin\theta\frac{\partial}{\partial\theta}\right)+\frac{1}{\sin^2\theta}\frac{\partial^2}{\partial\phi^2}\right]R(r)\,\Theta(\theta)\,\Phi(\phi)$$

$$= -\hbar^2 R\left[\frac{\Phi}{\sin\theta}\frac{d}{d\theta}\left(\sin\theta\frac{d\Theta}{d\theta}\right)+\frac{\Theta}{\sin^2\theta}\frac{d^2\Phi}{d\phi^2}\right] \quad \text{... (16)}$$

Since

$$\Phi = Ae^{im_l\phi},\ m_l = 0, \pm 1, \pm 2, \ldots$$

hence

$$\frac{d^2\Phi}{d\phi^2} = (im_l)^2 Ae^{im_l\phi} = -m_l^2\Phi$$

Substituting above into eqn. (16), one gets

$$L^2\varphi = -\hbar^2 R\,\Phi\left[\frac{1}{\sin\theta}\frac{d}{d\theta}\left(\sin\theta\frac{d\Theta}{d\theta}\right)-\frac{m_l^2}{\sin^2\theta}\Theta\right] \quad \text{... (17)}$$

Since Θ is solution of equation,

$$\frac{1}{\sin\theta}\frac{d}{d\theta}\left(\sin\theta\frac{d\Theta}{d\theta}\right)+\left\{-\frac{m_l^2}{\sin^2\theta}\Theta+l\,(l+1)\,\Theta\right\}=0$$

Therefore,

$$\frac{1}{\sin\theta}\frac{d}{d\theta}\left(\sin\theta\frac{d\Theta}{d\theta}\right)-\frac{m_l^2}{\sin^2\theta}\Theta=-\,l\,(l+1)\,\Theta$$

Substituting above into eqn. (17), one gets

$$L^2\,\varphi = l(l+1)\,\hbar^2\,R\,\Theta\,\Phi = l(l+1)\,\hbar^2\,\varphi \quad \text{... (18)}$$

Note that writing

$$\varphi = \Theta(\theta)\,\Phi(\phi) = Y_l^{m_l}\,(\theta,\phi)$$

would have been sufficient for the discussion as r does not appear in any of the differential operators in L^2 or as L^2 is function of θ and ϕ, the eigen functions also should be functions of θ and ϕ only. Thus

$$L^2\,Y_l^{m_l}\,(\theta,\phi) = l(l+1)\,\hbar^2\,Y_l^{m_l}\,(\theta,\phi) \quad \text{... (19)}$$

is the eigen value equation of L^2. Therefore, $l(l+1)\,\hbar^2$ is the eigen value of L^2 with, of course $l = 0, 1, 2, \ldots$ or the only possible values which L^2 can have in a spherically symmetric potential $V(\vec{r}) = V(r)$ are $l(l+1)\,\hbar^2$ and $Y_l^{m_l}\,(\theta,\phi)$ is the eigen function of L^2. Accordingly l is given the name orbital quantum number.

8.3 EIGEN VALUES AND EIGEN FUNCTIONS OF L_z

Operating L_z on $\varphi\,(r,\theta,\phi)$, we get

$$L_z\,\varphi\,(r,\theta,\phi) = -\,i\hbar\,\frac{\partial\varphi}{\partial\phi}$$

$$= -i\hbar\,R(r)\,\Theta\,(\theta)\,d\,\Phi\,(\phi)/d\phi$$

Since

$$\Phi = Ae^{im_l\phi},\ m_l = 0, \pm 1, \pm 2, \ldots$$

Therefore,

$$L_z\,\varphi\,(r,\theta,\phi) = -\,i\hbar\,R\,(r)\,\Theta\,(\theta)\,im_l\,Ae^{im_l\phi}$$

$$= m_l\,\hbar\,\varphi(r,\theta,\phi)$$

The above equation actually amounts to

$$L_z e^{im_l\phi} = m_l\,\hbar\,e^{im_l\phi}$$

Since L_z and L^2 commute, they will have simultaneous eigen functions (since L_x, L_y and L_z do not commute, it can be shown that the spherical harmonics $Y_l^{m_l}$ are not eigen functions of L_x and L_y. That the eigen functions of L_z are much simpler than those of L_x and L_y is the result of the fact that the z-axis was chosen as the axis of the polar coordinate system). Thus without any loss of generality (as L_z does not contain any differential operator which depends upon θ), $Y_l^{m_l}(\theta, \phi)$ can be taken as eigen function of L_z.

Therefore

$$L_z Y_l^{m_l}(\theta, \phi) = m_l\hbar\, Y_l^{m_l}(\theta, \phi) \qquad \text{... (20)}$$

The above equation (actually the eigen value equation for L_z) shows that the eigen values of L_z are $m_l\hbar$. Thus the measurement of L_z can yield as its result only the values $0, \pm\hbar, \pm 2\hbar, \ldots$. Since the z-axis points in an arbitrary chosen direction, it must be true that the angular momentum about any axis is quantized (called space quantization of angular momentum) and can upon measurement only reveal one of these discrete values.

The integral m_l is given the name magnetic quantum number because of the part played by this number in describing the effect of a uniform magnetic field $\vec{B}$ on a charged particle moving in a central force field in which the potential energy depends on a distance r of the moving particle from the fixed point, usually the coordinate origin.

A circulating electron is a current loop and has a magnetic field like that of a magnetic dipole. The magnetic moment of a current loop is given by

$$\vec{\mu}_l = -\frac{e}{2m}\vec{L} \qquad \text{... (21)}$$

where the quantity $-\dfrac{e}{2m}$ is called the gyromagnetic ratio and the minus sign signifies that $\vec{\mu}_l$ is in a direction opposite to $\vec{L}$. It may be pointed out that though the above expression for the magnetic moment of an orbital electron has been obtained by a classical calculation, quantum mechanics yields the same result.

In the presence of magnetic field $\vec{B}$, the magnetic potential energy of the dipole (current loop) is given by

$$V_m = -\vec{\mu}_l \cdot \vec{B} = \frac{e}{2m} LB\cos\theta,$$

θ being the angle between $\vec{\mu}_l$ and $\vec{B}$. If z-axis is chosen parallel to the field $\vec{B}$,

$$V_m = \frac{e}{2m} L_z B \qquad \text{... (22)}$$

The Schrödinger equation can then be written as

$$\left[\frac{p^2}{2m} + V(r) + \frac{e}{2m}L_z B\right]\varphi = E'\varphi \qquad \text{... (23)}$$

Using the eigen value eqn. (20) for L_z, we get

$$E' = E + \frac{e}{2m} B\, m_l \hbar$$

$$= E + m_l \mu_B B \qquad \text{... (24)}$$

which shows that the energy eigen values are changed in the presence of magnetic field. This indicates the splitting of energy levels (or spectral lines) in the presence of magnetic field and is called Zeeman effect which is clear cut confirmation of the space quantization. The quantity $\mu_B = e\hbar/2m$ is called the Bohr magneton. The above equation does indicate origin of the term magnetic quantum number for m_l. Note that the above expression (24) would be modified as soon as we associate spin (or spin angular momentum) with the electron.

The possible values of m_l for a given value of l range from $-l$ to $+l$ through 0, so that the number of possible orientations of $\vec{L}$ in a magnetic field is $(2l + 1)$. When $l = 0$, that is, $m_l = 0$, L_z can have only single value 0 which means $\vec{L}$ is $\perp$ to $\vec{B}$. When $l = 1$, L_z may be $-\hbar$, 0 or $\hbar$; when $l = 2$, L_z may be $-2\hbar$, $-\hbar$, 0, $+\hbar$ or $+2\hbar$; and so on.

The space quantization of $\vec{L}$ is shown in Fig. 8.2. An atom with certain value of m_l will assume the corresponding orientation of its orbital angular momentum $\vec{L}$ relative to the external magnetic field. Note that $\vec{L}$ can never be parallel or antiparallel to $\vec{B}$ as L_z is always smaller than the magnitude $\sqrt{l(l+1)}\,\hbar$ of $\vec{L}$.

Now let us see, why is only one component of $\vec{L}$ quantized? The answer is simple. If $\vec{L}$ is fixed in space, say in z-direction, then the electron would be confined to a definite plane xy at all the times. Therefore, its z-coordinate will be zero. That amounts to infinite uncertainty in p_z ($\Delta_z \Delta p_z \approx \hbar/2$) which is impossible if electron is part of an atom. Hence $\vec{L}$ precesses about z-axis (or direction of $\vec{L}$ is constantly changing) making a cone (see Fig. 8.3) such

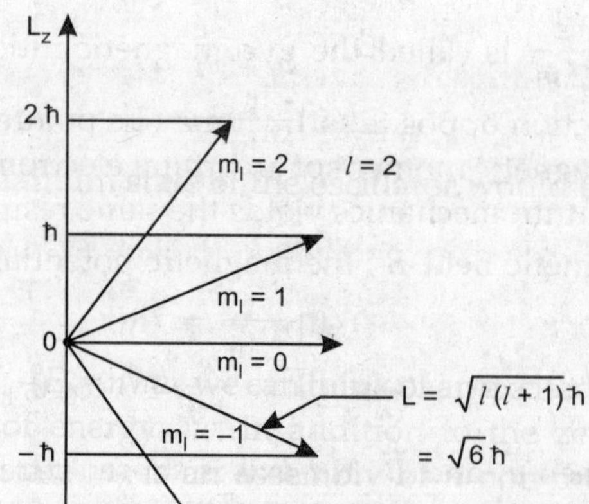

Fig. 8.2 *Space quantization of the orbital angular momentum L-case l = 2. Thus there are $2l + 1 = 5$ possible values of m_l with each value corresponding to different orientation relative to z-axis*

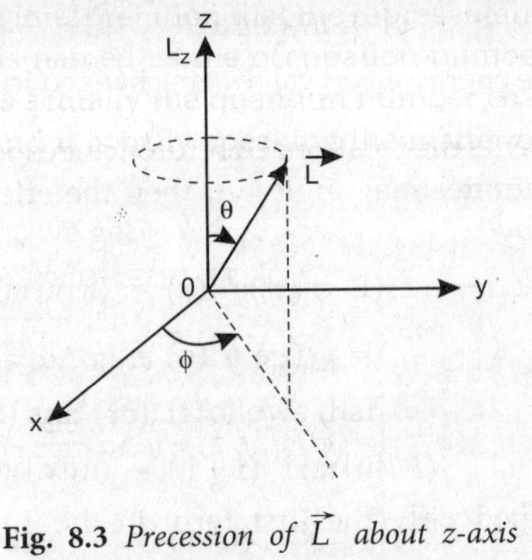

Fig. 8.3 *Precession of $\vec{L}$ about z-axis*

that its projection $L_z = m_l \hbar$ always and the average values of L_x and L_y are zero. Therefore only z-component of $\vec{L}$ will be quantized.

8.4 ORBITAL ANGULAR MOMENTUM AND ROTATIONS

We will now show that the angular momentum is related to the rotations of a system. Consider the case of a change of representation induced by the rotation of the coordinate axes, from $S : (oxyz)$ to $S' : (ox'\, y'\, z')$. To begin with let us see the effect of rotation about the oz-axis through an infinitesimal angle $\Delta\alpha$. The point P whose coordinates in S are $\vec{r}$, will have coordinates $\vec{r}\,'$ w.r.t. S'. The two coordinates are connected through the relations (see Fig. 8.4).

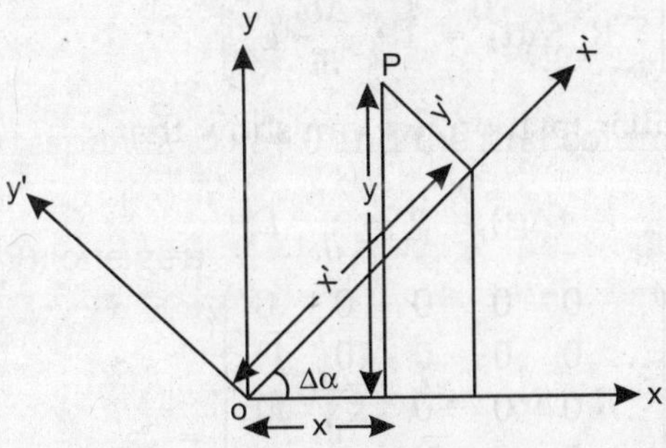

Fig. 8.4 *Rotation about the z-axis through an infinitesimal angle* $\Delta\alpha$

$$x' = x \cos \Delta\alpha + y \sin \Delta\alpha$$

$$y' = -x \sin \Delta\alpha + y \cos \Delta\alpha$$

$$z' = z \qquad \text{... (25)}$$

Since $\Delta\alpha$ is infinitesimal, we may write

$$x' = x + y\,\Delta\alpha$$

$$y' = y - x\,\Delta\alpha \qquad \text{.. (26)}$$

$$z' = z$$

Thus, if $R_z(\Delta\alpha)$ represents the rotational operator corresponding to a rotation about the *oz*-axis through an infinitesimal angle $\Delta\alpha$, then the effect of $R_z(\Delta\alpha)$ on arbitrary wave function φ is given by

$$\begin{aligned} R_z(\Delta\alpha)\,\varphi(x, y, z) &= \varphi(x', y' z') \\ &= \varphi(x + y\,\Delta\alpha,\, y - x\,\Delta\alpha,\, z) \\ &= \varphi(x, y, z) + y\,\Delta\alpha\,\frac{\partial\varphi}{\partial x} - x\,\Delta\alpha\,\frac{\partial\varphi}{\partial y} \end{aligned}$$

where we have retained only the first term in the Taylor expansion as $\Delta\alpha$ is infinitesimal. Therefore

$$R_z(\Delta\alpha)\,\varphi(x, y, z) = \left[1 + \Delta\alpha\left(y\frac{\partial}{\partial x} - x\frac{\partial}{\partial y}\right)\right]\varphi(x, y, z)$$

Now since

$$L_z = xp_y - yp_x = i\hbar\left(y\frac{\partial}{\partial x} - x\frac{\partial}{\partial y}\right),$$

$$R_z(\Delta\alpha)\,\varphi(x, y, z) = \left[1 + \frac{\Delta\alpha}{i\hbar}L_z\right]\varphi(x, y, z)$$

Since $\varphi(x, y, z)$ is an arbitrary wave function, we conclude that

$$R_z(\Delta\alpha) = 1 + \frac{\Delta\alpha}{i\hbar}L_z \qquad \text{... (27)}$$

Proceeding in the similar fashion, we can show that

$$R_x(\Delta\alpha) = 1 + \frac{\Delta\alpha}{i\hbar}L_x \qquad \text{... (28)}$$

and

$$R_y(\Delta\alpha) = 1 + \frac{\Delta\alpha}{i\hbar}L_y \qquad \text{... (29)}$$

Thus the operator L_x, L_y and L_z are called generators of the infinitesimal rotations about the three coordinate axes. More generally, if the rotation is about an arbitrary axis along the unit vector $\hat{n}$, we will have

$$R_n(\Delta\alpha) = 1 + \frac{\Delta\alpha}{i\hbar}\vec{L}\,.\,\hat{n} \qquad \text{... (30)}$$

In order to find $R_z(\alpha)$ for some finite rotation α about the *oz*-axis, we note that the effect of a small increase in the magnitude of α from α to $\alpha + \Delta\alpha$ is to follow the finite rotation $R_z(\alpha)$ by the infinitesimal rotation $R_z(\Delta\alpha)$ to give

$$R_z(\alpha + \Delta\alpha) = R_z(\Delta\alpha)\, R_z(\alpha) = \left[1 + \frac{\Delta\alpha}{i\hbar} L_z\right] R_z(\alpha)$$

Therefore,

$$Lt_{\Delta\alpha \to 0} = \frac{R_z(\alpha + \Delta\alpha) - R_z(\alpha)}{\Delta\alpha} = \frac{dR_z(\alpha)}{d\alpha} = -\frac{i}{\hbar} L_z R_z(\alpha)$$

The integration of the above equation yields,

$$R_z(\alpha) = e^{-i/\hbar\ L_z \alpha} \qquad \text{... (31)}$$

where we have incorporated the boundary condition, $R_z(0) = 1$.

As in the case of infinitesimal rotations, one can generalize the above result to a finite rotation about an arbitrary axis along the unit vector $\hat{n}$. Thus

$$R_n(\alpha) = e^{-1/\hbar\ \alpha \vec{L}.\hat{n}} \qquad \text{... (32)}$$

8.5 PAULI THEORY OF ELECTRON SPIN

So far we have considered the electron to be a mass point or point particle possessing three degrees of freedom associated with its three coordinates x, y and z. Consequently, the quantum theory that we have developed is based on the assumption that the state of the particle (or an electron state) at a given time, is characterized by a wave function $\varphi(x, y, z)$ which depends only on the spatial coordinates x, y, and z. Based on this theory as we would see that it is not possible to explain or interpret for instance the following observations.

(*i*) In the Stern and Gerlach experiment[1], a beam of neutral silver atoms sent through a non-uniform (or inhomogeneous) magnetic field is split into two components with equal and opposite deflections from the undeviated beam. Since the outermost single electron in these atoms in the ground state. (The silver atom has 47 electrons, the electronic configuration is, $4d^{10}5s^1$) has zero angular momentum (the electron is in the s-state which corresponds to $l = 0$. Therefore, $m_l = 0$ as m_l can take all the values from $-l$ to $+l$ through 0), therefore the magnetic quantum number m_l must be equal to zero and no deflection at all is expected (as it is proportional to the magnetic moment μ_z or L_z and $L_z = m_l \hbar = 0$). The observed deviation, however corresponds to a magnetic moment of one Bohr magneton ($\mu_B = e\hbar/2m$).

(ii) When an atom is placed in a uniform magnetic field, each of its spectral lines splits into a certain odd number of equidistant lines, the interval being proportional to the magnetic field, which is called (normal) Zeeman effect. This experimental observation can, however be explained qualitatively with the theory developed so far. But in the case of atoms with odd atomic number (in particular for the H-atom), the levels are divided into an even number of Zeeman levels (called anomalous Zeeman effect). This observation cannot be

[1] G. Stern and W. Gerlach, Z. Physik 8, 110 (1922)

explained based on the theory developed so far as this number should always be odd, being equal to $2l + 1$ with l being an integer (0, 1, 2, ...).

In order to resolve the preceding difficulties, two Dutch graduate students, Georage Uhlenbeck and Samuel Goudsmit[2] proposed the following hypothesis which consists of two parts

(*i*) that an electron has an intinsic angular momentum (not associated with the orbital motion) called spin or spin angular momentum,

$$\vec{S} = \frac{1}{2}\hbar$$

and

(*ii*) associated with this angular momentum is an intrinsic magnetic moment.

$$\vec{\mu}_s = \frac{e}{m}\vec{S}\left(= \frac{e}{m} \times \frac{1}{2}\hbar = \frac{e\hbar}{2\,m} = \mu_B\right).$$

Also that this magnetic moment can have only two values (projections),

$$\frac{e}{m} \times \pm\frac{1}{2}\hbar = \pm\frac{e\hbar}{2m} = \pm\mu_B$$

Note that the above is more precisely stated hypothesis by Pauli[3]. Also note that the coefficient of proportionality between the angular momentum and the magnetic moment is twice as large as in eqn. (21). In other words it simply means that the spin gyromagnetic ratio is twice the orbital gyromagnetic ratio.

In the early history of electron spin hypothesis it was thought that the electron behaves like a sphere rotating about its own axis (that is classical picture of spin). Assuming such a picture, that is, by assuming that the electron e circulates in a spherical region of radius r, one gets

$$\mu_s = iA = \left(\frac{ev}{2\pi r}\right)\pi r^2$$

which on equating this to the observed value of the magnetic moment , that is, $\mu_s = e\hbar/2m$, leads to

$$v = \frac{\hbar}{mr}$$

By putting $r = e^2/4\pi\epsilon_0 mv^2$ obtained from the classical theory of electron ($mv^2/r = e^2/4\pi\epsilon_0 r^2$) into above, one gets

$v/c = (e^2/4\pi\epsilon_0 \hbar c) = 137$ (the fine structure constant)

which implies that the linear (peripheral or equitorial) velocity of the electron is higher than the velocity of light, which contradicts the requirements of the special theory of relatively. That is, the classical model is unable in predicting the correct magnetic moment. Indeed it has been shown by Dirac's relativistic quantum mechanics that

---

[2]G. Unlebeck and S. Goudsmit, Naturwissenschafen, 13, 953 (1925)

[3]W. Pauli, Z. Physik 36, 336 (1926)

associated with the electron is a spin angular momentum $\vec{S} = \frac{1}{2}\hbar$ (the conservation of the total angular momentum in a central force field necessitates the introduction of spin angular momentum) and hence the magnetic moment. We can thus say that the spin has no classical analogue (it does not depend upon $\vec{r}$ and $\vec{p}$ variables).

The wave function of the electron, if the phenomenon related to the spin have also to be described, depends besides on the cartesian coordinates x, y, z also on the spin coordinate. This new coordinate is taken as the eigen value of S_z (z-component of the spin angular momentum). The eigen value of S_z is assumed to be $m_s\,\hbar$ (eigen value equation,

$$S_z\,\chi_s^{m_s} = m_s\,\hbar\,\chi_s^{m_s})$$

in analogy with $m_l\,\hbar$ in the case of L_z, m_s $(= s, s-1, \ldots, -s)$ is called the spin magnetic quantum number and the subscript s $\left(=\frac{1}{2}\right)$, the spin quantum number. Note that, the orbital quantum number can have values 0, 1, 2,... whereas s is fixed for a particle (The proton and neutron are also spin $\frac{1}{2}$ particles but their gyromagnetic ratios are different from that of the electron. At present we know particles of spin 0, $\frac{1}{2}$, 1, $\frac{3}{2}$, 2, ... upto higher value such as $\frac{11}{2}$). Thus we may write φ for a spin $\frac{1}{2}$ particle as

$$\varphi = \varphi\,(x, y, z, m_s),\; m_s = \pm\frac{1}{2} \qquad \text{... (33)}$$

It may be mentioned here that the way orbital angular momentum vector $\vec{L}$ can have $(2l + 1)$ orientations, the spin angular momentum $\vec{S}$ can have $2s + 1 = 2$ orientations, specified by $m_s = +\frac{1}{2}$ (spin up) and $m_s = -\frac{1}{2}$ (spin down) in the magentic field.

It is useful to write the above wave function as (1 × 2 matrix)

$$\varphi = \begin{pmatrix} \varphi_{m_s = 1/2}\,(x, y, z) \\ \varphi_{m_s = -1/2}\,(x, y, z) \end{pmatrix} \qquad \text{... (34)}$$

which describes (actually the two wave functions) the state of an electron at each point in space. This wave function is called a (Pauli) spinor. $\varphi_{1/2}$ and $\varphi_{-1/2}$ are components of the spinor. One must note that the wave function would have four components (as we would see) when the requirements of the relativity are also met.

In analogy with the spinless case (or orbital case) the wave equation of an electron with spin is assumed to be of the form, $i\hbar\;\partial\varphi/\partial t = H\varphi$ or the Pauli wave function (or spinor) satisfies the equation,

$$i\hbar\,\frac{\partial}{\partial t}\begin{pmatrix} \varphi_{\frac{1}{2}} \\ \varphi_{-\frac{1}{2}} \end{pmatrix} = H\begin{pmatrix} \varphi_{\frac{1}{2}} \\ \varphi_{-\frac{1}{2}} \end{pmatrix} \qquad \text{... (35)}$$

called Pauli wave equation. The form of the Hamiltonian H will be obtained while discussing the motion of Dirac particle in electromagnetic field and then taking its non-relativistic limit. Note that the normalization condition in this case would be read as

$$\int \psi^{+}{}_{t}(\vec{r})\,\varphi(\vec{r})\,d\tau = \int [\,|\,\varphi_{1/2}(\vec{r})\,|^2 + |\varphi_{-1/2}(\vec{r})|^2]\;\; d\tau = 1$$

8.6 PAULI SPIN MATRICES

Let us assume (or if we want to avoid Dirac's relativistic treatment) that the components of the spin angular momentum satisfy the same commutation relations as obeyed by L_x, L_y and L_z (see problem 8.1), that is,

$$[S_x, S_y] = i\hbar\, S_z,\ [S_y, S_z] = i\hbar\, S_x,\ [S_z, S_x] = i\hbar\, S_y \qquad \text{... (36)}$$

We now define a new vector operator $\vec{\sigma}$, called Pauli spin operator by

$$\vec{S} = \frac{1}{2}\hbar\,\vec{\sigma} \qquad \text{... (37)}$$

Which is equivalent to

$$S_x = \frac{1}{2}\hbar\sigma_s,\quad S_y = \frac{1}{2}\hbar\,\sigma_y,\quad S_z = \frac{1}{2}\hbar\,\sigma_z \qquad \text{... (38)}$$

Since the eigen values of S_z (or any component of $\vec{S}$) are $\pm\frac{1}{2}\hbar$, the eigen values of σ_z (or any other component of $\vec{\sigma}$) from the above equation would be $\pm$ 1. The components of $\vec{\sigma}$ in the light of eqns. (36) and (38) would be subjected to the following commutation relations,

$$[\sigma_x, \sigma_y] = 2i\sigma_z,\ [\sigma_y, \sigma_z] = 2i\sigma_x,\ [\sigma_z, \sigma_x] = 2i\sigma_{\mathrm{y}} \qquad \text{... (39)}$$

(The above commutation relations are sometimes combined symbolically in a single equation, $\vec{\sigma}\times\vec{\sigma} = 2i\vec{\sigma}$)

At the same time there is the following identity.

$$\sigma_x^2\,\sigma_z - \sigma_z\sigma_x^2 = \sigma_x\,(\sigma_x\sigma_z - \sigma_z\sigma_x) + (\sigma_x\sigma_z - \sigma_z\sigma_x)\sigma_x$$

where we have added and subtracted the term $\sigma_x\sigma_z\sigma_x$. The above identity together with eqn. (39) yields,

$$\begin{aligned}\sigma_x^2\sigma_1 - \sigma_z\sigma_x^2 &= \sigma_{\mathrm{x}}\,(-2i\,\sigma_y) + (-2i\,\sigma_y)\,\sigma_x \\ &= -\,2i\,(\sigma_x\sigma_y + \sigma_y\sigma_x)\end{aligned} \qquad \text{... (40)}$$

Since according to our hypothesis, the eigen values of σ_z (or any other component of $\vec{\sigma}$) are $\pm$ 1, σ_x^2 (or σ_y^2 or σ_z^2) will have only one eigen value equal to unity. This means, σ_x^2 (or σ_y^2 or σ_z^2) is not an operator but simply a number which commutes with any operator, σ_z inclusive. Therefore as the L.H.S. of eqn. (40) vanishes, the R.H.S. must vanish. Together with similar equalities for other components, this yields.

$$\begin{aligned}\sigma_x\sigma_y + \sigma_y\sigma_x &= 0 \\ \sigma_y\sigma_z + \sigma_z\sigma_y &= 0 \\ \sigma_z\sigma_x + \sigma_x\sigma_z &= 0\end{aligned} \quad \text{... (41)}$$

which are called the anticommutation relations (or the operators in this case are said to anticommute)

Now from eqns. (39) and (41) after adding and subtracting, one obtains

$$\sigma_x\,\sigma_y = -\,\sigma_y\,\sigma_x = i\sigma_z \quad \text{... (42)}$$

Similarly we can show that

$$\sigma_y\,\sigma_z = -\,\sigma_z\,\sigma_y = i\sigma_x \quad \text{... (43)}$$

and

$$\sigma_z\,\sigma_x = -\sigma_x\,\sigma_z = i\sigma_y \quad \text{... (44)}$$

Also we know that

$$\sigma_x^2 = \sigma_y^2 = \sigma_z^2 = I \quad \text{... (45)}$$

where I would mean a unit operator or a unit matrix.

We now look for explicit matrix representation of the σ' s. Since each σ has two eigen values, 2 × 2 matrices may be expected to serve the purpose. We begin by associating with σ_z, the simplest 2 × 2 matrix having eigen values 1 and –1, that is

$$\sigma_z = \begin{pmatrix} 1 & 0 \\ 0 & -1 \end{pmatrix} \quad \text{... (46)}$$

In that case the eigen values of σ_x and σ_y cannot be measured accurately and simultaneously (because of the anticommutation relations) and hence let us represent them by the following 2 × 2 matrices,

$$\sigma_x = \begin{pmatrix} a_{11} & a_{12} \\ a_{21} & a_{22} \end{pmatrix},\ \sigma_y = \begin{pmatrix} b_{11} & b_{12} \\ b_{21} & b_{22} \end{pmatrix} \quad \text{... (47)}$$

Using eqns. (41), (46) and (47), one gets

$$\sigma_z\sigma_x + \sigma_x\sigma_z = 2\begin{pmatrix} a_{11} & 0 \\ 0 & -a_{22} \end{pmatrix} = 0 = \begin{pmatrix} 0 & 0 \\ 0 & 0 \end{pmatrix},$$

the null matrix. It implies

$$a_{11} = a_{22} = 0 \quad \text{... (48)}$$

Similarly, one can show that

$$\sigma_y\sigma_z + \sigma_z\sigma_y = 2\begin{pmatrix} b_{11} & 0 \\ 0 & -b_{22} \end{pmatrix} = 0 = \begin{pmatrix} 0 & 0 \\ 0 & 0 \end{pmatrix}$$

which implies, $\quad b_{11} = b_{22} = 0 \quad$ ... (49*a*)

Thus relations (47) can take the form,

$$\sigma_x = \begin{pmatrix} 0 & a_{12} \\ a_{21} & 0 \end{pmatrix}, \sigma_y = \begin{pmatrix} 0 & b_{12} \\ b_{21} & 0 \end{pmatrix} \qquad \text{... (47')}$$

Now using eqns. (45) and (47′), one gets

$$\sigma_x^2 = \begin{pmatrix} a_{12}a_{21} & 0 \\ 0 & a_{21}a_{12} \end{pmatrix} = I \equiv \begin{pmatrix} 1 & 0 \\ 0 & 1 \end{pmatrix}$$

which amounts to

$$a_{12}a_{21} = |a_{12}|^2 = 1 \qquad \text{... (49b)}$$

as σ_x (and also σ_y and σ_z) is Hermitian operator with real eigen values and hence is self adjoint.

Similarly one can show that

$$\sigma_y^2 = \begin{pmatrix} b_{12}b_{21} & 0 \\ 0 & b_{21}b_{12} \end{pmatrix} = I \equiv \begin{pmatrix} 1 & 0 \\ 0 & 1 \end{pmatrix}$$

and hence

$$b_{12}b_{21} = |b_{12}|^2 = 1 \qquad \text{... (50)}$$

From eqns. (44) and (47′), one gets

$$\begin{pmatrix} 1 & 0 \\ 0 & -1 \end{pmatrix}\begin{pmatrix} 0 & a_{12} \\ a_{21} & 0 \end{pmatrix} = \begin{pmatrix} 0 & a_{12} \\ -a_{21} & 0 \end{pmatrix} = i\begin{pmatrix} 0 & b_{12} \\ b_{21} & 0 \end{pmatrix}$$

which implies

$$a_{12} = ib_{12} \qquad \text{... (51a)}$$

and

$$-a_{21} = i\,b_{21} \quad \text{or} \quad b_{21} = i\,a_{21} \qquad \text{... (51b)}$$

From eqn. (49*b*), the simplest possibility is to let

$$a_{12} = a_{21} = 1$$

In that case, from eqn. (51), one gets

$$b_{12} = -i, \qquad b_{21} = i,$$

therefore, σ_x, σ_y and σ_z will have the forms

$$\sigma_x = \begin{pmatrix} 0 & 1 \\ 1 & 0 \end{pmatrix}, \sigma_y = \begin{pmatrix} 0 & -i \\ i & 0 \end{pmatrix} \text{ and } \sigma_z = \begin{pmatrix} 1 & 0 \\ 0 & -1 \end{pmatrix} \qquad \text{... (52)}$$

The above matrices are called Pauli spin matrices.

8.7 THE SPIN EIGEN FUNCTIONS

The eigen value equation for σ_z may be written as

$$\sigma_z f = \pm f \qquad \text{... (53)}$$

where f is called the eigen function and ± 1, the eigen values of σ_z.

In the matrix form, the above equation corresponding to the eigen value +1 reads

$$\begin{pmatrix} 1 & 0 \\ 0 & -1 \end{pmatrix}\begin{pmatrix} a \\ b \end{pmatrix} = +1\begin{pmatrix} a \\ b \end{pmatrix} \quad \text{... (54)}$$

and corresponding to the eigen value – 1 reads

$$\begin{pmatrix} 1 & 0 \\ 0 & -1 \end{pmatrix}\begin{pmatrix} a \\ b \end{pmatrix} = -1\begin{pmatrix} a \\ b \end{pmatrix} \quad \text{... (55)}$$

Thus, eqn. (54) leads to

$$\begin{pmatrix} a \\ -b \end{pmatrix} = \begin{pmatrix} a \\ b \end{pmatrix}$$

which implies $-b = b = 0$ and similarly eqn. (55) leads to

$$\begin{pmatrix} a \\ -b \end{pmatrix} = \begin{pmatrix} -a \\ -b \end{pmatrix}$$

which implies $a = -a = 0$.

Therefore, the eigen functions corresponding to the eigen values +1 and –1 in the case of the electron, would be given by

$$f_+ = \begin{pmatrix} a \\ 0 \end{pmatrix}, \; f_- = \begin{pmatrix} 0 \\ b \end{pmatrix} \quad \text{... (56)}$$

which imply that f_+ and f_- are eigen functions, respectively corresponding to $m_s = +\frac{1}{2}$ or spin up state (or spin parallel to z-axis) and $m_s = -\frac{1}{2}$ or spin down state (or spin antiparallel to z-axis).

In an ensemble of spin $\frac{1}{2}$ particles, it is unlikely that all the particles are spin up or all are spin down. It is more likely that some of the particles are spin up and the others are spin down. Therefore, the spin wave function χ_s corresponding to an ensemble or a general spin state χ_s can be represented by a linear combination of f_+ and f_-, that is

$$\chi_s = \begin{pmatrix} a \\ 0 \end{pmatrix} + \begin{pmatrix} 0 \\ b \end{pmatrix} = \begin{pmatrix} a \\ b \end{pmatrix} \quad \text{... (57)}$$

From the normalization condition,

$$\chi_s^\dagger \chi_s = (a^* \; b^*)\begin{pmatrix} a \\ b \end{pmatrix} = |a|^2 + |b|^2 = 1 \quad \text{... (58)}$$

where $\chi_s^\dagger$ represents the transpose complex conjugate. The two spinors are said to be orthogonal if the above product is zero.

In accordance with the usual interpretation of wave functions, $|a|^2$ is the probability of finding the particle with spin up and $|b|^2$ represents the probability of finding it with spin down. The total probability of course must be unity which is ensured by the normalization as in eqn. (58). Since $|a|^2 = 1$ would mean, the particle is found in the spin up state, the simplest possibility, therefore is to let $a = 1$ for the spin up and similarly $b = 1$ for the spin down state. Hence the eigen functions for spin up and spin down states, would be given by

$$\alpha = \begin{pmatrix} 1 \\ 0 \end{pmatrix}; \ \beta = \begin{pmatrix} 0 \\ 1 \end{pmatrix} \quad \text{... (59)}$$

which are often called pure states or basic spinors as compared to φ of eqn. (34). They are also denoted by $|z\uparrow|$ and $|z\downarrow|$ states. It may be mentioned here that α and β are two othogonal spinors and they are normalized, as

$$\alpha^\dagger \alpha = \beta^\dagger \beta = 1$$

8.8 RAISING AND LOWERING OPERATORS

$$\sigma_+ = \sigma_x + i\sigma_y = \begin{pmatrix} 0 & 1 \\ 1 & 0 \end{pmatrix} + i \begin{pmatrix} 0 & -i \\ i & 0 \end{pmatrix}$$

$$= 2\begin{pmatrix} 0 & 1 \\ 0 & 0 \end{pmatrix},$$

therefore,

$$S_+ = S_x + iS_y = \frac{1}{2}\hbar\,(\sigma_x + i\sigma_y) = \hbar\begin{pmatrix} 0 & 1 \\ 0 & 0 \end{pmatrix} \quad \text{... (60)}$$

Similarly

$$\sigma_- = \sigma_x - i\sigma_y = 2\begin{pmatrix} 0 & 0 \\ 1 & 0 \end{pmatrix},$$

and therefore,

$$S_- = S_x - iS_y = \frac{1}{2}\hbar\,(\sigma_x - i\sigma_y) = \hbar\begin{pmatrix} 0 & 0 \\ 1 & 0 \end{pmatrix} \quad \text{... (61)}$$

Now

$$S_+\beta = \hbar\begin{pmatrix} 0 & 1 \\ 0 & 0 \end{pmatrix}\begin{pmatrix} 0 \\ 1 \end{pmatrix} = \hbar\begin{pmatrix} 1 \\ 0 \end{pmatrix} = \hbar\alpha \quad \text{... (62)}$$

$$S_-\alpha = \hbar\begin{pmatrix} 0 & 0 \\ 1 & 0 \end{pmatrix}\begin{pmatrix} 1 \\ 0 \end{pmatrix} = \hbar\begin{pmatrix} 0 \\ 1 \end{pmatrix} = \hbar\beta \quad \text{... (63)}$$

$$S_+\alpha = S_-\beta = 0 \quad \text{... (64)}$$

Thus S_+ flips the spin from down to up and S_- does the opposite. S_+ and S_- annihilate the states with highest and lowest values of m_s respectively as expected. Therefore S_+ and S_- are termed as the raising and lowering operators.

Problems

8.1 (*a*) Show that the commutator of L_x with L_y is

$$[L_x, L_y] = i\hbar\, L_z$$

(*b*) Find commutator of L_x with L^2

8.2 (*a*) Show that

$$L^2 = \frac{1}{2}(L_+ L_- + L_- L_+) + L_z^2$$

where

$$L_+ = L_x + iL_y \quad \text{and} \quad L_- = L_x - iL_y$$

are called ladder operators.

(*b*) Show that

$$[L_z, L_+] = i\hbar\, L_+ \quad \text{and} \quad [L_z, L_-] = -i\hbar\, L_-$$

(*c*) Show that

$$[L_+, L_-] = 2\hbar\, L_z$$

(*d*) Show that

$$L^2 = L_+ L_- + L_z^2 - \hbar\, L_z$$

(*e*) Show that

$$[L^2, L_\pm] = 0$$

8.3 Obtain differential operator forms of L_+ and L_- and then show that

$$L^2 = \frac{1}{2}(L_+ L_- + L_- L_+) + L_z^2$$

$$= -\hbar^2\left[\frac{1}{\sin\theta}\frac{\partial}{\partial\theta}\left(\sin\theta\frac{\partial}{\partial\theta}\right) + \frac{1}{\sin^2\theta}\frac{\partial^2}{\partial\phi^2}\right]$$

8.4 Show that

$$L_+ Y_l^{m_l} = [l(l+1) - m_l(m_l+1)]^{1/2}\, \hbar\, Y_l^{m_l+1}$$

and

$$L_- Y_l^{m_l} = [l(l+1) - m_l(m_l-1)]^{1/2}\, \hbar\, Y_l^{m_l-1}$$

8.5 Using the form of the rotational operator corresponding to the rotation of the coordinate axes about *z*-axis through an infinitesimal angle $\Delta\alpha$, that is

$$R_z(\Delta\alpha) = 1 + \frac{\Delta\alpha}{i\hbar} L_z,$$

obtain the differential operator representation of L_z in spherical polar coordinates.

8.6 (*a*) If σ_x, σ_y and σ_z represent the Pauli spin matrices and $\vec{A}$ and $\vec{B}$ any two constant vectors (or operators) show that

$$(\sigma . \vec{A})(\vec{\sigma} . \vec{B}) = \vec{A} . \vec{B} + i\vec{\sigma} . (\vec{A} \times \vec{B})$$

(*b*) From above, show that

$$(\vec{\sigma} . \hat{n})^2 = 1$$

where $\hat{n}$ is an arbitrary unit vector.

8.7 A sample of a certain element is placed in a 0.3 weber/m^2 magnetic field and is suitably excited. How far apart are the Zeeman components of the 4500 Å spectral line of this element?

8.8 In analogy with the result that L_x, L_y, L_z are said to be generators of infinitesimal rotation of the coordinates system about the coordinate axes x, y, z and finally the rotation operator corresponding to the finite rotation about an arbitrary axis along the unit vector $\hat{n}$ is

$$^{(L)}R_{\hat{n}}(\alpha) = e^{-i/\hbar \ \alpha \vec{L}.\hat{n}},$$

one can assume that the components of spin $\vec{S}$ generate infinitesimal rotations of the coordinate system about the coordinate axes and finally

$$^{(S)}R_{\hat{n}}(\alpha) = e^{-i/\hbar \ \alpha \vec{S}.\hat{n}} = e^{-\frac{i\alpha}{2}\vec{\sigma}.\hat{n}} \text{ for spin } \frac{1}{2},$$

thereby show that

$$^{(S)}R_{\hat{n}}(\alpha) = \cos\alpha/2 - i\,\vec{\sigma}\,.\,\hat{n}\,\sin\alpha/2$$

8.9 For an eigen state of L^2 and L_z (or for a state for which L^2 and L_z are well defined or for the state whose angular part is $Y_l^{m_l}(\theta, \phi)$, show that

$$< L_x > = < L_y > = 0$$

and

$$< L_x^2 > = < L_y^2 > = \frac{1}{2}\hbar^2 [l(l+1) - m_l^2]$$

Solutions

8.1 (*a*) $\vec{L} = \vec{r} \times \vec{p} = \begin{vmatrix} \hat{x} & \hat{y} & \hat{z} \\ x & y & z \\ p_x & p_y & p_z \end{vmatrix},$

Hence

$$\begin{aligned}(L_x, L_y) &= [yp_z - zp_y, zp_x - xp_z] \\ &= [yp_z, zp_x] - [yp_z, xp_z] - [zp_y, zp_x] + [zp_y, xp_z] \\ &= (yp_z\, zp_x - zp_x\, yp_z) - (yp_z\, xp_z - xp_z\, yp_z) \\ &\quad - (zp_y\, zp_x - zp_x\, zp_y) + (zp_y\, xp_z - xp_z\, zp_y) \\ &= yp_x(p_z z - zp_z) - 0 - 0 + p_y x(zp_z - p_z z) \\ &= -yp_x[z, p_z] + p_y x[z, p_z] \\ &= [z, p_z](xp_y - yp_x) \\ &= i\hbar L_z\end{aligned}$$

where use has been made of the results that $[z, p_z]\,\varphi = i\,\psi$ as φ is an arbitray wave function and $p_z \to -i\hbar\frac{\partial}{\partial z}$.

Analogous calculations yield the other two commutators and we finally obtain

$$[L_x, L_y] = i\hbar L_z,\ [L_y, L_z] = i\hbar L_x \text{ and } [L_z, L_x] = i\hbar L_y \quad \text{... (8.1)}$$

which indicates that the components of the orbital angular momentum operator do not commute among themselves and hence cannot be measured simultaneously and accurately.

The above result can be generalized to a system of n particles. The total orbital angular momentum of such a system is,

$$\vec{L} = \sum_{i=1}^{n} \vec{L}i \quad \text{... (8.2)}$$

with

$$\vec{L}i = \vec{r}_i \times \vec{p}_i \quad \text{... (8.3)}$$

Now each of the individual angular momenta $\vec{L}_i$ satisfies the commutation relations (8.1) and commutes with $\vec{L}_j$ when $j \neq i$.

(*b*) $[L_x, L^2] = \left[L_x, L_x^2 + L_y^2 + L_z^2\right]$

$$= [L_x, L_x^2] + [L_x, L_y L_y] + [L_x, L_z L_z]$$

$$= 0 + [L_x, L_y]\,L_y + L_y\,[L_x, L_y] + [L_x, L_z]L_z + L_z\,[L_x, L_z]$$

$$= i\hbar\, L_z L_y + L_y\, i\hbar\, L_z - i\hbar\, L_y L_z - i\hbar\, L_z L_y$$

$$= 0$$

Analogous calculations yield the other two commutators and we finally obtain

$$[L_x, L^2] = [L_y, L^2] = [L_z, L^2] = 0 \quad \text{... (8.4)}$$

Thus the components of orbital angular momentum operator commute with the square of the orbital angular momentum operator. That is, one component of $\vec{L}$ and L^2 can be measured simultaneously and accurately.

8.2 (*a*) $L_+ L_- = (L_x + iL_y)(L_x - iL_y)$

$$= L_x^2 + L_y^2 + i\,L_y L_x - i\,L_x L_y$$

and

$$L_- L_+ = (L_x - i\,L_y)(L_x + i\,L_y)$$

$$= L_x^2 + L_y^2 - i\,L_y\,L_x + i\,L_x\,L_y$$

Therefore

$$L^2 = L_x^2 + L_y^2 + L_z^2$$

or

$$L^2 = \frac{1}{2}(L_+L_- + L_-L_+) + L_z^2 \qquad \text{... (8.5)}$$

(*c*) $[L_+, L_-] = [L_x + i\,L_y, L_x - i\,L_y]$

$$= [L_x, L_x] - i[L_x, L_y] + i[L_y, L_x] + [L_y, L_y]$$

$$= 0 - i \times i\hbar\; L_z + i \times i\hbar\; L_z + 0$$

$$= 2\,\hbar\,L_z \qquad \text{... (8.6)}$$

(*d*) $L_+ L_- = L_x^2 + L_y^2 - i(L_x L_y - L_y L_x)$

$$= L_x^2 + L_y^2 + \hbar L_z$$

Hence

$$L_+ L_- + L_z^2 = L_x^2 + L_y^2 + L_z^2 + \hbar L_z$$

$$= L^2 + \hbar\,L_z$$

or

$$L^2 = L_+ L_- + L_z^2 - \hbar\,L_z \qquad \text{... (8.7)}$$

8.3 Using eqns. (12) and (13),

$$L_+ = L_x + i\,L_y$$

$$= i\hbar\left(\sin\phi\frac{\partial}{\partial\theta} + \cot\theta\cos\phi\frac{\partial}{\partial\phi}\right)$$

$$+ \hbar\left(\cos\phi\frac{\partial}{\partial\theta} - \cot\theta\sin\phi\frac{\partial}{\partial\phi}\right)$$

$$= \hbar\,(\cos\phi + i\sin\phi)\frac{\partial}{\partial\theta} + i\,\hbar\cot\theta\,(\cos\phi + i\sin\phi)\frac{\partial}{\partial\phi}$$

$$= \hbar\,e^{i\theta}\left(\frac{\partial}{\partial\theta} + i\cot\theta\frac{\partial}{\partial\phi}\right) \text{ and} \qquad \text{... (8.8)}$$

$$L_- = L_x - iL_y$$

$$= i\,\hbar\left(\sin\phi\frac{\partial}{\partial\theta} + \cot\theta\cos\phi\frac{\partial}{\partial\phi}\right)$$

$$- \hbar\left(\cos\phi\frac{\partial}{\partial\theta} - \cot\theta\sin\phi\frac{\partial}{\partial\phi}\right)$$

$$= -\,\hbar\,(\cos\phi - i\sin\phi)\frac{\partial}{\partial\theta} + i\,\hbar\cot\theta\,(\cos\phi - i\sin\phi)\frac{\partial}{\partial\phi}$$

$$= -\,\hbar\,e^{-i\theta}\left(\frac{\partial}{\partial\theta} - i\cot\theta\frac{\partial}{\partial\phi}\right) \qquad \text{... (8.9)}$$

Thus

$$L_+ L_- \varphi = -\hbar^2 e^{i\phi}\left(\frac{\partial}{\partial\theta} + i\cot\theta\frac{\partial}{\partial\phi}\right) e^{-i\phi}\left(\frac{\partial}{\partial\theta} - i\cot\theta\frac{\partial}{\partial\phi}\right)\varphi$$

$$= -\hbar^2\left[\frac{\partial^2\varphi}{\partial\theta^2} - i(-\text{cosec}^2\theta)\frac{\partial\varphi}{\partial\phi} + e^{i\phi}\, i\cot\theta\,(-i)\, e^{-i\phi}\frac{\partial\varphi}{\partial\theta}\right.$$

$$\left. + e^{i\phi}\cot^2\theta\,(-i)\, e^{-i\phi}\frac{\partial\varphi}{\partial\phi} + e^{i\phi}\cot^2\theta\, e^{-i\phi}\frac{\partial^2\varphi}{\partial\phi^2}\right]$$

$$= -\hbar^2\left[\frac{\partial^2\varphi}{\partial\theta^2} + i\,\text{cosec}^2\frac{\partial\varphi}{\partial\theta} + \cot\theta\frac{\partial\varphi}{\partial\theta^2}\right.$$

$$\left. - i\cot^2\theta\frac{\partial\varphi}{\partial\phi} + \cot^2\theta\frac{\partial^2\varphi}{\partial\phi^2}\right]$$

$$= -\hbar^2\left[\frac{\partial^2\varphi}{\partial\theta^2} + i\frac{\partial\varphi}{\partial\phi} + \cot\theta\frac{\partial\varphi}{\partial\theta} + \cot^2\theta\frac{\partial^2\varphi}{\partial\phi^2}\right]$$

Therefore

$$L_+ L_- = -\hbar^2\left[\frac{\partial^2}{\partial\theta^2} + i\frac{\partial}{\partial\phi} + \cot\theta\frac{\partial}{\partial\theta} + \cot^2\theta\frac{\partial^2}{\partial\phi^2}\right] \qquad \text{... (8.10)}$$

Similarly

$$L_- L_+ = -\hbar^2\left[\frac{\partial^2}{\partial\theta^2} + i\frac{\partial}{\partial\phi} + \cot\theta\frac{\partial}{\partial\theta} + \cot^2\theta\frac{\partial^2}{\partial\phi^2}\right] \qquad \text{... (8.11)}$$

Since

$$L_z^2 = -\hbar^2\frac{\partial^2}{\partial\phi^2},$$

Hence

$$L^2 = \frac{1}{2}(L_+ L_- + L_- L_+) + L_z^2$$

$$= -\hbar^2\left[\frac{1}{\sin\theta}\frac{\partial}{\partial\theta}\left(\sin\theta\frac{\partial}{\partial\theta}\right) + \frac{1}{\sin^2\theta}\frac{\partial^2}{\partial\phi^2}\right] \qquad \text{... (8.12)}$$

8.4 $L_z L_+ = L_z(L_x + iL_y)$

$= L_z L_x + i L_z L_y$

which using eqn. (8.1) leads to

$$L_z L_+ = L_x L_z + i\hbar L_y + i(L_y L_z - i\hbar L_x)$$

$$= (L_x + iL_y)L_z + \hbar(L_x + iL_y)$$

$$= L_+ L_z + \hbar L_+$$

$$= L_+(L_z + \hbar) \qquad \text{... (8.13)}$$

(which also amounts to $[L_z, L_+] = \hbar\, L_+$)

Similarly, it can be shown that

$$L_z L_- = L_- (L_z - \hbar) \qquad \text{... (8.14)}$$

(which also amounts to $[L_z, L_-] = -\hbar\, L_-$)

Now operating L_z and $L_+ \; Y_l^{m_l}$ and making use of eqn. (8.13) leads to

$$\begin{aligned} L_z (L_+ Y_l^{m_l}) &= L_+ (L_z + \hbar)\, Y_l^{m_l} \\ &= L_+ (m_l + 1)\, \hbar\, Y_l^{m_l} \\ &= (m_l + 1)\, \hbar\, (L_+ Y_l^{m_l}) \qquad \text{... (8.15)} \end{aligned}$$

Similarly, it can be shown that

$$L_z (L_- Y_l^{m_l}) = (m_l - 1)\, \hbar\, (L_- Y_l^{m_l}) \qquad \text{... (8.16)}$$

Thus L_+ and L_- are step up and step down operators with respect to the eigen value of L_z. The $L_+ \; Y_l^{m_l}$ is an eigen function of L_z with eigen value $(m_l + 1)\, \hbar$, that is one greater and $L_- \; Y_l^{m_l}$ an eigen function of L_z with eigen value $(m_l - 1)\, \hbar$, that is, one unit less than the eigen value of L_z corresponding to the eigen function $Y_l^{m_l}$. That is why the operators L_+ and L_- are also called ladder operators as the successive application of L_+ (or L_-) creates a ladder of eigen states of L_z. Thus, one may write

$$L_+ Y_l^{m_l} = C_+ Y_l^{m_l + 1} \qquad \text{... (8.17)}$$

and

$$L_- Y_l^{m_l} = C_- Y_l^{m_l - 1} \qquad \text{... (8.18)}$$

where C_+ and C_- are constants which can be found. To find them let us multiply both sides of eqn. (8.17) with its complex conjugate and then integrating over the entire space, we get

$$\langle L_+ Y_l^{m_l} \mid L_+ Y_l^{m_l} \rangle = |C_+|^2 \langle Y_l^{m_l} \mid Y_l^{m_l} \rangle = |C_+|^2 \qquad \text{... (8.19)}$$

where we have used the requirement that the spherical harmonics are normalized.

Using the definition of L_+, the L.H.S. of eqn. (8.19) can be written as

$$\left\langle (L_x + iL_y)\, Y_l^{m_l} \mid L_x \mid Y_l^{m_l} \right\rangle + i \left\langle (L_x + iL_y)\, Y_l^{m_l} \mid L_y \mid Y_l^{m_l} \right\rangle$$

Since the operators L_x and L_y are Hermitian, the above integral takes the form,

$$\begin{aligned} \left\langle Y_l^{m_l} \mid (L_x - iL_y)\, L_x \mid Y_l^{m_l} \right\rangle + i \left\langle Y_l^{m_l} \mid (L_x - iL_y)\, L_y \mid Y_l^{m_l} \right\rangle \\ = \left\langle Y_l^{m_l} \middle| L_x^2 + L_y^2 + i\,(L_x L_y - L_y L_x) \middle| Y_l^{m_l} \right\rangle \\ = \left\langle Y_l^{m_l} \middle| L_x^2 + L_y^2 - \hbar\, L_z \middle| Y_l^{m_l} \right\rangle \end{aligned}$$

$$= \langle Y_l^{m_l} | L_x^2 + L_y^2 + L_z^2 - L_z^2 - \hbar L_z | Y_l^{m_l} \rangle$$

$$= \langle Y_l^{m_l} | L^2 - L_z^2 - \hbar L_z | Y_l^{m_l} \rangle$$

$$= [l\ (l+1)\hbar^2 - m_l^2 \hbar^2 m_l] \langle Y_l^{m_l} \mid Y_l^{m_l} \rangle$$

Once again using the requirement that the spherical harmonics are normalized and substituting above into eqn. (8.19), we get

$$|C_+|^2 = [l\,(l+1) - m_l\,(m_l+1)]\,\hbar^2 \qquad \text{... (8.20)}$$

Similarly one can show that

$$|C_-|^2 = [l\,(l+1) - m_l\,(m_l-1)]\,\hbar^2 \qquad \text{... (8.21)}$$

Therefore

$$L_+ Y_l^{m_l} = [l\,(l+1) - m_l\,(m_l+1)]^{1/2}\,\hbar\, Y_l^{m_l+1} \qquad \text{... (8.22)}$$

and

$$L_- Y_l^{m_l} = [l\,(l+1) - m_l\,(m_l-1)]^{1/2}\,\hbar\, Y_l^{m_l-1} \qquad \text{... (8.23)}$$

8.5 The transformation of the spherical polar coordinates in this case are given by (see Fig. 8.5)

$$r \to r,\ \theta \to \theta,\ \phi \to \phi - \Delta\alpha$$

Thus

$$R_z(\Delta\alpha)\,\varphi\,(r, \theta, \phi) = \varphi\,(r, \theta, \phi - \Delta\alpha)$$

Using eqn. (27) and the Taylor series expansion, we get

$$\left[1 + \frac{\Delta\alpha}{i\hbar} L_z\right] \varphi\,(r, \theta, \phi) = \varphi\,(r, \theta, \phi) - \Delta\alpha \frac{\partial \varphi}{\partial \phi}$$

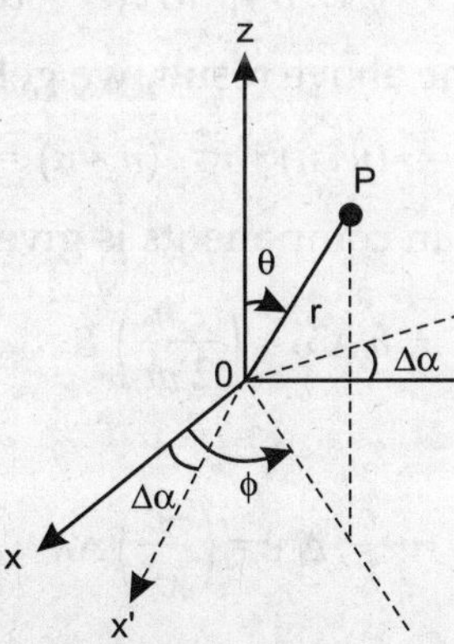

Fig. 8.5 *Transformations of the spherical polar coordinates due to the rotation through an infinitesimal angle Δα about z-axis*

Cancelling first term on both the sides and since φ is arbitrary, we obtain the following differential operator representation of L_z,

$$L_z = -i\hbar\frac{\partial}{\partial\phi}$$

8.6 (*a*)

$$(\vec{\sigma}\,.\,\vec{A})(\vec{\sigma}\,.\,\vec{B}) = (\sigma_x A_x + \sigma_y A_y + \sigma_z A_z)(\sigma_x B_x + \sigma_y B_y + \sigma_z B_z)$$

$$= \sigma_x^2 A_x B_x + \sigma_y^2 A_y B_y + \sigma_z^2 A_z B_z$$

$$+ \sigma_x\sigma_y A_x B_y + \sigma_x\sigma_z A_x B_z + \sigma_y\sigma_x A_y B_x$$

$$+ \sigma_y\sigma_z A_y B_z + \sigma_z\sigma_x A_z B_x + \sigma_z\sigma_y A_z B_y$$

Using the following properties of Pauli spin matrices,

$$\sigma_x^2 = \sigma_y^2 = \sigma_z^2 = I$$

and

$$\sigma_x\sigma_y = -\sigma_y\sigma_x = i\sigma_z$$

$$\sigma_y\sigma_z = -\sigma_z\sigma_y = i\sigma_x$$

$$\sigma_z\sigma_x = -\sigma_x\sigma_z = i\sigma_y$$

we get

$$(\vec{\sigma}\,.\,\vec{A})(\vec{\sigma}\,.\,\vec{B}) = A_x B_x + A_y B_y + A_z B_z - \sigma_y\sigma_x (A_x B_y - A_y B_x)$$

$$-\sigma_x\sigma_z (A_z B_x - A_x B_z) - \sigma_z\sigma_y (A_y B_z - A_z B_y)$$

$$= \vec{A}\,.\,\vec{B} + i\sigma_z (A_x B_y - A_y B_x) + i\sigma_y (A_z B_x - A_x B_z)$$

$$+ i\sigma_x (A_y B_z - A_z B_y)$$

$$= (\vec{A}\,.\,\vec{B}) + i\vec{\sigma}\,.(\vec{A}\times\vec{B})$$

(*b*) Taking $\vec{A} = \vec{B} = \hat{n}$ in the above result, we get

$$(\vec{\sigma}\,.\,\hat{n})^2 = (\hat{n}\,.\,\hat{n}) + i\vec{\sigma}\,.\,(\hat{n}\times\hat{n}) = 1$$

8.7 The separation of the Zeeman components is given by

$$E' - E = h\,\Delta\upsilon = \left(\frac{e\hbar}{2m}\right)B$$

Since

$$\upsilon = \frac{c}{\lambda}, \Delta\upsilon = \left(\frac{-c}{\lambda^2}\right)\Delta\lambda$$

(the minus sign signifies the fact that if υ increases λ would decrease). Therefore,

$$\Delta\lambda = \lambda^2\frac{\Delta\upsilon}{c}$$

$$= e\,B\,\frac{\lambda^2}{4\,mc\pi}$$

$$= 0.0283\ \text{Å}$$

8.8 Using definition of the exponential of an operator

$$^{(1/2)}R_{\hat{n}}(\alpha) = 1 - i\,\alpha\,\vec{\sigma}\,.\,\frac{\hat{n}}{2} + \left(\frac{-i\,\alpha}{2}\right)^2 \frac{(\vec{\sigma}\,.\,\hat{n})^2}{2!} + \ldots$$

$$+\left(\frac{-i\,\alpha}{2}\right)^r \frac{(\vec{\sigma}\,.\,\hat{n})^r}{r!} + \ldots \qquad \ldots (8.24)$$

Applying the identity

$$(\vec{\sigma}\,.\,\vec{A})\,(\vec{\sigma}\,.\,\vec{B}) = (\vec{A}\,.\,\vec{B}) + i\,\vec{\sigma}\,.\,(\vec{A}\times\vec{B}),$$

we get

$$(\vec{\sigma}\,.\,\hat{n})\,(\vec{\sigma}\,.\,\hat{n}) = (\vec{\sigma}\,.\,\hat{n})^2 = 1,$$

which leads to

$$(\vec{\sigma}\,.\,\hat{n})^r = \begin{Bmatrix} 1 & \text{for } r \text{ even} \\ \vec{\sigma}\,.\,\hat{n} & \text{for } r \text{ odd} \end{Bmatrix}$$

Consequently, if we group together the even and the odd terms, the expression (8.24) can be written as

$$^{(1/2)}R_{\hat{n}}(\alpha) = \left[1 - \frac{(\alpha/2)^2}{2!} + \ldots + (-1)^p\,\frac{(\alpha/2)^{2p}}{(2p)!} + \ldots\right]$$

$$-i\,\vec{\sigma}\,.\,\hat{n}\left[\frac{\alpha}{2} - \frac{(\alpha/2)^3}{3!} + \ldots + (-1)^p\,\frac{(\alpha/2)^{2p+1}}{(2p+1)!} + \ldots\right]$$

$$= \cos\frac{\alpha}{2} - i\vec{\sigma}\,.\,\hat{n}\,\sin\frac{\alpha}{2}$$

8.9 $Y_l^{m_l}(\theta, \phi)$ being simultaneous eigen function of L^2 and L_z,

$$L^2\,Y_l^{m_l}(\theta,\phi) = l(l+1)\,\hbar^2\,Y_l^{m_l}(\theta,\phi)$$

and

$$L_z\,Y_l^{m_l}(\theta,\phi) = m_l\,\hbar\,Y_l^{m_l}(\theta,\phi)$$

Since $[L_x, L_z] \neq 0$, L_x (and also L_y) cannot be measured simultaneously and accurately. Therefore

$$< L_x > = \left\langle Y_l^{m_l}\middle|\frac{1}{2}(L_+ + L_-)\middle|Y_l^{m_l}\right\rangle$$

$$= \frac{1}{2}\langle Y_l^{m_l}|L_+|Y_l^{m_l}\rangle + \frac{1}{2}\langle Y_l^{m_l}|L_-|Y_l^{m_l}\rangle$$

Chapter 8

Using eqns. (8.22), (8.23) and orthonormality property of the spherical harmonics,

$$< L_x > = \frac{1}{2}[l(l+1) - m_l(m_l+1)]^{1/2}\hbar\left\langle Y_l^{m_l} \mid Y_l^{m_l+1}\right\rangle$$

$$+ \frac{1}{2}[l(l+1) - m_l(m_l-1)]^{1/2}\hbar\left\langle Y_l^{m_l} \mid Y_l^{m_l-1}\right\rangle$$

$$= 0 + 0 = 0$$

Similarly, one can show that

$$< L_y > = \langle (L_+ - L_-)/2i \rangle = 0$$

Now

$$\langle L_x^2 \rangle = \left\langle Y_l^{m_l}\left|\left\{\frac{1}{2}(L_+ + L_-)\right\}^2\right| Y_l^{m_l}\right\rangle$$

$$= \frac{1}{4}[\langle Y_l^{m_l}|L_+L_+|Y_l^{m_l}\rangle + \langle Y_l^{m_l}|L_+L_-|Y_l^{m_l}\rangle$$

$$+ \langle Y_l^{m_l}|L_-L_+|Y_l\rangle + \langle Y_l^{m_l} \mid L_-L_- \mid Y_l^{m_l}\rangle]$$

Again using eqns. (8.22), (8.23) and orthonormality property of the spherical harmonics.

$$\langle L_x^2 \rangle = \frac{1}{4}\Big[\{l(l+1) - m_l(m_l+1)\}^{1/2}\hbar\left\langle Y_l^{m_l}|L_+|Y_l^{m_l+1}\right\rangle$$

$$+ \{l(l+1) - m_l(m_l-1)^{1/2}\hbar\left\langle Y_l^{m_l}|L_+|Y_l^{m_l-1}\right\rangle$$

$$+ \{l(l+1) - m_l(m_l+1)\}^{1/2}\hbar\left\langle Y_l^{m_l}|L_-|Y_l^{m_l+1}\right\rangle$$

$$+ \{l(l+1) - m_l(m_l-1)\}^{1/2}\hbar\left\langle Y_l^{m_l}|L_-|Y_l^{m_l-1}\right\rangle\Big]$$

$$= \frac{1}{4}[\{l(l+1) - m_l(m_l+1)\}^{1/2}\{l(l+1)$$

$$- (m_l+1)(m_l+1+1)\}^{1/2}\hbar^2\left\langle Y_l^{m_l} \mid Y_l^{m_l+2}\right\rangle$$

$$+ \{l(l+1) - m_l(m_l+1)\}^{1/2}$$

$$\{l(l+1) - (m_l-1)(m_l-1+1)\}^{1/2}\hbar^2\langle Y_l^{m_l} \mid Y_l^{m_l}\rangle$$

$$+ \{l(l+1) - m_l(m_l-1)\}^{1/2}$$

$$\{l(l+1) - (m_l+1)(m_l+1-1)^{1/2}\hbar^2\langle Y_l^{m_l} \mid Y_l^{m_l}\rangle$$

$$+ \{l(l+1) - m_l(m_l-1)\}^{1/2}$$

$$\{\, l\,(l+1) - (m_l - 1)\,(m_l - 1 - 1)\}^{1/2}\, \hbar^2 \left\langle Y_l^{m_l} \mid Y_l^{m_l - 2} \right\rangle$$

$$= \frac{1}{4}\,\hbar^2 [0 + \{l\,(l+1) - m_l(m_l - 1)\}^{1/2} \{l\,(l+1) - m_l(m_l - 1)\}^{1/2}$$

$$+ \{l(l+1) - m_l(m_l + 1)\}^{1/2}\, \{l(l+1) - m_l(m_l + 1)\}^{1/2} + 0]$$

$$= \frac{1}{4}\,\hbar^2\,[\{l\,(l+1) - m_l(m_l - 1)\} + \{l\,(l+1) - m_l(m_l + 1)\}]$$

$$= \frac{1}{2}\,\hbar^2\,[l\,(l+1) - m_l^2]$$

Similarly it can be shown that

$$\left\langle L_y^2 \right\rangle = \frac{1}{2}\,\hbar^2\,[l\,(l+1) - m_l^2]$$

❑❑❑

Chapter 9

ANGULAR MOMENTUM: GENERAL

9.1 EIGEN VALUES (OR EIGEN VALUE SPECTRUM) OF J^2 AND J_Z

In the last chapter, we have shown that the components of quantum mechanical angular momentum operator $\vec{L}$ satisfy the following commutation relations (problem 8.1*a*)

$$
\begin{aligned}
[L_x, L_y] &= i\hbar L_z \\
[L_y, L_z] &= i\hbar L_x \qquad \text{... (1)} \\
[L_z, L_x] &= i\hbar L_y
\end{aligned}
$$

It was also shown (Problem 8.1*b*) that the operator $L^2 = L_x^2 + L_y^2 + L_z^2$ which is square of the orbital angular momentum operator $\vec{L}$ commutes with L_x, L_y and L_z *i.e.*,

$$[L^2, L_x] = [L^2, L_y] = [L^2, L_z] = 0 \qquad \text{... (2)}$$

We shall now adopt a more general point of view and define an angular momentum operator $\vec{J}$ (called general angular momentum operator) and would assume that the commutation relations (1) and (2) are also satisfied by $\vec{J}$, that is

$$
\begin{aligned}
[J_x, J_y] &= i\hbar J_z \\
[J_y, J_z] &= i\hbar J_x \qquad \text{... (3)} \\
[J_z, J_x] &= i\hbar J_y
\end{aligned}
$$

and

$$[J^2, J_x] = [J^2, J_y] = [J^2, J_z] = 0 \qquad \text{... (4)}$$

which imply that it is impossible to measure simultaneously and accurately the three components of the angular momentum. However, J^2 and any component of $\vec{J}$ are compatible (or can be measured simultaneously and accurately).

In the derivation to be presented here, we will use the Dirac notation for the sake of convenience. Since the operators J^2 and J_z commute, we can construct a complete set of orthonormal eigen kets to be denoted by $|\lambda m\rangle$ (In Schrödinger notation $|\lambda m\rangle \rightarrow \varphi_{\lambda m}$) which are simultaneous eigen kets of J^2 and J_z belonging to the eigen values $\lambda\hbar^2$ and $m\hbar$, respectively. Therefore

$$J^2\,|\lambda m\rangle = \lambda\hbar^2\,|\lambda m\rangle \qquad \text{... (5)}$$

$$J_z|\lambda m\rangle = m\,\hbar\,|\lambda m\rangle \qquad \text{... (6)}$$

Instead of using the components J_x and J_y of the angular momentum operator $\vec{J}$, it is more convenient to introduce the following linear combinations,

$$J_+ = J_x + i\,J_y \qquad \text{... (7)}$$

and

$$J_- = J_x - i\,J_y \qquad \text{... (8)}$$

the so called ladder operators. Note that the operators J_+ and J_- are not Hermitian ($J_+^\dagger \neq J_+$). Instead, they are adjoint of each other. That is,

$$J_+^\dagger = J_-,\; J_-^\dagger = J_+ \qquad \text{... (9)}$$

Using eqns. (3) and (7), one leads to

$$\begin{aligned}[J_z, J_+] &= [J_z, J_x + i\,J_y] \\ &= [J_z, J_x] + i\,[J_z, J_y] \\ &= i\,\hbar\,J_y + i\,(-i\,\hbar\,J_x) \\ &= \hbar\,J_+ \end{aligned} \qquad \text{... (10)}$$

Similarly, from eqns. (3) and (8),

$$[J_z, J_-] = -\,\hbar\,J_- \qquad \text{... (11)}$$

Let us calculate the products J_+J_- and J_-J_+. Using eqn. (3), we find

$$\begin{aligned}J_+J_- &= (J_x + i\,J_y)\,(J_x - i\,J_y) \\ &= J_x^2 + J_y^2 - i\,(J_x\,J_y - J_y\,J_x) \\ &= J_x^2 + J_y^2 + J_z^2 - J_z^2 + \hbar\,J_z \\ &= J^2 - J_z^2 + \hbar\,J_z \end{aligned} \qquad \text{... (12)}$$

and similarly

$$J_-J_+ = J^2 - J_z^2 - \hbar\,J_z \qquad \text{... (13)}$$

Adding eqns. (12) and (13), one gets

$$J^2 = \frac{1}{2}\,(J_+J_- + J_-J_+) + J_z^2 \qquad \text{... (14)}$$

Subtracting (13) from eqn. (12), one gets

$$[J_+, J_-] = 2\hbar J_z \qquad \text{... (15)}$$

Using eqn. (14), one can show that

$$[J^2, J_+] = \left[\frac{1}{2}(J_+J_- + J_-J_+) + J_z^2, J_+\right]$$

$$= \frac{1}{2}[J_+J_-, J_+] + \frac{1}{2}[J_-J_+, J_+] + [J_z^2, J_+]$$

which on the application of the identity

$$[AB, C] = A[B, C] + [A, C]B$$

and then using eqns. (15) and (10) gives

$$[J^2, J_+] = \frac{1}{2}J_+[J_-, J_+] + \frac{1}{2}[J_+, J_+]J_-$$

$$+ \frac{1}{2}J_-[J_+, J_+] + \frac{1}{2}[J_-, J_+]J_+$$

$$+ J_z[J_z, J_+] + [J_z, J_+]J_z$$

$$= \frac{1}{2}J_+(-2\hbar J_z) + 0 + 0 + \frac{1}{2}(-2\hbar J_z)J_+$$

$$+ J_z(\hbar J_+) + (\hbar J_+)J_z$$

$$= 0 \qquad \text{... (16)}$$

Similarly, we can show that

$$[J^2, J_-] = 0 \qquad \text{... (17)}$$

We shall first show that, $\lambda \geq m^2$. Using eqn. (9),

$$\langle \lambda m | J_+J_- + J_-J_+ | \lambda m\rangle = \langle \lambda m | J_+J_+^\dagger + J_+^\dagger J_+ | \lambda m\rangle$$

which is a non-negative quantity (if one operator is row matrix, the other would be a column matrix with complex elements, that is the products of the type aa^*, which are positive definite)

Using eqn. (14), the above result reduces to

$$2\langle \lambda m | J^2 - J_z^2 | \lambda m\rangle \geq 0$$

which further using eqns. (5) and (6), gives

$$\lambda - m^2 \geq 0 \text{ (or } \lambda \geq m^2) \qquad \text{... (18)}$$

We do not know the values which λ and m are permitted to take. Therefore, let us first determine what these are? To do so, we consider the effect of the ladder operators on the ket $|\lambda m\rangle$. Using eqns. (10) and (6), one gets

$$J_z(J_+|\lambda m\rangle) = (\hbar J_+ + J_+ J_z)|\lambda m\rangle$$

$$= (m+1)\,\hbar\,(J_+|\lambda m\rangle) \quad \text{... (19)}$$

Now using eqns. (16) and (17) and then eqn. (5), one gets

$$J^2 (J_\pm|\lambda m\rangle) = J_\pm J^2|\lambda m\rangle$$

$$= \lambda\hbar^2 (J_\pm|\lambda m\rangle) \quad \text{... (20)}$$

Eqns. (19) and (20) imply that $J_+ |\lambda m\rangle$ is a simultaneous eigen ket of the operators J^2 and J_z belonging to the eigen values $\lambda\hbar^2$ and $(m+1)\hbar$, provided $J_+|\lambda m\rangle$ is not a null ket. Similarly, using eqns. (11) and (6), we can show that

$$J_z(J_-|\lambda m\rangle) = (m-1)\,\hbar(J_-|\lambda m\rangle) \quad \text{... (21)}$$

Eqns. (20) and (21) imply that $J_-|\lambda m\rangle$ is a simultaneous eigen ket of J^2 and J_z belonging to the eigen values $\lambda\hbar^2$ and $(m-1)\,\hbar$, respectively, provided it is not a null ket.

Comparing eqns. (19) and (20) with eqns. (6) and (5), we conclude that $J_+|\lambda m\rangle$ is nothing but $|\lambda, m+1\rangle$ apart from a possible normalization constant. Similarly $J_-|\lambda m\rangle$ is proportional to $|\lambda, m-1\rangle$. Thus

$$J_+|\lambda m\rangle = C_+\,\hbar\,|\lambda, m+1\rangle \quad \text{... (22)}$$

and

$$J_-|\lambda m\rangle = C_-\,\hbar\,|\lambda, m-1\rangle \quad \text{... (23)}$$

where C_+ and C_- are the normalization constants to be determined. The above equations thus justify the name ladder operators for J_+ and J_-. J_+ is an operator that raises the eigen ket to the next higher value of m and thus is also termed as raising operator and similarly J_- is a lowering operator. In the similar manner we may argue to show that $J_+|\lambda, m+1\rangle$ and $J_-|\lambda, m-1\rangle$ are proportional to $|\lambda, m+2\rangle$ and $|\lambda, m-2\rangle$.

This, however, cannot go on indefinitely since for a given ket $|\lambda m\rangle$, the value of m is limited to $m^2 \le \lambda$. Hence there exists a maximum value of m, say $m = j$ for any given λ for which

$$J_+|\lambda\, j\rangle = 0 \quad \text{... (24)}$$

as $J_+\,|\lambda\, j\rangle$ would be proportional to $|\lambda, j+1\rangle$ and as λ does not change by the application of J_+.

Multiplying above from the left by J_- and then using eqn. (13), we get

$$(J^2 - J_z^2 - \hbar\, J_z)|\lambda\, j\rangle = 0$$

which using eqns. (5) and (6) and then cancelling $\hbar^2$ throughout, amounts to

$$(\lambda - j^2 - j)|\lambda\, j\rangle = 0$$

and since $|\lambda\, j\rangle \neq 0$ (that is not a null ket), the number

$$\lambda - j^2 - j = 0$$

or the eigen value of J^2 is connected to the maximum value of m through

$$\lambda = j\,(j+1) \qquad \text{... (25)}$$

The lowest value of m, say $m = j'$ can be determined by requiring that

$$J_-|\lambda\, j'\rangle = 0 \qquad \text{... (26)}$$

as $J_-|\lambda\, j'\rangle$ would be proportional to $|\lambda, j'-1\rangle$

Multiplying above from the left by J_+ and then using eqn. (2), finally leads to

$$\lambda - j'^2 + j' = 0$$

or

$$\lambda = j'\,(j'-1) \qquad \text{... (27)}$$

Equating (25) to (26), leads to $j' = -j$ or $j + 1$. The later solution is unacceptable as it makes the minimum value of $m = j'$ greater than the maximum. Thus the minimum value of m is $-j$. Therefore, if we start with the state $|\lambda\, j\rangle$ having the maximum value of j for m, and then reduce the value of m by unity at a time by repeated application of J_- on $|\lambda\, j\rangle$, we should in the end arrive at the minimum value $-j$ (the repeated application of J_+ on $|\lambda, -j\rangle$ increases the value of m by unity at a time and in the end, we arrive at the maximum value j). Thus for given λ (or j), the possible values of m are

$$m = -j,\ -(j-1),\ \ldots\ j-1,\ j \qquad \text{... (28)}$$

We have seen that the difference between any two values of m must be an integer. We, therefore arrive at the result that the difference between the maximum and the minimum values of m which is $2j$, must be a non-negative integer. Thus j is necessarily an integer or half integer (half integer means half of an odd integer) or the allowed (possible) values of j are determined to be

$$j = 0,\ \frac{1}{2},\ 1,\ \frac{3}{2},\ \ldots \qquad \text{... (29)}$$

Now for each value of j, the m values go from $-j$ to $+j$ in steps of unity. Thus there are $(2j+1)$ eigen kets corresponding to the different eigen values of J_z but to same eigen value $\lambda\hbar^2$ $[= j\,(j+1)\hbar^2]$ of J^2. We label these eigen kets of J^2 and J_z from now on as $|jm\rangle$ rather than by $|\lambda m\rangle$ as has been done so far. That is,

$$|\lambda m\rangle = |jm\rangle;\ m = -j,\ -(j-1),\ \ldots,\ j-1,\ j \qquad \text{... (30)}$$

The kets $|jm\rangle$, of course satisfy the orthonormality condition

$$\langle j'\, m' \mid jm\rangle = \delta_{jj'}\,\delta_{mm'} \qquad \text{... (31)}$$

Note that the eigen value spectrum has thus been obtained in a remarkable simple fashion starting from the commutation rules. Moreover, half integral values of j have automatically emerged from this general treatment. As we already know, only integral values are possibles in the case of orbital angular momentum and the half integral values when the spin is involved.

9.2 DETERMINATION OF THE CONSTANTS $C_{\pm}$ AND MATRIX ELEMENTS OF J_+ AND J_-

Starting from eqn. (22), let us define

$$|P\rangle = J_+ |jm\rangle = C_+ \hbar |j, m+1\rangle \qquad \text{... (32)}$$

then

$$\langle P| = \langle jm| J_+^\dagger = C_+^* \hbar \langle j, m+1|$$

and since $J_+^\dagger = J_-$,

$$\langle P| = \langle jm| J_- = C_+^* \hbar \langle j, m+1| \qquad \text{... (33)}$$

Thus taking the scalar product of $\langle P|$ with (32) and then using the orthonormality relation

$$\begin{aligned}\langle P \mid P\rangle &= \langle jm \mid J_- J_+ \mid jm\rangle \\ &= |C_+|^2 \hbar^2 \langle j, m+1 \mid j, m+1\rangle \\ &= |C_+|^2 \hbar^2\end{aligned} \qquad \text{... (34)}$$

But using eqns. (13), (5), (6) and finally (31), one gets

$$\begin{aligned}\langle jm| J_- J_+ |jm\rangle &= \langle jm| J^2 - J_z^2 - \hbar J_z | jm\rangle \\ &= (\lambda \hbar^2 - m^2 \hbar^2 - m \hbar^2) \langle jm \mid jm\rangle\end{aligned}$$

or

$$\langle jm| J_- J_+ |jm\rangle = (\lambda \hbar^2 - m^2 \hbar^2 - m\hbar^2) \qquad \text{... (35)}$$

Comparing eqns. (34) and (35), and then using eqn. (25), we get,

$$C_+ = [(j-m)(j+m+1)]^{1/2} \qquad \text{... (36)}$$

Putting above into eqn. (22),

$$J_+ |jm\rangle = [(j-m)(j+m+1)]^{1/2} \hbar \mid j, m+1\rangle \qquad \text{... (37)}$$

and therefore

$$\langle j, m+1| j_+ |jm\rangle = [(j-m)(j+m+1)]^{1/2} \hbar \qquad \text{... (38)}$$

In the similar manner, we can show that

$$C_- = [(j+m)(j-m+1)]^{1/2} \qquad \text{... (39)}$$

and hence from eqn. (23),

$$J_- |jm\rangle = [(j+m)(j-m+1)]^{1/2} \hbar |j, m-1\rangle \qquad \text{... (40)}$$

Thus

$$\langle j, m-1| J_- |jm\rangle = [(j+m)(j-m+1)]^{1/2} \hbar \qquad \text{... (41)}$$

All other matrix elements of J_+ and J_- vanish. The matrix elements of J_x and J_y follow from

$$J_x = \frac{1}{2}(J_+ + J_-),\ J_y = \frac{1}{2i}(J_+ - J_-) \quad \text{... (42)}$$

Note that we have considered the matrix representation of the components of the angular momentum operator in a basis (or coordinate system) in which J^2 and J_z are diagonal, *i.e.*, we have chosen as basis vectors (or states) the simultaneous eigen states of J^2 and J_z.

9.3 ANGULAR MOMENTUM MATRICES CORRESPONDING TO $j = 1/2$ IN THE $|jm\rangle$ BASIS

For $j = \frac{1}{2}$, we have $m = \frac{1}{2}$ and $-\frac{1}{2}$ as m varies from $-j$ to $+j$. Therefore, we have two basis vectors (or states) which we denote by $|1\rangle$ and $|2\rangle$.

$$|1\rangle = |j, m\rangle = \left|\frac{1}{2}, \frac{1}{2}\right\rangle \text{ and } |2\rangle = |j, m\rangle = \left|\frac{1}{2}, -\frac{1}{2}\right\rangle$$

Using eqns. (5), (6), (25) and (31), one gets

$$(J^2)_{11} = \langle 1|J^2|1\rangle = \left\langle \frac{1}{2}, \frac{1}{2}\right| J^2 \left|\frac{1}{2}, \frac{1}{2}\right\rangle$$

$$= \frac{1}{2} \times \left(\frac{1}{2} + 1\right)\hbar^2 \left\langle \frac{1}{2}, \frac{1}{2} \middle| \frac{1}{2}, \frac{1}{2}\right\rangle$$

$$= \frac{3}{4}\hbar^2$$

$$(J^2)_{22} = \langle 2|J^2|2\rangle = \left\langle \frac{1}{2}, -\frac{1}{2}\right| J^2 \left|\frac{1}{2}, -\frac{1}{2}\right\rangle$$

$$= \frac{3}{4}\hbar^2$$

and all other elements of J^2 are zero, J^2 being diagonal.

Similarly,

$$(J_z)_{11} = \langle 1|J_z|1\rangle = \left\langle \frac{1}{2}, \frac{1}{2}\right| J_z \left|\frac{1}{2}, \frac{1}{2}\right\rangle$$

$$= \frac{1}{2}\hbar$$

and

$$(J_z)_{22} = \langle 2|J_z|2\rangle = \left\langle \frac{1}{2}, -\frac{1}{2}\right| J_z \left|\frac{1}{2}, -\frac{1}{2}\right\rangle$$

$$= -\frac{1}{2}\hbar$$

All other matrix elements of J_z would be zero, J_z being diagonal. Let us now calculate the matrix elements of J_+ and J_- using eqns. (38) and (41), that is,

$$\langle j, m+1 | J_+ | jm \rangle = [(j-m)(j+m+1)]^{1/2}\, \hbar \qquad \text{... (38)}$$

and

$$\langle j, m-1 | J_- | jm \rangle = [(j+m)(j-m+1)]^{1/2}\, \hbar \qquad \text{... (41)}$$

we get,

$$(J_+)_{12} = \left\langle \frac{1}{2}, \frac{1}{2} | J_+ | \frac{1}{2}, -\frac{1}{2} \right\rangle$$

$$= \left[\left\{\frac{1}{2} - \left(-\frac{1}{2}\right)\right\}\left(\frac{1}{2} - \frac{1}{2} + 1\right)\right]^{1/2} \hbar$$

$$= \hbar$$

$$(J_+)_{11} = (J_+)_{22} = (J_+)_{21} = 0$$

$$(J_-)_{21} = \left\langle \frac{1}{2}, -\frac{1}{2} | J_- | \frac{1}{2}, \frac{1}{2} \right\rangle$$

$$= \left[\left(\frac{1}{2} + \frac{1}{2}\right)\left(\frac{1}{2} - \frac{1}{2} + 1\right)\right]^{1/2} \hbar$$

$$= \hbar$$

$$(J_-)_{11} = (J_-)_{22} = (J_-)_{12} = 0$$

Thereafter using the relations,

$$J_x = \frac{1}{2}(J_+ + J_-), \qquad \text{... (42)}$$

and

$$J_y = \frac{1}{2i}(J_+ - J_-)$$

one gets,

$$(J_x)_{11} = \frac{1}{2}[(J_+)_{11} + (J_-)_{11}] = 0 = (J_x)_{22}$$

$$(J_x)_{12} = \frac{1}{2}(\hbar + 0) = \frac{1}{2}\hbar = (J_x)_{21}$$

Similarly,

$$(J_y)_{11} = \frac{1}{2i}[(J_+)_{11} - (J_-)_{11}] = 0 = (J_y)_{22}$$

$$(J_y)_{12} = \frac{1}{2i}(\hbar - 0) = -i\,\hbar/2$$

$$(J_y)_{21} = \frac{1}{2i}(0 - \hbar) = i\hbar/2$$

Thus the angular momentum matrices for $j = \frac{1}{2}$ are

$$J_x = \frac{1}{2}\hbar\begin{pmatrix}0 & 1\\ 1 & 0\end{pmatrix}, J_y = \frac{1}{2}\hbar\begin{pmatrix}0 & -i\\ i & 0\end{pmatrix} \quad \text{... (43)}$$

$$J_z = \frac{1}{2}\hbar\begin{pmatrix}1 & 0\\ 0 & -1\end{pmatrix}, J^2 = \frac{3}{4}\hbar\begin{pmatrix}1 & 0\\ 0 & 1\end{pmatrix}$$

We can write

$$\vec{J} = \frac{1}{2}\hbar\vec{\sigma} \quad \text{... (44)}$$

where

$$\sigma_x = \begin{pmatrix}0 & 1\\ 1 & 0\end{pmatrix}, \sigma_y = \begin{pmatrix}0 & -i\\ i & 0\end{pmatrix} \text{ and } \sigma_z = \begin{pmatrix}1 & 0\\ 0 & -1\end{pmatrix}$$

are known as the Pauli spin matrices. Therefore, $j = \frac{1}{2}$ corresponds to the intrinsic angular momentum or spin of a particle (electron).

PROBLEMS

9.1 Find out the angular momentum matrices J^2, J_+, J_-, J_x, J_y and J_z corresponding to $j = 0$ in the representation in which J^2 and J_z are diagonal.

9.2 Find out the angular momentum matrices J^2, J_+, J_-, J_x, J_y and J_z corresponding to $j = 1$ in a representation in which J^2 and J_z are diagonal.

9.3 Find out J^2, J_+, J_-, and J_z matrices corresponding to $j = 3/2$ in a representation in which J^2 and J_z are diagonal.

9.4 For an eigen state of J^2 and J_z (or for a state for which J^2 and J_z are diagonal or are well defined), show that

$$\langle J_x \rangle = \langle J_y \rangle = 0$$

and

$$\langle J_x^2 \rangle = \langle J_y^2 \rangle = \frac{1}{2}\hbar\{j(j+1) - m^2\}$$

SOLUTIONS

9.1 For $j = 0$, m would have only one value zero. Therefore all the matrices, $J^2, J_z, J_+, J_-, J_x, J_y$, using the following equations,

$$\langle jm | J^2 | jm \rangle = j(j+1)\hbar^2$$

$$\langle jm | J_z | jm \rangle = m\hbar$$

$$\langle j, m+1 | J_+ | j\, m\rangle = [(j-m)(j+m+1)]^{1/2}\, \hbar$$

$$\langle j, m-1 | J_- | j\, m\rangle = [(j+m)(j-m+1)]^{1/2}\, \hbar$$

$$J_x = \frac{1}{2}(J_+ + J_-),\; J_y = \frac{1}{2i}[J_+ - J_-]$$

lead to

$$J^2 = J_+ = J_- = J_x = J_y = J_z = 0$$

9.2 For $j = 1$, we have $m = 1, 0, -1$ as m can take all the values from $-j$ to $+j$. The three basis states which are simultaneous eigen states of J^2 and J_z are

$$|1\rangle = |jm\rangle = |1, 1\rangle, |2\rangle = |1, 0\rangle \text{ and } |3\rangle = |1, -1\rangle$$

Using eqns. (5), (6), (25) and (31),

$$(J^2)_{11} = \langle 1 | J^2 | 1\rangle = \langle 1, 1 | J^2 | 1, 1\rangle = 1(1+1)\,\hbar^2 = 2\hbar^2 = (J^2)_{22} = (J^2)_{33}$$

$$(J_z)_{11} = \langle 1, 1 | J_z | 1, 1\rangle = 1\,\hbar$$

$$(J_z)_{22} = \langle 1, 0 | J_z | 1, 0\rangle = 0\,\hbar$$

$$(J_z)_{33} = \langle 1, -1 | J_z | 1, -1\rangle = -1\,\hbar$$

All other matrix elements (off diagonal elements) of J^2 and J_z are zero as both are diagonal in this representation.

Now using eqns. (38) and (41), that is

$$\langle j, m+1 | J_+ | jm\rangle = [(j-m)(j+m+1)]^{1/2}\, \hbar$$

and

$$\langle j, m-1 | J_- | jm\rangle = [(j+m)(j-m+1)]^{1/2}\, \hbar,$$

we get

$$(J_+)_{11} = \langle 1 | J_+ | 1\rangle = \langle 1, 1 | J_+ | 1, 1\rangle = 0 = (J_+)_{22} = (J_+)_{33} =$$

$$(J_+)_{13} = (J_+)_{21} = (J_+)_{31} = (J_+)_{32}$$

$$(J_+)_{12} = \langle 1, 1 | J_+ | 1, 0\rangle = [(1-0)(1+0+1)]^{1/2} = \sqrt{2}\,\hbar = (J_+)_{23}$$

Similarly,

$$(J_-)_{11} = (J_-)_{22} = (J_-)_{33} = (J_-)_{12} = (J_-)_{13} = (J_-)_{23} = (J_-)_{31} = 0$$

$$(J_-)_{21} = \langle 1, 0 | J_- | 1, 1\rangle = [(1+1)(1-1+1)]^{1/2}\hbar = \sqrt{2}\,\hbar = (J_-)_{32}$$

Now, using the relations,

$$J_x = \frac{1}{2}(J_+ + J_-) \quad \text{and} \quad J_y = \frac{1}{2i}(J_+ - J_-),$$

we get

$$(J_x)_{11} = \frac{1}{2}[(J_+)_{11} + (J_-)_{11}] = 0 = (J_x)_{13} = (J_x)_{22} = (J_x)_{31} = (J_x)_{33}$$

$$(J_x)_{12} = \frac{1}{2}[(J_+)_{12} + (J_-)_{12}] = \frac{1}{2}(\sqrt{2}\,\hbar + 0) = \sqrt{2}\,\hbar =$$

$$(J_x)_{21} = (J_x)_{23} = (J_x)_{32}$$

and

$$(J_y)_{11} = \frac{1}{2i}[(J_+)_{11} - (J_-)_{11}] = 0 = (J_y)_{22} = (J_y)_{33} = (J_y)_{43} = (J_y)_{31}$$

$$(J_y)_{12} = \frac{1}{2i}[(J_+)_{12} + (J_-)_{12}] = \frac{1}{2i}(\sqrt{2}\,\hbar - 0) = -i\hbar/\sqrt{2}$$

$$(J_y)_{21} = i\hbar/\sqrt{2}, (J_y)_{23} = -i\hbar/\sqrt{2}, (J_y)_{32} = i\hbar/\sqrt{2}$$

Thus,

$$J^2 = \hbar^2\begin{pmatrix}2 & 0 & 0\\ 0 & 2 & 0\\ 0 & 0 & 2\end{pmatrix}, J_+ = \sqrt{2}\,\hbar\begin{pmatrix}0 & 1 & 0\\ 0 & 0 & 1\\ 0 & 0 & 0\end{pmatrix},$$

$$J_- = \sqrt{2}\,\hbar\begin{pmatrix}0 & 0 & 0\\ 1 & 0 & 0\\ 0 & 1 & 0\end{pmatrix}, J_x = \hbar/\sqrt{2}\begin{pmatrix}0 & 1 & 0\\ 1 & 0 & 1\\ 0 & 1 & 0\end{pmatrix},$$

$$J_y = i\hbar/\sqrt{2}\begin{pmatrix}0 & -1 & 0\\ 1 & 0 & -1\\ 0 & 1 & 0\end{pmatrix}, J_z = \hbar\begin{pmatrix}1 & 0 & 0\\ 0 & 0 & 0\\ 0 & 0 & -1\end{pmatrix}$$

9.3 For $j = \frac{3}{2}$, we have $m = \frac{3}{2}, \frac{1}{2}, -\frac{1}{2}, -\frac{3}{2}$ as m can take all the values from $+j$ to $-j$.

The four basis states which are simultaneous eigen states of J^2 and J_z are

$$|1\rangle = \langle j, m\rangle = \left|\frac{3}{2}, \frac{3}{2}\right\rangle, |2\rangle = \left|\frac{3}{2}, \frac{1}{2}\right\rangle, |3\rangle$$

$$= \left|\frac{3}{2}, -\frac{1}{2}\right\rangle \text{ and } |4\rangle = \left|\frac{3}{2}, -\frac{3}{2}\right\rangle$$

Using eqns. (5), (6), (25) and (31),

$$(J^2)_{11} = \langle 1|J^2|1\rangle = \left\langle\frac{3}{2}, \frac{3}{2}\right|J^2\left|\frac{3}{2}, \frac{3}{2}\right\rangle = \frac{3}{2}\left(\frac{3}{2}+1\right)\hbar^2 = \frac{15}{4}\hbar^2$$

$$= (J^2)_{22} = (J^2)_{33} = (J^2)_{44}$$

$$(J_z)_{11} = \langle 1|J_z|1\rangle = \left\langle\frac{3}{2}, \frac{3}{2}\right|J_z\left|\frac{3}{2}, \frac{3}{2}\right\rangle = \frac{3}{2}\hbar$$

$$(J_z)_{22} = \frac{1}{2}\hbar, (J_z)_{33} = -\frac{1}{2}\hbar \text{ and } (J_z)_{44} = -\frac{3}{2}\hbar$$

All other matrix elements of J^2 and J_z would be zero as in this representation J^2 and J_z are diagonal.

Now using eqns. (38) and (41), that is

$$\langle j, m+1 | J_+ | j\, m\rangle = [(j-m)(j+m+1)]^{1/2}\,\hbar,$$

and

$$\langle j, m-1 | J_- | j\, m\rangle = [(j+m)(j-m+1)]^{1/2}\,\hbar,$$

we get

$$(J_+)_{11} = \langle 1|J_+|1\rangle = \left\langle \frac{3}{2}, \frac{3}{2}\,|J_+|\, \frac{3}{2}, \frac{3}{2}\right\rangle = 0 = (J_+)_{22} = (J_+)_{33} = (J_+)_{44}$$

$$(J_+)_{12} = \langle 1|J_+|2\rangle = \left\langle \frac{3}{2}, \frac{3}{2}\,|J_+|\, \frac{3}{2}, \frac{1}{2}\right\rangle$$

$$= \left[\left(\frac{3}{2} - \frac{1}{2}\right)\left(\frac{3}{2} + \frac{1}{2} + 1\right)\right]^{1/2}\hbar = \sqrt{3}\,\hbar$$

$$(J_+)_{13} = \langle 1|J_+|3\rangle = \left\langle \frac{3}{2}, \frac{3}{2}\,|J_+|\, \frac{3}{2}, -\frac{1}{2}\right\rangle = 0 = (J_+)_{14} = (J_+)_{21}$$

$$(J_+)_{23} = \langle 2|J_+|3\rangle = \left\langle \frac{3}{2}, \frac{1}{2}\,|J_+|\, \frac{3}{2}, -\frac{1}{2}\right\rangle$$

$$= \left[\left(\frac{3}{2} + \frac{1}{2}\right)\left(\frac{3}{2} - \frac{1}{2} + 1\right)\right]^{1/2}\hbar = 2\,\hbar$$

$$(J_+)_{31} = (J_+)_{32} = (J_+)_{41} = (J_+)_{42} = (J_+)_{43}$$

$$(J_+)_{34} = \langle 3|J_+|4\rangle = \left\langle \frac{3}{2}, -\frac{1}{2}\,|J_+|\, \frac{3}{2}, -\frac{3}{2}\right\rangle$$

$$= \left[\left(\frac{3}{2} + \frac{3}{2}\right)\left(\frac{3}{2} - \frac{3}{2} + 1\right)\right]^{1/2}\hbar = \sqrt{3}\,\hbar$$

Similarly, one can show that

$$(J_-)_{11} = (J_-)_{22} = (J_-)_{33} = (J_-)_{44} = 0$$

$$(J_-)_{12} = (J_-)_{13} = (J_-)_{14} = (J_-)_{31} = (J_-)_{34} = (J_-)_{41} = (J_-)_{42} = 0$$

$$(J_-)_{21} = \sqrt{3}\,\hbar, (J_-)_{32} = 2\,\hbar, (J_-)_{43} = \sqrt{3}\,\hbar$$

Thus

$$J^2 = \frac{15}{4}\hbar^2\begin{pmatrix}1&0&0&0\\0&1&0&0\\0&0&1&0\\0&0&0&1\end{pmatrix}, J_z = \hbar\begin{pmatrix}\frac{3}{2}&0&0&0\\0&\frac{1}{2}&0&0\\0&0&-\frac{1}{2}&0\\0&0&0&-\frac{3}{2}\end{pmatrix}$$

$$J_+ = \hbar\begin{pmatrix}0&\sqrt{3}&0&0\\0&0&2&0\\0&0&0&\sqrt{3}\\0&0&0&0\end{pmatrix}, J_- = \hbar\begin{pmatrix}0&0&0&0\\\sqrt{3}&0&0&0\\0&2&0&0\\0&0&\sqrt{3}&0\end{pmatrix}$$

9.4 The state $|jm\rangle$ being a simultaneous eigen state of J^2 and J_z;

$$J^2\,|jm\rangle = j(j+1)\,\hbar^2\,|jm\rangle$$

and

$$J_z\,|jm\rangle = m\,\hbar\,|jm\rangle$$

Since, $[J_x, J_y] \neq 0$ (that is, J_x and J_y do not commute), J_x and J_y cannot be measured simultaneously and accurately or will not be well defined. Therefore,

$$\langle J_x\rangle = \langle jm\,|\,J_x\,|\,jm\rangle$$

$$= \frac{1}{2}\langle jm\,|\,J_+|jm\rangle + \frac{1}{2}\langle jm\,|\,J_-\,|\,jm\rangle$$

Using eqns. (38) and (41),

$$\langle J_x\rangle = 0$$

Similarly, it can be shown that $\langle J_y\rangle = 0$

Now

$$\langle J_x^2\rangle = \left\langle jm\left|\left\{\frac{1}{2}(J_+ + J_-)\right\}^2\right|jm\right\rangle$$

$$= \frac{1}{4}[\langle jm|\ J_+\,J_+|jm\rangle + \langle jm\,|\,J_+\,J_-|jm\rangle$$

$$+ \langle jm\,|\,J_-\,J_+|jm\rangle + \langle jm\,|\,J_-\,J_-|jm\rangle$$

Using eqns. (31), (37) and (40),

$$\langle J_x^2\rangle = \frac{1}{4}[\{(j-m)\,(j+m+1)\}^{1/2}\,\hbar\,\langle jm|\,J_+|j, m+1\rangle$$

$$+ \{(j+m)\,(j-m+1)\}^{1/2}\,\hbar\,\langle jm\,|\,J_+|\,j, m-1\rangle$$

$$+ \{(j-m)\,(j+m+1)\}^{1/2}\,\hbar\,\langle jm\,|\,J_-|j, m+1\rangle$$

$$+ \{(j+m)(j-m+1)\}^{1/2}\, \hbar \langle jm | J_- | j, m-1 \rangle]$$

Now using eqns. (31), (38) and (41), one gets

$$\langle J_x^2 \rangle = \frac{1}{4} [0 + \{(j+m)(j-m+1)\}^{1/2}\, \hbar \{(j - \overline{m-1})(j + \overline{m-1} + 1)\}^{1/2} \hbar$$

$$+ \{(j-m)(j+m+1)\}^{1/2}\, \hbar \{(j + \overline{m+1})(j - \overline{m+1} + 1)\}^{1/2}\, \hbar + 0]$$

$$= \frac{1}{4} \hbar^2 [(j+m)(j-m+1) + (j-m)(j+m+1)]$$

$$= \frac{1}{4} \hbar^2 [j(j-m+1+j-m+1) + m(j-m+1-j-m-1)]$$

$$= \frac{1}{2} \hbar^2 [j(j+1) - m^2]$$

Similarly, it can be shown that

$$\langle J_y^2 \rangle = \frac{1}{2} \hbar^2 [j(j+1) - m^2]$$

□□□

Chapter 10

ADDITION OF ANGULAR MOMENTA

10.1 CLEBSCH–GORDAN COEFFICIENTS

Consider a physical system consisting of two non-interacting subsystems (particles). Let $\vec{J_1}$ and $\vec{J_2}$ denote the angular momentum operators corresponding to the subsystems 1 and 2, respectively. The total (or resultant) angular momentum operator $\vec{J}$ of the system under consideration is defined as

$$\vec{J} = \vec{J_1} + \vec{J_2} \qquad \text{... (1)}$$

because angular momenta are vectors and would be added vectorially. Note that if the system consists of one particle only, in that case $\vec{J_1}$ may represent the orbital angular momentum $\vec{L}$ and $\vec{J_2}$, the spin angular momentum $\vec{S}$ of the same particle.

The components of $\vec{J_1}$ and $\vec{J_2}$ satisfy the commutation relations (eqn. (3) of the last chapter),

$$[J_{1x}, J_{1y}] = i\hbar J_{1z} \qquad \text{... (2)}$$

$$[J_{2x}, J_{2y}] = i\hbar J_{2z}$$

In the similar fashion, the commutation relations for the other components of $\vec{J_1}$ and $\vec{J_2}$ can be written. Also we have

$$[J_1^2, J_{1x}] = 0 = [J_2^2, J_{2x}] \text{ etc.} \qquad \text{... (3)}$$

We also assume that the components of $\vec{J_1}$ and $\vec{J_2}$ (being independent of each other) commute with each other, *i.e.*,

$$[J_{1x}, J_{2x}] = 0 = [J_{1x}, J_{2y}] \text{ etc.} \qquad \text{... (4)}$$

Now using the relation,

$$\vec{J} \times \vec{J} = \begin{vmatrix} x & y & z \\ J_x & J_y & J_z \\ J_x & J_y & J_z \end{vmatrix}$$

we can show that

$$(\vec{J} \times \vec{J})_x = J_y J_z - J_z J_y$$

$$= (J_{1y} + J_{2y})(J_{1z} + J_{2z}) - (J_{1z} + J_{2z})(J_{1y} + J_{2y})$$

$$= (J_{1y} J_{1z} - J_{1z} J_{1y}) + (J_{2y} J_{2z} - J_{2z} J_{2y})$$

$$= [J_{1y}, J_{1z}] + [J_{2y}, J_{2z}]$$

$$= i\hbar J_{1x} + i\hbar J_{2x}$$

or $$[J_y, J_z] = i\hbar J_x \quad \text{... (5)}$$

(or the commutation relations in general can be written as $\vec{J} \times \vec{J} = i\hbar \vec{J}$)

The above commutation relation implies that the components of $\vec{J}$ satisfy the same commutation relations as that satisfied by an angular momentum operator. Therefore,

$$[J^2, J_z] = 0 \text{ etc.} \quad \text{... (6)}$$

where

$$J^2 = (\vec{J_1} + \vec{J_2}) \cdot (\vec{J_1} + \vec{J_2})$$

$$= J_1^2 + J_2^2 + 2\vec{J_1} \cdot \vec{J_2} \quad \text{... (7)}$$

in which we have used the relation $\vec{J_1} \cdot \vec{J_2} = \vec{J_2} \cdot \vec{J_1}$ as components of $\vec{J_1}$ commute with the components of $\vec{J_2}$.

Let $|j_1 m_1\rangle$ be the simultaneous eigen vectors of J_1^2 and J_{1z}, that is

$$J_1^2 |j_1 m_1\rangle = j_1 (j_1 + 1)\hbar^2 |j_1 m_1\rangle$$

and $$J_{1z} |j_1 m_1\rangle = m_1 \hbar |j_1 m_1\rangle \quad \text{... (8a)}$$

Similarly, let $|j_2 m_2\rangle$ represent the simultaneous eigen vectors of J_2^2 and J_{2z}, that is

$$J_2^2 |j_2 m_2\rangle = j_2 (j_2 + 1)\hbar^2 |j_2 m_2\rangle$$

and $$J_{2z} |j_2 m_2\rangle = m_2 \hbar |j_2 m_2\rangle \quad \text{... (8b)}$$

Now since the operators J_1^2, J_2^2, J_{1z} and J_{2z} commute with each other, $|j_1, j_2, m_1, m_2\rangle$ (commonly denoted by $\varphi(j_1, j_2, m_1, m_2) \equiv \varphi(m_1, m_2)$ as j_1 and j_2 are fixed and are called φ functions), therefore represent a set of simultaneous eigen vectors of J_1^2, J_2^2, J_{1z} and J_{2z}. Further since for the given values of j_1 and j_2, m_1 can take $(2j_1 + 1)$ values and m_2 can take $(2j_2 + 1)$ values respectively there will be $(2j + 1)(2j_2 + 1)$ possible eigen vectors.

Now since the operators J_1^2, J_2^2, J^2 and J_z also commute with each other, we can construct eigen vectors $|j_1 j_2, j, m\rangle$ (commonly denoted by $\phi(j_1, j_2, j, m) \equiv \phi(j, m)$) and called ϕ functions) which are simultaneous eigen vectors of the operators J_1^2, J_2^2, J^2 and J_z belonging to the eigen values $j_1(j_1 + 1)\hbar^2$, $j_2(j_2 + 1)\hbar^2$, $j(j + 1)\hbar^2$ and $m\hbar$, respectively.

Before we proceed further it must be noted that the set of states denoted by $|j_1, j_2, j, m\rangle$ or simply by $|jm\rangle$ (as j_1 and j_2 are fixed) and the set of states denoted by

$|j_1, j_2, m_1, m_2\rangle$ or simply by $|m_1, m_2\rangle$ are different bases for the same space—the space of states where in the two angular momenta have values j_1 and j_2, respectively. That is, the total number of ϕ function (for the given values of j_1 and j_2) will be same as the number of φ functions. Thus, we can choose either the φ functions or the ϕ functions as the base states and can represent each of the φ functions as a linear combination of ϕ functions and vice versa.

Let us now find out the possible values of j for the given values of j_1 and j_2. For doing so let us assume that $j_1 \geq j_2$. Since the largest values of m_1 and m_2 are the numbers j_1 and j_2, the maximum (or largest) value of j would be

$$j_{max} = j_1 + j_2 = m_{max} = (m_1 + m_2)_{max}$$

which is necessarily the largest value of m too. There will thus be only one such function which will correspond to $m = j_1 + j_2$ (or $m_1 = j_1, m_2 = j_2$),

$$\phi\,(j_1, j_2, j_1 + j_2, j_1 + j_2) = \varphi\,(j_1, j_2, j_1, j_2) \quad \text{... (9)}$$

The next lower (or possible) value of m would be $j_1 + j_2 - 1$ which will give rise to two $\varphi\,(m_1, m_2)$ vectors (provided neither j_1 nor j_2 is zero), one with $m_1 = j_1$, $m_2 = j_2 - 1$ and other $m_1 = j_1 - 1$, $m_2 = j_2$, that is

$$\varphi\,(j_1, j_2 - 1) \text{ and } \varphi\,(j_1 - 1, j_2) \quad \text{... (10)}$$

and two $\phi\,(j, m)$ vectors as m can take the value $j_1 + j_2 - 1$ when $j = j_1 + j_2$ and also when $j = j_1 + j_2 - 1$, that is

$$\phi\,(j_1 + j_2, j_1 + j_2 - 1) \text{ and } \phi\,(j_1 + j_2 - 1, j_1 + j_2 - 1) \quad \text{... (11)}$$

Further for $m = j_1 + j_2 - 2$, we have three φ vectors (or states),

$$\varphi\,(j_1 - 2, j_2), \varphi\,(j_1 - 1, j_2 - 1) \text{ and } \varphi\,(j_1, j_2 - 2) \quad \text{... (12)}$$

and three ϕ vectors (as m can take the value $j_1 + j_2 - 2$ when $j = j_1 + j_2$, when $j = j_1 + j_2 - 1$ and also when $j = j_1 + j_2 - 2$)

$$\phi\,(j_1 + j_2, j_1 + j_2 - 2), \phi\,(j_1 + j_2 - 1, j_1 + j_2 - 2)$$
$$\text{and } \phi\,(j_1 + j_2 - 2, j_1 + j_2 - 2) \quad \text{... (13)}$$

In general for $m = j_1 + j_2 - k$, we will have following $(k + 1)$, φ vectors

$$\varphi\,(j_1 - k, j_2), \varphi\,(j_1 - k - 1, j_2 - 1), \ldots, \varphi\,(j_1, j_2 - k) \quad \text{... (14)}$$

and the following ϕ vectors

$$\phi\,(j_1 + j_2, j_1 + j_2 - k), \phi\,(j_1 + j_2 - 1, j_1 + j_2 - k), \ldots,$$
$$\phi\,(j_1 + j_2 - k, j_1 + j_2 - k) \quad \text{... (15)}$$

Since $j_1 \geq j_2$ and the smallest value of m_2 is $-j_2$, the largest value of k from eqn. (14) would correspond to

$$j_2 - k_{max} = -j_2$$

or

$$k_{max} = 2\,j_2$$

Thus the smallest value of j from eqn. (15) would be

$$j_{min} = j_1 + j_2 - k_{max} = j_1 + j_2 - 2j_2 = j_1 - j_2 \quad \text{... (16)}$$

Hence allowed values of j are

$$j_1 + j_2, j_1 + j_2 - 1, \ldots, |j_1 - j_2| \quad \text{... (17)}$$

Therefore, either all three quantum numbers j_1, j_2 and j are integers or two of them are half integers and one is an integer. Since for each value of j there would be $(2j + 1)$ values of m, the total number of ϕ functions or the total number of independent vectors $|jm\rangle$ for given j_1 and j_2 or the total number of eigen states would be

$$\sum_{j = j_{min} = |j_1 - j_2|}^{j = j_{max} = j_1 + j_2} (2j + 1) \quad \text{... (18)}$$

To calculate the above sum, put $j = j_1 - j_2 + i$. The sum then reduces to

$$\begin{aligned}\sum_{i=0}^{2j_2} [2(j_1 - j_2 + i) + 1] &= [2(j_1 - j_2) + 1] \sum_{i=0}^{2j_2} 1 + \sum_{i=0}^{2j_2} 2i \\ &= [2(j_1 - j_2) + 1](2j_2 + 1) + 2 \times 2j_2 (2j_2 + 1)/2 \\ &= (2j_1 + 1)(2j_2 + 1) \quad \text{... (19)}\end{aligned}$$

which as expected is same as the number of φ functions.

It may, however be mentioned that the representation defined by the vectors $|j_1 j_2 m_1 m_2\rangle$ (or $|m_1 m_2\rangle$) is called the uncoupled representation while the vectors $|j_1 j_2 jm\rangle$ (or $|jm\rangle$) define the coupled representation.

Now since $|jm\rangle$ and $|m_1 m_2\rangle$ are merely different bases in the same Hilbert space, they should be related by unitary transformation, Thus

$$|jm\rangle = \sum_{m_1 m_2} U^{j_1 j_2}_{m_1 m_2 : jm} |m_1 m_2\rangle \quad \text{... (20)}$$

where $U^{j_1 j_2}_{i:j}$ is the ijth element of the unitary matrix $U^{j_1 j_2}$ that transforms the basis $|m_1 m_2\rangle$ to the basis $|jm\rangle$. Note that the summation is to be carried out only over m_1 and m_2 as j_1 and j_2 are assumed to have fixed values.

Using the closure (or completeness) property of the basis $|m_1 m_2\rangle$ (that is eqns. (16) and (17) of sec 6.5), we can write

$$|j_1 j_2 jm\rangle = \sum_{m_1 = -j_1}^{+j_1} \sum_{m_2 = -j_2}^{+j_2} \langle j_1 j_2 m_1 m_2 | j_1 j_2 jm | j_1 j_2 m_1 m_2\rangle$$

or in the present notation where we have omitted j_1 and j_2, the above equation reads

$$|jm\rangle = \sum_{m_1 m_2} \langle m_1 m_2 | jm\rangle |m_1 m_2\rangle \quad \text{... (21)}$$

On comparing eqns. (18) and (19), we see that

$$U^{j_1 j_2}_{m_1 m_2 : jm} \equiv \langle j_1\, j_2\, m_1\, m_2 \mid j_1\, j_2\, jm\rangle$$

$$\equiv \langle m_1\, m_2 \mid jm\rangle \qquad \text{... (22)}$$

The above matrix elements are denoted by $C^{j_1 j_2 j}_{m_1 m_2 m}$ and are called Clebsch – Gordan, C, Wigner, vector coupling or vector addition coefficients. Eqn. (20) thus becomes

$$|\, jm\rangle = \sum_{m_1 m_2} C^{j_1 j_2 j}_{m_1 m_2 m} \,|\, m_1\, m_2\rangle \qquad \text{... (23)}$$

The above equation could be regarded as the defining equation for the C-coefficients. The problem of adding angular momenta is thus equivalent to the problem of finding C-coefficients.

Applying the operator $J_z = J_{1z} + J_{2z}$ on both the sides of the above equation and then using the eigen value equations,

$$J_z \,|\, jm\rangle = m\,\hbar \,|\, jm\rangle$$

and

$$(J_{1z} + J_{2z}) \,|\, m_1\, m_2\rangle = (m_1 + m_2)\, \hbar \,|\, m_1\, m_2\rangle,$$

we get

$$J_z \,|\, jm\rangle = (J_{1z} + J_{2z}) \,|\, jm\rangle = \sum_{m_1 m_2} C^{j_1 j_2 j}_{m_1 m_2 m} (J_{1z} + J_{2z}) \,|\, m_1\, m_2\rangle$$

or

$$m\,\hbar \sum_{m_1 m_2} C^{j_1 j_2 j}_{m_1 m_2 m} \,|\, m_1\, m_2\rangle = \sum_{m_1 m_2} (m_1 + m_2)\, \hbar\, C^{j_1 j_2 j}_{m_1 m_2 m} \,|\, m_1\, m_2\rangle$$

or

$$\sum_{m_1 m_2} (m - m_1 - m_2)\, C^{j_1 j_2 j}_{m_1 m_2 m} \,|\, m_1 m_2\rangle = 0$$

Since the vectors $|\, m_1\, m_2\rangle$ are linearly independent, in fact orthogonal, the above condition implies that

$$C^{j_1 j_2 j}_{m_1 m_2 m} \neq 0,\ m = m_1 + m_2 \qquad \text{... (24)}$$

The above condition is referred to as the selection rule. Due to this the double sum in eqn. (23) reduces to the single sum:

$$|\, jm\rangle = \sum_{m_1} \langle m_1,\, m - m_1 \,|\, jm\rangle \,|\, m_1,\, m - m_1\rangle \qquad \text{... (25)}$$

10.2 CLEBSCH—GORDAN COEFFICIENTS FOR $j_1 = j_2 = 1/2$ – AN EXAMPLE

For $j_1 = j_2 = \frac{1}{2}$, m_1 and m_2 both can take the values $+\frac{1}{2}$ and $-\frac{1}{2}$. Eventually one gets four $\varphi\,(m_1\, m_2)$ functions,

$$\varphi\left(\frac{1}{2}, \frac{1}{2}\right), \varphi\left(\frac{1}{2}, -\frac{1}{2}\right), \varphi\left(-\frac{1}{2}, \frac{1}{2}\right) \text{ and } \varphi\left(-\frac{1}{2}, -\frac{1}{2}\right)$$

j in this case can take all the values from $j_1 + j_2$ to $|j_1 - j_2|$, that is 1 and 0 and hence three $\phi\ (jm)$ functions corresponding to $j = 1$ and one corresponding to $j = 0$, that is

$$\phi\ (1, 1), \phi\ (1, 0), \phi\ (1, -1) \text{ and } \phi\ (0, 0)$$

Now

$$\phi\ (1, 1) = \varphi\left(\frac{1}{2}, \frac{1}{2}\right) \quad \text{... (26)}$$

as $\varphi\left(\frac{1}{2}, \frac{1}{2}\right)$ is the only state for which $m_1 + m_2 = m = 1$ and similarly

$$\phi\ (1, -1) = \varphi\left(-\frac{1}{2}, -\frac{1}{2}\right) \quad \text{... (27)}$$

as $\varphi\left(-\frac{1}{2}, -\frac{1}{2}\right)$ is the only state for which $m_1 + m_2 = m = -1$.

The remaining states $\phi\ (1, 0)$ and $\phi\ (0, 0)$ are given by

$$\phi\ (1, 0) = c_1\ \varphi\left(\frac{1}{2}, -\frac{1}{2}\right) + c_2 \varphi\left(-\frac{1}{2}, \frac{1}{2}\right)$$

and

$$\phi\ (0, 0) = c_3\ \varphi\left(\frac{1}{2}, -\frac{1}{2}\right) + c_4 \varphi\left(-\frac{1}{2}, \frac{1}{2}\right)$$

as ϕ functions can be represented as linear combinations of φ functions (and vice versa). Also as only the above two φ functions can give $m_1 + m_2 = m = 0$.

It may, however be noted that in setting the above equations we have effectively used the fact that all those coefficients for which $m \neq m_1 + m_2$ are zero.

Now using the known relation,

$$J_- \phi(jm) = [(j + m)\ (j - m + 1)]^{1/2}\ \phi\ (j, m - 1)$$

one gets

$$J_- \phi(1, 1) = \sqrt{2}\ \phi(1, 0)$$

where we have taken $j = 1$, $m = 1$. And hence

$$\phi\ (1, 0) = \frac{1}{\sqrt{2}} J_- \phi(1, 1)$$

Use of eqn. (26) gives,

$$\phi\ (1, 0) = \frac{1}{\sqrt{2}} (J_{1-} + J_{2-})\ \varphi\left(\frac{1}{2}, \frac{1}{2}\right)$$

$$= \frac{1}{\sqrt{2}} J_{1-} \varphi\left(\frac{1}{2}, \frac{1}{2}\right) + \frac{1}{\sqrt{2}} J_{2-} \varphi\left(\frac{1}{2}, \frac{1}{2}\right)$$

Once again using the relation,

$$J_-|jm\rangle = [(j+m)(j-m+1)]^{1/2}|j, m-1\rangle$$

and the fact that in one case when J_{1-} operates on φ, m_1 decreases by 1 and in the second case when J_{2-} operates on φ, m_2 decreases by 1. Also using the values of $j = j_1 = \frac{1}{2}$, $m = m_1 = \frac{1}{2}$ in one case and $j = j_2 = \frac{1}{2}$, $m = m_2 = \frac{1}{2}$ in the other case, we get

$$\phi(1, 0) = \frac{1}{\sqrt{2}}\left[\varphi\left(\frac{1}{2}, -\frac{1}{2}\right) + \varphi\left(-\frac{1}{2}, \frac{1}{2}\right)\right] \quad \text{... (28)}$$

For ϕ (1, 0), to be orthogonal to ϕ (0, 0) {eqn. (31) of sec. 9.1 that is, $\langle j'm' \mid jm\rangle = \delta_{jj'}$ $\delta mm'$},

$$\int \phi^*(1, 0)\, \phi(0, 0)\, d\tau = 0$$

or

$$\int \frac{1}{\sqrt{2}}\left[\varphi\left(\frac{1}{2}, -\frac{1}{2}\right) + \varphi\left(-\frac{1}{2}, \frac{1}{2}\right)\right]^* \left[c_3\varphi\left(\frac{1}{2}, -\frac{1}{2}\right) + c_4\,\varphi\left(-\frac{1}{2}, \frac{1}{2}\right)\right] d\tau = 0$$

Using the orthonormality property of φ functions, we get

$$\frac{1}{\sqrt{2}}(c_3 + c_4) = 0 \quad \text{or} \quad c_4 = -c_3$$

Thus

$$\phi(0, 0) = c_3\left[\varphi\left(\frac{1}{2}, -\frac{1}{2}\right) - \varphi\left(-\frac{1}{2}, \frac{1}{2}\right)\right]$$

Further using the orthonormality property of ϕ (0, 0) and then using, the orthonormality of φ functions, we get

$$c_3^2 + c_3^2 = 2\,c_3^2 = 1$$

or

$$c_3 = \frac{1}{\sqrt{2}}$$

Thus

$$\phi(0, 0) = \frac{1}{\sqrt{2}}\left[\varphi\left(\frac{1}{2}, -\frac{1}{2}\right) - \varphi\left(-\frac{1}{2}, \frac{1}{2}\right)\right] \quad \text{... (29)}$$

Note that we can not get ϕ (0, 0) by using ladder operators on ϕ (1, 0) as J_+ would lead to ϕ (1, 1) and J_- to ϕ (1, -1) and similarly on ϕ (1, –1) as J_+ would lead ϕ (1, 0) and J_- to ϕ (1, –2) = 0. We can thus write, the transformation from uncoupled to coupled representation by the matrix equation,

$$\begin{pmatrix} \phi(1,1) \\ \phi(1,0) \\ \phi(1,-1) \\ \phi(0,0) \end{pmatrix} = \begin{pmatrix} 1 & 0 & 0 & 0 \\ 0 & \frac{1}{\sqrt{2}} & \frac{1}{\sqrt{2}} & 0 \\ 0 & 0 & 0 & 1 \\ 0 & \frac{1}{\sqrt{2}} & -\frac{1}{\sqrt{2}} & 0 \end{pmatrix} \begin{pmatrix} \varphi\left(\frac{1}{2},\frac{1}{2}\right) \\ \varphi\left(\frac{1}{2},-\frac{1}{2}\right) \\ \varphi\left(-\frac{1}{2},\frac{1}{2}\right) \\ \varphi\left(-\frac{1}{2},-\frac{1}{2}\right) \end{pmatrix}$$

The matrix elements are the Clebsch-Gordan coefficients. It may be noted that the case discussed above is equivalent to a system of two spin $-\frac{1}{2}$ particles (electrons or silver atoms in the ground state-for example). Thus $j_1 = s_1 = \frac{1}{2}$ and $j_2 = s_2 = \frac{1}{2}$. The total spin s of the system thus can take the values 1 and 0. Corresponding to $s = 1$, m can take three values 1, 0 and –1. And corresponding to $s = 0$, m can take the value 0 only. The values $m = 1$ and –1 are not degenerate. Only one eigen vector corresponds to each of them: both spins up for the first one $\left(m_1 = \frac{1}{2}, m_2 = \frac{1}{2}\right)$ and both spins down for the second one $\left(m_1 = -\frac{1}{2}, m_2 = -\frac{1}{2}\right)$. on the other hand, m = 0 is two fold degenerate: two orthogonal eigen vectors $\left|\frac{1}{2}, -\frac{1}{2}\right\rangle$ and $\left|-\frac{1}{2}, \frac{1}{2}\right\rangle$ are associated with it. Any linear combination of these two eigen vectors is an eigen state of S_z, with eigen value 0. Corresponding to $s = 0$, there is only one state ϕ (0, 0) and is therefore known as singlet state. For $s = 1$, there are three states, $\phi(1, 1)$ $\phi(1, 0)$ and $\phi(1, -1)$ and $s = 1$ constitutes a triplet state. It may also be noted that the triplet states are symmetric with respect to exchange of two spins (see expressions (26), (27) and (28) whereas the singlet state is antisymmetric (see expression (29) by exchanging m_1 and m_2).

10.3 THE CONCEPT OF ISOSPIN

We call a nuclear constituent, neutron as well as proton, a nucleon. The proton is positively charged while the neutron has no charge. The mass of proton is almost equal to that of neutron. These two particles are, therefore regarded as two charge states of the nucleon. The quantum number which characterizes the charge state is known as isospin (or isotopic spin or isobaric spin) and is denoted by $\tau = \frac{1}{2}$ so that the neutron and proton correspond to the two possible values of τ_z (z-component of τ), namely $\tau_z = \pm\frac{1}{2}$ (one may see the analog with the spin angular momentum $S = \frac{\hbar}{2}$ where $S_z = +\frac{\hbar}{2}$ and $-\frac{\hbar}{2}$ correspond to spin up and spin down states, respectively). The charge e is related to τ_z as $e = \frac{1}{2} - \tau_z$.

Note that the above case is similar to $j_1 = \frac{1}{2}$ and $j_2 = \frac{1}{2}$. Assuming that the entire theory of angular momenta (commutation rules, addition) can be applied to isospin also, though physically isospin and angular momentum are quite different, we have four possible charge states.

$$\phi(1, 1) = \varphi\left(\frac{1}{2}, \frac{1}{2}\right) \qquad \text{... (30)}$$

$$\phi(1, 0) = \frac{1}{\sqrt{2}}\left[\varphi\left(\frac{1}{2}, -\frac{1}{2}\right) + \varphi\left(-\frac{1}{2}, \frac{1}{2}\right)\right] \qquad \text{... (31)}$$

$$\phi(1, -1) = \varphi\left(-\frac{1}{2}, -\frac{1}{2}\right) \qquad \text{... (32)}$$

$$\phi(0, 0) = \frac{1}{\sqrt{2}}\left[\varphi\left(\frac{1}{2}, -\frac{1}{2}\right) - \varphi\left(-\frac{1}{2}, \frac{1}{2}\right)\right] \qquad \text{... (33)}$$

Clearly $\phi(1, 1) = \varphi\left(\frac{1}{2}, \frac{1}{2}\right)$ corresponds to two neutrons as $\tau_{1z} = \tau_{2z} = \frac{1}{2}$ and similarly $\phi(1, -1) = \varphi\left(-\frac{1}{2}, -\frac{1}{2}\right)$ corresponds to two protons; both $\phi(1, 0)$ and $\phi(0, 0)$ $\left\{\tau_{1z} + \tau_{2z} = \frac{1}{2} - \frac{1}{2} = 0\right\}$ correspond to one neutron and one proton. However $\phi(1, 0)$ is symmetric function (see expression (31)) and $\phi(0, 0)$ is antisymmetric (see expression (33)) with respect to the exchange of τ_{1z} and τ_{2z}.

PROBLEMS

10.1 Calculate from the first principle, the Clebsch-Gordan coefficients for $j_1 = 1$ and $j_2 = \frac{1}{2}$. Use the calculated coefficients to write the wave functions for $2p$—states ($j_1 = l = 1$, $j_2 = s = \frac{1}{2}$) of an electron in hydrogen atom.

10.2 Calculate the possible values of the quantum numbers j and m for hydrogen atom in the ground state and as well as in the 2p-states.

10.3 Calculate from the first principle, the Clebsch-Gordan coefficients for $j_1 = 1$ and $j_2 = 1$. The case corresponds to a two particle system in which both orbital angular momenta are equal to 1. Since each of the two particles is then in a p-state, this is said to be a p^2 configuration.

SOLUTIONS

10.1 Corresponding to $j_1 = 1$, m_1 can take the values 1, 0 and –1. Similarly, corresponding to $j_2 = \frac{1}{2}$, m_2 can take the values $\frac{1}{2}$ and $-\frac{1}{2}$. The possible values of j would therefore be $\frac{3}{2}$ and $\frac{1}{2}$ (all values from $j_1 + j_2$ to $|j_1 - j_2|$).

Thus the following six $\varphi\ (m_1, m_2)$ functions would be possible,

$$\varphi\left(1,\frac{1}{2}\right), \varphi\left(1,-\frac{1}{2}\right), \varphi\left(0,\frac{1}{2}\right), \varphi\left(0,-\frac{1}{2}\right), \varphi\left(-1,\frac{1}{2}\right)$$

$$\text{and } \varphi\left(-1,-\frac{1}{2}\right)$$

Since the allowed values of m in this case are (all values from $+j$ to $-j$) $\frac{3}{2}, \frac{1}{2}, -\frac{1}{2}, -\frac{3}{2}$, and $\frac{1}{2}, -\frac{1}{2}$, the possible six $\phi\ (j, m)$ functions would be

$\phi\left(\frac{3}{2},\frac{3}{2}\right), \phi\left(\frac{3}{2},\frac{1}{2}\right), \phi\left(\frac{3}{2},-\frac{1}{2}\right)$ and $\phi\left(\frac{3}{2},-\frac{3}{2}\right)$ corresponding to $j = \frac{3}{2}$ and $\phi\left(\frac{1}{2},\frac{1}{2}\right)$ and $\phi\left(\frac{1}{2},-\frac{1}{2}\right)$ corresponding to $j = \frac{1}{2}$.

Now

$$\phi\left(\frac{3}{2},\frac{3}{2}\right) = \varphi\left(1,\frac{1}{2}\right) \qquad \text{... (10.1)}$$

as $\varphi\left(1,\frac{1}{2}\right)$ is the only state for which $m = m_1 + m_2 = \frac{3}{2}$ and similarly,

$$\phi\left(\frac{3}{2},-\frac{3}{2}\right) = \varphi\left(-1,-\frac{1}{2}\right) \qquad \text{... (10.2)}$$

as $\varphi\left(-1,-\frac{1}{2}\right)$ is the only state for which $m = m_1 + m_2 = -\frac{3}{2}$.

Now $\phi\left(\frac{1}{2},\frac{1}{2}\right)$ can be written as linear combination of $\varphi\left(1,-\frac{1}{2}\right)$ and $\varphi\left(0,\frac{1}{2}\right)$ as only these φ functions can give $m = m_1 + m_2 = \frac{1}{2}$. Thus

$$\phi\left(\frac{1}{2},\frac{1}{2}\right) = c_1\,\varphi\left(1,-\frac{1}{2}\right) + c_2\varphi\left(0,\frac{1}{2}\right) \qquad \text{... (10.3)}$$

Similarly the remaining states can be written as

$$\phi\left(\frac{3}{2},\frac{1}{2}\right) = c_3\,\varphi\left(1,-\frac{1}{2}\right) + c_4\,\varphi\left(0,\frac{1}{2}\right) \qquad \text{... (10.4)}$$

$$\phi\left(\frac{3}{2},-\frac{1}{2}\right) = c_5\,\varphi\left(0,-\frac{1}{2}\right) + c_6\varphi\left(-1,\frac{1}{2}\right) \qquad \text{... (10.5)}$$

and

$$\phi\left(\frac{1}{2}, -\frac{1}{2}\right) = c_7\varphi\left(0, -\frac{1}{2}\right) + c_8\varphi\left(-1, \frac{1}{2}\right) \quad \text{... (10.6)}$$

Now using the relation,

$$J_-\,\phi(jm) = [(j+m)\,(j-m+1)]^{1/2}\,\phi\,(j, m-1),$$

one gets

$$J_-\,\phi\left(\frac{3}{2}, \frac{3}{2}\right) = \sqrt{3}\,\phi\left(\frac{3}{2}, \frac{1}{2}\right)$$

where we have used $j = \frac{3}{2}$, $m = \frac{3}{2}$.

From above,

$$\phi\left(\frac{3}{2}, \frac{1}{2}\right) = \frac{1}{\sqrt{3}}\,J_-\,\phi\left(\frac{3}{2}, \frac{3}{2}\right)$$

Further using eqn. (10.1), we get

$$\phi\left(\frac{3}{2}, \frac{1}{2}\right) = \frac{1}{\sqrt{3}}\,J_-\,\varphi\left(1, \frac{1}{2}\right)$$

$$= \frac{1}{\sqrt{3}}\,(J_{1-} + J_{2-})\,\varphi\left(1, \frac{1}{2}\right)$$

When J_{1-} operates on $\varphi\left(1, \frac{1}{2}\right)$, m_1 decreases by 1 and $j = j_1 = 1$, $m = m_1 = 1$ in this case. Similarly when J_{2-} operates on $\varphi\left(1, \frac{1}{2}\right)$, m_2 decreases by 1 and $j = j_2 = \frac{1}{2}$, $m = m_2 = \frac{1}{2}$ in this case. Note that the relation

$$J_-\,|\,jm\rangle = [(j+m)\,(j-m+1)]^{1/2}\,|\,j, m-1\rangle$$

has been used once again. Therefore

$$\phi\left(\frac{3}{2}, \frac{1}{2}\right) = \frac{1}{\sqrt{3}}\left[\sqrt{2}\,\varphi\left(0, \frac{1}{2}\right) + \varphi\left(1, -\frac{1}{2}\right)\right]$$

$$= \frac{1}{\sqrt{3}}\,\varphi\left(1, -\frac{1}{2}\right) + \sqrt{\frac{2}{3}}\,\varphi\left(0, \frac{1}{2}\right) \quad \text{... (10.7)}$$

Now using the relation,

$$J_+\,\phi(jm) = [(j-m)\,(j+m+1)]^{1/2}\,\phi(j, m+1),$$

one gets

$$J_+\,\phi\left(\frac{3}{2}, -\frac{3}{2}\right) = \sqrt{3}\,\phi\left(\frac{3}{2}, -\frac{1}{2}\right)$$

or

$$\phi\left(\frac{3}{2}, -\frac{1}{2}\right) = \frac{1}{\sqrt{3}} J_+ \phi\left(\frac{3}{2}, -\frac{3}{2}\right)$$

Further using eqn. (10.2), we get

$$\phi\left(\frac{3}{2}, -\frac{1}{2}\right) = \frac{1}{\sqrt{3}} (J_{1+} + J_{2+})\, \varphi\left(-1, -\frac{1}{2}\right)$$

Since

$$J_{1+}\varphi\left(-1, -\frac{1}{2}\right) = \sqrt{2}\, \varphi\left(0, -\frac{1}{2}\right) \text{ as } j = j_1 = 1 \text{ and } m = m_1 = -\frac{1}{2}$$

in this case and also

$$J_{2+}\varphi\left(-1, -\frac{1}{2}\right) = \varphi\left(-1, \frac{1}{2}\right) \text{ as } j = j_2 = \frac{1}{2},\ m = m_2 = -\frac{1}{2}$$

in this case, one finds

$$\phi\left(\frac{3}{2}, -\frac{1}{2}\right) = \sqrt{\frac{2}{3}}\, \varphi\left(0, -\frac{1}{2}\right) + \frac{1}{\sqrt{3}}\, \varphi\left(-1, \frac{1}{2}\right) \quad \text{... (10.8)}$$

For $\phi\left(\frac{1}{2}, \frac{1}{2}\right)$ to be orthogonal to $\phi\left(\frac{3}{2}, \frac{1}{2}\right)$, the use of eqns. (10.3) and (10.7) leads to (of course using the orthonormality property of φ functions),

$$\frac{c_1}{\sqrt{3}} + c_2\sqrt{2/3} = 0$$

or

$$c_2 = -\frac{c_1}{\sqrt{2}}$$

Thus,

$$\phi\left(\frac{1}{2}, \frac{1}{2}\right) = c_1\left[\varphi\left(1, -\frac{1}{2}\right) - \frac{1}{\sqrt{2}}\, \varphi\left(0, \frac{1}{2}\right)\right]$$

Using the normality property of ϕ and the orthonormality of φ, we get

$$c_1^2 + \frac{c_1^2}{2} = 1$$

or

$$c_1 = \sqrt{\frac{2}{3}}$$

Thus,

$$\phi\left(\frac{1}{2}, \frac{1}{2}\right) = \sqrt{\frac{2}{3}}\, \varphi\left(1, -\frac{1}{2}\right) - \frac{1}{\sqrt{3}}\, \varphi\left(0, \frac{1}{2}\right) \quad \text{... (10.9)}$$

Using again J_- to calculate $\phi\left(\frac{1}{2}, -\frac{1}{2}\right)$, we get

$$\phi\left(\frac{1}{2}, -\frac{1}{2}\right) = J_-\, \phi\left(\frac{1}{2}, \frac{1}{2}\right)$$

Use of eqn. (9), leads to

$$\phi\left(\frac{1}{2},-\frac{1}{2}\right) = J_-\left[\sqrt{\frac{2}{3}}\,\varphi\left(1,-\frac{1}{2}\right)-\frac{1}{\sqrt{3}}\,\varphi\left(0,\frac{1}{2}\right)\right]$$

$$= \sqrt{\frac{2}{3}}\,(J_{1-}+J_{2-})\,\varphi\left(1,-\frac{1}{2}\right)-\frac{1}{\sqrt{3}}\,(J_{1-}+J_{2-})\,\varphi\left(0,\frac{1}{2}\right)$$

$$= \sqrt{\frac{2}{3}}\,J_{1-}\,\varphi\left(1,-\frac{1}{2}\right)+\sqrt{\frac{2}{3}}\,J_{2-}\,\varphi\left(1,-\frac{1}{2}\right)$$

$$-\frac{1}{\sqrt{3}}\,J_{1-}\,\varphi\left(0,\frac{1}{2}\right)-\frac{1}{\sqrt{3}}\,J_{2-}\,\varphi\left(0,\frac{1}{2}\right)$$

$$= \sqrt{\frac{2}{3}}\,\sqrt{2}\,\varphi\left(0,-\frac{1}{2}\right)+\sqrt{\frac{2}{3}}\times 0-\frac{1}{\sqrt{3}}\,\sqrt{2}\,\varphi\left(-1,\frac{1}{2}\right)$$

$$-\frac{1}{\sqrt{3}}\,\varphi\left(0,-\frac{1}{2}\right)$$

or

$$\phi\left(\frac{1}{2},-\frac{1}{2}\right) = \frac{1}{\sqrt{3}}\,\varphi\left(0,-\frac{1}{2}\right)-\sqrt{\frac{2}{3}}\,\varphi\left(-1,\frac{1}{2}\right) \qquad \text{... (10.10)}$$

Thus the transformation from uncoupled to coupled representation can be written as

$$\begin{pmatrix} \phi\left(\frac{3}{2},\frac{3}{2}\right) \\ \phi\left(\frac{3}{2},\frac{1}{2}\right) \\ \phi\left(\frac{3}{2},-\frac{1}{2}\right) \\ \phi\left(\frac{3}{2},-\frac{3}{2}\right) \\ \phi\left(\frac{1}{2},\frac{1}{2}\right) \\ \phi\left(\frac{1}{2},-\frac{1}{2}\right) \end{pmatrix} = \begin{pmatrix} 1 & 0 & 0 & 0 & 0 & 0 \\ 0 & \frac{1}{\sqrt{3}} & \sqrt{\frac{2}{3}} & 0 & 0 & 0 \\ 0 & 0 & 0 & 0 & \sqrt{\frac{2}{3}} & \frac{1}{\sqrt{3}} \\ 0 & 0 & 0 & 1 & 0 & 0 \\ 0 & \sqrt{\frac{2}{3}} & \frac{-1}{\sqrt{3}} & 0 & 0 & 0 \\ 0 & 0 & 0 & 0 & \frac{1}{\sqrt{3}} & -\sqrt{\frac{2}{3}} \end{pmatrix} \begin{pmatrix} \varphi\left(1,\frac{1}{2}\right) \\ \varphi\left(1,-\frac{1}{2}\right) \\ \varphi\left(0,\frac{1}{2}\right) \\ \varphi\left(-1,-\frac{1}{2}\right) \\ \varphi\left(0,-\frac{1}{2}\right) \\ \varphi\left(-1,\frac{1}{2}\right) \end{pmatrix}$$

The matrix elements are the Clebsch-Gordan coefficients. The wave functions for the $2p$-states of hydrogen atom are given by the eqns. (10.1), (10.2) and (10.7) to (10.10).

where

$$\varphi(m_1 = m_l,\, m_2 = m_s) = R_{nl}\,(r)\,Y_l^{m_l}\,(\theta,\varphi)\,\chi_{m_s},\ \text{that is}$$

$$\varphi\left(1,\frac{1}{2}\right) = R_{21}\,(r)\,Y_1^1\,\alpha,$$

$$\varphi\left(0, \frac{1}{2}\right) = R_{21}(r)\, Y_1^0\, \alpha,$$

$$\varphi\left(-1, \frac{1}{2}\right) = R_{21}(r)\, Y_1^{-1}\, \alpha,$$

$$\varphi\left(1, -\frac{1}{2}\right) = R_{21}(r)\, Y_1^1\, \beta,$$

$$\varphi\left(0, -\frac{1}{2}\right) = R_{21}(r)\, Y_1^0\, \beta,$$

$$\varphi\left(-1, -\frac{1}{2}\right) = R_{21}(r)\, Y_1^{-1}\, \beta,$$

where $\alpha = |\uparrow\rangle = \begin{pmatrix}1\\0\end{pmatrix}$ and $\beta = |\downarrow\rangle = \begin{pmatrix}0\\1\end{pmatrix}$ represent the spin up and spin down states, respectively.

10.2 For the ground state of hydrogen atom, $j_1 = l = 0, j_2 = s = \frac{1}{2}$ (spin of the electron).

Therefore the allowed values of j are $j_1 + j_2 = l + \frac{1}{2} = 0 + \frac{1}{2} = \frac{1}{2}$ to $|j_1 - j_2| = \left|0 - \frac{1}{2}\right| = \frac{1}{2}$, that is $j = \frac{1}{2}$ is the only allowed value of j in the case of hydrogen atom in the ground state. m can thus have the values $\frac{1}{2}$ and $-\frac{1}{2}$. The magnitude of total angular momentum $\vec{J}$ would, therefore be

$$\{j(j+1)\}^{1/2}\, \hbar = \frac{\sqrt{3}}{2}\hbar$$

and its z-component

$$J_z = S_z = m_s \hbar$$

that is, either $\frac{1}{2}\hbar$ or $-\frac{1}{2}\hbar$

When the hydrogen atom is in the $2p$–state, $j_1 = l = 1, j_2 = s = \frac{1}{2}$

Then j can have values $j = l + s = 1 + \frac{1}{2} = \frac{3}{2}$ to $|j_1 - j_2| = \left|1 - \frac{1}{2}\right| = \frac{1}{2}$. That is, the allowed values of j would be $\frac{3}{2}$ and $\frac{1}{2}$.

For $j = \frac{3}{2}$, m can have values $\frac{3}{2}, \frac{1}{2}, -\frac{1}{2}, -\frac{3}{2}$ and for $j = \frac{1}{2}$, m can have values $\frac{1}{2}$ and $-\frac{1}{2}$. Thus the possible values of quantum numbers (j, m) are

$$\left(\frac{3}{2}, \frac{3}{2}\right), \left(\frac{3}{2}, \frac{1}{2}\right), \left(\frac{3}{2}, -\frac{1}{2}\right), \left(\frac{3}{2}, -\frac{3}{2}\right), \left(\frac{1}{2}, \frac{1}{2}\right) \text{ and } \left(\frac{1}{2}, -\frac{1}{2}\right)$$

10.3 For $j_1 = j_2 = 1$, m_1 can take the values 1, 0 and – 1 and similarly m_2 can also take the values 1, 0 and –1. The possible values of j would therefore be (from $j_1 + j_2$ to $|j_1 - j_2|$) 2, 1 and 0. Thus there would be nine $\varphi\ (m_1\ m_2)$ functions, *viz.*

$$\varphi\,(1, 1), \varphi\,(1, 0), \varphi\,(1, -1), \varphi\,(0, 1), \varphi\,(0, 0), \varphi\,(0, -1)$$

$$\varphi\,(-1, 1), \varphi\,(-1, 0) \text{ and } \varphi\,(-1, -1)$$

Since the allowed values of m (all values from j to $-j$) in this case are 2, 1, 0, –1, –2 (corresponding to $j = 2$), 1, 0, –1 (corresponding to $j = 1$) and 0 (corresponding to $j = 0$), one gets nine $\phi\ (jm)$ functions, *viz.*

$$\phi\,(2, 2), \phi\,(2, 1), \phi\,(2, 0), \phi\,(2, -1), \phi\,(2, -2), \phi\,(1, 1),$$
$$\phi\,(1, 0), \phi\,(1, -1) \text{ and } \phi\,(0, 0).$$

Now

$$\phi\,(2, 2) = \psi\,(1, 1) \qquad \text{... (10.11)}$$

as $\varphi\,(1, 1)$ is the only state for which $m = m_1 + m_2 = 2$ and similarly

$$\phi\,(2, -2) = \varphi\,(-1, -1) \qquad \text{... (10.12)}$$

as $\varphi\,(-1, -1)$ is the only state with $m = m_1 + m_2 = -2$.

Applying J_- to (10.11), we get

$$\begin{aligned}\phi\,(2, 1) &= \frac{1}{2} J_- \phi\,(2, 2) \\ &= \frac{1}{2} (J_{1-} + J_{2-})\, \phi\,(1, 1) \\ &= \frac{1}{2} [\sqrt{2}\, \varphi\,(0, 1) + \sqrt{2}\, \varphi\,(1, 0)] \\ &= \frac{1}{\sqrt{2}} [\varphi\,(0, 1) + \varphi\,(1, 0)] \qquad \text{... (10.13)}\end{aligned}$$

To get $\phi\,(2, 0)$ we apply J_- to $\phi\,(2, 1)$ which results in

$$J_- \phi\,(2, 1) = \sqrt{6}\, \phi\,(2, 0)$$

or

$$\begin{aligned}\phi\,(2, 0) &= \frac{1}{\sqrt{6}} J_- \phi\,(2, 1) \\ &= \frac{1}{\sqrt{6}} \times \frac{1}{\sqrt{2}} J_- [\varphi\,(0, 1) + \varphi\,(1, 0)] \\ &= \frac{1}{\sqrt{12}} [(J_{1-} + J_{2-})\, \varphi\,(0, 1) + (J_{1-} + J_{2-})\, \varphi\,(1, 0)] \\ &= \frac{1}{\sqrt{12}} [\sqrt{2}\, \varphi\,(-1, 1) + \sqrt{2}\, \varphi\,(0, 0)]\end{aligned}$$

$$+ \sqrt{2}\, \varphi (0, 0) + \sqrt{2}\, \varphi (1, -1)]$$

$$= \frac{1}{\sqrt{6}} [\varphi (1, -1) + 2\, \varphi (0, 0) + \varphi (-1, 1)] \quad ... (10.14)$$

Since the eigen vector ϕ (2, – 1) is linear combination of eigen vectors φ (0, –1) and φ (–1, 0) and φ (–1, 0), $\phi (2, -1) = a_2\, \varphi (0, -1) + b_2\, \varphi (-1, 0)$

Using the fact that ϕ (2, – 1) is orthogonal to ϕ (2, 1) and then using the normality of ϕ (2, –1), one gets (of course the condition that φ functions are orthonormal has been used)

$$\phi (2, -1) = \frac{1}{\sqrt{2}} [\varphi (0, -1) + \varphi (-1, 0)] \quad ... (10.15)$$

Similarly one can write,

$$\phi (1, 1) = a_3\, \varphi (1, 0) + b_3\, \varphi (0, 1)$$

Using the fact that ϕ (1, 1) is orthogonal to ϕ (2, 1) and the normality condition of ϕ (1, 1), we finally get

$$\phi (1, 1) = \frac{1}{\sqrt{2}} [\varphi (1, 0) - \varphi (0, 1)] \quad ... (10.16)$$

Application of J_- once again enables one to find ϕ (1, 0) and ϕ (1, –1). Thus

$$\phi (1, 0) = \frac{1}{\sqrt{2}} [\varphi (1, -1) - \varphi (-1, 1)] \quad ... (10.17)$$

and

$$\phi (1, -1) = \frac{1}{\sqrt{2}} [\varphi (0, -1) - \varphi (-1, 0)] \quad ... (10.18)$$

It may be noted that the expansion (10.17) does not contain the eigen vector φ (0, 0), though it also corresponds to $m = 0$. Actually the corresponding Clebsch-Gordan coefficient happens to be zero.

We are now left with the calculation of the eigen vector ϕ (0, 0), which is linear combination of the three eigen vectors for which

$$m = m_1 + m_2 = 0,$$

$$\phi(0, 0) = a_4\, \varphi (1, -1) + b_4\, \varphi (0, 0) + c_4\, \varphi (-1, 1)$$

Using the fact that ϕ (0, 0) must be orthogonal to ϕ (2, 0) and then using the normality of f (0, 0), we get

$$\phi (0, 0) = \frac{1}{\sqrt{3}} [\varphi (1, -1) - \varphi (0, 0) + \varphi (-1, 1)] \quad ... (10.19)$$

We can thus represent the transformation from uncoupled to coupled representation by the matrix equation,

$$\begin{pmatrix} \phi(2,2) \\ \phi(2,1) \\ \phi(2,0) \\ \phi(2,-1) \\ \phi(2,-2) \\ \phi(1,1) \\ \phi(1,0) \\ \phi(1,-1) \\ \phi(0,0) \end{pmatrix} = \begin{pmatrix} 1 & 0 & 0 & 0 & 0 & 0 & 0 & 0 & 0 \\ 0 & \frac{1}{\sqrt{2}} & 0 & \frac{1}{\sqrt{2}} & 0 & 0 & 0 & 0 & 0 \\ 0 & 0 & \frac{1}{\sqrt{6}} & 0 & \frac{2}{\sqrt{6}} & 0 & \frac{1}{\sqrt{6}} & 0 & 0 \\ 0 & 0 & 0 & 0 & 0 & \frac{1}{\sqrt{2}} & 0 & \frac{1}{\sqrt{2}} & 0 \\ 0 & 0 & 0 & 0 & 0 & 0 & 0 & 0 & 1 \\ 0 & \frac{1}{\sqrt{2}} & 0 & -\frac{1}{\sqrt{2}} & 0 & 0 & 0 & 0 & 0 \\ 0 & 0 & \frac{1}{\sqrt{2}} & 0 & 0 & 0 & -\frac{1}{\sqrt{2}} & 0 & 0 \\ 0 & 0 & 0 & 0 & 0 & \frac{1}{\sqrt{2}} & 0 & -\frac{1}{\sqrt{2}} & 0 \\ 0 & 0 & \frac{1}{\sqrt{3}} & 0 & -1 & 0 & 0 & \frac{1}{\sqrt{3}} & 0 \end{pmatrix} \begin{pmatrix} \varphi(1,1) \\ \varphi(1,0) \\ \varphi(1,-1) \\ \varphi(0,1) \\ \varphi(0,0) \\ \varphi(0,-1) \\ \varphi(-1,1) \\ \varphi(-1,0) \\ \varphi(-1,-1) \end{pmatrix}$$

The matrix elements are the Clebsch-Gordan coefficients.

❑❑❑

Chapter 11

Invariance, Symmetry and Conservation Laws

11.1 SYMMETRIES AND CONSERVATION LAWS

The physical observables which during the course of motion of a physical system do not change with time, are called constants of motion (or integrals of motion) and the principles expressing their constancy in time are referred to as the conservation laws. We would now show that the constancy of a physical observable implies the invariance of the Hamiltonian of system under a certain symmetry operation.

In doing so we would mainly confine our discussion to geometrical symmetries (also called space-time symmetries) which are of special importance and the associated conservation laws. In this case, we would deal with the symmetries arising from the homogeneity of space and time (homogenous means, possesses translational symmetry), isotropy of space (isotropic means, possesses rotational symmetry)and from the inversion of space and time. The other type of symmetries, called dynamical symmetries are associated with particular features of the interaction involved. For example, the $2l + 1$ fold degeneracy of the energy level associated with a particular angular momentum quantum number l is a feature of motion in any centrally symmetric field for which the potential energy V depends only on the radial distance **r** from some point. Several other symmetries which relate to molecules, crystals, etc. would also be omitted here. We would actually confine ourselves to a single particle or equivalently to a pair of particles in a centre of mass (com) coordinate system.

Let us first of all establish the relationship between a conservation law and a symmetry operation and then discuss each of the symmetry operations in detail.

The statement that some transformation does not change the Hamiltonian (or the Hamiltonian is invariant under the transformation) means that, if we make this transformation on the function $\hat{H}\,\varphi$, the result is same as if we make it on the function φ and then apply the operator $\hat{H}$. Mathematically this can be written as follows.

Let $\hat{U}$ be the operator which effects the transformation in question. Then we have

$$\hat{U}\,(\hat{H}\,\varphi) = \hat{H}\,(\hat{U}\,\varphi) \qquad \text{... (1)}$$

In other words, the invariance of $\hat{H}$ with respect to a transformation realized by the operator $\hat{U}$ reduces to the condition that $\hat{U}$ commutes with the Hamiltonian, that is

$$\hat{U}\hat{H} = \hat{H}\hat{U} \quad \text{... (2)}$$

or

$$\hat{U}\hat{H}\hat{U}^{\dagger} = \hat{H}UU^{\dagger} = \hat{H} \quad \text{... (3)}$$

which represents a unitary transformation. The operator $\hat{U}$, called the unitary operator may be written as

$$\hat{U} = e^{i\in\hat{A}} \quad \text{... (4)}$$

where $\hat{A}$ is Hermitian and $\in$ is a real parameter. Since $\hat{A}^{\dagger} = \hat{A}$, we have $U^{\dagger} = e^{-i\in\hat{A}^{\dagger}}$ $= e^{-i\in\hat{A}} = \hat{U}^{-1}$, the definition of a unitary operator.

Using the definition of the exponential of an operator, the condition (2), that $\hat{U}$ commutes with $\hat{H}$ reduces to

$$\hat{A}\hat{H} = \hat{H}\hat{A}$$

or

$$[\hat{A}, \hat{H}] = 0 \quad \text{... (5)}$$

From eqn. (10) chapter 6, it amounts to

$$d\langle\hat{A}\rangle \mid dt = 0$$

or in other words, an observable is a constant of motion if the corresponding operator commutes with the Hamiltonian. Thus we have a result that if the observable corresponding to the operator $\hat{A}$ is conserved during the motion of the system, then the Hamiltonian of the system is invariant under the unitary transformation generated by $\hat{A}$. Such transformations which leave the Hamiltonian invariant are called symmetry transformations. As a result, a conservation law invariably implies the existence of a symmetry transformation for the system.

11.2 SYMMETRY AND DEGENERACY

From eqn. (2), we see that, if $|u_k\rangle$ is an eigen vector of $\hat{H}$ belonging to the eigen value E_k, then

$$\hat{H}\hat{U}\,|u_k\rangle = \hat{U}\hat{H}\,|u_k\rangle$$

$$= E_k\,\hat{U}\,|u_k\rangle \quad \text{... (6)}$$

which amounts to the fact that $\hat{U}\,|u_k\rangle$ is also an eigen vector of H belonging to the same eigen value E_k. If $\hat{U}\,|u_k\rangle$ is linearly independent of $|u_k\rangle$, then energy eigen value E_k is degenerate. Thus degeneracy of the energy eigen values is another consequence of the invariance of the Hamiltonian under a symmetry operation.

11.3 SPACE-TIME SYMMETRIES

Let us now study the integrals of motion arising from the homogeneity of space, isotropy of space (that is the space has the same properties in all directions) and from the homogeneity of time. It is necessary to point out here that the reproducibility of an experimental result at different locations, orientations and times truly speaking stemps from the above properties of space and time.

(*a*) **Homogeneity (or Uniformity) of Space—Conservation of Linear Momentum.** The uniformity of space means that the properties of a closed system, that is, a system which is not subjected to the action of any external agent or a system that is acted upon by constant external forces, do not change under any parallel displacement of the system as a whole. Since in quantum mechanics the properties of a system are determined by its Hamiltonian, uniformity of space must imply that the Hamiltanian is invariant when the system suffers a parallel displacement over any distance. A displacement in space can be described either as a displacement of the coordinate system but the physical system remaining fixed or as a displacement of the physical system but the coordinate system remaining fixed. The first one is put under the head passive point of view and the later under the head active point of view. Now, a displacement of the physical system through the vector **r**, is equivalent to a displacement of the coordinate system through –**r**. Accordingly any of the view points could be chosen.

Further any finite displacement can be constructed out of infinitesimal (infinitely small) displacements. It is thus sufficient to consider the invariance of the Hamiltonian under an infinitesimal displacement ∂r which signifies a transformation under which the radius vector $\mathbf{r}_i$ (i being the number of particle) of all the particles receive the same increment $\partial \mathbf{r} : \mathbf{r}_i \to \mathbf{r}_{i'} + \partial \mathbf{r}$. This amounts to same as the displacement of the coordinate system by $-\partial \mathbf{r}$. Thus an arbitrary function $\varphi\left(\vec{r}_1, \vec{r}_2, ...\right)$ of the coordinates of the particles under such a transformation becomes a function,

$$\begin{aligned}\varphi(\mathbf{r}_1 - \partial\mathbf{r}, \mathbf{r}_2 - \partial\mathbf{r}, ...) &= \varphi(\vec{r}_1, \vec{r}_2, ...) - \partial\mathbf{r}.\, \Sigma_i \vec{\nabla}_i\, \varphi \\ &= (1 - \partial\mathbf{r}.\, \Sigma_i \vec{\nabla}_i)\, \varphi(\mathbf{r}_1, \mathbf{r}_2, ...) \quad ...(7)\end{aligned}$$

where Taylor series expansion of $\varphi(\mathbf{r} - \partial\mathbf{r})$ around **r** has been used and where $\vec{\nabla}_i$ denotes the operator of differentiation with respect to $\mathbf{r}_i$. The operator $(1 - \partial\mathbf{r}\, .\, \Sigma_i\, \vec{\nabla}_i)$ can be called the infinitesimal displacement operator or the unitary operator corresponding to the displacement $\partial\mathbf{r}$ and would be given by

$$\hat{U}(\partial\mathbf{r}) = 1 - \partial\mathbf{r}\, .\, \Sigma_i\, \vec{\nabla}_i = 1 - \frac{i}{h}(\partial\mathbf{r}\, .\, \hat{p}) \quad ...(8)$$

where, $\hat{p} = \Sigma_i\, \hat{p}_i\ (\hat{p}_i = -i\,\hbar\,\vec{\nabla}_i)$ is the sum operator of the momenta of all the particles of the system or is called the total linear momentum of the system. Since a finite displacement **r** could be thought as succession of infinitesimal displacements, the unitary operator corresponding to a displacement over a finite distance **r**, can be written as

$$\hat{U}\,(\mathbf{r}) = e^{-i\vec{r}\,.\,\hat{p}/\hbar} \qquad \text{... (9)}$$

in analogy with eqns. (27) and (31) of section 8.4.

Invariance of the Hamiltonian under such a transformation requires (see eqns. (2), (4) and (5)) that the operator $\hat{U}$ (**r**) and hence the operator $\hat{p}$ commutes with $\hat{H}$ or $[\hat{p}, \hat{H}] = 0$. This implies that the total linear momentum of the system is conserved. In other words, conservation of the linear momentum of a physical system is consequence of the translational invariance of the Hamiltonian of the system.

(b) Homogeneity (or Uniformity) of Time-Conservation of Energy

As the time is a uniform quantity, the Hamiltonian of any closed system will not depend explicitly on time (experiment performed at different times would lead to the same experimental result). In that case $\frac{\partial \hat{H}}{\partial t} = 0$ and hence form eqn. (10) chapter 6, we have

$$\frac{d\langle \hat{H}\rangle}{dt} = \frac{1}{i\hbar}\langle[\hat{H}, \hat{H}]\rangle = 0$$

which amounts to

$$\frac{d\langle \hat{H}\rangle}{dt} = \frac{dE}{dt} = 0$$

or the energy of the system is conserved. Thus the homogeneity of time leads to the energy conservation law.

(c) Isotropy of Space-Conservation of Angular Momentum

Since the space has the property of isotropy (all directions in it are equivalent), the Hamiltonian of a closed system cannot change when the system rotates as a whole through an arbitrary angle about an arbitrary axis. It is sufficient to require the fulfilment of this condition for an infinitesimal rotation and there from determine the required operator.

Consider an infinitesimal rotation characterized by a vector $\partial\vec{\phi}$, the length (or magnitude) of which is equal to the angle $\partial\phi$ over which the system is rotated and the direction of which is along the axis defined by the unit vector **n** about which the rotation takes place. The changes $\partial\mathbf{r}_i$ (in the radius vectors $\mathbf{r}_i$ of the particles) in such a rotation are

$$\partial\mathbf{r}_i = \partial\vec{\phi} \times \mathbf{r}_i \qquad \text{... (10)}$$

Retaining only the first order terms, an arbitrary function $\varphi\,(\mathbf{r}_1, \mathbf{r}_2, \ldots)$ under such a transformation goes into the function

$$\begin{aligned}\varphi\,(\mathbf{r}_1 - \partial\mathbf{r}_1, \mathbf{r}_2 - \partial\mathbf{r}_2, \ldots) &= \varphi\,(\mathbf{r}_1, \mathbf{r}_2, \ldots) - \Sigma_i\,\partial\mathbf{r}_i \times \vec{\nabla}_i\,\varphi \\ &= \varphi\,(\mathbf{r}_1, \mathbf{r}_2, \ldots) - \partial\vec{\phi}\,.\,\Sigma_i\,\mathbf{r}_i \times \vec{\nabla}_i\,\varphi\end{aligned}$$

$$= (1 - \partial\vec{\phi} \,.\, \Sigma_i \, \mathbf{r}_i \times \vec{\nabla}_i) \, \varphi (\mathbf{r}_1, r_2, ...)$$

where the identify **a** . **b** × **c** = **a** × **b** . **c** has been utilized.

From above, it follows that

$$\hat{R} (\partial\vec{\phi}) = 1 - \partial\vec{\phi} \,.\, \Sigma_i \, \mathbf{r}_i \times \vec{\nabla}_i$$

$$= 1 - \frac{i}{\hbar} \partial\vec{\phi} \,.\, \hat{L} \qquad ... (11)$$

is the operator for an infinitesimal rotation $\partial\vec{\phi}$ and the operator $\hat{L}$ corresponds to the total angular momentum of the system. Each of the terms $- i \hbar \, \mathbf{r}_i \times \vec{\nabla}_i$, corresponds to the angular momentum of an individual particle.

We can construct from the operators of infinitesimal rotation around an axis defined by the unit vector **n**, the operator of rotation around the same axis over a finite angle ϕ using eqn. (32) of sec. 8.4. Thus

$$\hat{R}_n (\phi) = e^{-i/\hbar \, \phi \, (\hat{L} \,.\, \mathbf{n})} \qquad ... (12)$$

Since the total angular momentum operator $\hat{L}$ is Hermitian, $\hat{R}_n (\phi)$ is unitary. Thus the unitary operator corresponding to a rotation of the physical system through an angle ϕ about the axis **n**, would be given by

$$\hat{U} (\mathbf{n}, \phi) = \hat{R}_n (\phi) = e^{- i/\hbar \, \phi \, (\hat{L} \,.\, \mathbf{n})} \qquad ... (13)$$

The invariance of the Hamiltonian under rotations in space requires that the operator $\hat{U} (\mathbf{n}, \phi)$ and hence the operator $\hat{L}$ commutes with the Hamiltonian, that is $(\hat{L}, \hat{H}) = 0$. This implies that the total angular momentum of the system is conserved. In other words, the conservation of the angular momentum is consequence of the rotational invariance of the Hamiltonian of the system.

11.4 CONTINUOUS AND DISCRETE TRANSFORMATIONS

The above consideration of translations (displacements in space and time) and rotations (in space) refers to a class of continuous transformations since we can realize them through multiple consecutive applications of infinitesimal transformations. The invariance of the Hamiltonian under these transformations leads (as we have seen) to the conservation laws for linear momentum, energy and angular momentum which correspond to the conservation laws of classical mechanics. Symmetry conditions may lead not only to continuous transformations but also to discrete transformations (associated with inversion of space and time) which can not be reduced to infinitesimal transformations. Invariance of Hamilton's function under such transformations does not lead to a conservation law in classical mechanics. However, in quantum mechanics, there is no essential difference between continuous and discrete transformations. Therefore, discrete transformations in quantum mechanics also lead to conservation laws. Before we proceed further it may be pointed out that the continuous symmetries have universal validity, that is, they hold good for all the four known classes of interactions (strong, electromagnetic, weak and gravitational). But it is known that the space inversion invariance is not valid in the case of weak interactions (β decay of nuclei). Time reversal (or inversion) invariance is also found to be violated in the case of certain reactions such as the decay of neutral Kaons.

11.5 SPACE INVERSION (OR SPATIAL REFLECTION)—PARITY

Let us consider the discrete transformation under which the Hamiltonian remains invariant as a result of space inversion. This consists of simultaneous change in sign of all three spatial coordinates, that is

$$x \to -x, \quad y \to -y, \quad z \to -z \qquad \text{... (14)}$$

and amounts to the reversal of the direction of each coordinate axis. A right handed coordinate system then becomes left handed and vice versa (see Fig. 11.1). Since these two types of coordinate systems are mirror images of each other, a physical system which is invariant under space inversion is said to possess reflection symmetry or lacks chirality, that is handedness. This means that the phenomena associated with the physical system will look the same irrespective of the handedness of the coordinate system in which they are observed. In other words, there is parity between the system and its mirror image. For this very reason, the space invasion is also called parity operation.

Let us denote by $\hat{P}$ (for parity) an inversion operator whose effect on a wave function ϕ (**r**) is to change the sign of the coordinates,

$$\hat{P}\varphi(\mathbf{r}) = \varphi(-\mathbf{r}) \quad .$$

If we want to find the eigen values P of the inversion operator $\hat{P}$, we must solve the equation,

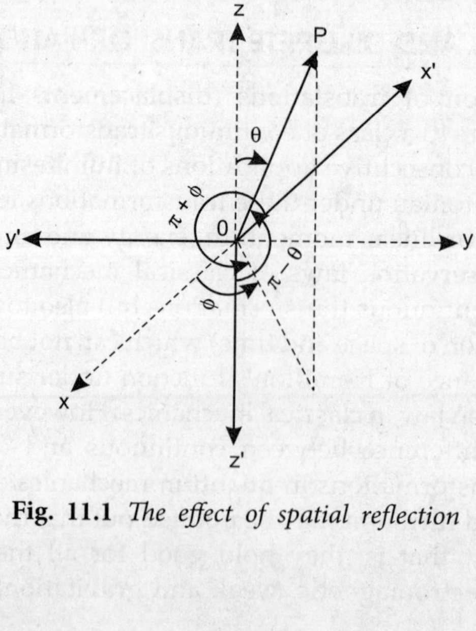

Fig. 11.1 *The effect of spatial reflection*

$$\hat{P}\,\varphi(\mathbf{r}) = P\,\varphi(\mathbf{r}) \qquad \text{... (15)}$$

Since

$$\hat{P}\,\hat{P}\,\varphi(\mathbf{r}) = \hat{P}\,P\,\varphi(\mathbf{r}) = P^2\,\varphi(\mathbf{r})$$

and

$$\hat{P}\,\hat{P}\,\varphi(\mathbf{r}) = \hat{P}\,\varphi(-\mathbf{r}) = \varphi(\mathbf{r}),$$

$$P^2\,\varphi(\mathbf{r}) = \varphi(\mathbf{r})$$

which amounts to $P^2 = 1$ or $P = \pm 1$. Thus the eigen values of the inversion operator are $\pm$ 1. Eqn. (15) can, therefore be written as

$$\hat{P}\,\varphi(\mathbf{r}) = \pm\,\varphi(\mathbf{r}) \qquad \text{... (16)}$$

Therefore, the eigen functions of the inversion operator are either unchanged or change in sign when acted upon by this operator. In the first case, the wave function is said to be even (and the corresponding state to be even parity state) and in the second case it is said to be odd (and the corresponding state to be odd parity state).

The invariance of Hamiltonian of a closed system under inversion (that is the fact that $\hat{P}\,\hat{H} = \hat{H}\,\hat{P}$ or the operators $\hat{P}$ and $\hat{H}$ commute) thus expresses the law of conservation of parity: if the state of a closed system has a definite parity (that is if it is even or odd) then this parity is conserved in the course of time. It is necessary to mention here that this refers to the non-relativistic theory. However, there exist interactions in nature, falling in the realm of relativistic theory as we will point out which violate the conservation of parity.

The parity conservation law is satisfied to a very high degree of accuracy in all phenomenon determined by nuclear and electromagnetic interactions. Prior to 1956, it was assumed that this law was a universal law of nature. However, it was established later on by Lee and Yang[1] and by Wu and co-workers[2] that in the β-decay of atomic nuclei an asymmetry was observed which made it possible to distinguish between left handedness and right handedness. This phenomenon showed that for weak interaction which determines this decay process, the symmetry between left and right is violated, that is the invariance under spatial inversion is violated. The parity conservation law is thus violated. This means that phenomenon mediated by weak interactions, such as β-decay, can be distinguished from their mirror images.

11.6 TIME REVERSAL (OR REFLECTION) AND KRAMERS' THEOREM

Time reversal is the operation in which $t \to -t$ (that is when future and past are interchanged). With respect to such an operation, the physical quantities can be divided into two classes. The physical quantities such as coordinates of a point, the total energy, kinetic energy (quantities which contain t in even powers) which do not change under time reversal, belong to the first class. To the second class belong

[1]T.D. Lee and C.N. Yang, Phys. Rev. 104, 254 (1956)

[2]C.S. Wu et al., Phys. Rev. 105, 1413 (1957)

physical quantities such as velocity, momentum, angular momentum, and all other physical quantities which contain time in odd powers.

Let us consider the time dependent Schrödinger equation

$$\hat{H}\,\psi(\mathbf{r}, t) = i\hbar\,\partial\,\psi(\mathbf{r}, t)/\partial t \qquad \text{... (17a)}$$

which determined the change with time of some state $\psi(\mathbf{r}, t)$. Under the substitution $t \to -t$, the above equation becomes

$$\hat{H}\,\psi\,(\mathbf{r}, -t) = -i\hbar\,\partial\psi(\mathbf{r}, -t)/\partial t \qquad \text{... (17b)}$$

as the operator $\hat{H}$ is invariant under time reversal.

The Schrödinger eqn. (17) is thus not invariant under time reversal.

Let us denote by ψ_{rev} $(\mathbf{r}, t)$, the wave function of the state which we get from the state $\Psi(\mathbf{r}, t)$ by time reversal and therefore called the time reversed wave function. Let us find the time reversal operator $\hat{T}$ which changes the wave function $\Psi(\mathbf{r}, t)$ into Ψ_{rev} $(\mathbf{r}, t)$. By definition, the function Ψ_{rev} $(\mathbf{r}, t)$ satisfies the Schrödinger equation,

$$\hat{H}\,\psi_{\text{rev}}(\mathbf{r}, t) = i\hbar\,\partial\psi_{\text{rev}}(\mathbf{r}, t)/\partial(-t)$$

or

$$\hat{H}\,\Psi_{\text{rev}}(\mathbf{r}, t) = -\,i\hbar\,\partial\psi_{\text{rev}}(\mathbf{r}, t)/\partial t \qquad \text{... (18)}$$

Let us consider the conjugate complex of the eqn. (17 *a*),

$$\hat{H}^*\Psi^*(\mathbf{r}, t) = H\,\Psi^*(\mathbf{r}, t) = -\,i\hbar\,\partial\Psi^*(\mathbf{r}, t)/\partial t \qquad \text{... (19)}$$

if H is real.

If there exists a unitary operator $\hat{U}$ satisfying the conditions.

$$\hat{U}\hat{H} = \hat{H}\hat{U}, \hat{U}^{+}\hat{U} = 1, \qquad \text{... (20)}$$

we find by operating from the left on both sides of eqn. (19) with $\hat{U}$,

$$\hat{U}\hat{H}\,\psi^*(\mathbf{r}, t) = \hat{H}\hat{U}\,\Psi^*(\vec{r}, t) = -i\hbar\,\partial\hat{U}\,\Psi^*(\mathbf{r}, t)/\partial t$$

Comparing above equation with eqn. (18), we find that

$$\Psi_{\text{rev}}(\mathbf{r}, t) = \hat{T}\,\Psi\,(\mathbf{r}, t) = \hat{U}\,\Psi^*(\mathbf{r}, t) = \hat{U}\hat{K}_c\,\Psi(\mathbf{r}, t) \qquad \text{... (21)}$$

where $\hat{K}_c$ is the operator of complex conjugation and its property that $\hat{K}_c\Psi = \Psi^*$ has been used. Thus the time reversal operator $\hat{T}$ transforming the function $\psi(\mathbf{r}, t)$ into ψ_{rev} $(\mathbf{r}, t)$, that is $\hat{T}\,\psi(\mathbf{r}, t) = \psi_{\text{rev}}(\mathbf{r}, t) = \psi^*(\mathbf{r}, -t)$ (see problem 11.4), is of the form

$$\hat{T} = \hat{U}\,\hat{K}_c \qquad \text{... (22)}$$

The complex conjugation operator $\hat{K}_c$ is an antilinear (or antiunitary) operator since its action upon the function $a_1\,\psi_1 + a_2\,\psi_2$ is characterized by the equation,

$$\hat{K}_c\,(a_1\,\psi_1 + a_2\psi_2) = a_1^*\,\hat{K}_c\,\psi_1 + a_2^*\,\hat{K}_c\,\psi_2 \qquad \text{... (23)}$$

as $\qquad \hat{K}_c a_i = a_i^*\,\hat{K}_c$

The product of a unitary and antiunitary operator is an antiunitary operator; the time reversal operator $\hat{T}$ of eqn. (22) is, therefore an antiunitary operator. The explicit form of the time reversal operator, however depends upon the form of the Hamiltonian and on the representation in which the wave function is given. Let us now consider a few examples.

11.7 $\hat{T}$ FOR ZERO SPIN PARTICLE

The physical significance of $\hat{T}$ as the time reversal operator requires that, while the coordinate operator remains unchanged, all velocities must be reversed (momentum and angular momentum operators change their signs). Hence, when

$$\hat{r} = \mathbf{r} \text{ and } \hat{p} = -i\hbar\vec{\nabla},$$

$$\hat{r}_{\text{rev}} = \hat{T}\hat{r}\hat{T}^{-1} = \hat{r} \quad \text{... (24a)}$$

$$\hat{p}_{rev} = \hat{T}\hat{p}\hat{T}^{-1} = -\hat{p} \quad \text{... (24b)}$$

$$\hat{L}_{rev} = \hat{T}\hat{L}\hat{T}^{-1} = -\hat{L} \quad \text{... (24c)}$$

For a spinless particle, in the coordinate representation in which $\hat{r}$ = r is real and $\hat{p} = -i\hbar\vec{\nabla}$ is imaginary, it is clear that all the requirements (24) are satisfied with

$$\hat{U} = 1, \hat{T} = \hat{K}_c \quad \text{... (25)}$$

as

$$\hat{r}_{\text{rev}} = \hat{T}\hat{r}\hat{T}^{-1} = \hat{K}_c\hat{r}\hat{K}_c^{-1} = \hat{r},$$

$$\hat{p}_{rev} = \hat{T}\hat{p}\hat{T}^{-1} = \hat{K}_c(-i\hbar\vec{\nabla})\hat{K}_c^{-1} = i\hbar\vec{\nabla} = -\hat{p}$$

$$\hat{L}_{rev} = \hat{T}\hat{L}\hat{T}^{-1} = \hat{K}_c(\hat{r}\times\hat{p})\hat{K}_c^{-1} = -\hat{L}$$

and as i goes to $-i$ under the operation $\hat{K}_c$. Therefore $\hat{T}$ is a complex conjugation operator.

In the momentum representation, $\hat{p} = \vec{p}$ is real and $\hat{r} = i\hbar\vec{\nabla}_p$ is imaginary ($x = i\hbar\, \partial/\partial p_x$, $y = i\hbar\, \partial/\partial p_y$, $z = i\hbar\, \partial/\partial p_z$). Equations (28), therefore require $\hat{U}$ to be an operator that changes $\hat{p}$ to $-\hat{p}$: $\hat{U}\hat{p}\hat{U}^{-1} = -\hat{p}$ and

$$\hat{T}\psi(\vec{p}, t) = \psi^*(-\vec{p}, -t)$$

It may be mentioned here that, although the time reversal operator $\hat{T}$ commutes with the Hamiltonian, that is, $\hat{T}\hat{H} - \hat{H}\hat{T} = 0$ (or $\hat{H}$ is invariant under time reversal), it is not a constant of motion as

$$i\hbar\, d\langle\hat{A}\rangle/dt = \langle\hat{A}\hat{H} - \hat{H}\hat{A}\rangle$$

holds only for linear operators.

11.8 $\hat{T}$ FOR A NON-ZERO SPIN PARTICLE

In the case of particles with non-zero spin, we expect in analogy with eqn. (24c) that the spin angular momentum operator $\hat{S}$ should change sign under time reversal:

$$\hat{T}\hat{S}\hat{T}^{-1} = -\hat{S} \quad \text{... (24d)}$$

(in general $\hat{T}\hat{J}\hat{T}^{-1} = -\hat{J}$ where $\hat{J}$ is the total angular momentum operator).

Since the explicit form for $\hat{T}$ actually depends on the representation chosen, let us choose the coordinate representation for $\hat{r}$ and $\hat{p}$ and the standard representation for the angular momentum, that is

$$\hat{S} = \frac{1}{2}\hbar\,\hat{\sigma} \qquad \text{... (26a)}$$

where

$$\hat{\sigma}_x = \begin{pmatrix} 0 & 1 \\ 1 & 0 \end{pmatrix}, \hat{\sigma}_y = \begin{pmatrix} 0 & -i \\ i & 0 \end{pmatrix} \text{ and } \hat{\sigma}_z = \begin{pmatrix} 1 & 0 \\ 0 & -1 \end{pmatrix} \qquad \text{... (26b)}$$

are the Pauli spin matrices. Then $\hat{r}$, $\hat{S}_x$, $\hat{S}_z$ are real whereas $\hat{p}$ and $\hat{S}_y$ are imaginary. (Note that the particular case of spin half particles has been considered). This amounts to

$$\hat{K}_c\hat{S}_x = \hat{S}_x\hat{K}_c, \hat{K}_c\hat{S}_y = -\hat{S}_y\hat{K}_c \text{ and } \hat{K}_c\hat{S}_z = \hat{S}_z\hat{K}_c$$

On putting $\hat{T} = \hat{U}\hat{K}_c$ in eqn. (24*d*), one gets

$$\hat{T}\,\hat{S}_x\hat{T}^{-1} = \hat{U}\hat{K}_c\hat{S}_x\,(\hat{U}\hat{K}_c)^{-1} = \hat{U}\hat{S}_x\hat{K}_c\hat{U}^{-1}\hat{K}_c^{-1}$$

or

$$\hat{U}\hat{S}_x\hat{U}^{-1} = -\hat{S}_x \qquad \text{... (27a)}$$

$$\hat{T}\hat{S}_y\hat{T}^{-1} = \hat{U}\hat{K}_c\hat{S}_y\,(\hat{U}\hat{K}_c)^{-1} = -\,\hat{U}\hat{S}_y\hat{K}_c\hat{U}^{-1}\hat{K}_c^{-1}$$

or

$$-\,\hat{U}\hat{S}_y\hat{U}^{-1} = -\,\hat{S}_y$$

or

$$\hat{U}\hat{S}_y\hat{U}^{-1} = \hat{S}_y \qquad \text{... (27b)}$$

and similarly

$$\hat{U}\hat{S}_z\hat{U}^{-1} = -\,\hat{S}_z \qquad \text{... (27c)}$$

That is $\hat{U}$, the unitary part of $\hat{T}$ leaves $\hat{S}_y$ invariant but changes the signs of $\hat{S}_x$ and $\hat{S}_z$. U should, therefore represent a rotation of the coordinate system through an angle π about the y-axis. From sec. 8.4, we know that the operator for such a rotation is $e^{i\pi\hat{S}_y/\hbar}$ ($S_y/\hbar$ being the generator of rotation about the y-axis). The positive sign in the exponent is because of the fact that $\hat{U}$ represents rotation of the coordinate system rather than of the physical system. Thus

$$\hat{T} = e^{i\pi\hat{S}_y/\hbar}\,\hat{K}_c \qquad \text{... (28)}$$

For spin $-\frac{1}{2}$, the above equation reduces to

$$\hat{T} = i\hat{\sigma}_y\,\hat{K}_c \qquad \text{... (29)}$$

Since $i\,\hat{S}_y$ or $i\hat{\sigma}_y$ is real, $\hat{K}_c$ commutes with it ($e^{i\pi\hat{S}_y/\hbar} = 1 + i\,\pi\,\hat{S}_y/\hbar. + ...$ and hence when $\hat{K}_c$ operates on $i\,\hat{S}_y$, it does not change). Using this fact ($\hat{T} = e^{i\pi\hat{S}_y/\hbar}\,\hat{K}_c = \hat{K}_c\,e^{i\pi\hat{S}_y/\hbar}$) and the relation $\hat{K}_c^2 = 1\,(\hat{K}_c\,a = a^*, \hat{K}_c\hat{K}_c a = \hat{K}_c a^* = a)$, we obtain in general

$$\hat{T}^2 = e^{2i\pi\hat{S}_y/\hbar} \qquad \text{... (30)}$$

and for the spin $-\frac{1}{2}$ particles

$$\hat{T}^2 = i\hat{\sigma}_y \hat{K}_c i\hat{\sigma}_y \hat{K}_c = i\hat{\sigma}_y \hat{K}_c \hat{K}_c i\hat{\sigma}_y = -\hat{\sigma}_y^2 = -1 \quad \text{... (31)}$$

Note that the same result holds for a system of particles since each particle contributes its own factor –1 to $\hat{T}^2$. If the system consists of n spin – $\frac{1}{2}$ particles, the time reversal operator is obtained from eqn. (29) by the simple generalization,

$$\hat{T}_n = i^n \hat{\sigma}_{1y} \hat{\sigma}_{2y} \hat{\sigma}_{3y} ... \hat{\sigma}_{ny} \hat{K}_c \quad \text{... (29}a\text{)}$$

By applying the time reversal operator twice corresponds to

$$\hat{T}_n^2 = (-1)^n \quad \text{... (32}a\text{)}$$

Thus $\hat{T}_n^2$ has eigen value +1 for n even or for integral spin particles and -1 for half odd-integral spin particles or for n odd. From eqn. (17a) and from eqns. (11.7) and (11.8), it is evident that if the Hamiltonian $\hat{H}$ is time reversal invariant and if $\psi(\mathbf{r}, t)$ is an energy eigen function, $\hat{T}\ \psi\ (\mathbf{r}, t)$ is also an energy eigen function with the same energy eigen value. As already mentioned $\hat{T}\ \psi(\mathbf{r}, t) = \psi^*\ (\mathbf{r}, -t)$ is referred to as the time reversed solution of the Schrödinger equation.

We first suppose that there is no degeneracy. Then $\hat{T}\ \Psi_n\ (\mathbf{r}, t)$ represents the same state, so that

$$\hat{T}_n \Psi (\mathbf{r}, t) = c\, \psi\, (\mathbf{r}, t) \quad \text{... (33)}$$

where c is some number (or scalar). Operating both sides of the above equation with the time reversal operator, we get

$$\hat{T}_n^2 \psi\, (\mathbf{r}, t) = \hat{T}_n \hat{T}_n \psi\, (\mathbf{r}, t) = \hat{T}_n [c\, \psi\, (\mathbf{r}, t)] = c^* \{\hat{T}_n\, \psi\, (\mathbf{r}, t)\}$$

as $\hat{T}_n$ is antiunitary operator. Further from above and eqn. (33).

$$\hat{T}_n^2 \Psi\, (\mathbf{r}, t) = c^* c\, \psi(\mathbf{r}, t) = |c|^2\, \psi\, (\mathbf{r}, t) \quad \text{... (34)}$$

If $\hat{T}_n^2 = +1$. (or $(-1)^n = +1$, that is if number of particles in the system is even), $|c|^2 = 1$, which is a possible solution, that is there is no degeneracy.

But if $\hat{T}_n^2 = -1$ (that is, if the number of particles in the system is odd or if the total spin has a half odd-integral value), there is no number c which satisfies $|c|^2 = -1$, which means, there is some degeneracy. In fact $\hat{T}_n\ \psi\ (\mathbf{r}, t)$ is orthogonal to $\psi\ (\mathbf{r}, t)$.

Using antiunitary property of $\hat{T}_n$, that is

$$\langle \hat{T}_n \Phi \mid \hat{T}_n \psi \rangle = \langle \psi \mid \Phi \rangle$$

and taking $\Phi = \hat{T}_n \psi$, we get

$$\langle \hat{T}_n^2 \psi \mid \hat{T}_n \psi \rangle = \langle \psi \mid \hat{T}_n \psi \rangle$$

Since $\hat{T}_n^2 = -1$, the left hand side of the above equation is negative of the right hand side. Hence both sides must be zero of $\hat{T}_n\ \psi(\mathbf{r}, t)$ is orthogonal to $\psi(\mathbf{r}, t)$.

Thus, every state $\psi(\mathbf{r}, t)$ would be degenerate with its time reversed state, $\hat{T}_n \psi\ (\mathbf{r}, t)$. In other words, every eigen value would be two fold degenerate. This is referred to as Kramers, degeneracy (or Kramers' theorem). A state of an atom with an odd number of electrons, provides an example of Kramers' degeneracy. The degeneracy would be present even in the presence of an electric field (as the potential $V(\mathbf{r}) = -E.\mathbf{r}$ commutes with $\hat{T}$ or $\mathbf{r}$ does not change sign on time reversal), but would be lifted by a velocity dependent field (velocity changes sign on time reversal) such as the magnetic field.

PROBLEMS

11.1 Show that, if the Hamiltonian is an even function of x, that is invariant under space inversion, the eigen functions of the Hamiltonian fall into two classes.

11.2 Show that the angular momentum eigen functions with even l have even parity and those with odd l have odd parity.

11.3 Using the concept of parity, show that the dipole transition $2s \to 1s$ in the case of hydrogen atom is forbidden in the dipole approximation.

11.4 Show that

$$\hat{T}\,\psi(\mathbf{r}, t) \equiv \psi_{rev}(\mathbf{r}, t) \neq \psi(\mathbf{r}, -t)$$

but is equal to $\psi^*(\mathbf{r}, -t)$

11.5 Show that in the case of spin $-\frac{1}{2}$ particle, $\hat{T} = i\,\hat{\sigma}_y\,\hat{K}_c$ where $\hat{\sigma}_y$ is the Pauli spin operator and $\hat{K}_c$ is the complex conjugation operator.

SOLUTIONS

11.1 Consider the case of the dimensional harmonic oscillator in which

$$\hat{H}(-x) = \hat{H}(x) \qquad \text{... (11.1)}$$

The Schrödinger equation in the above case can be written as

$$\hat{H}(x)\,\varphi(x) = E\,\varphi(x) \qquad \text{... (11.2)}$$

Changing x to $-x$ everywhere, we get

$$\hat{H}(-x)\,\varphi(-x) = E\,\varphi(-x)$$

which by using eqn. (11.1) becomes

$$\hat{H}(x)\,\varphi(-x) = E\,\varphi(-x) \qquad \text{... (11.3)}$$

that is, $\varphi(-x)$ is also an eigen function of $\hat{H}$ with the same eigen value E. Now since for each energy eigen value there is only one eigen function or the energy eigen functions are non-degenerate in this case, $\varphi(-x)$ must be same as $\varphi(x)$ except for the multiplication factor, that is

$$\varphi(-x) = \lambda\varphi(x)$$

Changing the sign of x in the above equation gives,

$$\varphi(x) = \lambda\varphi(-x) = \lambda^2\,\varphi(x)$$

or

$$\lambda^2 = 1,$$

which amounts to $\lambda = \pm 1$. Thus the eigen functions in this case divide into two classes. The eigen functions which belong to the class $\lambda = +1$ are said to have even parity and those with $\lambda = -1$ to have odd parity.

11.2 The inversion transformation ($x \to -x$, $y \to -y$ and $z \to -z$) in spherical polar coordinates, is the transformation (see Fig. 11.1),

$$\mathbf{r} \to \mathbf{r},\ \theta \to \pi - \theta,\ \phi \to \pi + \phi \quad \text{... (11.4)}$$

The dependence of the wave function of the particle on the angle is given by the spherical harmonic $Y_l^{m_l}(\theta, \phi)$ which apart from the constant (not important here) has the form $P_l^{m_l}(\cos\theta)\, e^{im_l\phi}$. When ϕ is replaced by $\pi + \phi$, the factor $e^{im_l\phi}$ is multiplied by $e^{i\pi m_l} = (-1)^{-m_l}$ and when θ is replaced by $\pi - \theta$, $P_l^{m_l}(\cos\theta)$ becomes $P_l^{m_l}(-\cos\theta) = (-1)^{l-m_l} P_l^{m_l}(\cos\theta)$. Thus the whole function is multiplied by $(-1)^l$ (independent of m_l), that is the parity of the state with given value of l is

$$P = (-1)^l \quad \text{... (11.5)}$$

That is, all states with even l are even parity states and all those with odd l are odd.

11.3 The relevant matrix element is given by

$$\mathbf{M}_{200-100} = \iiint_{-\infty}^{+\infty} \varphi_{200}(\mathbf{r})\, \mathbf{M}(\mathbf{r})\, \varphi_{100}(\mathbf{r})\, d\mathbf{r},\ d\mathbf{r} = dx\, dy\, dz \quad \text{... (11.6)}$$

where $\mathbf{M} = e\mathbf{r}$ is the electric dipole moment operator.

Changing the variable $\mathbf{r}$ to $-\mathbf{r}$ and using the results

$$\begin{aligned}\varphi_{100}(\mathbf{r}) &= c_1\, e^{-r/a_0} = \varphi_{100}(-\mathbf{r}),\ \varphi_{200}(\mathbf{r}) \\ &= c_2\,(2 - \mathbf{r}/a_0)\, e^{-r/2a_0} = \varphi_{200}(-\mathbf{r})\end{aligned}$$

and

$$\mathbf{M}(-\mathbf{r}) = -e\mathbf{r} = -\mathbf{M}(\mathbf{r}),$$

one gets

$$\begin{aligned}\mathbf{M}_{200-100} &= \iiint_{+\infty}^{-\infty} \varphi_{200}(-\mathbf{r})\, \mathbf{M}(-\mathbf{r})\, \varphi_{100}(-\mathbf{r})\, d(-\mathbf{r}) \\ &= \iiint_{+\infty}^{-\infty} \varphi_{200}(\mathbf{r})\, \mathbf{M}(\mathbf{r})\, \varphi_{100}(\mathbf{r})\, d\mathbf{r} \\ &= -\mathbf{M}_{200-100}\end{aligned}$$

where use has been made of eqn. (11.6). Therefore $\mathbf{M}_{200-100} = 0$. Thus $2s \to 1s$ dipole transition is forbidden as only those transitions are allowed for which $\mathbf{M} \neq 0$ (see section 13.6).

11.4 Replacing t by $-t$ in eqn. (17*a*),

$$\hat{H}\,\psi(\mathbf{r}, -t) = -i\hbar\,\partial\,\psi(\mathbf{r}, -t)/\partial t$$

and then taking the complex conjugate, we get

$$\hat{H}\,\psi^*(\mathbf{r}, -t) = i\hbar\,\partial\,\psi^*(\mathbf{r}, -t)/\partial t \qquad \text{... (11.7)}$$

Operating (17*a*) by $\hat{T}$, one gets

$$\hat{T}\hat{H}\,\psi(\mathbf{r}, t) = \hat{H}\hat{T}\,\psi(\mathbf{r}, t) = i\hbar\,\partial\hat{T}\,\psi(\mathbf{r}, t)/\partial t \qquad \text{... (11.8)}$$

Comparing eqns. (11.7) and (11.8), one gets the required result, that is

$$\hat{T}\,\psi(\mathbf{r}, t) \equiv \psi_{rev}(\mathbf{r}, t) = \psi^*(\mathbf{r} - t) \qquad \text{... (11.9)}$$

From the comparison of eqn. (11.7) with eqn. (17*a*), it may be noted that if $\psi(\mathbf{r}, t)$ is the solution of Schrödinger equation, then $\hat{T}\,\psi(\mathbf{r}, -t) = \psi^*(\mathbf{r}, -t)$ is also the solution. $\psi^*(\mathbf{r}, -t)$ is often referred to as the time reversed solution of the Schrödinger equation with respect to $\psi(\mathbf{r}, t)$. The behaviour of wave equation exhibited by eqn. (11.7) is called its invariance under time reversal.

❑❑❑

Chapter 12

APPROXIMATION METHODS

12.1 INTRODUCTION

So far we have considered the problems in which potential energy was either constant (for example, in square well potential and potential step problems) or had simple forms (like in harmonic oscillator, rigid rotator and hydrogen atom problems) and solved them exactly. If the potential energy does not have a simple form, the solution of Schrödinger equation even in one dimension, that is the solution of the differential equation,

$$i\hbar\, \partial \Psi(x, t)/\partial t = H\Psi(x, t) \qquad \text{... (1)}$$

becomes a complicated problem and it requires the use of approximation method. There exist several methods of approximation for tackling various types of problems in quantum mechanics. These methods could be broadly divided into two categories : those for time independent problems (where time independent Schrödinger equation is to be solved) and those for time dependent problems (where time dependent Schrödinger equation is to be solved). In this chapter, we would consider some of these approximation methods. In the first category, we would discuss the WKB approximation, the time independent perturbation theory and the variational method. In the second category, we would discuss the time dependent perturbation theory.

12.2 WKB APPROXIMATION[1]

This approximation or method is named after Wentzel, Kramers and Brillouin who have put forward its theory. It is also called semi-classical or quasi-classical since it deals with situations in which $\hbar$ is small (classical mechanics corresponds to $\hbar \to 0$). Still another name attached with the method is called phase integral method (why so

[1]G. Wentzel, Z. Physik 38, 518 (1926); H.A. Kramers, Z. Physik 39, 828 (1926); L. Brillouin, Comptes Rendus 183, 24 (1926).

would be evident later). The method is suited only to problems in one dimension or to problems which can be reduced to one dimensional form.

In the present case of time independent or stationary problems where Hamiltonian of the system is independent of time, $\Psi(x, t)$ can be expressed as a product of two functions, one being function of space coordinate x and the other function of time, that is $\Psi(x, t) = \varphi(x) f(t)$, we have to solve the time independent Schrödinger equation,

$$\frac{d^2\varphi(x)}{dx^2} + \frac{2m}{\hbar^2}[E - V(x)]\,\varphi(x) = 0 \qquad \text{... (2)}$$

The basic idea behind the WKB method is simple. If V is constant, that is independent of x, the above equation has solutions of the form $e^{\pm ikx}$ (or $e^{\pm ipx/\hbar}$). This suggests that if V or effectively p while no longer constant varies only slowly with x, we might try a solution of the form.

$$\varphi(x) = A_0\, e^{\pm iS(x)/\hbar},\ A_0 = \text{constant} \qquad \text{... (3)}$$

where $S(x)$ is not simply proportional to x.

Substituting $\varphi(x)$ from eqn. (3) into eqn. (2) and then multiplying the resulting equation by $-\hbar^2$, we get

$$\left(\frac{dS}{dx}\right)^2 - i\hbar\frac{d^2S}{dx^2} = 2m\,[E - V(x)] \qquad \text{... (4)}$$

which is nothing but the Hamilton- Jacobi equation[2] of classical mechanics for $\hbar \to 0$. That is, if we drop the second term on the left hand side of the above quantum mechanical equation, we get the well known Hamilton–Jacobi equation of classical mechanics. In other words the finite value of $\hbar$ is responsible for the difference between classical and quntum mechanics. Since $\hbar$ is a constant quantity, such a transition must be considered to be merely formal. It is justified only when the term in eqn. (4) containing $\hbar$ is small compared to the other term. To solve eqn. (4), we can, therefore try an expansion of $S(x)$ in powers of $\hbar$, that is, we can write,

$$S(x) = S_0(x) + \hbar\, S_1(x) + \hbar^2\, S_2(x) + \ldots$$

Since $\hbar$ is small, the terms containing $\hbar^2$ and higher powers of $\hbar$ can be neglected. Hence

$$S(x) = S_0(x) + \hbar\, S_1(x) \qquad \text{... (5)}$$

Note that the classical approximation consists in neglecting all powers of $\hbar$ higher than zero. Here terms upto first power of $\hbar$ are retained and hence the approximation is termed as semi-classical or corresponds to WKB approximation. After substituting S(x) from eqn. (5) into eqn. (4), we get

$$(S_0' + \hbar\, S_1')^2 - i\,\hbar\,(S_0'' + \hbar\, S_1'') = 2m\,\{E - V(x)\}$$

where prime denotes differentiation with respect to the coordinate x.

[2]H. Goldstein, "Classical Mechanics", 2nd ed. Sec. 10-1, Addison–Wesley, Philippines, 1980.

Equating the coefficients of $\hbar^0$ and $\hbar^1$ on both sides of the above equation, we get

$$\left(\frac{dS_0}{dx}\right)^2 = 2\text{m}\,\{E - V(x)\} \qquad \text{... (6a)}$$

and

$$2\frac{dS_0}{dx}\frac{dS_1}{dx} - i\frac{d^2S_0}{dx^2} = 0 \qquad \text{... (6b)}$$

From eqn. (6*a*), we have

$$\frac{dS_0}{dx} = \pm\,[2m\,\{E - V(x)\}]^{1/2}$$

or

$$S_0 = \pm\int [2m\,\{E - V(x)\}]^{1/2}\,dx \qquad \text{... (7a)}$$

Eqn. (6*b*) can be written as

$$\frac{dS_1}{dx} = \frac{i}{2}\frac{d^2S_0}{dx^2}\Big/\frac{dS_0}{dx} = \frac{i}{2}\frac{d}{dx}\left\{\ln\left(\frac{dS_0}{dx}\right)\right\}$$

Integrating, we find

$$S_1 = \frac{i}{2}\ln\frac{dS_0}{dx} = i\ln\left(\frac{dS_0}{dx}\right)^{1/2}$$

or

$$-iS_1 = \ln\left(\frac{dS_0}{dx}\right)^{1/2} \qquad \text{... (7b)}$$

Note that in eqns. (7) no constants of integration have been introduced as they only affect the normalization which is best accomplished after all the desired approximations have been made. Substituting for $S(x)$ from eqn. (5) into eqn. (2), we get

$$\varphi(x) = A_0\, e^{i\,(S_0 + \hbar\, S_1)/\hbar}$$

$$= A_0\, e^{iS_1}\, e^{iS_0/\hbar}$$

$$= \frac{A_0}{e^{-iS_1}}\, e^{iS_0/\hbar} \qquad \text{... (8)}$$

In the classical region of motion where $E - V(x)$ is everywhere non-negative, $[2m\,\{E - V(x)\}]^{1/2} = p(x)$ being the momentum of the particle of mass m moving in the x–direction under the action of the force $F = -dV/dx$, the solution (that is, eqn. (8)) using eqns. (7) becomes

$$\varphi\,(x) = A_0\, p^{-1/2}\, e^{+\,i/\hbar \int^{x} p\,(x')\,dx'}$$

The most general solution (or the WKB wave function) in the said region would, therefore, be given by

$$\varphi(x) = A_1 P^{-1/2} e^{i/\hbar \int^{x} p(x')dx'}$$

$$+B_1 p^{-1/2} e^{-i/\hbar \int^{x} p(x')dx'}, E > V(x) \quad \text{... (9)}$$

where A_1 and B_1 are arbitrary constants. The factor $p^{-1/2}$ in the above wave function has a simple interpretation. The probability of finding the particle at a point with coordinate between x and $x + dx$ is given by $|\varphi|^2 dx$, that is proportional to $1/p$ or inversely proportional to the velocity of classical particle. This is exactly what we should expect for a quasi-classical particle since in classical motion, the time spent by a particle in the segment dx is inversely proportional to the velocity of the particle.

In the region where $V(x) > E$ or where $p(x)$ is imaginary or where the classical motion is forbidden, we write $p(x) = i\ |p(x)|$. Substituting this into eqn. (9), the exponents become real and hence the most general solution (or the WKB wave function) in the non-classical region would be given by

$$\varphi(x) = A_2 |p|^{-1/2} e^{-1/\hbar \int^{x} |p(x')| dx'}$$

$$+ B_2 |P|^{-1/2} e^{1/\hbar \int^{x} |p(x')| dx'}, V(x) > E \quad \text{... (10)}$$

where A_2 and B_2 are arbitrary constants. Note that the lower limit for the integrals in eqns. (9) and (10) would be classical turning points as will be seen shortly.

One observes that the WKB solutions, that is eqn. (9) or (10) approach infinity as $p(x)$ approaches zero. That is, the solutions do not hold near the classical turning points (at the values of x where $E - V(x) = 0$ or $E = V(x)$) where the particle stops or comes to halt ($p(x) = 0$) and then begins to move in the opposite direction. This limitation on the applicability of the WKB solutions is actually the result of the approximation that the second term on the left hand side of eqn. (4) is taken to be small compared to the first term, that is we have

$$\hbar\,|\,S''/S'^2\,| \ll 1$$

$$|d(\hbar/S')/dx| \ll 1$$

In the first approximation (where we write $S(x) = S_0(x)$, we have from eqn. (7*a*), $S'(x) = S'_0(x) = p(x)$ so that the above condition can be written as

$$|d[\hbar/p(x)]/dx| \ll 1 \quad \text{... (11a)}$$

which can further be written as

$$d(\lambda/2\pi)/dx \ll 1 \quad \text{... (11b)}$$

where $\lambda(x) = h/p(x) = 2\pi\,\hbar/p(x)$ is the de Broglie wave length of the particle or is the wavelength corresponding to the value of momentum at point x. The condition (11*a*) can also be expressed as

$$\hbar\,|\{dp(x)/dx\}/p^2(x)| \ll 1$$

or

$$\lambda/2\pi\,|dp(x)/dx|/|p(x)| \ll 1 \qquad \text{... (11c)}$$

Therefore, WKB solutions are applicable in the regions where the wave length of the particle changes slowly with x (eqn. 11*b*). At the turning points λ (or $p(x)$) varies rapidly or $V(x)$ has steep behaviour. The above condition (eqn. (11*c*)) also amounts to the fact that for the applicability of the WKB approximation the fractional change of momentum over a de Broglie wave length of the particle be small.

The condition (11*a*) can still be written in another form by noting that

$$\frac{dp}{dx} = \frac{d}{dx}\,[2m\,\{E - V(x)\}]^{1/2}$$

$$= \frac{1}{2}\,[2m\,\{E - V(x)\}]^{1/2}\,2m\left(-\frac{dV}{dx}\right)$$

$$= mF/p(x)$$

where $F = -\dfrac{dV}{dx}$ is the classical force acting on the particle in the external field. In terms of this force, we find

$$m\hbar\,|F|/|p(x)|^3 \ll 1 \qquad \text{... (11d)}$$

that is, the WKB approximation becomes inapplicable if the momentum of the particle is too small, In particular, it is clearly inapplicable near turning points where $p(x) = 0$.

12.3 CONNECTION FORMULAE

Before we proceed further, it is necessary to point out that the WKB method can be used in the problems in which we have two regions where the WKB approximation is valid but which are separated by a classical turning point. See Fig. 12.1 in which a particle of energy E encounters a potential function $V(x)$. Region *I* ($E > V(x)$, classical region) and II ($E \langle\, V(x)$, non-classical region) are separated by a classical truning point a, that is, the point corresponding to $E = V(x)$.

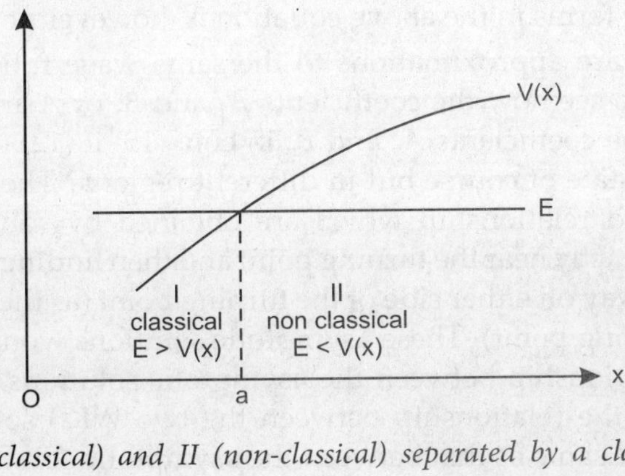

Fig. 12.1 *Region I (classical) and II (non-classical) separated by a classical turning point a*

It is also necessary to mention that the above method can be applied only if we can find a way to extend the WKB solution from one region to the other through the turning point or if we can obtain a connection between the WKB wave functions in the two regions. The equations connecting the WKB wave functions in the classical and in the non-classical regions are given the name connection formulae. Let us try to discuss the method for obtaining the same.

In the region I ($E \rangle V(x)$), the WKB wave function from eqn. (9) is given by

$$\varphi_1(x) = A_1 p^{-1/2} e^{i/\hbar \int_a^x p(x')\,dx'} + B_1 p^{-1/2} e^{-i/\hbar \int_a^x p(x')\,dx'} \quad \text{... (12)}$$

which can be written as

$$\varphi_1(x) = A\, p^{-1/2} \sin\left(\frac{1}{\hbar}\int_x^a p(x')\,dx' + \frac{\pi}{4}\right)$$

$$+ B\, p^{-1/2} \cos\left(\frac{1}{\hbar}\int_x^a p(x')\,dx' + \frac{\pi}{4}\right) \quad \text{... (13)}$$

where

$$A = (A_1 - iB_1)\, e^{i\pi/4} \quad \text{... (14a)}$$

$$B = -(A_1 + iB_1)\, e^{-i\pi/4} \quad \text{... (14b)}$$

The reason for writing $\varphi_I(x)$ in the peculiar form (13) would be clear later. In region II ($E < V(x)$), the WKB wave function from eqn. (12) using $p(x) = i\,|p(x)|$ would be given by

$$\varphi_{II}(x) = A_2 |p|^{-1/2} e^{-1/\hbar \int_a^x |p(x')|dx'} + B_2 |p|^{-1/2} e^{1/\hbar \int_a^x |p(x')|dx'} \quad \text{... (15)}$$

From eqns. (13) and (15), it is clear that the WKB wave function of the particle is oscillatory in the classical region and exponential in the non-classical region. Note that the first term in eqn. (15) decreases exponentially with increasing x whereas the second term increases exponentially. At a slightly later stage, we would see that the retention of both the terms in the above equation is, however not permissible.

Now φ_I and φ_{II} are approximations to the same wave function but in different regions. Let us try to see how the coefficients A_1 and B_1 or A and B in eqns. (12) and (13) are related to the coefficients A_2 and B_2 in eqn. (15) if (12) or (13) and (15) are to represent the same state of course but in different regions? The connection formulae provide the required relationship which are obtained by solving the Schrödinger equation (eqn. (2)) exactly near the turning point and then finding the asymptotic forms of the solution far away on either side of the turning point (as the WKB approximation is invalid at the turning point). These asymptotic solutions would resemble the WKB solutions. If the relationship between the asymptotic solutions in the two regions is known, we can get the relationship between the two WKB solutions assuming the relationship to be the same as that between the asymptotic solutions.

To achieve the goal, let us assume (as $V(x)$ varies slowly with x) that $V(x)$ is linear in the neighbourhood of the turning point a : That is, we can write (retaining only the first two terms of the expansion),

$$V(x) \underset{x \sim a}{\simeq} V(a) + (x-a)\left.\frac{dV}{dx}\right|_{x=a} = E + C(x-a) \qquad \text{... (16)}$$

(as $x = a$ corresponds to the turning point and at the turning point $E = V(x) = V(a)$) where

$$C = \left.\frac{dV}{dx}\right|_{x=a} = -\left(-\left.\frac{dV}{dx}\right|_{x=a}\right) = -F_0,\ F_0 = -\left.\frac{dV}{dx}\right|_{x=a} \qquad \text{... (17)}$$

is slope of the potential curve at $x = a$. Thus, the problem in this region is one of the movement in a homogenous field (in an electric field of intensity ε, this force is $F_0 = e\varepsilon$, where e is the charge on the particle)

Use of eqn. (16) gives

$$p^2(x) = 2m\{E - V(x)\} \underset{x \sim a}{\simeq} -2m\,C(x-a) \qquad \text{... (18)}$$

Substituting above into eqn. (2) leads to

$$\frac{d^2Z}{d\xi^2} - \xi Z = 0 \qquad \text{... (19)}$$

where

$$\xi = (2m\,C/\hbar^2)^{1/3}(x-a) \qquad \text{... (20)}$$

and

$$Z(\xi) = \varphi(x)$$

Note that the regions I and II and the turning point correspond, respectively to $\xi < 0$, $\xi > 0$ and $\xi = 0$. The solution of eqn. (19) are known as the Airy functions and are given by (linearly independent solutions)

$$Z_1 = (\xi/3\pi^2)^{1/2} K_{1/2}(\zeta) \qquad \text{... (21a)}$$

and

$$Z_2 = (\xi/3)^{1/2}[I_{1/3}(\zeta) + I_{-1/3}(\zeta)] \qquad \text{... (21b)}$$

in which $\zeta = \frac{2}{3}(\xi)^{3/2}$ and I and K are the modified Bessel functions of the first and second kinds, respectively.

We are interested only in the asymptotic forms of Z_1 and Z_2. These are given by,

$$Z_1 \underset{\xi \ll 0}{\simeq} (-\pi^2\xi)^{-1/4}\,\text{Sin}\left\{\frac{2}{3}(-\xi)^{3/2} + \frac{\pi}{4}\right\} \qquad \text{... (22a)}$$

$$Z_1 \underset{\xi \gg 0}{\simeq} \frac{1}{2}(\pi^2\xi)^{-1/4} e^{-2/3\,\xi^{3/2}} \qquad \text{... (22b)}$$

$$Z_2 \underset{\xi \ll 0}{\simeq} (-\pi^2\xi)^{-1/4}\,\text{Cos}\left\{\frac{2}{3}(-\xi)^{3/2} + \frac{\pi}{4}\right\} \qquad \text{... (23a)}$$

$$Z_2 \underset{\xi \gg 0}{\simeq} (\pi^2\xi)^{-1/4} e^{2/3\,\xi^{3/2}} \qquad \text{... (23b)}$$

Now for ξ negative (that is, for $x < a$),

$$\frac{2}{3}(-\xi)^{3/2} = \int_0^{\xi} (-\xi')^{1/2} d(-\xi')$$

After substituting for ξ' from eqn. (20), we get

$$\frac{2}{3}(-\xi)^{3/2} = \int_x^a \left(2m\frac{C}{\hbar^2}\right)^{1/6} (a-x')^{1/2}\left(2m\frac{C}{\hbar^2}\right)^{1/3} dx'$$

$$= \int_x^a \left[\left(2m\frac{C}{\hbar^2}\right)(a-x')\right]^{\frac{1}{2}} dx'$$

$$= \frac{1}{\hbar}\int_x^a p(x')\,dx' \qquad \text{... (24a)}$$

where use of eqn. (18) has been made. Also,

$$(-\xi)^{1/4} = \left(2m\frac{C}{\hbar^2}\right)^{1/12} (a-x)^{1/4}$$

Using eqn. (20) and then eqn. (18), we get

$$(-\xi)^{1/4} = (2m\,C\,\hbar)^{-1/6}\{2m\,C\,(a-x)\}^{1/4}$$

$$= (2m\,C\,\hbar)^{-1/6}\,p(x)^{1/2} \qquad \text{... (25a)}$$

Similarly, for ξ positive (or $x > a$), it can be shown that

$$\frac{2}{3}\xi^{3/2} = \frac{1}{\hbar}\int_a^x |p(x')|\,dx' \qquad \text{... (24b)}$$

and

$$\xi^{1/4} = (2m\,C\,\hbar)^{-1/6}\,|p(x)|^{1/2} \qquad \text{... (25b)}$$

Substituting from eqns. (24*a* – 25*b*) into eqns. (22*a* – 23*b*), we get the asymptotic forms of Z_1 and Z_2 given by

$$Z_1 = \varphi_1^{OSC} \underset{x \ll a}{\backsim} \alpha\,p^{-1/2}\sin\left(\frac{1}{\hbar}\int_x^a p(x')\,dx' + \frac{\pi}{4}\right) \qquad \text{... (26a)}$$

$$Z_1 = \varphi_1^{\exp} \underset{x \gg a}{\backsim} \frac{\alpha|p|^{-1/2}}{2} e^{-\frac{1}{\hbar}\int_a^x |p(x')|\,dx'} \qquad \text{... (26b)}$$

$$Z_2 = \varphi_2^{OSC} \underset{x \ll a}{\backsim} \alpha|p|^{-1/2}\cos\left(\frac{1}{\hbar}\int_x^a p(x')\,dx' + \frac{\pi}{4}\right) \qquad \text{... (27a)}$$

$$Z_2 = \varphi_2^{\exp} \underset{x \gg a}{\backsim} \alpha|p|^{-1/2} e^{1/\hbar\int_a^x |p(x')|\,dx'} \qquad \text{... (27b)}$$

where

$$\alpha = (2m\, C\, \hbar/\pi^3)^{1/6} \qquad \text{... (28)}$$

Since φ_1^{exp} and φ_1^{OSC} (similarly of course φ_2^{exp} and φ_2^{OSC} are approximations to the same wave function, φ_1^{exp} (or φ_2^{exp}) is the continuation into the non-classical region of φ_1^{OSC} (or φ_2^{OSC}) the classical region. Thus, the connection between the approximate forms of the wave functions in the classical and in the non-classical regions are given by the formulae.

Classical region (Oscillatory)	Non-classical region (exponential)

$$\alpha\, p^{-1/2} \sin\left(\frac{1}{\hbar}\int_x^a p(x')\,dx' + \frac{\pi}{4}\right) \leftrightarrow \frac{\alpha|p|^{-1/2}}{2}\, e^{-1/\hbar \int_a^x |p(x')|\,dx'} \qquad \text{... (29a)}$$

$$\alpha\, p^{-1/2} \cos\left(\frac{1}{\hbar}\int_x^a p(x')\,dx' + \frac{\pi}{4}\right) \leftrightarrow \alpha|p|^{-1/2}\, e^{1/\hbar \int_x^a |p(x')|\,dx'} \qquad \text{... (29b)}$$

We note from the above that a wave function that is represented by the Sine function in the classical region becomes a decreasing exponential in the non-classical region whereas an increasing exponential in the non-classical region corresponds to the Cosine function in the classical region. We may also note that the constant multiplying the decreasing exponential is half that multiplying the Sine function whereas the constant multiplying the increasing exponential is same as that multiplying the Cosine function.

The wave function of the physical system would be a general solution of the Airy equation (eqn. (19)) and thus would be linear combination of φ_1 and φ_2. Therefore

$$\varphi^{\text{OSC}}(x) \underset{x \ll a}{\simeq} a\, p^{-1/2} \sin\left(\frac{1}{\hbar}\int_x^a p(x')\,dx' + \frac{\pi}{4}\right) + a\, p^{-1/2} \cos\left(\frac{1}{\hbar}\int_x^a p(x')\,dx' + \frac{\pi}{4}\right) \qquad \text{... (30a)}$$

and

$$\varphi^{\text{exp}}(x) \underset{x \gg a}{\simeq} \frac{\alpha|p|^{-1/2}}{2}\, e^{-1/\hbar \int_a^x |p(x')|\,dx'} + \alpha|p|^{-1/2}\, e^{1/\hbar \int_a^x |P(x')|\,dx'} \qquad \text{... (30b)}$$

Comparing the above equations with eqns. (13) and (15), that is, with φ_{I} and φ_{II}, we see that φ^{osc} is the WKB wave function φ_1 in the classical region and φ^{exp} is the WKB wave function φ_{II} in the non-classical region. We also infer that

$$Ap^{-1/2} = \alpha p^{-1/2},\ Bp^{-1/2} = \alpha p^{-1/2} \text{ or } A = \alpha = B$$

and

$$A_2|p|^{-1/2} = \alpha\,|p|^{-1/2},\ B_2\,|p|^{-1/2} = \alpha\,|p|^{-1/2} \text{ or } A_2 = \frac{\alpha}{2},\ B_2 = \alpha$$

which amounts to

$$A_2 = \frac{\alpha}{2} = \frac{A}{2}$$

$$B_2 = \alpha = B = \left(\frac{2m\,C\hbar}{\pi^3}\right)^{1/6} = \left(\frac{2m\,\hbar}{\pi^3}\frac{dV}{dx}\bigg|_{x=a}\right)^{1/6} \quad \text{... (31)}$$

in which value of C has been substituted.

Thus eqns. (29*a*, *b*) give the connection between the WKB wave functions in the classical and in the non-classical regions, respectively. The said equations are, therefore the connection formulae of the WKB approximation.

12.4 ANOTHER FORM OF VALIDITY CONDITION

From eqn. (18)

$$|p|^2 = 2mC\,|x - a|$$

Differentiating above with respect to x, we get,

$$|p|\left|\frac{dp}{dx}\right| = mC$$

which on substitution into the validity condition (11*c*) lead to

$$\frac{\lambda}{2\pi}\left|\frac{dp}{dx}\right|/|p| = \frac{\lambda}{2\pi}\frac{mC}{|p|^2} = \frac{\lambda\, mC}{2\,\pi\, 2\, mC\,|x-a|} \ll 1$$

or

$$|x - a| \gg \frac{\lambda}{4\,\pi} \quad \text{... (32)}$$

That is, the WKB approximation is applicable at distances (values of x) from the turning point satisfying the above inequality, where λ is the wavelength corresponding to the value of momentum at x. From above it can further be concluded that for WKB method to be applicable, it is necessary that the separation between turning points be atleast several de Broglie wavelengths, otherwise it cannot be applied (see the case in Fig. 12.2). Note that the derivation in sec, 12.3 is in accordance with the restriction (32). Eqns. (26*a* – 27*b*) are written only for x very far from the turning point.

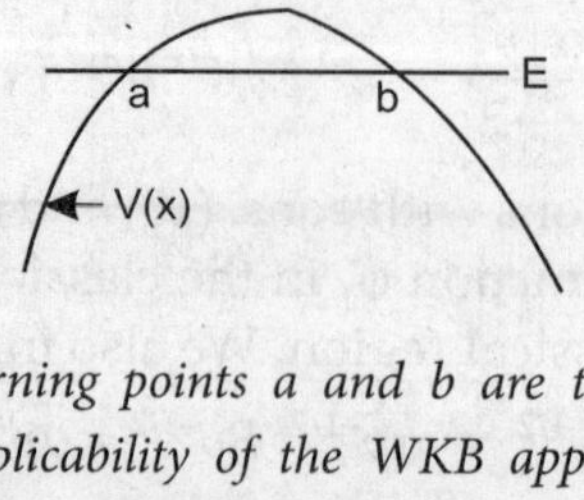

Fig. 12.2 *The two turning points a and b are too near to each other for the applicability of the WKB approximation*

12.5 APPLICATION TO BOUND STATE

Let us apply WKB method for the determination of energy levels and wave functions of a particle of mass m moving in one-dimensional potential well displayed in Fig. 12.3. There are three regions separated by two turning points a and b determined by the condition.

$$V(a) = V(b) = E$$

In the regions I and III (non-classical regions), we can use the functions (10) of the WKB approximation. The wave functions in these regions which are exponentially decreasing when we go away from the turning points have the forms (since the system is bound, the wave function should go to zero for $x \to \pm\infty$).

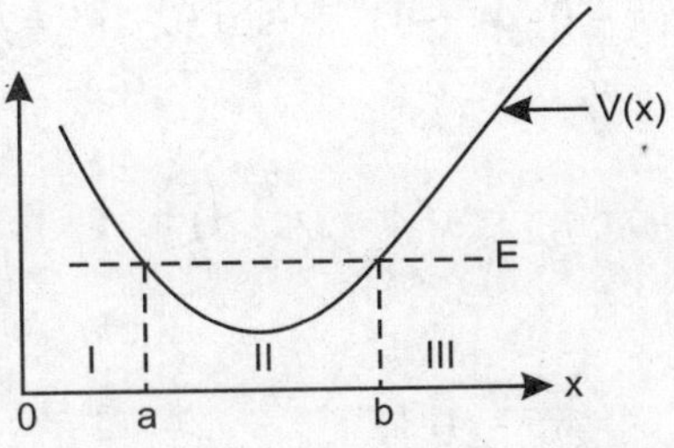

Fig. 12.3 *Potential well showing three regions separated by two turning points*

$$\varphi_I(x) = A_2|p|^{-1/2} e^{-1/\hbar \int_x^a |p(x')|dx'}, x < a \quad \text{... (33)}$$

and

$$\varphi_{III}(x) = A_3 \,|p|^{-1/2} e^{-1/\hbar \int_b^x |p(x')|dx'}, x > b \quad \text{... (34)}$$

where A_2 and A_3 are arbitrary constants.

According to the connection formula (29*a*) the wave function (oscillatory solution) in the region II (classical) would be given by

$$\varphi_{II}(x) = 2A_2 p^{-1/2} \sin\left(\frac{1}{\hbar}\int_a^x p(x')\,dx' + \frac{\pi}{4}\right), a \le x \le b \quad \text{... (35a)}$$

$$= 2A_3 p^{-1/2} \sin\left(\frac{1}{\hbar}\int_x^b p(x')\,dx' + \frac{\pi}{4}\right), a \le x \le b \quad \text{... (35b)}$$

as the constant factor in place of $\frac{\alpha}{2}$ in the non-classical region is α in the classical region.

Since

$$\int_a^x p(x')\,dx' = \int_a^b p(x')\,dx' - \int_x^b p(x')\,dx',$$

we can use the substitution,

$$\sin\left(\frac{1}{\hbar}\int_a^x p(x')\,dx' + \frac{\pi}{4}\right) = \sin\left\{\left(\frac{1}{\hbar}\int_a^b p(x')\,dx' + \frac{\pi}{2}\right) - \left(\frac{1}{\hbar}\int_x^b p(x')\,dx' + \frac{\pi}{4}\right)\right\}$$

In eqn. (35*a*) and then equating the result to (35*b*), one gets

$$A_2 \sin\left\{\left(\frac{1}{\hbar}\int_a^b p(x')\,dx' + \frac{\pi}{2}\right) - \left(\frac{1}{\hbar}\int_x^b p(x')\,dx' + \frac{\pi}{4}\right)\right\}$$

$$= A_3 \sin\left(\frac{1}{\hbar}\int_x^b p(x')\,dx' + \frac{\pi}{4}\right)$$

Comparing above with the identity,

$$\sin(n'\pi - \theta) = (-1)^{n'-1}\sin\theta, \quad n' = 1, 2, 3, \ldots \qquad \ldots (36)$$

one gets

$$\frac{1}{\hbar}\int_a^b p(x)\,dx + \frac{\pi}{2} = n'\pi = (n+1)\,\pi,\ n = 0, 1, 2, \ldots \qquad \ldots (37a)$$

and

$$\frac{A_3}{A_2} = (-1)^n \qquad \ldots (37b)$$

If we introduce the phase integral (called so because it is equal to the area enclosed by the curve representing the motion in the phase space formed by $p(x)$ and (x)

$$\oint p(x)\,dx = 2\int_a^b p(x)\,dx$$

along the path from a to b and back from b to a, that is integral over one complete period of the classical motion, the eqn. (37*a*), can be written as

$$\frac{2\pi}{h}\int_a^b p(x)\,dx + \frac{\pi}{2} = \frac{\pi}{h}\oint p(x)\,dx + \frac{\pi}{2} = (n+1)\,\pi$$

or

$$\oint p(x)\,dx = \left(n + \frac{1}{2}\right)h, \quad n = 0, 1, 2, \ldots \qquad \ldots (38)$$

Since the term on the left hand side of the above equation represents the phase integral, the WKB method is also referred to as the phase integral method. If the potential is known as a function of x, the said integral (which contains $p(x) = \sqrt{2m\{E - V(x)\}}$ can be evaluated, yielding $E(= E_n)$ in terms of n and the parameters of the potential. Thus n designates the different energy levels of the system. In other words, eqn. (38) determines the stationary states of the particle in the WKB approximation. It corresponds to the Bohr-Sommerfeld quantization rule in the old quantum theory except for the fact that the integral quantum numbers are replaced by odd half-integral quantum numbers. The term $\frac{h}{2}$ (which is absent in the Bohr-Sommerfeld formula) brings eqn. (38) in better agreement with the exact result.

Since the derivation of eqn. (38) assumed that the turning points were so well separated that we could employ the asymptotic solutions (22*a* – 23*b*) for the Airy functions, the condition for the validity of eqn. (38) would be $n >> 1$. It may be mentioned here that though the quantization rule (38) is really applicable only for large n but in the cases of energy levels in a Coulomb field and those of a harmonic oscillator, it gives for the function E_n an expression which is the exact one.

The wave function φ decreases exponentially outside the interval a to b, but inside that interval, that is $a \le x \le b$, function (eqn. (35*a*))

$$\varphi(x) = 2A_2 p^{-1/2} \sin\left(\frac{1}{\hbar}\int_a^x p(x')\,dx' + \frac{\pi}{4}\right)$$

oscillates and hence can be taken as the approximate wave function of the system. The phase of the Sine function (eqn. (35*a*)) varies from $\frac{\pi}{4}$ at $x = a$ to $\left(n + \frac{3}{4}\right)\pi$ at $x = b$ (use eqns. (35*a*) and (38*a*)). Thus the Sine function vanishes n times (or the function has n zeros or nodes) in the range a to b. As has been shown earlier, the number n is large in the WKB approximation. In that case, the Sine function oscillates rapidly in the interval a to b so that the square of the Sine function can be approximated to its average value $\frac{1}{2}$ when we determine the constant A_2 from the condition that the wave function is normalized. Then

$$1 = \int_a^b |\varphi(x)|^2\,dx = 4|A_2|^2 \int_a^b \frac{dx}{2p(x)}$$

If we bear in mind that

$$m\int_a^b \frac{dx}{p(x)} = \frac{T}{2}$$

is the time it takes the particle to move through the interval a to b, we can introduce the angular frequency $\omega = \frac{2\pi}{T}$ of the periodic motion of the particle. Hence

$$1 = |A_2|^2 \frac{2\pi}{m\omega}$$

or

$$A_2 = \left(\frac{m\omega}{2\pi}\right)^{1/2} \quad \text{... (39)}$$

Therefore, the normalized wave function would be given by

$$\varphi(x) = \left(\frac{2m\omega}{\pi p}\right)^{1/2} \sin\left(\frac{1}{\hbar}\int_a^b p(x')\,dx' + \frac{\pi}{4}\right) \quad \text{... (40)}$$

12.6 WKB SOLUTION OF THE RADIAL WAVE EQUATION

We would show here that in case a three dimensional problem can be reduced to one dimensional form, the WKB method can be applied. The radial equation for a particle in a central force field described by $V(\mathbf{r}) = -e^2/\mathbf{r}$ (the case of hydrogen atom) is given by

$$\frac{d^2 u(r)}{dr^2} + \frac{2m}{\hbar^2}\left[E - V(r) - \frac{l(l+1)\hbar^2}{2mr^2}\right]u(r) = 0 \qquad \text{... (41)}$$

where $u(\mathbf{r}) = \mathbf{r}\,\mathbf{R}(\mathbf{r})$ with $\mathbf{R}(\mathbf{r})$ being the radial wave function.

Note that though the above equation looks like eqn. (2) but these two differ because $0 \le \mathbf{r} \le \infty$ in eqn. (41) whereas $-\infty \le x \le \infty$ in the case of eqn. (2). To remove this difference, let us change the variable in eqn. (41) to x so that $\mathbf{r} = 0$ corresponds to $x = -\infty$ and $\mathbf{r} = \infty$ corresponds to $x = \infty$. This is achieved if we introduce the transformation,

$$\mathbf{r} = e^x \qquad \text{... (42)}$$

The eqn. (41) then takes the form,

$$\frac{d^2 u}{dx^2} - \frac{du}{dx} + \frac{2m}{\hbar^2}\left[E - V(e^x) - \frac{l(l+1)\hbar^2}{2m}e^{-2x}\right]e^{2x}u = 0 \qquad \text{... (43)}$$

To bring above equation into the form that of eqn. (2) or in the form of one dimensional Schrödinger equation, we introduce a new variable

$$u(x) = w(x)\,e^{-x/2} \qquad \text{... (44)}$$

Then $w(x)$ satisfies the equation of the desired form

$$\frac{d^2 w(x)}{dx^2} + \frac{2m}{\hbar^2}\left[E - V(e^x) - \frac{\hbar^2}{2m}\left(l + \frac{1}{2}\right)^2 e^{-2x}\right]e^{2x}w(x) = 0 \qquad \text{... (45)}$$

The results of WKB approximation can now be used. The quantization condition (38) in the light of the above equation can be expressed as

$$\int_a^b \left[2m\left\{E - V(e^x) - \left(l + \frac{1}{2}\right)^2 \frac{\hbar^2}{2m}e^{-2x}\right\}e^{2x}\right]^{1/2} dx = \left(n + \frac{1}{2}\right)\pi\hbar,\ n = 0, 1, 2, \ldots$$

Transforming $\hbar$ to the L.H.S of the above equation, we get

$$\int_a^b \left[\frac{2m}{\hbar^2}\left\{E - V(e^x) - \left(l + \frac{1}{2}\right)^2 \frac{\hbar^2}{2m}e^{-2x}\right\}e^{2x}\right]^{1/2} dx = \left(n + \frac{1}{2}\right)\pi$$

As $\mathbf{r} = e^x$ (or $d\mathbf{r} = e^x\,dx$) and as $V(\mathbf{r}) = -e^2/\mathbf{r}$, after substituting these in the above equation, one gets

$$\int_{r_1}^{r_2} \left[\frac{2m}{\hbar^2}\left\{E + e^2/r - \left(l + \frac{1}{2}\right)^2 \frac{\hbar^2}{2mr^2}\right\}\right]^{1/2} d\mathbf{r} = \left(n + \frac{1}{2}\right),\ n_{\mathbf{r}} = 0, 1, 2, \ldots \qquad \text{... (46)}$$

which is in the form of the original variable $\mathbf{r}$. Also note that the n has been replaced by $n_{\mathbf{r}}$, as n finally would represent the total quantum number.

Writing

$$\xi = \frac{2m e^2 r}{\hbar^2},$$

the following term takes the form,

$$\frac{e^2}{r} - \left(l + \frac{1}{2}\right)^2 \hbar^2 / 2mr^2 = \frac{e^2\, 2me^2}{\xi \hbar^2} - \left(l + \frac{1}{2}\right)^2 \hbar^2 \frac{4\, m^2\, e^4}{2m\, \xi^2\, \hbar^4}$$

$$= \frac{2me^4}{\hbar^2}\left[\frac{1}{\xi} - \left(l + \frac{1}{2}\right)^2 / \xi^2\right]$$

and in the present case of bound state problem, that is, for $E \langle 0$, writing

$$\lambda^2 = -\frac{2\, me^4}{\hbar^2\, E} \qquad \text{... (47)}$$

eqn. (46) finally becomes

$$\int_{\xi_1 = a}^{\xi_2 = b} \left[-\frac{1}{\lambda^2} + \frac{1}{\xi} - \left(l + \frac{1}{2}\right)^2 / \xi^2\right]^{1/2} d\xi = \left(n_r + \frac{1}{2}\right)\pi$$

which can be written as

$$\int_{\xi_1 = a}^{\xi_2 = b} \frac{[(\xi - a)\,(b - \xi)]^{1/2}}{\lambda \xi}\, d\xi = \left(n_r + \frac{1}{2}\right)\pi \qquad \text{... (48)}$$

where a and b are the turning points or are solutions of integrand = 0 or $E - V = 0$, that is

$$-\xi^2 + \lambda^2 \xi - \lambda^2 \left(l + \frac{1}{2}\right)^2 = 0$$

Comparing above equation with

$$(\xi - a)\,(b - \xi) = 0$$

or

$$-\xi^2 + (a + b)\,\xi - ab = 0,$$

one gets

$$a + b = \lambda^2,\; ab = \lambda^2 \left(l + \frac{1}{2}\right)^2 \qquad \text{... (49)}$$

Using the result,

$$\int_a^b \frac{[(x - a)\,(b - x)]^{1/2}}{x}\, dx = \frac{\pi}{2}\,[a + b - 2\,(ab)^{1/2}],$$

from eqn. (48), we get (also using eqn. (49)

$$\frac{1}{\lambda} \times \frac{\pi}{2}\left[\lambda^2 - 2\lambda\left(l + \frac{1}{2}\right)\right] = \left(n_r + \frac{1}{2}\right)\pi$$

which gives,

$$\lambda = 2\,(n_r + l + 1)$$

Squaring both the sides and then using eqn. (47),

$$\lambda^2 = -\frac{2me^4}{\hbar^2 E} = 4\,(n_r + l + 1)^2$$

from which we conclude that the energy levels in the case of hydrogen atom are given by

$$E = E_n = -\frac{me^4}{2\hbar^2 n^2} \qquad \dots (50)$$

which is exact result with $n = n_r + l + 1$, the total or principal quantum number.

12.7 APPLICATION TO BARRIER PENETRATION

The penetration problem in the case of a square potential barrier has already been discussed and solved exactly. For a barrier of arbitrary shape, the Schrödinger equation usually cannot be solved exactly and the WKB approximation is often suited for the problem.

A potential barrier such as mentioned above is shown in Fig. 12.4. Particles of energy E less than the height of the potential barrier are incident from the left of the barrier. At the classical turning point a, some of the particles would be reflected and rest would pass on to the classically forbidden region ($a \le x \le b$) defined by $E < V(x)$. Of these, some would be reflected back at

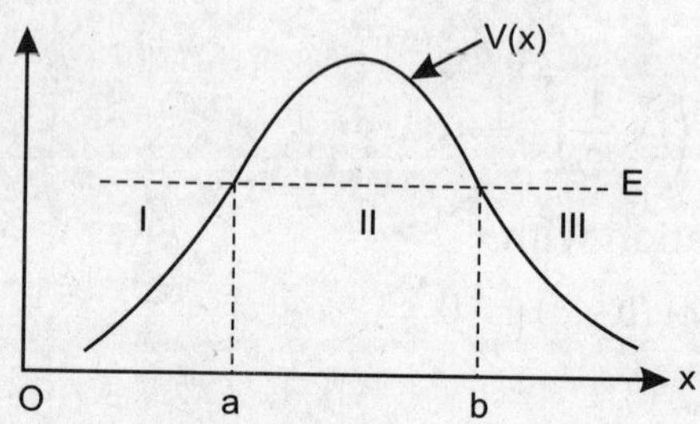

Fig. 12.4 *Potential barrier of arbitrary shape*

the other turning point b but the others would escape to the classically allowed region to the right of the potential barrier, that is $x > b$. Thus there is possibility that a certain fraction of the total number of incident particles would be transmitted through the potential barrier. The ratio of the flux of transmitted particles to that of the incident particles is called the transmission coefficient of the potential barrier.

Let us now obtain an expression for the transmission coefficient in terms of the parameters of the potential barrier such as height, width and shape using the WKB approximation. To do so, we divide the whole region of motion of particle into three parts I, II and III as shown in Fig. 12.4. These are defined respectively, by $x \le a$, $a \le x \le b$ and $x \ge b$. According to eqn. (9), the WKB wave function in region I ($E > V(x)$) is given by

$$\varphi_1(x) = A_1 p^{-1/2} e^{i/\hbar \int_a^x p(x')\,dx'} + B_1 p^{-1/2} e^{-i/\hbar \int_a^x p(x')dx'}, x < a \quad \text{... (51)}$$

Note that the first term on the R.H.S. of the above equation represents the incident particles travelling from left to right whereas the second term represents the particles reflected at the turning point a and travelling from right to left. Since in region III, there is only a transmitted wave (as there is no discontinuity of potential), the WKB solution in the said region would be given by

$$\varphi_{III}(x) = A_3 p^{-1/2} e^{i/\hbar \int_b^x p(x')dx'}, x > b \quad \text{... (52)}$$

Now using the definitions of flux J and density p (= $|\varphi|^2$) and also the fact that the momenta of incident and transmitted particles are the same (same medium hence same k), the expression for the transmission coefficient T would be given by

$$T = \frac{|J_{tr}|}{|J_{\text{inc}}|} = \frac{|\rho_{tr} V_{tr}|^2}{|\rho_{\text{inc}} V_{\text{inc}}|^2} = \frac{|\varphi_{tr} p_{tr}^{1/2}|^2}{|\phi_{inc}\, p_{inc}^{1/2}|^2}$$

$$= \frac{|\varphi_{\text{III}} p^{1/2}|^2}{|\phi_{\text{inc}} p^{1/2}|^2} = \left|\frac{A_3}{A_1}\right|^2 \quad \text{... (53)}$$

Therefore if we can determine A_1 in terms of A_3, T can be calculated from the above expression. For doing so, let us rewrite eqn. (52), using the relation

$$e^{i\phi} = \cos\phi + i\sin\phi \text{ as}$$

$$\varphi_{\text{III}}(x) = A\, p^{-1/2}\left[\cos\left(\frac{1}{\hbar}\int_b^x p(x')\,dx' + \frac{\pi}{4}\right)\right.$$

$$\left. + i\sin\left(\frac{1}{\hbar}\int_b^x p(x')\,dx' + \frac{\pi}{4}\right)\right], x > b$$

where

$$A = A_3 e^{-i\pi/4} = A_3\left(\cos\frac{\pi}{4} - i\sin\frac{\pi}{4}\right) = \frac{(1-i)}{\sqrt{2}} A_3 \quad \text{... (54)}$$

Now using the connection formulae (29a, b), we have the following WKB solution in region II,

$$\varphi_{\text{II}}(x) = A|p|^{-1/2} e^{1/\hbar \int_x^b |p(x')|dx'}$$

$$+ \frac{iA|p|^{-1/2}}{2} e^{-1/\hbar \int_x^b |p(x')|dx'}, a < x < b \quad \text{... (55a)}$$

writing,

$$\int_x^b p(x')\,dx' = \int_a^b p(x')\,dx' - \int_a^x p(x')\,dx'$$

and defining

$$\Theta = e^{1/\hbar \int_a^b |p(x')|\,dx'} \quad \text{... (56)}$$

where Θ measures the height ($|p(x)| = [2m\{V(x) - E\}]^{1/2}$) and width $(b - a)$ of the barrier, eqn. (55*a*) takes the form,

$$\varphi_{II}(x) = A\,|p|^{-1/2}\left[\Theta\, e^{-1/\hbar \int_a^x |p(x')|\,dx'} + \frac{i}{2\Theta} e^{1/\hbar \int_a^x |p(x')|\,dx'}\right],\ a < x < b \quad \text{... (55b)}$$

In order to obtain the solution in region I, we once again apply the connection formulae (29*a*, *b*) and get

$$\varphi_I(x) = Ap^{-1/2}\left[2\,\Theta \sin\left(\frac{1}{\hbar}\int_x^a p(x')\,dx' + \frac{\pi}{4}\right) + \frac{i}{2\Theta}\cos\left(\frac{1}{\hbar}\int_x^a p(x')\,dx' + \frac{\pi}{4}\right)\right]$$

Which on using eqn. (54) reduces to

$$\varphi_1(x) = A_3\, e^{-i\pi/4} p^{-1/2}\left[2\,\Theta \sin\left(\frac{1}{\hbar}\int_x^a p(x')\,dx' + \frac{\pi}{4}\right) + \frac{i}{2\,\Theta}\cos\left(\frac{1}{\hbar}\int_x^a p(x')\,dx' + \frac{\pi}{4}\right)\right],\ x < a \quad \text{... (57)}$$

Now

$$A_3\, e^{-i\pi/4} p^{-1/2}\left[\left(\Theta + \frac{1}{4\Theta}\right)\left\{\cos\left(\frac{1}{\hbar}\int_a^x p(x')\,dx' + \frac{\pi}{4}\right) + i\sin\left(\frac{1}{\hbar}\int_a^x p(x')\,dx' + \frac{\pi}{4}\right)\right\} + \left(\Theta - \frac{1}{4\Theta}\right)\left\{\cos\left(\frac{1}{\hbar}\int_a^x p(x')\,dx' + \frac{\pi}{4}\right) - i\sin\left(\frac{1}{\hbar}\int_a^x p(x')\,dx' + \frac{\pi}{4}\right)\right\}\right]$$

$$= A_3\, e^{-i\pi/4} p^{-1/2}\left[\cos\left(\frac{1}{\hbar}\int_a^x p(x')\,dx' + \frac{\pi}{4}\right)\left\{\Theta + \frac{1}{4\Theta} + \Theta - \frac{1}{4\Theta}\right\} + i\sin\left(\frac{1}{\hbar}\int_a^x p(x')\,dx' + \frac{\pi}{4}\right)\left\{\Theta + \frac{1}{4\Theta} - \Theta + \frac{1}{4\Theta}\right\}\right]$$

$$= A_3 e^{-i\pi/4} p^{-1/2}\left[2\Theta\cos\left(\frac{1}{\hbar}\int_a^x p(x')\,dx' + \frac{\pi}{4}\right) + \frac{i}{2\Theta}\sin\left(\frac{1}{\hbar}\int_a^x p(x')\,dx' + \frac{\pi}{4}\right)\right]$$

$$= A_3\, e^{-i\pi/4} p^{-1/2}\left[2\,\Theta \cos\left(-\frac{1}{\hbar}\int_x^a p(x')\,dx' + \frac{\pi}{4}\right) + \frac{i}{2\,\Theta}\sin\left(-\frac{1}{\hbar}\int_x^a p(x')\,dx' + \frac{\pi}{4}\right)\right]$$

$$= A_3\, e^{-i\pi/4}\, p^{-1/2}\left[2\,\Theta\,\sin\left(\frac{\pi}{2} + \frac{1}{\hbar}\int_x^a p(x')\,dx' - \frac{\pi}{4}\right) + \frac{i}{2\Theta}\cos\left(\frac{\pi}{2} + \frac{1}{\hbar}\int_x^a p(x')dx' - \frac{\pi}{4}\right)\right]$$

$$= A_3\, e^{-i\pi/4} p^{-1/2}\left[2\,\Theta \sin\left(\frac{1}{\hbar}\int_x^a p(x')\,dx' + \frac{\pi}{4}\right) + \frac{i}{2\,\Theta}\cos\left(\frac{1}{\hbar}\int_x^a p(x')\,dx' + \frac{\pi}{4}\right)\right]$$

where we have used the relations,

$$\sin\left(\frac{\pi}{2} - \theta\right) = \cos\theta \text{ and } \cos\left(\frac{\pi}{2} - \theta\right) = \sin\theta$$

The eqn. (57), therefore amounts to,

$$\varphi_1(x) = A_3\, p^{-1/2}\left[\left(\Theta + \frac{1}{4\Theta}\right) e^{1/\hbar \int_a^x p(x')\,dx'} - i\left(\Theta - \frac{1}{4\Theta}\right) e^{-1/\hbar \int_a^x p(x')\,dx'}\right],\ x < a \qquad \text{... (58)}$$

where once again the use of relation $e^{i\phi} = \cos\phi + i\sin\phi$ and that $-i = e^{-i\pi/2}$ have been made.

Comparing eqns. (51) and (58), we get

$$A_1 = \left(\Theta + \frac{1}{4\Theta}\right) A_3 \qquad \text{... (59a)}$$

$$B_1 = -i\left(\Theta - \frac{1}{4\Theta}\right) A_3 \qquad \text{... (59b)}$$

The transmission coefficient T(or probability for penetration of the barrier) would, therefore be given by

$$T = \left|\frac{A_3}{A_1}\right|^2 = \left(\Theta + \frac{1}{4\Theta}\right)^{-2} \qquad \text{... (60)}$$

Note that in case one is interested, one can calculate the reflection coefficient **R** $(= |B_1/A_1|^2)$ by using eqns. (59a) and (59b). From eqn. (56),

$$\Theta = e^{\int_a^b 2\pi/\lambda\, dx}$$

where we have used the de Broglie relation $\frac{|p|}{\hbar} = \frac{2\pi}{\lambda}$.

For WKB approximation to be valid, the width of the barrier $b - a$ should be several times the de Broglie wavelength λ. Hence

$$\Theta \approx e^{\text{several times } \lambda/\lambda} >> 1$$

In the light of the above, one can neglect $\frac{1}{4\Theta}$ compared to Θ in eqn. (60). After doing so, we get

$$T = \frac{1}{\Theta^2} = e^{-2/\hbar \int_a^b |p(x)|\, dx} = e^{-2/\hbar \int_a^b [2m\{V(x)-E\}]^{1/2}\, dx} \quad \text{... (61)}$$

From the above expression, we may note that increasing either the width or the height of the barrier, decreases the probability for the penetration of the barrier. Also note from the same expression that the transmission coefficient decreases rapidly with increasing particle mass. For example, when we increase the mass from that of an electron to that of a proton, the transparency of the barrier decreases by a factor $e^{\sqrt{1840}} \approx 10^{18}$.

12.8 TIME INDEPENDENT PERTURBATION THEORY

It has already been pointed out earlier that very few problems in quantum mechanics can be solved exactly. For example, in the case of harmonic oscillator problem if over and above the potential energy function $V(x) = \frac{1}{2} kx^2$, there exists a term of the type $v(x) = \lambda x^4$, it becomes difficult to solve the problem exactly. One then uses approximation methods. One of these methods to be used here is called stationary or time independent perturbation method[3] (usually referred to as the Rayleigh-Schrödinger perturbation theory) and is applicable to the cases in which the additional potential energy term is only slightly different from that of the well solved problem.

The Hamiltonian H of the system being studied is accordingly

$$H = H^{(0)} + H' \quad \text{... (62)}$$

where $H^{(0)}$ is time independent and is called unperturbed Hamiltonian. $\lambda H'$ is known as perturbation (or small correction) and is also time independent in this case. The dimension less parameter $\lambda(<<1)$ real and measures the size or strength of the perturbation. The eigen values $E_n^{(0)}$ and the orthonormal eigen functions $\varphi_n^{(0)}$ of the unperturbed Hamiltonian $H^{(0)}$ are taken as known, that is the exact solutions of the equation.

$$H^{(0)} \varphi_n^{(0)} = E_n^{(0)} \varphi_n^{(0)} \quad \text{... (63)}$$

are known.

Our aim here is to develop a perturbation method so that we can solve the Schrödinger equation.

$$H\varphi_n = (H^{(0)} + \lambda H')\, \varphi_n = E_n\, \varphi_n \quad \text{... (64)}$$

and can obtain approximate expressions for the eigen values E_n and the eigen functions φ_n of the perturbed Hamiltonian H. λ being small, the eigen values and the eigen functions of the perturbed H will not be very much different from that of the

[3]E. Schrödinger, Ann. Physik 80, 437 (1926)

unperturbed Hamiltonian $H^{(0)}$. Therefore, let us express the eigen values and the eigen functions of H as a power series in λ, keeping only a finite number of terms (often two) in these expansions, that is

$$E_n = E_n^{(0)} + \lambda E_n^{(1)} + \lambda^2 E_n^{(2)} + \dots \quad \dots (65a)$$

$$\varphi_n = \varphi_n^{(0)} + \lambda \varphi_n^{(1)} + \lambda^2 \varphi_n^{(2)} + \dots \quad \dots (65b)$$

such that the coefficients of the first power of λ gives first order correction and the second power of λ gives the second order correction and so on. Substituting the expansions given by eqns. (65) into eqn. (64), one obtains

$$(H^{(0)} + \lambda H')(\varphi_n^{(0)} + \lambda \varphi_n^{(1)} + \lambda^2 \varphi_n^{(2)} + \dots)$$
$$= (E_n^{(0)} + \lambda E_n^{(1)} + \lambda^2 E_n^{(2)} + \dots)(\varphi_n^{(0)} + \lambda\varphi_n^{(1)} + \lambda^2 \varphi_n^{(2)} + \dots)$$

which is equivalent to

$$H^{(0)} \varphi_n^{(0)} + \lambda (H' \varphi_n^{(0)} + H^{(0)} \varphi_n^{(1)}) + \lambda^2 (H' \varphi_n^{(1)} + H^{(0)} \varphi_n^{(2)}) + \dots$$
$$= E_n^{(0)} \varphi_n^{(0)} + \lambda (E_n^{(1)} \varphi_n^{(0)} + E_n^{(0)} \varphi_n^{(1)})$$
$$+ \lambda^2 (E_n^{(0)} \varphi_n^{(2)} + E_n^{(1)} \varphi_n^{(1)} + E_n^{(2)} \varphi_n^{(0)}) + \dots \quad \dots (66)$$

In order that the above equation is satisfied for all values of λ, the coefficients of equal powers of λ on either side of the said equation must be same. Thus we obtain for zeroth order terms in λ;

$$H^{(0)} \varphi_n^{(0)} = E_n^{(0)} \varphi_n^{(0)} \quad \dots (67)$$

for first order terms;

$$H^{(0)} \varphi_n^{(1)} + H' \varphi_n^{(0)} = E_n^{(0)} \varphi_n^{(1)} + E_n^{(1)} \varphi_n^{(0)} \quad \dots (68)$$

for second order terms;

$$H^{(0)} \varphi_n^{(2)} + H' \varphi_n^{(1)} = E_n^{(0)} \varphi_n^{(2)} + E_n^{(1)} \varphi_n^{(1)} + E_n^{(2)} \varphi_n^{(0)} \quad \dots (69)$$

Note that the eqn. (67) has already been solved. If eqn. (68) can be solved, one obtains $\varphi_n^{(1)}$ and $E_n^{(1)}$. Similarly the solution of eqn. (69) gives $\varphi_n^{(2)}$ and $E_n^{(2)}$.

Let us consider the effect of perturbation $\lambda H'$ on the nth energy state which will be assumed to be non-degenerate for the time being, that is corresponding to the eigen value $E_n^{(0)}$, there corresponds a single eigen function.

12.9 FIRST ORDER PERTURBATION

Eqn. (68) can be written in the form

$$(H^{(0)} - E_n^{(0)}) \varphi_n^{(1)} + (H' - E_n^{(1)}) \varphi_n^{(0)} = 0 \quad \dots (68a)$$

Let us assume that all the eigen functions $\varphi_n^{(0)}$ of $H^{(0)}$ form a complete (or closed) set, that is, we may expand $\varphi_n^{(1)}$ in terms of the unperturbed eigen functions $\varphi_m^{(0)}$,

$$\varphi_n^{(1)} = \sum_m a_m^{(1)} \varphi_m^{(0)} \quad \dots (70)$$

where the superscript (1) refers to the fact that we are considering first order perturbation and the summation represents sum over all the states. Substituting $\varphi_n^{(1)}$ from eqn. (70) into eqn. (68*a*), one obtains

$$\sum_m a_m^{(1)} \{H^{(0)} - E_n^{(0)}\} \varphi_m^{(0)} + (H' - E_n^{(1)}) \varphi_n^{(0)} = 0$$

Use of eqn. (63), reduces the above equation into the form,

$$\sum_m a_m^{(1)} \{E_m^{(0)} - E_n^{(0)}\} \varphi_m^{(0)} + (H' - E_n^{(1)}) \varphi_n^{(0)} = 0 \qquad \text{... (71)}$$

Multiplying the above equation from the left by $\varphi_k^{(0)*}$ and integrating the resulting expression over the entire space, we get

$$\sum_m a_m^{(1)} \{E_m^{(0)} - E_n^{(0)}\} \delta_{km} + H'_{kn} - E_n^{(1)} \delta_{kn} = 0 \qquad \text{... (72)}$$

where we have used the orthonormality condition,

$$\int \varphi_k^{(0)*} \varphi_m^{(0)} \, d\tau = \delta_{km} \qquad \text{... (73a)}$$

and the definition that the matrix element of H' between the kth and nth state,

$$H'_{kn} = \int \varphi_k^{(0)*} H' \varphi_n^{(0)} \, d\tau \qquad \text{... (73b)}$$

Further making use of the fact that the kronecker delta function $\delta_{km} = 1$ for $k = m$ and equal to zero for $k \neq m$, eqn. (72) leads to

$$a_k^{(1)} \{E_k^{(0)} - E_n^{(0)}\} + H'_{kn} - E_n^{(1)} \delta_{kn} = 0 \qquad \text{... (74)}$$

To find $E_n^{(1)}$, the first order correction to the nth energy state, one puts $k = n$ in the above equation and gets

$$E_n^{(1)} = H'_{nn} = \int \varphi_n^{(0)*} H' \varphi_n^{(0)} \, d\tau \qquad \text{... (75)}$$

which is equal to the quantum mechanical mean value of the perturbation taken in the unperturbed state $\varphi_n^{(0)}$. Truly speaking $\lambda E_n^{(1)} = \lambda H'_{nn}$ is the first order correction according to our notation.

Let us now proceed for the calculation of first order correction to the wave function of the nth energy state, that is $\varphi_n^{(1)} \left(= \sum_m a_m^{(1)} \varphi_m^{(0)}\right)$ so that we can finally evaluate φ_n $(= \varphi_n^0 + \lambda \varphi_n^{(1)})$. If $k \neq n$ and if $E_k^{(0)} \neq E_n^{(0)}$, that is if the states are non-degenerate, from eqn. (74), one obtains

$$a_k^{(1)} = H'_{kn} / (E_n^{(0)} - E_k^{(0)}) \qquad \text{... (76)}$$

From the above equation, all the coefficients $a_k^{(1)}$ except $k = n$, that is $a_n^{(1)}$ can be calculated. To evaluate $a_n^{(1)}$, let us require that the perturbed wave function φ_n be normalized, that is

$$\int \varphi_n^* \varphi_n \, d\tau = 1 \qquad \text{... (77a)}$$

which amounts to

$$\int (\varphi_n^{(0)} + \lambda\varphi_n^{(1)})^* (\varphi_n^{(0)} + \lambda\varphi_n^{(1)}) \, d\tau = 1 \qquad \text{... (77b)}$$

where the terms of the order λ^2 and higher have been neglected as presently we are interested in the calculation of first order correction. Note that the above condition simply implies that

$$\int \varphi_n^{(0)*} \varphi_n^{(1)} \, d\tau = 0 \qquad \text{... (78)}$$

that is, $\varphi_n^{(1)}$ is orthogonal to $\varphi_n^{(0)}$ which of course is normalized.

Putting the value of $\varphi_n^{(1)}$ from eqn. (70) into eqn. (77*b*), we get

$$\int \left(\varphi_n^{(0)} + \lambda \sum_m a_m^{(1)} \varphi_m^{(0)}\right)^* \left(\varphi_n^{(0)} + \lambda \sum_m a_m^{(1)} \varphi_m^{(0)}\right) d\tau = 1$$

which reduces to

$$\int \varphi_n^{(0)*} \varphi_n^{(0)} \, d\tau + \lambda \left(\sum_m a_m^{(1)*} \int \varphi_m^{(0)*} \varphi_n^{(0)} \, d\tau + \sum_m a_m^{(1)} \int \varphi_n^{(0)*} \varphi_m^{(0)} \, d\tau\right) = 1$$

where once again we have neglected the higher order terms. The above equation further reads

$$1 + \lambda \left(\sum_m a_m^{(1)*} \delta_{mn} + \sum_m a_m^{(1)} \delta_{nm}\right) = 1$$

which effectively means

$$a_n^{(1)*} + a_n^{(1)} = 0 \qquad \text{... (79)}$$

that is, $a_n^{(1)}$ must be purely imaginary, say $a_n^{(1)} = i\alpha$.

Now the corrected (upto first order in perturbation) or the perturbed wave function is given by

$$\varphi_n = \varphi_n^{(0)} + \lambda \sum_k a_k^{(1)} \varphi_k^{(0)}$$

$$= \varphi_n^{(0)} + \lambda a_n^{(1)} \varphi_n^{(0)} + \lambda \Sigma' a_k^{(1)} \varphi_k^{(0)}$$

where we have separated the term corresponding to $k = n$ from the summation. Prime indicates the exclusion of the term corresponding to $k = n$ from the summation.

After substituting the value of $a_n^{(1)}$ in the above equation, one obtains.

$$\varphi_n = \varphi_n^{(0)} + \lambda \, i \, \alpha \, \varphi_n^{(0)} + \lambda \Sigma' a_k^{(1)} \varphi_k^{(0)}$$

$$= \varphi_n^{(0)} (1 + i \alpha \lambda) + \lambda \Sigma' a_k^{(1)} \varphi_k^{(0)}$$

Now using the expansion $e^{i\alpha\lambda} = 1 + i\alpha\lambda$, where higher order terms in λ have been neglected as it is small, one gets

$$\varphi_n = \varphi_n^{(0)} e^{i\alpha\lambda} + \lambda \Sigma' a_k^{(1)} \varphi_k^{(0)}$$

Since in the physical problems, we are interested only in the absolute values which in the case of $e^{i\alpha\lambda}$ is 1, therefore

$$\varphi_n = \varphi_n^{(0)} + \lambda \Sigma' a_k^{(1)} \varphi_k^{(0)}$$

which effectively amounts to $a_n^{(1)} = 0$.

Now substituting the value of $a_k^{(1)}$ from eqn. (76) into the above equation, we get the required result.

$$\varphi_n = \varphi_n^{(0)} + \Sigma' \lambda H'_{kn} / (E_n^{(0)} - E_k^{(0)}) \varphi_k^{(0)} \quad \text{... (80)}$$

12.10 SECOND ORDER PERTURBATION

Eqn. (69) can be written in the form

$$(H^{(0)} - E_n^{(0)}) \varphi_n^{(2)} + (H' - E_n^{(1)}) \varphi_n^{(1)} - E_n^{(2)} \varphi_n^{(0)} = 0 \quad \text{... (69a)}$$

Further expanding $\varphi_n^{(2)}$ in terms of the unperturbed eigen functions $\varphi_m^{(0)}$ of $H^{(0)}$ as given by the expression

$$\varphi_n^{(2)} = \sum_m a_m^{(2)} \varphi_m^{(0)} \quad \text{... (81)}$$

where the superscript (2) of course refers to the fact that we are considering second order perturbation and then substituting the expansions (70) and (81) into eqn. (69*a*), we get

$$\sum_m a_m^{(2)} (H^{(0)} - E_n^{(0)}) \varphi_m^{(0)} + \sum_m a_m^{(1)} (H' - E_n^{(1)}) \varphi_m^{(0)} - E_n^{(2)} \varphi_n^{(0)} = 0$$

Making use of eqn. (63) and then multiplying the above equation by $\varphi_k^{(0)*}$ from the left and integrating the resulting expression over the entire space, we get

$$\sum_m a_m^{(2)} (E_m^{(0)} - E_n^{(0)}) \delta_{km} + \sum_m a_m^{(1)} (H'_{km} - E_n^{(1)} \delta_{km}) - E_n^{(2)} \delta_{kn} = 0$$

which amounts to

$$a_k^{(2)} (E_k^{(0)} - E_n^{(0)}) + \sum_m a_m^{(1)} H'_{km} - a_k^{(1)} E_n^{(1)} - E_n^{(2)} \delta_{kn} = 0 \quad \text{... (82)}$$

where we have used the ortho normality relation (73*a*), the definition of H'_{km} from eqn. (73*b*) and also the property of kronecker delta function.

$E_n^{(2)}$, the second order correction to the *n*th energy state can be determined by putting $k = n$ in eqn. (82) which leads to

$$E_n^{(2)} = \sum_m a_m^{(1)} H'_{nm} - E_n^{(1)} a_n^{(1)}$$

There after putting the value of $E_n^{(1)}$ from eqn. (75), one gets

$$E_n^{(2)} = \sum_m a_m^{(1)} H'_{nm} - H'_{nn} a_n^{(1)}$$

On separating the term corresponding to $n = m$ from the summation, one obtains

$$E_n^{(2)} = \Sigma' a_m^{(1)} H'_{nm} + a_n^{(1)} H'_{nn} - a_n^{(1)} H'_{nn}$$

where prime over the summation indicates $m \neq n$. Cancelling second and third terms in the above expression and then replacing m by k (both being dummy indices), one gets

$$E_n^{(2)} = \Sigma' a_k^{(1)} H'_{nk}$$

Further substituting the value of $a_k^{(1)}$ from eqn. (76) into the above equation, one gets

$$E_n^{(2)} = \Sigma' H'_{nk} H'_{kn} / (E_n^{(0)} - E_k^{(0)})$$

Since H' is hermitian (self adjoint), $H'_{nk} = (H'_{kn})^*$, the transposed complex conjugate. The second order correction to the nth energy state would, therefore be given by

$$E_n^{(2)} = \Sigma' |H'_{kn}|^2 / (E_n^{(0)} - E_k^{(0)}) \quad \text{... (83)}$$

which according to our notations is $\lambda^2 E_n^{(2)}$. Also note that the second order correction to the energy of ground state ($n = 0$) is always negative as $E_0^{(0)} < E_k^{(0)}$ for all $k \neq 0$.

Let us now try to find the second order correction to the wave function of the nth energy state, that is $\psi_n^{(2)}$ $(= \sum_m a_m^{(2)} \varphi_m^{(0)})$. From eqn. (82) $k \neq n$ and for $E_n^{(0)} \neq E_k^{(0)}$, one find

$$a_k^{(2)} = \frac{1}{\left(E_n^{(0)} - E_k^{(0)}\right)} \left(\sum_m a_m^{(1)} H'_{km} - a_k^{(1)} E_n^{(1)} \right) \quad \text{... (84)}$$

which gives all the coefficients $a_k^{(2)}$ except $k = n$, that is $a_n^{(2)}$. To find $a_n^{(2)}$, we again use the condition that the perturbed wave function φ_n be normalized, that is

$$\int \varphi_n^* \varphi_n \, d\tau = 1 \quad \text{... (85)}$$

which amounts to

$$\int (\varphi_n^{(0)} + \lambda \varphi_n^{(1)} + \lambda^2 \varphi_n^{(2)})^* (\varphi_n^{(0)} + \lambda \varphi_n^{(1)} + \lambda^2 \varphi_n^{(2)}) d\tau = 1$$

where terms only upto λ^2 have been retained.

Using the expansions (70) and (81), the above equation reduces to

$$\int \left(\varphi_n^{(0)*} + \lambda \sum_m a_m^{(1)*} \varphi_m^{(0)*} + \lambda^2 \sum_m a_m^{(2)*} \varphi_m^{(0)*} \right) \times \left(\varphi_n^{(0)} + \lambda \sum_m a_m^{(1)} \varphi_m^{(0)} + \lambda^2 \sum_m a_m^{(2)} \varphi_m^{(0)} \right) d\tau = 1$$

or

$$1+\lambda\left(\sum_m a_m^{(1)*}\,\delta_{mn}+\sum_m a_m^{(1)}\,\delta_{nm}\right)+\lambda^2\left(\sum_m a_m^{(2)*}\,\delta_{mn}+\sum_m a_m^{(2)}\,\delta_{nm}+\sum_{mm'} a_m^{(1)*}\,a_{m'}^{(1)}\,\delta_{mm'}\right)=1$$

Cancelling 1 on both sides of the above equation and then making use of the property of the kronecker delta function, one gets

$$\lambda\,(a_n^{(1)*}+a_n^{(1)})+\lambda^2\left(a_n^{(2)*}+a_n^{(2)}+\sum_m |a_m^{(1)}|^2\right)=0$$

Further making use of eqn. (79), one gets

$$a_n^{(2)*}+a_n^{(2)} \;=\; -\,\Sigma\,|a_m^{(1)}|^2=-\,\Sigma\,|a_k^{(1)}|^2$$

as m and k dummy indices.

The imaginary part on the left hand side of the above equation, of course vanishes or can be made to vanish as before.

The real part of $a_n^{(2)}$ would, therefore be given by

$$\mathbf{Re}\;a_n^{(2)} \;=\; -\frac{1}{2}\,\Sigma|\,a_k^{(1)}\,|^2=-\frac{1}{2}\,\Sigma'\,|\,a_k^{(1)}\,|^2 \qquad \text{... (86)}$$

as $a_n^{(1)}$ is zero, Further

$$\varphi_n^{(2)} \;=\; \Sigma\,a_m^{(2)}\,\varphi_m^{(0)}$$

$$=\; \Sigma'\,a_k^{(2)}\,\varphi_k^{(0)}+a_n^{(2)}\,\varphi_n^{(0)}$$

where the summation has been split into two parts. After substituting the values of $a_k^{(2)}$ and $a_n^{(2)}$ from eqns. (84) and (86), respectively, one gets

$$\varphi_n^{(2)} \;=\; \Sigma'\,\frac{1}{\left(E_n^{(0)}-E_k^{(0)}\right)}\left(\sum_m{}' a_m^{(1)}\,H'_{km}-E_n^{(1)}\,a_k^{(1)}\right)\varphi_k^{(0)}-\frac{1}{2}\,\Sigma'\,|\,a_k^{(1)}\,|^2\;\varphi_n^{(0)}$$

Note that the excluded term corresponding to $m = n$ from the summation over m is zero as $a_n^{(1)} = 0$. Now substituting the values of $a_m^{(1)}$ and $a_k^{(1)}$ from eqn. (76) into above equation, one obtains the following expression for the second order correction to the wave function of the nth state,

$$\varphi_n^{(2)} \;=\; \sum_k{}'\left[\left\{\sum_m{}'\frac{H'_{mn}}{\left(E_n^{(0)}-E_m^{(0)}\right)}\frac{H'_{km}}{\left(E_n^{(0)}-E_k^{(0)}\right)}-\frac{H'_{nn}\,H'_{kn}}{\left(E_n^{(0)}-E_k^{(0)}\right)^2}\right\}\varphi_k^{(0)}-\frac{1}{2}\,\frac{|H'_{kn}|^2}{\left(E_n^{(0)}-E_k^{(0)}\right)^2}\,\varphi_n^{(0)}\right] \qquad \text{... (87)}$$

Note that the said second order correction is $\lambda^2\varphi_n^{(2)}$ according to our notation.

12.11 TIME INDEPENDENT PERTURBATION THEORY (DEGENERATE CASE)

For first order correction to the wave function in the case of nth energy state being non-degenerate, we obtained

$$\lambda\varphi_n^{(1)} = \sum_k{}' \lambda H'_{kn} / \left(E_n^{(0)} - E_k^{(0)}\right) \varphi_k^{(0)}$$

and similarly for the second order correction to the eigen value of nth energy state, we obtained

$$\lambda^2 E_n^{(2)} = \sum_k{}' |\lambda H'_{kn}|^2 / \left(E_n^{(0)} - E_k^{(0)}\right)$$

In both the above expressions $\left(E_n^{(0)} - E_k^{(0)}\right)$ appears in the denominator and the summations are for all values of k except $k = n$. But if there exists an $E_n^{(0)}$ equal to $E_k^{(0)}$ for $k \neq n$ (that is, $E_n^{(0)}$ and $E_k^{(0)}$ are degenerate or in other words there exist two eigen functions corresponding to the same energy eigen value), both the above corrections diverge (result infinity) and the earlier theory then fails. Therefore, we need to develop a perpturbation theory for the degenerate states.

For the sake of simplicity or clarity, we would consider an unperturbed state which is two fold degenerate. Let $\varphi_1^{(0)}$ and $\varphi_2^{(0)}$ be two orthonormal unperturbed eigen functions belonging to the eigen values $E_1^{(0)}$ and $E_2^{(0)}$ where $E_1^{(0)} = E_2^{(0)} = E^{(0)}$ say.

The eigen value equations (or Schrödinger equations) for the unperturbed Hamiltonian $H^{(0)}$ would be given by

$$H^{(0)} \varphi_1^{(0)} = E_1^{(0)} \varphi_1^{(0)} \quad \text{... (88a)}$$

and

$$H^{(0)} \varphi_2^{(0)} = E_2^{(0)} \varphi_2^{(0)} \quad \text{... (88b)}$$

We have already shown in section 2.11 that for a degenerate state, a linear combination of the unperturbed wave functions is also a solution of $H^{(0)}$ with the same energy eigen value. Thus

$$\varphi^{(0)} = c_1 \varphi_1^{(0)} + c_2 \varphi_2^{(0)} \quad \text{... (89a)}$$

satisfies the eigen value equation

$$H^{(0)} \varphi^{(0)} = E^{(0)} \varphi^{(0)} \quad \text{... (89b)}$$

The Schrödinger equation for the perturbed system will be given by

$$H \varphi = E \varphi \quad \text{... (90)}$$

where H, E and φ represent perturbed Hamiltonian, perturbed energy and perturbed eigen function, respectively.

For the first order perturbation calculations, we write

$$(H^{(0)} + \lambda H')(\varphi^{(0)} + \lambda\varphi^{(1)}) = (E^{(0)} + \lambda E^{(1)})(\varphi^{(0)} + \lambda\varphi^{(1)}) \quad \text{... (91)}$$

where the perturbed Hamiltonian H has been broken into two parts, $\lambda H'$ being perturbation and where we have again assumed that the perturbed energy and the perturbed eigen function can be expanded in terms of a power series in λ ($\lambda << 1$ and real).

If the above equation has to be valid for all values of λ, the coefficients of the equal powers of λ on both sides of the above equation must be equal. Equating the coefficients of λ^0 and λ^1 on both the sides, we get

$$H^{(0)}\,\varphi^{(0)} = E^{(0)}\,\varphi^{(0)}, \qquad \text{... (89b)}$$

and

$$(H^{(0)} - E^{(0)})\,\varphi^{(1)} + (H' - E^{(1)})\,\varphi^{(0)} = 0 \qquad \text{... (92)}$$

Again expanding $\varphi^{(1)}$ in terms of the orthonormal set,

$$\varphi^{(1)} = \sum_m a_m^{(1)}\,\varphi_m^{(0)} \qquad \text{... (93)}$$

and substituting above into eqn. (92), we obtain

$$\sum_m a_m^{(1)}\,(H^{(0)} - E^{(0)})\,\varphi_m^{(0)} + (H' - E^{(1)})\,\varphi^{(0)} = 0$$

Using the eigen value eqn. (88) for the unperturbed Hamiltonian, that is

$$H^{(0)}\,\varphi_m^{(0)} = E_m^{(0)}\,\varphi_m^{(0)},$$

one gets

$$\sum_m a_m^{(1)}\,(E_m^{(0)} - E^{(0)})\,\varphi_m^{(0)} + (H' - E^{(1)})\,\varphi^{(0)} = 0$$

Substituting the value of $\varphi^{(0)}$ from eqn. (89a) into above equation, one gets

$$\sum_m a_m^{(1)}\,(E_m^{(0)} - E^{(0)})\,\varphi_m^{(0)} + (H' - E^{(1)})\,(c_1\,\varphi_1^{(0)} + c_2\,\varphi_2^{(0)}) = 0$$

Multiplying above equation by $\varphi_k^{(0)*}$ from the left and then integrating the resulting expression over the entire space, we get

$$\sum_m a_m^{(1)}\,(E_m^{(0)} - E^{(0)})\,\delta_{km} + c_1\,(H'_{k1} - E^{(1)}\,\delta_{k1}) + c_2\,(H'_{k2} - E^{(1)}\,\delta_{k2}) = 0$$

or

$$a_k^{(1)}\left(E_k^{(0)} - E^{(0)}\right) + c_1\,(H'_{k1} - E^{(1)}\,\delta_{k1}) + c_2\,(H'_{k2} - E^{(1)}\,\delta_{k2}) = 0 \qquad \text{... (94)}$$

where δ_{k1} and δ_{k2} denote the so called Kronecker delta symbols.

Taking $k = 1$, $E_k^{(0)} - E^{(0)} = E_1^{(0)} - E^{(0)} = 0$ and similarly for $k = 2$, $E_k^{(0)} - E^{(0)} = E_2^{(0)} - E^{(0)} = 0$. Thus corresponding to $k = 1$ and $k = 2$ from eqn. (94). one obtains the following two simultaneous linear homogenous equations.

$$c_1\,(H'_{11} - E^{(1)}) + c_2\,H'_{12} = 0$$

and

$$c_1 H'_{21} + c_2 (H'_{22} - E^{(1)}) = 0$$

or

$$c_1 (\lambda H'_{11} - \lambda E^{(1)}) + c_2 \lambda H'_{12} = 0 \qquad \text{... (95a)}$$

and

$$c_1 \lambda H'_{21} + c_2 (\lambda H'_{22} - \lambda E^{(1)}) = 0 \qquad \text{... (95b)}$$

For non-vanishing solutions of c_1 and c_2, the determinant of their coefficients must be zero, that is

$$\begin{vmatrix} \lambda H'_{11} - \lambda E^{(1)} & \lambda H'_{12} \\ \lambda H'_{21} & \lambda H'_{22} - \lambda E^{(1)} \end{vmatrix} = 0 \qquad \text{... (96a)}$$

In general for a g-fold degenerate state, we will obtain a set of g simultaneous linear homogeneous equations and finally for non-vanishing solutions of $c_1, c_2, \ldots, c_g$.

$$\begin{vmatrix} \lambda H'_{11} - \lambda E^{(1)} & \lambda H'_{12} & \lambda H'_{13} \ldots \lambda H_{1g} \\ \lambda H'_{21} & \lambda H'_{22} - \lambda E^{(1)} & \ldots \\ \vdots & & \\ \lambda H'_{g1} & \ldots & \lambda H'_{gg} - \lambda E^{(1)} \end{vmatrix} = 0 \qquad \text{... (96b)}$$

Eqn. (96*a*) (also of course 96*b*) is known as the secular equation. It (eqn. 96*a*) is quadratic in $\lambda E^{(1)}$. Therefore, one gets two values of $\lambda E^{(1)}$, *viz.*

$$\lambda E^{(1)}_{12} = \frac{1}{2} \lambda (H'_{11} + H'_{22}) \pm \frac{1}{2} [(\lambda H'_{11} - \lambda H'_{22})^2 + 4 |\lambda H'_{12}|^2]^{1/2}$$

Thus the degenerate level is split into two levels,

$$E^{(0)} + \lambda E^{(1)}_1 \text{ and } E^{(0)} + \lambda E^{(1)}_2$$

as a result of perturbation. That means the degeneracy is lifted (removed) in the first order or in other words the levels are no more degenerate. If both the roots of the secular equation happen to be same, that is, if $\lambda E^{(1)}_1 = \lambda E^{(1)}_2$ in that case the degeneracy is not removed in the first order and one has then to proceed for the calculations of the second order perturbation which of course can be applied by the usual method.

Let us see whether we can write the secular eqn. (96*a*) in terms of the perturbed Hamiltonian and the perturbed energy so that the perturbed energy can be determined directly.

$$\begin{aligned} \lambda H'_{11} &= \int \varphi_1^{(0)*} \lambda H' \varphi_1^{(0)} \, d\tau \\ &= \int \varphi_1^{(0)*} (H - H^{(0)}) \varphi_1^{(0)} \, d\tau \\ &= \int \varphi_1^{(0)*} H \varphi_1^{(0)} \, d\tau - \int \varphi_1^{(0)*} H^{(0)} \varphi_1^{(0)} \, d\tau \\ &= H_{11} - E^{(0)} \int \varphi_1^{(0)*} \varphi_1^{(0)} \, d\tau \\ &= H_{11} - E^{(0)} \end{aligned}$$

Thus

$$\lambda H'_{11} - \lambda E^{(1)} = H_{11} - E^{(0)} - \lambda E^{(1)} = H_{11} - E$$

Similarly

$$\lambda H'_{12} = H_{12}, \lambda H'_{21} = H_{21}, \lambda H'_{22} - \lambda E^{(1)} = H_{22} - E$$

Substituting above into eqn. (96a), one gets

$$\begin{vmatrix} H_{11} - E & H_{12} \\ H_{21} & H_{22} - E \end{vmatrix} = 0 \quad \text{... (97)}$$

which is quadratic in E. Thus the two roots of E corrected upto first order in perturbation can be found directly.

Let us now find the correct zero order wave functions associated with the energy levels, $E^{(0)} + \lambda E_1^{(1)}$ and $E^{(0)} + \lambda E_2^{(1)}$. Putting $\lambda E^{(1)} = \lambda E_1^{(1)}$ into eqn. (95), we can find the ratio of c_1 and c_2. Further using the normalization condition, $c_1^2 + c_2^2 = 1$, we can determine c_1 and c_2. Substituting these values of c_1 and c_2 into eqn. (89*a*), we can determine the normalized unperturbed eigen function (or correct zeroth order wave function) $\varphi^{(0)}$ associated with the level $E^{(0)} + \lambda E_1^{(1)}$. Similarly, putting $\lambda E^{(1)} = \lambda E_2^{(1)}$ into eqn. (95), we can again find the ratio of c_1 and c_2. Using the normalization condition, $c_1^2 + c_2^2 = 1$, we can find c_1 and c_2. These values of c_1 and c_2 can be substituted into eqn. (89*a*) and one can obtain the normalized unperturbed zeroth order eigen function associated with the level $E^{(0)} + \lambda E_2^{(1)}$.

The method outlined here would be more clear when we would determine the correct unperturbed eigen functions in the zeroth approximation in the case of $n = 2$ state of hydrogen atom under the action of uniform electric field in the z-direction.

12.12 VARIATIONAL METHOD (OR RITZ METHOD[4])

In the case of perturbation theory, it was clearly pointed out that the method is applicable only to the problems in which the Hamiltonian of the system can be broken into two parts, *viz.*

$$H = H^{(0)} + \lambda H'$$

where $H^{(0)}$ is the unperturbed Hamiltonian for which the exact solution is known and $\lambda H'$ is the perturbation or small addition.

Suppose we do not know the exact solution for $H^{(0)}$, that is there may not be any closely related problem which can be solved exactly, the time independent perturbation theory then cannot be applied. The said theory also cannot be applied to the problem where the exact solution of the unperturbed.

Hamiltonian is known but the second order perturbation may be extremely difficult to calculate. Example-the ground state energy of Helium atom (or two electron atom).

For such a system differing substantially from that subject to the exact treatment or the one for which the second order perturbation calculations are extremely difficult, it

[4]J. Ritz, Reine u. angew. Math. 135, 1 (1908)

is some times possible to write a trial function of the system on the basis of intelligent guess (by using physical criteria taking the symmetry of the problem into account). One then uses a method called variational method by which the ground state energy of the system can be found out approximately. The method can, however be applied for the investigation of the excited states as well.

Let φ_n denote the complete orthonormal set of eigen functions of the Hamiltonian H, that is

$$H\varphi_n = E_n \varphi_n \qquad \dots (98)$$

One can then expand any arbitrary function Φ (which is assumed to be an approximation to the correct wave function of the system under consideration) as the linear combination of φ_n, that is

$$\Phi = \sum_{n=0}^{\infty} a_n \varphi_n \qquad \dots (99)$$

The expectation value of H in the state Φ is given by

$$E_\Phi = \int \Phi^* H \Phi \, d\tau \Big/ \int \Phi^* \Phi \, d\tau \qquad \dots (100)$$

On substituting Φ from eqn. (99) into eqn. (100), one obtains

$$E_\Phi = \sum_n \sum_m a_n^* a_m \int \varphi_n^* H \varphi_m \, d\tau \Big/ \sum_n \sum_m a_n^* a_m \int \varphi_n^* \varphi_m \, d\tau$$

which on using eqn. (98) and the orthonormality relation for φ_n functions becomes

$$E_\Phi = \sum_n \sum_m E_m a_n^* a_m \delta_{nm} \Big/ \sum_n \sum_m a_n^* a_m \delta_{nm}$$

$$= \sum_{n=0}^{\infty} |a_n|^2 E_n \Big/ \sum_{n=0}^{\infty} |a_n|^2 \qquad \dots (101)$$

Subtracting E_0, the ground state energy of the system from both sides of the above equation, we have

$$E_\Phi - E_0 = \sum_{n=0}^{\infty} |a_n|^2 (E_n - E_o) \Big/ \sum_{n=0}^{\infty} |a_n|^2$$

Now $E_n \geq E_0$ (or $E_n - E_0 \geq 0$) for all n because by definition E_0 is the ground state energy and that $\sum_{n=0}^{\infty} |a_n|^2 = 1$ (if Φ is assumed to be normalized) and in general $|a_n|^2$ is positive being absolute square of a quantity, we must have from the above equation

$$E_\Phi - E_0 > 0 \text{ or } E_\Phi > E_0 \qquad \dots (102)$$

This important inequality can be used to obtain an upper limit or upper bound (the term used would be clear at a slightly later stage) to the ground state energy E_0. The variational method consists in evaluating E_Φ by choosing a trial function Φ which depends on number of parameters λ_i ($i = 1, 2, \dots p$) and the parameters are varied (or chosen) so as to make E_Φ a minimum by solving the p equations,

$$\frac{\partial E_\Phi}{\partial \lambda_i} = 0, \ (i = 1, 2, \ldots, p) \qquad \ldots (103)$$

The minimum value of E_Φ so determined would be close to the actual ground state energy (or equal to the upper bound or limit to E_0 or always larger than E_0 or it can never be smaller than E_0) if the trial function has the form closely resembling that of the actual ground state wave function Φ_0 (say). In the case in which the chosen trial function, happens to be exactly same as the ground state wave function, $(E_\Phi)_{min} = E_0$ or the equality condition in eqn. (102) holds good. The following example would make this point more clear.

12.13 THE HYDROGEN ATOM—AN EXAMPLE

In order to illustrate the use of variational method, we would apply the method to the hydrogen atom problem. That is, to the problem in which a particle of mass m is bound by a potential $V(\mathbf{r}) = -e^2/\mathbf{r}$. Choosing a trial function $\Phi = e^{-\alpha r}$, let us get a good limit on the lowest energy value where α is a parameter which is to be varied.

We will calculate the expectation value of H in the state Φ, that is

$$E_\Phi = \int \Phi^* H \Phi \, d\tau / \int \Phi^* \Phi \, d\tau = \langle \Phi | H | \Phi \rangle / \langle \Phi | \Phi \rangle \qquad \ldots (104)$$

and set $\frac{\partial E_\Phi}{\partial \alpha} = 0$ which will give us the value of α for which E_Φ is minimum (one may start with $\Phi = Ae^{-\alpha r}$ and find the normalization constant A using the normalization condition, $\int \Phi^* \Phi \; d\tau$. In that case the starting equation would be, $E_\Phi = \int \Phi^* H \Phi \, d\tau$ with Φ being the normalized wave function).

Now

$$\langle \Phi | \Phi \rangle = \int_0^\infty \int_0^\pi \int_0^{2\pi} e^{-2\alpha r} \, \mathbf{r}^2 \sin\theta \, d\theta \, d\phi \, d\mathbf{r}$$

$$= 4\pi \int_0^\infty e^{-2\alpha r} \, r^2 \, dr$$

the expression which can be written directly in this case as the wave function is independent of θ and ϕ and hence $d\tau$ = volume element between $\mathbf{r}$ and $\mathbf{r} + d\mathbf{r}$, that is equal to $4\pi \mathbf{r}^2 d\mathbf{r}$. Putting the value of the integral, one gets

$$\langle \Phi | \Phi \rangle = 4\pi \times \frac{2}{(2\alpha)^3} = \frac{\pi}{\alpha^3} \qquad \ldots (105)$$

Similarly

$$\langle \Phi | H | \Phi \rangle = \langle \Phi | -\hbar^2/2m \, \nabla^2 + V(r) | \Phi \rangle$$

$$= -\hbar^2/2m \, \langle \Phi | \nabla^2 | \Phi \rangle + \langle \Phi | V(r) | \Phi \rangle \qquad \ldots (106)$$

Now

$$\langle \Phi | V(r) | \Phi \rangle = 4\pi \int_0^\infty e^{-\alpha r} (-e^2/r) \, e^{-\alpha r} \, r^2 \, dr$$

$$= -4\pi e^2 \int_0^\infty e^{-2\alpha r}\, r dr$$

$$= -4\pi e^2 \frac{1}{(2\alpha)^2}$$

$$= -\frac{\pi e^2}{\alpha^2} \quad \text{... (107a)}$$

and

$$\langle \Phi \mid \nabla^2 \mid \Phi \rangle = \left\langle \Phi \left| \frac{1}{r^2} \frac{d}{dr}\left(r^2 \frac{d}{dr} \right) \right| \Phi \right\rangle$$

as Φ depends upon **r** only and hence the radial part of ∇^2 has been taken. Simplifying above,

$$\langle \Phi \mid \nabla^2 \mid \Phi \rangle = \left\langle \Phi \left| \frac{d^2}{dr^2} + \frac{2}{r}\frac{d}{dr} \right| \Phi \right\rangle$$

$$= 4\pi \int_0^\infty e^{-\alpha r}\, \alpha^2 e^{-\alpha r}\, r^2\, dr + 4\pi \int_0^\infty e^{-\alpha r} (-2\alpha)\, e^{-\alpha r}\, r dr$$

$$= 4\pi\alpha^2 \frac{2}{(2\alpha)^3} - 8\pi\alpha \frac{1}{(2\alpha)^2}$$

$$= -\frac{\pi}{\alpha} \quad \text{... (107b)}$$

Substituting values from eqns. (105) and (107) into eqn. (104), one gets

$$E_\Phi(\alpha) = \frac{\hbar^2}{2m}\alpha^2 - e^2\alpha \quad \text{... (108)}$$

Let us minimize $E_\Phi(\alpha)$ with respect to α, that is equating

$$dE_\Phi / d\alpha = \frac{2\alpha\hbar^2}{2m} - e^2 = 0$$

which gives,

$$\alpha = \frac{me^2}{\hbar^2} = \frac{1}{a_0} \quad \text{... (109)}$$

where a_0 is called Bohr's radius.

Substituting the value of α from the above equation into eqn. (108), one gets

$$(E_\Phi)_{\min} = \frac{\hbar^2}{2ma_0^2} - \frac{e^2}{a_0}$$

Further substituting $\frac{\hbar^2}{m}$ from eqn. (109) into above, one gets

$$(E_\Phi)_{\min} = -\frac{e^2}{2a_0} \quad \text{... (110)}$$

which is the exact result for the ground state energy of hydrogen atom as it indeed should be, because we had started with correct form of Φ (normalized ground state wave function using eqns. (105) and (109) is $\Phi_0 = \varphi_0 = (\pi a_0^3)^{-1/2} e^{-r/a_0}$, that is the value of Φ for that value of α for which $(E_\Phi)_{min} = E_0$)

12.14 THE GROUND STATE OF HELIUM ATOM – AN APPLICATION

The Hamiltonian for such a system (see Fig. 12.5) with nuclear charge Ze would be given by

$$H = -\frac{\hbar^2}{2m}\nabla_1^2 - \frac{Ze^2}{r_1} - \frac{\hbar^2}{2m}\nabla_2^2 - \frac{Ze^2}{r_2} + \frac{e^2}{r_{12}} \quad \text{... (111)}$$

where $\mathbf{r}_{12} = |\vec{r}_1 - \vec{r}_2|$, $\vec{r}_1$ and $\vec{r}_2$ represent the position vectors of the two electrons with nucleus at the origin of coordinate system.

The trial function Φ for the ground state of helium atom is chosen to be the product of two hydrogen atom wave functions with the variational parameter λ written instead of Z (= 2), the atomic number or true nuclear charge. This is based on the fact that we know that the electrons in atom repel each other (due to the repulsive term $e^2/\mathbf{r}_{12}$) and hence one expects that each electron is held to the atom a little more loosely (or nuclear attraction on the electron would reduce) than if the other electron were not present or the effective nuclear charge (λe) is less than the true nuclear charge Ze. Thus

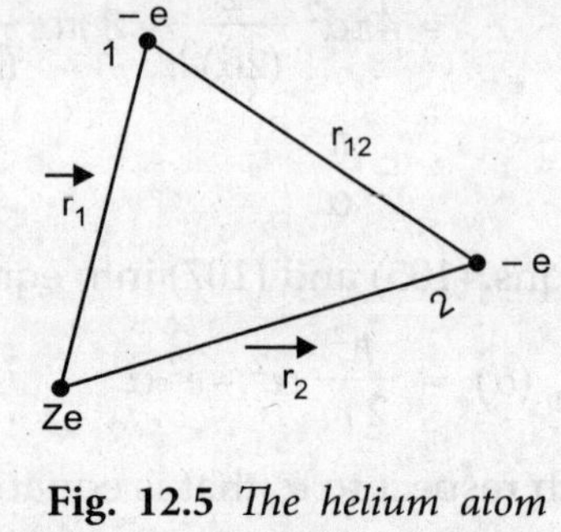

Fig. 12.5 *The helium atom*

$$\Phi(\vec{r}_1, \vec{r}_2) = \frac{\lambda^3}{\pi a_0^3} e^{-\lambda(r_1 + r_2)/a_0} \quad \text{... (112)}$$

The expectation value of H in the state Φ is given by

$$E_\Phi = \langle \Phi \mid H \mid \Phi \rangle$$

as Φ is already normalized. Note that Z rather than λ appears in the Hamiltonian of the said system since we cannot change the actual charge on the nucleus.

Putting the value of H in the above equation, one gets

$$E_\Phi = \left\langle \Phi \left| -\frac{\hbar^2}{2m}\nabla_1^2 - \frac{Ze^2}{r_1} - \frac{\hbar^2}{2m}\nabla_2^2 - \frac{Ze^2}{r_2} + \frac{e^2}{r_{12}} \right| \Phi \right\rangle \quad \text{... (113)}$$

Now

$$\left\langle \Phi \left| -\frac{\hbar^2}{2m} \nabla_1^2 \right| \Phi \right\rangle = -\frac{\hbar^2}{2\,m} \frac{\lambda^6}{\pi^2 a_0^6} \int_{r_1=0}^{\infty} \int_{r_2=0}^{\infty} e^{-\lambda(r_1+r_2)/a_o}$$

$$\left(\frac{\partial^2}{\partial r_1^2} + \frac{2}{r_1} \frac{\partial}{\partial r_1} \right) e^{-\lambda(r_1+r_2)/a_0} (4\pi r_1^2\, dr_1)(4\pi r_2^2\, dr_2)$$

$$= -\frac{\hbar^2}{2\,m} \frac{\lambda^6}{\pi^2 a_0^6} 16\,\pi^2 \int_0^{\infty} e^{-2\lambda r_2 / a_0} r_2^2\, dr_2 \times$$

$$\int_0^{\infty} e^{-\lambda r_1 / a_0} \left(\frac{d^2}{dr_1^2} + \frac{2}{r_1} \frac{d}{dr_1} \right) e^{-\lambda r_1 / a_0} r_1^2\, dr_1$$

$$= -\frac{\hbar^2}{2\,m} \frac{16\,\lambda^6}{a_0^6} \frac{2}{(2\,\lambda/a_o)^3} \int_0^{\infty} \left(\frac{\lambda^2}{a_0^2} - \frac{2\,\lambda}{a_o r_1} \right) e^{-2\lambda r_1/a_o} r_1^2\, dr_1$$

$$= -\frac{\hbar^2}{2\,m} \frac{4\lambda^3}{a_0^3} \left[\frac{\lambda^2}{a_0^2} \frac{2}{(2\lambda/a_0)^3} - \frac{2\lambda}{a_0} \frac{1}{(2\lambda/a_0)^2} \right]$$

$$= \frac{\lambda^2 e^2}{2a_0} \qquad \text{... (114a)}$$

where use been made of the relation, $a_0 = \dfrac{\hbar^2}{me^2}$.

Similarly, we can show that

$$\left\langle \Phi \left| -\frac{\hbar^2}{2m} \nabla_2^2 \right| \Phi \right\rangle = \frac{\lambda^2 e^2}{2\,a_0} \qquad \text{... (114b)}$$

Now

$$\left\langle \Phi \left| -\frac{Ze^2}{r_1} \right| \Phi \right\rangle = -\frac{Ze^2 \lambda^6}{\pi^2 a_0^6} \times 16\,\pi^2 \int_0^{\infty} r_1\, e^{-2\lambda r_1/a_0}\, dr_1 \int_0^{\infty} r_2^2\, e^{-2\lambda r_2/a_0}\, dr_2$$

$$= -\frac{Ze^2 \lambda}{a_0} \qquad \text{... (114c)}$$

Similarly,

$$\left\langle \Phi \left| -\frac{Ze^2}{r_2} \right| \Phi \right\rangle = -\frac{Ze^2 \lambda}{a_0} \qquad \text{... (114d)}$$

Now

$$\left\langle \Phi \left| \frac{e^2}{r_{12}} \right| \Phi \right\rangle = \frac{\lambda^6}{\pi^2 a_0^6} \int\!\!\int e^{-\lambda(r_1+r_2)/a_0} \frac{e^2}{r_{12}} e^{-\lambda(r_1+r_2)/a_o}\, d\mathbf{r_1}\, d\mathbf{r_2}$$

$$= \frac{\lambda^6 e^2}{\pi^2 a_0^6} \int\int \frac{1}{r_{12}} e^{-2\lambda(r_1 + r_2)/a_0} d\mathbf{r}_1 \, d\mathbf{r}_2 \quad \text{... (114e)}$$

Consider first the integration over r_2, keeping r_1 fixed. The relevant part of the integral is

$$\int \frac{1}{r_{12}} e^{-2\lambda r_2/a_0} d\mathbf{r}_2 = \int_0^\infty r_2^2 \, dr_2 \int d\Omega_2 \, e^{-2\lambda r_2/a_0} \frac{1}{r_{12}} \quad \text{... (114f)}$$

where we have introduced polar coordinates (r_2, θ_2, ϕ_2) for r_2 and $d\Omega_2$ is the element of solid angle $\sin\theta_2 \, d\theta_2 \, d\phi_2$. To evaluate above, we will use the expansion,

$$\frac{1}{r_{12}} = \frac{1}{|\mathbf{r}_1 - \mathbf{r}_2|} = \frac{1}{r_1} \sum_{l=0}^{\infty} \left(\frac{r_2}{r_1}\right)^l P_l(\cos\theta), \ \mathbf{r}_1 \geq \mathbf{r}_2$$

$$= \frac{1}{r_2} \sum_{l=0}^{\infty} \left(\frac{r_1}{r_2}\right)^l P_l(\cos\theta), \ \mathbf{r}_1 > \mathbf{r}_2$$

where $P_l(\cos\theta)$ is the Legendre's polynomial and θ is the angle between r_1 and r_2. Also choose the polar axis to be along the direction of r_1 so that the polar angle θ_2 is just the angle θ. The eqn. (114*f*), then becomes

$$\int_0^{r_1} r_2^2 \, dr_2 \int d\Omega_2 \, e^{-2\lambda r_2/a_0} \left[\frac{1}{r_1} \sum_{l=0}^{\infty} \left(\frac{r_2}{r_1}\right)^l P_l(\cos\theta)\right.$$

$$\left. + \int_{r_1}^{\infty} r_2^2 \, dr_2 \int d\Omega_2 \, e^{-2\lambda r_2/a_0} \frac{1}{r_2} \sum_{l=0}^{\infty} \left(\frac{r_1}{r_2}\right)^l P_l(\cos\theta)\right]$$

Since $\int P_l(\cos\theta) \, d\Omega_2 = 4\pi \, \delta_{l0}$, only the term with $l = 0$ survives in each integral. Thus the above equation reduces to

$$\int_0^{r_1} r_2^2 \, e^{-2\pi r_2/a_0} dr_2 \frac{1}{r_1} 4\pi + \int_{r_1}^{\infty} dr_2 \, r_2 \, e^{-2\lambda r_2/a_0} 4\pi$$

$$= -\frac{\pi a_0^2}{\lambda^2}\left(1 + \frac{a_0}{\lambda r_1}\right) e^{-2\lambda r_1/a_0} + \frac{\pi a_0^3}{\lambda^3 r_1}$$

where integration has been done by parts.

Substituting above into eqn. (114e), one gets

$$\left\langle \Phi \left| \frac{e^2}{r_{12}} \right| \Phi \right\rangle = \frac{\lambda^6 e^2}{\pi^2 a_0^6} \int \left[-\frac{\pi a_0^2}{\lambda^2}\left(1 + \frac{a_0}{\lambda r_1}\right) e^{-2\lambda r_1/a_0} + \frac{\pi a_0^3}{\lambda^3 r_1}\right] \times e^{-2\lambda r_1/a_0} \, d\mathbf{r}_1$$

$$= \frac{\lambda^6 e^2}{\pi^2 a_0^6} \left[-\frac{\pi a_0^2}{\lambda^2} \int_0^\infty e^{-4\lambda r_1/a_0} 4\pi r_1^2 \, dr_1 - \frac{\pi a_0^3}{\lambda^3} \int_0^\infty e^{-4\lambda r_1/a_0} 4\pi \, r_1 \, dr_1\right.$$

$$+ \frac{\pi a_0^3}{\lambda^3} \int_0^\infty e^{-2\lambda r_1/a_0} \, 4\pi \, r_1 \, dr_1 \Bigg]$$

$$= \frac{4\lambda^6 e^2}{a_0^6} \left[-\frac{a_0^2}{\lambda^2} \frac{2}{(4\lambda/a_0)^3} - \frac{a_0^3}{\lambda^3} \frac{1}{(4\lambda/a_0)^2} + \frac{a_0^3}{\lambda^3} \frac{1}{(2\lambda/a_0)^2} \right]$$

$$= \frac{5}{8} \frac{\lambda e^2}{a_0} \qquad \text{... (114g)}$$

Substituting values from eqns. (114) into eqn. (113), one gets

$$E_\Phi = \frac{\lambda^2 e^2}{2a_0} + \frac{\lambda^2 e^2}{2a_0} - \frac{Ze^2\lambda}{a_0} - \frac{Ze^2 \lambda}{a_0} + \frac{5}{8} \frac{\lambda e^2}{a_0}$$

$$= \left(\lambda^2 - 2\lambda Z + \frac{5\lambda}{8} \right) \frac{e^2}{a_0} \qquad \text{... (115)}$$

The minimum value of E_Φ (or the ground state energy of the helium atom) can be found from the condition, $\frac{dE_\Phi}{d\lambda} = 0$, which leads to

$$\lambda = Z - \frac{5}{16} \left(= 2 - \frac{5}{16} = \frac{27}{16} \right) \qquad \text{... (116)}$$

This λ (or λe) value corresponding to the energy minimum is called effective nuclear charge and comes out to be less than $2e$ as expected. Putting the above value into eqn. (115), we have

$$(E_\Phi)_{\min} = -\left(Z - \frac{5}{16} \right)^2 \frac{e^2}{a_0}$$

$$= -\left(2 - \frac{5}{16} \right)^2 \frac{e^2}{a_0}$$

$$= -2.85 \frac{e^2}{a_0} \qquad \text{... (117)}$$

Note that the estimate is better than the result obtained from the first order perturbation theory. The experimental value for the minimum energy required to remove both the electrons from the helium atom is $2.904 \frac{e^2}{a_0}$ which is very close to the value found above. If we use the result $\frac{e^2}{2 a_0} = 13.6$ eV, the above result, reads,

$$(E_\Phi)_{\min} = -76.2 \text{ eV}.$$

12.15 APPLICATION TO EXCITED STATES

If we denote by φ_0 the wave function of the ground state, the evaluation of the energy E_1 of the first excited state reduces to solving the variational problem,

$$E_1 = \text{Min} \int \Phi_1^* H \Phi_1 \, d\tau \quad \text{... (119)}$$

Under the conditions that

$$\int \Phi_1^* \varphi_1 \, d\tau = 1 \quad \text{... (120a)}$$

and

$$\int \Phi_1^* \varphi_0 \, d\tau = 0 \quad \text{... (120 b)}$$

(in case Φ_1 is not normalized one has to find minimum of $\int \Phi_1^* H \Phi_1 \, d\tau / \Phi_1^* \Phi_1 \, d\tau$)

To prove above, we can proceed as for the case of the ground state but keeping in mind that due to the orthogonality condition (120*b*) in the expansion of Φ_1 in terms of the eigen functions of H, there is no term with

$$\varphi_0 \left(\Phi_1 = \sum_{n=0}^{\infty} b_n \varphi_{n'} \int \Phi_1^* \varphi_0 \, d\tau = \int \sum_n b_n^* \varphi_n^* \varphi_0 \, d\tau \right.$$

$$\left. = 0 \text{ or } \sum_n b_n^* \delta_{n0} = 0 \text{ or } b_0 = 0 \right)$$

that is,

$$\Phi_1 = \sum_{n=1}^{\infty} b_n \varphi_{n'} \sum_{n=1}^{\infty} |b_n|^2 = 1 \quad \text{... (121)}$$

The evaluation of the energy of the second excited level reduces to the solution of the variational problem

$$E_2 = \text{Min} \int \Phi_2^* H \Phi_2 \, d\tau \quad \text{... (122)}$$

with the conditions that

$$\int \Phi_2^* \Phi_2 \, d\tau = 1 \quad \text{... (123a)}$$

and

$$\int \Phi_2^* \varphi_1 \, d\tau = \int \Phi_2^* \varphi_0 \, d\tau = 0 \quad \text{... (123b)}$$

The evaluation of the energy of the third excited level reduces to the solution of a varitional problem with additional conditions. The variational problem thus becomes increasingly complicated for the evaluation of higher excited levels.

12.16 ONE DIMENSIONAL HARMONIC OSCILLATOR – A SIMPLE EXAMPLE

The Hamiltonian in this case is given by

$$H = -\frac{\hbar^2}{2m} \frac{d^2}{dx^2} + \frac{1}{2} m\omega^2 x^2 \quad \text{... (124)}$$

For choosing the trial function, we must keep in mind that the wave function must vanish as $x \to \pm\infty$. Also, the ground state wave function must not have nodes. We put, therefore

$$\varphi_0 = Ae^{-\alpha x^2/2} \qquad \text{... (125)}$$

To find energy and the wave function of the first excited state, we must take a trial function Φ_1 orthogonal to φ_0. The simplest function satisfying this condition $\left(\int \Phi_1^* \varphi_0 \, dx = 0\right)$ is

$$\Phi_1 = Bx\, e^{-\beta x^2/2}$$

From the normalization condition, we find

$$\int_{-\infty}^{+\infty} \Phi_1^* \Phi_1 \, dx = B^2 \int_{-\infty}^{+\infty} x^2 e^{-\beta x^2} dx = 1$$

or $$B^2 \frac{2}{2(2\beta)^1}\left(\frac{\pi}{\beta}\right)^{1/2} = 1$$

Therefore,

$$B = \left(\frac{4\beta^3}{\pi}\right)^{1/4} \qquad \text{... (127)}$$

Let us now find,

$$\int_{-\infty}^{+\infty} \Phi_1^* H \Phi_1 \, dx = B^2 \int_{-\infty}^{+\infty} xe^{-\beta x^2/2}\left[-\frac{\hbar^2}{2m}\frac{d^2}{dx^2} + \frac{1}{2} m\omega^2 x^2\right] xe^{-\beta x^2/2} dx$$

$$= -\frac{\hbar^2}{2m} B^2 \left[\int_{-\infty}^{+\infty} \beta^2 x^4 e^{-\beta x^2} dx - 3\beta \int_{-\infty}^{+\infty} x^2 e^{-\beta x^2} dx\right]$$

$$+ \frac{1}{2} m\omega^2 B^2 \int_{-\infty}^{+\infty} x^4 e^{-\beta x^2} dx$$

$$= -\frac{\hbar^2}{2m} B^2 \left[\beta^2\, 2 \times \frac{1\cdot 3}{2(2\beta)^2}\left(\frac{\pi}{\beta}\right)^{1/2} - \frac{3\beta \times 2 \times 1}{2(2\beta)^1}\left(\frac{\pi}{\beta}\right)^{1/2}\right]$$

$$+ \frac{1}{2} m\omega^2 B^2 \times 2 \times \frac{1\cdot 3}{2(2\beta)^2}\left(\frac{\pi}{\beta}\right)^{1/2}$$

$$= \frac{3}{4}\left[\frac{\beta\hbar^2}{m} + \frac{m\omega^2}{\beta}\right] \qquad \text{... (128)}$$

From the condition that the above is minimum, we find $\beta = \beta_0 = \dfrac{m\omega}{\hbar}$.

Substituting this value into eqn. (128), we find

$$E_1 = \frac{3}{2}\hbar\omega \qquad \text{... (129)}$$

and hence the wave function of the first excited state

$$\Phi_1 = \left(\frac{2}{\pi^{1/2}}\right)^{1/2} \left(\frac{m\omega}{\hbar}\right)^{3/4} x e^{-m\omega x^2/2\hbar} = \varphi_1 \quad \text{... (130)}$$

Here once again, we find exactly the energy E_1 and the associated eigen state because the trial function chosen happens to be the correct wave function.

12.17 TIME DEPENDENT PERTURBATION THEORY

While describing the time independent perturbation theory, it was made clear that the theory would be applicable only in the cases in which the total Hamiltonian could be broken into two parts, the large part describing a system for which the Schrödinger equation can be solved exactly whereas the small part can be treated as perturbation. As an example of the non-degenerate case, one dimensional harmonic oscillator of charge e in the presence of an electric field ε in the x-direction (perturbation $\lambda H' = -e\varepsilon x$) was discussed. All the energy levels in the said case were found to be shifted equally (by $-e^2\varepsilon^2/2m\omega^2$). The effect of perturbation in the case of degenerate systems was also considered. As an example, a particle in an infinitely deep potential well in side a cube of dimension L in the presence of perturbation, λx^2 was discussed. In the case of first excited state which is three fold degenerate, the degeneracy was shown to be lifted partially.

Now we would go over to the cases in which the total Hamiltonian can of course be broken up into two parts as

$$H = H^{(0)} + \lambda H' \quad \text{... (131)}$$

but where $H^{(0)}$ is the time independent un-perturbed Hamiltonian while the part $\lambda H'$ is a small time dependent perturbation ($\lambda \ll 1$, is a real dimensionless parameter) arising due to the interaction of external agency with the system under consideration. Interaction of electromagnetic radiation with the atom or an atom irradiated by light, is an example.

Since the perturbation is time dependent, the time behaviour of wave functions of the perturbated system, is determined by solving the time dependent Schrödinger equation

$$H\Psi(\mathbf{r}, t) = (H^{(0)} + \lambda H')\Psi(\mathbf{r}, t) = i\hbar\,\partial\Psi(\mathbf{r}, t)/\partial t \quad \text{... (132)}$$

Like in the time independent case, let us express the perturbed wave function $\psi(\mathbf{r}, t)$ as an expansion in the wave functions $\Psi_n^{(0)}(\mathbf{r}, t)$ of the unperturbed time dependent Schrödinger equation, that is

$$\Psi(\mathbf{r}, t) = \sum_n a_n(t)\,\Psi_n^{(0)}(\mathbf{r}, t) = \sum_n a_n(t)\,\varphi_n^{(0)}(\mathbf{r})\,e^{-iE_n^{(0)}t/\hbar} \quad \text{... (133)}$$

where we have used the time dependence of $\Psi_n^{(0)}(\mathbf{r}, t)$. $\varphi_n^{(0)}(\mathbf{r})$ and $\Psi_n^{(0)}(\mathbf{r}, t)$ describe the states of the unperturbed system at time $t = 0$ and at a later time t, respectively. Note that the expansion coefficients $a_n(t)$ here are taken to be function of time. This method of treating time dependent problem is known as method of variation of constants (why is it called so would be clear at a slightly later stage) and was developed by Dirac[5]. The wave function is taken to be normalized at all times such that

$$\begin{aligned} 1 &= \int \Psi^*(\mathbf{r},t)\,\Psi(\mathbf{r},t)\,d\tau \\ &= \sum_{nn'} a_n^*(t)\,a_{n'}(t) \int \varphi_n^{(0)*}(\mathbf{r})\,\varphi_{n'}^{(0)}(\mathbf{r})\,e^{-i/\hbar\,(E_{n'}^{(0)}-E_n^{(0)})t}\,d\tau \\ &= \sum_{nn'} a_n^*(t)\,a_{n'}(t)\,e^{-i/\hbar\,(E_{n'}^{(0)}-E_n^{(0)})t}\,\delta_{nn'} \\ &= \sum_n |a_n(t)|^2 \end{aligned} \qquad \text{... (134)}$$

where $|a_n(t)|^2$ can be interpreted as the probability of finding the system in the nth state at time t.

The Schrödinger equation for the unperturbed Hamiltonian $H^{(0)}$ for the nth quantum state is given by

$$H^{(0)}\,\Psi_n^{(0)}(\mathbf{r},t) = i\hbar\,\partial\Psi_n^{(0)}(\mathbf{r},t)/\partial t \qquad \text{... (135a)}$$

Similarly $\varphi_n^{(0)}(\mathbf{r})$ and $E_n^{(0)}$ denote the eigen functions and eigen values, respectively of the time independent Hamiltonian $H^{(0)}$, that is

$$H^{(0)}\,\varphi_n^{(0)}(\mathbf{r}) = E_n^{(0)}\,\varphi_n^{(0)}(\mathbf{r}) \qquad \text{... (135b)}$$

Substituting eqn. (133) into eqn. (132), one gets

$$\sum_n a_n(t)\,(H^{(0)}+\lambda H')\,\varphi_n^{(0)}\,e^{-i/\hbar\,E_n^{(0)}t} = i\hbar\frac{\partial}{\partial t}\left[\sum_n a_n(t)\,\varphi_n^{(0)}\,e^{-i/\hbar\,E_n^{(0)}t}\right]$$

Using the relation (135b) and then cancelling the term

$$\sum_n a_n(t)\,E_n^{(0)}\,\varphi_n^{(0)}\,e^{-iE_n^{(0)}t/\hbar}$$

on both the sides, one gets

$$i\hbar\sum_n \dot{a}_n(t)\,\varphi_n^{(0)}\,e^{-iE_n^{(0)}t/\hbar} = \sum_n a_n(t)\,\lambda H'\,\varphi_n^{(0)}\,e^{-E_n^{(0)}t/\hbar}$$

Multiplying above equation by $\varphi_f^{(0)*}$ from the left, integrating over the entire space and using the orthonormality property of the eigen functions $\varphi_n^{(0)}$, one gets

$$i\hbar\sum_n \dot{a}_n(t)\,\delta_{fn}\,e^{-i/\hbar\,E_n^{(0)}t} = \sum_n a_n(t)\,\lambda\,H'_{fn}\,e^{-i/\hbar\,E_n^{(0)}t} \qquad \text{... (136a)}$$

where

$$H'_{fn} = \int \varphi_f^{(0)*}(\mathbf{r})\,H'(\mathbf{r},t)\,\varphi_n^{(0)}(\mathbf{r})\,d\tau$$

represents the matrix element of the perturbation between the states $\varphi_f^{(0)}$ and $\varphi_n^{(0)}$. Using the definition of kronecker delta function, eqn. (136a) leads to

[5]P.A.M. Dirac, Proc. Roy. Soc. Lond. A112, 661 (1926)

$$\dot{a}_f(t) = (i\hbar)^{-1} \sum_n a_n(t)\, \lambda\, H'_{fn}\, e^{i\omega_{fn} t}, \qquad \text{... (136}b\text{)}$$

in which $\omega_{fn} = \left(E_f^{(0)} - E_n^{(0)}\right)/\hbar$ denotes the Bohr's frequency of transition corresponding to the pair of unperturbed levels $E_f^{(0)}$ and $E_n^{(0)}$.

The set of equations (136) enables us to determine the time dependence of the coefficients $a_f(t)$ and consequently dependence of wave function on time and hence the set is completely equivalent to the Schrödinger eqn. (132). Also note that the set of eqns. (136b) is exact as no approximation has been made so far. From this set, we observe that if the perturbation $\lambda H'$ were absent all the $a_f(t)$ would be constants. The perturbation $\lambda H'$ makes them vary with time. For this very reason the method is often referred to as the method of variation of constants.

Since $\lambda H'$ is supposed to be very small, it is reasonable to express $a_f(t)$ as a power series in λ;

$$a_f(t) = a_f^{(0)}(t) + \lambda a_f^{(1)}(t) + \lambda^2 a_f^{(2)}(t) + \ldots \qquad \text{... (137)}$$

It is here that the idea of perturbation method enters in. Superscripts denote the order. Substituting the values of $a_f(t)$ and $a_n(t)$ from the above equation into eqn. (136b) and equating the coefficients of λ^r on both sides of the resulting equation, we get:

(*i*) for $r = 0$:

$$\dot{a}_f^{(0)}(t) = 0 \qquad \text{... (138}a\text{)}$$

(*ii*) for $r = 1$:

$$\dot{a}_f^{(1)}(t) = (i\hbar)^{-1} \sum_n a_n^{(0)}(t)\, H'_{fn}\, e^{i\omega_{fn} t}, \qquad \text{... (138}b\text{)}$$

and

(*iii*) for $r = 2$:

$$\dot{a}_f^{(2)}(t) = (i\hbar)^{-1} \sum_n a_n^{(1)}(t)\, H'_{fn}\, e^{i\omega_{fn} t}, \qquad \text{... (138}c\text{)}$$

From the relation (138*a*), one gets

$$a_f^{(0)}(t) = \text{constant.}$$

or in other words $a_f^{(0)}(t)$ does not depend upon time. Therefore

$$a_f^{(0)}(t) = a_f^{(0)}(0) \qquad \text{... (139)}$$

At $t = 0$ when the perturbation has not yet had a chance to act, $a_f^{(1)}(t), a_f^{(2)}(t), \ldots$ must necessarily be zero. Accordingly using eqns. (137) and (139), one gets

$$a_f(0) = a_f^{(0)}(0) \qquad \text{... (139}a\text{)}$$

12.18 FIRST ORDER PERTURBATION (OR SOLUTION TO FIRST ORDER IN λ)

After making use of eqn. (139), eqn. (138*b*) results in

$$\dot{a}_f^{(1)}(t) = (i\hbar)^{-1}\sum_n a_n^{(0)}(0)\,H'_{fn}\,e^{i\omega_{fn}t} \quad \text{... (140)}$$

In many practical applications, the initial conditions are such that at $t = 0$, that is before the perturbation is applied (or turned on or radiation begins arriving) the system (e.g. an atom) is described by one of the unperturbed eigen functions, say $\varphi_I^{(0)}$. This would mean that our initial conditions are

$$a_f^{(0)}(0) = \delta_{fI} \quad \text{... (139b)}$$

(if $I \neq f$, $a_f^{(0)}(0) = a_f(0) = 0$ as the system is in the state $\varphi_I^{(0)}$ at $t = 0$. Actually $|a_f(0)|^2$ is the probability that he system is found in the fth state at $t = 0$). Making use of the above equation, eqn. (140) results in

$$\dot{a}_f^{(1)}(t) = (i\hbar)^{-1}\sum_n \delta_{nI}\,H'_{fn}\,e^{i\omega_{fn}t},$$

Further using the definition of Kronecker delta function, one gets

$$\dot{a}_f^{(1)}(t) = (i\hbar)^{-1}\,H'_{fI}\,e^{i\omega_{fI}t},$$

and hence

$$a_f^{(1)}(t) = (i\hbar)^{-1}\int_0^t H'_{fI}\,e^{i\omega_{fI}t}\,dt + a \text{ constant of integration}$$

which will be $a_f^{(1)}(0)$ and is actually zero because at $t = 0$, there is no perturbation. Therefore

$$a_f^{(1)}(t) = (i\hbar)^{-1}\int_0^t H'_{fI}\,e^{i\omega_{fI}t}\,dt \quad \text{... (141)}$$

which is basis for the first order time dependent perturbation theory.

12.19 TRANSITIONS TO A CONTINUUM – CONSTANT PERTURBATION

We now consider the case of a perturbation $\lambda H'$ which acts from time $0 - t$ and remains constant during this period. Such a perturbation is called constant perturbation. In this case H'_{fI} can be pulled outside the integral in eqn. (141) which then reduces to

$$a_f^{(1)}(t) = (i\hbar)^{-1}H'_{fI}\left.\frac{e^{i\omega_{fI}t}}{i\omega_{fI}}\right|_0^t$$

$$= -\frac{H'_{fI}}{\hbar}\frac{(e^{i\omega_{fI}t} - 1)}{\omega_{fI}} \quad \text{... (142)}$$

Note that $a_f^{(1)}(t)$ would be zero if $\lambda H' = 0$ or if there were no perturbation. That is, it is due to the nonzero value (or presence) of the perturbation that $a_f^{(1)}(t) \neq 0$ and hence transition to the state f takes place.

Let us now find the probability, $|a_f^{(1)}(t)|^2$ (correspondingly, $a_f^{(1)}(t)$ is called first order transition amplitude) that the transition to such a state has occured. From eqn. (142).

$$|a_f^{(1)}(t)|^2 = \frac{|H'_{fl}|^2}{\hbar^2} \frac{(e^{-\omega_{fl}t}-1)(e^{i\omega_{fl}t}-1)}{\omega_{fl}^2}$$

$$= \frac{|H'_{fl}|^2}{\hbar^2\omega_{fl}^2}[1+1-(e^{i\omega_{fl}t}+e^{-i\omega_{fl}t})]$$

$$= \frac{|H'_{fl}|^2}{\hbar^2\omega_{fl}^2}(2-2\cos\omega_{fl}t)$$

$$= \frac{|H'_{fl}|^2}{\hbar^2}\sin^2\omega_{fl}\,t/2\Big/\left(\frac{\omega_{fl}}{2}\right)^2 \quad \text{... (143)}$$

Because of the presence of the factor ω_{fl}^2 in the denominator of the above equation, it is evident that the probability of finding the system in the state f is small unless the energy of this state is close to the energy of the initial state, that is $E_1^{(0)}$. Let us, therefore assume that the levels form a cluster around $E_t^{(0)}$ with a range $\pm\,\Delta E_f^{(0)}$ (such a case arises, for example, in the scattering problem). Under this condition, the total transition probability into the continuum of final states would be given by

$$W = \sum_f{}' |a_f^{(1)}(t)|^2$$

where the prime indicates $f \neq I$. The use of eqn. (143), therefore leads to

$$W = \sum_f{}' \frac{|H'_{fl}|^2}{\hbar^2(\omega_{fl}/2)^2} \quad \sin^2\omega_{fl}^{t/2}$$

Since $\Delta E_f^{(0)}$ is small enough, the variation of H'_{fl} with f with in this interval is negligible. Hence the total transition probability into the continuum of final states f would be given by

$$W = \frac{|H'_{fl}|^2}{\hbar^2} \quad \sum_f{}' \sin^2\omega_{fl}\frac{t}{2}\Big/\left(\frac{\omega_{fl}}{2}\right)^2 \quad \text{... (144)}$$

As the level spacing is close enough (being cluster), the summation can be replaced by integration by defining a density function $\rho\left(E_f^{(0)}\right)$, the so called density of states around the energy $E_f^{(0)}$ such that $\rho\left(E_f^{(0)}\right)dE_f^{(0)}$ is the number of states within the energy range $dE_f^{(0)}$.Therefore, from the above equation,

$$W = \frac{|H'_{fl}|^2}{\hbar^2}\int \sin^2\omega_{fl}\frac{t}{2}\Big/\left(\frac{W_{fl}}{2}\right)^2 \rho\left(E_f^{(0)}\right)dE_f^{(0)}$$

Since, $$\omega_{fl} = \frac{\left(E_f^{(0)}-E_1^{(0)}\right)}{\hbar}, d\,\omega_{fl} = \frac{dE_f^{(0)}}{\hbar}$$

as $E_l^{(0)}$ is discrete. Thus

$$W = \frac{|H'_{fl}|^2}{\hbar} \int \rho\left(E_f^{(0)}\right) \sin^2 \omega_{fl}\, t/2 \Big/ \left(\frac{\omega_{fl}}{2}\right)^2 d\,\omega_{fl}$$

We further assume that $\rho\left(E_f^{(0)}\right)$, like H'_{fl} varies so little with in the interval $\Delta E_f^{(0)}$ that it can be treated as constant and hence can be taken outside the integral. As a result

$$W = \frac{|H'_{fl}|^2}{\hbar} \rho\left(E_f^{(0)}\right) \int \sin^2 \omega_{fl} \frac{t}{2} \Big/ \left(\frac{\omega_{fl}}{2}\right)^2 d\omega_{fl} \qquad \text{... (145)}$$

Fig. 12.6 shows the dependence of the function $\sin^2 \omega_{fl} \frac{t}{2} \Big/ \left(\frac{\omega_{fl}}{2}\right)^2$ on ω_{fl} for a given t. Note that for $\omega_{fl} \to 0$ (or $E_f^{(0)}$ very near to $E_1^{(0)}$),

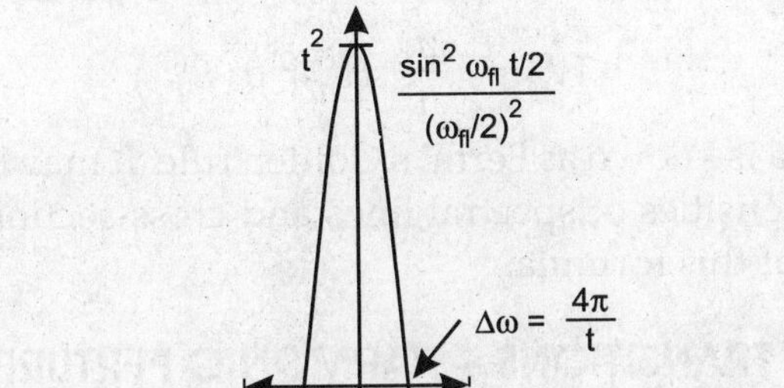

Fig. 12.6 *The transition probability as a function of ω_{fl} for fixed t*

$$\sin^2 \omega_{fl} \frac{t}{2} \Big/ \left(\frac{\omega_{fl}}{2}\right)^2 = \sin^2 \omega_{fl} \frac{t}{2} \Big/ \left(\frac{\omega_{fl}\, t}{2}\right)^2 t^2 = t^2$$

as $\sin\theta$ goes to θ for small θ. That is the height of the central peak at $\omega_{fl} = 0$, is t^2. $\Delta\omega$, the resonant width approximately defined as the distance between the first two zeros in $\sin^2 \omega_{fl} \frac{t}{2} \Big/ \left(\frac{\omega_{fl}}{2}\right)^2$ about $\omega_{fl} = 0$, being $\frac{4\pi}{t}$, decreases inversely as t (see Fig. 12.6). It means area under the curve (which is nothing but the total transition probability into the continuum of final states *f*) is proportional to *t*. Let us now try to find an explicit expression for the said probability. Since most of the contribution to the integral in eqn. (145) is from the first maximum (compare the curve with the single slit diffraction pattern), there is no harm taking the limits of integration from $-\infty$ to $+\infty$. Also multiplying the numerator and denominator by t^2 and then putting

$$x = \omega_{fI}\frac{t}{2}, dx = \frac{t}{2}\, d\omega_{fI}$$

into eqn. (145), one gets

$$W = \frac{|H'_{fI}|^2}{\hbar}\rho\left(E_f^{(0)}\right)\int_{-\infty}^{+\infty}\frac{\sin^2 x}{x^2}\frac{2t^2}{t}\, dx$$

Using the result $\int_{-\infty}^{+\infty}\frac{\sin^2 x}{x^2}\, dx = \pi$, the above equation reduces to

$$W = \frac{2\pi}{\hbar}\, t\,|H'_{fI}|^2\,\rho\left(E_f^{(0)}\right) \qquad \text{... (146)}$$

that is, it increases linearly with t. Thus we can say that for a transition from the initial state I to any of the final states f with energy nearly equal to $E_I^{(0)}$, the transition probability per unit time, W_{fI} (called transition rate) would be given by

$$W_{fI} = \frac{2\pi}{\hbar}|H'_{fI}|^2\,\rho\left(E_f^{(0)}\right) \qquad \text{... (146)}$$

The above formula is known as Fermi's Golden rule. It may be pointed out here that the calculations of intensities of spectral lines and cross-sections in various processes are based on the use of this formula.

12.20 RADIATIVE TRANSITIONS – HARMONIC PERTURBATION

Let us now consider the situation in which the perturbation is an explicit function of time (say, harmonic function) and is supposed to act over a time interval $0 - t$. It can be represented by

$$H'(\mathbf{r}, t) = 2H'(\mathbf{r})\cos\omega t = H'(\mathbf{r})\,(e^{i\omega t} + e^{-i\omega t}) \qquad \text{... (148)}$$

where $H'(\mathbf{r})$, also written as H'^0, is a time independent operator.

Suppose we consider the harmonic perturbation which results when atoms or molecules are exposed to the electromagnetic radiation*. The above equation in that case would imply an oscillating electric field associated with the electromagnetic radiation of frequency ω. $2H'^0$ will be the amplitude of ε. Substituting the above equation into eqn. (141), one immediately obtains the following expression for the first order transition amplitude

$$a_f^{(1)}(t) = (i\hbar)^{-1}\, H'^0_{fI}\int_0^t (e^{i\omega t} + e^{-i\omega t})\,e^{i\omega_{fI}t}\, dt,$$

*The electromagnetic field of the radiation is represented by an oscillating electric field ε and an oscillating magnetic field $H \perp$ to each other and $\perp$ to the direction of propagation of the waves. The interaction of the electromagnetic radiation with electrons or with any charged particles of the matter will be due, mainly to the electric field. This is because of the fact that the force exerted on a particle of charge e by the electric field is eε whereas the force on the same particle by a magnetic field is evH/c, where v is the velocity of the particle and c that of light. Since in most of the cases $v << c$, the magnetic contribution is much smaller than the electric contribution of the interaction.

where H'^0_{fl} has been taken outside the integral because it is not a function of time. Integrating the above expression, we obtain

$$a_f^{(1)}(t) = -i/\hbar\, H'^0_{fl} \int_0^t [e^{i(\omega_{fl}+\omega)t} + e^{i(\omega_{fl}-\omega)t}]dt$$

$$= \frac{-H'^0_{fl}}{\hbar}\left[\frac{(e^{i(\omega_{fl}+\omega)t}-1)}{(\omega_{fl}+\omega)} + \frac{(e^{i(\omega_{fl}-\omega)t}-1)}{(\omega_{fl}-\omega)}\right] \quad \text{... (149)}$$

From the above equation, it is evident that when the angular frequency ω of the radiation equals ω_{fl} ($\omega \sim \omega_{fl}$), the term that contains the factor $\omega_{fl} - \omega$ in the denominator would predominate. As a result, one can neglect the other term and find the probability of transition as found earlier in section 12.18. The only difference between the $a_f^{(1)}(t)$ here and in eqn. (141) would be that we have $\omega_{fl} - \omega$ in the present case instead of ω_{fl} in the earlier case and thus the expression for the probability $|a_f^{(1)}(t)|^2$ will be same as that obtained earlier with ω_{fl} replaced by $\omega_{fl} - \omega$. That means, only those transitions in which

$$\omega_{fl} - \omega = 0 \text{ or } E_f^{(0)} - E_I^{(0)} = \hbar\omega$$

can take place with appreciable probability. This effectively means that the perturbation (148) can induce transitions from $E_I^{(0)}$ only to level $E_f^{(0)}$ whose energy is higher by just $\hbar\omega$. Such a transition may be described as absorption of energy $\hbar\omega$ from the external agency by the system and is put under the head absorption (or stimulated absorption or induced absorption).

Similarly when

$$\omega_{fl} + \omega = 0 \text{ or } E_f^{(0)} - E_I^{(0)} = -\hbar\omega$$

the term containing $\omega_{fl} + \omega$ in the denominator of eqn. (149) would predominate and hence the other term can be neglected. That means in this case $E_f^{(0)}$ is lower than $E_I^{(0)}$ $\left(E_f^{(0)} < E_I^{(0)}\right)$ and the energy $\hbar\omega$ given away by the system to the perturbing (external) agency. The case corresponds to emission (or stimulated emission).

PROBLEMS

12.1 Show that the energy eigen values of a one dimensional harmonic oscillator calculated with the help of WKB method match with the exact results.

12.2 Using WKB method, calculate the transmission coefficient for the triangular barrier for which the potential energy variation is given by

$$V(x) = V_0(1 - |x|/a), \quad |x| < a$$

$$= 0, \quad |x| > a$$

12.3 Calculate the transmission coefficient of the parabolic potential barrier given by

$$V(x) = V_0(1 - x^2/a^2), -a \le x \le +a \text{ (or } |x| \le a)$$
$$= 0, \qquad a \le x \le -a \text{ (or } |x| \ge a)$$

using WKB method.

12.4 Using WKB method, calculate the transmission coefficient for the potential barrier displayed in Fig. 12.7 and mathematically represented by

$$V(x) = 0, x < 0$$
$$= V_0 - e\varepsilon x, x > 0$$

where ε is the external electric field. Note that the problem corresponds to the cold emission of electrons from a metal.

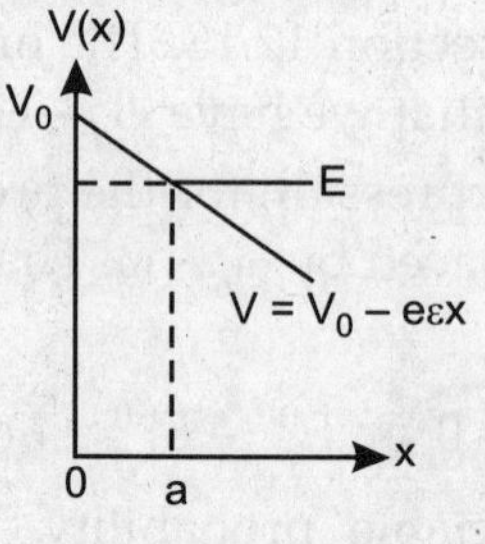

Fig. 12.7 *Shape of potential barrier in the presence of electric field* ε

12.5 In the case of penetration through the potential barrier discussed in sec 12.7, show that if $[2m(V_0 - E)a^2/\hbar^2]^{1/2} >> 1$ where V_0 is the height and a is the width of the barrier, the transmission coefficient is given by

$$T \approx e^{-2\{2m(V_0 - E)\}^{1/2} a/\hbar}$$

12.6 Find the probability that a particle will emerge from a centrally symmetric potential well with

$$V(\mathbf{r}) = -V_0, x < r_0$$
$$= A/x, x > r_0$$

shown in the Fig. 12.8 Note that the problem corresponds to the potential energy variation as seen by an α-partiele. Once the particle is formed inside the nucleus, it cannot escape unless it penetrates the surrounding Coulomb barrier. $x = \mathbf{r}_0$ and $\mathbf{r}_1$ are the turning points.

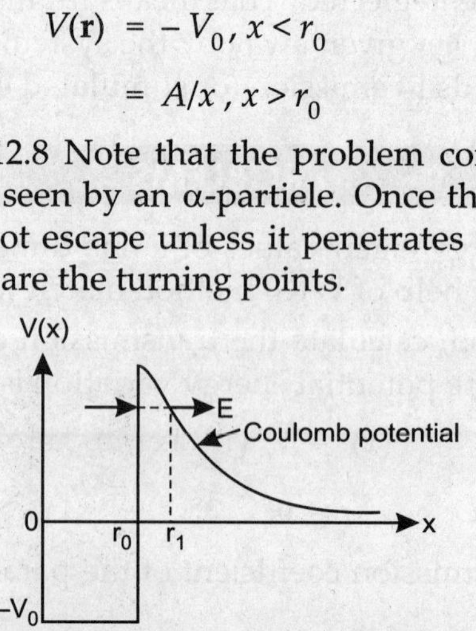

Fig. 12.8 *The potential energy variation as seen by the α-particle*

12.7 A 1 kg ball rolls slowly on a level track, then encounters a hump, 1 m long and 0.1m high as shown in Fig. 12.9. If the kinetic energy of the ball is neglected, find the probability that it can overcome the hump using WKB method.

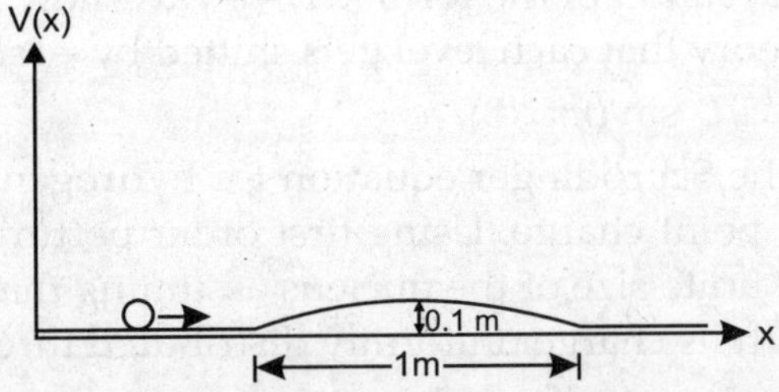

Fig. 12.9 *Ball on track*

12.8 Consider one dimensional harmonic oscillator with anharmonic perturbation λx^4, (see Fig. 12.10). Find first order correction to the ground state energy level using perturbation theory.

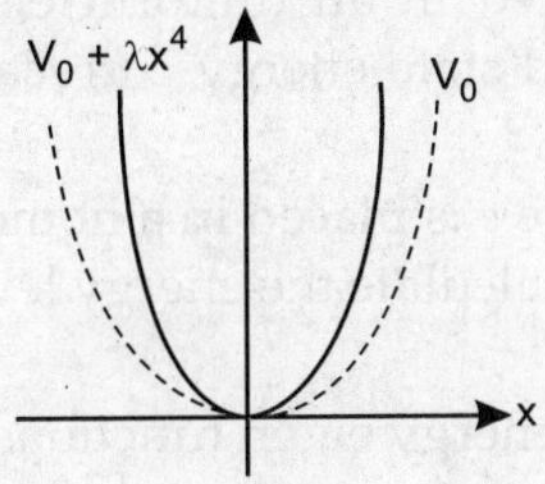

Fig. 12.10 *Anharmonic oscillator with* $\lambda H' = \lambda x^4$ *and* $V_0 = \dfrac{1}{2} m \omega^2 x^2$

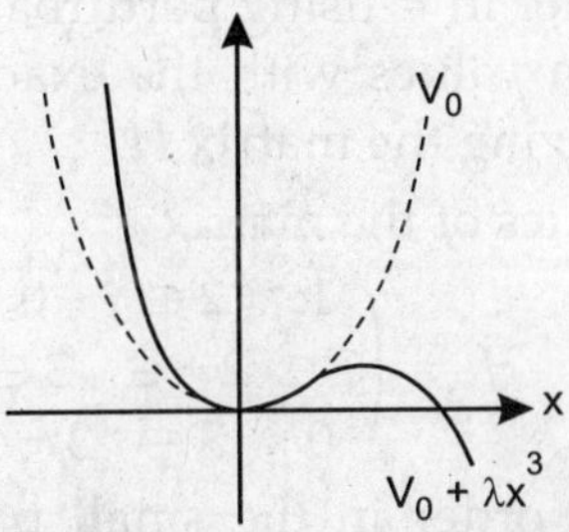

Fig. 12.11 *Anharmonic oscillator with* $\lambda H' = \lambda x^3$

12.9 Consider one dimensional harmonic oscillator with perturbation λx^3 (See Fig. 12.11). Find first order correction to the *n*th energy state.

12.10 Consider one dimensional harmonic oscillator with perturbation λx^4. Find *n*th energy state and the *n*th energy eigen function corrected upto the first order in perturbation.

12.11 Consider an electron inside an infinitely deep one dimensional potential well given by

$$V(x) = 0 \text{ for } 0 < x < L$$
$$= \infty \text{ for } 0 > x > L$$

Assuming perturbation of the form $\lambda H' = - e\varepsilon x$, show by using the first order perturbation theory that each level gets shifted by $- e\varepsilon L/2$.
Given, $\varphi_n(x) = \sqrt{2/L} \sin(n\pi x/L)$.

12.12 While solving the Schrödinger equation for hydrogen atom one assumes the nucleus to be a point charge. Using first order perturbation theory, calculate the effect of the finite size of the nucleus assuming the nucleus to be a sphere of radius $\mathbf{R}_N$ with its charge uniformly distributed throughout its volume.

12.13 Consider a linear harmonic oscillator of charge e perturbed by an electric field ε in the x-direction. Find energy levels corrected upto second order using perturbation theory. Given

$$\varphi_n^{(0)} = \left(\frac{\alpha}{\pi}\right)^{1/4} 2^{-n/2} (n!)^{-1/2} H_n(\sqrt{\alpha}\, x) e^{-\alpha x^2/2}, \alpha = m\omega/\hbar.$$

12.14 Hydrogen atom is placed in an uniform electric field ε applied along the z-direction. Find ground state energy corrected upto the second order using perturbation theory.

12.15 A plane rotator of charge e is placed in a homogeneous electric field ε parallel to the plane of rotator. Calculate the energy levels corrected upto second order in perturbation.

12.16 Find energy levels and energy eigen functions of the matrix

$$H = \begin{pmatrix} 1+\in & \in \\ \in & -1+\in \end{pmatrix}, \in \ll 1$$

corrected upto first order in $\in$ using perturbation theory. Compare the above evaluated energy eigen values with the exact ones, that is with the results obtained by diagnonalizing the matrix H.

12.17 Work out the eigen values of the matrix

$$H = \begin{pmatrix} 1 & 2\in & 0 \\ 2\in & 2+\in & 3\in \\ 0 & 3\in & 3+\in \end{pmatrix}$$

corrected upto second order in the small parameter $\in$ using perturbation theory.

12.18 Find eigen values of the matrix

$$H = \begin{pmatrix} 1 & \in \\ \in & 3 \end{pmatrix}, \in \ll 1$$

corrected upto second order in perturbation. Compare the results with the exact ones.

12.19 Consider a particle in an infinitely deep potential well inside a cube of dimension L described by the potential energy variation.

$$V(x, y, z) = 0, 0 < x < L, 0 < y < L, 0 < z < L$$
$$= \infty \text{ elsewhere}$$

Consider a perturbation λx^2. Apply first order perturbation theory to the first excited state of the particle. Also find the corresponding eigen vectors.

12.20 A hydrogen atom in the $n = 2$ energy state is subjected to an electric field ε in the z-direction. Find the energy corrections using perturbation theory. (The splitting of energy levels or spectral lines in the presence of electric field is called Stark effect). One can also find the corresponding eigen vectors.

12.21 Consider two coupled identical harmonic oscillators with the Hamiltonian

$$H = H^{(0)} + \lambda H'$$

where

$$H^{(0)} = \frac{1}{2m}(p_{x_1}^2 + p_{x_2}^2) + \frac{1}{2} m\omega^2 (x_1^2 + x_2^2)$$

and the perturbation

$$\lambda H' = \lambda x_1 x_2, \lambda << 1$$

Calculate the ground state and first excited state energies corrected upto first order in perturbation. Also find the zeroth order corrected eigen functions corresponding to each member of the first excited state.

12.22 Consider two dimensional harmonic oscillator with the unperturbed Hamiltonian

$$H^{(0)} = \frac{p_x^2}{2m} + \frac{p_y^2}{2m} + \frac{1}{2} m\omega^2 (x^2 + y^2)$$

which has a level with energy $E = 2\hbar\,\omega$ and is two fold degenerate. The oscillator is subjected to a perturbation $\lambda H' = \lambda m\omega^2 xy$, $\lambda << 1$. Show by using the first order perturbation theory that the level would split into a doublet. Also find the zeroth order eigen functions associated with the each member of the doublet.

12.23 For a hydrogen atom (neglecting the effects due to spin angular momentum of electron) placed in a weak magnetic field in the z-direction, show using perturbation theory that s states ($l = 0$) are not split whereas the p states ($l = 1$) are split into three states separated by the energy interval $\mu_B B$. This is known as the normal Zeeman effect (The splitting of atomic levels when the atom is placed in a uniform weak magnetic field).

12.24 Find the eigen values of the matrix

$$H = \begin{pmatrix} 1+\epsilon & \epsilon \\ \epsilon & 1+\epsilon \end{pmatrix}$$

corrected upto first order in perturbation ($\epsilon << 1$). Also find the corresponding eigen vectors. Compare with the eigen values obtained by the exact theory.

12.25 Find the eigen values of the matrix

$$H = \begin{pmatrix} 1 & \in & 0 \\ \in & 1 & 0 \\ 0 & 0 & 2+\in \end{pmatrix}$$

corrected upto first order in perturbation ($\in << 1$). Also obtain the corresponding eigen vectors.

12.26 The neutron-proton short range interaction is described by the potential energy variation

$$V(r) = -V_0, \ r < r_0$$
$$= 0, \ r > r_0$$

displayed in Fig. 5.9. Using the trial function $\Phi = e^{-\alpha/r_0}$ derive a range depth ($\mathbf{r}_0$, V_0) relation for the given potential and hence binding energy of the deuteron using variational technique.

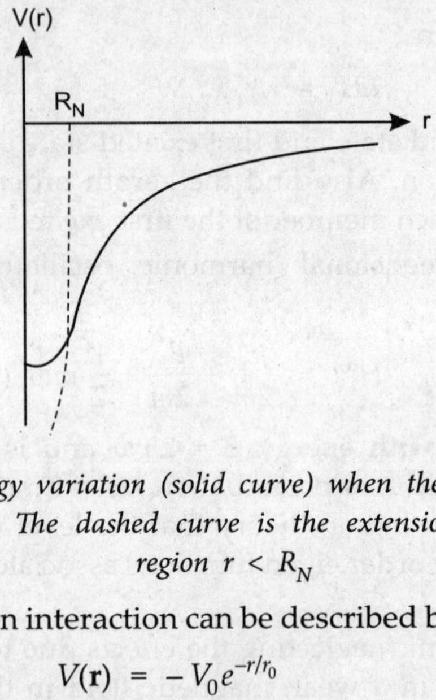

Fig. 12.12 *The potential energy variation (solid curve) when the effect of finite size of nucleus has been taken into account. The dashed curve is the extension of the 1/r variation in the region $r < R_N$*

12.27 The neutron–proton interaction can be described by the exponential potential,

$$V(\mathbf{r}) = -V_0 e^{-r/r_0}$$

where $\dfrac{\hbar^2}{\mu V_0 r_0^2} = \dfrac{3}{4}$. Assuming a trial function of the form $\Phi = e^{-\alpha r}$, carry out a variational analysis to obtain the ground state energy of the deuteron.

12.28 Obtain an approximate value for the lowest energy of the H-atom ($V(r) = -e^2/r$ in this case) using the variational technique starting with the trial function,

$$\Phi = e^{-\alpha (r/a_0)^2}$$

which actually represents the ground state wave function for the three dimensional harmonic oscillator.

12.29 The force between neutron and proton can be described, to a good approximation, by a potential energy function of the form

$$V(r) = -V_0\, e^{-r/r_0}/r/r_0$$

where V_0 is the strength of potential and r_0 the range. The above potential is called Yukawa potential. Using a trial function, $\Phi = e^{-\alpha r/r_0}$, calculate the binding energy of deuteron using variational method.

12.30 Choosing a trial function, $\Phi = A(1 + \alpha r)\, e^{-\alpha r}$, show by using the variational method that the ground state energy is given by

$$E_0 = -\frac{e^2}{2.07\, a_0}$$

12.31 A hydrogen atom in its ground state is placed between the plates of a condenser. A voltage pulse is applied to the condenser so as to produce a homogenous electric field that has the time dependence.

$$E = 0, t < 0 : E = E_0\, e^{-t/\tau}, t > 0$$

Find first order transition probability that the atom is in the 2s ($n = 2$, $l = 0$, $m = 0$) state after a long time. Also find the probability that it is in the $2p_z$ state. Note that the probabilities to the orbitals $2p_x$ and $2p_y$ come out to be zero.

12.32 A hydrogen atom in its ground state is placed in a homogeneous electric field applied in the z-direction with time dependence

$$E = \frac{E_0}{\pi} \frac{\tau}{(t^2 + \tau^2)}$$

Find the first order transition probability that the atom is in the $2p_z$ state after a long time.

SOLUTIONS

12.1 The quantization condition is given by

$$\oint p(x)\, dx = \left(n + \frac{1}{2}\right) h$$

or

$$2\int_a^b p(x)\, dx = \left(n + \frac{1}{2}\right) h, n = 0, 1, 2, \ldots \quad \ldots (12.1)$$

where a and b represent the classical turning points and are given by

$$E = V(x) = \frac{1}{2} m\omega^2 x^2$$

or

$$x = a = -\left(\frac{2E}{m\omega^2}\right)^{1/2}$$

and

$$x = b = +\left(\frac{2E}{m\omega^2}\right)^{1/2}$$

Therefore,

$$\int_a^b p(x)\,dx = \int_a^b \left[2m\left(E - \frac{1}{2}m\omega^2 x^2\right)\right]^{1/2} dx$$

$$= (2mE)^{1/2}\int_a^b \left(1 - \frac{m\omega^2}{2E}x^2\right)^{1/2} dx$$

After making the substitution,

$$\left(\frac{m\omega^2}{2E}\right)^{1/2} x = t$$

or

$$dx = dt\left(\frac{2E}{m\omega^2}\right)^{1/2}$$

we get,

$$\int_a^b p(x)\,dx = \frac{2E}{\omega}\int_{-1}^{+1}(1-t^2)^{1/2}\,dt$$

$$= \frac{2E}{\omega} \times \frac{\pi}{2}$$

$$= \frac{\pi E}{\omega}$$

Substituting above into eqn. (12.1), we get

$$\frac{2\pi E}{\omega} = \left(n + \frac{1}{2}\right)h$$

or

$$E = E_n = \left(n + \frac{1}{2}\right)\hbar\,\omega,\ n = 0, 1, 2, \ldots$$

which is same as that obtained from the exact theory.

12.2 The transmission coefficient in this case is given by

$$T \approx e^{-2/\hbar \int_{x_1}^{x_2} |2m\{V(x) - E\}|^{1/2}\,dx}$$

where x_1 and x_2 are the turning points of motion and are given by

$$E = V(x) = V_0(1 - |x|/a)$$

or

$$|x| = (V_0 - E)\,a/V_0 \text{ (or say } x = \pm b)$$

Therefore,

$$T \approx e^{-\frac{2}{\hbar}\int_{-b}^{+b}(2m)^{1/2}(V_0 - V_0|x|/a - E)^{1/2}dx}$$

$$\approx e^{-\frac{4}{\hbar}\int_{0}^{b}(2m)^{1/2}(V_0 - E - V_0|x|/a)^{1/2}dx}$$

$$\approx e^{-\frac{4}{\hbar}(2m)^{1/2}\frac{2}{3}(V_0 - E - V_0|x|/a)^{3/2}\Big|_0^b(-a/V_0)}$$

$$\approx e^{-\frac{8}{3\hbar}(2m)^{1/2}\left[\left\{V_0 - E - \frac{V_0}{a}(V_0 - E)\frac{a}{V_0}\right\}^{3/2} - (V_0 - E)^{3/2}\right]\left(-\frac{a}{V_0}\right)}$$

$$\approx e^{-\frac{8a}{3\hbar V_0}(2m)^{1/2}(V_0 - E)^{3/2}}$$

12.3 For the given problem, the transmission coefficient is given by

$$T \approx e^{-\frac{2}{\hbar}\int_{x_1}^{x_2}|2m\{V(x) - E\}|^{1/2}dx}$$

where x_1 and x_2 are the turning points of the classical motion of the particle and are given by

$$E = V(x) = V_0(1 - x^2/a^2)$$

or

$$x^2 = a^2(V_0 - E)/V_0$$

or

$$x = \pm a[(V_0 - E)/V_0]^{1/2} \text{ (or } x = \pm b)$$

Therefore,

$$T \approx e^{-\frac{2}{\hbar}\int_{-b}^{+b}(2m)^{1/2}\left(V_0 - \frac{V_0 x^2}{a^2} - E\right)^{1/2}dx}$$

$$\approx e^{-\frac{2}{\hbar}(2mV_0)^{1/2}(1 - E/V_0)^{1/2}\int_{-b}^{+b}\left[1 - \frac{1}{(1 - E/V_0)}\frac{x^2}{a^2}\right]^{1/2}dx}$$

Making the substitution,

$$\frac{x}{b} = \frac{x}{a(1 - E/V_0)^{1/2}} = t$$

or

$$dx = (1 - E/V_0)^{1/2}a\,dt,$$

the above equation results in

$$T \approx e^{-\frac{2}{\hbar}(2mV_0a^2)^{1/2}(1 - E/V_0)\int_{-1}^{+1}(1 - t^2)^{1/2}dt}$$

Using the result

$$\int_{-1}^{+1}(1 - t^2)^{1/2}dt = \frac{\pi}{2}$$

we get,

$$T \approx e^{-(2m\pi^2 V_0 a^2/\hbar^2)^{1/2}(1-E/V_0)}$$

12.4 For present problem, the transmission coefficient is given by

$$T \approx e^{-\frac{2}{\hbar}\int_{x_1}^{x_2}[2m\{V(x)-E\}]^{1/2}dx}$$

where x_1 and x_2 are the turning points of motion. Taking $x = x_1 = 0$, the other turning point is given by

$$E = V(x) = V_0 - e\varepsilon x_2$$

or

$$x_2 = a = (V_0 - E)/e\varepsilon$$

Therefore,

$$T \approx e^{-\frac{2}{\hbar}\int_0^a[2m(V_0-e\varepsilon x-E)]^{1/2}dx}$$

$$\approx e^{-\frac{2}{\hbar}(2m)^{1/2}\frac{2}{3}(V_0-E-e\varepsilon x)^{3/2}\times 1/e\varepsilon\Big|_0^a}$$

$$\approx e^{-\frac{2}{\hbar}(2m)^{1/2}\frac{2}{3}\times -1/e\varepsilon\left[\{V_0-E-e\varepsilon(V_0-E)/e\varepsilon\}^{3/2}-(V_0-E)^{3/2}\right]}$$

$$\approx e^{-4(2m)^{1/2}(V_0-E)^{3/2}/3e\varepsilon}$$

12.6 As soon as an α-particle comes out of nucleus (the remaining charge on the nucleus becomes $(Z-2)\,e$), it experiences Coulombian repulsive force represented by the potential energy function,

$$V(x) = \frac{A}{x},\; A = 2(Z-2)\,e^2/4\pi\epsilon_0 \qquad \text{... (12.2)}$$

The transmission coefficient, therefore would be given by (note that the fact that the centrally symmetric problem reduces to one dimensional form, has been used here).

$$T \approx e^{-\frac{2}{\hbar}\int_{r_0}^{r_1}[2m\{V(x)-E\}]^{1/2}dx} \qquad \text{... (12.3)}$$

where $\mathbf{r}_0$ and $\mathbf{r}_1$ are the turning points of the motion, $\mathbf{r}_0$ corresponds to the range of the nuclear force whereas the other turning point $\mathbf{r}_1$ would be given by

$$E = V(x = \mathbf{r}_1) = \frac{A}{r_1} \qquad \text{... (12.4)}$$

The expression for T can be written as

$$T \approx e^{-\frac{2}{\hbar}\int_{r_0}^{r_1}\left[2m\left(\frac{A}{x}-E\right)\right]^{1/2}dx}$$

which on making use of eqn. (12.4) becomes

$$T \approx e^{-\frac{2}{\hbar}(2mE)^{1/2}\int_{r_0}^{r_1}\left(\frac{r_1}{x}-1\right)^{1/2}dx}$$

Making the substitution,

$$x = r_1 \cos^2\theta$$

or

$$dx = r_1\, 2\cos\theta\,(-\sin\theta\, d\theta)$$

and also changing the limits to θ (at $x = r_0$, $\theta_0 = \cos^{-1}(r_0/r_1)$ and at $x = r_1$, $\theta = 0$), one gets

$$T \approx e^{-2(2mE/\hbar^2)^{1/2} r_1 \int_0^{\theta_0} 2\left(\frac{1}{\cos^2\theta}-1\right)^{1/2}\sin\theta\cos\theta\, d\theta}$$

Since $\left(\frac{1}{\cos^2\theta}-1\right)^{1/2} = \frac{\sin\theta}{\cos\theta}$, the above expression becomes

$$T \approx e^{-2(2mE/\hbar^2)^{1/2} r_1 \int_0^{\theta_0} 2\sin^2\theta\, d\theta}$$

$$\approx e^{-2(2mE/\hbar^2)^{1/2} r_1 \int_0^{\theta_0}(1-\cos 2\theta)\, d\theta}$$

$$\approx e^{-2(2mE/\hbar^2)^{1/2} r_1 (\theta_0 - \sin\theta_0\cos\theta_0)}$$

$$\approx e^{-(2mE/\hbar^2)^{1/2} r_1 (2\theta_0 - \sin 2\theta_0)} \qquad \text{... (12.5)}$$

Since far away from the nucleus the energy of the α –particle is totally kinetic,

$$E = \frac{1}{2}mv^2 = A/r_1$$

Substituting above into eqn. (12.5), the transmission coefficient would be given by

$$T \approx e^{-\frac{2A}{\hbar v}(2\theta_0 - \sin 2\theta_0)} \qquad \text{... (12.6)}$$

As it has been already mentioned, the above discussed case refers to nothing but the α-decay. It may also be mentioned here that in case the α-particle strikes the barrier from the right and appears in the nucleus, the result would again be given by the expression (12.6) and is referred to as the Gamow factor.

12.7 The probability that the ball can overcome the hump (or the transmission coefficient) would be given by

$$T \approx e^{-\frac{2}{\hbar}\int_0^L |2m\{V(x)-E\}|^{1/2}dx}$$

$$\approx e^{-\frac{2}{\hbar}\int_0^L |2m\,V(x)|^{1/2}dx}$$

where E has been taken to be zero. Rest of the steps are same as that in problem 3.10 with $V(x) = mgh \sin \pi x/L$.

12.8 The ground state wave function in the present case is given by

$$\varphi_0^{(0)}(x) = \left(\frac{\alpha}{\pi}\right)^{1/4} e^{-\alpha x^2/2}, \; \alpha = m\omega/\hbar$$

The first order correction to the ground state energy level would, therefore be given by

$$\lambda E_0^{(1)} = \lambda H'_{00} = \int_{-\infty}^{+\infty} \varphi_0^{(0)*}(x)\, \lambda x^4\, \varphi_0^{(0)}(x)\, dx$$

$$= \lambda \left(\frac{\alpha}{\pi}\right)^{1/2} \int_{-\infty}^{+\infty} e^{-\alpha x^2} x^4\, dx$$

$$= \lambda \left(\frac{\alpha}{\pi}\right)^{1/2} \frac{3}{4} \left(\frac{\pi}{\alpha^5}\right)^{1/2}$$

$$= \frac{3}{4} \frac{\lambda}{\alpha^2}$$

$$= \frac{3}{4} \lambda \left(\frac{\hbar}{m\omega}\right)^2$$

Therefore the total ground state energy corrected upto first order would to given by

$$E_0 = \frac{1}{2}\hbar\omega + \frac{3}{4}\lambda(\hbar/m\omega)^2$$

12.9 The wave function of the nth energy state in the case of one dimensional harmonic oscillator is given by

$$\varphi_n^{(0)} = \left(\frac{\alpha}{\pi}\right)^{1/4} 2^{-n/2} (n!)^{-1/2} H_n(\sqrt{\alpha}\, x)\, e^{-\alpha x^2/2}, \; \alpha = m\omega/\hbar$$

The first order correction to the nth energy state would, therefore be given by

$$\lambda E_n^{(1)} = \lambda H'_{nn} = \lambda \int_{-\infty}^{+\infty} \varphi_n^{(0)*}(x)\, x^3\, \varphi_n^{(0)}(x)\, dx$$

Since x^3 is an odd function and $\varphi_n^{(0)*}(x)\,\varphi_n^{(0)}(x)$ is an even function of x, the above integral would give zero. Hence the first order correction due to the perturbation λx^3 would be zero.

12.10 The first order correction to the nth energy state would be given by

$$\lambda E_n^{(1)} = \lambda H'_{nn} = \lambda \int_{-\infty}^{+\infty} \varphi_n^{(0)*}(x)\, x^4\, \varphi_n^{(0)}(x)\, dx$$

Substituting the value of $\varphi_n^{(0)}(x)$ from the above problem with

$$\xi = \sqrt{\alpha}\, x \text{ (or } d\xi = \sqrt{\alpha}\, dx)$$

one gets

$$\lambda E_n^{(1)} = \frac{\lambda}{(\pi^{1/2}\, n!\, 2^n\, \alpha^2)} \int_{-\infty}^{+\infty} e^{-\xi^2} |H_n(\xi)|^2 \xi^4\, d\xi \quad \text{... (12.7)}$$

The recursion relation for Hermite polynomials given by

$$\xi H_n(\xi) = \frac{1}{2} H_{n+1}(\xi) + n\, H_{n-1}(\xi) \quad \text{... (12.8)}$$

Therefore,

$$\xi^2 H_n(\xi) = \frac{1}{2}\, \xi H_{n+1}(\xi) + n\, \xi H_{n-1}(\xi) \quad \text{... (12.9)}$$

Using the recursion relation (12.8) for $\xi H_{n+1}(\xi)$ and similarly for $\xi H_{n-1}(\xi)$, that is

$$\xi H_{n+1}(\xi) = \frac{1}{2} H_{n+2}(\xi) + (n+1) H_n(\xi)$$

and

$$\xi H_{n-1}(\xi) = \frac{1}{2} H_n(\xi) + (n-1) H_{n-2}(\xi),$$

and after substituting the above into eqn. (12.9), gets

$$\begin{aligned}\xi^2 H_n(\xi) &= \frac{1}{2}\left[\frac{1}{2} H_{n+2}(\xi) + (n+1) H_n(\xi)\right] \\ &\quad + n\left[\frac{1}{2} H_n(\xi) + (n-1) H_{n-2}(\xi)\right] \\ &= \frac{1}{4} H_{n+2}(\xi) + \left(n + \frac{1}{2}\right) H_n(\xi) \\ &\quad + n(n-1) H_{n-2}(\xi) \end{aligned} \quad \text{... (12.10)}$$

Substituting the above equation into eqn. (12.7), one gets

$$\lambda E_n^{(1)} = \frac{\lambda}{(\pi^{1/2}\, n!\, 2^n\, \alpha^2)} \int_{-\infty}^{+\infty} e^{-\xi^2} \left[\frac{1}{4} H_{n+2}(\xi) + \left(n + \frac{1}{2}\right) H_n(\xi) + n(n-1) H_{n-2}(\xi)\right]^2 d\xi$$

Using the known relation,

$$\begin{aligned}\int_{-\infty}^{+\infty} e^{-\xi^2} H_n(\xi) H_m(\xi)\, d\xi &= 0, \; m \neq n \\ &= 2^n\, n!\, \pi^{1/2}, \; m = n\end{aligned}$$

one gets,

$$\lambda E_n^{(1)} = \frac{\lambda}{\alpha^2}\left[\frac{1}{16}2^{n+2}(n+2)!+\left(n+\frac{1}{2}\right)^2 2^n n!\right.$$

$$\left.+n^2(n-1)^2 2^{n-2}(n-2)!\right]/(2^n n!)$$

$$= 3\lambda(2n^2+2n+1)/4\alpha^2$$

Thus the energy of the nth state of one dimensional harmonic oscillator corrected upto first order would be given by

$$E_n = E_n^{(0)} + \lambda E_n^{(1)}$$

$$= \left(n+\frac{1}{2}\right)\hbar\omega + 3\lambda\hbar^2\omega^2(2n^2+2n+1)/4k^2,$$

$$n = 0, 1, 2, \ldots$$

The energy eigen function corrected upto first order in perturbation would be given by

$$\varphi_n = \varphi_n^{(0)} + \sum_k{}' \int \lambda H'_{kn}/(E_n^{(0)} - E_k^{(0)})\,\varphi_k^{(0)} \qquad \ldots (12.11)$$

Now

$$\sum_k{}' \lambda H'_{kn} = \lambda \sum_k{}' \int \varphi_k^{(0)*}(x)\,x^4\,\varphi_n^{(0)}(x)\,dx$$

$$= \lambda \sum_k{}' \int \varphi_k^{(0)*}(x)\,x^2x^2\,\varphi_n^{(0)}(x)\,dx \qquad \ldots (12.12)$$

The recursion relation for the harmonic oscillator wave functions is given by

$$x\varphi_n^{(0)} = \frac{1}{\alpha^{1/2}}\left[\frac{(n+1)^{1/2}}{2^{1/2}}\varphi_{n+1}^{(0)} + \left(\frac{n}{2}\right)^{1/2}\varphi_{n-1}^{(0)}\right] \qquad \ldots (12.13)$$

Therefore

$$x^2\varphi_n^{(0)} = \frac{1}{\alpha^{1/2}}\left[\frac{(n+1)^{1/2}}{2^{1/2}}x\,\varphi_{n+1}^{(0)} + \left(\frac{n}{2}\right)^{1/2}x\,\varphi_{n-1}^{(0)}\right]$$

Using the recursion relation (12.13) for $x\varphi_{n+1}^{(0)}$ and for $x\varphi_{n-1}^{(0)}$, one gets

$$x^2\varphi_n^{(0)} = \frac{1}{\alpha}\left[\left\{\frac{(n+1)(n+2)}{4}\right\}^{1/2}\varphi_{n+2}^{(0)} + \frac{(2n+1)}{2}\varphi_n^{(0)}\right.$$

$$\left.+\left\{\frac{n(n-1)}{4}\right\}^{1/2}\varphi_{n-2}^{(0)}\right]$$

Substituting above into eqn. (12.12),

$$\lambda\sum_k{}' H'_{kn} = \frac{\lambda}{\alpha^2}\sum_k{}'\int\left\{\frac{\sqrt{(k+1)(k+2)}}{2}\varphi_{k+2}^{(0)*} + \frac{(2k+1)}{2}\varphi_k^{(0)*}\right.$$

$$\left.+\frac{\sqrt{k(k-1)}}{2}\varphi_{k-2}^{(0)*}\right\} \times \left\{\frac{\sqrt{(n+1)(n+2)}}{2}\varphi_{n+2}^{(0)}\right.$$

$$\left.+\frac{(2n+1)}{2}\varphi_n^{(0)} + \frac{\sqrt{n(n-1)}}{2}\varphi_{n-2}^{(0)}\right\}dx$$

$$= \frac{\lambda}{\alpha^2}\left[\frac{\sqrt{(k+1)(k+2)}}{2}\frac{(2n+1)}{2}\delta_{k+2,n}\right.$$

$$+\frac{\sqrt{(k+1)(k+2)n(n-1)}}{4}\delta_{k+2,n-2}$$

$$+\frac{(2k+1)}{2}\frac{\sqrt{(n+1)(n+2)}}{2}\delta_{k,n+2}$$

$$+\frac{(2k+1)}{2}\frac{\sqrt{n(n-1)}}{2}\delta_{k,n-2}$$

$$+\frac{\sqrt{k(k-1)(n+1)(n+2)}}{4}\delta_{k-2,n+2}$$

$$\left.+\frac{(2n+1)}{2}\frac{\sqrt{k(k-1)}}{2}\delta_{k-2,n}\right]$$

where terms corresponding to $k = n$ have not been taken. Further substituting above in (12.11)

$$\varphi_n = \varphi_n^{(0)} + \frac{\lambda}{\alpha^2}\left[\frac{(2n+1)}{2}\frac{\sqrt{(n-2+1)(n-2+2)}}{2}\right.$$

$$\frac{1}{(E_n^{(0)} - E_{n-2}^{(0)})}\varphi_{n-2}^{(0)}$$

$$+\frac{\sqrt{(n-4+1)(n-4+2)n(n-1)}}{4}\frac{1}{(E_n^{(0)} - E_{n-4}^{(0)})}\varphi_{n-4}^{(0)}$$

$$+\frac{\{2(n+2)+1\}}{2}\frac{\sqrt{(n+1)(n+2)}}{2}\frac{1}{(E_n^{(0)} - E_{n+2}^{(0)})}\varphi_{n+2}^{(0)}$$

$$+\frac{\{2(n-2)+1\}}{2}\frac{\sqrt{n(n-1)}}{2}\frac{1}{(E_n^{(0)}\ 0\ E_{n-2}^{(0)})}\varphi_{n-2}^{(0)}$$

$$+\frac{\sqrt{(n+4)(n+4-1)(n+1)(n+2)}}{4}\frac{1}{(E_n^{(0)} - E_{n+4}^{(0)})}\varphi_{n+4}^{(0)}$$

$$\left.+\frac{\sqrt{(n+2)(n+2-1)}}{2}\frac{(2n+1)}{2}\frac{1}{(E_n^{(0)} - E_{n+2}^{(0)})}\varphi_{n+2}^{(0)}\right]$$

$$= \varphi_n^{(0)} + \frac{\lambda}{\alpha^2}\left[\frac{n(n-1)}{4}(2n-1)\frac{1}{2\hbar\omega}\varphi_{n-2}^{(0)}\right.$$

$$+\frac{\sqrt{n(n-1)(n-2)(n-3)}}{4}\frac{1}{4\hbar\omega}\varphi_{n-4}^{(0)}$$

$$-\frac{\sqrt{(n+1)(n+2)}}{2}(2n+3)\frac{1}{2\hbar\omega}\varphi_{n+2}^{(0)}$$

$$- \frac{\sqrt{(n+1)(n+2)(n+3)(n+4)}}{4} \frac{1}{4\,\hbar\omega} \varphi_{n+4}^{(0)} \Bigg]$$

$$= \varphi_n^{(0)} + \frac{\lambda\hbar\omega}{k^2} \Bigg[\sqrt{n(n-1)} \frac{(2n-1)}{4} \varphi_{n-2}^{(0)}$$

$$+ \sqrt{n(n-1)(n-2)(n-3)} \frac{1}{16} \varphi_{n-4}^{(0)}$$

$$- \sqrt{(n+1)(n+2)} \frac{(2n+3)}{4} \varphi_{n+2}^{(0)}$$

$$- \sqrt{(n+1)(n+2)(n+3)(n+4)} \frac{1}{16} \varphi_{n+4}^{(0)} \Bigg]$$

12.11 The first order correction to the nth energy state in the present case is given by

$$\lambda E_n^{(1)} = \lambda H'_{nn} = \lambda \int \varphi_n^{(0)*} H' \varphi_n^{(0)}\, dx$$

$$= -2e\varepsilon/L \int_0^L x \sin^2 n\pi x/L\, dx$$

$$= -2e\varepsilon/L \int_0^L x(1-\cos 2\pi x/L)/2\, dx$$

Integral of the part gives $-2e\varepsilon/L\; x^2/2 \,|_0^L = -e\varepsilon\, L/2$ whereas integrating second function by parts leads to zero. Thus

$$\lambda E_n^{(1)} = -e\,\varepsilon L/2$$

which is independent of n. Thus each energy level gets shifted by $-e\,\varepsilon L/2$.

12.12 Since the charge Ze of the nucleus is assumed to be distributed uniformly throughout its volume, the charge density is

$$Ze \Big/ \left(\frac{4}{3}\pi R_N^3\right)$$

Thus the force acting on the electron, directed towards the centre of the sphere (found with the help of Guass's theorem) is given by

$$F(r) = Ze \Big/ \left(\frac{4}{3}\pi R_N^3\right)\left(\frac{4}{3}\pi r^3\right)\frac{(-e)}{r^2},\; r \le R_N$$

$$= -\frac{Ze^2}{r^2},\; r \ge R_N$$

The potential energy of the electron being at a distance r from the centre of the nucleus (see Fig. 12.12) would be given by

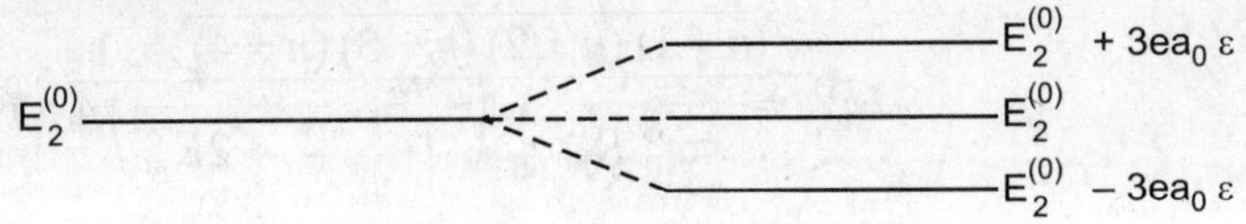

Fig. 12.13 *The splitting of the n = 2 state of the hydrogen atom in an electric field ε*

$$V(r) = -\frac{Ze^2}{R_N}\left(\frac{3}{2} - \frac{r^2}{2R_N^2}\right), \qquad r \le R_N$$

$$= -\frac{Ze^2}{r}, \qquad r \ge R_N$$

where we have used the relation,

$$F(r) = -\frac{dV}{dr}\left(V(r) = \frac{Ze^2 r^2}{2R_N^3} + c\right), \quad \text{for } r \le R_N.$$

The constant c is found by equating the R.H.S. to the potential energy,

$V(r = R_N) = -\dfrac{Ze^2}{R_N}$ which gives $c = -\dfrac{3}{2}\dfrac{Ze^2}{R_N}$.

By substituting the potential energy, $-\dfrac{Ze^2}{r}$ corresponding to the point like nucleus, the perturbation due to the finite size of the nucleus can be determined to be

$$\lambda H' = -\frac{Ze^2}{R_N}\left(\frac{3}{2} - \frac{r^2}{2R_N^2}\right) + \frac{Ze^2}{r}$$

$$= -\frac{Ze^2}{R_N}\left(\frac{3}{2} - \frac{r^2}{2R_N^2} - \frac{R_N}{r}\right), \; r \le R_N$$

$$= -\frac{Ze^2}{r} + \frac{Ze^2}{r} = 0, \quad r \ge R_N$$

Let us now calculate the first order correction $\lambda E_n^{(1)}$ for the 1*s* electron orbit, that is ground state. the corresponding wave function $\varphi_{100} = \varphi^{(0)}$ is given by

$$\varphi^{(0)} = \left(\frac{Z^3}{\pi a_0^3}\right)^{1/2} e^{-Zr/a_0}$$

Therefore the first order correction in the ground state energy would be given by

$$\lambda E_1^{(1)} = \int \varphi^{(0)*}\, \lambda H'\, \varphi^{(0)}\, d\tau$$

$$= \frac{Z^4 e^2}{\pi a_0^3 R_N} \int_0^{R_N} \left(\frac{R_N}{r} - \frac{3}{2} + \frac{r^2}{2R_N^2}\right) e^{-2Zr/a_0}\, 4\pi r^2\, dr$$

Since the radius of the nucleus is much smaller than the Bohr radius, that is $R_N \ll a_0$, we may assume the wave function to remain constant over the region of integration. Thus we may replace $\varphi^{(0)}$ by its value at $r = 0$. Therefore

$$\lambda E_1^{(1)} = \frac{Z^4 e^2}{\pi a_0^3 R_N} \int_0^{R_N} \left(\frac{R_N}{r} - \frac{3}{2} + \frac{r^2}{2R_N^2} \right) 4\pi r^2 \, dr$$

$$= \frac{4Z^4 e^2}{a_0^3 R_N} \left[\frac{R_N r^2}{2} - \frac{3}{2}\frac{r^3}{3} + \frac{1}{2R_N^2}\frac{r^5}{5} \right]_0^{R_N}$$

$$= \frac{2}{5} \frac{Z^4 e^2}{a_0^3} R_N^2$$

The energy of an electron moving in the 1s orbit around the nucleus (point like) is

$$E_1^{(0)} = -\frac{Z^2 e^2}{2 a_0}$$

Thus the relative energy change is

$$\eta = \frac{\lambda E_1^{(1)}}{|E_1^{(0)}|}$$

$$= \frac{4 Z^2 R_N^2}{5 a_0^2}$$

For lead atom, taken as an example, the relative energy change $\eta \simeq 10^{-4}$. Thus the difference is very small as compared to the energy calculated by supposing the nucleus to be a point charge. However, the spectroscopic investigations can reveal even this energy correction. For the isotopes of a given element where the atomic number Z is the same but the atomic weight and therefore the nuclear radii, are different, the energy correction would be different and due to this the spectral lines will not coincide completely. This phenomenon is called isotope shift.

12.13 The perturbation in this case is given by

$$\lambda H' = -e\varepsilon x$$

The first order correction to the nth energy state would therefore be given by

$$\lambda E_n^{(1)} = \lambda H'_{nn} = -e\,\varepsilon \int_{-\infty}^{+\infty} \varphi_n^{(0)*} \, x \, \varphi_n^{(0)} \, dx$$

$$= -\frac{e\varepsilon}{\alpha} \left(\frac{\alpha}{\pi} \right)^{1/2} 2^{-n} (n!)^{-1} \int_{-\infty}^{+\infty} |H_n(\xi)|^2 e^{-\xi^2} \xi d\xi, \; \xi = \sqrt{\alpha}\, x$$

The integrand being an odd function of ξ, the integral vanishes. Thereby implying that the first order correction $\lambda E_n^{(1)}$ is zero. The second order correction to the nth energy state is given by

$$\lambda^2 E_n^{(2)} = \sum_k{}' \frac{|\lambda H'_{kn}|^2}{\left(E_n^{(0)} - E_k^{(0)} \right)}$$

To evaluate $\lambda H'_{kn}$, let us note that the recursion relation for the harmonic oscillator wave function is given by

$$x\,\varphi_n^{(0)} = \frac{1}{\alpha^{1/2}}\left[\sqrt{\frac{(n+1)}{2}}\,\varphi_{n+1}^{(0)}(x) + \sqrt{\frac{n}{2}}\,\varphi_{n-1}^{(0)}(x)\right] \quad \ldots (12.13)$$

Therefore,

$$\lambda H'_{kn} = -e\,\varepsilon \int \varphi_n^{(0)*}\, x\, \varphi_n^{(0)}\, dx$$

After substituting the value of $x\varphi_n^{(0)}$ from eqn. (12.13) gives

$$\lambda H'_{kn} = -\frac{e\varepsilon}{\alpha^{1/2}}\left[\int_{-\infty}^{+\infty} \varphi_k^{(0)*}\sqrt{\frac{(n+1)}{2}}\,\varphi_{n+1}^{(0)}\,dx + \int_{-\infty}^{+\infty} \varphi_k^{(0)*}\sqrt{\frac{n}{2}}\,\varphi_{n-1}^{(0)}\,dx\right]$$

$$= -\frac{e\varepsilon}{\alpha^{1/2}}\left[\sqrt{\frac{(n+1)}{2}}\,\delta_{k,n+1} + \sqrt{\frac{n}{2}}\,\delta_{k,n-1}\right]$$

and hence

$$\lambda^2 E_n^{(2)} = \frac{e^2\varepsilon^2}{(m\omega/\hbar)}\left[\frac{(n+1)/2}{(E_n^{(0)} - E_{n+1}^{0})} + \frac{n/2}{(E_n^{(0)} - E_{n-1}^{(0)})}\right]$$

Using the result for the energy levels of the linear harmonic oscillator, that is

$$E_n^{(0)} = \left(n + \frac{1}{2}\right)\hbar\,\omega, \qquad n = 0, 1, 2, \ldots$$

one gets

$$\lambda^2 E_n^{(2)} = \left(\frac{e^2\varepsilon^2\hbar}{2m\omega}\right)\left[-\frac{(n+1)}{\hbar\omega} + \frac{n}{\hbar\omega}\right] = -\frac{e^2\varepsilon^2}{2\,m\omega^2}$$

which is independent of n. Thus all the energy levels are equally shifted in this case.

12.14 The perturbation in the present case is given by

$$\lambda H' = -e\,\varepsilon z = -e\varepsilon \mathbf{r}\cos\theta$$

where use has been made of the spherical polar coordinates and the unperturbed energy levels are given by

$$E_n^{(0)} = -\frac{e^2}{(2n^2 a_0)}$$

where $n = 1$ corresponds to the ground state energy level.

The wave function of the ground state of the hydrogen atom (or 1s orbital) is given by

$$\varphi_{100}^{(0)} = \varphi_1^{(0)}\ (\text{say}) = (\pi a_0^3)^{-1/2}\, e^{-r/a_0}$$

Therefore, the first order correction to the ground state energy level would be given by

$$\lambda E_1^{(1)} = \lambda H'_{11} = -e\,\varepsilon \int \varphi_1^{(0)*}\, r\cos\theta\, \varphi_1^{(0)}\, d\tau$$

$$= -\frac{e\varepsilon}{\pi a_0^3}\int_0^\infty e^{-2r/a_0} r^3\, dr \int_0^\pi \sin\theta\cos\theta\, d\theta \int_0^{2\pi} d\phi$$

which is equal to zero as $\int_0^\pi \sin\theta\cos\theta\, d\theta = 0$

Thus there is no first order effect on the ground state of hydrogen atom. The given perturbation may cause second order effect due to the mixing of $\varphi_1^{(0)}$ with certain appropriate excited states. The excited states which lie closest to $\varphi_1^{(0)}$ in energy are $\varphi_{nlm_l}^{(0)} = \varphi_{200}^{(0)}$ (2s), $\varphi_{211}^{(0)}$ $(2p_y)$, $\varphi_{210}^{(0)}$ $(2p_z)$, $\varphi_{21-1}^{(0)}$ $(2p_x)$, the $n = 2$ state is, therefore four fold degenerate. Of these four states, only the interaction between 1s and $2p_z$ orbitals ($\varphi_{210}^{(0)} = (32\,\pi a_0^3)^{-1/2}\, r\cos\theta/a_0\, e^{-r/2a_0}$ denoted by $\varphi_2^{(0)}$) are non-vanishing while the other ineractions vanish because the angular parts of the integral are zero. Thus second order correction to the ground state energy is given by

$$\lambda^2 E_1^{(2)} = \sum_k{}' \frac{|\lambda H'_{kn}|^2}{\left(E_n^{(0)} - E_k^{(0)}\right)}$$

$$= \frac{e^2\varepsilon^2 |H'_{21}|^2}{(E_1^{(0)} - E_2^{(0)})}$$

To evaluate above, let us first find H'_{21} which is given by

$$H'_{21} = \int \varphi_2^{(0)*}\, r\cos\theta\, \varphi_1^{(0)}\, d\tau$$

$$= \frac{1}{4\sqrt{2}\,\pi a_0^4}\int_0^\infty e^{-3r/2a_0} r^4 dr \int_0^\pi \sin\theta\cos^2\theta\, d\theta \int_0^{2\pi} d\phi$$

Using the results,

$$\int_0^\infty e^{-ar} r^n\, dr = \frac{n!}{a^{n+1}}$$

$$\int_0^\pi \sin\theta\cos^2\theta\, d\theta = -\cos^3\theta\,|_0^\pi / 3 = \frac{2}{3}\ and \int_0^{2\pi} d\phi = 2\pi,$$

one gets

$$H'_{21} = a_0\left(\frac{2}{3}\right)^5 \frac{2^3}{2^{1/2}}$$

Therefore,

$$\lambda^2 E_1^{(2)} = -\frac{\varepsilon^2 a_0^3\, 2^{18}}{3^{11}}$$

12.15 The perturbation in this case is,

$$\lambda H' = -e\,\varepsilon a \cos\phi$$

where ϕ is the plane polar angle, that is the azimuth and a is the radius of the circular orbit in which the mass point revolves around another fixed point. The normalized unperturbed wave functions in the said case are given by

$$\varphi_n^{(0)} = (2\pi)^{-1/2} e^{in\phi}, \quad n = 0, \pm 1, \pm 2, \ldots$$

Further the unperturbed energy eigen values are given by

$$E_n^{(0)} = \frac{n^2\hbar^2}{2I} \quad \ldots (12.14)$$

The first order correction to the nth energy state is given by

$$\lambda E_n^{(1)} = \lambda H'_{nn} = \int \varphi_n^{(0)*} \lambda H' \varphi_n^{(0)}\, d\phi$$

$$= -\frac{\varepsilon e a}{2\pi} \int_0^{2\pi} e^{-in\phi} \cos\phi\, e^{in\phi} d\phi$$

$$= -\frac{\varepsilon e a}{2\pi} . \sin\phi \Big|_0^{2\pi}$$

$$= 0,$$

that is, the first order energy correction is zero.

To calculate second order correction to the nth energy state, that is $\lambda^2 E_n^{(2)}$, we have to first find $\lambda H'_{kn}$ which is given by

$$\lambda H'_{kn} = -\frac{\varepsilon e a}{2\pi} \int_0^{2\pi} e^{-ik\phi} \cos\phi\, e^{in\phi} d\phi$$

$$= -\frac{\varepsilon e a}{2\pi} \int_0^{2\pi} e^{i(n-k)\phi} \frac{(e^{i\phi} + e^{-i\phi})}{2} d\phi$$

$$= -\frac{\varepsilon e a}{4\pi} \int_0^{2\pi} e^{i(n-k+1)\phi} d\phi - e\varepsilon \frac{a}{4\pi} \int_0^{2\pi} e^{i(n-k-1)\phi} d\phi$$

$$= -\frac{\varepsilon e a}{2} (\delta_{k,n+1} + \delta_{k,n-1})$$

where δ is the kronecker symbol.

Therefore,

$$\lambda^2 E_n^{(2)} = \sum_k{}' \frac{|\lambda H'_{kn}|^2}{\left(E_n^{(0)} - E_k^{(0)}\right)}$$

$$= \frac{e^2\varepsilon^2 a^2}{4} \left[\left\{\frac{1}{(E_n^{(0)} - E_{n+1}^{(0)})}\right\} + \left\{\frac{1}{(E_n^{(0)} - E_{n-1}^{(0)})}\right\}\right]$$

Using equation (12.14), we get

$$\lambda^2 E_n^{(2)} = \frac{Ie^2 \varepsilon^2 a^2}{\hbar^2} \frac{1}{(4n^2 - 1)} \quad \text{... (12.15)}$$

Thus the energy levels corrected upto second order are

$$E_n = E_n^{(0)} + \lambda E_n^{(1)} + \lambda^2 E_n^{(2)}$$

$$= \frac{\hbar^2 n^2}{2I} + 0 + \frac{Ie^2 \varepsilon^2 a^2}{\hbar^2} \frac{1}{(4n^2 - 1)}$$

From eqn. (12.15), it may be noted that the second order correction to the ground state ($n = 0$) being negative, the external field reduces the energy of the rotator. For the excited states ($n \geq 1$), the said correction is positive, the perturbation therefore, lifts the energy levels.

12.16 Let us write the given matrix as $H = H^{(0)} + \in H'$, where

$$H^{(0)} = \begin{pmatrix} 1 & 0 \\ 0 & -1 \end{pmatrix}$$

represents the unperturbed matrix (or Hamiltonian) and

$$\in H' = \in \begin{pmatrix} 1 & 1 \\ 1 & 1 \end{pmatrix}$$

the perturbation. The eigen values of the unperturbed matrix are +1 and –1 and the corresponding eigen functions (or eigen vectors) are

$$\varphi_1^{(0)} = \begin{pmatrix} 1 \\ 0 \end{pmatrix} \text{ and } \varphi_2^{(0)} = \begin{pmatrix} 0 \\ 1 \end{pmatrix}$$

Now $\in H'_{11} = \in$ and $\in H'_{22} = \in$, therefore the eigen values corrected upto first order in perturbation would be

$$E_1 = 1 + \in H'_{11} = 1 + \in \text{ and } E_2 = -1 + \in H'_{22} = -1 + \in.$$

The eigen functions corrected upto first order in $\in$ would be given by

$$\varphi_1 = \varphi_1^{(0)} + \frac{\in H'_{21}}{(E_1^{(0)} - E_2^{(0)})} \varphi_2^{(0)}$$

$$= \varphi_1^{(0)} + \frac{\in}{\{1 - (-1)\}} \varphi_2^{(0)}$$

$$= \begin{pmatrix} 1 \\ 0 \end{pmatrix} + \frac{\in}{2} \begin{pmatrix} 0 \\ 1 \end{pmatrix}$$

$$= \begin{pmatrix} 1 \\ \frac{\in}{2} \end{pmatrix}$$

and similarly

$$\varphi_2 = \varphi_2^{(0)} + \frac{\in H'_{12}}{(E_2^{(0)} - E_1^{(0)})}\varphi_1^{(0)}$$

$$= \varphi_2^{(0)} + \frac{\in}{(-1-1)}\varphi_{(1)}^{(0)}$$

$$= \begin{pmatrix} 0 \\ 1 \end{pmatrix} - \frac{\in}{2}\begin{pmatrix} 1 \\ 0 \end{pmatrix}$$

$$= \begin{pmatrix} -\frac{\in}{2} \\ 1 \end{pmatrix}$$

Now the exact eigen values of H can be found from

$$\begin{vmatrix} 1+\in-\lambda & \in \\ \in & -1+\in-\lambda \end{vmatrix} = 0$$

or

$$\lambda = \in \pm (1+\in^2)^{1/2}$$

$$= \in \pm \left(1 + \frac{\in^2}{2}\right)$$

Hence

$$\lambda_1 = 1 + \in + \frac{\in^2}{2} \text{ and } \lambda_2 = -1 + \in - \frac{\in^2}{2}$$

which corrected upto the first order in $\in$ are same as E_1 and E_2.

12.17 Let us write the given matrix as

$$H = H^{(0)} + \in H'$$

where

$$H^{(0)} = \begin{pmatrix} 1 & 0 & 0 \\ 0 & 2 & 0 \\ 0 & 0 & 3 \end{pmatrix}$$

represents the unperturbed matrix and

$$\in H' = \in \begin{pmatrix} 0 & 2 & 0 \\ 2 & 1 & 3 \\ 0 & 3 & 2 \end{pmatrix}$$

the perturbation, $\in \ll 1$. The eigen values of $H^{(0)}$ are 1, 2 and 3. The corresponding eigen functions are

$$\varphi_1^{(0)} = \begin{pmatrix} 1 \\ 0 \\ 0 \end{pmatrix}, \varphi_2^{(0)} = \begin{pmatrix} 0 \\ 1 \\ 0 \end{pmatrix}, \varphi_3^{(0)} = \begin{pmatrix} 0 \\ 0 \\ 1 \end{pmatrix}$$

Now the eigen values corrected upto second order are given by

$$E_n = E_n^{(0)} + \in H'_{nn} + \sum_k{}' \frac{|\in H'_{kn}|^2}{\left(E_n^{(0)} - E_k^{(0)}\right)}$$

Therefore,

$$E_1 = 1 + \in H'_{nn} + \frac{|\in H'_{21}|^2}{(E_1^{(0)} - E_1^{(0)})} + \frac{|\in H'_{31}|^2}{(E_1^{(0)} - E_3^{(0)})}$$

$$= 1 + 0 + \frac{(2\in)^2}{(1-2)} + 0$$

$$= 1 - 4\in^2,$$

$$E_2 = 2 + \in H'_{22} + \frac{|\in H'_{12}|^2}{(E_2^{(0)} - E_1^{(0)})} + \frac{|\in H'_{32}|^2}{(E_2^{(0)} - E_3^{(0)})}$$

$$= 2 + \in + \frac{(2\in)^2}{(2-1)} + \frac{(3\in)^2}{(2-3)}$$

$$= 2 + \in - 5\in^2$$

$$E_3 = 3 + \in H'_{33} + \frac{|\in H'_{13}|^2}{(E_3^{(0)} - E_1^{(0)})} + \frac{|\in H'_{23}|^2}{(E_3^{(0)} - E_2^{(0)})}$$

$$= 3 + 2\in + 0 + \frac{(3\in)^2}{(3-2)}$$

$$= 3 + 2\in + 9\in^2$$

12.19 The energy eigen values in the present case are given by

$$E^{(0)}_{n_x, n_y, n_z} = \frac{\pi^2 \hbar^2}{2mL^2}\left(n_x^2 + n_y^2 + n_z^2\right), n_x, n_y, n_z = 1, 2, 3, \ldots$$

Thus the energy eigen value corresponding to the first excited state for which n_x, n_y or n_z is to be taken equal 2, is given by

$$E^{(0)}_{211} = \frac{\pi^2 \hbar^2}{2mL^2}(2^2 + 1^2 + 1^2) = \frac{6\pi^2 \hbar^2}{2\,mL^2}$$

The said state is 3-fold degenerate because for the given eigen value, it has three linearly independent wave function $\varphi^{(0)}_{n_x, n_y, n_z}$ energy given by

$$\varphi^{(0)}_{211} = \varphi^{(0)}_1 = \left(\frac{8}{L^3}\right)^{1/2} \sin n_x \pi\, x/L \sin n_y \pi\, y/L \sin n_z\, \pi\, z/L$$

$$= \left(\frac{8}{L^3}\right)^{1/2} \sin 2\pi\, x/L \sin \pi\, y/L \sin \pi\, z/L$$

$$\varphi^{(0)}_{121} = \varphi^{(0)}_2 = \left(\frac{8}{L^3}\right)^{1/2} \sin \pi\, x/L \sin 2\pi\, y/L \sin \pi\, z/L$$

$$\varphi^{(0)}_{112} = \varphi^{(0)}_3 = \left(\frac{8}{L^3}\right)^{1/2} \sin \pi\, x/L \;\sin \pi\, y/L \;\sin^2 \pi\, z/L$$

Now since

$$\lambda H'_{kn} = \int \varphi_k^{(0)*} \lambda H' \varphi_n^{(0)} \, d\tau$$

$$= \lambda \int_0^L \int_0^L \int_0^L \varphi_k^{(0)} x^2 \varphi_n^{(0)} \, dx \, dy \, dz$$

One can find $\lambda H'_{11}$ and can show that it is equal to

$$\lambda L^2 \left(\frac{1}{3} - \frac{1}{8\pi^2}\right) = g_1 \quad \text{(say)},$$

$$\lambda H'_{22} = \lambda H'_{33} = \lambda L^2 \left(\frac{1}{3} - \frac{1}{2\pi^2}\right) = g_2 \quad \text{(say)},$$

One can also show that $\lambda H'_{12}, \lambda H'_{13}, \ldots$ are all zero due to the orthonormality of wave functions. Therefore, the secular equation takes the form,

$$\begin{vmatrix} \lambda H'_{11} - \lambda E^{(1)} & 0 & 0 \\ 0 & \lambda H'_{22} - \lambda E^{(1)} & 0 \\ 0 & 0 & \lambda H'_{33} - \lambda E^{(1)} \end{vmatrix}$$

$$= \begin{vmatrix} g_1 - \lambda E^{(1)} & 0 & 0 \\ 0 & g_2 - \lambda E^{(1)} & 0 \\ 0 & 0 & g_2 - \lambda E^{(1)} \end{vmatrix} = 0$$

The roots of above equation are g_1, g_2 and g_2. Hence the degeneracy is partially lifted and the perturbed energy levels are

$$E^{(0)} + g_1,\ E^{(0)} + g_2 \text{ and } E^{(0)} + g_2$$

The correct zeroth order wave function in this case is given by

$$\varphi = c_1 \varphi_1^{(0)} + c_2 \varphi_2^{(0)} + c_3 \varphi_3^{(0)}$$

In order to calculate above, we have to solve the following three equations for different values of $\lambda E^{(1)}$,

$$c_1 (g_1 - \lambda E^{(1)}) + c_2 \times 0 + c_3 \times 0 = 0$$

$$c_1 \times 0 + c_2 (g_2 - \lambda E^{(1)}) + c_3 \times 0 = 0$$

$$c_1 \times 0 + c_2 \times 0 + c_3 (g_2 - \lambda E^{(1)}) = 0$$

For $\lambda E^{(1)} = g_1$, $c_2 = c_3 = 0$, $c_1 \neq 0$. Thus the zeroth order wave function for which energy correction $\lambda E^{(1)} = g_1$ is

$$\varphi_1 = c_1 \varphi_1^{(0)} + c_2 \varphi_2^{(0)} + c_3 \varphi_3^{(0)}$$

$$= c_1 \varphi_1^{(0)}$$

Now using the normalization condition that φ_1 is normalized,

$$c_1^2 + c_2^2 + c_3^2 = c_1^2 = 1$$

Hence

$$\varphi_1 = \varphi_1^{(0)}$$

Now for $\lambda E^{(1)} = g_2$, $c_1 = 0$, $c_2 \neq 0$, $c_3 \neq 0$. Therefore, the zeroth order wave function for which energy correction $\lambda E^{(1)} = g_2$ is

$$\varphi_2 = c_1 \varphi_1^{(0)} + c_2 \varphi_2^{(0)} + c_3 \varphi_3^{(0)}$$

$$= c_2 \varphi_2^{(0)} + c_3 \varphi_3^{(0)}$$

where c_2 and c_3 are connected by the normalization requirement,

$$c_1^2 + c_2^2 + c_3^2 = c_2^2 + c_3^2 = 1.$$

12.20 The Hamiltonian for the perturbed system can be written as

$$H = H^{(0)} + \lambda H' = H^{(0)} - e\,\varepsilon z$$

where $H^{(0)}$ is the Hamiltonian for the unperturbed hydrogen atom and $\lambda H' = -e\,\varepsilon z$ is the perturbation. The unperturbed wave function φ_{nlm_1} for $n = 2$ state (l can take all the values from $0 - (n - 1)$, that is 0 and 1 and similarly m_l can taken all the values from $-l, \ldots, 0, \ldots + l$, that is 0 and $-1, 0, +1$) represented by

$$\varphi_{200} = \varphi_1^{(0)} = (32\,\pi a_0^3)^{-1/2} \left(2 - \frac{r}{a_0}\right) e^{-r/2a_0},$$

$2s$ term corresponding to $l = 0$.

$$\varphi_{211} = \varphi_2^{(0)} = (32\,\pi a_0^3)^{-1/2} \frac{r}{a_0} \sin\theta \sin\phi\, e^{-r/2a_0},$$

$2p_y$ term corresponding to $l = 1$.

$$\varphi_{210} = \varphi_3^{(0)} = (32\,\pi a_0^3)^{-1/2} \frac{r}{a_0} \cos\theta\, e^{-r/2a_0}.$$

$2p_z$ term corresponding to $l = 1$.

$$\varphi_{21-1} = \varphi_4^{(0)} = (32\,\pi a_0^3)^{-1/2} \frac{r}{a_0} \sin\theta \cos\phi\, e^{-r/2a_0},$$

$2p_x$ term corresponding to $l = 1$.

Since the $n = 2$ level is four fold degenerate (energy depends upon the value of n only), we have to solve the following secular equation,

$$\begin{vmatrix} \lambda H'_{11} - \lambda E^{(1)} & \lambda H'_{12} & \lambda H'_{13} & \lambda H'_{14} \\ \lambda H'_{21} & \lambda H'_{22} - \lambda E^{(1)} & \lambda H'_{23} & \lambda H'_{24} \\ \lambda H'_{31} & \lambda H'_{32} & \lambda H'_{33} - \lambda E^{(1)} & \lambda H'_{34} \\ \lambda H'_{41} & \lambda H'_{42} & \lambda H'_{43} & \lambda H'_{44} - \lambda E^{(1)} \end{vmatrix} = 0$$

in which

$$\lambda H'_{11} = -e\varepsilon\int |\varphi_1^{(0)}|^2 r\cos\theta\, r^2 \sin\theta\, d\theta\, d\phi\, dr$$

$$= -\frac{e\varepsilon}{32\,\pi a_0^4}\int_0^\infty\left(2-\frac{r}{a_0}\right)e^{-r/a_0}\, r^4\, dr \int_0^\pi \sin^2\theta\cos\theta\, d\theta \int_0^{2\pi}\sin\phi\, d\phi$$

$$= 0$$

as $\int_0^{2\pi}\sin\phi\, d\phi = 0$

$$\lambda H'_{12} = -e\varepsilon\int \psi_1^{(0)*}\, r\cos\theta\,\psi_2^{(0)}\, r^2\sin\theta\, d\theta\, d\phi\, dr$$

$$= -\frac{e\varepsilon}{32\,\pi a_0^4}\int_0^\infty\left(2-\frac{r}{a_0}\right)e^{-r/a_0}\, dr \int_0^\pi \sin^2\theta\cos\theta\, d\theta \int_0^{2\pi}\sin\phi\, d\phi$$

$$= 0$$

as $\int_0^{2\pi}\sin\phi\, d\phi = 0$

$$\lambda H'_{13} = -e\varepsilon\int \varphi_1^{(0)*}\, r\cos\theta\,\varphi_3^{(0)}\, r^2\sin\theta\, d\theta\, d\phi\, dr$$

$$= -\frac{e\varepsilon}{32\,\pi\, a_0^4}\int_0^\infty\left(2-\frac{r}{a_0}\right)e^{-r/a_0}\, r^4\, dr \int_0^\pi \sin\theta\cos^2\theta\, d\theta \int_0^{2\pi} d\phi$$

in which

$$\int_0^\pi \cos^2\theta\sin\theta\, d\theta = -\frac{1}{3}[\cos^3\theta]_0^\pi = \frac{2}{3},\ \int_0^{2\pi} d\phi = 2\pi$$

and

$$\int_0^\infty\left(2-\frac{r}{a_0}\right)e^{-r/a_0}\, r^4\, dr = 2a_0^5\int_0^\infty e^{-r/a_0}\left(\frac{r}{a_0}\right)^4 d\left(\frac{r}{a_0}\right) - a_0^5\int_0^\infty e^{-r/a_0}\left(\frac{r}{a_0}\right)^5 d\left(\frac{r}{a_0}\right)$$

$$= 2a_0^4\, 4! - a_0^5\, 5!$$

where use has been made of the integral,

$$\int_0^\infty e^{-x}x^n\, dx = n!$$

Thus

$$\lambda H'_{13} = -\frac{e\varepsilon}{32\,\pi\, a_0^4}(-72\, a_0^5)\frac{2}{3}\times 2\pi = 3\, e\,\varepsilon a_0$$

Similarly, it can be shown that $\lambda H'_{31} = 3\,e\varepsilon a_0$ and $\lambda H'_{14} = \lambda H'_{21} = \lambda H'_{22} = \lambda H'_{23} = \lambda H'_{24} = \lambda H'_{32} = \lambda H'_{33} = \lambda H'_{34} = \lambda H'_{41} = \lambda H'_{42} = \lambda H'_{43} = \lambda H'_{44} = 0$

Substituting all these values in the secular equation (or determinant), we get

$$\begin{vmatrix} -\lambda E_2^{(1)} & 0 & 3e\varepsilon a_0 & 0 \\ 0 & -\lambda E_2^{(1)} & 0 & 0 \\ 3e\varepsilon a_0 & 0 & -\lambda E_2^{(1)} & 0 \\ 0 & 0 & 0 & -\lambda E_2^{(1)} \end{vmatrix} = 0$$

The four roots of the above equation are given as $\lambda E_2^{(1)} = 0, 0, 3e\ \varepsilon a_0, -3e\ \varepsilon a_0$ and the corresponding perturbed energy eigen values are given as

$$E_2^{(0)}, E_2^{(0)}, E_2^{(0)} + 3e\ \varepsilon a_0, E_2^{(0)} - 3\,e\,\varepsilon\,a_0.$$

we see that splitting is proportional to the strength of the electric field (hence the name linear stark effect). A diagrammatic representation of the effect is given in Fig. 12.13. Note that the $n = 2$ level is split into three of which the middle one is still doubly degenerate.

The correct zeroth order wave function in this case is given by

$$\varphi = c_1\,\varphi_1^{(0)} + c_2\varphi_2^{(0)} + c_3\varphi_3^{(0)} + c_4\varphi_4^{(0)}$$

In order to calculate above, we have to solve the following four equations for the different values of $\lambda E_2^{(1)}$,

$$c_1(-\lambda E_2^{(1)}) + c_2 \times 0 + c_3 \times 3e\varepsilon a_0 + c_4 \times 0 = 0$$

$$c_1 \times 0 + c_2(-\lambda E_2^{(1)}) + c_3 \times 0 + c_4 \times 0 = 0$$

$$c_1 \times 3e\varepsilon a_0 + c_2 \times 0 + c_3(-\lambda E_2^{(1)}) + c_4 \times 0 = 0$$

$$c_1 \times 0 + c_2 \times 0 + c_3 \times 0 + c_4(-\lambda E_2^{(1)}) = 0$$

For $\lambda E_2^{(1)} = 3\,e\varepsilon a_0, c_1 = c_3, c_2 = c_4 = 0$. Therefore, the zeroth order wave function foe which the energy correction is $3\,e\varepsilon a_0$, would be given by

$$\varphi_1 = c_1\varphi_1^{(0)} + c_3\,\varphi_3^{(0)}$$

Using the normalization condition (that is φ_1 is normalized), $c_1^2 + c_3^2 = 1, c_1 = c_3 = \frac{1}{\sqrt{2}}$. Thus the required wave function is

$$\varphi_1 = \frac{1}{\sqrt{2}}(\varphi_1^{(0)} + \varphi_3^{(0)})$$

Similarly the zeroth order wave function for which the energy correction is $-3\,e\,\varepsilon\,a_0$ would be given by

$$\varphi_2 = \frac{1}{\sqrt{2}}(\varphi_1^{(0)} - \varphi_3^{(0)})$$

When $\lambda E_2^{(1)} = 0$, $c_1 = c_3 = 0$ and c_2 and c_4 are indeterminate. Thus φ_3 and $\varphi_4 = c_2\,\varphi_2^{(0)} + c_4\varphi_4^{(0)}$, where c_2 and c_4 are connected by the normalization requirement, $c_2^2 + c_4^2 = 1$.

12.21 The normalized eigen functions of $H^{(0)}$ are given by

$$\varphi_{n_1 n_2}^{(0)}(x_1, x_2) = \varphi_{n_1}^{(0)}(x_1)\,\varphi_{n_2}^{(0)}(x_2)$$

$$= N_{n_1} N_{n_2} H_{n_1}(\sqrt{\alpha}\, x_1) H_{n_2}(\sqrt{\alpha}\, x_2)\, e^{-\alpha x_1^2/2}\, e^{-\alpha x_2^2/2}$$

$$\alpha = \frac{m\omega}{\hbar}, \quad n_1, n_2 = 0, 1, 2, \ldots$$

where

$$N_{n_1} N_{n_2} = \left(\frac{\alpha}{\pi}\right)^{1/2} \left(\frac{1}{2^{n_1} n_1!} \times \frac{1}{2^{n_2} n_2!}\right)^{1/2}$$

The corresponding eigen values are

$$E_{n_1 n_2}^{(0)} = E_{n_1}^{(0)} + E_{n_2}^{(0)}$$

$$= \left(n_1 + \frac{1}{2}\right)\hbar\,\omega + \left(n_2 + \frac{1}{2}\right)\hbar\,\omega$$

$$= (n_1 + n_2 + 1)\,\hbar\,\omega$$

For the ground state ($n_1 = n_2 = 0$) which is non-degenerate

$$\varphi_{00}^{(0)}(x_1, x_2) = \varphi_0^{(0)} = \left(\frac{\alpha}{\pi}\right)^{1/2} e^{-\alpha(x_1^2 + x_2^2)/2}$$

and

$$E_{00}^{(0)} = \hbar\,\omega$$

Therefore, the first order correction to the ground state energy would be given by

$$\lambda E_{00}^{(1)} = \lambda H'_{00} = \lambda \frac{\alpha}{\pi} \int_{-\infty}^{+\infty}\int_{-\infty}^{+\infty} e^{-\alpha(x_1^2 + x_2^2)/2} x_1\, x_2\, dx_1\, dx_2$$

$$= 0$$

as integrand is odd function of x_1 and as well as of x_2. Thus, the first order correction to the ground state is zero and is not shifted.

Now the first excited state is doubly (or two fold) degenerate ($n_1 = 1$, $n_2 = 0$ and $n_1 = 0$, $n_2 = 1$), the corresponding eigen value being $2\,\hbar\,\omega$. The wave functions for the said state are

$$\varphi_{10}^{(0)} = \varphi_1^{(0)} = \left(\frac{\alpha}{2\pi}\right)^{1/2} 2\sqrt{\alpha}\, x_1\, e^{-\alpha(x_1^2 + x_2^2)/2}$$

$$\varphi_{01}^{(0)} = \varphi_2^{(0)} = \left(\frac{\alpha}{2\pi}\right)^{1/2} 2\sqrt{\alpha}\, x_2\, e^{-\alpha(x_1^2 + x_2^2)/2}$$

For calculating the first order corrections to the first excited state, we have to solve the following secular equation,

$$\begin{vmatrix} \lambda H'_{11} - \lambda E_1^{(1)} & \lambda H'_{12} \\ \lambda H'_{21} & \lambda H'_{22} - \lambda E_1^{(1)} \end{vmatrix} = 0$$

Now

$$\lambda H'_{11} = \frac{\lambda\alpha}{2\pi} 4\alpha \int_{-\infty}^{+\infty} e^{-\alpha x_1^2} x_1^3\, dx_1 \int_{-\infty}^{+\infty} e^{-\alpha x_2^2} x_2\, dx_2 = 0$$

as the integrands are odd functions of x_1 and x_2, respectively. Similarly $\lambda H'_{22} = 0$.

$$\lambda H'_{12} = \frac{\lambda\alpha}{2\pi} \int_{-\infty}^{+\infty} e^{-\alpha x_1^2} x_1 2\sqrt{\alpha}\, x_1\, dx_1 \int_{-\infty}^{+\infty} e^{-\alpha x_2^2} x_2 2\sqrt{\alpha}\, x_2\, dx_2$$

$$= \frac{\lambda\alpha}{2\pi}\frac{1}{\alpha^2} \int_{-\infty}^{+\infty} 2e^{-\xi_1^2}\xi_1^2\, d\xi_1 \int_{-\infty}^{+\infty} 2\, e^{-\xi_2^2}\xi_2^2\, d\xi_2$$

$$= \frac{\lambda}{2\alpha}$$

where we have used the result,

$$\int_{-\infty}^{+\infty} x^2 e^{-x^2}\, dx = \frac{\sqrt{\pi}}{2}.$$

Similarly

$$\lambda H'_{21} = \frac{\lambda}{2\alpha}$$

Substituting all the above values in the secular equation and solving, one gets

$$\lambda E_1^{(1)} = \pm \frac{\lambda\hbar}{2\, m\omega}$$

Thus the first excited state splits up into two energy levels.

$$2\,\hbar\,\omega + \frac{\lambda\hbar}{2\, m\omega} \quad \text{and} \quad 2\hbar\,\omega - \frac{\lambda\,\hbar}{2m\omega}$$

The corresponding zeroth order corrected wave functions are given by

$$\frac{1}{\sqrt{2}}(\varphi_1^{(0)} + \varphi_2^{(0)}) \quad \text{and} \quad \frac{1}{\sqrt{2}}(\varphi_1^{(0)} - \varphi_2^{(0)})$$

12.22 In the given problem, the unpeturbed Hamiltonian is equivalent to the sum of the Hamiltonians of two one dimensional harmonic oscillators, therefore the solution of the said problem is same as that of the earlier problem with x_1 replaced by x and with x_2 replaced by y and similarly λ replaced by $\lambda\, m\omega^2$. Thus the normalized eigen function of $H^{(0)}$ would be given by

$$\varphi^{(0)}_{n_x n_y}(x, y) = \varphi^{(0)}_{n_x}(x)\,\varphi^{(0)}_{n_y}(y)$$

$$= N_{n_x} N_{n_y} H_{n_x}(\sqrt{\alpha}\, x)\, H_{n_y}(\sqrt{\alpha}\, y)\, e^{-\alpha x^2/2}\, e^{-\alpha y^2/y},$$

$$n_x, n_y = 0, 1, 2, ...$$

where

$$N_{n_x} N_{n_y} = \left(\frac{\alpha}{\pi}\right)^{1/2}\left(\frac{1}{2^{n_x} n_x!}\,\frac{1}{2^{n_y} n_y!}\right)^{1/2}$$

The corresponding eigen values are given by

$$E^{(0)}_{n_x n_y} = E^{(0)}_{n_x} + E^{(0)}_{n_y} = (n_x + n_y + 1)\,\hbar\,\omega$$

The first excited state corresponds to $n_x = 1$, $n_y = 0$ and $n_x = 0$ and $n_y = 1$ and is thus doubly degenerate with energy $2\,\hbar\omega$. The wave functions for the said state would be given by $\varphi^{(0)}_{10} = \varphi^{(0)}_1$ and $\varphi^{(0)}_{01} = \varphi^{(0)}_2$. The first order energy corrections to the first excited state would be $\dfrac{\pm\,\lambda m\omega^2\,\hbar}{2\,m\omega} = \dfrac{\pm\,\lambda\hbar\omega}{2}$. Thus the first excited state splits into two energy levels.

$$2\,\hbar\omega + \frac{\lambda\hbar\omega}{2} \quad \text{and} \quad 2\,\hbar\omega - \frac{\lambda\hbar\omega}{2}$$

The corresponding zeroth order corrected eigen functions would be given by

$$\frac{1}{\sqrt{2}}(\varphi^{(0)}_1 + \varphi^{(0)}_2) \quad \text{and} \quad \frac{1}{\sqrt{2}}(\varphi^{(0)}_1 - \varphi^{(0)}_2)$$

12.23 The effect of the magnetic field is to change the momentum from $\mathbf{p}$ to $(\mathbf{p} - e/c\,\mathbf{A})$ where $\mathbf{A}$ is the vector potential related to the magnetic induction $\mathbf{B}$ by the relation.

$$\mathbf{B} = \vec{\nabla} \times \mathbf{A}$$

Now,

$$\mathbf{B} \times \mathbf{r} = \vec{\nabla} \times \mathbf{A} \times \mathbf{r} = (\vec{\nabla} \cdot \mathbf{r})\,\mathbf{A} - (\mathbf{A} \cdot \vec{\nabla})\,\mathbf{r}$$ and using the relations

$$\vec{\nabla} \cdot \mathbf{r} = \frac{\partial x}{\partial x} + \frac{\partial y}{\partial y} + \frac{\partial z}{\partial z} = 3$$

and

$$\begin{aligned}(\mathbf{A} \cdot \vec{\nabla})\mathbf{r} &= \left(\frac{A_x \partial}{\partial x} + \frac{A_y \partial}{\partial y} + \frac{A_z \partial}{\partial z}\right)(\hat{x}x + \hat{y}y + \hat{z}z) \\ &= (A_x \hat{x} + A_y \hat{y} + A_z \hat{z}) \\ &= \mathbf{A}\end{aligned}$$

we get

$$\mathbf{B} \times \mathbf{r} = 3\mathbf{A} - \mathbf{A} = 2\mathbf{A}$$

Therefore,

$$\mathbf{A} = \frac{1}{2}(\mathbf{B} \times \mathbf{r})$$

The Hamiltonian of hydrogen atom in the presence of magnetic induction **B** is given by

$$\begin{aligned}\mathrm{H} &= \frac{1}{2m}\left(\mathbf{p} - \frac{e}{c}\mathbf{A}\right)^2 + V(r) \\ &= \frac{p^2}{2m} + V(r) - \frac{e}{2mc}(\mathbf{p} \cdot \mathbf{A} + \mathbf{A} \cdot \mathbf{p}) + \frac{e^2}{2mc^2}A^2 \\ &= H^{(0)} + \lambda H' + \mu H''\end{aligned}$$

where

$$H^{(0)} = \frac{p^2}{2m} + V(r),$$

is the unperturbed Hamiltonian

$$\lambda H' = -\frac{e}{2mc}(\mathbf{p} \cdot \mathbf{A} + \mathbf{A} \cdot \mathbf{p})$$

and

$$\mu H'' = \frac{e^2}{2mc^2}A^2$$

For the weak fields, the second order perturbation term $\mu H''$ containing A^2 may be neglected and hence the perturbed Hamiltonian takes the form

$$H = H^{(0)} + \lambda H'$$

Now,

$$\begin{aligned}(\mathbf{p} \cdot \mathbf{A} + \mathbf{A} \cdot \mathbf{p})\varphi &= (-i\hbar\vec{\nabla} \cdot \mathbf{A} - i\hbar\mathbf{A} \cdot \vec{\nabla})\varphi \\ &= -i\hbar\vec{\nabla} \cdot \mathbf{A}\varphi - i\hbar\, \mathbf{A} \cdot \vec{\nabla}\varphi\end{aligned}$$

Using the vector identity,

$$\vec{\nabla}\,.\,\mathbf{A}\varphi = (\vec{\nabla}.\,\mathbf{A})\,\varphi + \mathbf{A}\,.\,\vec{\nabla}\varphi$$

$$(\mathbf{p}\,.\,\mathbf{A} + \mathbf{A}\,.\,\mathbf{p})\varphi = -i\hbar\,(\vec{\nabla}\,.\,\vec{\mathbf{A}})\,\varphi - i\hbar\mathbf{A}\,.\,\vec{\nabla}\,\varphi - i\hbar\mathbf{A}.\,\vec{\nabla}\varphi$$

$$= -i\hbar\,(\vec{\nabla}\,.\,\mathbf{A})\,\varphi - 2i\hbar\mathbf{A}\,.\,\vec{\nabla}\varphi$$

Since

$$\vec{\nabla}\,.\,\mathbf{A} = \frac{1}{2}\vec{\nabla}\,.\,(\mathbf{B}\times\mathbf{r}) = \frac{1}{2}[\mathbf{r}.\,(\vec{\nabla}\times\mathbf{B}) - \mathbf{B}.\,(\vec{\nabla}\times\mathbf{r})]$$

$$= 0$$

as $\vec{\nabla}\times\mathbf{r} = 0$ and as the field is uniform (independent of position), $\vec{\nabla}\times\mathbf{B} = 0$, we get

$$\mathbf{p}\,.\,\mathbf{A} + \mathbf{A}\,.\,\mathbf{p} = -2i\hbar\,\mathbf{A}\,.\,\vec{\nabla} = 2\mathbf{A}\,.\,\mathbf{p}$$

Therefore,

$$\lambda H' = -\frac{e}{2mc}\times 2\mathbf{A}\,.\,\mathbf{p} = -\frac{e}{mc}\,\mathbf{A}\,.\,\mathbf{p}$$

$$= -\frac{e}{2mc}\,(\mathbf{B}\times\mathbf{r})\,.\,\mathbf{p}$$

$$= -\frac{e}{2mc}\mathbf{B}.\,(\mathbf{r}\times\mathbf{p})$$

where we have used the vector identity for $\mathbf{p}\,.\,(\mathbf{B}\times\mathbf{r})$ and have set $\vec{\nabla}\times\mathbf{B} = 0$. Thus

$$\lambda H' = -\frac{e}{2mc}\mathbf{B}\,.\,\mathbf{L} = -\mu_0\,.\,\mathbf{B}$$

where $\mathbf{L} = \mathbf{r}\times\mathbf{p}$ represents the orbital angular momentum and $\vec{\mu}_0 = \frac{e\vec{L}}{2mc}$ can be interpreted as the magnetic moment associated with the orbital motion of the charged particle (the subscript 0 signifies correspondence with the orbital motion)

By using **B** along the *z*-direction, $\mathbf{B}\,.\,\mathbf{L} = BL_z$ or $\mu H' = -\frac{e}{2mc}BL_z$. Therefore, the first order correction to the *n*th energy state $\left(\varphi^{(0)}_{nlm_l} = \varphi^{(0)}_n \text{ say}\right)$ would be given by

$$\lambda E^{(1)}_n = \lambda H'_{nn} = \int \varphi^{(0)*}_n\,\lambda H'\,\varphi^{(0)}_n\,d\tau$$

$$= -\frac{e}{2mc}B\int \varphi^{(0)*}_n L_z \varphi^{(0)}_n\,d\tau$$

$$= -\frac{e}{2mc}B\,m_l\hbar\int \varphi^{(0)*}_n\,\varphi^{(0)}_n\,d\tau$$

$$= -\frac{e\hbar}{2mc}m_l B$$

$$= -\mu_B\,m_l\,B$$

where $\mu_B = \dfrac{e\hbar}{2mc}$ is called the Bohr magneton and the magnetic quantum number m_l can take all the values from $-l$ to $+l$ that is, total $(2l + 1)$ values. For s state, $l = 0$ so $m_l = 0$. Thus s state is not split. For p state $l = 1$ so m_l can take $(2l + 1)$ values. Thus p state is split into three components corresponding to $m_l = -1$, 0 and $+1$. The Fig. 12.14 shows the splittings of 1s, 2s and 2p levels in the presence of magnetic field. Thus the degenerate states are split by magnetic field into $(2l + 1)$ distinct components which are equally spaced and are odd in number. In many spectra, an even number of Zeeman levels is observed (anomalous Zeeman effect) in place of predicted odd number. This requires inclusion of spin in the theory.

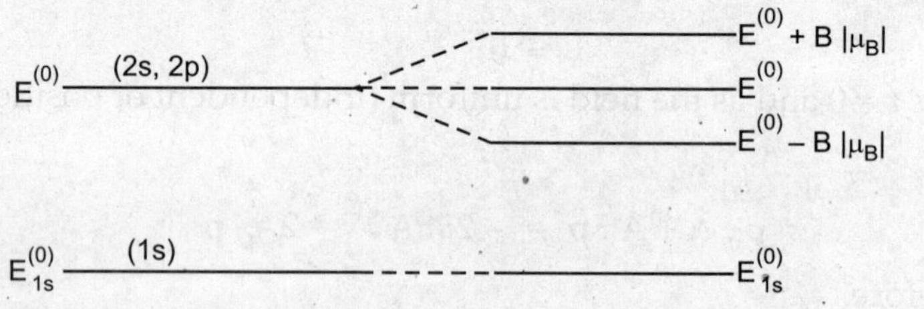

Fig. 12.14. *Zeeman effect in hydrogen atom (neglecting spin of the electron)*

12.24 Let us write,

$$H = H^{(0)} + \in H'$$

where

$$H^{(0)} = \begin{pmatrix} 1 & 0 \\ 0 & 1 \end{pmatrix} \text{ and } \in H' = \in \begin{pmatrix} 1 & 1 \\ 1 & 1 \end{pmatrix}.$$

The unperturbed matrix (or Hamiltonian) $H^{(0)}$ has only one eigen value + 1 which is two fold degenerate. The corresponding eigen vectors are

$$\varphi_1^{(0)} = \begin{pmatrix} 1 \\ 0 \end{pmatrix} \quad \text{and} \quad \varphi_2^{(0)} = \begin{pmatrix} 0 \\ 1 \end{pmatrix}$$

Now $\in H'_{11} = \in H'_{12} = \in H'_{21} = \in H'_{22} = \in$, therefore the secular equation would be given by

$$\begin{vmatrix} \in H'_{11} - \in E^{(1)} & \in H'_{12} \\ \in H'_{21} & \in H'_{22} - \in E^{(1)} \end{vmatrix} = \begin{vmatrix} \in - \in E^{(1)} & \in \\ \in & \in - \in E^{(1)} \end{vmatrix} = 0$$

which gives, $\in E^{(1)} = 0, 2\in$. Therefore, the eigen values corrected upto first order would be $1 + 0 = 1$ and $1 + 2\in$. Thus the degeneracy is removed in the first order.

Since for a degenerate state the linear combination of the two unperturbed eigen functions is also an eigen function, we have

$$\varphi = c_1 \varphi_1^{(0)} + c_2 \varphi_2^{(0)}$$

To evaluate above, we know that

$$c_1 (\in H'_{11} - \in E^{(1)}) + c_2 \in H'_{12} = 0$$

$$c_1 \in H'_{21} + c_2 (\in H'_{22} - \in E^{(1)}) = 0$$

which for $\in E^{(1)} = 0$, gives $c_1 = -c_2$ and further using the normalization requirement that $\varphi = \varphi_2$ is normalized , that is $c_1^2 + c_2^2 = 1$, one gets

$$\varphi_1 = \frac{1}{\sqrt{2}}(\varphi_1^{(0)} - \varphi_2^{(0)}) = \frac{1}{\sqrt{2}}\begin{pmatrix}1\\-1\end{pmatrix}$$

Similarly for $\in E^{(1)} = 2\in$, $c_1 = c_2$ and further using the normalization requirement that $\varphi = \varphi_2$ is normalized, we get

$$\varphi_2 = \frac{1}{\sqrt{2}}(\varphi_1^{(0)} + \varphi_2^{(0)}) = \frac{1}{\sqrt{2}}\begin{pmatrix}1\\1\end{pmatrix}$$

For finding the eigen values of H exactly, we will have to solve the equation

$$\begin{vmatrix}1+\in-\lambda & \in \\ \in & 1+\in-\lambda\end{vmatrix} = 0$$

which gives,

$$(1+\in-\lambda)^2 - \in^2 = 0$$

or

$$\lambda = 1 + \in \mp \in = 1, 1 + 2\in$$

Thus the first order perturbation theory gives exact results in this case.

12.26 The expectation value of H in the state Φ is given by

$$E_\Phi = \frac{\langle\Phi|H|\Phi\rangle}{\langle\Phi|\Phi\rangle} \quad \text{... (12.16)}$$

Now

$$\langle\Phi|\Phi\rangle = \int e^{-2\alpha r/r_0} d\tau$$

$$= 4\pi\int_0^\infty e^{-2\alpha r/r_0} r^2 dr$$

$$= 4\pi\frac{2}{(2\alpha/r_0)^3} = \frac{\pi r_0^3}{\alpha^3} \quad \text{... (12.17)}$$

and

$$\langle\Phi|H|\Phi\rangle = \left\langle\Phi\left|-\frac{\hbar^2}{2\mu}\nabla^2 + V(r)\right|\Phi\right\rangle$$

$$= -\frac{\hbar^2}{2\mu}\langle\Phi|\nabla^2|\Phi\rangle + \langle\Phi|V(r)|\Phi\rangle$$

$$= -\frac{\hbar^2}{2\mu}\left\langle\Phi\left|\frac{d^2}{dr^2} + \frac{2}{r}\frac{d}{dr}\right|\Phi\right\rangle + \langle\Phi|V(r)|\Phi\rangle \quad \text{... (12.18)}$$

Let us calculate,

$$\langle\Phi|V(r)|\Phi\rangle = \int_0^\infty e^{-\alpha r/r_0} V(r) e^{-\alpha r/r_0} d\tau$$

$$= 4\pi\int_0^\infty e^{-2\alpha r/r_0} V\, r^2\, dr$$

Using the value of $V(r)$ in the present case

$$\langle\Phi|\, V(r)\, |\Phi\rangle = 4\pi\int e^{-2\alpha r/r_0}(-V_0)\, r^2 dr$$

Putting $\frac{r}{r_0} = x$, $dr = r_0\, dx$, the above equation becomes

$$\langle\Phi|\, V(r)\, |\Phi\rangle = -4\pi V_0 r_0^3 \int_0^1 e^{-2\alpha x} x^2 dx$$

$$= -4\pi V_0 r_0^3 \left[\frac{x^2 e^{-2\,\alpha x}}{(-2\alpha)}\Bigg|_0^1 - \int_0^1 \frac{2xe^{-2\alpha x}}{(-2\alpha)} dx \right]$$

$$= -4\pi V_0 r_0^3 \left[-\frac{e^{-2\alpha}}{2\alpha} + \frac{1}{\alpha}\left\{ \frac{xe^{-2\alpha x}}{(-2\alpha)}\Bigg|_0^1 - \int_0^1 \frac{e^{-2\alpha x}}{(-2\alpha)} dx \right\} \right]$$

$$= -4\pi V_0 r_0^3 \left[-\frac{e^{-2\alpha}}{2\alpha} - \frac{e^{-2\alpha}}{2\alpha^2} + \frac{1}{2\alpha^2} \frac{e^{-2\alpha x}}{(-2\alpha)}\Bigg|_0^1 \right]$$

$$= -4\pi V_0 r_0^3 \left[-\frac{e^{-2\alpha}}{2\alpha} - \frac{e^{-2\alpha}}{2\alpha^2} - \frac{1}{4\alpha^3}(e^{-2\alpha} - 1) \right] \quad \text{... (12.19)}$$

Now

$$\left\langle \Phi \left| \frac{d^2}{dr^2} + \frac{2}{r}\frac{d}{dr} \right| \Phi \right\rangle = \int_0^\infty e^{-\alpha r/r_0} \left(\frac{d^2}{dr^2} + \frac{2}{r}\frac{d}{dr} \right) e^{-\alpha r/r_0} d\tau$$

$$= 4\pi \int_0^\infty e^{-\alpha r/r_0} \left(\frac{\alpha^2}{r_0^2} - \frac{2\alpha}{rr_0} \right) e^{-\alpha r/r_0}\, r^2 dr$$

$$= \frac{4\,\pi\alpha^2}{r_0^2} \int_0^\infty e^{-2\,\alpha r/r_0}\, r^2\, dr - \frac{8\,\pi\alpha}{r_0} \int_0^\infty e^{-2\,\alpha r/r_0}\, rdr$$

$$= \frac{4\pi\alpha^2}{r_0^2} \frac{2}{(2\alpha/r_0)^3} - \frac{8\pi\alpha}{r_0} \frac{1}{(2\alpha/r_0)^2}$$

$$= -\frac{\pi r_0}{\alpha}$$

Therefore

$$-\frac{\hbar^2}{2\mu}\langle\Phi|\, \nabla^2\, |\Phi\rangle = \frac{\hbar^2}{2\mu}\frac{\pi r_0}{\alpha} \quad \text{... (12.20)}$$

Substituting eqns. (12.19) and (12.20) into eqn. (12.18) and then the resulting equation and (12.17) into (12.16), one gets

$$E_\Phi = \left[\frac{\hbar^2}{2\mu}\frac{\pi r_0}{\alpha} + 4\pi V_0 r_0^3 \left(\frac{1}{2\alpha} + \frac{1}{2\alpha^2} + \frac{1}{4\alpha^3} \right) e^{-2\alpha} - \frac{\pi V_0 r_0^3}{\alpha^3} \right] \Bigg/ \frac{\pi r_0^3}{\alpha^3}$$

or

$$\frac{E_\Phi}{V_0} = \lambda\alpha^2 + (1 + 2\alpha + 2\alpha^2)e^{-2\alpha} - 1 \qquad \text{... (12.21)}$$

where $\lambda = \dfrac{\hbar^2}{2\mu V_0 r_0^2}$ is a dimensionless quantity.

Minimizing E_Φ with respect to α, we get

$$\frac{dE_{\Phi(\alpha)}}{d\alpha} = 0 = 2\lambda\alpha - 4\alpha^2 e^{-2\alpha}$$

or

$$\lambda = 2\alpha e^{-2\alpha}$$

which is a transcendental equation for the variational parameter α. From this equation for different values of α, λ can be found out (that is, the range-depth relation). Putting these values in eqn. (12.21) minimum E_Φ (or binding energy of deuteron) can be found in terms of V_0.

12.27 The expectation value of H in the state Φ is given by

$$E_\Phi = \frac{\langle\Phi \mid H \mid \Phi\rangle}{\langle\Phi \mid \Phi\rangle}$$

Now

$$\langle\Phi \mid \Phi\rangle = 4\pi\int e^{-2\alpha r} r^2\, dr = \frac{\pi}{\alpha^3}$$

and

$$\langle\Phi \mid H \mid \Phi\rangle = -\frac{\hbar^2}{2\mu}\left\langle\Phi\left|\frac{d^2}{dr^2} + \frac{2}{r}\frac{d}{dr}\right|\Phi\right\rangle + \langle\Phi \mid V(r) \mid \Phi\rangle$$

out of which

$$\langle\Phi \mid V(r) \mid \Phi\rangle = -4\pi V_0\int_0^\infty e^{-(2\alpha + 1/r_0)r} r^2 dr$$

$$= -4\pi V_0 \frac{2}{(2\alpha + 1/r_0)^3}$$

$$= -\frac{8\pi V_0}{(2\alpha + 1/r_0)^3}$$

and

$$\left\langle\Phi\left|\frac{d^2}{dr^2} + \frac{2}{r}\frac{d}{dr}\right|\Phi\right\rangle = 4\pi\int_0^\infty e^{-\alpha r}\left(\alpha^2 - \frac{2\alpha}{r}\right)e^{-\alpha r} r^2\, dr$$

$$= 4\pi\alpha^2\frac{2}{(2\alpha)^3} - 8\pi\alpha\frac{1}{(2\alpha)^2}$$

$$= -\frac{\pi}{\alpha}$$

Hence

$$\langle \Phi \mid H \mid \Phi \rangle = \frac{\hbar^2 \pi}{2\mu\alpha} - \frac{8\pi V_0}{(2\alpha + 1/r_0)^3}$$

Thus

$$E_\Phi = \left[\frac{\hbar^2 \pi}{2\mu\alpha} - \frac{8\pi V_0}{(2\alpha + 1/r_0)^3}\right] \Big/ \frac{\pi}{\alpha^3}$$

$$= \frac{\hbar^2 \alpha^2}{2\,\mu} - \frac{8 V_0 \alpha^3}{(2\alpha + 1/r_0)^3}$$

Putting $\alpha = \dfrac{\beta}{r_0}$ to make $e^{-\alpha r}$, a dimensionless quantity, we get

$$\frac{E_\Phi}{V_0} = \frac{\hbar^2\,\beta^2}{2\,\mu V_0 r_0^2} - \frac{8\,\beta^3}{r_0^3\left(\dfrac{2\beta}{r_0} + \dfrac{1}{r_0}\right)^3}$$

$$= \frac{3}{8}\beta^2 - \frac{8\,\beta^2}{(2\beta + 1)^3} \quad \text{... (12.22)}$$

where we have substituted $\dfrac{\hbar^2}{\mu V_0 r_0^2} = \dfrac{3}{4}$

Now let us minimize E_Φ (β), that is put

$$\frac{dE_\Phi(\beta)}{d\beta} = 0 = \frac{3\beta}{4} - \frac{24\beta^2}{(2\,\beta + 1)^3} + \frac{48\beta^3}{(2\beta + 1)^4}$$

which on simplification gives

$$\frac{(2\beta + 1)^4}{\beta} = 32,$$

which is transcendental equation in β. It can be solved by numerical method to give $\beta = \dfrac{1}{2}$. Substituting this value of β in eqn. (12.22), one gets the value of ground state energy to be

$$E_\Phi = -\frac{V_0}{32}$$

12.28 The expectation value of H in the state Φ is given by

$$E_\Phi = \frac{\langle \Phi \mid H \mid \Phi \rangle}{\langle \Phi \mid \Phi \rangle}$$

Now

$$\langle \Phi \mid \Phi \rangle = 4\,\pi \int_0^\infty e^{-2\alpha (r/a_0)^2}\, r^2\, dr$$

$$= 4\pi \frac{1}{2\,(4\alpha/a_0^2)^1}\left(\frac{\pi}{2\alpha/a_0^2}\right)^{1/2}$$

$$= \left(\frac{\pi\, a_0^2}{2\,\alpha}\right)^{3/2}$$

and

$$\langle \Phi \mid H \mid \Phi \rangle = -\frac{\hbar^2}{2\,m}\langle \Phi \mid \nabla^2 \mid \Phi \rangle + \langle \Phi \mid V(r) \mid \Phi \rangle$$

out of which

$$\langle \Phi \mid V(r) \mid \Phi \rangle = 4\pi \int_0^\infty e^{-\alpha (r/a_0)^2}\left(-\frac{e^2}{r}\right) e^{-\alpha (r/a_0)^2} r^2\, dr$$

$$= -\,4\,\pi\, e^2 \frac{1}{2\,(2\alpha/a_0^2)^1}$$

$$= -\frac{\pi a_0^2\, e^2}{\alpha}$$

and

$$\langle \Phi \mid \nabla^2 \mid \Phi \rangle = \left\langle \Phi \left| \frac{d^2}{dr^2} + \frac{2}{r}\frac{d}{dr} \right| \Phi \right\rangle$$

$$= 4\pi \int_0^\infty e^{-\,\alpha (r/a_0)^2}\left(\frac{d^2}{dr^2} + \frac{2}{r}\frac{d}{dr}\right) e^{-\,\alpha (r/a_0)^2} r^2\, dr$$

$$= 4\pi \int_0^\infty e^{-\,\alpha (r/a_0)^2}\left(-\frac{2\alpha}{a_0^2} + \frac{4\alpha^2 r^2}{a_0^4} - \frac{4\alpha}{a_0^2}\right) e^{-\alpha (r/a_0)^2} r^2 dr$$

$$= -\frac{8\pi\alpha}{a_0^2}\int_0^\infty e^{-2\alpha (r/a_0)^2} r^2\, dr + \frac{16\,\pi\alpha^2}{a_0^4}$$

$$\int_0^\infty e^{-2\alpha (r/a_0)^2} r^4\, dr - \frac{16\,\pi\alpha}{a_0^2}\int_0^\infty e^{-2\alpha (r/a_0)^2} r^2 dr$$

$$= -\frac{24\,\pi\,\alpha}{a_0^2}\int_0^\infty e^{-2\,\alpha (r/a_0)^2} r^2\, dr + \frac{16\,\pi\alpha^2}{a_0^4}$$

$$\int_0^\infty e^{-2\alpha\, (r/a_0)^2} r^4\, dr$$

$$= -\frac{24\,\pi\alpha}{a_0^2}\,\frac{1}{2\left(2\times\dfrac{2\alpha}{a_0^2}\right)^1}\left(\frac{\pi}{2\alpha/a_0^2}\right)^{1/2}$$

$$+\frac{16\pi\alpha^2}{a_0^4}\frac{1.3}{2(2\times 2\,\alpha/a_0^2)^2}\left(\frac{\pi}{2\alpha/a_0^2}\right)^{1/2}$$

$$= -\frac{3\pi^{3/2}\,a_0}{2^{3/2}\,\alpha^{1/2}}$$

Therefore,

$$E_\Phi = \left[\frac{\hbar^2}{2\,m}\frac{3\pi^{3/2}\,a_0}{2^{3/2}\,\alpha^{1/2}} - \frac{\pi a_0^2\,e^2}{\alpha}\right]\Bigg/\left(\frac{\pi a_0^2}{2\alpha}\right)^{3/2}$$

On putting $\frac{\hbar^2}{me^2} = a_0$ and then simplifying

$$E_\Phi(\alpha) = \left[\frac{3\alpha}{2} - 2^{3/2}\left(\frac{\alpha}{\pi}\right)^{1/2}\right]\frac{e^2}{a_0}$$

Minimizing $E_\Phi(\alpha)$, that is putting $\frac{dE_\Phi\,(\alpha)}{d\,\alpha} = 0$, one gets $\alpha = \frac{8\pi}{9}$.
Substituting this value of α in the above expression the corresponding value of the lowest energy comes out to be

$$(E_\Phi)_{min} = -\frac{0.85\,e^2}{2\,a_0}$$

The above result is indicative of the fact that if the trial function chosen for the problem is poorly guessed the result obtained for the ground state energy would also be poor.

12.29 The expectation value of H in the state Φ is given by

$$E_\Phi = \frac{\langle\Phi \mid H \mid \Phi\rangle}{\langle\Phi \mid \Phi\rangle}$$

Now

$$\langle\Phi \mid \Phi\rangle = 4\pi\int e^{-2\alpha r/r_0}\,r^2\,dr = \frac{\pi r_0^3}{\alpha^3}$$

and

$$\langle\Phi \mid H \mid \Phi\rangle = \left\langle\Phi\left|-\frac{\hbar^2}{2\mu}\nabla^2 + V(r)\right|\Phi\right\rangle$$

where **r** is the neutron-proton separation and μ is the reduced mass and is one-half the mass of nucleon.

$$\left\langle\Phi\left|-\frac{\hbar^2}{2\mu}\nabla^2\right|\Phi\right\rangle = -\frac{\hbar^2}{2\,\mu}\int_0^\infty e^{-\alpha\,r/r_0}\left(\frac{d^2}{dr^2} + \frac{2}{r}\frac{d}{dr}\right)e^{-\alpha r/r_0}\,4\pi r^2\,dr$$

$$= -\frac{\hbar^2}{2\mu}\int_0^\infty e^{-\alpha r/r_0}\left(\frac{\alpha^2}{r_0^2} - \frac{2\alpha}{rr_0}\right)e^{-\alpha r/r_0}\,4\pi r^2\,dr$$

$$= \frac{\hbar^2}{2\mu}\frac{\pi r_0}{\alpha}$$

and

$$\langle \Phi \mid V(r) \mid \Phi \rangle = 4\pi \int_0^\infty e^{-\alpha r/r_0}\left(-\frac{V_0 e^{-r/r_0}}{r/r_0}\right)e^{-\alpha r/r_0} r^2\, dr$$

$$= -4\pi V_0 r_0 \int_0^\infty e^{-(2\alpha+1)/r_0}\, r\, dr$$

$$= -\frac{4\pi V_0 r_0^3}{(2\alpha+1)^2}$$

Therefore,

$$E_\Phi = \left[\frac{\hbar^2}{2\mu}\frac{\pi r_0}{\alpha} - \frac{4\pi r_0^3}{(2\alpha+1)^2}\right] \bigg/ \frac{\pi r_0^3}{\alpha^3}$$

$$= \frac{\hbar^2\alpha^2}{2\mu r_0^2} - \frac{4V_0 r_0^3}{(2\alpha+1)^2}$$

or

$$\frac{E_\Phi}{V_0} = \lambda\alpha^2 - \frac{4\alpha^3}{(2\alpha+1)^3} \qquad \text{... (12.23)}$$

where

$$\lambda = \frac{\hbar^2}{2\mu V_0 r_0^2}$$

Minimizing $E_\Phi(\alpha)$, that is putting $\frac{dE_\Phi}{d\alpha} = 0$, one finds that α must satisfy the condition,

$$\lambda = \lambda_0 = \frac{2\alpha(2\alpha+3)}{(2\alpha+1)^3} \qquad \text{... (12.24)}$$

Putting the above value of $\lambda = \lambda_0$ into eqn. (12.23), one gets the following equation for the lowest or binding energy of deuteron,

$$-\frac{E_\Phi}{V_0} = \frac{\alpha^2(2\alpha-1)}{(2\alpha+3)}$$

If the quantities V_0 and r_0 are given, that is, λ is given, the α value corresponding to the minimum of E_Φ can be found from eqn. (12.24).

This value of α is then substituted into the above equation and the required E_Φ can be determined.

12.31 The wave functions corresponding to 1*s*, 2*s* and $2p_z$ orbitals of the hydrogen atom are given by

$$\varphi_{n l m_l} = \varphi_{1s} = \varphi_{100} = \varphi_1 \text{ say} = (\pi a_0^3)^{-1/2}\, e^{-r/a_0}$$

$$= \varphi_{2s} = \varphi_{200} = \varphi_2 \text{ say} = (32\pi a_0^3)^{-1/2}\left(2 - \frac{r}{a_0}\right) e^{-r/2a_0}$$

$$= \varphi_{2p_z} = \varphi_{210} = \varphi_3 \text{ say} = (32\pi a_0^5)^{-1/2}\, r\cos\theta\, e^{-r/2a_0}$$

If the z-axis of the coordinate system is taken in the direction of the external field, then the perturbation

$$\lambda H' = -e\,E\,z = -e\,E_0\,e^{-t/\tau}\,r\cos\theta$$

Therefore,

$$H'_{fl} = H'_{21} = \int \varphi_2^*\,\lambda H'\,\varphi_1\,d\tau$$

$$= -eE_0 e^{-t/\tau}\,\frac{1}{\sqrt{32}\,\pi\,a_0^3}\int_0^\infty \left(2 - \frac{r}{a_0}\right) e^{-3r/2a_0}\,r^3\,dr$$

$$\int_0^\pi \cos\theta\sin\theta\,d\theta \int_0^{2\pi} d\phi$$

As $\int_0^\pi \sin\theta\cos\theta\,d\theta = \frac{1}{2}\int_0^\pi \sin 2\theta\,d\theta = 0,\ H'_{21} = 0$. As a result

$$a_f^{(1)}(t) = (i\hbar)^{-1}\int_0^t H'_{fl}\,e^{i\omega_{fl}t}\,dt = 0, \text{ which}$$

effectively means that the first order transition probability $|a_f^{(1)}(t)|^2$ that the atom is found in the 2s state is zero.

Now let us calculate the probability that the atom is in the $2p_z$ state (or φ_{210}) after a long time. To do so, we have to first find, $H'_{fl} = H'_{31}$ in this case, which is given by

$$H'_{31} = \int \varphi_3^*\,\lambda H'\,\varphi_1\,d\tau$$

$$= \frac{1}{\sqrt{32}\,\pi a_0^4}\,(-e\,E_0 e^{-t/\tau})\int_0^\infty e^{-3r/2a_0}\,r^4\,dr$$

$$\int_0^\pi \cos^2\theta\sin\theta\,d\theta \int_0^{2\pi} d\phi$$

Putting $\cos\theta = w,\ -\sin\theta\,d\theta = dw$,

$$\int_0^\pi \cos^2\theta\sin\theta\,d\theta = \int_{-1}^{+1} w^2\,dw = \frac{2}{3},$$

Therefore,

$$H'_{31} = \frac{1}{\sqrt{32}\,\pi\, a_0^4}(-eE_0 e^{-t/\tau})\frac{4!}{\left(\frac{3}{2a_0}\right)^5}\frac{2}{3}2\pi$$

$$= Ae^{-t/\tau},\ A = -\frac{eE_0 a_0}{\sqrt{2}}\frac{2^8}{3^5}$$

In the present case,

$$w_{fl} = w_{21} = \frac{(E_2 - E_1)}{\hbar} = -\frac{e^2}{2\,a_0\,\hbar}\left(\frac{1}{4}-1\right)$$

$$= \frac{3e^2}{8\,a_0\,\hbar}$$

To calculate the first order transition probability at $t \to \infty$, we have to first calculate $a_f^{(1)}\,(t \to \infty)$ which is given by

$$a_f^{(1)}\,(\infty) = (i\hbar)^{-1}\int_0^\infty H'_{fl}\, e^{iw_{fl}t}\, dt$$

$$= \frac{A}{i\hbar}\int_0^\infty e^{-t/\tau}\, e^{i3e^2/8\hbar a_0 t}\, dt$$

$$= \frac{A}{i\hbar}\int_0^\infty e^{-(1/\tau - i\,3e^2/8\,\hbar a_0)t}\, dt$$

$$= \frac{A}{i\hbar}\left.\frac{e^{-(1/\tau - i3\,e^2/8a_0\hbar)t}}{-(1/\tau - i3e^2/8a_0\hbar)}\right|_0^\infty$$

$$= \frac{A}{i\hbar}\frac{1}{(1/\tau - i3e^2/8\,\hbar\, a_0)}$$

Thus

$$|a_f^{(1)}\,(\infty)|^2 = \frac{A^2}{\hbar^2}\left(\frac{1}{\tau^2}+\frac{9e^4}{64\,a_0^2\,\hbar^2}\right)$$

12.32 The perturbation in the present case is given by

$$\lambda H' = -eE_z = -eE\,r\cos\theta$$

and the first order transition probability that the atom is in the $2p_z$ state is $|a_{31}\,(t)|^2$. To calculate this, we have to first find $a_{31}^{(1)}\,(t)$ or actually H'_{31} which is given by

$$H'_{31} = \int \varphi_3^*\,\lambda H'\,\varphi_1\, d\tau$$

$$= \int \varphi_{210}\,(-eEz)\,\varphi_{100}\, d\tau$$

Chapter 12

$$= -\frac{eEa_0\, 2^8}{3^5\, 2!}$$

Thus,

$$|a_{31}^{(1)}(t)|^2 = \left|(i\hbar)^{-1}\int H'_{31}\, e^{iw_{21}t}\, dt\right|^2$$

or

$$|a_{31}^{(1)}(\infty)|^2 = \frac{1}{\hbar^2}\frac{2^{15}}{3^{10}}\, e^2 a_0^2 E_0^2\, \frac{\tau^2}{\pi^2}\left|\int_0^\infty \frac{e^{iw_{21}t}}{(t^2+\tau^2)}\, dt\right|^2$$

Putting the value of the above integral which is equal to $\frac{\pi}{2\,\tau}\, e^{-w_{21}\,\tau}$, one gets

$$|a_{31}^{(1)}(\infty)|^2 = 2^{13}\,\frac{e^2 a_0^2\, E_0^2\, e^{-2w_{21}\tau}}{\hbar^2\, 3^{10}}$$

where

$$w_{21} = \frac{3e^2}{8a_0\,\hbar}$$

❑❑❑

Chapter 13

INTERACTION OF RADIATION[1] WITH MATTER—SEMI-CLASSICAL TREATMENT

13.1 INTRODUCTION

The time dependent perturbation theory developed in the last chapter would now be used to study the interaction of an atom (material particle) with electromagnetic field in detail. We shall, however confine ourselves to first order perturbation calculations only. Also the electromagnetic field would be treated classically and the atom with which the field interacts quantum mechanically. That is, atom is assumed to exist in discrete states. For this very reason, the theory falls under the category of semi-classical theories. We shall find that the said theory of atom-field interaction would enable us to calculate probabilities of induced absorption and induced emission but is insufficient to describe the phenomenon of spontaneous emission of radiation, that is, the emission in the absence of any external agency. The rate of spontaneous emission would be calculated using Einstein idea. It may be mentioned here that the fully quantum theory (in which electromagnetic field is also subjected to the rules of quantization) which we may not discuss here, describes the spontaneous emission automatically.

13.2 RADIATIVE TRANSITIONS IN ATOMS

The Hamiltonian of an atomic electron of mass m and charge e in the presence of an electromagnetic field is given by

$$H = \frac{1}{2m}\left(\mathbf{P} - \frac{e}{c}\mathbf{A}\right)^2 + e\phi + V \qquad \text{... (1)}$$

where $\mathbf{A}$ ($\mathbf{r}$, t) and ϕ ($\mathbf{r}$, t) are the vector and scalar potentials of the electromagnetic field, respectively at the position of the electron. $\mathbf{p}$ is the momentum of the electron

[1]For detailed discussion, see W. Heitler, The Quantum Theory of Radiation, 3rd ed., Oxford University Press, 1954.

situated at a distance r from the origin and bound to the origin by a central potential $V(r)$ (of electrostatic origin) created by nucleus assumed to be at rest at the origin.

Expanding (1), one gets

$$H = \frac{1}{2m}\left\{\left(\mathbf{p} - \frac{e}{c}\mathbf{A}\right).\left(\mathbf{p} - \frac{e}{c}\mathbf{A}\right)\right\} + e\phi + V$$

$$= \frac{1}{2m}\left\{p^2 - \frac{e}{c}(\mathbf{A}.\mathbf{p} + \mathbf{p}.\mathbf{A}) + \frac{e^2}{c^2}A^2\right\} + e\phi + V \quad \text{... (2a)}$$

Now

$$(\mathbf{A}.\mathbf{p} - \mathbf{p}.\mathbf{A})\varphi = (-i\hbar\,\mathbf{A}.\vec{\nabla} + i\hbar\vec{\nabla}.\mathbf{A})\varphi$$

$$= -i\hbar\mathbf{A}.\vec{\nabla}\varphi + i\hbar\,\vec{\nabla}.\mathbf{A}\varphi$$

$$= -i\hbar\,\mathbf{A}.\vec{\nabla}\varphi + i\hbar\,(\vec{\nabla}.\mathbf{A})\varphi + i\hbar\,\mathbf{A}.\vec{\nabla}\varphi$$

$$= i\hbar(\vec{\nabla}.\mathbf{A})\varphi$$

where φ is an arbitrary function and **A** and **p** are operators.

Thus $$\mathbf{A}.\mathbf{p} - \mathbf{p}.\mathbf{A} = i\hbar \,\text{div}\,\mathbf{A} \quad \text{... (2b)}$$

For a free wave (as the radiation field has no sources in the vicinity of atom), $\phi = 0$. Hence from Lorentz condition

$$\text{div}\,\mathbf{A} + \frac{1}{c}\frac{\partial\phi}{\partial t} = 0,$$

One gets

$$\text{div}\,\mathbf{A} = 0 \quad \text{... (2c)}$$

Substituting above into (2*b*), one gets

$$\mathbf{A}.\mathbf{p} = \mathbf{p}.\mathbf{A} \quad \text{... (2d)}$$

Further substituting (2*d*) into eqn. (2*a*), one gets

$$H = -\frac{\hbar^2}{2m}\nabla^2 + V - \frac{e}{mc}\mathbf{A}.\mathbf{p} + \frac{e^2}{2mc^2}A^2 \quad \text{... (3)}$$

One may assume field to be weak (which is a valid assumption as with ordinary light sources, the intensity of electromagnetic wave is sufficiently low) or effectively $|\mathbf{A}|$ to be small. In that case $\frac{e^2}{2mc^2}A^2$ can be neglected as compared to $\frac{e}{mc}\mathbf{A}.\mathbf{P}$. Also because of the reason that we are confining ourselves to the first order perturbation calculations, the said term can be neglected. In case we want to go to higher order of perturbation, the term $\frac{e^2}{2mc^2}A^2$ must be included. The Hamiltonian for atomic electron in the presence of electromagnetic field thus becomes

$$H = -\frac{\hbar^2}{2m}\nabla^2 + V - \frac{e}{mc}\mathbf{A}.\mathbf{p} \quad \text{... (4)}$$

Keeping above in mind, we can write the Hamiltonian for an atom (atomic number Z) in the presence of electromagnetic field as

$$H = \sum_{j=1}^{Z}\left(-\frac{\hbar^2}{2m}\nabla_j^2 + V\right) - \sum_{j=1}^{Z}\frac{e}{mc}\mathbf{A}(\mathbf{r}_j, t).\mathbf{p}_j \quad \text{... (5)}$$

in which first term is the Hamiltonian in the absence of electromagnetic field and the second term represents, the interaction Hamiltonian. $\mathbf{p}_j$ represents the momentum of the jth electron. $\mathbf{A}(\mathbf{r}_j, t)$ is the vector potential corresponding to the field at the position of the jth electron. The term V in this case contains the interaction (Coulombian and nonelectrical (nuclear forces)) between the different particles of the system.

Let us now write the total Hamiltonian of the system as

$$H = H^{(0)} + \lambda H' \quad \text{... (6)}$$

where

$$H^{(0)} = \sum_{j=1}^{Z}\left(-\frac{\hbar^2}{2m}\nabla_j^2 + V\right) \quad \text{... (6a)}$$

is the unperturbed (or atomic) Hamiltonian and

$$\lambda H' = -\sum_{j=1}^{Z}\frac{e}{mc}\mathbf{A}(\mathbf{r}_j, t).\mathbf{p}_j \quad \text{... (6b)}$$

is the interaction Hamiltonian (to be treated as perturbation) with the incident electromagnetic wave (or radiation).

According to the time dependent perturbation theory, the first order transition amplitudes at time t (from eqn. (140) of the section 12.17) are given by

$$i\hbar\dot{a}_k^{(1)}(t) = \sum_m H'_{km}\, a_m^{(0)}(0)e^{iw_{km}t} \quad \text{... (7a)}$$

Let us consider that the system is initially in the nth quantum state before the perturbation is applied. In that case

$$a_k^{(0)} = \delta_{kn} \quad \text{... (7b)}$$

(if $k \neq n$, $a^0(0) = 0$ as the system is in the state $\varphi_n^{(0)}$ at $t = 0$)

Using (7b), eqn. (7a) takes the form

$$i\hbar\dot{a}_k^{(1)}(t) = H'_{kn}\, e^{iw_{kn}t} \quad \text{... (7c)}$$

in which H'_{kn}, the matrix element of H' between kth and nth states, using eqn. (6b) would be given by

$$H'_{kn} = \int \varphi_k^{(0)*}\lambda H'\varphi_n^{(0)}\, d\tau$$

$$= -\sum_{j=1}^{Z} \frac{e}{mc} \int \varphi_k^{(0)*} \mathbf{A}(\mathbf{r}_j . t) . \mathbf{p}_j \varphi_n^{(0)} d\tau \qquad \text{... (8)}$$

where the value of $\lambda H'$ has been substituted from (*6b*).

In the present case (where $\phi = 0$; $\vec{\nabla} . \mathbf{A} = 0$), the vector potential satisfies the wave equation,

$$\nabla^2 \mathbf{A} - \frac{1}{c^2} \frac{\partial^2 \mathbf{A}}{\partial t^2} = 0 \qquad \text{... (9)}$$

The plane wave solution[2] of the above equation is given by

$$\mathbf{A}(\mathbf{r}, t) = \mathbf{A}_0 e^{i(\mathbf{k}.\mathbf{r} - wt)} + \mathbf{A}_0^* e^{-i(\mathbf{k}.\mathbf{r} - wt)} \qquad \text{... (10)}$$

in which $\mathbf{A}_0$ ($\mathbf{n}\,|\mathbf{A}_0|$ where $\mathbf{n}$ denotes the polarization vector) is a constant complex vector normal to the propagation vector $\mathbf{k}$ and $\mathbf{w} = c\,|\mathbf{k}|$.

Substituting above in eqn. (8), we obtained

$$H'_{kn} = H'^{a}_{kn} e^{-iwt} + H'^{e}_{kn} e^{iwt} \qquad \text{... (11)}$$

(the superscript *a* and *e* would finally correspond to absorption and emission) where

$$H'^{a}_{kn} = -\sum_{j=1}^{Z} \frac{e}{mc} \int \varphi_k^{(0)*} e^{i\mathbf{k}.\mathbf{r}} \mathbf{A}_0 . \mathbf{p}_j \varphi_n^{(0)} d\tau \qquad \text{... (12a)}$$

and

$$H'^{e}_{kn} = -\sum_{j=1}^{Z} \frac{e}{mc} \int \varphi_k^{(0)*} e^{-i\mathbf{k}.\mathbf{r}} \mathbf{A}_0^* . \mathbf{p}_j \varphi_n^{(0)} d\tau \qquad \text{... (12b)}$$

Substituting eqn. (11) into eqn. (*7b*), one obtains

$$i\hbar \dot{a}_k^{(1)}(t) = H'^{a}_{kn} e^{i(w_{kn} - w)t} + H'^{e}_{kn} e^{i(w_{kn} + w)t}$$

which on integration leads to the following expression for the first order transition amplitudes at time *t*,

$$a_k^{(1)}(t) = -\frac{H'^{a}_{kn}}{\hbar} - \frac{H'^{a}_{kn}}{\hbar} \frac{(e^{i(w_{kn} - w)t} - 1)}{(w_{kn} - w)} - \frac{H'^{e}_{kn}}{\hbar} \frac{(e^{i(w_{kn} + w)t} - 1)}{(w_{kn} + w)} \qquad \text{... (13)}$$

From above, it is clear that when frequency of radiation w is equal to one of the Bohr frequencies w_{kn} of the physical system under consideration, that is when $E_k^{(0)} = E_n^{(0)} + \hbar w$, the first term predominates and hence the second term compared to the first one can be neglected. When $w = -w_{kn}$, that is when $E_k^{(0)} = E_n^{(0)} - \hbar w$, the second term predominates and hence the first term can be neglected compared to the first one. The

[2]J.D. Jackson, classical Electrodynamics (John–wiley, N.Y. 1963) Sec. 12.5

first case, $E_k^{(0)} > E_n^{(0)}$ corresponds to induced or stimulated absorption and the second, $E_k^{(0)} < E_n^{(0)}$ corresponds to induced or stimulated emission.

13.3 ELECTRIC DIPOLE APPROXIMATION—CASE OF INDUCED ABSORPTION

In this case neglecting the second term compared to the first one (which predominates) in eqn. (13), the first order transition amplitudes at time t would be given by

$$a_k^{(1)}(t) = -H'^{a}_{kn} \frac{[e^{i(w_{kn}-w)t} - 1]}{\hbar\,(w_{kn} - w)}$$

Accordingly the probability of finding the system in the state k at time t would be

$$|a_k^{(1)}(t)|^2 = a_k^{(1)}(t)^* \, a_k^{(1)}(t)$$

$$= 4|H'^{a}_{kn}|^2 \sin^2 (w_{kn} - w)t/2 \,/\hbar^2 (w_{kn} - w)^2 \qquad \text{... (14)}$$

From above expression note that as long as the perturbation (incident radiation) contains only a single frequency, the transition probability oscillates with time as in eqn. (143), section 12.17 with w_{kn} (or w_{fl}) replaced by $(w_{kn} - w)$. In case the perturbation involves a spectrum of frequencies, the total transition probability $P_{n \to k}$ is found by adding up all the probabilities arising from different frequency components (the phases of different frequency components are unrelated to each other or these are incoherent). Thus

$$P_{n\to k} = \sum_{w} 4|H'^{a}_{kn}|^2 \sin^2(w_{kn} - w)\,t/2\,/\hbar^2 (w_{kn} - w)^2$$

For a continuous radiation (if frequencies present in the spectrum are closely spaced), the summation can be replaced by integration and thus

$$P_{n\to k} = \int_{\Delta w} \frac{4|H'^{a}_{kn}|^2}{\hbar^2} \frac{\sin^2(w_{kn} - w)\,t/2}{(w_{kn} - w)^2} dw$$

$$= \int_{\Delta w} \frac{4|H'^{a}_{kn}|}{\hbar^2} \frac{\sin^2(wkn - w)\,t/2}{(w_{kn} - w)^2} d(w - w_{kn}) \qquad \text{... (15)}$$

where Δw represents the range of frequencies present in the spectrum of incident radiation and the transitions being between discrete levels, w_{kn} is a fixed quantity and therefore, $dw_{kn} = 0$.

The plot of the time factor $\sin (w_{kn} - w)\, t/2\,/\,(w_{kn} - w)^2$ versus $(w_{kn} - w)$ would be same as that shown in Fig. 12.6 of the last chapter (w_{kn} replaced by $w_{kn} - w$). Since this factor has a pronounced peak (or first maximum) at $w = w_{kn}$ and also most of the contribution being from the said maximum, there is no harm taking the integration from $-\infty$ to $+\infty$ (as was done earlier in the case of eqn. (145) section 12.18) and also evaluating other terms at $w = w_{kn}$, one gets

$$P_{n\to k} = \frac{4|H'^{a}_{kn}|^2_{w=w_{kn}}}{\hbar^2}\int_{-\infty}^{+\infty}\frac{\sin^2(w_{kn}-w)t/2}{(w_{kn}-w)^2}d(w-w_{kn})$$

$$= \frac{4|H'^{a}_{kn}|^2_{w=w_{kn}}}{\hbar^2}\left(\frac{1}{2}\pi t\right)$$

$$= \frac{2\pi}{\hbar^2}|H'^{a}_{kn}|^2_{w=w_{kn}}$$

The transition probability per unit time for the upward transition (or the stimulated absorption) would therefore be given by

$$\frac{P_{n\to k}}{t} = \frac{2\pi}{\hbar^2}|H'^{a}_{kn}|^2 \qquad \text{... (16)}$$

For the visible light, $\lambda \simeq 10^{-5}$ cm falling on an atom of size, $r \simeq 10^{-8}$ cm. $\mathbf{k}\cdot\mathbf{r} \sim 10^{-3}\left(k = \frac{2\pi}{\lambda}\right)$. Therefore, $e^{i\mathbf{k}\cdot\mathbf{r}}$ is expanded in a power series

$$e^{i\mathbf{k}\cdot\mathbf{r}} = 1 + i\mathbf{k}\cdot\mathbf{r} + \frac{1}{2}(i\mathbf{k}\cdot\mathbf{r})^2 + \ldots \qquad \text{... (17)}$$

and only the first term which is unity is retained. This simplification is called the long wavelength approximation and the corresponding transition is called electric dipole transition (why is it called so would be evident after some time). Accordingly eqn. (12*a*) results in

$$H'^{a}_{kn} = -|\mathbf{A}_0|\sum_{j=1}^{Z}\frac{e}{mc}\int \varphi_k^{(0)*}\,\mathbf{n}\cdot\mathbf{p}_j\varphi_n^{(0)}\,d\tau \qquad \text{... (18)}$$

Now

$$\int \varphi_k^{(0)*}\,\mathbf{p}_j\varphi_n^{(0)}\,d\tau = (\mathbf{p}_j)_{kn}$$

where $(\mathbf{p}_j)_{kn}$ denotes the matrix element of the momentum of particle between *k*th and *n*th states.

Further making use of the Heisenberg equation of motion,

$$\frac{(\mathbf{p}_j)_{kn}}{m} = \frac{d(\mathbf{r}_j)_{kn}}{dt} = \frac{1}{i\hbar}[\mathbf{r}_j, H^{(0)}]_{kn}$$

and then using Hermiticity of $H^{(0)}$ and as well as the relation, $H^{(0)}\,\varphi_n^{(0)} = E_n^{(0)}\varphi_n^{(0)}$, one gets

$$\frac{(\mathbf{p}_j)_{kn}}{m} = i/\hbar\,(E_k^{(0)} - E_n^{(0)})(\mathbf{r}_j)_{kn}$$

$$= iw_{kn}\,(\mathbf{r}_j)_{kn}$$

Substituting above into eqn. (18), one gets

$$H'^{a}_{kn} = -\frac{i|\mathbf{A}_0|}{c} wkn \sum_{j=1}^{Z} \int \varphi_k^{(0)*}\ \mathbf{n}\,.\,e\mathbf{r}_j \varphi_n^{(0)}\, d\tau$$

$$= -\frac{i|\mathbf{A}_0|}{c} w_{kn} (\mathbf{n}\,.\,\mathbf{M})_{kn} \qquad \text{... (19)}$$

where $M = \sum_{j=1}^{Z} e\mathbf{r}_j$ denotes the total electric dipole moment of the atom (called electric dipole moment operator) and $(\mathbf{n}\,.\,\mathbf{M})_{kn}$ represents the matrix element of $n\,.\,M$ between the kth and nth states. As a result

$$|H'^{a}_{kn}|^2 = |\mathbf{A}_0|^2 |(\mathbf{n}\,.\,\mathbf{M})_{kn}|^2 \frac{w_{kn}^2}{c^2} \qquad \text{... (20)}$$

Now the intensity of the electromagnetic field is given by the Poynting vector

$$\mathbf{S} = \frac{c}{4\pi} (\mathbf{E} \times \mathbf{B}) \qquad \text{... (21)}$$

in which the electric field intensity vector

$$E = -\frac{1}{c}\frac{\partial A}{\partial t} \left(\text{actually } \mathbf{E} = -\vec{\nabla}\phi - \frac{1}{c}\frac{\partial A}{\partial t}\ \textit{but}\ \phi = 0 \text{ in this case}\right)$$

which using eqn. (10) takes the form

$$\mathbf{E} = \frac{iw}{c}\left[\mathrm{A}_0 e^{i(\mathbf{k}\,.\mathbf{r}\,-\,wt)} - \mathbf{A}_0^* \, e^{-i(\mathbf{k}.\mathbf{r}\,-\,wt)}\right] \qquad \text{... (22)}$$

and magnetic induction

$$\mathbf{B} = \vec{\nabla} \times \mathbf{A} = i\mathbf{k}\,[\mathbf{A}_0\, e^{i(\mathbf{k}\,.\mathbf{r}\,-\,wt)} - \mathbf{A}_0^* e^{-i(\mathbf{k}\,.\mathbf{r}\,-\,wt)}]$$

which using the above equation for **E** and the relation $w = ck$ becomes

$$\mathbf{B} = \left(\frac{\mathbf{k} \times \mathbf{E}}{k}\right) \qquad \text{... (23)}$$

Thus

$$\mathbf{S} = \frac{c}{4\pi} \mathbf{E} \times (\mathbf{k} \times \mathbf{E}) = \frac{c}{4\pi} \frac{\mathbf{k}\,(\mathbf{E}\,.\,\mathbf{E})}{k}$$

where use has been made of the vector identity, $\mathbf{a} \times (\mathbf{b} \times \mathbf{c}) = \mathbf{b}\,(\mathbf{a}\,.\,\mathbf{c}) - \mathbf{c}\,(\mathbf{a}\,.\,\mathbf{b})$ and the fact that $\mathbf{k}\,.\,\mathbf{E} = 0$ as the wave is transverse. The above equation, then using eqn. (22) reduces to

$$\mathbf{S} = \frac{\mathbf{k}}{k} \frac{w^2}{4\pi c}\left[2|\mathbf{A}_0|^2 - \mathrm{A}_0^2 e^{2i(\mathbf{k}\,.\mathbf{r}\,-\,wt)} - A_0^{*2}\, e^{-2i(\mathbf{k}\,.\mathbf{r}\,-\,wt)}\right]$$

The intensity associated with the plane wave (10) or average magnitude of **S** over a period is thus given by

$$I(w) = \frac{1}{(2\pi/w)}\int_0^{2\pi} \mathbf{S}dt = \frac{w^2|\mathbf{A}_0|^2}{2\pi c} \qquad \text{... (24)}$$

as on taking the time average, the oscillating terms give zero. Using the above equation at $w = w_{kn}$, (20) takes the form

$$|H'^{a}_{kn}|^2 = \frac{2\pi I(w_{kn})|(\mathbf{n}\,.\,\mathbf{M})_{kn}|^2}{c}$$

Accordingly the transition probability per unit time for the upward transition from eqn. (16) would be given by

$$\frac{P_{n\to k}}{t} = \frac{4\pi^2 I(w_{kn})|(\mathbf{n}\,.\,\mathbf{M})_{kn}|^2}{\hbar^2 c} \qquad \text{... (25)}$$

Further if the incident radiation is unpolarized, **n** can have any direction and hence $\mathbf{n}\,.\,\mathbf{M} = M\cos\theta$, where θ is the angle between the polarization vector n and the electric dipole moment vector **M**. Accordingly

$$\frac{P_{n\to k}}{t} = \frac{4\pi^2 I(w_{kn})|\mathbf{M}_{kn}|^2\cos^2\theta}{\hbar^2 c}$$

and thus the average (average over all the orientations, that is

$$\overline{\cos^2\theta} = \frac{1}{4\pi}\int_0^{2\pi}\int_0^{\pi}\cos^2\theta\sin\theta\,d\theta\,d\phi = \frac{1}{3}\Bigg)$$

rate of absorption, would be given by

$$w_{abs} = \frac{4\pi^2 I(w_{kn})|\mathbf{M}_{kn}|^2}{3\hbar^2 c} \qquad \text{... (26)}$$

Note that in the above equation, $I(w_{kn})$ is the characteristic of radiation whereas $\frac{4\pi^2|M_{kn}|^2}{3\hbar^2 c}$ is that of the absorbing system. Thus, for the transition induced by unpolarized field,

$$w_{abs} = B_{n\to k} I(w_{kn}) \qquad \text{... (27)}$$

where

$$B_{n\to k} = \frac{4\pi^2|M_{kn}|^2}{3\hbar^2 c} \qquad \text{... (27a)}$$

is called Einstein *B* coefficient for absorption.

13.4 RATE OF INDUCED EMISSION

In analogous to eqns. (16) and (27), it can be shown that the transition probability per unit time from state k to state n (that is, for downward transition) would be given by

$$w_{em} = \frac{2\pi}{\hbar^2}|H'^{e}_{kn}|^2_{w = w_{kn}} \quad \text{... (28)}$$

which finally for the transition induced by the unpolarized field, would take the form

$$w_{em} = B_{k \to n} I(w_{kn}) \quad \text{... (29)}$$

where

$$B_{k \to n} = \frac{4\pi^2 |\mathbf{M}_{kn}|^2}{3\hbar^2 c}$$

is called Einstein B coefficient for the induced emission (In case one considers quantum theory of radiation, the result obtained differs from the formula (29) for emission by an additional term which does not vanish in absence of the radiation field. However, the result obtained in the case of induced absorption is in agreement with eqn. (27).

From eqns. (27) and (29),

$$B_{n \to k} = B_{k \to n} \quad \text{... (30a)}$$

$$(w_{abs})_{n \to k} = (w_{em})_{k \to n} \quad \text{... (30b)}$$

Eqn. (30b) implies that the probability that the radiation field induces the transition form state n to k is same as the probability that the radiation field induces the inverse transition from state k to n. This is known as the principle of detailed balance. In spite of this, the intensity of stimulated emission between two atomic levels is found to be much less than that in the case of stimulated absorption. This is because of the greater initial number of atoms in the lower state. The stimulated emission can, however be made to predominate over the absorption process by achieving an inversion of the normal population of the two levels as in the case of masers and lasers (quantum mechanical light generators).

It is also necessary to mention that the transitions which we have been discussing here are called electric dipole transitions since only the matrix element of the electric dipole moment of the system is involved.

13.5 RATE OF SPONTANEOUS EMISSION

Here once again we do not use the Dirac's theory (in which the radiation is also treated quantum mechanically), instead to obtain the rate at which spontaneous emission takes place, we use Einstein[3] idea (since the theory developed in the above sections does not account for the spontaneous emission) which is based on the consideration of matter (atomic system) in equilibrium with radiation and assume that the resultant radiation obeys Planck's law.

[3]A. Einstein, Z. Physik 18, 121 (1917)

Let N_n be the number of atoms in the state n and the number in the state k is N_k. The probability of absorption per unit time (or rate of absorption) would be proportional to N_n and also to $I(w_{kn})$, the intensity of incident radiation and thus can be written as

$$N_n B_{n \to k} I(w_{kn}) \quad \text{... (31)}$$

where $B_{n \to k}$, the coefficient of proportionality is called Einstein B coefficient for induced absorption.

On the other hand when the atom is in the excited state, an incident signal of appropriate frequency triggers (or forces) the atom to emit radiation, the process called stimulated emission. The rate of this type of transition is

$$N_k B_{k \to n} I(w_{kn}) \quad \text{... (32)}$$

where $B_{k \to n}$ is called Einstein B coefficient of induced (or stimulated) emission.

At equilibrium, the number of transitions from the state n to the state k must equal the number of transitions from state k to state n by the principle of detailed balance. If there were no other transition mechanism (other than the one considered above) then $N_n B_{n \to k} I(w_{kn}) = N_k B_{k \to n} I(w_{kn})$ which in turn is equal to $N_k B_{n \to k} I(w_{kn})$ as $B_{n \to k} = B_{k \to n}$. However, Boltzmann's law states that at equilibrium

$$N_n/N_k = \frac{e^{-E_n^{(0)}/k_B T}}{e^{-E_k^{(0)}/k_B T}} = e^{(E_k^{(0)} - E_n^{(0)})/k_B T}$$

$$= e^{\hbar w_{kn}/k_B T} \quad \text{... (33)}$$

where k_B is Boltzmann's constant and therefore $N_n B_{n \to k} I(w_{kn})$ cannot be equal to $N_k B_{n \to k} I(w_{kn})$. To resolve this paradox, Einstein postulated the existence of spontaneous emission, an emission process which is independent of the incident radiation (that is a process in which an atom in the excited state emits radiation even in the absence of any incident radiation). The probability of spontaneous emission per unit time (or rate of spontaneous emission) is thus given as

$$N_k A_{k \to n} \quad \text{... (34)}$$

where $A_{k \to n}$, the coefficient of proportionality is called Einstein A coefficient of spontaneous emission.

With the inclusion of this process of emission the principle of detailed balance at equilibrium, gives

$$N_k B_{n \to k} I(w_{kn}) = N_k B_{n \to k} I(w_{kn}) + N_k A_{k \to n}$$

or

$$A_{k \to n} = B_{n \to k} I(w_{kn}) (N_n/N_k - 1) \quad \text{... (35)}$$

Using eqn. (33), the above equation reduces to

$$A_{k \to n} = (e^{\hbar w_{kn}/k_B T} - 1) B_{n \to k} I(w_{kn}) \quad \text{... (36)}$$

Now according to Planck's radiation law, the intensity of incident radiation in equilibrium with our system at temperature T, is given by

$$I(w) = \frac{\hbar w^3}{\pi^2 c^2} \frac{1}{(e^{\hbar w/k_B T} - 1)} \quad \text{... (37)}$$

(one must note that $I(w) = c\,\rho(w)$, where $\rho(w)\,dw$ represents the radiation energy per unit volume within the frequency interval between w and $w + dw$).

Substituting above (that is, the value of $I(w)$ at $w = w_{kn}$ from (37)) into eqn. (36), one gets

$$A_{k \to n} = \frac{B_{n \to k} \hbar w_{kn}^3}{\pi^2 c^2} \left(= \frac{B_{k \to n} \hbar w_{kn}^3}{\pi^2 c^2} \right) \quad \text{... (38)}$$

Thus we get the following relation between the Einstein A and B coefficients,

$$\frac{A_{k \to n}}{B_{n \to k}} = \frac{\hbar \omega_{kn}^3}{\pi^2 c^2} \quad \text{... (39)}$$

Further substituting the value of $B_{n \to k} = B_{k \to n} = \dfrac{4\pi^2 |\mathbf{M}_{kn}|^2}{3\hbar^2 c}$ from eqn. (27*a*) into the above equation, one gets

$$A_{k \to n} = \frac{\hbar w_{kn}^3 |\mathbf{M}_{kn}|^2}{3\hbar c^2} \quad \text{... (40)}$$

which represents nothing but the transition probability per unit time (see eqn. (34)) for spontaneous emission. The above equation shows that the said probability increases with the transition frequency and is thus very important at high frequencies.

13.6 ELECTRIC DIPOLE SELECTION RULES – CASE OF HYDROGEN ATOM

It is now clear that the transition probability per unit time for transitions between the states k and n is proportional to $|\mathbf{M}_{kn}|^2$. Thus only those transitions for which $M_{kn} \neq 0$ would be allowed and those for which $\mathbf{M}_{kn} = 0$ would be forbidden (of course only in the dipole approximation). If, as is almost always the case, the states under consideration have definite angular momenta, the conditions mentioned above give selection rules governing the change of angular momentum.

As an application of the semi-classical theory of radiation using dipole approximation, we would obtain the selection rules for the transitions in hydrogen atom. The eigen functions for hydrogen atom are given by

$$\varphi_{nlm_l} = N_{nlm_l} R_{nl}(r) P_l^{m_l}(\cos\theta)\, e^{im_l\phi} \quad \text{... (41)}$$

where $R_{nl}(r)$ represents the radial part and $P_l^{m_l}(\cos\theta)\, e^{im_l\phi} = Y_l^{m_l}(\theta, \phi)$ represents the angular part and is called spherical harmonics. $P_l^{m_l}(\cos\theta)$ denotes the Legendre's polynomial.

The electric dipole moment in the present case is given by

$$\mathbf{M} = e\mathbf{r} \quad \text{... (42)}$$

and hence the matrix element of **M** between the states $\varphi_{n'l'ml'}$ (final state) and φ_{nlml} (initial state) would be given by

$$(\mathbf{M})_{n'l'm_l'\,nlm_l} = N_{n'l'm_l'} N_{nlm_l} \int R^*_{n'l'}(r)\, Y_{l'}^{m_{l'}*}(\theta, \phi)\, e\mathbf{r}\, R_{nl}(r)\, y_l^{m_l}(\theta, \phi)\, d\tau$$

The above relation can further be written as

$$(\mathbf{M})_{n'l'\,m_l'\,nlm_l} = eN_{n'l'm_{l'}} N_{nlm_l} \int R^*_{n'l'}(r)\, Y_{l'}^{m_{l'}*}(\theta, \phi) \begin{vmatrix} r\sin\theta\cos\phi \\ r\sin\theta\sin\phi \\ r\sin\theta \end{vmatrix} \times$$

$$R_{nl}(r) Y_l^{m_l}, (\theta, \phi) r^2 \sin\theta\, d\theta\, d\phi\, dr$$

which amounts to three terms corresponding to x, y and z components of the electric dipole moment vector. Further

$$(\mathbf{M})_{n'l'm_{l'}\,nlm_l} = eN_{n'l'm_{l'}} N_{nlm_l} \int_0^\infty R^*_{n'l'}(r) r^3 Rn_l(r)\, dr \times \int_0^\pi P_{l'}^{m_{l'}}(\cos\theta)$$

$$\begin{vmatrix} \sin\theta \\ \sin\theta \\ \cos\theta \end{vmatrix} P_l^{m_l}(\cos\theta)\sin\theta\, d\theta \int_0^{2\pi} e^{i(m_l - m_{l'})\phi} \begin{vmatrix} \cos\phi \\ \sin\phi \\ 1 \end{vmatrix} d\phi \qquad \text{... (43)}$$

The radial part of the integral is of course non-vanishing and hence may be omitted for discussing selection rules. Let us first consider the case when the light is polarized in the z-direction. In that case the transition rate involves the matrix element of the z-component of the dipole moment vector. The ϕ part of the integration,

$$\int_0^{2\pi} e^{i(m_l - m_l)\phi} \times 1\, d\phi = 0,\ m_l \neq m_{l'} \qquad \text{... (44)}$$

$$\neq 0,\ m_l = m_{l'}$$

The θ part of the integration can then be written as

$$\int_0^\pi P_{l'}^{m_{l'}}(\cos\theta)\cos\theta\, P_l^{m_l}(\cos\theta)\sin\theta\, d\theta$$

which on making the substitution $w = \cos\theta$, takes the form,

$$\int_{-1}^{+1} wP_{l'}^{m_{l'}}(w) P_l^{m_l}(w)\, dw$$

Using the recursion relation for the Legendre's polynomials, that is,

$$wP_l^{m_l} = \frac{(l + m_l)}{(2l + 1)} P_{l-1}^{m_l}(w) + \frac{(l - m_l + 1)}{(2l + 1)} P_{l+1}^{m_l}(w),$$

the θ part of the integration becomes,

$$\int_{-1}^{+1} \frac{(l+m_l)}{(2l+1)} P_{l'}^{m_{l'}}(w) P_{l-1}^{m_l}(w)\, dw$$

$$+ \int_{-1}^{+1} \frac{(l-m_l+1)}{(2l+1)} P_{l'}^{m_l}(w) P_{l+1}^{m_l}(w)\, dw \qquad \text{... (45)}$$

Further using the orthonormality relation for the Legendre's polynomials, that is,

$$\int_{-1}^{+1} P_{l'}^{m_{l'}}(w) P_{l}^{m_l}(w)\, dw \neq 0,\ l' = l$$

the first term in (45) is non-vanishing when $l' = l - 1$ and the second is non-vanishing when $l' = l + 1$. Thus the matrix element of the *z*-component of dipole moment vector is non-vanishing when

$$m_l = m_{l'} \text{ or } \Delta m_l = 0$$

and $\quad l' = l - 1 \text{ and } l' = l + 1 \qquad$ ... (46)

or $\quad \Delta l = \pm 1$

The above conditions represent selection rules in the case of light polarized in the *z*-direction and since spin angular momenta are not involved, we have $\Delta j = \pm 1$ as well.

For light polarized in the *x*-direction, the matrix element of the *x*-component of the dipole moment vector would be involved. The ϕ part of the integral in this case would be given by

$$\int_0^{2\pi} e^{i(m_l - m_{l'})}\phi \cos\phi\, d\phi = \int_0^{2\pi} \frac{[e^{i(m_{l'} - m_l' + 1)\phi} + e^{i}(m_l - m_{l'} - 1)]}{2} d\phi \qquad \text{... (47)}$$

where we have used the relation, $\cos\phi = \dfrac{(e^{i\phi} + e^{-i\phi})}{2}$

The first part of the above integral is non-vanishing when $m_{l'} = m_l + 1$ and the second when $m_{l'} = m_l - 1$. Thus for the non-zero result

$$\Delta m_l = \pm 1 \qquad \text{... (48)}$$

The θ part of the integral in this case can be written as

$$\int_0^{\pi} P_{l'}^{m_{l'}}(\cos\theta) \sin\theta P_l^{m_l}(\cos\theta) \sin\theta\, d\theta$$

which on substituting $w = \cos\theta$ takes the form

$$\int_{-1}^{+1} P_{l'}^{m_{l'}}(w) (1 - w^2)^{1/2} P_l^{m_l}(w)\, dw \qquad \text{... (49)}$$

Using $m_{l'} = m_l + 1$, the above result takes the form

$$\int_{-1}^{+1} P_{l'}^{m_{l'}}(w)\,(1-w^2)^{1/2} P_l^{m_l}(w)\,dw$$

Now using recursion relation for the Legendre's polynomials, that is,

$$(1-w^2)^{1/2} P_l^{m_l}(w) = \frac{1}{(2l+1)}[P_{l+1}^{m_l}(w) - P_{l-1}^{m_{l+1}}(w)]$$

the θ part finally becomes,

$$\frac{1}{(2l+1)} \int_{-1}^{+1} P_{l'}^{m_l+1}(w)\left[P_{l+1}^{m_l+1}(w) - P_{l-1}^{m_l+1}(w)\right]dw \quad \text{... (50)}$$

Further using the orthonormality relation for Legendre's polynomials, we find that the first part of the above integral is non-zero when $l' = l + 1$ and the second when $l' = l - 1$.

Similarly when we put $m_{l'} = m_l - 1$ in eqn. (49), we get

$$\int_{-1}^{+1} P_{l'}^{m_l-1}(w)\,(1-w^2)^{1/2} P_l^{m_l}(w)\,dw$$

The use of recursion relation,

$$(l-w^2)^{1/2} P_l^{m_l}(w) = \frac{1}{(2l+1)}\Big[(l+m_l)(l+m_l-1)P_l^{m_l-1}(w)$$

$$-(l-m_l+1)(l-m_l+2)P_{l+1}^{m_l-1}(w)\Big]$$

reduces above into

$$\frac{1}{(2l+1)} \int_{-1}^{+1} P_l^{m_l-1}(w)\left[(l+m_l)(l+m_l-1)P_{l-1}^{m_l-2}(w) - (l - m_l + 1)(l - m_l + 2) P_{l+1}^{m_l-1}(w)\,dw\right]$$

and finally the use of orthonormality relation dictates that the first part of above integral would be non-vanishing when $l' = l - 1$ and the second part when $l' = l + 1$. Thus the non-zero value of the matrix element of the x-component of the dipole moment vector results, when

$$\Delta m_l = \pm 1 \quad \text{and} \quad \Delta l = \pm 1 \quad \text{... (51)}$$

If one proceeds in the similar fashions, it can be shown that one obtains the same result for the non-vanishing of the matrix element of the y-component of the dipole moment vector. It may be mentioned here that for the light polarized in the $x - y$ plane (plane polarized light) one has to find the matrix element of $x \pm iy$ or $\cos\phi \pm i \sin\phi$ or $e^{\pm i\phi}$ which can be evaluated easily with the outcome of the same conditions.

Since the radial part of eqn. (43) is never zero, Δn can take any value. Thus only those transitions (of course in the electric dipole approximation) would be allowed for which the conditions (51) are satisfied in the case of hydrogen atom or in general in the case of a single charged particle that moves in a central force field. Conditions (51) are called selection rules.

13.7 FORBIDDEN TRANSITIONS

Transitions which are forbidden by the selection rules of the dipole approximation may arise from the higher order terms in the expansion of $e^{ik\,.\,r}$ (or $e^{-ik\,.\,r}$). The nth term of the series is of the order $(kr)^{n-1}$. Thus each term is smaller than the preceding one by a factor of order $kr \sim 10^{-3}$ and the corresponding probability is smaller by $\sim (kr)^2$. Evidently, the forbidden lines (or transitions, called electric quadrupole and magnetic dipole) arising from the presence of even $i\mathbf{k}\,.\,\mathbf{r}(-\,i\mathbf{k}\,.\,\mathbf{r})$, that is the second term in the expansion of $e^{ik\,.\,\mathbf{r}}$ (or $e^{-\,ik\,.\,\mathbf{r}}$) are very much weaker (since the intensity of spectral line is proportional to energy absorbed or emitted per second) than those from the allowed dipole transitions.

PROBLEMS

13.1 Obtain the selection rules for dipole transitions in the case of one dimensional harmonic oscillator.

13.2 Find the probability per unit time for spontaneous radiative transitions from $2p$ state to $1s$ state of the hydrogen atom and hence show that the mean life time of the atom in $2p$ state is 1.6×10^{-9} sec.

13.3 The intensity of spectral line (radiation) is proportional to the energy emitted per second, that is $\hbar\, w_{k \to n}\, A_{k \to n}$. Show that the ratio of intensities of the first two Lyman lines of the hydrogen atom ($L_y\,\alpha$: $2p \to 1s$; $L_y\beta$: $3p \to 1s$) is 3.2.

SOLUTIONS

13.1 The wave function for one–dimensional harmonic oscillator is given by

$$\varphi_n^{(0)}(\xi) = N_n'\, H_n(\xi) e^{-\xi^2/z} \quad \text{... (13.1)}$$

where $\xi = \sqrt{\alpha}\, x$, $\alpha = \dfrac{mw}{\hbar}$ and H_n (ξ) is Hermite polynomial of degree n. N_n' represents the normalization constant.

The electric dipole moment in the present case (in which a particle of charge e executes harmonic motion about a charge-e at the origin) is given by, $M = ex$ and as a result,

$$M_{kn} = \int \varphi_k^* M \varphi_n^{(0)}\, dx$$

where n and k denote the initial and final oscillator states, respectively. Thus

$$M_{kn} = N_n'\, N_k' \int_{-\infty}^{+\infty} e^{-\xi^2} H_k(\xi)\, ex\, H_n(\xi)\, dx$$

$$= \frac{eN'_n N'_k}{\alpha} \int_{-\infty}^{+\infty} e^{-\xi^2} H_k(\xi) H_n d\xi$$

$$= \frac{eN'_n N'_k}{\alpha} \int_{-\infty}^{+\infty} e^{-\xi^2} H_k(\xi) [n H_{n-1}(\xi)] + \frac{1}{2} H_{n+1}(\xi)] d\xi \quad \text{... (13.2)}$$

where the use of recursion relation for Hermite polynomials has been made. The use of orthonormality relation reveals that the first term in eqn. (13.2) is non-zero only when $k = n - 1$ and similarly, the second term is non-zero when $k = n + 1$. Therefore

$$M_{kn} \neq 0 \text{ when } k - n = \pm 1 \text{ or } \Delta n = \pm 1 \quad \text{... (13.3)}$$

Accordingly, in the case of one-dimensional harmonic oscillator only those transitions for which $\Delta n = \pm 1$ holds good would be allowed in the dipole approximation and others would be forbidden. Thus only the fundamental frequency w is absorbed or emitted in this case.

13.2 The eigen functions for $2p$ and $1s$ states of the hydrogen atom are

$$\varphi_i = \varphi_{2p} = \varphi_{nlm_l} = \varphi_{21\,m_l} = R_{21}(r) Y_1^{m_l}(\theta, \phi)$$

$$= \left(\frac{1}{2a_0}\right)^{3/2} \frac{r}{\sqrt{3}\, a_0} e^{-r/2a_0} Y_1^{m_l}$$

where $m_l = +1, 0,$ and -1 and

$$\varphi_f = \varphi_{1s} = \psi_{nlm_l} = \varphi_{100} = R_{100}(r)\, Y_0^0(\theta, \phi)$$

$$= (\pi a_0^3)^{-1/2} e^{-r/a_0}$$

We want to calculate the matrix elements of the electric dipole moment, that is effectively, x_{fi}, y_{f_i} and z_{fi} where

$$x = r \sin\theta \cos\phi = \left(\frac{2\pi}{3}\right)^{1/2} r(-Y_1^1 + Y_1^{-1}),$$

$$y = r \sin\theta \sin\phi = i\left(\frac{2\pi}{3}\right)^{1/2} r(Y_1^1 + Y_1^{-1}),$$

and

$$z = r\cos\theta = \left(\frac{4\pi}{3}\right)^{1/2} r Y_1^0$$

as

$$Y_1^{\pm 1}(\theta, \phi) = \mp\left(\frac{3}{8\pi}\right)^{1/2} \sin\theta\, e^{\pm i\phi}$$

and

$$Y_1^0(\theta, \phi) = \left(\frac{3}{4\pi}\right)^{1/2} \cos\theta$$

Now

$$x_{fi} = \int \varphi_f^* x \varphi_i d\tau$$

$$= \frac{1}{(\pi a_0^3)^{1/2}} \frac{1}{(2a_0)^{3/2}} \frac{1}{\sqrt{3}\, a_0} \left(\frac{2\pi}{3}\right)^{1/2}$$

$$\int_0^\infty r^4 e^{-3r/2a_0} dr \int (-Y_1^1 + Y_1^{-1}) Y_1^{m_l} d\Omega$$

Using the relation,

$$Y_1^{-m_l} = (-1)^{m_l} Y_1^{m_l^*}$$

we get,

$$x_{f_i} = 1/6a_0^4 \frac{4!}{(3a_0/2)^5} \int (Y_1^{-1^*} - Y_1^{1^*}) Y_1^{m_l} d\Omega$$

$$= 4\left(\frac{2}{3}\right)^5 a_0 (\delta_{-1m_l} - \delta_{1m_l})$$

$$= -4\left(\frac{2}{3}\right)^5 a_0 \; for\; m_l = 1$$

$$= 4\left(\frac{2}{5}\right)^5 a_0 \text{ for } m_l = -1$$

$$= 0 \text{ for } m_l = 0$$

Similarly,

$$y_{f_i} = \int \varphi_f^* y \varphi_i d\tau$$

$$= \frac{i}{6a_0^4} \int_0^\infty r^4 e^{-3r/2a_0} dr \int (Y_1^1 + Y_1^{-1}) Y_1^{m_l} d\Omega$$

$$= -4i\left(\frac{2}{3}\right)^5 a_0 \int (-Y_1^{-1^*} - Y_1^{1^*}) Y_1^{m_l} d\Omega$$

$$= -4i\left(\frac{2}{3}\right)^5 a_0 (\delta_{-1\, m_l} + \delta_l m_l)$$

$$= -4i\left(\frac{2}{5}\right)^5 a_0 \; for\; m_l = \pm 1$$

$$= 0 \text{ for } m_l = 0$$

and
$$z_{f_i} = \int \varphi_f^* z \varphi_i d\tau$$
$$= 4\sqrt{2}\left(\frac{2}{3}\right)^5 a_0 \int Y_1^0 Y_1^{m_l} d\Omega$$
$$= 4\sqrt{2}\left(\frac{2}{3}\right)^5 a_0 \delta_0 \delta_0 me$$
$$= 0 \text{ for } m_l = \pm 1$$
$$= 4\sqrt{2}\left(\frac{2}{3}\right)^5 \text{ for } m_l = 0$$

Now rate of spontaneous transition from state $i \to f$,

$$A_{i\to f} = \frac{4w_{if}^3}{3\hbar c^3} |M_{fi}|^2$$
$$= \frac{4w_{if}^3}{3\hbar c^3} \left|\int \varphi_f^* er\, \varphi_i d\tau\right|^2$$
$$= \frac{4w_{21}^3 e^2}{3\hbar c^3} |\mathbf{r}_{fi}|^2$$
$$= \frac{4w_{21}^3 e^2}{3\hbar c^3} (|x_{fi}|^2 + |y_{fi}|^2 + |z_{fi}|^2)$$

in which,

$$w_{21} = \frac{(E_2^{(0)} - E_1^{(0)})}{\hbar} = \frac{\left(-\frac{1}{4}+1\right)}{\hbar} \frac{e^2}{2a_0} = \frac{3e^2}{8\hbar a_0}$$

Therefore,

$$A_{i\to f} = A_{2\to l} = \left(\frac{4}{9}\alpha\right)^4 \frac{c}{a_0} \simeq 6 \times 10^8 \text{ sec}^{-1}$$

for $m_l = 0, +1$ and for $m_l = -1$, that is, $A_{2\to 1}$ is independent of m_l value. $\alpha = \frac{e^2}{\hbar c} \simeq \frac{1}{137}$ is known as the fine structure constant.

The coefficient $A_{2\to 1}$ gives the probability per unit time for spontaneous emission from state 2 to state 1. Thus if there are N_2 atoms in state 2 and if $-dN_2$ represents the number of atoms undergoing spontaneous transitions in time dt, then

$$dN_2 = -A_{2\to 1} N_2\, dt$$

or

$$N_2(t) = N_2(0)\, e^{-A_{21}t}$$

Thus the mean life time of the state is given by

$$\tau = \frac{1}{A_{2\to 1}} \simeq 1.6 \times 10^{-9} \text{ sec.}$$

13.3 Since the intensity of the emitted radiation is proportional to the energy emitted per second, the required ratio is,

$$\frac{I_1}{I_2} = \frac{\hbar w_{21} A_{2\to 1}}{\hbar w_{31} A_{3\to 1}}$$

where

$$\hbar w_{21} = E_2^{(0)} - E_1^{(0)} = \left(-\frac{1}{4}+1\right)\frac{e^2}{2a_0} = \frac{3e^2}{8a_0}$$

$$\hbar w_{31} = E_3^{(0)} - E_1^{(0)} = \left(-\frac{1}{9}+1\right)\frac{e^2}{2a_0} = \frac{4e^2}{9a_0}$$

and

$$A_{2\to 1} = \left(\frac{4}{9}\alpha\right)^4 \frac{c}{a_0} \text{ form the last problem,}$$

$$A_{3\to 1} = \frac{4}{3}\alpha\frac{w_{31}^3}{c^2}\left|\int \varphi_{1s}^* \mathbf{r}\varphi_{3p}\, d\tau\right|^2$$

$$= \frac{4}{3}\alpha\frac{w_{31}^3}{c^2}\left(|x_{fi}|^2 + |y_{fi}|^2 + |z_{fi}|^2\right)$$

In which $\alpha = \frac{e^2}{\hbar c}$ is the fine structure constant.

Now

$$\varphi_i = \varphi_{3p} = R_{31}(r)\, Y_1^{m_l}(\theta, \phi)$$

$$= \left(\frac{2}{3a_0}\right)^{3/2} \frac{2}{9a_0}\left(r - \frac{r^2}{6a_0}\right) e^{-r/3a_0}\, Y_1^{m_l},$$

$$\varphi_f = \varphi_{1s} = (\pi a_0^3)^{-1/2}\, e^{-r/a_0}$$

and

$$x = \left(\frac{2\pi}{3}\right)^{1/2} r(-Y_1^1 + Y_1^{-1})$$

$$y = i\left(\frac{2\pi}{3}\right)^{1/2} r(Y_1^1 + Y_1^{-1})$$

$$z = \left(\frac{4\pi}{3}\right)^{1/2} rY_1^0$$

Therefore

$$x_{31} = \int \varphi_f^* \, x \, \varphi_i d\tau$$

$$= \frac{1}{a_0^4}\frac{4\sqrt{2}}{3^4}\int_0^\infty \left(r^4 - \frac{r^5}{6a_0}\right)e^{-4r/3a_0}dr\int(-Y_1^1 + Y_1^{-1})Y_1^{m_l}d\Omega$$

Now using the relation, $Y_1^{m_l} = (-1)^{m_l}\,Y_1^{m_l^*}$, one gets

$$x_{31} = \frac{1}{a_0^4}\frac{4\sqrt{2}}{3^4}\left[\frac{4!}{(4/3a_0)^5} - \frac{5!}{6a_0(4/3a_0)^6}\right]\int (Y_1^{-1^*} - Y_1^{1^*})\,Y_1^{m_l}d\Omega$$

$$= \frac{1}{a_0^4}\frac{8}{3^4}\frac{3^5}{4^5}a_0^5(24-15)(\delta_{-1m_l} - \delta_{1m_l})$$

$$= -8a_0\frac{3^3}{4^5}(\delta_{-1m_l} - \delta 1_{m_l})$$

Similarly

$$y_{31} = -8a_0 i\frac{3^3}{4^5}(\delta_{1m_l} + \delta_{-1m_l})$$

and

$$z_{31} = -8\sqrt{2}\,a_0\frac{3^3}{4^5}\delta_{0m_l}$$

Thus,

$$A_{3\to 1} = 128a_0^2\frac{3^6}{4^{10}}\frac{4}{3}\alpha\frac{w_{31}^3}{c^2}$$

$$= 128a_0^2\frac{3^6}{4^{10}}\frac{4}{3}\alpha\frac{\left(\frac{4}{9}\frac{e^2}{a_0}\right)^3}{\hbar^3c^2}$$

and hence on simplification,

$$\frac{I_1}{I_2} = \frac{4^4}{3^4} = 3.2$$

❑❑❑

Chapter 14

QUANTUM THEORY OF SCATTERING

14.1 INTRODUCTION

So far we have applied quantum mechanics to the bound state problems ($E < 0$ case). For example, the problem of hydrogen atom which consists of two particles, an electron and a proton, separated by a distance r and bound together by a Coulombian force $-\frac{e^2}{r^2}$ which extends out to $r = \infty$. The mutual potential energy of the two particles is, therefore $-\frac{e^2}{r^2}$. When we substitute this potential energy into the Schrödinger equation and apply the admissibility conditions on the wave function, we find that the energy levels form a discrete set.

Now we want to apply quantum mechanics to scattering problems or problems involving collision between atomic or nuclear particles. In this case, the total energy of the system is positive and continuous (not discrete) or each positive value of energy is an eigen value of the Schrödinger equation. This is specified in advance. We substitute both the specified total energy of the system and the potential energy function (the potential energy describing the interaction between the colliding particles) into the Schrödinger equation and determine the behaviour of the wave function of the two particles (at large separations, why so would be clear soon) in terms of energy.

In our study of scattering problem,

(*i*) We shall confine ourselves to the case of elastic scattering (which does not affect the internal structure of the colliding particles, also momentum before the scattering is equal to the momentum after the scattering) of incident particles by the target particles.

(*ii*) We shall assume that the incident and target particles have no spin, of course for the sake of simplicity.

(*iii*) We shall assume that the target is thin enough so that we can neglect the multiple scattering processes, that is, the processes during which a particular incident particle is scattered several times before leaving the target.

(*iv*) We shall also assume that the interaction between the incident and target particles can be described by a potential energy $V(\mathbf{r}_1 - \mathbf{r}_2)$, which depends only on the relative position, $\mathbf{r} = \mathbf{r} - \mathbf{r}_2$ of the particles. In the centre of mass reference frame, the problem then reduces to the study of the scattering of a single particle (of mass μ, called reduced mass of the system related to the masses m_1 and m_2 of the two particles by the formula $\mu = \frac{m_1 m_2}{(m_1 + m_2)}$) by the potential $V(r)$.

Before we proceed further, let us mention some of the scattering experiments which provide the following useful information,

(*i*) Rutherford's experiment on the scattering of α-particles by nuclei provided the experimental basis for the nuclear atom model that eventually led to the Bohr atom model and finally to quantum mechanics.

(*ii*) Scattering of nucleons of various energies from a nucleus gives information about the strength and range of nuclear forces also about structure of nucleus.

(*iii*) Scattering of high energy electrons from nuclei and as well as from nucleons has helped in determining the charge distribution in nucleus and even in a nucleon.

(*iv*) Scattering of neutrons from a nucleus reveals magnetic properties of the nucleus.

14.2 SCATTERING BY SPHERICALLY SYMMETRIC POTENTIAL

Spherically symmetric (or central) potential means, the potential energy depends only on the magnitude of the distance between the two particles, that is, $V = V(|\mathbf{r}_1 - \mathbf{r}_2|) = V(|\mathbf{r}|) = V(\mathbf{r})$. In that case, the time independent Schrödinger equation for a particle of mass $\mu \left(= \frac{m_1 m_2}{(m_1 + m_2)}\right)$, called the reduced mass. m_1 and m_2 are masses of the colliding particles) moving in a field described by the potential energy function V(r), is given by

$$\nabla^2 \varphi(\mathbf{r}) + \frac{2\mu}{h^2}[E - V(r)]\,\varphi(\mathbf{r}) = 0 \quad \text{... (1)}$$

where E represents the energy in the centre of mass system.

At large distances ($r \to \infty$) from the scattering centre (or target, the position of which is taken at the origin of the coordinate system which infact is the centre of mass of the two particle system) the form of the wave function φ(**r**), that is the solution of eqn. (1) should be such that, it must consist of part φ_{inc} corresponding to the parallel beam of incident particles and another part φ_{sc} representing the scattered particles moving radially outward from the scattering centre, that is

$$\varphi(\vec{r}) = \varphi(r, \theta, \phi) \xrightarrow[r \to \infty]{} \varphi_{inc} + \varphi_{sc} \quad \text{... (2)}$$

Called asymptotic form of the solution of Schrödinger eqn. (1).

For $r \to \infty$, the incident wave is described by $\varphi_{inc} = Ae^{ikz}$ (a plane wave) where we have assumed that the wave travels in the positive z-direction (If e^{ikz} is mutiplied by the time factor $e^{-i\omega t}$, it is seen that the resulting term represents a plane wave propagating from left to right along the z–axis). According to the interpretation of the unnormalizable wave functions, $|\varphi_{inc}|^2$ is to be understood as the number of incident particles per unit volume. If we normalize the wave function φ_{inc} such that the density of particles in the incident beam is one per unit volume, then $A = 1$. Thus

$$\varphi_{inc} = e^{ikz}, \qquad \text{... (3)}$$

Now since the scattering is assumed to be elastic, φ_{sc} must have the same propagation constant k (or the scattered particles must have the same momentum $\hbar k$ as the incident ones) but it must be a spherical wave as these particles move radially outward in all the directions, of course at large distances from the target. The direction of any scattered particle is indicated by polar angles (θ, ϕ) with the z-axis in the direction of incident beam taken as the polar axis (see Fig. 14.1). Then θ, called the angle of scattering is the angle between scattered and incident directions. These two directions together define the plane of scattering. The azimuthal angle ϕ specifies the orientation of this plane with respect to some reference plane containing the z-axis.

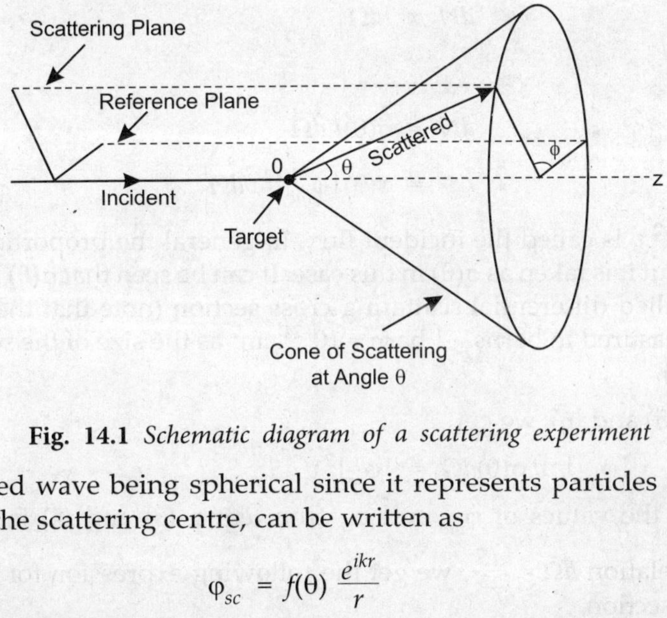

Fig. 14.1 *Schematic diagram of a scattering experiment*

The scattered wave being spherical since it represents particles moving radially outward from the scattering centre, can be written as

$$\varphi_{sc} = f(\theta)\,\frac{e^{ikr}}{r} \qquad \text{... (4)}$$

as in the case of a spherical wave **k** and **r** are in the same direction, hence $\mathbf{k} \cdot \mathbf{r} = kr$. $f(\theta)$ is called the scattering amptitude (why is it called so would be soon clear) and would in general be function of θ and ϕ, but for the present case, $f = f(\theta)$ as the potential is central (or has spherical symmetry). The number of particles which enter the collector or detector placed (far from the region under the influence of the potential) at (r, θ, ϕ) per unit time would be given by (see Fig. 14.2),

$$dN = |\varphi_{sc}|^2 v\, dS \qquad \text{... (5)}$$

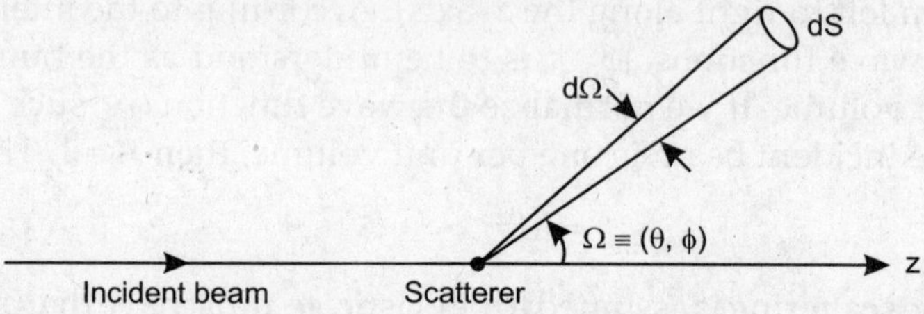

Fig. 14.2 *Scattering experiment showing particle scattered into the solid angle dΩ and collected by the detector*

Where v is the particle velocity, dS face area of the detector and $|\varphi_{sc}|^2 v$ represents flux of the scattered particles (number of particles per unit time per unit area).

The above number dN would be proportional to the intensity I (number of particles crossing unit area normal to the beam per unit time) of the incident beam and to the solid angle $d\Omega = \frac{dS}{r^2} = \sin\theta\, d\theta\, d\phi$ subtended by the collector at the centre of force. Thus

$$dN \propto I d\Omega$$

or

$$dN = \sigma(\theta) I\, d\Omega$$

$$= \sigma(\theta)|\varphi_{inc}|^2 v d\Omega \qquad \text{... (6)}$$

Where $|\varphi_{inc}|^2 v$ is called the incident flux. In general the proportionality factor is taken as $\sigma(\theta, \phi)$ but it is taken as $\sigma(\theta)$ in this case. It can be seen that $\sigma(\theta)$ has dimensions of area and is called differential scattering cross section (note that the cross–sections are generally measured in barns – 1 barn = 10^{-24} cm^2 as the size of the scattering centre is about 10^{-12}cm).

From eqns. (5) and (6), we get

$$|\varphi_{inc}|^2 v\, \sigma(\theta) d\Omega = |\varphi_{sc}|^2 v\, dS \qquad \text{... (7)}$$

Substituting the values of φ_{inc} and φ_{sc} from eqns. (3) and (4) into eqn. (7) and then using the relation $d\Omega = \frac{dS}{r^2}$, we get the following expression for the differential scattering cross-section,

$$\sigma(\theta) = \frac{d\sigma}{d\Omega} = |f(\theta)|^2 \qquad \text{... (8)}$$

In view of this relation, $f(\theta)$ is called the scattering amplitude. The total number of particles scattered in unit time can be obtained from eqn. (6) by integrating it over all directions and hence

$$\int dN = \int \sigma(\theta)\, I d\Omega$$

or

$$N_{scatt} = I \int \sigma(\theta)\, d\Omega = I\,\sigma \quad \text{... (9)}$$

where σ is called the total scattering cross-section and is given by

$$\sigma = \int \sigma(\theta)\, d\Omega = \int_0^{2\pi}\int_0^{\pi} \sigma(\theta) \sin\theta\, d\theta\, d\phi = 2\pi \int_0^{\pi} \sigma(\theta) \sin\theta d\theta \quad \text{... (9a)}$$

It is clear now that in order to calculate the scattering cross-section what we have to do is to determine the wave function φ (*r*) which has the following asymptotic form demanded by eqns. (2) to (4):

$$\varphi(\mathbf{r}) \xrightarrow[r \to \infty]{} e^{ikz} + f(\theta) \frac{e^{ikr}}{r} \quad \text{... (2a)}$$

The amplitude of scattered wave is then $f(\theta)$. $\sigma(\theta)$ and σ can then be determined from eqns. (8) and (9*a*).

14.3 METHOD OF PARTIAL WAVES

We know that in case the potential is central, eqn. (1), the Schrödinger equation can be separated into radial and angular parts. That is, the solution can be written in the form, $R_l(r)\, Y_l^{m_l}(\theta, \phi) = R_l(r)\, P_l(\cos\theta)\, e^{im_l\phi}$ where $Y_l^{m_l}(\theta, \phi)$ is called the spherical harmonics and $P_l(\cos\theta)$ Legendre polynomial of degree *l* and where $R_l(r)$, the radial wave functions satisfy the equation,

$$\frac{1}{r^2}\frac{d}{dr}\left(\frac{r^2 dR_l(r)}{dr}\right) + \frac{2\mu}{\hbar^2}\left[E - V(r) - \frac{l(l+1)\hbar^2}{2\mu r^2}\right] R_l(r) = 0 \quad \text{... (10)}$$

The general solution of equ, (1) can thus be written as linear combination of such solutions, that is

$$\varphi(\mathbf{r}) = \sum_{l=0}^{\infty} \sum_{m_l=-l}^{+l} R_l(r)\, P_l(\cos\theta)\, e^{im_l\phi} \quad \text{... (11)}$$

Since in the scattering problem, we have already assumed that the particle is travelling in the *z*–direction initially, the *z*–component of its angular momentum, $\mathbf{L}_z = (\mathbf{r} \times \mathbf{p})_z = m_l\hbar$, is zero (which implies $m_l = 0$). This continues to be zero as the central potential does not disturb the angular momentum (or as the angular momentum is conserved in such a field). In other words, this simply means that $m_l = 0$. Hence the general solution (11) becomes,

$$\varphi(\mathbf{r}) = \sum_{l=0}^{\infty} R_l(r)\, P_l(\cos\theta) \quad \text{... (12)}$$

The term corresponding to a particular l in the above series is called the lth partial wave. For different values of l (= 0, 1, 2, 3, . . .), we get $S, P, D, F, \ldots$ waves.

Before we proceed further, let us first consider the case in which $V(r) = 0$. that is, the free particle case. In the said case, eqn. (10) takes the form,

$$\frac{1}{r^2}\frac{d}{dr}\left(\frac{r^2 dR_l(r)}{dr}\right)+\left[k^2-\frac{l(l+1)}{r^2}\right]R_l(r)=0,\ k^2=\frac{2\mu E}{\hbar^2} \quad \ldots (13)$$

The solution of the above equation which resembles with the spherical Bessel's equation (for $k = 1$, the above equation exactly becomes the spherical Bessel's equation) is given by

$$R_l(r) = a_l j_l(kr) + b_l \eta_l(kr) \quad \ldots (14)$$

where $j_l(kr)$ and $\eta_l(kr)$ are called spherical Bessel and spherical—Neumann Functions, respectively and are given by

$$j_l(kr) = J_{l+1/2}(kr)\left(\frac{\pi}{2kr}\right)^{1/2} \quad \ldots (15a)$$

And

$$\eta_l(kr) = (-1)^{l+1} J_{-(l+1/2)}(kr)\left(\frac{\pi}{2kr}\right)^{1/2} \quad \ldots (15b)$$

Where J is ordinary Bessel function of half integer order.

Now,

$$j_l(kr) \xrightarrow[kr \to 0]{} \frac{(kr)^l}{(2l+1)!!} \quad \ldots (16a)$$

and

$$n_l(kr) \xrightarrow[kr \to 0]{} \frac{-(2l+1)!!}{(2l+1)}(kr)^{-l-1} \quad \ldots (16b)$$

In which

$$(2l+1)!! = 1.3.5 \ldots (2l+1)$$

We see from eqn. (16b), that $\eta_l(kr)$ is not finite at $r = 0$. Since the wave function has to be finite for all values of r, η_l is to be discarded. Hence from eqn. (14),

$$R_l(r) = \mathbf{a}_l j_l(kr) \quad \ldots (17)$$

The complete wave function in the case of free particle will, therefore be given by

$$\varphi(\mathbf{r}) = \sum_{l=0}^{\infty} \mathbf{a}_l j_l(kr) P_l(\cos\theta) \quad \ldots (18)$$

Now a plane wave (or a free particle moving in the positive z–direction) is represented by e^{ikz} (solution of the wave equation for a free particle). Therefore,

$$e^{ikz} = e^{ikr\cos\theta} = \sum_{l=0}^{\infty} a_l j_l(kr) P_l(\cos\theta) \quad \ldots (19)$$

Multiplying both sides of the above equation by $P_{l'}(\cos\theta)\sin\theta$ and then integrating the resulting expression from 0–π, we find

$$\int_0^{\pi} e^{ikr\cos\theta} P_{l'}(\cos\theta)\sin\theta\, d\theta = \sum_{l=0}^{\infty} a_l j_l(kr)\int_0^{\pi} P_l(\cos\theta)\, P_l(\cos\theta)\sin\theta d\theta$$

Or

$$\int_{-1}^{+1} e^{ikr}\cos\theta\, P_{l'}(\cos\theta)\, d(\cos\theta) = \sum_{l=0}^{\infty} a_l j_l(kr)\int_{-1}^{+1} P_l(\cos\theta)\, P_{l'}(\cos\theta)\, d(\cos\theta)$$

After substituting the value of integral on the L.H.S. of the above equation and after using the orthonormality relation for the Legendre polynomials on the R.H.S., we get

$$2i^{l'} j_l(kr) = \sum_{l=0}^{\infty} a_l\, j_l(kr)\frac{2}{(2l+1)}\delta l'$$

$$= a_{l'}\, j_{l'}(kr)\frac{2}{(2l'+1)}$$

And hence,

$$a_l = i^l(2l+1) \qquad \text{... (20)}$$

Substituting the value of a_l from the above equation into eqn. (9), the plane wave (or free particle) expansion becomes

$$e^{ikz} = e^{ikr\cos\theta} = \sum_{l=0}^{\infty} i^l(2l+1)\, j_l(kr)\, P_l(\cos\theta) \qquad \text{... (21)}$$

The above result is known as Bauer's formula.

Now in the presence of the scattering potential $V(r)$, the radial wave eqn. (10) can be rewritten as

$$\frac{1}{r^2}\frac{d}{dr}\left(\frac{r^2 dR_l(r)}{dr}\right)+\left[k^2-U(r)-\frac{l(l+1)}{r^2}\right]R_l(r)=0, \qquad \text{... (22)}$$

$$k^2 = \frac{2\mu E}{\hbar^2},\ U(r)=\frac{2\mu V(r)}{\hbar^2}.$$

Let us write down the solution of the above equation for r large. In most cases of practical interest, the scattering potential $V(r)$ is such that it vanishes for $r > R$, where R is small and would finally represent the range of the potential (Note that the Coulomb potential vanishes at r — ∞ and hence is to be dealt with separately). Thus the term, $l(l+1)$ should not be neglected in eqn. (22) while considering the solution for r large. Its solution, therefore for r large would be given by

$$R_l(k) \xrightarrow[r\to\infty]{} A_l\, j_l(kr) + B_l\, \eta_l(kr) \qquad \text{... (23)}$$

where

$$j_l(kr) \xrightarrow[r \to \infty]{} \frac{\sin\left(kr - \frac{l\pi}{2}\right)}{kr} \quad \text{... (24a)}$$

and

$$n_l(kr) \xrightarrow[r \to \infty]{} -\frac{\cos\left(kr - \frac{l\pi}{2}\right)}{kr} \quad \text{... (24b)}$$

The reason for excluding the spherical Neumann function is not valid here since the solution now is restricted to regions outside the range of potential which does not include $r = 0$. Accordingly, the asymptotic form of $R_l(r)$ or

$$R_l(r) \xrightarrow[r \to \infty]{} \frac{A_l}{kr} \sin\left(kr - \frac{l\pi}{2}\right) - \frac{B_l}{kr} \cos\left(kr - \frac{l\pi}{2}\right)$$

which can be written as

$$R_l(r) \xrightarrow[r \to \infty]{} \frac{C_l}{kr} \sin\left(kr - \frac{l\pi}{2} + \delta_l\right) \quad \text{... (25)}$$

Where C_l and δ_l are related to A_t and B_l by

$$\tan \delta_l = -\frac{B_l}{A_l} \text{ and } A_l = C_l \cos \delta_l, -B_l = C_l \sin \delta_j \quad \text{... (26a)}$$

and hence

$$C_l^2 = A_l^2 + B_l^2 \quad \text{... (26b)}$$

Thus the asymptotic form of the general solution of eqn. (1), using eqns. (12) and (25) would be given by

$$\varphi(\mathbf{r}) \xrightarrow[r \to \infty]{} \sum_{l=0}^{\infty} \frac{C_l}{kr} \sin\left(kr - \frac{l\pi}{2} + \delta_l\right) P_l(\cos\theta) \quad \text{... (27)}$$

We have now to choose the constants C_l in such a way that the above solution agrees with eqn. (2*a*).

To do this, we need the free particle expansion (21):

$$e^{ikz} = \sum_{l=0}^{\infty} i^l (2l+1)\, j_l(kr)\, P_l(\cos\theta)$$

$$\xrightarrow[r \to \infty]{} \sum_{l=0}^{\infty} i^l (2l+1) \frac{\sin(kr - l\pi/2)}{kr} P_l(\cos\theta) \quad \text{... (28)}$$

where we have put the asymptotic value of $j_l(kr)$.

Substituting above into eqn. (2*a*) and then equating the resulting equation to eqn. (27), we obtain

$$\sum_{l=0}^{\infty} i^l (2l+1) \frac{\sin (kr - l\pi/2)}{kr} P_l (\cos \theta) + f(\theta) \frac{e^{ikr}}{r}$$

$$= \sum_{l=0}^{\infty} C_l \frac{\sin (kr - l\pi_2 + \delta_l)}{kr} P_l (\cos \theta)$$

Further using the identity, $\sin x = \frac{(e^{ix} - e^{-ix})}{2i}$, we get from above,

$$e^{ikr} \left[2ik\, f(\theta) + \sum_{l=0}^{\infty} i^l (2l+1)\, e^{-il\pi/2} P_l(\cos\theta) \right] - e^{-ikr} \sum_{l=0}^{\infty} i^l (2l+1)\, e^{il\pi/2} P_l (\cos \theta)$$

$$= e^{ikr} \sum_{l=0}^{\infty} C_l\, e^{-il\pi/2} e^{i\delta_l} P_l (\cos\theta) - e^{-ikr} \sum_{l=0}^{\infty} C_l\, e^{il\pi/2} e^{-i\delta_l} P_l (\cos \theta) \quad \text{... (29)}$$

Comparing the coefficients of e^{-ikr} on both sides of the above equation, the constants C_l are determined to be

$$C_l = i^l (2l + 1)\, e^{i\delta_l} \quad \text{... (30)}$$

On substituting the value of C_l from above into eqn. (29) and then comparing the coefficients of e^{ikr} on both sides of the resulting equation, we get

$$2ikf(\theta) + \sum_{l=0}^{\infty} i^l (2l+1)\, e^{-il\pi/2} P_l (\cos \theta)$$

$$= \sum_{l=0}^{\infty} i^l (2l+1)\, e^{2i\delta_l} e^{-il\pi/2} P_l(\cos\theta)$$

which on simplification leads to

$$f(\theta) = \frac{1}{2ik} \sum_{l=0}^{\infty} i^l (2l+1)\, e^{-il\pi/2} (e^{2i\delta_l} - 1) P_l (\cos \theta) \quad \text{... (31)}$$

Making use of the fact that

$$i^l = (\cos \pi/2 + i \sin \pi/2)^l = (e^{i\pi/2})^l = e^{il\pi/2}$$

and further substituting this value into eqn. (31), one gets

$$f(\theta) = \frac{1}{2ik} \sum_{l=0}^{\infty} (2l+1)\, (e^{2i\delta_l} - 1) P_l(\cos\theta)$$

$$= \frac{1}{2ik} \sum_{i=0}^{\infty} (2l+1)\, e^{i\delta_l} (e^{i\delta_l} - e^{-i\delta_l}) P_l (\cos \theta)$$

$$= \frac{1}{k} \sum_{l=0}^{\infty} (2l+1)\, e^{i\delta_l} \sin \delta_l\, P_l(\cos\theta) \quad \text{... (32)}$$

which gives $f(\theta)$, the scattering amplitude as sum of contributions from partial waves with $l = 0$ to $l = \infty$. It also shows that $f(\theta)$ is independent of the azimuthal angle ϕ in this case. The above formula indeed means the solution of the scattering problem.

Using the above equation and eqn. (8), the differential scattering cross-section is given by

$$\sigma(\theta) = \frac{d\sigma}{d\Omega} = |f(\theta)|^2 = \frac{1}{k^2}\left|\sum_{l=0}^{\infty}(2l+1)e^{i\delta_l}\sin\delta_l P_l(\cos\theta)\right|^2$$

$$= \frac{1}{k^2}\sum_{l=0}^{\infty}\sum_{l'=0}^{\infty}(2l+1)\ (2l'+1)\ e^{i(\delta_l-\delta_{l'})}$$

$$\sin\delta_l \sin\delta_{l'} \times P_l(\cos\theta)\ P_{l'}(\cos\theta) \qquad \text{... (33)}$$

The total scattering cross-section using eqns. (9) and (33) is therefore given by

$$\sigma = \int \frac{d\sigma}{d\Omega}\, d\Omega = 2\pi \int_0^{\pi} \sigma(\theta)\sin\theta d\theta$$

$$= \frac{2\pi}{k^2}\int_0^{\pi}\sum_{l=0}^{\infty}\sum_{i=0}^{\infty}(2l+1)\ (2l'+1)e^{i(\delta_l-\delta_{l'})}$$

$$\sin\delta_l \sin\delta_{l'} \times P_l(\cos\theta)\ P_{l'}(\cos\theta)\sin\theta d\theta$$

Further using the orthonormality relation,

$$\int_0^{\pi} P_l(\cos\theta)\ P_{l'}(\cos\theta)\sin\theta d\theta = \frac{2}{(2l'+1)}\ \delta ll'$$

For the Legendre polynomials, $P_l(\cos\theta)$, one gets

$$\sigma = \frac{2\pi}{k^2}\sum_{l=0}^{\infty}(2l+1)^2\sin^2\delta_l\ \frac{2}{(2l+1)}$$

$$= \frac{4\pi}{k^2}\sum_{l=0}^{\infty}(2l+1)\sin^2\delta_l \qquad \text{... (34)}$$

It may be noted here that the cross terms (those with $l \neq l'$) representing interference between different partial waves which are present in the expression (33) for $\sigma(\theta)$, have vanished in the case of σ because of the orthonormality relation of Legendre polynomials. That means as far as the total scattering cross-section is concerned the different partial waves contribute independently.

It is necessary to point out here that the method developed above (also called partial wave analysis) is applicable in the cases in which the potential is central and is of finite range. We will see later on that the method is really useful when the energy of incident particles is rather low.

14.4 OPTICAL THEOREM

The forward scattering amplitude, f(at $\theta = 0$) from eqn. (32) is given by

$$f(0) = \frac{1}{k}\sum_{l=0}^{\infty}(2l+1)e^{i\delta_l}\sin\delta_l\, P_l\,(1)$$

Using the fact that, P_l (1) = 1 for all l values, one gets

$$f(0) = \frac{1}{k}\sum_{l=0}^{\infty}(2l+1)\sin\delta_l\,(\cos\delta_l + i\sin\delta_l)$$

From which

$$I_m f(0) = \frac{1}{k}\sum_{l=0}^{\infty}(2l+1)\,\sin^2\delta_l \quad \text{... (35)}$$

where I_m represents the imaginary part.

Comparing eqns. (34) and (35), one gets the following relation between the total scattering cross–section and the imaginary part of the amplitude of scattering through an angle zero,

$$\sigma = \frac{4\pi}{k} I_m f\,(0) \quad \text{... (36)}$$

The above relation (which is a special case of the more general relation involving σ_{abs}) is known as the optical theorem. The said relation is also sometimes referred to as the Bohr-Pierls—Placzek formula.

14.5 PHYSICAL MEANING OF δ_l

We have seen that the asymptotic solution of the radial part of the Schrödinger equation in the presence of scattering potential $V(r)$, is given by

$$R_l(r) \xrightarrow[r\to\infty]{} \frac{C_l}{kr}\sin\left(kr - \frac{l\pi}{2} + \delta_l\right) \quad \text{... (25)}$$

whereas the solution of the radial part of the free particle Schrödinger equation (that is for $V(r) = 0$) is given by

$$R_l(r) = a_l j_l\,(kr) \quad \text{... (17)}$$

which in the asymptotic region takes the form,

$$R_l\,(r) \xrightarrow[r\to\infty]{} \frac{a_l}{kr}\sin\left(kr - \frac{l\pi}{2}\right) \quad \text{... (37)}$$

By comparing eqns. (25) and (37), we conclude that δ_l is the difference in phase and is, therefore called phase shift of the lth partial wave. Also,

$$\delta_l = 0 \text{ for } V(r) = 0.$$

The total scattering cross = section given by eqn. (34) can be written as

$$\sigma = \sum_{l=0}^{\infty}\sigma_l \quad \text{... (38)}$$

where

$$\sigma_l = \frac{4\pi}{k^2}(2\,l+1)\,\sin^2\delta_l \quad \text{... (38}a\text{)}$$

is called the lth partial wave cross-section for scattering of particles with given orbital angular momentum l. Note that the phase shifts completely determine the scattering and if the scattering potential is such that, $\delta_l = s\pi$ ($s = 0, 1, 2, \ldots$) for all l, the total scattering cross-section, σ vanishes. This will be maximum for values of $\delta_l = \pm\frac{\pi}{2}, \pm\frac{3\pi}{2} \ldots$.

The angular momentum L of the particle in the incident beam is related to the linear momentum p of the particle and the impact parameter b (defined as the perpendicular distance between the scattering centre and the incident direction)—also called distance of closest approach, by the relation

$$L = pb$$

or $$\sqrt{l(l+1)}\,\hbar = \hbar\, kb$$

or $$\sqrt{l(l+1)} = kb \quad \ldots (39)$$

Let R be the range of scattering potential $V(r)$. This implies that the particles will only be scattered (or the waves would suffer phase shift) when

$$b < R$$

or $$kb < kR$$

Using eqn. (39), the above condition leads to

$$\sqrt{l(l+1)} < kR$$

or $$l << kR \quad \ldots (40)$$

From the above condition, it is very clear that smaller the value of kR, smaller will be the number of terms in the summations (32) or (34). For a given value of R, the number of terms in the said summations would depend on the value of k only. Since $p = \hbar k$, it follows that smaller the momentum (and therefore, the energy) of the particles in the incident beam, smaller will be the number of terms required to describe the scattering. For say $kR = 0.2$, the above condition implies, $l = 0$, that is only the S = wave contributes to the scattering.

14.6 SCATTERING LENGTH

If the energy is so low that the particles with $l = 0$ only are scattered, then from eqn. (32), we have

$$f(\theta) = \frac{1}{k}\, e^{i\delta_0} \sin \delta_0 \quad \ldots (41)$$

as $P_0 = 1$

The limiting value of energy for which the above equation holds good, is called zero energy. Thus the angular distribution of the scattered particles at zero energy, is independent of the scattering angle θ. In other words, the zero energy scattering is isotropic.

The negative of the scattering amplitude in the zero energy limit, is called (Fermi) scattering length and is denoted by a. Thus from eqn. (41),

$$a = Lt_{E\to 0}\,[-f(\theta)] = -\frac{1}{k}\,e^{i\delta_0}\sin\delta_0 \qquad \text{... (42)}$$

Therefore, the zero energy total scattering cross = section, using eqns. (41) and (42) would be given by

$$\sigma = \int_0^{2\pi}\int_0^{\pi}\left|\frac{1}{k}e^{i\delta_0}\sin\delta_0\right|^2 \sin\theta\, d\,\theta\, d\,\phi$$

$$= 4\,\pi\,a^2 \qquad \text{... (43)}$$

which can also be written as

$$\sigma = \sigma_0 = \frac{4\pi}{k^2}\sin^2\delta_0 \qquad \text{... (44)}$$

It may be pointed out here that in case the energy is higher, waves other than $l = 0$ would also contribute and hence $f(\theta)$ would no more be independent of θ but instead would be function of θ (eqn. (32)) and effectively the scattering would depend on the scattering angle.

14.7 DEPENDENCE OF δ_l ON V(r)

It is obvious from the section 14.5 that the phase shift δ_l depends upon the scattering potential $V(r)$ and of course $\delta_l = 0$ for $V(r) = 0$. Let us now try to find its explicit dependence on $V(r)$.

To do so let us make the substitution, $R_l(r) = \dfrac{f_l(r)}{r}$ in the radial part of the Schrödinger equation in the absence of the scattering potential $V(r)$, that is in eqn. (11) which then reduces to

$$\frac{d^2 f_l(r)}{dr^2} + \left[k^2 - \frac{l(l+1)}{r^2}\right] f_l(r) = 0$$

Dividing above equation by k^2 and then putting $\rho = kr$, we get

$$\frac{d^2 f_l(\rho)}{d\rho^2} + \left[1 - \frac{l(l+1)}{\rho^2}\right] f_l(\rho) = 0 \qquad \text{... (45)}$$

Similarly making the substitution $R_l(r) = \dfrac{F_l(r)}{r}$ in the radial part of the Schrödinger equation in the presence of the scattering potential $V(r)$, that is in eqn. (9) which then reduces to

$$\frac{d^2 F_l(r)}{dr^2} + \left[k^2 - U(r) - \frac{l(l+1)}{r^2}\right] F_l(r) = 0 \qquad \text{... (46)}$$

Further dividing the above equation by k^2 and then putting $\rho = kr$, we get

$$\frac{d^2Fl(\rho)}{d\rho^2}+\left[1-\frac{V(r)}{E}-\frac{l(l+1)}{\rho^2}\right]F_l(\rho)=0 \qquad \text{... (46a)}$$

Multiplying eqn. (45) by $F_l(\rho)$ and eqn. (46*a*) by $f_l(\rho)$, respectively and subtracting later from the former and finally integrating the resulting equation from 0 to ∞, we obtain

$$\int_0^\infty f_l\frac{d^2F_l}{d\rho^2}\,d\rho-\int_0^\infty F_l\frac{d^2f_l}{d\rho^2}\,d\rho=\int_0^\infty\frac{V(r)}{E}F_lf_l\,d\rho$$

Integrating the L.H.S. of the above equation by parts and then cancelling the equal and opposite terms, we have

$$\left(f_l\frac{dF_l}{d\rho}-F_l\frac{df_l}{d\rho}\right)\Big|_0^\infty=\int_0^\infty\frac{V(r)}{E}F_lf_l\,d\rho \qquad \text{... (47)}$$

Now recalling that both f_l and F_l vanish at $r = 0$, we get

$$\left(f_l\frac{dF_l}{d\rho}-F_l\frac{df_l}{d\rho}\right)_{\rho\to\infty}=\int_0^\infty\frac{V(r)}{E}F_lf_l\,d\rho \qquad \text{... (47a)}$$

The above relation is clearly independent of the normalization of f_l and F_l. Putting the asymptotic forms of f_l and F_l on the L.H.S. of the above equation, we get

$$r\sin\left(kr-\frac{l\pi}{2}\right)\Big/kr\,\frac{d}{d\rho}\left\{r\sin\left(kr-\frac{l\pi}{2}+\delta_l\right)\Big/kr\right\}$$

$$-r\sin\left(kr-\frac{l\pi}{2}+\delta_l\right)\Big/kr\,\frac{d}{d\rho}\left\{r\sin\left(kr-\frac{l\pi}{2}\right)\Big/kr\right\}$$

$$=\int_0^\infty\frac{V(r)}{E}F_lf_l\,d\rho \qquad \text{... (47b)}$$

The above equation on simplification reduces to

$$\sin\left(kr-\frac{l\pi}{2}\right)\cos\left(kr-\frac{l\pi}{2}+\delta_l\right)-\sin\left(kr-\frac{l\pi}{2}+\delta_l\right)\cos\left(kr-\frac{l\pi}{2}\right)$$

$$=k^2\int_0^\infty\frac{V(r)}{E}F_lf_l\,d\rho$$

which gives

$$\sin\delta_l=-\int_0^\infty U(r)\,F_l(kr)\,f_l(kr)\,d\rho$$

$$=-k\int_0^\infty U(r)\,F_l(kr)\,f_l(kr)\,dr \qquad \text{... (48)}$$

where we have put, $E = \frac{\hbar^2 k^2}{2\mu}$.

Note that in the above equation, the potential energy function appears explicity as a factor in the integrand and also implicity in the function $F_l\,(kr)$ which is solution of eqn. (46), the radial equation in the presence of $V(r)$ or $U(r)$.

14.8 DEPENDENCE OF δ_l ON ANGULAR MOMENTUM AND ENERGY

We will now show that if l is sufficiently large, the dependence of $F_l(kr)$ upon $V(r)$ can be neglected to a good approximation and hence $F_l(kr)$ in the integrand of eqn. (48) can be replaced by $f_l(kr)$. Let us see how to achieve this.

Eqn. (46) can be written as

$$\frac{d^2 F_l}{dr^2} + \frac{2\mu}{\hbar^2}\,[E - V_{eff}\,(r)]\,F_l = 0 \qquad \text{... (49)}$$

where $V_{eff}\,(r) = V(r) + \frac{l(l+1)\hbar^2}{2\mu r^2}$ is called the effective potential energy.

Eqn. (49), thus represents a particle of energy $E = \frac{\hbar^2 k^2}{2\mu}$ moving under the influence of potential $V_{eff}(r)$ or $V_{eff}(r)$ represents the potential energy function for the equivalent one-dimensional problem.

Fig. 14.3 Shows the variation of $V_{eff}\,(r)$ and $\frac{l(l+1)\hbar^2}{2\mu r^2}$ for a typical potential function $V(r)$. R_1 represents the classical turning point and is given by

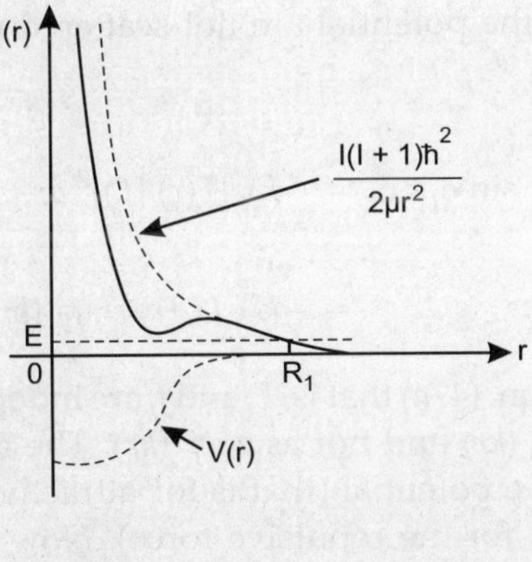

Fig. 14.3 *The effective potential energy V_{eff} (shown by solid line) is the sum of potential energy function $V(r)$ and the centrifugal term, $\frac{l(l+1)\hbar^2}{2\mu r^2}$. $r = R_1$ represents the classical turning point*

$$E = V_{eff}(R_l) = V(R_1) + \frac{l(l+1)\hbar^2}{2\mu R_1^2} \qquad \text{... (50)}$$

From the above expression, it is clear that for a given value of energy E, R_1 can be made as large as desired by choosing sufficiently large value of l. For $R_1 > R$, where R is the range of the potential, $V(r = R)$ is very small or in other words for R_l large or l large,

$$F_l = f_l \qquad \text{... (51)}$$

This effectively amounts to the condition

$$\lim_{l \to \infty} \delta l \approx 0 \qquad \text{... (52)}$$

Since for very small $V(r = R_1)$, $\sin \delta_l$ would be very small and hence δ_l.

Now, the condition $R_1 > R$ is equivalent to the requirement,

$$V(R_1) << \frac{l(l+1)\hbar^2}{2\mu R_1^2} \qquad \text{... (53)}$$

Therefore, from eqn. (50),

$$E = \frac{\hbar^2 k^2}{2\mu} = V_{eff}(R_1) = \frac{l(l+1)\hbar^2}{2\mu R_1^2}$$

or

$$l(l+1) \approx l^2 \approx k^2 R_1^2 >> k^2 R^2$$

as

$$R_1 > R. \text{ That is,}$$

$$l >> kR. \qquad \text{... (54)}$$

Thus, whenever the condition (54) is satisfied, the phase shift δ_l is small ($\delta_l \approx 0$). Classically this result corresponds to the fact that the particles with impact parameters larger than the range of the potential are not scattered (see eqn. (39)). The eqn. (48), then becomes

$$\sin \delta_l \approx \delta_l \approx -k \int_0^\infty U(\mathrm{r})\, [f_l)]^2\, dr$$

$$= -k \int_0^\infty U(r)\, r^2\, j_l^2(kr) dr \qquad \text{... (55)}$$

Note that in view of eqn. (47*a*) that is, F_l and f_l are independent of the normalization, f_l is taken as $r\, R_l(r) = r\, j_l(kr)$ and not as $a_l\, r j_l(kr)$. The above equation shows that δ_l is positive for the negative potential (that is for attractive force) and negative for the positive potential (that is for the repulsive force). Now the condition (54) would be satisfied by any non=zero value of l if k (and hence the energy) is small ($l >> kR$) would then correspond to $l = 1, 2, 3, \ldots$). Thus

$$\lim_{E \to 0} \delta_l = 0,\ l \neq 0 \qquad \text{... (56)}$$

Also when energy is small (or k small), phase shifts are given by eqn. (55). Substituting

$$j_l(kr) \xrightarrow[k \text{ or } \rho \to 0]{} \frac{\rho^l}{(2l+1)!!}, \quad (2l+1)!! = 1.3.5 \ldots (2l+1)$$

and also the value of k $\left(E = \frac{\hbar^2 k^2}{2\mu}\right)$ in eqn. (55), one gets

$$\delta_l \approx -\left(\frac{2\mu E}{\hbar^2}\right)^{1/2} \frac{2\mu}{\hbar^2} \int_0^\infty V(r)\, r^2 (kr)^{2l} \frac{1}{\{(2l+1)!!\}^2}\, dr, \text{ for } l \neq 0$$

$$= -(\alpha_l)^2 \left(\frac{2\mu}{\hbar^2}\right)^{l+3/2} E^{l+1/2} \int_0^\infty V(r)\, r^{2l+2}\, dr, \text{ for } l \neq 0, \ \alpha_l = \frac{1}{(2l+1)!!} \quad \ldots (57)$$

Thus the phase shifts vary as $E^{l+1/2}$ (or phase shifts are odd functions of k) for small values of E.

14.9 EFFECTIVE RANGE: SCATTERING BY A SQUARE WELL POTENTIAL

We have seen in the section 14.6 that the zero energy scattering can be characterized by just one parameter, the scattering length. As the energy is increased, more and more of the partial waves would begin to get scattered and in turn making scattering dependent both on energy and the scattering angle. If the energy is only slightly higher than what is termed as zero energy, the scattering would still be confined to the S-wave (and hence angle independent) but the energy dependence makes its appearance (see eqn. (57)). This energy dependence of the scattering cross-section at low energy can be described in terms of a parameter called effective range. We would illustrate this in the case of scattering by a short range attractive square well potential which is displayed in Fig. 14.4 and is represented mathematically by

$$V(r) = -V_0, \quad r < R \quad \ldots (58)$$

$$= 0, \quad r > R$$

We know that the radial part of Schrödinger equation in the presence of scattering potential $V(r)$ for $l = 0$, is given by

$$\frac{1}{r^2}\frac{d}{dr}\left(\frac{r^2 dR_0}{dr}\right) + [k^2 - U(r)]\, R_0 = 0 \quad \ldots (59a)$$

where $$k^2 = \frac{2\mu E}{\hbar^2}, \quad U(r) = \frac{2\mu V(r)}{\hbar^2} \quad \ldots (59b)$$

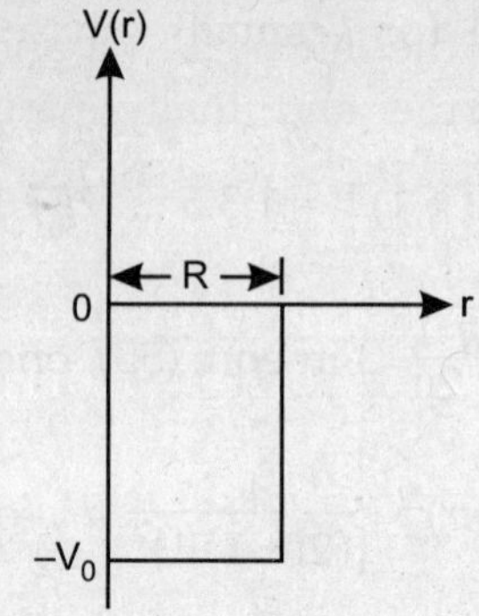

Fig. 14.4 *The square well potential*

After substituting $R_0(r) = \frac{F(r)}{r}$, the above equation takes the form,

$$\frac{d^2 F(r)}{dr^2} + k'^2 F(r) = 0, \ r < R \qquad \text{... (60a)}$$

with

$$k'^2 = k^2 + k_0^2, \ k_0^2 = \frac{2\mu V_0}{\hbar^2} \qquad \text{... (60b)}$$

Similarly putting $U(r) = 0$ in eqn. (59) and further substituting, $R_0^0(r) = \frac{f(r)}{r}$ where zero in the superscript of R represents radial solution in the absence of $V(r)$, we get

$$\frac{d^2 f(r)}{dr^2} + k^2 f(r) = 0, \quad r > R \qquad \text{... (61)}$$

Since we are interested to find the dependence of the cross-section on the energy, let us consider eqns. (60) and (61) for two different values of k, say k_1 and k_2. We have, therefore the following four equations,

$$\frac{d^2 F_1(r)}{dr^2} + (k_1^2 + k_0^2)\, F_1(r) = 0 \qquad \text{... (62a)}$$

$$\frac{d^2 F_2(r)}{dr^2} + (k_2^2 + k_0^2)\, F_2(r) = 0 \qquad \text{... (62b)}$$

and

$$\frac{d^2 f_1(r)}{dr^2} + k_1^2 f_1(r) = 0 \qquad \text{... (63a)}$$

$$\frac{d^2 f_2(r)}{dr^2} + k_2^2 f_2(r) = 0 \qquad \text{... (63b)}$$

Multiplying eqn. (62*a*) from the left by F_2 (*r*) and eqn. (62*b*) by F_1 (*r*) and then subtracting the later from the former and finally integrating the resulting expression over *r*, we get

$$\left(\frac{F_2 dF_1}{dr} - \frac{F_1 dF_2}{dr}\right)\Bigg|_0^\infty = (k_2^2 - k_1^2)\int_0^\infty F_1 F_2 \, dr \quad \text{... (64)}$$

Similarly from eqns. (63), we have

$$\left(\frac{f_2 df_1}{dr} - \frac{f_1 df_2}{dr}\right)\Bigg|_0^\infty = (k_2^2 - k_1^2)\int_0^\infty f_1 f_2 \, dr \quad \text{... (65)}$$

Further subtracting (65) from (64), one gets

$$\left(\frac{F_2 dF_1}{dr} - \frac{F_1 dF_2}{dr}\right)\Bigg|_0^\infty - \left(\frac{f_2 df_1}{dr} - \frac{f_1 df_2}{dr}\right)\Bigg|_0^\infty = (k_2^2 - k_1^2)\int_0^\infty (F_1 F_2 - f_1 f_2)\, dr \quad \text{... (66)}$$

Now $F_1(0) = F_2(0) = 0$ as $R_0(r) = \frac{F(r)}{r}$ and hence

$$\left(\frac{F_2 dF_1}{dr} - \frac{F_1 dF_2}{dr}\right)\Bigg|_{r=0} = 0$$

Also,

$$\left(\frac{F_2 dF_1}{dr} - \frac{F_1 dF_2}{dr}\right)\Bigg|_{r=\infty} = \left(\frac{f_2 df_1}{dr} - \frac{f_1 df_2}{dr}\right)\Bigg|_{r=\infty}$$

as for large values of *r*, *V* (*r*) would vanish in both the cases and effectively both the solutions *F*(*r*) and *f*(*r*) would coincide.

In the light of above, eqn. (66) reduces to

$$\left(\frac{f_2 df_1}{dr} - \frac{f_1 df_2}{dr}\right)\Bigg|_{r=0} = (k_2^2 - k_1^2)\int_0^\infty (F_1 F_2 - f_1 f_2)\, dr \quad \text{... (67)}$$

Solutions of eqns. (60) and (61) are given by (see solution 14.1)

$$F(r) = A \sin k' r, \quad r < R \quad \text{... (68a)}$$

and

$$f(r) = B \sin (kr + \delta), \quad r > R \quad \text{... (68b)}$$

Let us adopt the normalization, $f(0) = 1$

$$f(0) = 1 \quad \text{... (69)}$$

Since *f*(*r*) corresponds to the solution only outside the range of the potential, the value of *f*(*r*) at $r = 0$, that is *f*(0) which has been taken equal to 1 does not affect the value of the actual solution which is zero at $r = 0$. Thus from eqns. (68*b*) and (69), one gets

$$B = \frac{1}{\sin \delta} \quad \text{... (70)}$$

Therefore,

$$f(r) = \frac{\sin(kr + \delta)}{\sin \delta}, \; r > R \quad \text{... (71)}$$

From above equation, it follows that

$$\left.\frac{f_1(r)\, df_2}{dr}\right|_{r=0} = k_2 \cot \delta_2 \quad \text{... (72)}$$

so that

$$\left.\left(\frac{f_1 df_2}{dr} - \frac{f_2 df_1}{dr}\right)\right|_{r=0} = k_2 \cot \delta_2 - k_1 \cot \delta_1 \quad \text{... (73)}$$

Substituting above into eqn. (67), one gets

$$k_2 \cot \delta_2 - k_1 \cot \delta_1 = (k_2^2 - k_1^2) \int_0^\infty (f_1 f_2 - F_1 F_2)\, dr \quad \text{... (74)}$$

Consider two energies which differ only by an infinitesimal amount.
Then,

$$(k_2^2 - k_1^2) = \Delta(k^2)$$

and similarly,

$$k_2 \cot \delta_2 - k_1 \cot \delta_1 = \Delta(k \cot \delta)$$

Substituting these into eqn. (74) and then taking the limit, $\Delta(k^2) \to 0$, one gets

$$Lt_{\Delta(k^2)\to 0} \frac{\Delta(k \cot \delta)}{\Delta(k^2)} = \frac{d}{dk^2}(k \cot \delta) = \int_0^\infty (f^2 - F^2)\, dr \quad \text{... (75)}$$

as $f_1 = f_2 = f$ say and $F_1 = F_2 = F$ say in this situation

Integrating above over k^2 between limits 0 and k^2, one gets

$$(k \cot \delta)_k = (k \cot \delta)_{k=0} + \int_0^\infty dk^2 \int_0^\infty (f^2 - F^2)\, dr \quad \text{... (76)}$$

Now from eqn. (41) in the zero energy limit, we know that

$$f(\theta) = \frac{1}{k} e^{i\delta_0} \sin \delta_0$$

As δ_0 is small and also that in the expansion of $e^{i\delta_0}$, only the first term ($e^{i\delta_0} \simeq 1$) can be retained, the above expression can be written as

$$f(\theta) = \frac{\tan \delta_0}{k} = -a$$

where a is called the scattering length. Hence

$$(k \cot \delta)_{k=0} = -\frac{1}{a} \qquad \text{... (77)}$$

Note that the δ at $k \to 0$ is nothing but δ_0. Further making use of the above equation, the relation (76) takes the form

$$k \cot \delta_0 (k) = -\frac{1}{a} + \left[\int_0^\infty (f^2 - F^2)\, dr\right] k^2$$

$$= -\frac{1}{a} + \frac{1}{2} r_{eff} k^2 \qquad \text{... (78a)}$$

where

$$r_{eff} = 2\int_0^\infty (f^2 - F^2)\, dr \qquad \text{... (78b)}$$

which has dimension of length as $f^2 - F^2$ is dimensionless, is called the effective range of the square potential well. Also note that no doubt the energy is slightly higher than what is termed as the zero energy, the scattering would still be confined to *S*-wave or

$$k (\cot \delta)_k = k \cot \delta_0 (k)$$

The total scattering cross-section in this case would thus given by

$$\sigma = \frac{4\pi}{k^2} \sin^2 \delta_0(k) = \frac{4\pi}{k^2 \operatorname{cosec}^2 \delta_0(k)} = \frac{4\pi}{k^2 \{1 + \cot^2 \delta_0(k)\}}$$

$$= \frac{4\pi}{[k^2 + k^2 \cot^2 \delta_0(k)]} = \frac{4\pi}{\left[k^2 + \left(\frac{1}{2} r_{eff} k^2 - \frac{1}{a}\right)^2\right]} \qquad \text{... (79)}$$

where use has been made of eqn. (78*a*).

The above formula gives the energy dependence of the total *S*-wave scattering cross-section at low energies and shows that the scattering by a short range attractive potential can be described in terms of two parameters, the scattering length *a*, and the effective range, r_{eff}. Note that for $k \to 0$, expression (79) reduces to (43). Further note that the absolute value of *a* can be determined from measurements at zero energy (eqn. 43). Then the value of r_{eff} can be found by measuring the cross=section at slightly higher energies (eqn. (79)). Finally the sign of *a* (which truly speaking determines in the case of an attractive potential whether *a* bound state is possible) can be determined from eqn. (78*a*) provided the δ_0 measurement (eqn. (44)) is sufficiently accurate.

14.10 RAMSAUER—TOWNSEND EFFECT

If at some small value of *k* for which $kR \ll 1$ (which implies $l = 0$) and the scattering potential $V(r)$ be large enough such that $\delta_1 = \delta_0 = 180°$, then sin δ_0 (= sin 180°) become zero. As a result no scattering takes place (see eqn. (44)). This is the Ramsauer—

Townsend effect actually observed in the case of scattering of low energy electrons by rare gas atoms like, Argon, Kripton, Xenon and Neon. The scattering cross-section drops to extremely small value at about 0.7 eV incident or bombarding energy.

Let us try to explain the above observation. We know that the phase shift δ_l would be negative in the case of a repulsive potential ($V > 0$) which implies that the radial wave function in the presence of $V(r)$, that is, $R_l(r)$ would be pushed out in comparison with the force free wave function, $j_l(kr)$. Similarly in the case of an attractive potential ($V < 0$), the phase shift δ_l being positive, the radial wave function $R_l(r)$ would be pulled in as compared to $j_l(kr)$. These results are displayed in Fig 14.5 where $R_l(r)$ and $j_l(kr)$, the solutions in the presence of $V(r)$ and in the absence of $V(r)$, respectively are compared when they are arbitrarily made to start out at $r = 0$.

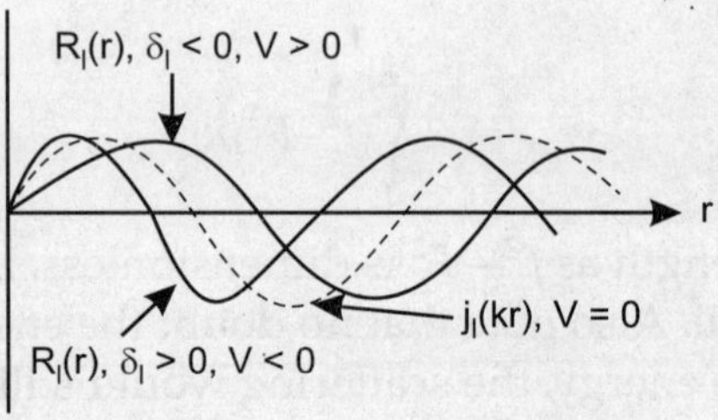

Fig. 14.5 *Effect of potential on the phase of the scattered wave*

Fig. 14.5 suggests that the attractive potential might be strong enough so that one of the partial waves is pulled in by just half a cycle and its phase shift is 180°. If $kR \ll 1$, then only $l = 0$ wave would contribute and hence the scattering cross-section would vanish.

We may note that in the case of attractive potential, $k'^2 = k^2 + k_0^2 \left(k_0^2 = \dfrac{2\mu V_0}{\hbar^2}\right)$ or $k < k'$ (see problem 14.1) whereas in the case of repulsive potential, $k'^2 = k^2 - k_0^2$ or $k > k'$. Hence in the later case waves other than $l = 0$ will also contribute and therefore $R - T$ effect cannot be observed even if $\delta_0 = 180°$.

Let us now try to find the condition for the occurrence of $R - T$ effect. From Problem 14.1,

$$\delta_0 = \tan^{-1}\left\{\frac{k}{k'}\tan k'R\right\} - kR$$

$$= \tan^{-1}\left\{\frac{k}{k'}\tan k'R\right\} - \tan^{-1}(\tan kR) \qquad \text{... (80)}$$

which is difference of two angles. Thus

$$\tan \delta_0 = \left(\frac{k}{k'}\tan k'R - \tan kR\right)\Big/\left(1 + \frac{k}{k'}\tan k'R \tan kR\right)$$

For $R - T$ effect to occur, $\delta_0 = 180°$. Therefore

$$\frac{k}{k'}\tan k'R = \tan kR \qquad \text{... (81)}$$

For $k \to 0$ in the present case, $k' = (k^2 + k_0^2)^{1/2} \to k_0$. Hence eqn. (81) leads to

$$\tan k_0 R = k_0 R, \; k_0^2 = \frac{2\mu V_0}{\hbar^2} \qquad \text{... (82)}$$

If V_0 and R are such that the above equation (the condition for $R-T$ effect to occur) is satisfied, then $\sigma_0 \to 0$. If eqn. (82) is nearly satisfied, then $\sigma_0 \to$ a small value. This thus explains the observed Ramsauer—Townsend effect.

Before we close the discussion on $R - T$ effect, let us point out that the field of the inert gas atoms (having closed shell configuration—the orbital electrons are compactly packed) decreases appreciably faster with distance than the field of any other atom, so that, to a first approximation, we can replace this field by a spherical potential well and use eqn. (14.11). (see problem 14.1) to evaluate the scattering cross-section for slow electrons. As the energy of the electrons is approximately 0.7 eV, eqn. (82) is approximately satisfied and thus $\sigma_0 \approx 0$.

14.11 RESONANCE SCATTERING

In this section, we would derive another expression for phase shifts in the case of short or finite range potentials. We would finally see that it would be more suitable than eqn. (57) for the discussion of scattering at low energies. We have already seen in section 14.3 that the solution of the radial part of the Schrödinger equation in the exterior region, that is, for $r > R$ ($U(r) = 0$) is given by

$$R_l(r) \xrightarrow[r \to \infty]{} A_l j_l(kr) + B_l n_l(kr)$$

$$= \frac{C_l}{kr} \sin\left(kr - \frac{l\pi}{2} + \delta_l\right) \qquad \text{... (25)}$$

where

$$\tan \delta_l = -\frac{B_l}{A_l} \text{ and } A_l = C_l \cos \delta_l, \; -B_l = C_l \sin \delta_l \qquad \text{... (26a)}$$

and hence

$$C_l^2 = A_l^2 + B_l^2 \qquad \text{... (26b)}$$

Suppose we do not know $R_l(r)$ in the interior region, that is, the region for which $r < R$. We of course know that it has to match with the exterior wave function at $r = R$. That is, if the logarithmic derivative

$$\frac{1}{R_l(r)} \frac{dR_l(r)}{dr} \to \gamma_l \qquad \text{... (83)}$$

as $r \to R$ from the interior region, the logarithmic derivative of the solution for the exterior region must also tend to γ_l as $r \to R$ which can be obtained from eqn. (25) and is given by

$$\gamma_l = k\frac{[A_l\, j_l'(kR) + B_l\, n_l'(kR)]}{[A_l\, j_l(kR) + B_l\, n_l(kR)]}$$

$$= k\left[j_l'(kR) + \frac{B_l}{A_l}n_l'(kR)\right]\Bigg/\left[j_l(kR) + \frac{B_l}{A_l}n_l(kR)\right]$$

$$= k\frac{[j_l'(kR) - \tan\delta_l n_l'(kR)]}{[j_l(kR) - \tan\delta_l n_l(kR)]}$$

From above,

$$\tan\delta_l = \frac{[k\, j_l'(kR) - \gamma_l\, j_l(kR)]}{[k\, n_l'(kR) - \gamma_l\, n_l(kR)]} \quad \text{... (84)}$$

which is the expression for the phase shift δ_l in terms of γ_l. Writing,

$$i\tan\delta_l = \frac{i\sin\delta_l}{\cos\delta_l}\left(= \frac{a}{b} = \{ik\, j_l'(kR) - i\gamma_l\, j_l(kR)\}/\{kn_l'(kR) - \gamma_l n_l(kR)\} = \frac{c}{d}\ say\right)$$

and using the relationship,

$$\frac{(a+b)}{(a-b)} = \frac{(c+d)}{(c-d)}$$

one obtains,

$$\frac{(i\sin\delta_l + \cos\delta_l)}{(i\sin\delta_l - \cos\delta_l)} = \frac{ik\, j_l'(kR) - i\gamma_l\, j_l(kR) + kn_l'(kR) - \gamma_l\, n_l(kR)}{ik\, j_l'(kR) - i\gamma_l\, j_l(kR) - kn_l'(kR) + \gamma_l\, n_l(kR)}$$

After multiplying numerator and denominator of the *R.H.S.* of the above equation by R and then dividing both by i, one gets

$$\frac{e^{i\delta_l}}{e^{-i\delta_l}} = e^{2i\delta_l}$$

$$= \frac{-kR\, j_l'(kR) + \beta_l\, j_l(kR) + ikRn_l'(kR) - i\beta_l\, n_l(kR)}{kR\, j_l'(kR) - \beta_l\, j_l(kR) + ikRn_l'(kR) - i\beta_l\, n_l(kR)},$$

$$\beta_l(k) = R\gamma_l(k) \quad \text{... (85)}$$

From above equation,

$$e^{2i\delta_l} = \frac{\beta_l(j_l - in_l) - kR\,(j_l' - in_l')}{kR\,(j_l' + in_l) - \beta\,(j_l + in_l)}$$

Or

$$e^{2\,i\delta_l} = \frac{(j_l - in_l)}{-(j_l + in_l)} \times \frac{\beta_l - kR\dfrac{(j_l' - in_l')}{(j_l - in_l)}}{\beta_l - kR\dfrac{(j_l' + in_l')}{(j_l + in_l)}}$$

$$= \frac{ij_l - i^2 n_l}{-ij_l - i^2 n_l} \times \frac{\beta_l - kR\dfrac{(j_l' - in_l')(j_l + in_l)}{(j_l - in_l)(j_l + in_l)}}{\beta_l - kR\dfrac{(j_l' + in_l')(j_l - in_l)}{(j_l + in_l)(j_l - in_l)}}$$

$$= \frac{n_l + ij_l}{n_l - ij_l} \times \frac{\beta_l - kR\dfrac{j_l j_l' + n_l n_l'}{j_l^2 + n_l^2} - ikR\dfrac{-j_l n_l' + n_l j_l'}{j_l^2 + n_l^2}}{\beta_l - kR\dfrac{j_l j_l' + n_l n_l'}{j_l^2 + n_l^2} + ikR\dfrac{j_l' n_l - n_l' j_l}{j_l^2 + n_l^2}} \quad \text{... (86)}$$

Putting

$$\Delta_l(\rho_0) = \rho_0 \frac{j_l j_l' + n_l n_l'}{j_l^2 + n_l^2}, \; \rho_0 = kR \quad \text{... (87a)}$$

And using the known relation,

$$j_l(\rho_0)\,\eta_l'(\rho_0) - j_l'(\rho_0)\,\eta_l(\rho_0) = \frac{1}{\rho_0^2} \quad \text{... (87b)}$$

and then putting

$$S_l(\rho_0) = \frac{1}{[\rho_0(j_l^2 + \eta_l^2)]} \quad \text{... (87c)}$$

Into eqn. (86), one gets

$$e^{2i\delta_l}, = \frac{n_l + ij_l}{n_l \; ij_l} \times \frac{\beta_l - \Delta_l + iS_l}{\beta_l - \Delta_l - iS_l}$$

Further multiplying numerator and denominator of the *R.H.S.* of the above equation by *i*, one gets

$$e^{2i\delta_l}, = \frac{j_l - in_l}{j_l + in_l} \times \frac{\beta_l - \Delta_l + iS_l}{\beta_l - \Delta_l - iS_l}$$

which can be written as

$$e^{2i\xi_l} = e^{2i(\xi_l + \zeta_l)} \quad \text{... (88)}$$

where

$$e^{2i\xi_l} = -\frac{j_l(\rho_0) - in_l(\rho_0)}{j_l(\rho_0) + in_l(\rho_0)} \quad \text{... (89a)}$$

and

$$e^{2i\xi_l} = \frac{\beta_l - \Delta_l + iS_l}{\beta_l - \Delta_l - iS_l} \qquad \text{... (89b)}$$

Writing

$$\frac{i \sin \xi_l}{\cos \xi_l} = \frac{ij_l(\rho_0)}{n_l(\rho_0)} \text{ as } \frac{a'}{b'} = \frac{c'}{d'} \text{ and}$$

further using the relation, $\frac{(a'+b')}{(a'-b')} = \frac{(c'+d')}{c'-d')}$, one gets back eqn. (89*a*).

This simply leads to the conclusion that eqn. (89*a*) is equivalent to the relation,

$$\tan \xi_l = \frac{j_l(\rho_0)}{\eta_l(\rho_0)} \qquad \text{... (90)}$$

which means, ξ_l is real and also that is does not depend upon the shape and depth of the potential, but only on its range. Similarly, it can be shown that eqn. (89*b*) is equivalent to the relation

$$\tan \zeta_l = \frac{S_l}{(\beta_l - \Delta_l)} \qquad \text{... (91)}$$

which shows that Δ_l, S_l and hence ζ_l are also real and at the same time depend upon the details of potential through β_l.

Eqn. (88) shows that the phase shift δ_l can be written as the sum of two terms :

$$\delta_l = \xi_l + \zeta_l \qquad \text{... (92)}$$

At low energies (k→0),

$$j_l(\rho_0) \xrightarrow[\rho_0 \to 0]{} \frac{(\rho_0)^l}{(2l+l)!!}$$

And

$$n_l(\rho_0) \xrightarrow[\rho_0 \to 0]{} -\frac{(2l-1)!!}{\rho_0^{l+1}}$$

Substituting above expression into eqn. (90), one gets

$$\tan \xi_{l\,kR \tilde{\to} 0} \frac{-(kR)^{2l+1}}{(2l+1)\{(2l-1)!!\}^2} \qquad \text{... (93)}$$

for all *l*.

Thus at low energy, ξ_l goes to zero as $E^{l+1/2}$ $\left(E = \frac{\hbar^2 k^2}{2\mu} or\ E\, \alpha k^2\right)$or

ξ_l is a slowly varying function of energy. Also we see from eqn. (89*b*) that $\zeta_l \to 0$ as $\beta_l \to \infty$. Therefore, ξ_l is the total phase shift (see eqn. (92)) whenever β_l is infinite. Now an infinite logarithmic derivative, $\beta_l \left(= \frac{R}{R_l(r)} \frac{dR_l(r)}{dr}\right)$at $r = R$ occurs for all l if the potentials represent a hard sphere core of radius R so that $R_l\ (r = R) = 0$ or if particle cannot penetrate the region. The angles ξ_l are, therefore often referred to as the hard sphere phase shifts.

Before we proceed further, let us first try to evaluate the energy dependence of β_l. Writing radial part of the Schrödinger equation in the presence of $V(r)$, that is eqn. (10) with energies E_1 and E_2 but corresponding to the same angular momentum, one gets the following two equations,

$$\left[\frac{1}{r^2}\frac{d}{dr}\left(r^2\frac{d}{dr}\right)+\frac{2\mu}{\hbar^2}\left\{E_1 - V(r) - \frac{l(l+1)\hbar^2}{2\mu r^2}\right\}\right] R_{l,k_1}(r) = 0 \quad \text{... (94a)}$$

and

$$\left[\frac{1}{r^2}\frac{d}{dr}\left(r^2\frac{d}{dr}\right)+\frac{2\mu}{\hbar^2}\left\{E_2 - V(r) - \frac{l(l+1)\hbar^2}{2\mu r^2}\right\}\right] R_{l,k_2}(r) = 0 \quad \text{... (94b)}$$

Multiplying (94*a*) by $R_l,\ k_2^{(r)}$ and (94*b*) by $R_l,\ k_l^{(r)}$ and then subtracting, one gets

$$\frac{1}{r^2}\left[\frac{d}{dr}\left(\frac{r^2 dR_{l,k_2}}{dr}\right)R_{l,k_l} - \frac{d}{dr}\left(\frac{r^2 dR_{l,k_l}}{dr}\right)R_{l,k_2}\right]$$

$$+\frac{2\mu}{\hbar^2}(E_2 - E_1)\, R_{l,k_1} R_{l,k_2} = 0$$

which on simplification gives.

$$\frac{d}{dr}\left(r^2 R_{l,k_1}\frac{dR_{l,k_2}}{dr} - r^2 R_{l,k_2}\frac{dR_{l,k_1}}{dr}\right) = -\frac{2\mu}{\hbar^2}(E_2 - E_1)\, R_{l,k_1} R_{l,k_2}\, r^2$$

Integrating above expression between the limits 0—R, one gets

$$\left\{r^2 R_{l,k_1}\frac{d}{dr}R_{l,k_2} - r^2 R_{l,k_2}\frac{d}{dr}R_{l,k_1}\right\}\Bigg|_0^R$$

$$= -\frac{2\mu}{\hbar^2}(E_2 - E_1)\int_0^R R_{l,k_1} R_{l,k_2}\, r^2\, dr$$

Or

$$R^2 R_{l,k_1} R_{l,k_2}\left[\frac{1}{R_{l,k_2}}\frac{dR_{l,k_2}}{dr} - \frac{1}{R_{l,k_1}}\frac{dR_{l,k_1}}{dr}\right]_{r=R}$$

$$= -\frac{2\mu}{\hbar^2}(E_2 - E_1)\int_0^R R_{l,k_1} R_{l,k_2}\, r^2\, dr$$

Dividing the above expression by the product $R_{l,k_1}(r)\, R_{l,k_2}(r)$, one finds that the two values of the logarithmic derivative are related by the formula,

$$\beta_l(E_2) - \beta_l(E_1) = \frac{2\mu(E_2 - E_1)\int_0^R r^2 R_{l,k_1}(r) R_{l,k_2}(r)\, dr}{\hbar^2 R\, R_{l,k_1}(R)\, R_{l,k_2}(R)} \quad \text{... (95)}$$

as $$\beta_l(E_2) = \frac{R}{R_{l,k_2}} \frac{dR_{l,k_2}}{dr}\bigg|_{r=R}$$

which shows that β_l decreases with increase of energy or is a decreasing function of energy.

Due to the presence of $\beta_l - \Delta_l$ in the denominator of eqn. (91), it is clear that a sudden increase in the value of tan ζ_l occurs for $\beta_l \simeq \Delta_l$ and if the energy is small enough ξ_l would then be negligible (see eqn. (93)) compared with ζ_l (which from eqn. (91) tends to $\pi/2$ at $\beta_l \approx \Delta_l$) at this energy. Hence from eqn. (92),

$$\delta_l \approx \zeta_l \qquad \text{... (96)}$$

and such a phenomenon is described as resonance in the lth partial wave. In this circumstance, the phase shifts corresponding to the partial waves other than the one for which $\beta_l \approx \Delta_l$ can be neglected because for the other partial waves the phase shift is ξ_l and at low energy it can be neglected. Then, the total scattering cross-section (or cross-section of the resonance scattering) is

$$\sigma \approx \sigma_l^{(\text{res})} = \frac{4\pi}{k^2}(2l+1)\sin^2\delta_l$$

$$= \frac{4\pi}{k^2}(2l+1)\sin^2\zeta_l$$

$$= \frac{4\pi}{k^2}\frac{(2l+1)\tan^2\zeta_l}{(1+\tan^2\zeta_l)} \qquad \text{... (97)}$$

Let $E_l^{(0)}$ be the energy at which $\beta_l = \Delta_l$. For E in the neighbourhood of $E_l^{(0)}$, let us assume that β_l is linear, that is expanding β_l in a Taylor series in the neighbourhood of $E_l^{(0)}$. we have

$$\beta_l(E) \approx \beta_l(E_l^{(0)}) + (E - E_l^{(0)})\left(\frac{\partial \beta_1}{\partial E}\right)_{E=E_l^{(0)}}$$

$$\approx \Delta_l + b_l(E - E_l^{(0)}) \qquad \text{... (98)}$$

as at $$E = E_l^{(0)},\ \beta_l = \Delta_l$$

Since β_l is a decreasing function of energy as shown above, b_l is negative (as differential shows increase). Substituting the value of β_l from the above equation into (91), one gets

$$\tan\zeta_l = \frac{-2(S_l/b_l)}{2(E_l^{(0)} - E)}$$

$$= \frac{\tau_l}{2\left(E_l^{(0)} - E\right)} \qquad \text{... (99)}$$

where

$$\tau_l = -2\,S_l/b_l$$

Chapter 14

By its definition, S_l is positive (see eqn. (87c)), hence the quantity τ_l defined above would also be positive. Substituting for tan ζ_l from eqn. (99) into eqn. (97), one gets

$$\sigma_l^{(res)}(E) = \frac{4\pi}{k^2}(2l+1)\left(\frac{\tau_l}{2}\right)^2 \Big/ \left\{(E - E_l^{(0)})^2 + \left(\frac{\tau_l}{2}\right)^2\right\} \quad \text{... (100)}$$

Fig. 14.6 shows the variation of $\sigma_l^{(res)}$ (E) with E. The cross-section has a sharp maximum centered around $E = E_l^{(0)}$ termed as the resonant energy. Such large variations in the cross-section, $\sigma_l^{(res)}$ (E) (or phase shifts as $\sigma_l^{(res)}$ (E) α $\sin^2 \delta_l$) are attributed to the fact that a certain partial wave is in resonance with the potential at small energies. The phenomenon is, therefore called resonance scattering. Naturally, it is the partial wave with $l > 0$ which is in resonance at non zero energy ($E = E_l^{(0)}$) in contrast with the S-wave case where the cross-section is highest $\left(\sigma_0 = \frac{4\pi}{k^2}\sin^2\delta_0\right)$ at zero energy (at $k \to 0$). From Fig. 14.6, $\frac{\tau_l}{2}$ is the value of $|E - E_l^{(0)}|$ for which $\sigma_l^{(res)}(E)$ falls to half of its peak value or τ_l is the width of the resonance peak. The relation (101) is known as the Breit–Wigner formula.

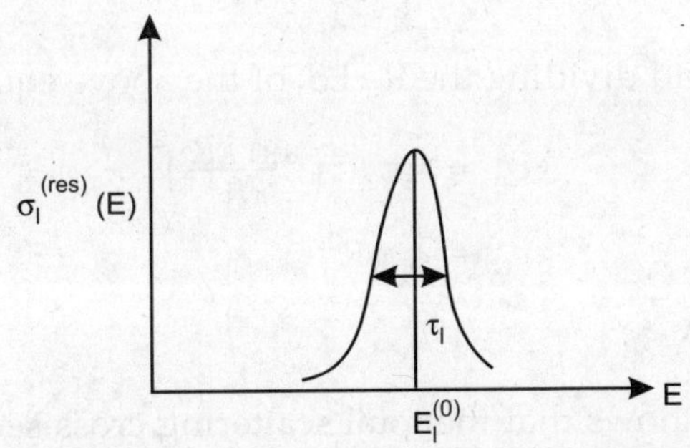

Fig. 14.6 *The cross-section of resonance scattering. $E_l^{(0)}$ is the resonance energy and τ_l the width of the resonance*

14.12 SCATTERING BY A PERFECTLY RIGID SPHERE POTENTIAL

Perfectly rigid (or hard) sphere (or infinitely repulsive) potential is given by

$$V(r) = \infty, r < R$$

$$= 0, r > R \quad \text{... (101)}$$

which is another example of a central potential. Let us study the problem under low as well as under high energy limits.

(i) Low energy limit

In this case (since $kR \ll 1$) only $l = 0$ partial wave would contribute. The solution of the radial part of the Schrödinger equation would be given by (see problem 14.1)

$$f_0(r) = B \sin(kr + \delta_0),\ r > R \qquad \text{... (14.5)}$$

Since the particle cannot penetrate the region $r < R$ (as the particle cannot have infinite energy), the wave function in that region must vanish. Thus

$$F_0(r) = 0,\ r < R \qquad \text{... (102)}$$

Let us now apply the boundary conditions that the wave function and its derivative must match at $r = R$ (in this case we cannot use the second boundary condition that the derivative, f_0' is continuous at $r = R$ because of the fact that V has infinite discontinuity over there). Thus applying only the first boundary condition, we get

$$B \sin(kR + \delta_0) = 0$$

or

$$\delta_0 = -kR \qquad \text{... (103)}$$

Accordingly the total scattering cross-section in the low energy limit would be given by

$$\sigma = \sigma_0 = \frac{4\pi}{k^2} \sin^2 \delta_0$$

$$= \frac{4\pi}{k^2} \sin^2 kR$$

After multiplying and dividing the R.H.S. of the above equation by R^2, we get

$$\sigma_0 = 4\pi R^2 \left(\frac{\sin kR}{kR}\right)^2$$

$$= 4\pi R^2 \qquad \text{... (104)}$$

as $\sin \frac{kR}{kR} \to 1$ for $kR \ll 1$.

The above relation shows that the total scattering cross-section at low energies is four times that of the classical cross-section presented by a sphere of radius R. Also the scattering is so tropic (that is independent of θ and hence spherically symmetric) and at the same time independent of energy.

(ii) High energy limit

In this limit since $kR \gg 1$, the partial waves other than $l = 0$ would also contribute. Accordingly the total scattering cross-section would be given by

$$\sigma = \sigma_t = \frac{4\pi}{k^2} \sum_{l=0}^{\infty} (2l + 1) \sin^2 \delta_l \qquad \text{... (105)}$$

From section 14.5, we know that δ_l would be appreciable only when $b < R$ or effectively if $l \ll kR$. In the circumstance, the above expression takes the form

$$\sigma_t = \frac{4\pi}{k^2} \sum_{l=0}^{kR} (2l + 1) \sin^2 \delta_l \qquad \text{... (106)}$$

The radial part of the Schrödinger equation in the present case takes the form

$$\frac{1}{r^2}\frac{d}{dr}\left(\frac{r^2 dR_l(r)}{dr}\right)+\left(k^2-\frac{l(l+1)}{r^2}\right)R_l(r)=0,\ r>R$$

The solution of above equation is,

$$R_l(r) = A_l j_l(kr) + B_l n_l(kr)$$

which in the asymptotic region ($r > R$) is given by

$$R_l(r) \xrightarrow[r\to\infty]{} \frac{A_l}{kr}\sin\left(kr-\frac{l\pi}{2}\right)-\frac{B_l}{kr}\cos\left(kr-\frac{l\pi}{2}\right)$$

$$= C_l/kr\left[\cos\delta_l \sin\left(kr-\frac{l\pi}{2}\right)+\sin\delta_l\cos\left(kr-\frac{l\pi}{2}\right)\right] \quad \text{... (107)}$$

According to what has been said before, $R_l(r)$ has to vanish on the surface of the sphere, that is, at $r = R$. Therefore

$$\cos\delta_l \sin\left(kR-\frac{l\pi}{2}\right)+\sin\delta_l\cos\left(kR-\frac{l\pi}{2}\right) = 0$$

or

$$\tan\delta_l = -\tan\left(kR-\frac{l\pi}{2}\right)$$

or

$$\delta_l = -kR+\frac{l\pi}{2} \quad \text{... (108)}$$

Thus the total scattering cross-section would be given by

$$\sigma_t = \frac{4\pi}{k^2}\sum_{l=0}^{kR}(2l+1)\sin^2\left(kR-\frac{l\pi}{2}\right)$$

As $kR \gg 1$, we have to sum over large number of terms. Therefore replacing the summation in the above equation by integration, we get

$$\sigma_t = \frac{4\pi}{k^2}\int_0^{kR}(2l+1)\sin^2\left(kR-\frac{l\pi}{2}\right)dl$$

To simplify the above integral, let us take the average value of $\sin^2\left(kR-\frac{l\pi}{2}\right)$ which is $\frac{1}{2}$. Thus

$$\sigma_t = \frac{4\pi}{k^2}\times\frac{1}{2}\int_0^{kR}(2l+1)\,dl$$

$$= \frac{2\pi}{k^2}(l^2+l)\Big|_0^{kR}$$

$$= 2\pi R^2 \quad \text{... (109)}$$

where we have neglected the term containing $1/k$ as k is large. Thus the total scattering cross-section in the high energy limit comes out to be twice that of the classical result.

14.13 THE INTEGRAL EQUATION METHOD

We have seen that the method of partial waves is limited to the cases of low energy scattering (that is, in the cases in which $kR << 1$) by short range (inspite of the fact that we considered the asymptotic solution, the term containing r^2 in the denominator was taken in to account) central potentials (the cases in which the wave function can be expanded in spherical harmonics). In the problems where kR is large, the method of partial waves becomes impracticable (as one has to sum over large number of l values).

We will now discuss an approximation method which is suitable for large energies and also where the potential need not be necessarily central. The Schrödinger equation to be solved for the relative motion of the particles involved in the scattering process is

$$\nabla^2 \varphi(\mathbf{r}) + \frac{2\mu}{\hbar^2}[E - V(\mathbf{r})]\varphi(\mathbf{r}) = 0$$

or

$$(-\nabla^2 - k^2)\,\varphi(\mathbf{r}) = -U(\mathbf{r})\,\varphi(\mathbf{r}) = F(\mathbf{r}) \text{ say} \quad \text{... (110)}$$

with

$$k^2 = \frac{2\mu E}{\hbar^2} \text{ and } U(\mathbf{r}) = \frac{2\mu V(\mathbf{r})}{\hbar^2}$$

The above inhomogeneous differential equation can be converted to an integral equation with the help of Green's functions. The technique is simple but instead of solving eqn. (110) as a special case, we would indicate a method for solving more general inhomogeneous partial differential equation,

$$(\hat{\Omega} - w_0)\,\phi(\mathbf{r}) = F(\mathbf{r}) \quad \text{... (111)}$$

where $\hat{\Omega}$ is a Hermitian operator, $F(\mathbf{r})$ is the inhomogeneity and w_0 is a number. The general solution of the above equation would consist of a particular solution $\phi(\mathbf{r})$ and a complementary solution $\chi(r)$. The later is solution of the homogeneous equation,

$$(\hat{\Omega} - w_0)\,\chi(\mathbf{r}) = 0 \quad \text{... (111a)}$$

that is, eqn. (111) without the right hand side. $\phi(\mathbf{r})$ is the solution of the complete equation,

$$(\hat{\Omega} - w_0)\,\phi(\mathbf{r}) = F(\mathbf{r}) \quad \text{... (111b)}$$

$\hat{\Omega}$ being a Hermitian operator, defines a complete orthonormal set of eigen functions $\varphi_w(\mathbf{r})$ with real eigen values w. Thus

$$\hat{\Omega}\varphi_w(\mathbf{r}) = \omega\,\varphi_w(\mathbf{r}) \text{ (eigen value equation)} \quad \text{... (112a)}$$

$$\int \varphi_w^*(\mathbf{r})\,\varphi_{\omega'}(\mathbf{r})\,d\mathbf{r} = \delta(w - w') \quad \text{... (112b)}$$

which defines δ – function normalization for the eigen function of a continuous spectrum and

$$\int \varphi_w^*(\mathbf{r})\,\varphi_w(\mathbf{r}')\,dw = \delta(\mathbf{r} - \mathbf{r}') \text{ (completeness)} \quad \text{... (112c)}$$

where it is assumed that the eigen values w form a continuous set and δ refers to the Dirac delta function.

Let us expand the particular function $\phi(\mathbf{r})$ in terms of $\varphi_w(\mathbf{r})$ (as any arbitrary function can be expanded in terms of an orthonormal set), that is, write

$$\phi(\vec{\mathbf{r}}) = \int A_w \varphi_w(\mathbf{r})\, dw \qquad \text{... (113)}$$

Substituting above into (111*b*), one gets

$$\int A_w(\hat{\Omega} - w_o)\varphi_w(\mathbf{r})\, dw = F(\mathbf{r})$$

Using eqn. (112*a*), the above equation becomes

$$\int A_w(w - w_o)\varphi_w(\mathbf{r})\, dw = F(\mathbf{r}) \qquad \text{... (114)}$$

Multiplying the above equation by $\varphi^*_{w'}(\mathbf{r})$ and then integrating the resulting expression over entire space, one gets

$$\iint A_w(w - w_0)\varphi^*_{w'}(\mathbf{r})\varphi_w(\mathbf{r})\, dw\, dr = \int \varphi^*_{w'}(\mathbf{r})F(\mathbf{r})\, dr$$

Further using (112*b*), one gets

$$\int A_w(w - w_0)\delta(w - w')\, dw = \int \varphi^*_{w'}(\mathbf{r})\, F(\mathbf{r})\, d\mathbf{r}$$

Therefore,

$$A_{w'} = \int \frac{\varphi^*_{w'}(\mathbf{r})F(\mathbf{r})\, d\mathbf{r}}{(w' - w_0)}$$

or

$$A_{w'} = \int \frac{\varphi^*_{w'}(\mathbf{r}')\, F(\mathbf{r}')\, d\mathbf{r}'}{(w' - w_0)} \qquad \text{... (115)}$$

Substituting above into eqn. (113), the solution of eqn. (111*b*) takes the form

$$\phi(\mathbf{r}) = \int A_{w'}\varphi_{w'}(\mathbf{r})\, dw'$$

$$= \iint \frac{\varphi^*_{w'}(\mathbf{r}')\varphi_{w'}(\mathbf{r})F(\mathbf{r}')\, d\mathbf{r}'\, dw'}{(w' - w_0)}$$

$$= \int G_{w_0}(\mathbf{r}, \mathbf{r}')F(\mathbf{r}')\, d\mathbf{r}' \qquad \text{... (116)}$$

where the function,

$$G_{w_0}(\mathbf{r}, \mathbf{r}') = \int \frac{\varphi^*_{w'}(\mathbf{r}')\varphi_{w'}(r)\, dw'}{(w' - w_0)} \qquad \text{... (116a)}$$

is called the Green function for the operator $\hat{\Omega}$, and number w_0.

Comparison of eqns. (110*b*) and (111*b*) indicate that the operator $\hat{\Omega}$ can be taken as $-\nabla^2$ and the number w_0 as k^2 ($-\nabla^2$ multiplied by $\frac{\hbar^2}{2\mu}$ is the free particle Hamiltonian). That is, the Green function (116*a*) is to be evaluated when the operator $\hat{\Omega}$ is the Hamiltonian operator for a free particle.

A normalized eigen function $\varphi_{k'}(\mathbf{r})$ of the free particle Hamiltonian or of the operator $-\nabla^2$ is given by $(\varphi_{k'}(\mathbf{r}))$ satisfies ,

$$(\nabla^2 + k'^2)\ \varphi_{k'}(\mathbf{r}) = 0$$

which is the Schrödinger equation for a free particle)

$$\varphi_{k'}(\mathbf{r}) = (2\pi)^{-3/2}\ e^{i.\mathbf{k}'.\mathbf{r}} \qquad \text{... (117)}$$

where $\mathbf{k}'$ is any wave vector of magnitude k'. Thus the required Green function is (after substituting the above into eqn. (116*a*)),

$$G_k(\mathbf{r}, \mathbf{r}') = (2\pi)^{-3} \int \frac{\mathbf{e}^{\vec{i}.\vec{\mathbf{k}'}.\vec{\rho}}}{k'^2 - k^2}\, d\mathbf{k}',\ \rho = \mathbf{r} - \mathbf{r}' \qquad \text{... (118)}$$

The integrand in the above equation being scalar, the integral is independent of the coordinate system in which it is evaluated. We perform the $\mathbf{k}'$ ($d\mathbf{k}' = dk'_x\, dk'_y\, dk'_z$) integration in the spherical polar coordinates with polar axis (z-axis) in the direction of the vector $\vec{\rho} = \mathbf{r} - \mathbf{r}'$ (see Fig. 14.7). Then

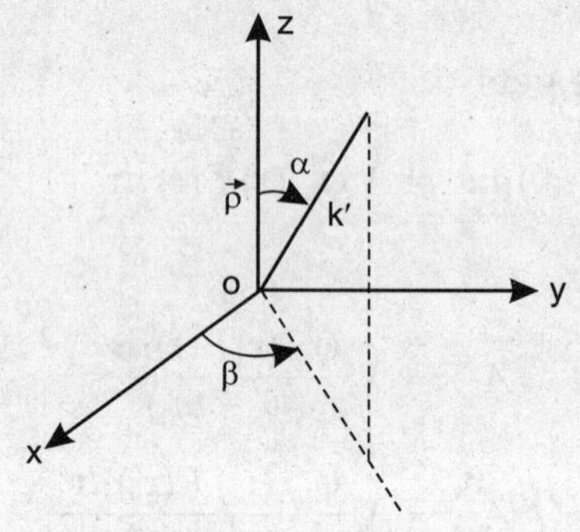

Fig. 14.7 $\vec{k'}$ *integration in the spherical polar coordinates*

$$G_k(\mathbf{r}, \mathbf{r}') = (2\pi)^{-3} \int_0^{\infty}\int_0^{\pi}\int_0^{2\pi} \frac{e^{-ik'\rho\cos\alpha}}{k'^2 - k^2} k'^2 \sin dk'\, d\alpha\, d\beta$$

$$= (2\pi)^{-2} \int_0^{\infty}\int_0^{\pi} \frac{e^{ik'\rho\cos\alpha}}{k'^2 - k^2} k'^2 \sin\alpha\, dk'\, d\alpha$$

$$= (2\pi)^{-2} \int_0^{\infty}\int_{-1}^{+1} \frac{e^{ik'\rho\cos\alpha}}{k'^2 - k^2} k'^2\, dk'\, d(\cos\alpha)$$

$$= (2\pi)^{-2} \int_0^{\infty} \left. \frac{e^{ik'\rho\cos\alpha}}{ik'\rho} \right|_{-1}^{+1} \frac{k'^2\, dk'}{k'^2 - k^2}$$

$$= (2\pi)^{-2} \int_0^{\infty} \frac{e^{ik'\rho} - e^{-ik'\rho}}{ik'\rho} \frac{k'^2 dk'}{k'^2 - k^2}$$

Alter multiplying and dividing the integrand by 2 and then using the relation, $\sin x = \dfrac{(e^{ix} - e^{-ix})}{2i}$, one gets

$$G_k(\mathbf{r}, \mathbf{r}') = \frac{2}{4\pi^2\rho} \int_0^{\infty} \frac{k' \sin k'\rho}{k'^2 - k^2} dk'$$

$$= \frac{1}{4\pi^2\rho}\int_{-\infty}^{\infty}\frac{k'\sin k'\rho}{k'^2-k^2}dk'$$

as the integrand is an even function of k'. There being singularities at the points $k' = \pm k$, the above integral can be evaluated by the method of residue which results in

$$G_k(\mathbf{r}, \mathbf{r}') = \frac{1}{4\pi^2\rho}(\pi e^{ik\rho})$$

$$= \frac{1}{4\pi}\frac{e^{ik\,|\,\mathbf{r}-\mathbf{r}'\,|}}{|\,\mathbf{r}-\mathbf{r}'\,|} \qquad \text{... (119)}$$

Substituting above into eqn. (116), one gets

$$\phi(r) = \frac{1}{4\pi}\int\frac{e^{ik|\mathbf{r}-\mathbf{r}'|}}{|\vec{\mathbf{r}}-\vec{\mathbf{r}}'|}F(\mathbf{r}')\,d\mathbf{r}' \qquad \text{... (120)}$$

The general solution of eqn. (111) is, therefore given by

$$\Phi(\mathbf{r}) = \chi(\mathbf{r}) + \phi\,(\mathbf{r}) \qquad \text{... (121a)}$$

Note that the $\chi(r)$ is a free particle solution, $e^{i\mathbf{k}_0\cdot\mathbf{r}}$ ($\mathbf{k}_0$ is wave vector of magnitude k for the incident free particle) but keeping in mind the coordinate system, we introduced in Fig. 14.1 for the discussion of scattering, we prefer to use a plane wave proceeding in the z-direction, that is, we set $\chi(\mathbf{r}) = e^{ikz}$.

Thus

$$\Phi(\mathbf{r}) = e^{ikz} + \frac{1}{4\pi}\int\frac{e^{ik\,|\,\mathbf{r}-\mathbf{r}'\,|}}{|\,\mathbf{r}-\mathbf{r}'\,|}F(\mathbf{r}')\,d\mathbf{r}' \qquad \text{... (121b)}$$

Applying the above procedure to eqn. (110), we get

$$\varphi(\mathbf{r}) = e^{ikz} - \frac{1}{4\pi}\int\frac{e^{ik|\mathbf{r}-\mathbf{r}'|}}{|\mathbf{r}-\mathbf{r}'|}\varphi(\mathbf{r}')\,U(\mathbf{r}')\,d\mathbf{r}' \qquad \text{... (122)}$$

where the value of $F(\mathbf{r}')$ has been substituted from eqn. (110).

From the above equation, we observe that the potential function $U(\mathbf{r}')$ or $V(\mathbf{r}')$ is a factor in the integrand. Hence regions of $\mathbf{r}'$ where $V(\mathbf{r}')$ is very small do not contribute appreciably to the integral. Therefore, though the integration is in principle to be carried over all $\mathbf{r}'$, it extends effectively only over the region of $\mathbf{r}'$ near the origin (see Fig. 14.8) where $V(\mathbf{r}')$ is large. This means that in considering the asymptotic form of eqn. (122), that is for $\mathbf{r} \to \infty$, we can take $|\mathbf{r}| >> |\mathbf{r}'|$ in the second term (the first term on the R.H.S. has already the desired form). Thus

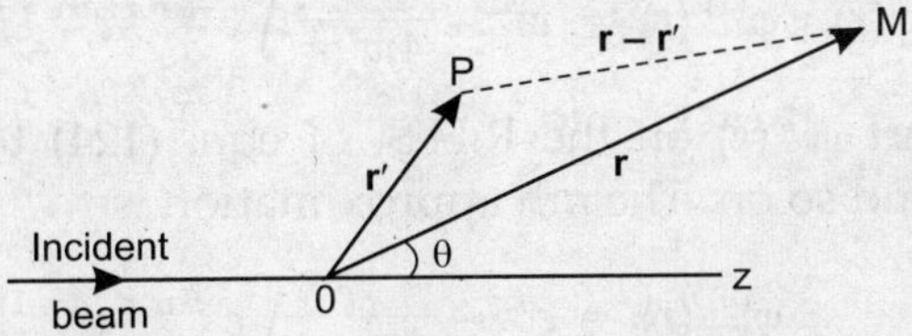

Fig. 14.8 *Approximate calculation of the distance* $|\mathbf{r} - \mathbf{r}'|$ *between a point M very far from 0 and a point P situated in the region of influence of the potential*

$$|\mathbf{r} - \mathbf{r}'| = (r^2 - 2\mathbf{r} \cdot \mathbf{r}' + r'^2)^{1/2}$$

$$= r\left(1 - \frac{2\vec{n} \cdot \vec{r}'}{r} + r'^2/r^2\right)^{1/2} \xrightarrow[r \to \infty]{} \mathbf{r}\left(1 - \frac{2\mathbf{n} \cdot \mathbf{r}'}{r}\right)^{1/2}$$

$$= r(1 - \mathbf{n} \cdot \mathbf{r}'/r) \qquad \text{... (123 } a)$$

where **n** is the unit vector along **r** and where the higher order terms have been neglected in the binomial expansion.

Similarly,

$$\frac{1}{|\mathbf{r} - \mathbf{r}'|} = \frac{1}{\{(r^2 - 2\mathbf{r} \cdot \mathbf{r}' + r'^2)^{1/2}\}} \xrightarrow[r \to \infty]{} \frac{\left(1 + \mathbf{n} \cdot \frac{\mathbf{r}'}{\mathbf{r}}\right)}{r}$$

$$= \frac{1}{r} \qquad \text{... (123 } b)$$

Thus the asymptotic form of $\varphi(\mathbf{r})$ from eqn. (122) using eqns. (123) would be given by

$$\varphi(\mathbf{r}) \xrightarrow[r \to \infty]{} e^{ikz} - \frac{1}{4\pi} \frac{e^{ikr}}{r} \int e^{-ik\mathbf{n} \cdot \mathbf{r}'} \varphi(\mathbf{r}') U(\mathbf{r}') d\mathbf{r}' \qquad \text{... (124)}$$

which implies that the second term is the product of an out going spherical wave e^{ikr}/r and some angular amplitude independent of r.

Comparing above equation with eqn. (2a) in the general case of $V(r) \neq V(r)$, we get

$$f(\theta, \phi) = -\frac{1}{4\pi} \int e^{-ik\mathbf{n} \cdot \mathbf{r}'} \varphi(\mathbf{r}') U(\mathbf{r}') d\mathbf{r}' \qquad \text{... (125)}$$

Note that the normalization constant while writing the free particle soloution in eqns. (121) and (122) was taken to be 1 so that it could be compared with eqn. (2a).

14.14 THE BORN APPROXIMATION[1]

In the above section, we have converted the differential eqn. (110) into an integral eqn. (124) with the help of Green functions. We will now use a well known method of successive approximations (or method of iteration) to solve the said integral equation. In the zeroth approximation, we neglect the integral term in eqn. (124) and set $\varphi(\mathbf{r}) = \varphi^0(\mathbf{r}) = e^{ikz}$. Then we use this zeroth approximation on the R.H.S. of eqn. (124) to compute the first approximation (called first Born approximation) $\varphi^{(1)}(\mathbf{r})$. Thus we have

$$\varphi(\mathbf{r}) \approx \varphi^{(1)}(\mathbf{r}) = e^{ikz} - \frac{1}{4\pi} \frac{e^{ikr}}{r} \int e^{-ik\mathbf{n} \cdot \mathbf{r}'} e^{ikz'} U(\mathbf{r}') d\mathbf{r}' \qquad \text{... (126)}$$

Next we may insert $\varphi^{(1)}(\mathbf{r})$ on the R.H.S. of eqn. (124) to compute the second approximation $\varphi^{(2)}(\mathbf{r})$ and so on. The nth approximation is

$$\varphi^{(n)}(\mathbf{r}) = e^{ikz} - \frac{1}{4\pi} \frac{e^{ikr}}{r} \int e^{-ik\mathbf{n} \cdot \mathbf{r}'} \varphi^{(n-1)}(\mathbf{r}') U(\mathbf{r}') d\mathbf{r}' \qquad \text{... (127)}$$

[1] M.Born, Z. Physik 38, 803 (1926)

In the light of what has been said above, the scattering amplitude in the first Born approximation (hence forth to be called Born Approximation), thus would be given by

$$f(\theta, \phi) \approx \frac{1}{4\pi}\int e^{-ik\mathbf{n}.\mathbf{r}'} e^{ikz'} U(\mathbf{r}')d\mathbf{r}' \quad \text{... (128)}$$

Higher Born approximations for the scattering amplitude are obtained in an analogous fashion, but as a rule these are expressions which are too clumsy to discuss in generality. Thus we shall concentrate on the first Born approximation (128).

Let $\mathbf{k}_0$ be a vector of magnitude k that has the direction of the incident beam (polar axis), $\mathbf{k}$ is the vector of magnitude k for the scattered particle and θ is the angle of scattering ($|\mathbf{k}_0| = |\mathbf{k}| = k$ as the scattering is elastic). We can write

$$e^{ikz'} = e^{i\mathbf{k}_0 \cdot \mathbf{r}'}$$

Substituting above equation into (128), we get

$$f(\theta, \phi) = -\frac{1}{4\pi}\int e^{i(\mathbf{k}_0 - \mathbf{k}).\mathbf{r}'} U(\mathbf{r}')\, d\mathbf{r}'$$

$$= \frac{1}{4\pi}\int e^{i\mathbf{k}.\mathbf{r}'} U(\mathbf{r}')d\mathbf{r}' \quad \text{... (129)}$$

where we have defined $\mathbf{K} = \mathbf{k}_0 - \mathbf{k}$ (see Fig. 14.9). The magnitude of $\mathbf{K}$ is $2k \sin \theta/2$ $[\{(\mathbf{k}_0 - \mathbf{k}) \cdot (\mathbf{k}_0 - \mathbf{k})\}^{1/2} = (k^2 + k^2 - 2k^2 \cos \theta)^{1/2} = \{2k^2 (1 - \cos \theta)\}^{1/2} = \{2k^2 (1 - 1 + 2 \sin^2 \theta/2)\}^{1/2}]$. Its physical significance is that $\hbar\mathbf{K}$ is the momentum transferred to the particle in its encounter with the potential ($\mathbf{K}$ is also termed as the momentum transfer vector). The expression (129) shows that in the first Born approximation, the scattering amplitude is simply the Fourier transform of the scattering potential.

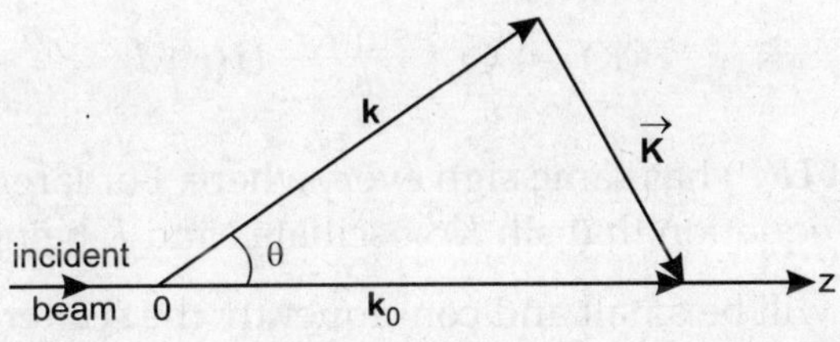

Fig. 14.9 *Incident wave vector k_0, scattered wave vector k and transferred wave vector* **K**

The differential scattering cross-section in the Born approximation would, thus be given by

$$\sigma(\theta, \phi) = |f(\theta, \phi)|^2 = \left(\frac{1}{4\pi}\right)^2 \left|\int e^{i\mathbf{K}.\mathbf{r}'} U(\mathbf{r}')\, d\mathbf{r}'\right|^2 \quad \text{... (130)}$$

The expression (129) simplifies greatly for a central force, that is if $V(\mathbf{r}) = V(r)$. In the majority of practical situations, we deal with central forces, so that it is worth while to work out this case in some detail. The scattering amplitude in this case would be given by

$$f(\theta) = \frac{-1}{4\pi}\int e^{i\mathbf{K}\cdot\mathbf{r}'} U(\mathbf{r}')d\mathbf{r}'$$

as $\vec{\mathbf{K}}$ depends upon θ. The integrand being scalar, the integral will be independent of the coordinate system in which it is evaluated. Performing the integration in the spherical polar coordinate system, we get

$$f(\theta) = \frac{-1}{4\pi}\int_0^{\infty}\int_0^{\pi}\int_0^{2\pi} e^{iKr'\cos\alpha'} U(r')\, r'^2 \sin\alpha'\, dr'\, d\alpha'\, d\beta'$$

where (r', α', β') define the position of particle w.r.t. the K-axis as the polar axis. Thus

$$f(\theta) = -\frac{1}{2}\int_0^{\infty} r' U(r') \left(\frac{e^{iKr'\cos\alpha}}{iK}\right)\Bigg|_{-1}^{+1} dr'$$

$$= \frac{-1}{K}\int_0^{\infty} r' \sin Kr'\, U(r')\, dr' \qquad \text{... (131)}$$

Hence the differential scattering cross-section will be given by

$$\sigma(\theta) = |f(\theta)|^2 = \frac{1}{K^2}\left|\int_0^{\infty} r' \sin Kr'\, U(r')\, dr'\right|^2 \qquad \text{... (132)}$$

From the above expression, we find that the $\sigma(\theta)$ is independent of ϕ (as expected) and depends on θ through the combination $K = 2k \text{ Sin } \theta/2$ and not on the momentum of incident particle or on scattering angle individually.

14.15 FORM FACTOR

Multiplying and dividing the integrand of eqn. (131) by r', it can be written as

$$f(K) = \int_0^{\infty} r'^2 \frac{\sin Kr'}{Kr'} U(r')\, dr' \qquad \text{... (133)}$$

Let us assume that $U(r')$ has same sign everywhere. For large values of K, it can be noted from the above equation that $\sin Kr'$ oscillates and K being in the denominator, the total term $\frac{\sin Kr'}{Kr'}$ will be small and consequently the scattering amplitude will be small. On the other hand for $K = 0$ (that is, $\mathbf{k}_0 \,||\, \mathbf{k}$ which amounts to $\theta = 0$), the scattering amplitude would be largest $\left(\text{for small } K, \frac{\sin Kr'}{Kr'} \to 1\right)$ and is given by

$$f(0) = \int_0^{\infty} r'^2 U(r')\, dr' \qquad \text{... (134)}$$

The ratio $\frac{f(K)}{f(0)}$ is called the form factor. A study of the form factor reveals the difference in scattering amplitude from a point scatterer and that from a spreaded one. It is easy to see that, in general, the effect of spreading (or finiteness which amounts to large θ or large K) is to reduce the scattering.

14.16 VALIDITY OF BORN APPROXIMATION

In obtaining eqn. (126), we have replaced $\varphi(\mathbf{r}')$ in the integral term by $e^{ikz'}$. This is possible (or Born approximation would be valid) only if the second term (or the scattered wave) can be neglected in comparison with the first term, that is, if

$$|\phi(\mathbf{r})| << |e^{ikz}| = 1 \qquad \text{... (135)}$$

where

$$\phi(\mathbf{r}) = \frac{-1}{4\pi}\int \frac{e^{ik|\mathbf{r}-\mathbf{r}'|}}{|\mathbf{r}-\mathbf{r}'|} e^{ikz'}\, U(\mathbf{r}')\, d\mathbf{r}' \qquad \text{... (120)}$$

which depends upon $U(\mathbf{r}')$ or effectively on $V(\mathbf{r}')$ which is appreciable near the origin for a short range potential. That is, the value of $\phi(\mathbf{r})$ would be largest near the origin (or the centre of scattering potential, that is at $\mathbf{r} = 0$). Thus the validity condition becomes

$$|\phi(0)| << 1 \qquad \text{... (136)}$$

From eqn. (120) in the case of spherically symmetric potential, that is, for $U(\mathbf{r}') = U(r')$, we get

$$\phi(0) = \frac{-1}{4\pi}\int_0^\infty\int_0^\pi\int_0^{2\pi} \frac{e^{ikr'}}{r'} e^{ikz'}\, U(r')\, r'^2 \sin\theta'\, dr'\, d\theta'\, d\phi'$$

$$= \frac{-1}{4\pi}\int_0^\infty\int_0^\pi\int_0^{2\pi} \frac{e^{ikr'}}{r'} e^{ikr'\cos\theta'}\, U(r')\, r'^2 \sin\theta'\, dr'\, d\theta'\, d\phi'$$

On integrating over the polar angles (θ', ϕ') of $\mathbf{r}'$, we get

$$\phi(0) = \frac{-1}{2}\int_0^\infty \frac{e^{ikr'\cos\theta'}}{ikr'}\Bigg|_1^{+1} U(r')\frac{e^{ikr'}}{r'} r'^2\, dr'$$

$$= \frac{-1}{2ik}\int_0^\infty U(r')\,(e^{2ikr'} - 1)\, dr' \qquad \text{... (137)}$$

The validity condition using eqns. (136) and (137) and the relation

$U(r') = 2\mu \dfrac{V(r')}{\hbar^2}$, thus becomes

$$\frac{\mu}{\hbar^2 k}\left|\int_0^\infty V(r')\,(e^{2ikr'} - 1)\, dr'\right| << 1 \qquad \text{... (138)}$$

a result which is likely to hold good for weak potentials and high energies or essentially a small V/k ratio is required for the reliability (or validity) of the Born approximation.

14.17 TRANSFORMATION FROM CENTRE OF MASS TO LABORATORY SYSTEM

We have developed the entire scattering theory in the Centre of mass (COM) coordinate system or that means the cross-sections are calculated in COM. Our next task is how to obtain these in the laboratory frame in which the measurements are actually made. For this let us denote the scattering and azimuthal angles in the laboratory and COM systems by (θ_0, ϕ_0) and (θ, ϕ), respectively and find the relation between them.

Fig. 14.10 (*a*) shows the laboratory system in which a particle (called bombarding particle) of mass m_1 is moving with velocity v before collision. m_2 is target and is initially at rest. As the particle m_1 moves towards m_2, the COM moves in the same direction with velocity V determined by the condition of the conservation of momentum.

$$V = \frac{(m_1\, v + 0)}{(m_1 + m_2)} \qquad \text{... (139)}$$

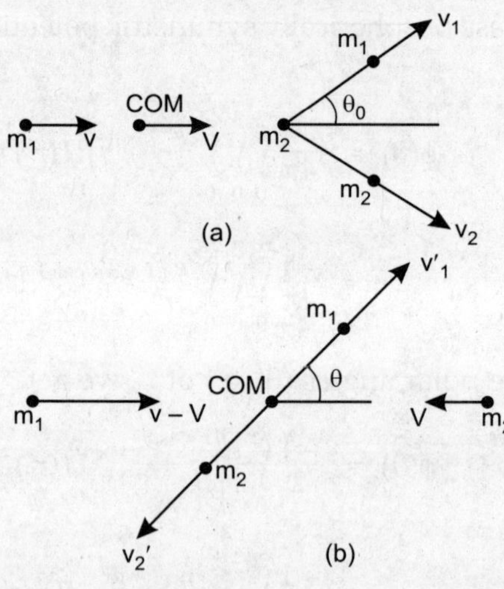

Fig. 14.10 (*a*) *Laboratory system (in which the target particle in initially at rest) (b) Centre of mass system (a frame in which COM of the two collinding particles is at rest or a frame which moves with a uniform velocity V relative to laboratory). The total momentum is always zero*

Fig. 14 .10(*b*) shows the COM system in which the momemtum is always zero and hence after collision the particles move in the opposite directions with equal momentum.

From the velocity addition formula (if we add **V** to COM system, it reduces to laboratory system), we know

$$\mathbf{v}_1 = \mathbf{v'}_1 + \mathbf{V}$$

Writing above in components form,

$$v_1 \cos\theta_0 = v_1' \cos\theta + V \qquad \text{... (140)}$$

$$v_1 \cos\theta_0 = v_1' \sin\theta \qquad \text{... (141)}$$

where v_1 and v_1' are the speeds of m_1 after scattering in laboratory and COM coordinate systems, respectively.

From eqns. (140) and (141) after division (or after eliminating v_1), we get

$$\tan\theta_0 = \frac{v_1' \sin\theta}{(v_1' \cos\theta + V)} = \frac{\sin\theta}{(\gamma + \cos\theta)}, \gamma = \frac{V}{v_1'} \quad \text{... (142)}$$

The initial kinetic energy in the COM system is given by

$$T_i = \frac{1}{2}m_1(v - V)^2 + \frac{1}{2}m_2 V^2$$

which after putting the value of v from eqn. (139) becomes

$$T_i = \frac{1}{2}\frac{m_2}{m_1}(m_1 + m_2) V^2 \quad \text{... (143)}$$

The final kinetic energy in the COM system is given by

$$T_f = \frac{1}{2}m_1 v_1'^2 + \frac{1}{2}m_2 v_2'^2$$

Further using the condition that the COM is always stationary in the frame called COM system, that is,

$$m_1 v_1' = m_2 v_2'$$

or

$$v_2' = \frac{m_1}{m_2} v_1',$$

One gets

$$T_f = \frac{1}{2}\frac{m_1}{m_2}(m_1 + m_2) v_1'^2 \quad \text{... (144)}$$

Now the scattering being elastic, that is, $T_i = T_f$, one gets

$$\gamma = \frac{V}{v_1'} = \frac{m_1}{m_2} \quad \text{... (145)}$$

From eqns. (142) and (145), it may be noted that for $m_1 = m_2$ that is, for scattering of identical particles (which is almost true for scattering of neutrons by protons), $\gamma = 1$ and hence $\theta_0 = \theta/2$. Thus as θ varies from $\theta - \pi$, θ_0 ranges from $0 - \pi/2$. In other words, as viewed in the laboratory, there is no back scattering of neutrons from hydrogen.

Now the number of particles which are scattered into a small solid angle $d\Omega_0$ about (θ_0, ϕ_0) in the laboratory system (as $\sigma_0(\theta_0, \phi_0)\, d\Omega_0$ is proportional to the said number) is equal to $\sigma_0(\theta_0, \phi_0)\, d\Omega_0$. These same particles appear with in the solid angle $d\Omega$ about (θ, ϕ) in the COM system and in terms of the differential scattering cross-section in this frame, their number is equal to $\sigma(\theta, \phi) d\Omega$. Thus

$$\sigma_0(\theta_0, \phi_0)\, d\Omega_0 = \sigma(\theta, \phi)\, d\Omega$$

or

$$\sigma_0(\theta_0, \phi_0) \sin\theta_0\, d\theta_0\, d\phi_0 = \sigma(\theta, \phi) \sin\theta\, d\theta\, d\phi \quad \text{... (146a)}$$

The azimuthal angles in the two frames are, however, equal, that is, $\phi = \phi_0$. Hence from the above equation

$$\sigma_0(\theta_0, \phi_0) = \frac{\sin\theta}{\sin\theta_0}\frac{d\theta}{d\theta_0}\sigma(\theta, \phi) \quad \text{... (146}b\text{)}$$

From (142) on differentiation, we get

$$\sec^2\theta_0\, d\theta_0 = \frac{(1+\gamma\cos\theta)}{(\gamma+\cos\theta)^2}d\theta$$

Substituting $\frac{d\theta}{d\theta_0}$ from above into eqn. (146*b*), we get

$$\sigma_0(\theta_0, \phi_0) = \frac{\sin\theta}{\sin\theta_0}\frac{(\gamma+\cos\theta)^2}{(1+\gamma\cos\theta)}\sec^2\theta_0\,\sigma(\theta, \phi)$$

$$= \frac{\sin\theta}{\sin\theta_0}\frac{(\gamma+\cos\theta)^2}{(1+\gamma\cos\theta)}\left[1+\frac{\sin^2\theta}{(\gamma+\cos\theta)^2}\right]\sigma(\theta, \phi)$$

where use has been made of eqn. (142). On simplification, one gets

$$\sigma_0(\theta_0, \phi_0) = \frac{\sin\theta}{\sin\theta_0}\frac{(1+\gamma^2+2\gamma\cos\theta)}{(1+\gamma\cos\theta)}\sigma(\theta, \phi) \quad \text{... (147)}$$

From eqn. (142), using Fig. 14.11,

$$\sin\theta_0 = \frac{\sin\theta}{(1+\gamma^2+2\gamma\cos\theta)^{1/2}} \quad \text{... (148)}$$

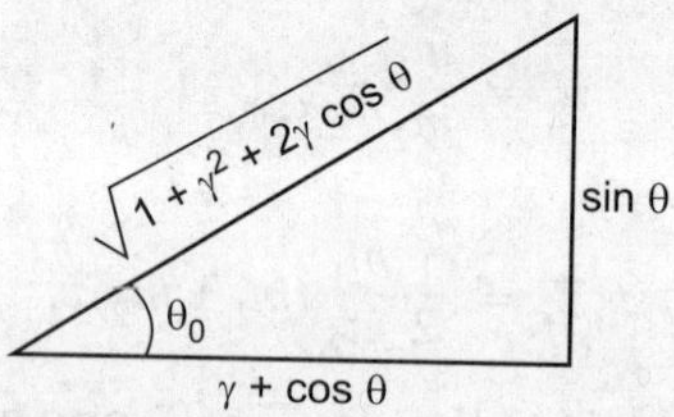

Fig. 14.11 *Relation between θ and θ_0 from eqn. (142)*

Substituting the value of $\frac{\sin\theta}{\sin\theta_0}$ from above into eqn. (147), we obtain the following relation between the differential scattering cross-sections in the two frames,

$$\sigma_0(\theta_0, \phi_0) = \frac{(1+2\gamma\cos\theta+\gamma^2)^{3/2}}{(1+\gamma\cos\theta)}\sigma(\theta, \phi) \quad \text{... (149)}$$

We must mention that it is $\sigma_0(\theta_0, \phi_0)$ which is measured and $\sigma(\theta, \phi)$ is obtained from the theory. Relation between θ and θ_0 can be found for known $\gamma = \left(\frac{m_1}{m_2}\right)$ from eqn. (142). Though the relations between angles and between differential scattering cross-sections in the two frames are complicated, it is clear from eqn. (146a) that the total scattering cross-section is same in both the coordinate systems

$$\left(\sigma_0 = \iint \sigma_0(\theta_0, \phi_0)\right) d\Omega_0 = \iint \sigma(\theta, \phi)\, d\Omega = \sigma)$$

as the total number of collisions is independent of the manner in which the process is described.

14.18 SCATTERING BY SPHERICAL SQUARE WELL POTENTIAL USING BORN APPROXIMATION

An attractive square well potential is represented by

$$V(r) = -V_0, \; r < R \qquad \ldots (58)$$

$$= 0, \; r > R$$

The scattering amptitude in the present context is given by

$$f(\theta) = -\frac{1}{K}\int_0^\infty rU(r)\sin Kr\,dr, \; K = 2k\sin\theta/2 \qquad \ldots (131)$$

which on substituting, $U(r) = \dfrac{2\mu V(r)}{\hbar^2}$ gives

$$f(\theta) = \frac{-2\mu}{\hbar^2 K}\int_0^\infty rV(r)\sin Kr\,dr \qquad \ldots (131a)$$

Further, the use of eqn. (58) leads to

$$f(\theta) = \frac{2\mu V_0}{\hbar^2 K}\int_0^\infty r\sin Kr\,dr$$

which on integrating by parts, gives

$$f(\theta) = \frac{2\mu V_0}{\hbar^2 K^3}(\sin KR - KR\cos KR) \qquad \ldots (150)$$

The differential scattering cross-section, therefore would be given by

$$\sigma(\theta) = |f(\theta)|^2 = \left(\frac{2\mu V_0 R^3}{\hbar^2}\right)^2 g(x) \qquad \ldots (151a)$$

where

$$g(x) = \left(\frac{\sin x - x\cos x}{x^3}\right)^2, \; x = KR = 2kR\sin\theta/2 \qquad \ldots (151b)$$

Further, the total scattering cross-section would be given by

$$\sigma_t = \left(\frac{2\mu V_0 R^3}{\hbar^2}\right)^2 \int_0^\pi\int_0^{2\pi}\left(\frac{\sin x - x\cos x}{x^3}\right)^2 \sin\theta\,d\theta\,d\phi$$

Now since,

$$x = 2kR\sin\theta/2,$$

$$dx = 2kR\cos\theta/2\,\frac{1}{2}d\theta \text{ and } xdx = (kR)^2\sin\theta\,d\theta$$

therefore,

$$\sigma_t = \frac{32\pi\mu^2 V_0^2 R^6}{\hbar^4}\gamma(y) \qquad \ldots (152a)$$

where

$$\gamma(y) = \frac{1}{y^2}\int_0^y \frac{(\sin x - x\cos x)^2}{x^5}dx$$

$$= \frac{1}{4y^2}\left(1 - \frac{1}{y^2} + \frac{\sin 2y}{y^3} - \frac{\sin^2 y}{y^4}\right), y = 2kR \quad \text{... (152 b)}$$

In the low energy limit ($kR << 1$),

$$g(x) \xrightarrow[x \to 0]{} \left[\frac{\left\{\left(x - \frac{x^3}{3!}\right) - x\left(1 - \frac{x^2}{2!}\right)\right\}}{x^3}\right] = \frac{1}{9} \quad \text{... (153)}$$

or g (0) = $\frac{1}{9}$. Fig. 14.12 shows the plot of function, $\frac{g(x)}{g(0)} = 9\,g(x)$ versus x. As θ increases, x increases (since K increases) and therefore $g(x)$ decreases.

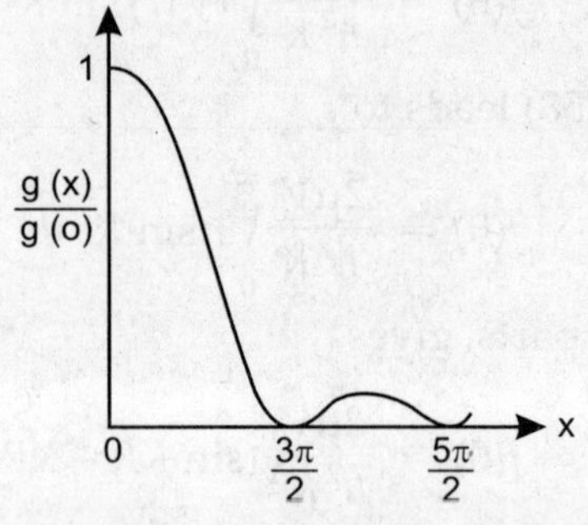

Fig. 14.12 *Variation of* $\frac{g(x)}{g(0)} = 9\ g(x)$ *with x. At x = 0,* $\frac{g(x)}{g(0)}$ *is unity but falls off rapidly and at x ~ 4, it is zero*

At $x = \frac{3\pi}{2} \simeq 4$, $\frac{g(x)}{g(0)} = 9\,g(x)$ is zero. The differential scattering cross-section using eqns. (151) and (153) would be given by

$$\sigma_{diff.} \xrightarrow[k\ small]{} \left(\frac{2\mu V_0 R^3}{3\hbar^2}\right)^2 \quad \text{... (154)}$$

which infers that the scattering is isotropic (independent of scattering angle) at low energies.

Now for y → 0, $\gamma(0) = \frac{1}{18}$.

Fig. 14.13 shows the plot of $\frac{\gamma(y)}{\gamma(0)} = 18\,\gamma(y)$ versus y. The total cross-section at low energies, that is, at $y \to 0$ from eqn. (152) would, therefore be given by

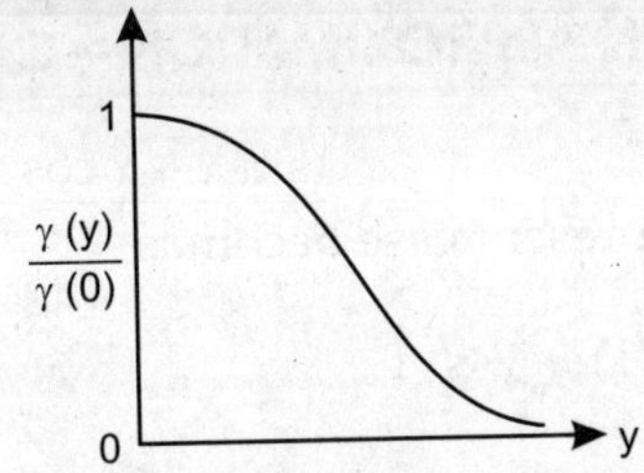

Fig. 14.13 *Variation of the function* $\frac{\gamma(y)}{\gamma(0)}$ *with y*

$$\sigma_t = \frac{32\pi\mu^2 V_0^2 R^6}{\hbar^4} \times \frac{1}{18}$$

$$= \frac{16\pi\mu^2 V_0^2 R^6}{9\hbar^4} \qquad \text{... (155)}$$

In the high energy limit ($kR >> 1$), x would increase much faster than $\sin x$ as the maximum value of $\sin x$ is 1. Hence

$$g(x) = \left(\frac{-\cos x}{x^2}\right)^2$$

Using above and eqn. (151*a*),

$$\sigma(\theta) \xrightarrow[k \text{ large}]{} \left(\frac{2\mu V_0 R^3}{\hbar^2}\right)^2 \frac{\cos^2(2kR\sin\theta/2)}{(2kR\sin\theta/2)^4}$$

$$= \left(\frac{\mu V_0 R}{2\hbar^2 k^2}\right)^2 \cos^2(2kR\sin\theta/2)\,\text{cosec}^4\theta/2$$

which fluctuates with θ. To get rid of the term $\cos^2(2kR\sin\theta/2)$ let us take its average value $\frac{1}{2}$. Thus

$$\sigma(\theta) = \frac{1}{2}\left(\frac{V_0 R}{4E}\right)^2 \text{cosec}^4\theta/2 \qquad \text{... (156)}$$

where $E = \frac{\hbar^2 k^2}{2\mu}$ is the bombarding energy in the COM system. The result indicates that the scattering shows a strong maximum in the forward direction ($\theta = 0$) at high energies.

Now for y large,

$$\gamma(y) \to \frac{1}{4y^2} = \frac{1}{16k^2R^2}$$

Using above and eqn. (152*a*), the total scattering cross-section at large energies would be given by

$$\sigma_t = \frac{32\pi\mu^2 V_0^2 R^6}{\hbar^4} \times \frac{1}{16k^2R^2}$$

$$= \frac{\mu\pi V_0^2 R^4}{\hbar^2 E} \qquad \text{... (157)}$$

14.19 VALIDITY OF BORN APPROXIMATION—CASE OF SQUARE WELL POTENTIAL

The validity condition in the present case becomes

$$\frac{\mu V_0}{\hbar^2 k}\left|\int_0^R (e^{2ikr}-1)\,dr\right| << 1$$

or $$\frac{\mu V_0}{\hbar^2 k}\left|\left[\frac{e^{2ikr}}{2ik}-r\right]_0^R\right| << 1$$

which can be written as

$$\frac{\mu V_0}{2\,\hbar^2 k^2}\left|e^{iy}-1-iy\right| << 1$$

or $$\frac{\mu V_0}{2\,\hbar^2 k^2}\left|(\cos y - 1 + i(\sin y - y)\right| << 1$$

Multiplying above by its complex conjugate and then taking the square root, one gets

$$\frac{\mu V_0}{2\,\hbar^2 k^2}\,[(\cos y - 1)^2 + (\sin y - y)^2]^{1/2} << 1 \qquad \text{... (158)}$$

In the low energy limit, that is for y small,

$$[(\cos y - 1)^2 + (\sin y - y)^2]^{1/2} \to \left[\left(1-\frac{y^2}{2!}-1\right)^2+\left(y-\frac{y^3}{3!}-y\right)^2\right]^{1/2}$$

$$\to \left(\frac{y^4}{4}+\frac{y^6}{36}\right)^{1/2}$$

$$\to \frac{y^2}{2}$$

where the term $\frac{y^6}{36}$ has been neglected in comparison to $\frac{y^4}{4}$. Substituting above into eqn. (158), one gets

$$\frac{\mu V_0 R^2}{\hbar^2} << 1 \qquad \text{... (159)}$$

The above inequality shows that if the potential is strong enough to bind a particle, the Born approximation then cannot be applied in the low energy limit or the Born approximation would be valid if the potential is too weak to bind a particle.

Further consider the case of $n - p$ scattering where the relation between depth and range of the potential is given by

$$\frac{\mu V_0 R^2}{\hbar^2} \approx 1$$

Thus from eqn. (159), it follows that the Born approximation is not applicable in the case of $n - p$ scattering at low energies.

In the high energy limit, that is, for y large, $[(\cos y - 1)^2 + (\sin y - y)^2]^{1/2} \to y$ as $\cos y$ for large y can be neglected as compared to 1. In the second term $\sin y$ can maximum be 1 and hence can be neglected as compared to y. Thus we are left with $(1 + y^2)^{1/2}$. Again 1 can be neglected in comparison to y and therefore finally we get y. As a result from eqn. (158), we get

$$\frac{\mu V_0}{2\hbar^2 k^2} \times y << 1$$

or

$$\frac{\mu V_0 R}{\hbar^2 k} << 1, \quad \text{... (160)}$$

The condition which is generally satisfied at high energies for weak potentials. Thus the Born approximation can safely be used in the high energy limit. Hence it supplements the partial wave analysis which is most useful in the case of low energies, that is, for $kR << 1$.

14.20 SCATTERING BY SCREENED COULOMB POTENTIAL

As an another application of the Born approximation, let us consider the scattering of electron by a neutral atom (also termed as scattering by screened Coulomb potential). If the nucleus is considered as a point charge, the interaction potential between the electron at a distance r from the nucleus is given by

$$V(r) = -\frac{Ze^2}{r}$$

where Z is the atomic number.

The effect of the orbital electrons which screen the nucleus from the incident electron, called "screening effect" can be accommodated by introducing an exponential factor, e^{-rla} where a is called the screening parameter (or atomic radius or radius of the atomic electron cloud that screens the nucleus). Therefore the screened Coulomb potential can be represented by

$$V(r) = -\frac{Ze^2}{r} e^{-r/a} \left(= -\frac{V_0 e^{-r/a}}{r} \right) \quad \text{... (161)}$$

The scattering amplitude in the Born approximation, is given by

$$f(\theta) = -\frac{2\mu}{\hbar^2 K} \int_0^\infty r\, V(r) \sin Kr\, dr, \; K = 2k \sin \theta/2 \quad \text{... (131a)}$$

From which

$$\sigma(\theta) = |f(\theta)|^2 = \left(\frac{2\mu}{\hbar^2 K}\right)^2 \left| \int_0^\infty r \sin Kr\, V(r)\, dr \right|^2$$

After substituting $V(r)$ from eqn. (161) into above, we get

$$\sigma(\theta) = \left(\frac{2\mu V_0}{\hbar^2 K}\right)^2 \left| \int_0^\infty e^{-r/a} \sin Kr\, dr \right|^2 \quad \text{... (162)}$$

Take,

$$I = \int_0^\infty e^{-r/a} \sin Kr\, dr$$

$$= \int_0^\infty \frac{[e^{-r(1/a - ik)} - e^{-r(1/a+ik)}]}{2i} dr$$

$$= \frac{1}{2i}\left[\frac{e^{-r(1/a - ik)}}{-(1/a - iK)} - \frac{e^{-r(1/a+iK)}}{-(1/a+iK)}\right]_0^\infty$$

$$= \frac{1}{2i}\left[\frac{1}{(1/a - iK)} - \frac{1}{(1/a+iK)}\right]$$

$$= \frac{K}{\left(\frac{1}{a^2}+K^2\right)} \quad \text{... (162a)}$$

Note that the above result can also be obtained by doing integration by parts. Doing so, one gets $\int_0^\infty e^{-r/a} \cos Kr\, dr$. Again integrating it by parts, one gets back I. Finally one can find I. Substituting (162*a*) into eqn. (162), one gets

$$\sigma(\theta) = \frac{4\mu^2 V_0^2}{\hbar^4} \frac{1}{\left(\frac{1}{a^2}+K^2\right)^2} \quad \text{... (163)}$$

Further note that the sign of V_0 is of no importance in the scattering problem, atleast in the Born approximation (that means, had we started even with the potential $V(r) = \frac{V_0 e^{-r/a}}{r}$, we would have got the same result). The above result shows that for a given θ, $\sigma(\theta)$ is a decreasing function of energy or $\sigma(\theta)$ for a given energy (that is, for fixed k) depends on the scattering angle. In particular, the cross-section in the forward direction ($\theta = 0$) is larger than the cross-section in the backward direction ($\theta = \pi$). It may be pointed out here that for a pure Coulomb potential, $a \to \infty$ and hence $\frac{1}{a^2}$ can be neglected in eqn. (163). Finally putting, $K = 2\,k \sin \theta/2$ and also replacing V_0 by Ze^2 and then e by $Z'e$ (that is, in the case of scattering of charged particles by atoms) in (163), one gets

$$\sigma(\theta) = \frac{ZZ'e^4}{16E^2} \operatorname{cosec}^4 \theta/2,\ E = \frac{\hbar^2 k^2}{2\mu} \quad \text{... (163a)}$$

which is nothing but Rutherford's formula. Of course, the way in which we have obtained it, does not constitute the proof. The theory we have used is not applicable to the Coulomb potential (as it is long range). However, it is interesting to note that the Born approximation for the given potential (eqn. (161)) gives precisely the Rutherford's formula for the limiting situation where the range of potential ('*a*' finally would be defined so) approaches infinity.

Before we proceed further, let us point out that in the discussion above, we have effectively considered a potential of the form,

$$V(r) = \frac{V_0 e^{-r/a}}{r} \qquad \text{... (161a)}$$

where V_0 and a, are real constants with 'a' positive. The potential is attractive or repulsive depending on whether V_0, is negative or positive. Its range is characterized by the distance 'a' since as the Fig. 14.14 shows, $V(r)$ is practically zero when r exceeds $2a$ or $3a$. The potential (161) bears the name of Yukawa who had the idea of associating it with nuclear forces whose range is of the order of Fermi. To explain the origin of this potential,

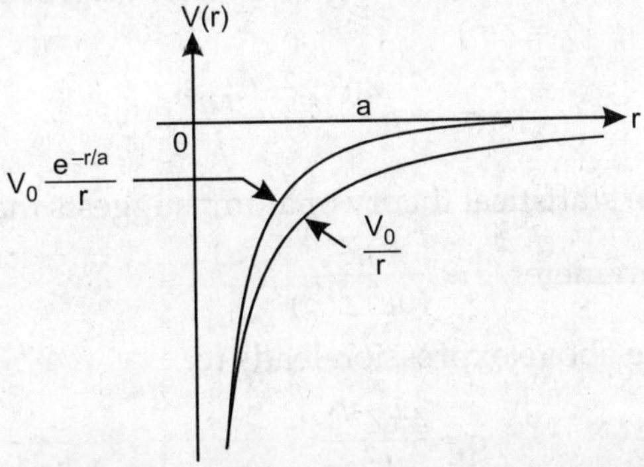

Fig. 14.14 *Yukawa and Coulomb potentials. The presence of the term $e^{-r/a}$ causes the Yukawa potential to approach zero much more rapidly when $r >> a$*

Yukawa was led to predict the existence of π-meson, which was indeed later discovered. As noticed earlier, for $a \to \infty$ this potential becomes the Coulomb potential which can thus be considered to be a Yukawa potential of infinite range.

For large energies of the relative motion ($Ka >> 1$), the particle will penetrate the atom so that the screening electrons are relatively ineffective. 1 can be neglected in comparison to K^2a^2 in eqn. (163). Hence the differential scattering cross-section on substituting $K = 2k \sin \theta/2$ and also replacing V_0 by Ze^2 and finally e by $Z'e$, one gets

$$\sigma(\theta) = \frac{ZZ'e^4}{16E^2} \operatorname{cosec}^4 \theta/2 \qquad \text{... (163b)}$$

which is nothing but the Rutherford's formula for the scattering of charged particles by atoms as expected.

The total scattering cross-section using eqn. (163), would be given by

$$\sigma_t = \int_0^{\pi}\int_0^{2\pi} \sigma(\theta) \sin\theta \, d\theta \, d\phi$$

$$= 2\pi \left(\frac{4\mu^2 V_0^2 a^4}{\hbar^4}\right) \int_0^{\pi} \frac{1}{(K^2a^2+1)} \sin\theta \, d\theta$$

Now since $K = 2k \sin \theta/2$,

$$\sigma_t = \frac{2\pi}{k^2}\left(\frac{4\mu^2 V_0^2 a^4}{\hbar^4}\right)\int_0^{2k} \frac{kdK}{(K^2a^2+1)^2}$$

$$= \frac{2\pi}{k^2}\left(\frac{4\mu^2 V_0^2 a^4}{\hbar^4}\right)\left[\frac{-1}{2(K^2a^2+1)a^2}\right]_0^{2k}$$

$$= \frac{4\mu^2 V_0^2}{\hbar^4}\frac{4\pi a^4}{(4k^2a^2+1)} \qquad \text{... (164)}$$

Since we are considering the case of high energies, 1 can be neglected in comparison to $4k^2a^2$ and thus (putting $V_0 = Ze^2$),

$$\sigma_t = \frac{4\mu^2 Z^2 e^4}{\hbar^4}\frac{\pi a^2}{k^2}$$

The Thomas—Fermi statistical theory of atom[2] suggests that for moderately heavy atoms, the screening parameter, $a \approx \frac{\hbar^2}{(\mu e^2 Z^{1/3})}$.

Using this value, the above expression leads to

$$\sigma_t = \frac{4\pi Z^{4/3}}{k^2} \qquad \text{... (165)}$$

14.21 VALIDITY OF BORN APPROXIMATION — CASE OF SCREENED COULOMB POTENTIAL

Substituting the value of $V(r)$ from eqn. (161), the validity condition in the present case takes the form,

$$\frac{\pi Z e^2}{\hbar^2 k}\left|\int_0^\infty \frac{(e^{2ikr}-1)e^{-a'r}}{r}dr\right| << 1,\ a' = \frac{1}{a} \qquad \text{... (166)}$$

To evaluate the integral,

$$I = \int_0^\infty \frac{(e^{2ikr}-1)e^{-a'r}}{r}dr, \qquad \text{... (167)}$$

We first differentiate it *w.r.t.*, the parameter a' and get

$$\frac{\partial I}{\partial a'} = -\int_0^\infty e^{-a'r}\,(e^{2ikr}-1)\,dr$$

$$= -\int_0^\infty [e^{-(a'-2ik)r} - e^{-a'r}]\,dr$$

[2] L.H. Thomas, Proc. Cambridge Phil. Soc. 23, 542 (1927); E. Fermi, Z. Physik 48, 73 (1928); 49, 550 (1928).

Using the result, $\int_0^\infty e^{-a'x}\,dx = \frac{1}{a'}$, one gets,

$$\frac{\partial I}{\partial a'} = -\left[\frac{1}{(a'-2ik)} - \frac{1}{a'}\right]$$

Integrating the above equation over a', we find

$$I = -\int\left[\frac{1}{(a'-2ik)} - \frac{1}{a'}\right] da' + C$$

$$= -\ln\frac{(a'-2ik)}{a'} + C$$

$$= -\ln(1-2ika) + C$$

where C is a constant of integration.

From (167), for $a \to 0$, $I \to 0$ so we must choose $C = 0$. Thus

$$I = -\ln(1-2\,ika)$$

Substituting above into eqn. (166), the validity condition becomes

$$\frac{\mu Ze^2}{\hbar^2 k}\,|\ln(1-2\,ika| << 1$$

or

$$\frac{\mu Ze^2}{\hbar^2 k}[\{\ln(1+4k^2a^2)^{1/2}\}^2 + (\tan^{-1}2ka)^2]^{1/2} << 1 \qquad \text{... (168)}$$

where the relation, $\ln(x-iy) = \ln(x^2+y^2)^{1/2} - i\tan^{-1}\frac{y}{x}$ has been used.

In the low energy limit ($ka << 1$), one can take, $\tan^{-1} 2ka \approx 2ka$. In the remaining term in eqn. (168), $4k^2a^2$ can be neglected in comparison to 1 and finally ln 1 = 0. Thus the validity condition becomes,

$$\frac{uZe^2}{\hbar^2 k}\,2ka << 1$$

Substituting,

$$a = \frac{\hbar^2}{\mu e^2 Z^{1/3}}$$

one gets

$$2\,Z^{2/3} << 1 \qquad \text{... (169)}$$

Since $Z \geq 1$, the above condition is never satisfied. Thus Born approximation cannot be used for the scattering of slow electrons by atoms.

In the high energy limit ($ka >> 1$), one can take $\tan^{-1} 2ka \approx \pi/2$ (as this is the maximum value) which being small compared to ln $2ka$ (1 has already been neglected in comparison to $4k^2a^2$) can be neglected. Also finally not even writing the big term *ln* $2ka$ in the multiplication, the validity condition from eqn. (168) becomes,

$$\frac{\mu Z e^2}{\hbar^2 k} << 1$$

Using the relation, $\frac{\hbar k}{\mu} = v$ for the relative velocity of the colliding particles, the above condition takes the form,

$$\frac{Ze^2}{\hbar v} = \frac{Ze^2}{\hbar c} \times \frac{c}{v} << 1$$

Using the value of $\frac{e^2}{\hbar c} = \frac{1}{137}$ for the fine structure constant and since the energy is large $\frac{c}{v} \approx 1$, one gets

$$\frac{Z}{137} << 1 \qquad \text{... (170)}$$

Thus the Born approximation becomes poor for heavier elements at large energies. That is, in order to use the Born approximation, we need incident particles of high energy and scatterer of low atomic number. Even for $Z = 10$ and energy 10 keV ($v = 6 \times 10^9$ cm/sec), $\frac{Z}{137} \times \frac{c}{v} \simeq 0.4$, which is not much less than 1.

PROBLEMS

14.1 Calculate the total S–wave scattering cross-section by a short range attractive spherical square well potential mathematically represented by

$$V(r) = -V_0, \ r < R$$
$$= 0, \ r > R$$

14.2 Determine the least value of V_0R^2 for a three dimensional square well potential mathematically represented by

$$V(r) = -V_0, \ r < R$$
$$= 0, r > R$$

in order that the scattering cross-section is zero at zero bombarding energy.

14.3 Determine the scattering cross-section for slow particles ($kR << 1$) by a spherical potential hump V_0 ($E < V_0$) and radius R, that is by a repulsive potential given by

$$V(r) = V_0, \ r < R$$
$$= 0, \ r > R$$

14.4 In the case of S–wave scattering (energy in this case being slightly bigger than what is termed as zero energy) by an attractive square well potential, show that

$$r_{eff} = R\left[1 - \frac{1}{k_0^2 R^2}\left(\frac{R}{a}\right) - \frac{1}{3}\left(\frac{R}{a}\right)^2\right]$$

where $k_0^2 = \dfrac{2\mu V_0}{\hbar^2}$. a and R are called scattering length and range of the potential, respectively.

14.5 Given,

$$\tan \delta_l = \frac{[k\, j_l'(kR) - \gamma_l\, j_l\,(kR)]}{[k\, n_l'(kR) - \gamma_l\, n_l\,(kR)]}$$

where $\gamma_l \left(= \dfrac{1}{R_l(r)} \dfrac{dR_l(r)}{dr}\bigg|_{r=R} \right)$ represents the logarithmic derivative of the radial wave function at the boundary $r = R$. Show that

$$\delta_0 = -kR$$

for low energy scattering by a hard sphere of radius R.

14.6 Show that

$$e^{i\delta_l} \sin \delta = e^{2i\xi l}\left[\frac{S_l}{(\beta_l - \Delta_l - iS_l)} + e^{-i\xi_l} \sin \xi_l\right]$$

where terms have their usual meaning. Finally obtain the relation

$$\sigma_l^{(res)}(E) = \frac{4\pi}{k^2}(2l+1)\frac{\left(\dfrac{\tau_l}{2}\right)^2}{\left[(E - E_l^{(0)})^2 + \left(\dfrac{\tau_l}{2}\right)^2\right]}$$

14.7 Using the results for ξ_l, Δ_l, S_l, β_l and δ_l, obtain expression for σ_0 for $E \to 0$ in the case of scattering by a spherical square well potential.

14.8 In the case of 3d square well potential described by

$$V(r) = -V_0,\ r < R$$
$$= 0,\ r > R$$

find δ_0 and δ_1. Also find the corresponding scattering cross-sections, that is, σ_0 and σ_1.

14.9 In the case of scattering by a perfectly rigid sphere potential, show that

$$\tan \delta_l = \frac{j_l(kR)}{n_l(kR)}$$

R being the range of potential. Further show that for

$$l = 0,\ \delta_0 = -kR$$

14.10 If the potential energy for scattering of an electron by an atom (or scattering by screened Coulomb potential) can be represented by $V(r) = -\dfrac{Ze^2}{r} e^{-r/a}$, where a is called shielding radius (or screening parameter), show that in the Born approximation, the scattering amplitude is given by

$$f(\theta) = \frac{2\mu Z e^2 a^2}{\{\hbar^2 (K^2 a^2 + 1)\}}$$

14.11 Calculate the Rutherford scattering cross-section starting from the expression for the scattering amplitude in Born approximation.

14.12 Using Born approximation, find the differential scattering cross-section from an exponential potential, $V(r) = -V_0 e^{-r/a}$, where V_0 and a are constants characterizing the strength and range of the potential, respectively. Also find total scattering cross-section. What is the validity criterion in this case?

14.13 Using Born approximation, determine the differential and total scattering cross-sections from an exponential potential, $V(r) = V_0 e^{-\alpha r^2}$, called Gaussian potential.

14.14 The scattering of fast (high energy) electrons by an atom of atomic number Z can be represented to a good approximation as the scattering due to a spherically symmetric potential of the form,

$$V(r) = -\frac{Ze^2}{r} + Ze^2 \int \frac{\rho(\vec{\mathbf{r}}')}{|\mathbf{r} - \mathbf{r}'|} dr'$$

where the first term represents the potential due to nucleus and the second, potential due to the cloud of electrons, $\rho(\mathbf{r}')$ is the density of electron cloud at $\mathbf{r}'$, $\mathbf{r}$ is the position of the scattered electron. Derive an expression for the scattering amplitude in the given case using Born approximation.

14.15 The electronic density for hydrogen atom in the ground state is given by, $\rho(r) = |\varphi_{100}|^2 = \frac{1}{\pi a_0^3} e^{-2r/a_0}$, which is spherically symmetric and where a_0 is the Bohr radius. Show that the atomic scattering (or atomic form factor) in the said case is,

$$F(K) = \frac{1}{\left(1 + \frac{K^2 a_0^2}{4}\right)^2}$$

Calculate $\sigma(\theta)$ from the above. Finally calculate the total scattering cross-section and show that for $ka_0 >> 1$,

$$\sigma_t \simeq \frac{7\pi}{3k^2}$$

14.16 Considering the scattering of or–particles by a potential represented by

$$V(r) = -V_0, \quad r < a$$
$$= \frac{2Ze^2}{r}, \quad r > a,$$

obtain an expression for the scattering cross-section in this case (which affectively amounts to scattering of α-particles by nuclei) using Born approximation.

14.17 Using first order time dependent perturbation theory, derive an expression for the Born approximation scattering cross–section.

SOLUTIONS

14.1 The radial wave equation corresponding to the S (that is, $l = 0$) — wave is given by

$$\frac{1}{r^2}\frac{d}{dr}\left(\frac{r^2 dR_0(r)}{dr}\right) + \{k^2 - U(r)\}\, R_0(r) = 0, \qquad \text{... (14.1}a\text{)}$$

where
$$k^2 = \frac{2\mu E}{\hbar^2},\; U(r) = \frac{2\mu V(r)}{\hbar^2} \qquad \text{... (14.1}b\text{)}$$

and where we have taken the term $\frac{l(l+1)}{r^2}$ equal to zero in the given case

Substituting, $R_0(r) = \frac{F_0(r)}{r}$ in eqn. (14.1a), one gets,

$$\frac{d^2F_0(r)}{dr^2} + \{k^2 - U(r)\}\, F_0(r) = 0$$

or

$$\frac{d^2F_0(r)}{dr^2} + k'^2 F_0(r) = 0,\; r < R \qquad \text{... (14.2}a\text{)}$$

where

$$k'^2 = k^2 - U(r) = k^2 + \frac{2\mu V_0}{\hbar^2} = k^2 + k_0^2,\; k_0^2 = \frac{2\mu V_0}{\hbar^2} \qquad \text{... (14.2}b\text{)}$$

Similarly substituting, $R_0(r) = \frac{f_0(r)}{r}$ in eqn. (14.1a), one gets,

$$\frac{d^2f_0(r)}{dr^2} + k^2 f_0(r) = 0,\; r > R \qquad \text{... (14.3)}$$

The solution of eqn. (14.2a) would be given by

$$F_0(r) = A \sin k' r,\; r < R \qquad \text{... (14.4)}$$

where the cosine solution has been discarded otherwise $F_0\,(= rR_0)$ which has to vanish at $r = 0$ would not vanish at the said point (origin), whereas the solution of eqn. (14.3) would be

$$f_0(r) = B \sin(kr + \delta_0),\quad r > R \qquad \text{... (14.5)}$$

Which is actually the asymptotic form of the solution and included both $j_0(kr)$ and $\eta_0(kr)$ or in other words $f_0(r)$ would be a linear combination of the functions sin kr and cos kr (as the region of the solution does not contain the point $r = 0$ or origin) as both of them could be solutions of eqn. (14.3).

At the boundary of the well, that is, at $r = R$, the wave functions and their derivatives must match. Hence

$$B \sin(kR + \delta_0) = A \sin k' R$$

and

$$Bk \cos (kR + \delta_0) = Ak' \cos k' R$$

Dividing former by the later, we obtain

$$\tan (kR + \delta_0) = \frac{k}{k'} \tan k'R \quad \text{... (14.6)}$$

or

$$\delta_0 = \tan^{-1}\left\{\frac{k}{k'} \tan k'R\right\} - kR \quad \text{... (14.7)}$$

Since we are interested in the S–wave scattering cross-section (that is corresponding to $l = 0$), this means

$$kR << 1 \quad \text{... (14.8)}$$

as $l = 0$ from the condition $l << kR$ amounts to $kR << 1$.

Under the above condition (that is, when k is very small), eqn. (14.7) reduces to

$$\delta_0 = \frac{k}{k'} \tan k'R - kR \quad \text{... (14.9)}$$

Now for small energies ($k \rightarrow 0$) of the relative motion and deep potential well (V_0 large), we have

$$k' = (k^2 + k_0^2)^{1/2} \approx k_0$$

Therefore,

$$\delta_0 = \frac{k}{k_0} \tan k_0 R - kR$$

$$\delta_0 = kR\left[\frac{\tan k_0 R}{k_0 R} - 1\right] \quad \text{... (14.10)}$$

Note that k being small, δ_0 is small. Thus the cross-section for scattering by a deep spherical potential well for small energies of relative motion would be given by

$$\sigma = \sigma_0 = \frac{4\pi}{k^2} \sin^2 \delta_0$$

$$= \frac{4\pi}{k^2} \delta_0^2$$

$$= \frac{4\pi}{k^2} k^2 R^2 \left(\frac{\tan k_0 R}{k_0 R} - 1\right)^2$$

$$= 4\pi R^2 \left(\frac{\tan k_0 R}{k_0 R} - 1\right)^2, \; k_0^2 = \frac{2\mu V_0}{\hbar^2} \quad \text{... (14.11)}$$

Which is non-vanishing isotropic (independent of θ) S-wave scattering cross-section (also called zero energy scattering cross-section).

Fig 14.15 shows a graph of the zero energy scattering cross-section derived from eqn. (14.11) as a function of well depth V_0 for a fixed

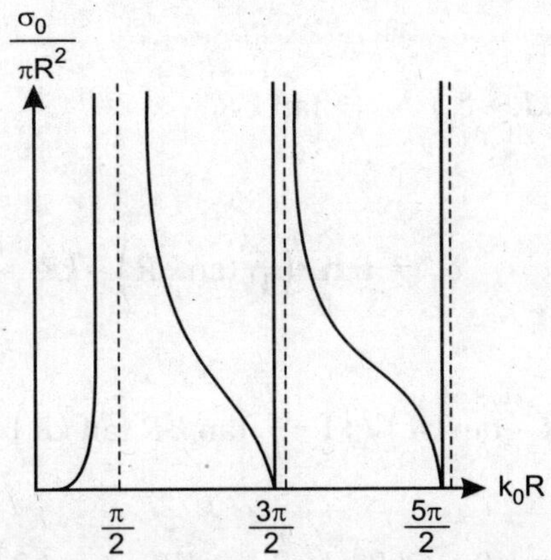

Fig. 14.15 *Zero energy cross-section (in units of πR^2) for an attractive spherical potential well as a function of well depth ($k_0R = (2\mu V_0 R^2//\hbar^2)^{1/2}$. R is range of the potential*

range R. If the well is shallow (that is, if V_0 is small), the cross-section is nearly zero.

One notices that when slow particles are scattered by the given potential and whenever k_0R is approximately equal to $\left(n+\frac{1}{2}\right)\pi$, n being an integer, the scattering cross-section reaches its maximum (infinity), the resonance value.

14.2 The radial part of the Schrödinger equation in the present case ($l = 0$) would take the forms (see problem 14.1)

$$\frac{d^2F_0(r)}{dr^2} + k'^2 F_0(r) = 0, \; r < R \qquad \text{... (14.2a)}$$

where

$$k'^2 = k^2 + k_0^2, \quad k^2 = \frac{2\mu E}{\hbar^2} \text{ and } k_0^2 = \frac{2\mu V_0}{\hbar^2} \qquad \text{... (14.2b)}$$

and

$$\frac{d^2f_0(r)}{dr^2} + k^2 f_0(r) = 0, \; r > R \qquad \text{... (14.3)}$$

The solutions of the above equations would be given by (see details in problem 14.1),

$$F_0(r) = A \sin k'r, \; r < R \qquad \text{... (14.4)}$$

$$f_0(r) = B \sin (kr + \delta_0), \; r > R \qquad \text{... (14.5)}$$

The continuity condition at $r = R$ demands that the logarithmic derivatives of (14.4) and (14.5) should be equal that is,

$$\frac{k'A \cos k'R}{A \sin k'R} = k\frac{B \cos (kR + \delta_0)}{B \sin (kR + \delta_0)}$$

or

$$\tan (kR + \delta_0) = \frac{k}{k'} \tan k'R \quad \text{... (14.6)}$$

or

$$\delta_0 = \tan^{-1}\left(\frac{k}{k'} \tan k'R\right) - kR \quad \text{... (14.7)}$$

Therefore,

$$\tan \delta_0 = \left(\frac{k}{k'} \tan k'R - \tan kR\right)\Big/\left(1 + \frac{k}{k'} \tan k'R \tan kR\right)$$

or

$$k \cot \delta_0 = k\left(1 + \frac{k}{k'} \tan k'R \tan kR\right)\Big/\left(\frac{k}{k'} \tan k'R - \tan kR\right) \quad \text{... (14.12)}$$

Further using the definition of the scattering length (eqn. 77) and then using the above equation,

$$-\frac{1}{a} = \underset{k \to 0}{Lt}\ k \cot \delta_0 = \frac{k_0}{\tan k_0 R - k_0 R} \quad \text{... (14.13)}$$

or

$$a = \left(R - \frac{\tan k_0 R}{k_0}\right) \quad \text{... (14.13a)}$$

The scattering cross-section at zero bombarding energy is given by

$$\sigma = 4 \pi a^2 \quad \text{... (43)}$$

which would be zero if $a = 0$. This implies that

$$a = R - \frac{\tan k_0 R}{k_0} = 0$$

or

$$\tan k_0 R = k_0 R \quad \text{... (14.14)}$$

The solution of the above transcendental equation is (the point of intersection in Fig. 14.16) given by

$$k_0 R \simeq \frac{3\pi}{2}$$

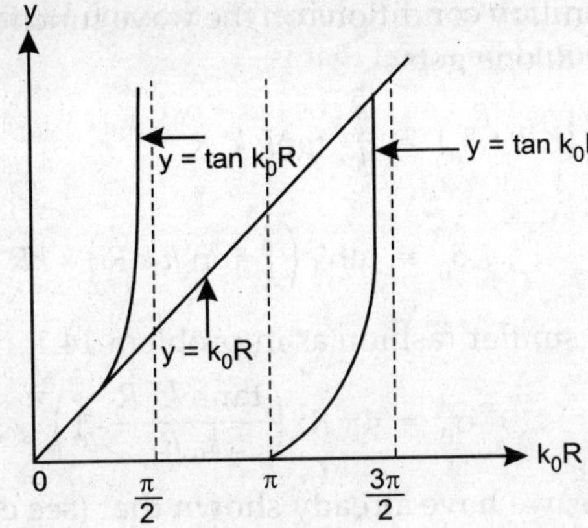

Fig. 14.16 *The point of intersection gives the solution of the transcendental equation tan* $k_0R = k_0R$

or $$\left(\frac{2\mu V_0}{\hbar^2}\right)^{1/2} R \simeq \frac{3\pi}{2}$$

or $$V_0R^2 \simeq \frac{3\pi^2}{4} \times \frac{\hbar^2}{2\mu}$$

14.3 In the present case since $kR \ll 1$, only the $l = 0$ wave would contribute. Hence, the radial part of the Schrödinger equation would be given by

$$\frac{1}{r^2}\frac{d}{dr}\left(r^2\frac{dR_0(r)}{dr}\right) + |k^2 - U(r)|\, R_0(r) = 0 \quad \text{... (14.1a)}$$

where $$k^2 = \frac{2\mu E}{\hbar^2},\ U(r) = \frac{2\mu V(r)}{\hbar^2} \quad \text{... (14.1b)}$$

Substituting $R_0(r) = \dfrac{F_0(r)}{r}$ in eqn. (14.1*a*) one gets

$$\frac{d^2F_0(r)}{dr^2} - k'^2 F_0(r) = 0,\ r < R \quad \text{... (14.15a)}$$

where $$k'^2 = U(r) - k^2 = k_0^2 - k^2,\ k_0^2 = \frac{2\mu V_0}{\hbar^2} \quad \text{... (14.15b)}$$

Similarly substituting $R_0(r) = \dfrac{f_0(r)}{r}$ in eqn. (14.1*a*), one gets

$$\frac{d^2f_0(r)}{dr^2} + k^2 f_0(r) = 0,\ r > R \quad \text{... (14.3)}$$

The solutions of eqns. (14.15*a*) and (14.3) arc given by

$$F_0(r) = A \sin h\, k'r,\ r < R \quad \text{... (14.4a)}$$

and

$$f_0(r) = B \sin (kr + \delta_0),\ r > R \quad \text{... (14.5)}$$

Applying the boundary conditions on the wave functions and their derivatives must match at $r = R$, one gets

$$\tan(KR + \delta_0) = \frac{k}{k'} \tanh k' R$$

or

$$\delta_0 = \tan^{-1}\left(\frac{k}{k'} \tan h\, k'R\right) - kR \quad \text{... (14.16)}$$

Proceeding in the similar fashion as in problem 14.1, one gets

$$\sigma_0 = 4\pi R^2 \left(\frac{\tanh k_0 R}{k_0 R} - 1\right)^2$$

14.4 In the present case, we have already shown that (see eqn. 14.12)

where
$$k'^2 = k^2 + k_0^2,\ k^2 = \frac{2\mu E}{\hbar^2} \text{ and } k_0^2 = \frac{2\mu V_0}{\hbar^2} \quad \text{... (14.2}b\text{)}$$

Now

$$k' = (k^2_, + k_0^2)^{1/2} = k_0\left(1 + \frac{k^2}{k_0^2}\right)^{1/2}$$

$$= k_0\left(1 + \frac{k^2}{2k_0^2}\right)$$

$$= k_0 + \frac{k^2}{2k_0} \quad \text{... (14.17)}$$

Using the Taylor series expansion about k_0R

$$\left\{f(x) = f(a + x - a) = f(a) + (x - a)\ \frac{df(x)}{dx}\bigg|_{x-a=0} + \ldots\right\}$$

and retaining only upto second term, we get

$$\tan k' R = \tan\{k_0R + (k^2 + k_0^2)^{1/2} R - k_0 R\}$$

$$\simeq k_0R + \frac{(k^2 R)}{2\,k_0} \sec^2 (k^2 + k_0^2)^{1/2} R\bigg|_{k^2 \to 0}$$

$$= \tan k_0R + k^2\left[\frac{R}{2k_0}(1 + \tan^2 k_0 R)\right] \quad \text{... (14.18)}$$

Further

$$\tan kR = \frac{\sin kR}{\cos kR} \underset{R \to 0}{\simeq} \left[kR - \frac{(kR)^3}{3!\ldots} + \ldots\right]\left[1 + \frac{(kR)^2}{2!\ldots} \ldots\right]$$

$$= kR + \frac{(kR)^3}{3} + \ldots \quad \text{... (14.19)}$$

Substituting from eqns. (14.17) - (14.19) into eqn. (14.12), we get

$$k \cot \delta_0 = k \,.\, \frac{\left(k_0 + \frac{k^2}{2k_0}\right) + k\left\{\tan k_0 R + \frac{(kR)^2}{2k_0 R}(1 + \tan^2 k_0 R)\right\}\left\{kR + \frac{(kR)^3}{3}\right\}}{k\left\{\tan k_0 R + \frac{k^2 R}{2k_0}(1 + \tan^2 k_0 R)\right\} - \left(k_0 + \frac{k^2}{2k_0}\right)\left\{kR + \frac{(kR)^3}{3}\right\}}$$

Cancelling k in the numerator and in the denominator of the R.H.S. of the above equation,

$$k \cot \delta_0 = \frac{\left(k_0 + \frac{k^2}{2k_0}\right) + k^2 R\left\{1 + \frac{(kR)^2}{3}\right\}\left\{\tan k_0 R + \frac{(kR)^2}{2k_0 R}(1 + \tan^2 k_0 R\right\}}{\tan k_0 R + \frac{k^2 R}{2k_0}(1 + \tan^2 k_0 R) - R\left(k_0 + \frac{k^2}{2k_0}\right)\left\{1 + \frac{(kR)^2}{3}\right\}}$$

$$= \frac{k_0 + k^2\left(\frac{1}{2k_0} + R \tan k_0 R\right)}{(\tan k_0 R - k_0 R) + k^2\left(\frac{R}{2k_0}\tan^2 k_0 R - \frac{k_0 R^3}{3}\right)}$$

$$= \frac{k_0 + k^2\left(\frac{1}{2k_0} + R \tan k_0 R\right)}{(\tan k_0 R - k_0 R)\left[1 + \frac{k^2}{(\tan k_0 R - k_0 R)}\left(\frac{R}{2k_0}\tan^2 k_0 R - \frac{k_0 R^3}{3}\right)\right]}$$

$$= \frac{k_0}{(\tan k_0 R - k_0 R)}\left\{1 + \frac{k^2}{k_0}\left(\frac{1}{2k_0} + R \tan k_0 R\right)\right\}$$

$$\left\{1 - \frac{k^2}{\tan k_0 R - k_0 R}\left(\frac{R}{2k_0}\tan^2 k_0 R - \frac{k_0 R^3}{3}\right)\right\}$$

$$= \frac{k_0}{(\tan k_0 R - k_0 R)}$$

$$\left[1 + k^2\left\{\frac{1}{2k_0^2} + \frac{R}{k_0}\tan k_0 R - \frac{1}{(\tan k_0 R - k_0 R)}\left(\frac{R}{2k_0}\tan^2 k_0 R - \frac{k_0 R^3}{3}\right)\right\}\right]$$

where terms of the order higher than k^2 have been neglected. Using the definition of scattering length, that is,

$$\underset{k \to 0}{Lt}\; k \cot \delta_0 = -\frac{1}{a} = \frac{k_0}{(\tan k_0 R - k_0 R)} \qquad \text{... (14.13)}$$

from which

$$k_0(R - a) = \tan k_0 R \qquad \text{... (14.20)}$$

The above equation leads to

$$k \cot \delta_0 = -\frac{1}{a}\left[1+k^2\left\{\frac{1}{2k_0}+R(R-a)+\frac{1}{k_0 a}\left(\frac{Rk_0}{2}R-a^2-\frac{k_0 R^3}{3}\right)\right\}\right]$$

$$= -\frac{1}{a}-\frac{k^2}{2a}\left[\frac{1}{k_0^2}-2Ra+\frac{1}{a}\left(R^3+Ra^2-\frac{2R^3}{3}\right)\right]$$

$$= -\frac{1}{a}-\frac{k^2}{2}\left[\frac{1}{ak_0^2}-2R+\frac{1}{a^2}\left(\frac{R^3}{3}+Ra^2\right)\right]$$

$$= -\frac{1}{a}-\frac{1}{2}k^2 R\left(\frac{1}{aRk_0^2}-1+\frac{R^2}{3a^2}\right)$$

$$= -\frac{1}{a}+\frac{1}{2}k^2 \mathrm{R}\left[1-\frac{1}{k_0^2 R^2}\left(\frac{R}{a}\right)-\frac{1}{3}\left(\frac{R}{a}\right)^2\right] \quad \text{... (14.21)}$$

Comparing above equation with (78*a*), one gets

$$r_{eff} = R\left[1-\frac{1}{k_0^2 R^2}\left(\frac{R}{a}\right)-\frac{1}{3}\left(\frac{R}{a}\right)^2\right]$$

14.5 Since $V(r) = \infty$, for $r \le R$, all $R_l(r)$ would vanish at $r = \mathrm{R}$. Thus $\gamma_l = \infty$ which amounts to

$$\tan \delta_l = \frac{j_l(kR)}{n_l(kR)}$$

$$= J_{l+1/2}(kR)\left(\frac{\pi}{2kR}\right)^{1/2} \Big/ (-1)^{l+1} J_{-(l+1/2)}(kR)\left(\frac{\pi}{2kR}\right)^{1/2}$$

where J is ordinary Bessel function of half odd integer order and where we have divided the numerator and denominator of the R.H.S. of the given equation by γ_l.

For the present case of low energy scattering ($kR \ll 1$) only $l = 0$ wave would, contribute. Thus

$$\tan \delta_0 = -\frac{J_{1/2}(kR)}{J_{-1/2}(kR)}$$

$$= -\left(\frac{2}{\pi kR}\right)^{1/2} \sin kR \Big/ \left(\frac{2}{\pi kR}\right)^{1/2} \cos kR$$

$$= -\tan kR$$

which means,

$$\delta_0 = -kR$$

14.6 $$e^{i\delta_l}\sin\delta_l = e^{i\delta_l}\frac{(e^{i\delta_l}-e^{-i\delta_l})}{2i} = \frac{(e^{2i\delta_l}-1)}{2i}$$

Using eqn. (88), the above equation takes the form

$$e^{i\delta_l}\sin\delta_l = \frac{(e^{2i\xi_l}e^{2i\zeta_l}-1)}{2i}$$

Further using eqn. (89*b*), we get

$$e^{i\delta_l}\sin\delta_l = \frac{e^{2i\xi_l}}{2i}\frac{(\beta_l-\Delta_l+iS_l)}{(\beta_l-\Delta_l-iS_l)}$$

$$= \frac{e^{2i\delta_l}}{2i}\left[\frac{(\beta_l-\Delta_l-iS_l+2iS_l)}{(\beta_l-\Delta_l-iS_l)} - e^{-2i\xi_l}\right]$$

$$= \frac{e^{2i\xi_l}}{2i}\left[1+\frac{2iS_l}{(\beta_l-\Delta_l-iS_l)} - e^{-2i\xi_l}\right]$$

$$= e^{2i\xi_l}\left[\frac{S_l}{(\beta_l-\Delta_l-iS_l)}+\frac{(l-e^{-2i\xi_l})}{2i}\right]$$

$$= e^{2i\xi_l}\left[\frac{S_l}{(\beta_l-\Delta_l-iS_l)}+e^{-i\xi_l}\sin\xi_l\right]$$

This is particularly handy form, since only this combination enters into the scattering amplitude expression (32).

Now

$$\frac{S_l}{(\beta_l-\Delta_l-iS_l)} = \left(\frac{S_l}{-b_l}\right)\Big/\left\{\frac{(\Delta_l-\beta_l)}{b_l}-i\left(\frac{S_l}{-b_l}\right)\right\}$$

$$= \frac{\tau_l/2}{\left(E_l^{(0)}-E\right)-i\tau_l/2}$$

where use has been made of eqns. (98) and (100). Thus

$$e^{i\delta_l}\sin\delta_l = e^{2i\xi_l}\left[\frac{\tau_l/2}{\left\{\left(E_l^{(0)}-E\right)-i\tau_l/2\right\}}+e^{-i\xi_l}\sin\xi_l\right]$$

Using above and eqn. (32), the partial scattering amplitude $f_l(\theta)$ would be given by

$$f_l(\theta) = \frac{(2l+1)}{k}e^{2i\xi_l}\left[\frac{(\tau_l/2)}{\left\{\left(E_l^{(0)}-E\right)-i\tau_l/2\right\}}+e^{-i\xi_l}\sin\xi_l\right]p_l(\cos\theta)$$

The above expression gives a clear separation of the resonant part of the amplitude from the non-resonant part, which depends only on ξ_l. If, as is the case for low energies, ξ_l, the hard sphere phase shifts are negligible, the resonance term dominates and we obtain for pure resonance differential cross-section

$$\sigma_l^{(res)}(\theta) = \frac{(2l+1)}{k^2} \frac{(\tau_l/2)^2}{\left\{\left(E - E_l^{(0)}\right)^2 + (\tau_l/2)^2\right\}} P_l^2(\cos\theta)$$

Integrating above over the angles, we obtain the resonant total cross-section.

$$\sigma_l^{(res)}(E) = \frac{4\pi}{k^2}(2l+1) \frac{(\tau_l/2)^2}{\left\{\left(E - E_l^{(0)}\right)^2 + (\tau_l/2)^2\right\}} \quad \text{... (101)}$$

14.7 Since the solution inside the given well is given by

$$R_l(r) = A_l j_l(k'r),\ r < R,\ k'^2 = \frac{2\mu(E + V_0)}{\hbar^2} \quad \text{... (14.22)}$$

Therefore,

$$\gamma_l = \frac{1}{R_l(r)} \frac{dR_l(r)}{dr}\bigg|_{r=R}$$

$$= \frac{1}{j_l(k'r)} \frac{dj_l(k'r)}{dr}\bigg|_{r=R}$$

or

$$\beta_l = \frac{R}{j_l(k'r)} \frac{dj_l(k'r)}{dr}\bigg|_{r=R}$$

In the low energy limit ($E \to 0$ or $k \to 0$) only $l = 0$ (S) – wave would contribute for computation of the total scattering cross-section. Hence

$$\beta_0 = \frac{R}{j_0(k'r)} \frac{dj_0(k'r)}{dr}\bigg|_{r=R}$$

Since,

$$j_0(k'r) = \frac{\sin k'r}{k'r},$$

Hence

$$\beta_0 = \frac{R}{\left(\dfrac{\sin k'R}{k'R}\right)} \left[\frac{d}{dr}\left(\frac{\sin k'r}{k'r}\right)\right]_{r=R}$$

$$= \frac{k'R^2}{\sin k'R}\left[\frac{(k'rk'\cos k'r - \sin k'r\, k')}{k'^2 r^2}\right]_{r=R}$$

$$= k'R\cot k'R - 1$$

Further, since

$$j_0(kr) = \frac{\sin kr}{kr}, \; \eta_0(kr) = -\frac{\cos kr}{kr},$$

Hence

$$j_0^2 + \eta_0^2 = \frac{\sin^2 kr}{k^2 r^2} + \frac{\cos^2 kr}{k^2 r^2}$$

$$= \frac{1}{k^2 r^2}$$

which gives

$$\Delta_0(kR) = kR\left[\frac{j_0(kr)\,j_0'(kr) + \eta_0(kr)\,\eta_0'(kr)}{j_0^2(kr) + \eta_0^2(kr)}\right]_{r=R}$$

$$= kR\left[\frac{1}{\left(\frac{1}{k^2 r^2}\right)}\left\{-\frac{\sin^2 kr}{k^3 r^3} - \frac{\cos^2 kr}{k^3 r^3}\right\}\right]_{r=R}$$

$$= -1$$

Now

$$\tan \xi_0 = \frac{j_0(kR)}{\eta_0(kR)} = -\tan kR$$

or

$$\xi_0 = -kR$$

Further

$$S_0(kR) = \frac{1}{kR}\frac{1}{\{j_0^2(kR) + \eta_0^2(kR)\}}$$

$$= kR$$

and

$$\tan \zeta_0 = \frac{S_0}{(\beta_0 - \Delta_0)} = \frac{kR}{\{(k'R\cot k'R - 1) - (-1)\}}$$

$$= \frac{k}{k'}\tan k'R$$

Hence

$$\delta_0 = \xi_0 + \zeta_0$$

$$= -kR + \tan^{-1}\left(\frac{k}{k'}\tan k'R\right)$$

$$= -kR + \frac{k}{k'}\tan k'R$$

as k is small. Finally $k' \xrightarrow[k \to 0]{} k_0$, hence

$$\delta_0 = kR\left[\frac{\tan k_0 R}{k_0 R} - 1\right],\ k_0^2 = \frac{2\mu V_0}{\hbar^2}$$

Thus *S*–wave scattering cross-section,

$$\sigma_0 = \frac{4\pi}{k^2}\sin^2\delta_0 = \frac{4\pi}{k^2}\delta_0^2 = 4\pi R^2\left[\frac{\tan k_0 R}{k_0 R} - 1\right]^2$$

14.8 In the given case, the interior solution is given by

$$R_l(r) = A_l j_l(k'r),\ r < R,\ k'^2 = \frac{2\mu(E + V_0)}{\hbar^2} \quad \text{... (14.22a)}$$

and hence the logarithmic derivative at $r = R$, would be given by

$$\gamma_l = \frac{1}{R_l(r)}\frac{dR_l(r)}{dr}\bigg|_{r=R}$$

$$= k'\frac{j_l'(k'R)}{j_l(k'R)} \quad \text{... (14.23)}$$

From eqn. (84), the phase shifts are given by

$$\tan\delta_l = \frac{[k\, j_l'(kR) - \gamma_l\, j_l(kR)]}{[k\, n_l'(kR) - \gamma_l\, n_l(kR)]}$$

Putting the volue of γ_l from eqn. (14.23) into above, we get

$$\tan\delta_l = \frac{[k\, j_l'(kR)\, j_l(k'R) - k'\, j_l'(k'R)\, j_l(kR)]}{[kn_l'(kR)\, j_l(k'R) - k'\, j_l'(k'R)\, \eta_l(kR)]} \quad \text{... (14.24)}$$

which is an exact expression.

Fig. 14.17 shows the variation of δ_l (for $l = 0$ and 1) as a function of k. The total cross-sections computed from (Using $\sigma_l = \frac{(2l+1)}{k^2}\sin^2\delta_l = \frac{(2l+1)}{k^2}\frac{\tan^2\delta_l}{(1+\tan^2\delta_l)}$) this are displayed in Fig. 14.18. For very low bombarding energies ($kR << 1$) the cross-section of the *P*–wave (and that of all other waves belonging to higher l or high orbital quantum numbers) is seen to be negligible small compared to the cross-section of the *S*–wave. For very high bombarding energy ($kR >> 1$), the cross-Section of every partial wave tends to zero. In the range of moderate energies a very interesting phenomenon can be observed.

The phase shift of *P*–wave increases sharply near $kR = 0.7$ or the total cross-section corresponding to *P*–wave,

$$\sigma_1 = \frac{4\pi}{k^2}(2 \times 1 + 1)\sin^2\delta_1$$

takes its maximum value, $\frac{12}{k^2}$. This produces a sharp maximum seen In Fig. 14.18. The occurrence of sharp maximum at a non-zero energy is a feature of resonance scattering in any partial wave with $l > 0$, in contrast with the S–wave case where the cross-section is highest at zero energy.

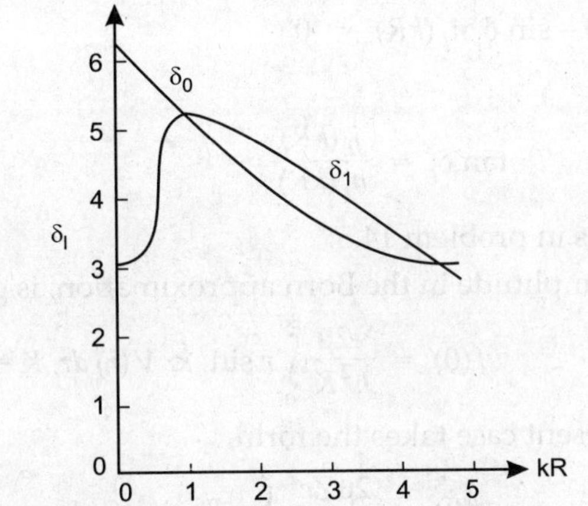

Fig. 14.17 *The phase shifts of S and P waves as function of k for scattering from spherical potential well with* $k_0R = (2\mu\ V_0/\hbar^2)^{1/2}\ R = 2.5$

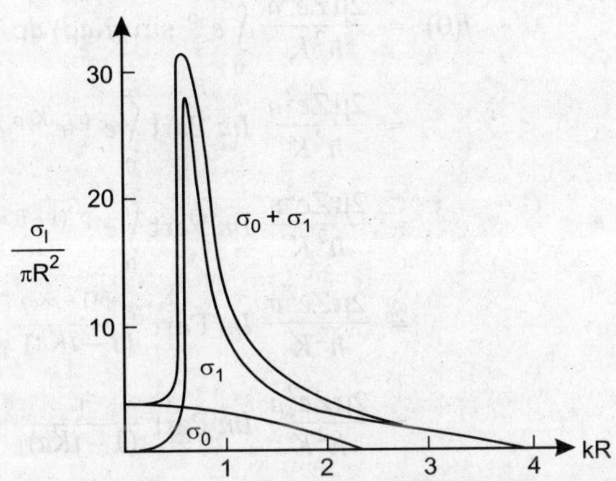

Fig. 14.18 *The partial cross-section for S and P waves corresponding to phase shifts of Fig. 14.17 as function of kR. The cross-sections are given in units of* πR^2

14.9 The solution of the radial part of the Schrödinger equation in the exterior region is given by

$$R_l(r) = A_l\, j_l\,(kr) + B_l\, n_l\,(kr)$$

$$= C_l\,[\cos\delta_l\, j_l\,(kr) - \sin\delta_l\, n_l\,(kr)],$$

where

$$A_l = C_l \cos\delta_l,\ - B_l = C_l \sin\delta_l,\ C_l^2 = A_l^2 + B_l^2$$

Due to the continuity of the wave function at $r = R$ and due to the fact that the particle cannot penetrate in the region in which its energy is infinity,

$$R_l(r) = 0 \text{ at } r = R$$

or

$$\cos\delta_l j_l(kR) - \sin\delta_1 n_l(kR) = 0$$

or

$$\tan\delta_l = \frac{j_l(kR)}{n_l(kR)}$$

For rest see steps in problem 14.5.

14.10 The scattering amplitude in the Born approximation, is given by

$$f(\theta) = \frac{-2\mu}{\hbar^2 K}\int_0^\infty r \sin\ kr\, V(r)\, dr,\ K = 2k\sin\frac{\theta}{2} \qquad \text{... (131}a\text{)}$$

which in the present case takes the form,

$$f(\theta) = \frac{2\mu Ze^2}{\hbar^2 K}\int_0^\infty e^{-r/a}\sin Kr\, dr$$

Putting $\frac{r}{a} = \rho$ or $dr = ad\rho$, the above equation becomes,

$$f(\theta) = \frac{2\mu Ze^2 a}{\hbar^2 K}\int_0^\infty e^{-\rho}\sin(Ka\rho)\, d\rho$$

$$= \frac{2\mu Ze^2 a}{\hbar^2 K}\ Im\ \text{Part}\int_0^\infty e^{-\rho}\, e^{iKa\rho}\, d\rho$$

$$= \frac{2\mu Ze^2 a}{\hbar^2 K}\ Im\ \text{Part}\int_0^\infty e^{-\rho\,(1-iKa)}\, d\rho$$

$$= \frac{2\mu Ze^2 a}{\hbar^2 K}\ Im\ \text{Part}\left.\frac{e^{-\rho(1-iKa)}}{(1-iKa)}\right|_0^\infty$$

$$= \frac{2\mu Ze^2 a}{\hbar^2 K}\ Im\ \text{Part}\frac{1}{(1-iKa)}$$

$$= \frac{2\mu Ze^2 a}{\hbar^2 K}\ Im\ \text{Part}\frac{(1+iKa)}{(1+K^2a^2)}$$

$$= \frac{2\mu Ze^2 a^2}{\{\hbar^2(1+K^2a^2)\}}$$

14.11 The Coulomb potential due to the charges Ze and $Z'e$ separated by a distance r, is given by

$$V(r) = ZZ'\frac{e^2}{r}$$

The scattering amplitude in the Born approximation, is given by

$$f(\theta) = -\frac{2\mu}{\hbar^2 K}\int_0^\infty r\sin Kr\, V(r)\, dr, K = 2k\sin\frac{\theta}{2} \qquad \text{... (131}a\text{)}$$

which in the present case becomes

$$f(\theta) = -2\mu ZZ' \frac{e^2}{\hbar^2 K} \int_0^\infty \sin Kr\, dr$$

The integral appearing on the R.H.S. of the above expression is not absolutely convergent but can be integrated by inserting a convergent factor $e^{-\eta r}$ (with $\eta \geq 0$). Thus

$$f(\theta) = \lim_{n \to 0} \left(\frac{-2\mu ZZ' e^2}{\hbar^2 K} \right) \int_0^\infty \sin Kr\, e^{-\eta r}\, dr$$

$$= \lim_{n \to 0} \left(\frac{-2\mu ZZ' e^2}{\hbar^2 K} \right) Im \text{ Part} \int_0^\infty e^{-(\eta - iK) r}\, dr$$

$$= \lim_{n \to 0} \left(\frac{-2\mu ZZ' e^2}{\hbar^2 K} \right) Im \text{ Part} \frac{1}{(\eta - iK)}$$

$$= \lim_{n \to 0} \left(\frac{-2\mu ZZ' e^2}{\hbar 2 K} \right) Im \text{ Part} \frac{(\eta + iK)}{(\eta^2 + K^2)}$$

$$= -2\mu ZZ' \frac{e^2}{\hbar^2 K^2}$$

Hence the differential scattering cross-section (or Rutherford scattering cross-section),

$$\sigma(\theta) = |f(\theta)|^2 = \left(\frac{\mu ZZ' e^2}{2\hbar^2 K^2} \right)^2 \text{cosec}^4\, \theta/2$$

14.12 The scattering amplitude in the Born approximation is given by

$$f(\theta) = -\frac{2\mu}{\hbar^2 K} \int_0^\infty r \sin Kr\, V(r)\, dr, \; K = 2k \sin\frac{\theta}{2} \qquad \text{... (131a)}$$

Substituting for $V(r)$ in the present case, one gets

$$f(\theta) = \frac{2\mu V_0}{\hbar^2 K} \int_0^\infty r \sin Kr\, e^{-r/a}\, dr$$

$$= \frac{2\mu V_0}{\hbar^2 K} Im \text{ Part} \int_0^\infty r\, e^{iKr}\, e^{-r/a}\, dr$$

$$= \frac{2\mu V_0}{\hbar^2 K} Im \text{ Part} \int_0^\infty r\, e^{-(1/a - iK) r}\, dr$$

$$= \frac{2\mu V_0}{\hbar^2 K} Im \text{ Part} \left[\left. \frac{re^{-(1/a - iK)r}}{-\left(\frac{1}{a} - iK\right)} \right|_0^\infty - \int_0^\infty \frac{e^{-(1/a - iK)r}}{-\left(\frac{1}{a} - iK\right)} dr \right]$$

$$= \frac{2\mu V_0}{\hbar^2 K}\, Im \text{ Part } \left.\frac{e^{-(1/a - iK)r}}{-\left(\frac{1}{a} - iK\right)^2}\right|_0^\infty$$

$$= \frac{2\mu V_0}{\hbar^2 K}\, Im \text{ Part } \frac{1}{\left(\frac{1}{a} - iK\right)^2}$$

$$= \frac{2\mu V_0}{\hbar^2 K}\, Im \text{ Part } \frac{1}{\left(\frac{1}{a^2} - K^2 - \frac{2\,iK}{a}\right)}$$

$$= \frac{2\mu V_0}{\hbar^2 K}\, Im \text{ Part } \frac{\left\{\left(\frac{1}{a^2} - K^2\right) + \frac{2\,iK}{a}\right\}}{\left\{\left(\frac{1}{a^2} - K^2\right)^2 + \frac{4K^2}{a^2}\right\}}$$

$$= \frac{2\mu V_0}{\hbar^2 K}\frac{2K}{a} \Bigg/ \left\{\left(\frac{1}{a^2} - K^2\right)^2 + \frac{4K^2}{a^2}\right\}$$

$$= \frac{4\mu V_0}{\hbar^2}\frac{1}{\left(\frac{1}{a^2} + K^2\right)^2}$$

Thus,

$$\sigma(\theta) = |f(\theta)|^2 = \left(\frac{4\mu V_0}{\hbar^2 a}\right)^2 \frac{1}{\left(\frac{1}{a^2} + K^2\right)^4}$$

From above, the total scattering cross-section would be given by

$$\sigma_t = \left(\frac{4\mu V_0}{\hbar^2 a}\right)^2 \int_0^\pi \int_0^{2\pi} 1 \Bigg/ \left(\frac{1}{a^2} + K^2\right)^4 \sin\theta\, d\theta\, d\phi$$

$$= \left(\frac{4\mu\, V_0}{\hbar^2\, a}\right)^2 2\pi \int_0^\pi \frac{1}{\left(\frac{1}{a^2} + K^2\right)^4} \sin\theta\, d\theta$$

Since $K = 2k \sin\theta/2$, $dK = k\cos\theta/2\, d\theta$ and hence $Kdk = k^2 \sin\theta d\theta$. Thus

$$\sigma_t = \left(\frac{4\mu V_0}{\hbar^2 a}\right)^2 \frac{2\pi}{k^2} \int_0^{2k} \frac{KdK}{\left(\frac{1}{a^2} + K^2\right)^4}$$

$$= \left(\frac{4\mu V_0}{\hbar^2 a}\right)^2 \frac{2\pi}{k^2} \left[\left.\frac{-1}{6\left(\frac{1}{a^2} + K^2\right)^3}\right|_0^{2k}\right]$$

$$= \left(\frac{4\mu V_0 a^2}{\hbar^2}\right)^2 \frac{\pi}{3k^2}\left[1 - \frac{1}{(4\,k^2\,a^2 + 1)^3}\right]$$

The validity condition in the present case takes the form

$$\frac{\mu V_0}{\hbar^2 k}\left|\int_0^\infty e^{-r/a}\,(e^{2ikr}-1)\,dr\right| << 1$$

or

$$\frac{\mu V_0}{h^2 k}\left|\int_0^\infty [e^{-(1/a-2ik)r} - e^{-r/a}]\,dr\right| << 1$$

or

$$\frac{\mu V_0}{\hbar^2 k}\left|\left[\frac{e^{-(1/a-2ik)r}}{-\left(\frac{1}{a}-2iK\right)} - \frac{e^{-r/a}}{\frac{-1}{a}}\right]_0^\infty\right| << 1$$

or

$$\frac{\mu V_0 a}{\hbar^2 k}\left|\frac{2\,ika}{(1-2\,ika)}\right| << 1$$

or

$$\frac{\mu V_0 a}{\hbar^2 k}\left|\frac{2\,ika\,(1+2\,ika)}{(1+4\,k^2 a^2)^{1/2}}\right| << 1$$

or

$$\frac{2\mu V_0 a^2}{\hbar^2}\;\frac{1}{(1+4\,k^2 a^2)^{1/2}} << 1$$

In the low energy limit ($ka << 1$), one gets

$$\frac{2\mu V_0 a^2}{\hbar^2} << 1$$

that is, the Born approximation would be applicable if the potential is too weak to bind the particle.

In the high energy limit ($ka >> 1$), one gets

$$\frac{\mu V_0 a}{\hbar^2 k} << 1$$

or

$$\frac{V_0 a}{\hbar v} << 1$$

where v is the relative velocity of the colliding particles. The above condition is generally satisfied for weak potential at high energies.

14.13 The Born scattering amplitude in the present case takes the form,

$$f(\theta) = -\frac{2\mu V_0}{\hbar^2 K}\int_0^\infty r\,\sin Kr\,e^{-\alpha r^2}\,dr$$

$$= -\frac{2\mu V_0}{\hbar^2 K}\,Im\text{ Part}\int_0^\infty r\,e^{iKr}\,e^{-\alpha r^2}\,dr$$

$$= -\frac{2\mu V_0}{\hbar^2 K}\,e^{-K^2/4\alpha}\,Im\text{ part}\int_0^\infty e^{-\alpha\,(r-iK/2\,\alpha)^2}\,rdr$$

Putting $$x = r - \frac{iK}{2\alpha}\left(\text{or } r = x + \frac{iK}{2\alpha}\right),\ dx = dr, \text{ one gets}$$

$$f(\theta) = -\frac{2\mu V_0}{\hbar^2 K} e^{-K^2/4\alpha}\, Im \text{ Part } (I_1 + I_2)$$

in which

$$I_1 = \int_{-iK/2\alpha}^{\infty} x e^{-\alpha x^2}\, dx$$

Put, $$\alpha x^2 = y,\ \alpha 2x\, dx = dy, \text{ hence}$$

$$I_1 = \frac{1}{2\alpha}\int_{-k^2/4\alpha}^{\infty} e^{-y}\, dy = \frac{1}{2\alpha}\frac{e^{-y}}{-1}\Big|_{-K^2/4\alpha}^{\infty} = \text{a real function}$$

and

$$I_2 = \frac{iK}{2\alpha}\int_{-iK/2\alpha}^{\infty} e^{-\alpha x^2}\, dx$$

$$= \frac{iK}{2\alpha}\left[\int_{-iK/2\alpha}^{0} e^{-\alpha x^2}\, dx + \int_{0}^{\infty} e^{-\alpha x^2}\, dx\right]$$

$$= I_3 + I_4$$

in which

$$I_3 = \frac{iK}{2\alpha}\int_{-iK/2\alpha}^{0} e^{-\alpha x^2}\, dx$$

and

$$I_4 = \frac{iK}{2\alpha}\int_{0}^{\infty} e^{-\alpha x^2} dx$$

Putting $$x = \frac{iK}{2\alpha} y$$

$$I_3 = \frac{iK}{2\alpha}\frac{iK}{2\alpha}\int_{-1}^{0} e^{K^2/4\alpha y^2}\, dy = \text{a real function}$$

Thus,

$$f(\theta) = -\frac{2\mu V_0}{\hbar^2 K} e^{-K^2/4\alpha}\, Im \text{ Part } \frac{iK}{2\alpha}\int_{0}^{\infty} e^{-\alpha x^2}\, dx$$

$$= -\frac{2\mu V_0}{\hbar^2 K} e^{-K^2/4\alpha}\, \frac{K}{2\alpha}\frac{1}{2}\left(\frac{\pi}{\alpha}\right)^{1/2}$$

$$= -\frac{\mu V_0}{2\hbar^2} e^{-K^2/4\alpha}\left(\frac{\pi}{\alpha^3}\right)^{1/2}$$

The differential scattering cross-section, therefore would be given by

$$\sigma(\theta) = |f(\theta)|^2 = \frac{\pi\mu^2 V_0^2}{4\hbar^4\alpha^3} e^{-K^2/2\alpha}$$

Thus, the total scattering cross-section,

$$\sigma_t = \frac{\pi\mu^2 V_0^2}{4\hbar^4\alpha^3} \int_0^{\pi}\int_0^{2\pi} e^{-K^2/2\alpha} \sin\theta \, d\theta \, d\phi$$

$$= \frac{\pi^2 \mu^2 V_0^2}{2\hbar^4 \alpha^3} \int_0^{\pi} e^{-K^2/2\alpha} \sin\theta \, d\theta$$

Since, $K = 2k \sin \theta/2, \; Kdk = k^2 \sin\theta \, d\theta$, hence

$$\sigma_t = \frac{\pi^2 \mu^2 V_0^2}{2\hbar^4 \alpha^3 k^2} \int_0^{2k} e^{-K^2/2\alpha} \, KdK$$

Putting $K^2 = t, \; 2K \, dK = dt$, one gets

$$\sigma_t = \frac{\pi^2 \mu^2 V_0^2}{4\hbar^4 \alpha^3 k^2} \int_0^{4k^2} e^{-t/2\alpha} \, dt$$

$$= \frac{\pi^2 \mu^2 V_0^2}{4\hbar^4 \alpha^3 k^2} \left[(-2\alpha) e^{-t/2\alpha} \Big|_0^{4k^2}\right]$$

$$= \frac{\pi^2 \mu^2 V_0^2}{4\hbar^4 \alpha^3 k^2} (1 - e^{-2k^2/\alpha})$$

14.14 The given potential for the sake of convenience can be expressed as (see Fig. 14.19)

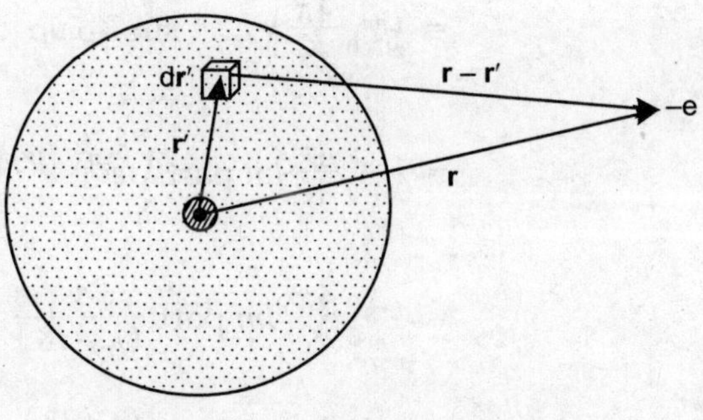

Fig. 14.19 *The vectors* **r** *and* **r'**

$$V(r) = -Ze^2 \int \frac{\delta(\mathbf{r'}) - \rho(\mathbf{r'})}{|\mathbf{r} - \mathbf{r'}|} d\mathbf{r'}$$

where $\delta(\mathbf{r'})$ is the three dimensional Dirac delta function and $\int \delta(\mathbf{r'}) \, d\mathbf{r'} = 1$ at $\mathbf{r'} = 0$. $\{\delta(\mathbf{r'}) - \rho(\mathbf{r'})\}$ is the total charge density (nuclear plus electronic).

The Born scattering amplitude in the present case would, therefore, be given by

$$f(\theta) = \frac{\mu Z e^2}{2\pi\hbar^2} \int\int \frac{\delta(\mathbf{r'}) - \rho(\mathbf{r'})}{|\mathbf{r} - \mathbf{r'}|} e^{i\vec{K}.r} \, d\mathbf{r'} \, d\mathbf{r}, \; K = 2k \sin \theta/2$$

The above expression can be written as

$$f(\theta) = \frac{\mu Z e^2}{2\pi\hbar^2} \int \{\delta(\mathbf{r}') - \rho(\mathbf{r}')\} e^{i\vec{\mathbf{K}}.\mathbf{r}'} d\mathbf{r}' \int \frac{e^{i\vec{\mathbf{K}}.(\mathbf{r}-\mathbf{r}')}}{|\mathbf{r}-\mathbf{r}|} d\vec{\mathbf{r}}$$

Out of which the integral,

$$I = \int \frac{e^{i\vec{\mathbf{K}}.(\mathbf{r}-\mathbf{r}')}}{|\mathbf{r}-\mathbf{r}'|} d\mathbf{r} = \int \frac{e^{i\vec{\mathbf{K}}.\vec{\rho}}}{\rho} d\rho \, , \rho = \mathbf{r} - \mathbf{r}'$$

Where the point **r'** has been taken as origin in the integration over **r**, is strictly not convergent, but can be integrated by inserting a convergent factor $e^{-\lambda\rho}$. Thus, we have

$$I = \lim_{\lambda\to 0} \int_0^\infty e^{-\lambda\rho} \int_0^\pi \int_0^{2\pi} \frac{e^{iK\rho\cos\theta'}}{\rho} \rho^2 \, d\rho \sin\theta' \, d\theta' \, d\phi'$$

$$= \lim_{\lambda\to 0} 2\pi \int_0^\infty e^{-\lambda\rho} \int_{-1}^{+1} e^{iK\rho\cos'}\theta' \rho d\rho d(\cos\theta')$$

$$= \lim_{\lambda\to 0} 2\pi \int_0^\infty e^{-\lambda\rho} \frac{e^{iK\rho\cos\theta'}}{iK\rho}\Bigg|_{-1}^{+1} \rho \, d\rho$$

$$= \lim_{\lambda\to 0} 2\pi \int_0^\infty e^{-\lambda\rho} \frac{(e^{iK\rho} - e^{-eK\rho})}{iK} d\rho$$

$$= \lim_{\lambda\to 0} \frac{4\pi}{K} \int_0^\infty e^{-\lambda\rho} \sin K\rho \, d\rho$$

$$= \lim_{\lambda\to 0} \frac{4\pi}{K} Im \text{ part} \int_0^\infty e^{-(\lambda - iK)\rho} d\rho$$

$$= \lim_{\lambda\to 0} \frac{4\pi}{K} Im \text{ part} \frac{1}{(\lambda - iK)}$$

$$= \lim_{\lambda\to 0} \frac{4\pi}{K} Im \text{ part} \frac{(\lambda + iK)}{(\lambda^2 + K^2)}$$

$$= \lim_{\lambda\to 0} \frac{4\pi}{(\lambda^2 + K^2)}$$

$$= \frac{4\pi}{K^2}$$

Thus,

$$f(\theta) = \frac{\mu Z e^2}{2\pi\hbar^2} \frac{4\pi}{K^2} \int \{\delta(\mathbf{r}') - \rho(\mathbf{r}')\} e^{i\vec{\mathbf{K}}.r'} d\vec{r}'$$

$$= \frac{2\mu Ze^2}{\hbar^2 K^2}\,[1 - F(K)]$$

where

$$F(K) = \int \rho(\mathbf{r}')\, e^{i\mathbf{K}.\mathbf{r}'}\, d\mathbf{r}'$$

$$= \int \rho(\mathbf{r})\, e^{i\mathbf{K}.\mathbf{r}}\, d\mathbf{r}$$

is called the atomic scattering factor (and is Fourier transform of the electronic density) which depends upon the electronic charge distribution and on the magnitude of the momentum transfer during the scattering.

Assuming the particle density to be spherically symmetric, the above expression can be written as

$$F(K) = \int \rho(r)\, e^{i\mathbf{K}.\mathbf{r}}\, d\mathbf{r}$$

$$= \int_0^\infty \int_0^\pi \int_0^{2\pi} \rho(r)\, e^{iKr\,\mathrm{Cos}\alpha}\, r^2\, dr \sin\alpha\, d\alpha\, d\beta$$

$$= 2\pi \int_0^\infty \rho(r)\, r^2\, dr \left.\frac{e^{iKr\cos\alpha}}{iKr}\right|_{-1}^{+1}$$

$$= \frac{4\pi}{K}\int_0^\infty r\rho(r) \sin Kr\, dr$$

14.15 The atomic scattering factor in the present case of spherically symmetric electronic density, is given by

$$F(K) = \frac{4\pi}{K}\int_0^\infty r\rho(r) \sin Kr\, dr,\; K = 2k \sin\theta/2$$

which in the present case, becomes

$$F(K) = \frac{4\pi}{Ka_0^3}\int_0^\infty e^{-2r/a_0} \sin Kr\, rdr$$

Putting
$$\frac{2r}{a_0} = x,\; dr = \frac{a_0}{2}\, dx$$

or

$$rdr = \left(\frac{a_0}{2}\right)^2 x\, dx, \text{ one gets}$$

$$F(K) = \frac{4}{Ka_0^3}\int_0^\infty e^{-x} \sin Ka_0\, x/2 \left(\frac{a_0}{2}\right)^2 x\, dx$$

$$= \frac{1}{Ka_0}\, Im \text{ Part} \int_0^\infty e^{-x\left(1 - iK\frac{a_0}{2}\right)}\, x\, dx$$

$$= \frac{1}{Ka_0}\, Im \text{ Part} \left[\frac{xe^{-x\left(1 - iK\frac{a_0}{2}\right)}}{-\left(1 - iK\frac{a_0}{2}\right)}\Bigg|_0^\infty - \int_0^\infty \frac{e^{-x\left(1 - iK\frac{a_0}{2}\right)}}{-\left(1 - iK\frac{a_0}{2}\right)} dx \right]$$

$$= -\frac{1}{Ka_0}\, Im \text{ Part} \frac{e^{-x\left(1 - \frac{iKa_0}{2}\right)}\Big|_0^\infty}{\left(1 - \frac{iKa_0}{2}\right)^2}$$

$$= \frac{1}{Ka_0}\, Im \text{ Part} \frac{1}{\left(1 - \frac{iKa_0}{2}\right)^2}$$

$$= \frac{1}{Ka_0}\, Im \text{ Part} \frac{1}{\left(1 - \frac{K^2 a_0^2}{4} - iKa_0\right)}$$

$$= \frac{1}{Ka_0}\, Im \text{ Part} \left(1 - \frac{K^2 a_0^2}{4} + iKa_0\right) \Bigg/ \left\{\left(1 - \frac{K^2 a_0^2}{4}\right)^2 + K^2 a_0^2\right\}$$

$$= \frac{1}{\left\{\left(1 - \frac{K^2 a_0^2}{4}\right)^2 + K^2 a_0^2\right\}}$$

$$= \frac{1}{\left(1 + \frac{K^2 a_0^2}{4}\right)^2}$$

The scattering amplitude is, therefore, given by

$$f(\theta) = \frac{2\mu e^2}{\hbar^2 K^2}[1 - F(K)]$$

$$= \frac{2\mu e^2}{\hbar^2 K^2}\left[1 - \frac{1}{\left(1 + \frac{K^2 a_0^2}{4}\right)^2}\right]$$

$$= \frac{2\mu e^2}{\hbar^2 K^2} \frac{(K^4 a_0^2 + 8K^2 a_0^2)^2}{(4 + K^2 a_0^2)^2}$$

Thus, the differential scattering cross-section would be given by

$$\sigma(\theta) = |f(\theta)|^2 = \left(\frac{2\mu e^2}{\hbar^2 K^2}\right)^2 \frac{(K^4 a_0^4 + 8K^2 a_0^2)^2}{(4 + K^2 a_0^2)^4}$$

$$= 4\, a_0^2 \frac{(8 + K^2 a_0^2)^2}{(4 + K^2 a_0^2)^4},\ a_0 = \frac{\hbar^2}{\mu e^2}$$

The total scattering cross-section using above would be given by

$$\sigma_t = 2\pi \int_0^{\pi} \sigma(\theta) \sin\theta \, d\theta$$

$$= \frac{2\pi}{k^2} \int_0^{2k} \sigma(K) \, KdK$$

$$= \frac{2\pi}{k^2} 4 a_0^2 \int_0^{2k} \frac{(8+K^2a_0^2)^2}{(4+K^2a_0^2)^4} KdK$$

Putting $y = Ka_0$ and then writing, $(8 + y^2)^2 = [4 + (4 + y^2)]^2 = 16 + 8(4 + y^2)^2$ $(4 + y^2)^2$, one gets

$$\sigma_t = \frac{8\pi}{k^2} \int_0^{2ka_0} \left[\frac{16\, ydy}{(4+y^2)^4} + \frac{8\, ydy}{(4+y^2)^3} + \frac{ydy}{(4+y^2)^2} \right]$$

$$= -\frac{8\pi}{k^2} \left[\frac{8}{3(4+y^2)^3} + \frac{4}{3(4+y^2)^2} + \frac{1}{2(4+y^2)} \right]_0^{2ka_0}$$

$$= \frac{8\pi}{k^2}\frac{7}{24} - \frac{8\pi}{k^2} \left[\frac{8}{3(4+4k^2a_0^2)^3} + \frac{4}{2(4+4k^2a_0^2)^2} + \frac{1}{2(4+4k^2a_0^2)} \right]$$

For $ka_0 >> 1$, the second term can be neglected compared to the first (as 1 can be neglected compared to $k^2a_0^2$ in the denominator of each term. Then the second term is left with powers of k^4 and higher powers). Thus

$$\sigma_t = \frac{7\pi}{3k^2}$$

14.16 The given potential is displayed in Fig. 14.20. Thus the Born scattering amplitude in the present case is given by

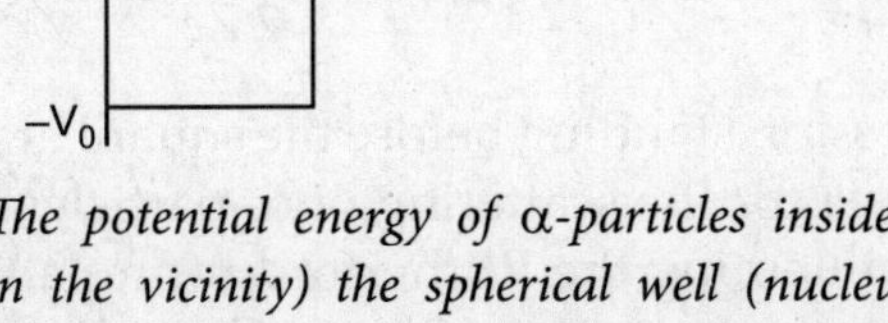

Fig. 14.20 *The potential energy of α-particles inside and outside (in the vicinity) the spherical well (nucleus)*

$$f(\theta) = -\frac{2\mu}{\hbar^2 K} \int_0^{\infty} r \sin Kr \, V(r) \, dr$$

$$= -\frac{2\mu}{\hbar^2 K} \left[\int_0^{a} -V_0 \, r \sin Kr \, dr + \int_a^{\infty} \frac{2Ze^2}{r} r \sin Kr \, dr \right.$$

$$= \frac{2\mu V_0}{\hbar^2 K}\int_0^a r \sin Kr\, dr - \frac{4\,\mu Ze^2}{\hbar^2 K}\int_0^\infty \sin Kr\, dr$$

$$= \frac{2\mu V_0}{\hbar^2 K}\,(\sin Ka - Ka \cos Ka) - \frac{4\,\mu Ze^2}{\hbar^2\, K}\int_a^\infty \sin Kr\, dr$$

where the first term has been obtained by evaluating the integral by parts. The value of the integral in the second term is indeterminate. This can be evaluated by introducing a screening or convergent factor $e^{-\eta r}$ ($\eta \geq 0$). Thus,

$$\int_a^\infty \sin Kr\, dr = \underset{\eta\to 0}{Lim}\int_a^\infty \sin Kr\, e^{-\eta r}\, dr$$

$$= \lim_{\eta\to 0}\; Im \text{ Part}\int_a^\infty e^{-(\eta - iK)r}\, dr$$

$$= \lim_{\eta\to 0} Im \text{ Part}\left.\frac{e^{-(\eta - iK)r}}{-(\eta - iK)}\right|_a^\infty$$

$$= \lim_{\eta\to 0} Im \text{ Part}\,\frac{e^{-(\eta - iK)a}}{(\eta - iK)}$$

$$= \lim_{\eta\to 0} Im \text{ Part}\,\frac{(\eta + iK)}{(\eta^2 + K^2)}\, e^{-(\eta - iK)a}$$

$$= \frac{\cos Ka}{K}$$

Therefore,

$$f(\theta) = \frac{2\mu V_0}{\hbar^2 K^3}\,(\sin Ka - Ka \cos \text{Ka}) - \frac{4\,\mu Ze^2}{\hbar^2\, K}\,\frac{\cos Ka}{K}$$

$$= -\frac{4\,\mu Ze^2}{\hbar^2\, K^2}\left[\cos Ka + \frac{V_0 a}{2Ze^2}\left(\cos Ka - \frac{\sin Ka}{Ka}\right)\right]$$

The differential scattering cross-section from above would be given by

$$\sigma(\theta) = \left(\frac{2\,\mu Ze^2}{2\hbar^2\, k^2}\right)^2 \text{cosec}^4\,\frac{\theta}{2}\left[\cos Ka + \frac{V_0 a}{2\, Ze^2}\left(\frac{\cos Ka - \sin Ka}{Ka}\right)\right]^2$$

Note that the expression standing before the square brackets is nothing but the Rutherford's formula for the scattering of α–particles by a point like nucleus. The expression augmenting the Rutherford formula takes into consideration the correction due to the finite size of the nucleus. For very small nuclear radii, that is, in the limiting case of $a \to 0$, the value of the expression

$$\left[\cos Ka + \frac{V_0 a}{2\, Ze^2}\left(\cos Ka - \frac{\sin Ka}{Ka}\right)\right]^2 \to 1$$

The other limiting case is represented by a very high bombarding energy of the α–particles, that is, $Ka >> 1$. In that case

$$\left[\cos Ka + \frac{V_0 a}{2Ze^2}\left(\cos Ka - \frac{\sin Ka}{Ka}\right)\right]^2$$

$$\rightarrow \left[\cos Ka\left(1 + \frac{V_0 a}{2\,Ze^2}\right)\right]^2$$

$$\approx \frac{1}{2}\left(1 + \frac{V_0 a}{2\,Ze^2}\right)^2$$

where $\cos^2 Ka$ is replaced by its average value $\frac{1}{2}$.

14.17 The scattering or collision process can be treated as a perturbation of the initial state corresponding to the free motion caused by time dependent interaction which is switched on at time $t = 0$ and is switched off at a later time $t = \tau$. If the potential which determines the energy of interaction between the colliding particles, is constant in time, then the time dependent perturbation caused by switching on and off can be represented by setting.

$$\begin{aligned} H'(t) &= V(\vec{r}, t) = 0, -\infty \le t < 0 \\ &= V(r), 0 \le t \le \tau \\ &= 0, \tau < t \le \infty \end{aligned}$$

We have to solve the Schrödinger equation,

$$\frac{i\hbar\, \partial\, \psi(\mathbf{r}, t)}{\partial t} = \{H^{(0)} + V(\mathbf{r}, t)\}\, \psi(\mathbf{r}, t)$$

in which $H^{(0)}$ the time independent part represents the free particle Hamiltonian and $V(r, t)$ the perturbation. We are interested in the quantum transitions (involving states of the continuous spectrum) from a given free initial state to some free final state caused by the time dependent potential.

Let us assume the whole system to be enclosed in a large cubical box of sides L. The initial and final states, describing the particle incident with momentum $\mathbf{k}_0$ and scattered with momentum $\mathbf{k}$, would be represented by the box normalized wave functions,

$$\varphi(\mathbf{r}) = L^{-3/2}\, e^{i\mathbf{k}_0.\mathbf{r}},\ \varphi_f(\mathbf{r}) = L^{-3/2}\, e^{i\mathbf{k}.\mathbf{r}}$$

where $|\mathbf{k}_0| = |\mathbf{k}| = k$

The matrix elements of the perturbation are

$$\begin{aligned} H'_{fi} &= \int \varphi^*_f(\mathbf{r})\, V(\mathbf{r})\, \varphi_i(\mathbf{r})\, d\vec{\mathbf{r}} \\ &= L^{-3} \int V(\mathbf{r})\, e^{i\vec{\mathbf{K}}.\vec{\mathbf{r}}}\, d\mathbf{r} \end{aligned}$$

where $\mathbf{K} = \mathbf{k}_0 - \mathbf{k}$, is momentum transfer vector.

We now want to determine the density of final states. In doing so we must remember that we are interested in transitions to states with momentum directions (direction of **k**) lying with in a solid angle $d\Omega = \sin\theta\, d\theta\, d\phi$ and magnitudes or lengths (of $\vec{\mathbf{k}}$) between k and $k + dk$.

The number of such states is equal to $\left(\frac{L}{2\pi}\right)^3 k^2\, dk\, d\Omega$. This is the number of states with energies between E_f and $E_f + dE_f$. Thus

$$\rho(E_f)\, dE_f = \left(\frac{L}{2\pi}\right)^3 k^2\, dk\, d\Omega$$

where

$$E_f = \frac{\hbar^2 k^2}{2\mu},\ dE_f = \frac{\hbar^2 k}{\mu}\, dk$$

From above the density of final states is given by

$$\rho(E_f) = \left(\frac{L}{2\pi}\right)^3 k^2\, dk\, d\Omega \frac{\mu}{\hbar^2 kdk} \hbar^2\, k\, \hbar^2$$

$$= \left(\frac{L}{2\pi}\right)^3 \frac{\mu k}{\hbar^2}\, d\Omega$$

Thus the probability per unit time, w_{fl} (the Fermi Golden rule) for scattering into the direction $d\Omega$ of particle whose incident wave function is φ_i,

$$w_{fi} = \frac{2\pi}{\hbar}\left(\frac{L}{2\pi}\right)^3 \frac{\mu k}{\hbar^2}\, dk \left| L^{-3} \int V(\vec{\mathbf{r}}) e^{i\vec{\mathbf{K}}.\vec{\mathbf{r}}}\, d\,\vec{\mathbf{r}} \right|^2$$

The particle flux (or incident flux) associated with this wave function φ_i is, $I_0 = |\varphi_i|^2\, v = L^{-3}\, \frac{\hbar k}{\mu}$. Since the differential scattering cross-Section is defined to be the number of scattered particles per unit time (or transition probability per unit time) per unit solid angle per unit incident flux,

$$\sigma(\theta, \phi) = \frac{w_{fi}}{d\Omega\, I_0}$$

$$= \left(\frac{\mu}{2\pi\hbar^2}\right)^2 \left| \int V(\mathbf{r}) e^{i\mathbf{K}.\mathbf{r}}\, d\vec{\mathbf{r}} \right|^2$$

The expression which coincides with what was obtained (see eqn. (130) section 14.14) from the Born approximation. Thus the first approximation of the time dependent approach gives the same result for the cross-section as does the first Born approximation in scattering theory. Also note that the size L of the box which was introduced to find $\rho(E_f)$ does not appear in the final result.

❑❑❑

Chapter 15

SYSTEMS OF IDENTICAL PARTICLES

15.1 IDENTICAL PARTICLES

Consider a hydrogen atom. It will always be possible to distinguish between the proton and the electron because of their different properties, But if we now shoot an electron at the hydrogen atom, will it be possible to distinguish between the incident and the atomic electron after scattering, that is which one reaches the recording device. Similarly in the scattering of a proton by a proton, the proton which comes to the detector could be either the projectile or the target proton, but it would be impossible to tell them apart. In classical theory, this is possible since we can follow each particle along its trajectory (one can even colour the two colliding billiard balls and distinguish them even after any collision). But in quantum mechanics where a collision (or scattering process) is represented by overlapping wave functions, the distinction will not be possible. This inability to tell apart the particles from one another in a system of identical particles (like electrons or protons in the above examples) is embodied in the principle of indistinguishability of identical particles. Quantum mechanics of a system of identical particles should be formulated in such a manner so as to be consistent with this principle. For example, in the scattering problem mentioned above, experiment does not distinguish between the projectile and the target. Therefore, in the quantum mechanical calculation of the scattering cross-section, no distinction should be permitted between the projectile and the target (we will see this in section 15.6).

15.2 SYMMETRY OF WAVE FUNCTIONS

In order to illustrate how the principle of indistinguishability influences the quantum mechanics of a system of identical particles, let us consider a simple case of a two particle system. The Schrödinger equation for a system of two identical particles (or say two atoms) can be written as

$$H(1, 2)\, \varphi(1, 2) = E\varphi(1, 2) \qquad \dots (1)$$

where $H(1, 2)$ stands for the Hamiltonian of the system with particle number 1 in the state α_1 (by state we mean quantum mechanical state) and the particle number 2 in the state α_2 (in general the symbol '1' denotes collectively the dynamical variables (position, momentum, spin etc.) for the first particle (or system) and the symbol '2' for the second particle).

Now the principle of indistinguishability reqnires that the physical observables of the system, and in particular the Hamiltonian, be invariant under the interchange of the two particles. That is,

$$H(1, 2) = H(2, 1) \quad \text{... (2)}$$

where $H(2, 1)$ represents the Hamiltonian of the system when particle number 2 is in the state α_1 and number 1 in the state α_2.

Before we proceed further, we must point out that if

$$H(1, 2) = H_1(1) + H_2(2) \quad \text{... (3)}$$

where either part depends on the variables of one system only, then the two systems are non-interacting.

If the systems 1 and 2 interact mutually, then $H(1, 2)$ cannot be separated as before; the presence of the interaction is usually manifested as an additional term $\lambda H'(1, 2)$ depending on the variables of both,

$$H(1, 2) = H_1(1) + H_2(2) + \lambda H'(1, 2) \quad \text{... (4)}$$

Thus, $\lambda H'(1, 2)$ is referred to as the interaction part of the Hamiltonian and $H_1(1) + H_2(2)$ as the free part.

Now on interchanging particles 1 and 2 in eqn. (1),

$$H(2, 1)\,\varphi(2, 1) = E\varphi(2, 1)$$

and then using eqn. (2), one gets

$$H(1, 2)\,\varphi(2, 1) = E\varphi(2, 1) \quad \text{... (5)}$$

Eqns. (1) and (5) show that if $\varphi(1, 2)$ is a possible state of the two particle system with a certain energy E, then $\varphi(2, 1)$ is also a possible state with the same energy. Hence the system is degenerate and any linear combination (or superposition).

$$\psi(1, 2) = A\,\varphi(1,2) + B\,\varphi(2, 1) \quad \text{... (6)}$$

is also an eigen function belonging to the same energy eigen value E.

Let us now consider the exchange (or permutation) operator $\hat{P}$ (also some times written as $\hat{P}_{12}$) which when operates on an arbitrary wave function $\varphi(1, 2)$ gives

$$\hat{P}\,\varphi(1, 2) = \varphi(2, 1) \quad \text{... (7)}$$

that is, it exchanges particles 1 and 2.

By further operating with $\hat{P}$ on both sides of the above eqnation, we get

$$\hat{P}\hat{P}\,\varphi(1, 2) = \hat{P}^2\,\varphi(1, 2) = \hat{P}\,\varphi(2, 1) = \varphi(1, 2)$$

which implies

$$\hat{P}^2 = 1$$

or

$$\hat{P} = \pm 1 \qquad \text{... (8)}$$

Thus eigen values of $\hat{P}$ are + 1 and – 1. Denoting the corresponding eigen functions by $\psi_S(1, 2)$ and $\psi_A(1, 2)$, we have

$$\hat{P}\,\psi_S(1,2) = 1\,\psi_S(1,2) \qquad \text{... (9a)}$$

and

$$\hat{P}\psi_A(1,2) = -1\,\psi_A(1,2) \qquad \text{... (9b)}$$

Using definition of the exchange operator $\hat{P}$, that is, eqns. (7) and (9*a*) one gets

$$\hat{P}\,\psi_S(1,2) = \psi_S(2,1) = 1\,\psi_S(1,2)$$

or

$$\psi_S(2,1) = \psi_S(1,2) \qquad \text{... (10a)}$$

Similarly from eqns. (7) and (9*b*),

$$\hat{P}\,\psi_A(1,2) = \psi_A(2,1) = -1\,\psi_A(1,2)$$

or

$$\psi_A(2,1) = -1\,\psi_A(1,2) \qquad \text{... (10b)}$$

We say that ψ_S is symmetric as interchange of particles produces no change in sign while ψ_A is antisymmetric as interchange of particles produced change in sign. Thus, the physical state of a system of two identical particles is represented by either ψ_S or ψ_A or these are the only wave functions that are consistent with the principle of indistinguishability of identical particles. For this it is necessary to show that these are the simultaneous eigen kets of H also (or these are solutions of the Schrödinger eqnation for the system two identical particles).

Applying $\hat{P}$ to $\hat{H}(1, 2)\,\varphi(1, 2)$ for arbitary φ, we obtain

$$\hat{P}\hat{H}(1,2)\,\varphi(1,2) = \hat{H}(2,1)\,\varphi(2,1)$$

Using eqn. (2),

$$\hat{P}\hat{H}(1,2)\,\varphi(1,2) = \hat{H}(1,2)\,\varphi(2,1)$$

$$= \hat{H}(1,2)\,\hat{P}\,\varphi(1,2)$$

where use of eqn. (7) has been made. Thus

$$[\hat{P}, \hat{H}] = 0 \qquad \text{... (11)}$$

which is the reqnired result (if two operators commute, they will have simultaneous eigen kets). Experiment shows that a system consisting of two electrons, or two protons is in all states described by antisymmetric wave functions only. A system consisting of two α-particles is always described by a symmetric function. Since $\hat{P}$ and $\hat{H}$ commute, the symmetry property with respect to interchange of two particles is thus an integral (or constant) of motion. Therefore, if a system is, say, in a symmetric state at some time, then it will continue to be in the symmetric state for all times.

Let us extend the above discussion to a system of N identical particles ($N > 2$) by defining an exchange operator $\hat{P}_{ij}$ for each pair of particles (i, j). What has been proved above in regard to $\hat{P}$ holds also for each $\hat{P}_{ij}$. In particular,

$$\hat{P}_{ij}\,\hat{H}\,(1, 2, \ldots\ldots, i, \ldots\ldots, j, \ldots\ldots, N) = \hat{H}\,(1, 2, \ldots\ldots, i, \ldots\ldots, j, \ldots\ldots, N)\,\hat{P}_{ij}$$

Since $\hat{P}_{ij}$ and $\hat{H}$ commute, any eigen function ψ of $\hat{H}$ belonging to a non-generate eigen value must also be an eigen function of $\hat{P}_{ij}$, that is,

$$\hat{P}_{ij}\,\psi(1, 2, \ldots\ldots, i, \ldots\ldots, j, \ldots\ldots, N) = \lambda\,\psi(1, 2, \ldots\ldots, i, \ldots\ldots, j, \ldots\ldots, N)$$

and finally $\hat{P}_{ij}^2 = \lambda^2 = 1$ or $\lambda = \pm 1$

Note that, we have written the eigen value simply as λ instead of λ_{ij} as it is actually independent of (i, j). This fact can easily be deduced from the observation that

$$\hat{P}_{kj}\,\hat{P}_{ik}\,\hat{P}_{ij} = \hat{P}_{ik}$$

Operating both sides of the above eqnation on ψ, one gets

$$\lambda_{kj}\,\lambda_{ik}\,\lambda_{ij} = \lambda_{ik}$$

Which on canceling λ_{ik} on both the sides, gives

$$\lambda_{kj}\,\lambda_{ij} = 1$$

Hence λ_{kj} and λ_{ij} must both be +1 or both be –1 (for arbitary i, k) showing that λ_{ij} is independent of i (as well as of j). If $\lambda = +1$, we say that ψ, the overall wave function of the N identical particles system is symmetric (it is symmetric under the interchange of any pair of particles) and denote it by ψ_S whereas if $\lambda = -1$, ψ is antisymmetric and is denoted by ψ_A. Thus we come to the conclusion that the wave function of a system of identical particles is symmetric or antisymmetric.

15.3 THE PAULI EXCLUSION PRINCIPLE

Let us consider two identical non-interacting particles (or systems). If 1 is in the eigen state φ_a (1) of $H(1)$ and particle 2 is in an eigen state $\varphi_b(2)$ of $H(2)$, where the subscripts a and b of course include the spin state i.e., they represent the quantum numbers n, l, m_l and m_S, then

$$\hat{H}(1)\,\varphi_a(1) = E_a\,\varphi_a(1),$$

$$\hat{H}(2)\,\varphi_b(2) = E_b\,\varphi_b(2),$$

Hence

$$\hat{H}(1,2)\,\varphi_a(1)\,\varphi_b(2) = \{\hat{H}(1) + \hat{H}(2)\}\,\varphi_a(1)\,\varphi_b(2)$$
$$= \hat{H}(1)\,\varphi_a(1)\,\varphi_b(2) + \hat{H}(2)\,\varphi_a(1)\,\varphi_b(2)$$

$$= E_a\, \varphi_a(1)\, \varphi_b(2) + E_b\, \varphi_a(1)\, \varphi_b(2)$$

$$= E\, \varphi_a(1)\, \varphi_b(2) \qquad \text{... (12)}$$

with $E = E_a + E_b$. Thus the eigen functions of the combined system consisting of two non-interacting subsystems are product of the eigen functions of the individual subsystems, while the energy eigen value E is sum of E_a and E_b. Two particles being identical, interchanging 1 and 2 in the above equation one gets

$$\hat{H}(2, 1)\, \varphi_a(2)\, \varphi_b(1) = (E_b + E_a)\, \varphi_a(2)\, \varphi_b(1)$$

Further using eqn. (2), one gets

$$\hat{H}\,(1, 2)\, \varphi_a(2)\, \varphi_b(1) = (E_a + E_b)\, \varphi_a(2)\, \varphi_b(1) \qquad \text{... (13)}$$

The general solution for the two particle system which correspond to energy, $E = E_a + E_b$ can, therefore be written as

$$\psi(1, 2) = \frac{1}{\sqrt{2}}\, [\varphi_a(1)\, \varphi_b(2) \pm \varphi_a(2)\, \varphi_b(1)] \qquad \text{... (14)}$$

where the factor $\frac{1}{\sqrt{2}}$ ensures the correct normalization. For $a = b$ (that is, it both the particles have the same set of quantum numbers n, l, m_l and m_s) ψ vanishes if it is antisymmetric (wave function with (–) negative sign in between) but does not vanish if ψ is symmetric (wave function with (+) positive sign in between). Hence the probability of two antisymmetric identical particles occupying identical states is zero. In other words it simply implies that, a given single particle state (say φ_a) cannot be occupied by more than one such particle. This is nothing but the statement of pauli exclusion principle. Thus the antisymmetric nature of the wave function leads automatically to the Pauli exclusion principle.

Before we proceed further, let us point out that the particles which are described by antisymmetric wave functions are known as Fermions and they are said to obey Fermi-Dirac statistics: examples of such particles are electrons, protons, neutrons, etc. On the other hand, the particles which are described by symmetric wave functions are known as Bosons and they are said to obey Bose-Einstein statistics: examples of such particles are photons, pions, etc. Pauli has proved that particles with half integral spins (1/2, 3/2, ...) are Fermions and particles with integral spins (0, 1, 2, ...) are Basons. In the case of system of N electrons occupying single particle states described by the wave functions $\varphi_{a_1}, \varphi_{a_2}, \ldots,$ and φ_{a_N}, we can represent the antisymmetric wave function of the system (generalization of eqn. (14) with (–) negative sign in between) by a determinant (called slater dominant)

$$\psi_A\,(1, 2, \ldots, N) = \frac{1}{(N!)^{1/2}} \begin{vmatrix} \varphi_{a_1}\;(1) & \varphi_{a_1}\;(2) \ldots\ldots & \varphi_{a_1}\;(N) \\ \varphi_{a_2}\;(1) & \varphi_{a_2}\;(2) \ldots\ldots & \varphi_{a_2}\;(N) \\ \vdots\;\;(1) & & \\ \varphi a_N\;(1) & \ldots\ldots & \varphi a_N\;(N) \end{vmatrix} \qquad \text{... (15)}$$

It is immediately clear from the above that if $\varphi_{a_1} = \varphi_{a_2}$, then the two rows would become the same and the determinant would vanish (note that $a_1, a_2, \ldots$ have been used in place of $a, b, \ldots$) and hence $\psi_A = 0$, meaning that no such state exists. That is, it is not possible for the two electrons to occupy the same state which is, of course the statement of the Pauli exclusion principle. For particles which are characterized by symmetric wave functions, there is no exclusion principle. For example, $\varphi_a(1)\,\varphi_a(2) \ldots \varphi_a(N)$ is a perfectly good symmetric wave function corresponding to all N particles having the same state or quantum numbers).

15.4 HEITLER – LONDON THEORY OF HYDROGEN MOLECULE

Lewis and Langmuir suggested (valence theory) that the chemical bond between two hydrogen atoms consisted of pair of electrons held jointly by the two atoms (that is, a covalent bond). Considering the interaction between two nuclei between two electrons and between electrons and nuclei, the binding energy was estimated to be – 0.36 eV whereas the experimental value stood at – 4.72 eV. To overcome this difficulty, Heitler and London[1] explained the formation of hydrogen molecule on the basis of quantum mechanics. The Hamiltonian of the said system is

$$-\frac{\hbar^2}{2m}\nabla_1^2 - \frac{\hbar^2}{2m}\nabla_2^2 - \frac{e^2}{r_{a_1}} - \frac{e^2}{r_{b_1}} - \frac{e^2}{r_{a_2}} - \frac{e^2}{r_{b_2}} + \frac{e^2}{r_{12}} + \frac{e^2}{R} \quad \ldots (16)$$

where the labels 1 and 2 refer to two electrons and the labels a and b to protons (or nuclei) and the distances between the pair of particles are as shown in Fig. 15.1.

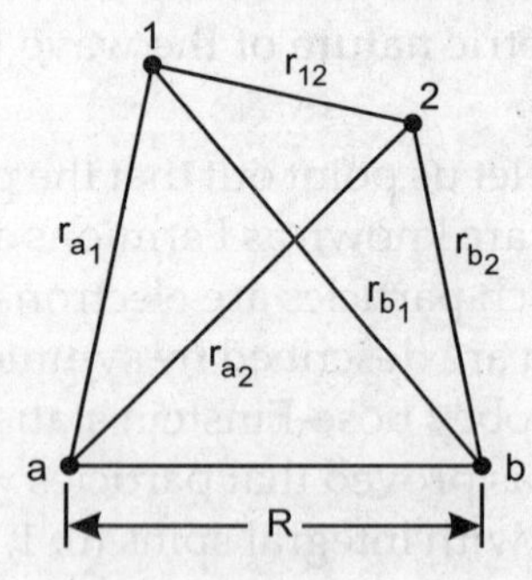

Fig. 15.1 *The coordinate system for a hydrogen molecule*

In the present case the electrons move around the nuclei which vibrate along the line joining them. The two motions are referred to as the electronic and nuclear motions, respectively. The electronic motion is rapid where as the nuclear motion is small. Therefore, the later may be regarded as stationary, that is, R may be taken to be a fixed quantity. This approximation is called Born-oppenheimer (or adiabatic) approximation and has been applied in writing the above Hamiltonian.

[1]W Heitler and F. London, Z. Physik 44, 455 (1927)

We shall assume that the atoms are sufficiently far from one another. In that case each atom has its own wave function (the hydrogen atom wave function) and thus the ground state wave function of the system (the wave function of the molecule in zeroth approximation) would be given by

$$\varphi_1^{(0)} = \varphi_{100}(r_{a_1})\,\varphi_{100}(r_{b_2}) \qquad \text{... (17}a\text{)}$$

where

$$\varphi_{100}(r_{a_1}) = (\pi a_0^3)^{-1/2}\, e^{-r_{a_1}/a_0} \text{ and } \varphi_{100}(r_{b_2}) = (\pi a_0^3)^{-1/2}\, e^{-r_{b_2}/a_0}$$

are the hydrogen atom ground state wave functions for electron 1 attached to nucleus a and for electron 2 attached to nucleus b, respectively.

Taking into account the fact that the electrons are identical (or indistinguishable) the system can also be described by the wave function,

$$\varphi_2^{(0)} = \varphi_{100}(r_{a_2})\,\varphi_{100}(r_{b_1}) \qquad \text{... (17}b\text{)}$$

Both the solutions (17) correspond to the same energy eigen value,

$$E^{(0)} = E_0 + E_0 = 2\,E_0 = 2\left(\frac{-e^2}{2a_0}\right) \qquad \text{... (18)}$$

Thus the energy eigen value is two fold degenerate (as corresponding to the single energy eigen value $E^{(0)}$, there are two eigen functions $\varphi_1^{(0)}$ and $\varphi_2^{(0)}$). This degeneracy arises from the exchange of two electrons being identical and hence is called exchange degeneracy.

The linear combination of the solutions (17), that is

$$\varphi_\pm^{(0)} = N_\pm\,(\varphi_1^{(0)} \pm \varphi_2^{(0)}) \qquad \text{... (19)}$$

would also be the solution of the said problem belonging to the same energy eigen value $E^{(0)}$, where $N_\pm$ is called the normalization constant. The solutions $\varphi_+^{(0)}$ and $\varphi_-^{(0)}$ are symmetric and antisymmetric with respect to the interchange of spatial coordinates because when the indices 1 and 2 are interchanged, $\varphi_+^{(0)}$ remains the same whereas $\varphi_-^{(0)}$ changes sign.

The normalization constant $N_\pm$ can be evaluated using the condition,

$$\iint |\varphi_\pm^{(0)}|^2\, d\tau_1\, d\tau_2 = 1$$

or $$N_\pm^2 \iint |\varphi_1^{(0)} \pm \varphi_2^{(0)}|^2\, d\tau_1\, d\tau_2 = 1$$

or $$N_\pm^2\left[\iint |\varphi_1^{(0)}|^2\, d\tau_1\, d\tau_2 + \iint |\varphi_2^{(0)}|^2\, d\tau_1\, d\tau_2 \pm 2\iint \varphi_1^{(0)}\,\varphi_2^{(0)}\, d\tau_1\, d\tau_2\right] = 1$$

or $$N_\pm^2\Big[\int |\varphi_{100}(r_{a_1})|^2\, d\tau_1 \int |\varphi_{100}(r_{b_2})|^2\, d\tau_2 + \int |\varphi_{100}(r_{b_1})|^2\, d\tau_1 \int |\varphi_{100}(r_{a_2})|^2\, d\tau_2$$
$$\pm 2\int \varphi_{100}(r_{a_1})\,\varphi_{100}(r_{b_1})\, d\tau_1 \int \varphi_{100}(r_{a_2})\,\varphi_{100}(r_{b_2})\, d\tau_2\Big] = 1$$

or $$N_\pm^2\left[1 + 1 \pm 2\int \varphi_{100}(r_{a_1})\,\varphi_{100}(r_{b_1})\, d\tau_1 \int \varphi_{100}(r_{a_2})\,\varphi_{100}(r_{b_2})\, d\tau_2\right] = 1$$

as the functions φ_{100} are already normalized.

Now both the integrals in the above eqnation being definite, are eqnal.

Thus, $$N_{\pm}^2 (2 \pm 2S^2) = 1$$

Where

$$S = \int \varphi_{100} (r_{a_1}) \varphi_{100}(r_{b_1})\, d\tau_1 = \int \varphi_{100} (r_{a_2}) \varphi_{100} (r_{b_2})\, d\tau_2$$

Thus,

$$\varphi_{\pm}^{(0)} = \frac{1}{(2 \pm 2S^2)^{1/2}} [\varphi_{100} (r_{a_1}) \varphi_{100} (r_{b_2}) \pm \varphi_{100} (r_{a_2}) \varphi_{100} (r_{b_1})] \quad \text{... (20)}$$

It may be noted that when R is very large in that case electron 1 cannot be associated with both the nuclei (and same is the case for electron 2) and hence S will be zero. When there is overlapping between the two wave functions, in that case, $S = 1$ as we cannot distinguish between $\varphi_{100} (r_{a_1})$ and $\varphi_{100} (r_{b_1})$. Thus S ranges from 0–1 and is called overlap integral.

The total Hamiltonian $\hat{H}$ (eqn. (16)) for the system can be written as the sum of unperturbed part (corresponding to the situation when the two atoms are sufficiently far from each other) and the perturbation which creeps when the atoms are brought closer, that is,

$$\hat{H} = \hat{H}^{(0)} + \lambda \hat{H}' \quad \text{... (21)}$$

where

$$\hat{H}^{(0)} = \left(\frac{-\hbar^2}{2m} \nabla_1^2 - \frac{e^2}{r_{a_1}} \right) + \left(\frac{-\hbar^2}{2m} \nabla_2^2 - \frac{e^2}{r_{b_2}} \right), \quad \text{... (21a)}$$

and

$$\lambda \hat{H}' = \frac{e^2}{R} + \frac{e^2}{r_{12}} - \frac{e^2}{r_{a_2}} - \frac{e^2}{r_{b_1}} \quad \text{... (21b)}$$

Let us now apply two fold degenerate perturbation theory to determine the energy correction (or finally perturbed energy) originating due to the interaction (or perturbation). To do so we have to solve the secular equation (or determinant)

$$\begin{vmatrix} H_{11} - E & H_{12} \\ H_{21} & H_{22} - E \end{vmatrix} = 0$$

It may be noted from eqn. (16) that, $H(1, 2) = H(2, 1)$. Thus the Hamiltonian is symmetric w.r.t. the interchange of the two elections. For very large values of R, we know that $\varphi_1^{(0)}$ and $\varphi_2^{(0)}$ both are solutions to the problem. This suggests that for small values of R, we should use the linear combinations of $\varphi_1^{(0)}$ and $\varphi_2^{(0)}$, that is $\varphi_+^{(0)}$ and $\varphi_-^{(0)}$.

Thus

$$H_{12} = \iint \varphi_+^{(0)} \hat{H} \varphi_-^{(0)}\, d\tau_1\, d\tau_2$$

Because of the indistinguishability of the particles, interchanging 1 and 2, we get

$$H_{12} = -\iint \varphi_+^{(0)} \hat{H}\varphi_-^{(0)}\, d\tau_1\, d\tau_2 \qquad \text{... (22)}$$

$$= -H_{12}$$

as (–) negative sign is introduced in $\varphi_-^{(0)}$ due to the interchange of 1 and 2. $\varphi_+^{(0)}$ and $\hat{H}$ are of course not affected. Thus $H_{12} = 0$ and similarly $H_{21} = 0$. Also note that the wave functions being real star is not used.

From the secular eqn. (22), thus the perturbed energies would be given by

$$E_+ = H_{11} = \iint \varphi_+^{(0)} \hat{H}\varphi_+^{(0)}\, d\tau_1\, d\tau_2$$

and

$$E_- = H_{22} = \iint \varphi_-^{(0)} \hat{H}\varphi_-^{(0)}\, d\tau_1\, d\tau_2$$

or in general,

$$E_\pm = \int \varphi_\pm^{(0)} \hat{H}\varphi_\pm^{(0)}\, d\tau_1\, d\tau_2$$

$$= N_\pm^2 \iint (\varphi_1^{(0)} \pm \varphi_2^{(0)})\, \hat{H}\, (\varphi_1^{(0)} \pm \varphi_2^{(0)})\, d\tau_1\, d\tau_2$$

$$= \frac{1}{(2 \pm 2S^2)} \iint (\varphi_1^{(0)} \pm \varphi_2^{(0)}) \left[\frac{-\hbar^2}{2m}\nabla_1^2 - \frac{\hbar^2}{2m}\nabla_2^2 - \frac{e^2}{r_{a_1}} - \frac{e^2}{r_{b_2}}\right.$$

$$\left. - \frac{e^2}{r_{a_2}} - \frac{e^2}{r_{b_1}} + \frac{e^2}{R} + \frac{e^2}{r_{12}}\right] (\varphi_1^{(0)} \pm \varphi_2^{(0)})\, d\tau_1\, d\tau_2 \qquad \text{... (23)}$$

Schrödinger eqnations for the wave functions $\varphi_1^{(0)}$ and $\varphi_2^{(0)}$ are given by

$$\left(\frac{-\hbar^2}{2m}\nabla_1^2 - \frac{\hbar^2}{2m}\nabla_2^2 - \frac{e^2}{r_{a_1}} - \frac{e^2}{r_{b_2}}\right)\varphi_1^{(0)} = 2E_0\, \varphi_1^{(0)} \qquad \text{... (24)}$$

and

$$\left(\frac{-\hbar^2}{2m}\nabla_1^2 - \frac{\hbar^2}{2m}\nabla_2^2 - \frac{e^2}{r_{a_2}} - \frac{e^2}{r_{b_1}}\right)\varphi_2^{(0)} = 2E_0\, \varphi_2^{(0)} \qquad \text{... (24a)}$$

Using above eqnations, (23) leads to

$$E_\pm = \frac{1}{(2 \pm 2S^2)} \iint (\varphi_1^{(0)} \pm \varphi_2^{0}) \left[\left(2E_0 - \frac{e^2}{r_{a_2}} - \frac{e^2}{r_{b_1}} + \frac{e^2}{R} + \frac{e^2}{r_{12}}\right)\varphi_1^{(0)}\right.$$

$$\left. \pm \left(2E_0 - \frac{e^2}{r_{a_1}} - \frac{e^2}{r_{b_2}} + \frac{e^2}{R} + \frac{e^2}{r_{12}}\right)\varphi_2^{(0)}\right] d\tau_1\, d\tau_2$$

$$= \frac{1}{(2 \pm 2S^2)} \left[2E_0 \iint (\varphi_1^{(0)} \pm \varphi_2^{(0)})(\varphi_1^{(0)} \pm \varphi_2^{(0)})\, d\tau_1\, d\tau_2\right.$$

$$+ \iint \varphi_1^{(0)} \left(\frac{-e^2}{r_{a_2}} - \frac{e^2}{r_{b_1}} + \frac{e^2}{R} + \frac{e^2}{r_{12}}\right)\varphi_1^{(0)}\, d\tau_1\, d\tau_2$$

$$+\iint \varphi_2^{(0)}\left(\frac{-e^2}{r_{a_1}}-\frac{e^2}{r_{b_2}}+\frac{e^2}{R}+\frac{e^2}{r_{12}}\right)\varphi_2^{(0)}\,d\tau_1\,d\tau_2$$

$$\pm\iint \varphi_1^{(0)}\left(\frac{-e^2}{r_{a_1}}-\frac{e^2}{r_{b_2}}+\frac{e^2}{R}+\frac{e^2}{r_{12}}\right)\varphi_2^{(0)}\,d\tau_1\,d\tau_2$$

$$\pm\iint \varphi_1^{(0)}\left(\frac{-e^2}{r_{a_2}}-\frac{e^2}{r_{b_1}}+\frac{e^2}{R}+\frac{e^2}{r_{12}}\right)\varphi_1^{(0)}\,d\tau_1\,d\tau_2$$

Interchanging 1 and 2, the second and third and similarly forth and fifth integrals in the above expression turn out to be the same. Thus

$$E_\pm = \frac{1}{(2\pm 2S^2)}\left[2E_0(2\pm 2S^2)+2\iint (\varphi_1^{(0)})^2\right]\left(\frac{-e^2}{r_{a_2}}-\frac{e^2}{r_{b_1}}+\frac{e^2}{R}+\frac{e^2}{r_{12}}\right)$$

$$d\tau_1\,d\tau_2\pm 2\iint \varphi_2^{(0)}\varphi_1^{(0)}\left(\frac{-e^2}{r_{a_2}}-\frac{e^2}{r_{b_1}}+\frac{e^2}{R}+\frac{e^2}{r_{12}}\right)d\tau_1\,d\tau_2\Bigg]$$

$$= 2E_0+\frac{2\iint (\varphi_1^{(0)})\frac{e^2}{R}\,d\tau_1\,d\tau_2\pm 2\iint \varphi_2^{(0)}\varphi_1^{(0)}\frac{e^2}{R}\,d\tau_1\,d\tau_2}{(2\pm 2S^2)}+\frac{2C\pm 2J}{2\pm 2S^2}$$

$$= 2E_0+\frac{(2\pm 2S^2)\frac{e^2}{R}}{(2\pm 2S^2)}+\frac{2C\pm 2J}{(2\pm 2S^2)}$$

$$= \frac{-e^2}{a_0}+\frac{e^2}{R}+\frac{(C\pm J)}{1\pm S^2} \qquad \text{... (25)}$$

where

$$C = \iint (\varphi_1^{(0)})^2\left(\frac{-e^2}{r_{a_2}}-\frac{e^2}{r_{b_1}}+\frac{e^2}{r_{12}}\right)d\tau_1\,d\tau_2$$

$$= \iint (\varphi_{100}(r_{a_1}))^2(\varphi_{100}(r_{b_2}))^2\left(\frac{-e^2}{r_{b_1}}-\frac{e^2}{r_{a_2}}+\frac{e^2}{r_{12}}\right)d\tau_1\,d\tau_2$$

$$= -\int (\varphi_{100}(r_{a_1}))^2\frac{e^2}{r_{b_1}}\,d\tau_1\int (\varphi_{100}(\varphi_{b_2}))^2\,d\tau_2$$

$$-\int (\varphi_{100}(r_{b_2}))^2\frac{e^2}{r_{a_2}}\,d\tau_2\int (\varphi_{100}(r_{a_1}))^2\,d\tau_1$$

$$+\iint (\varphi_{100}(r_{a_1}))^2\frac{e^2}{r_{12}}(\varphi_{100}(r_{b_2}))^2\,d\tau_1\,d\tau_2$$

$$= -\int (\varphi_{100}(r_{a_1}))^2\frac{e^2}{r_{b_1}}\,d\tau_1-\int (\varphi_{100}(r_{b_2}))^2\frac{e^2}{r_{a_2}}\,d\tau_2$$

$$+\iint (\varphi_{100}(r_{a_1}))^2 \frac{e^2}{r_{12}} (\varphi_{100}(r_{b_2}))^2 \, d\tau_1 \, d\tau_2 \quad \text{... (26)}$$

and

$$J = \iint \varphi_2^{(0)} \varphi_1^{(0)} \left(\frac{e^2}{r_{12}} - \frac{e^2}{r_{b_1}} - \frac{e^2}{r_{a_2}} \right) d\tau_1 \, d\tau_2$$

$$= \iint \varphi_{100}(r_{a_2}) \varphi_{100}(r_{b_1}) \varphi_{100}(r_{a_1}) \varphi_{100}(r_{b_2}) \left(\frac{e^2}{r_{12}} - \frac{e^2}{r_{b_1}} - \frac{e^2}{r_{a_2}} \right) d\tau_1 \, d\tau_2$$

$$= \iint \varphi_{100}(r_{a_1}) \varphi_{100}(r_{b_2}) \frac{e^2}{r_{12}} \varphi_{100}(r_{a_2}) \varphi_{100}(r_{b_1}) \, d\tau_1 \, d\tau_2$$

$$- S \int \varphi_{100}(r_{a_1}) \frac{e^2}{r_{b_1}} \varphi_{100}(r_{b_1}) \, d\tau_1 - S \int \varphi_{100}(r_{b_2}) \frac{e^2}{r_{a_2}} \varphi_{100}(r_{a_2}) \, d\tau_2 \quad \text{... (27)}$$

where C and J are called Coulomb and exchange integrals, respectively. C is called the Coulomb integral because the first term in eqn. (26) represents Coulomb interaction between the nucleus b and electron 1 which corresponds to an electron density, $e\{\varphi_{100}(r_{a_1})]^2$. The second term similarly represents the interaction between the electron 2 and nucleus a. Numerically, these two integrals are eqnal. The third term gives the Coulomb interaction between an electron in 1s orbital on nucleus a with an electron in 1s orbital on nucleus b. J is obtained because of the interchange of particles being identical (indistinguishable) and hence represents a non-classical interaction, called exchange interaction or integral.

The integrals C, J and S can be evaluated exactly but the expressions are cumbersome involving various transcendental eqnations and hence we do not present them here. Curves representing E_+ and E_- as functions of R/a_0 are shown in Fig. 15.2. It is seen that

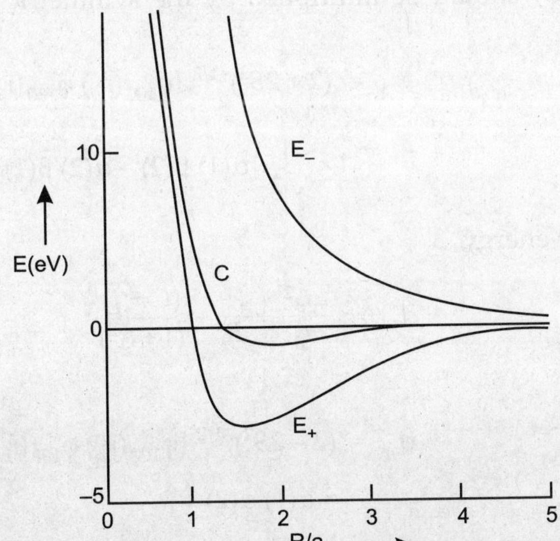

Fig. 15.2 *Energy curves for the hydrogen molecule as function of R/a_0*

E_- corresponds to repulsion at all the distances, their being no eqnilibrium position of the nuclei. The E_+ curve corresponds to attraction of the two hydrogen atoms with the formation of a stable molecule, the eqnilibrium value for R/a_0 being 1.5. (The middle curve in which we only consider the Coulomb interactions, gives the eqnilibrium value, R/a_0 =1.9 which corresponds to the binding energy – 0.36 eV). This corresponds to the binding energy – 3.14 eV (or the energy of dissociation of the molecule into atoms is eqnal to 3.14 eV) which is somewhat smaller than the correct value – 4.72 eV corresponding to R/a_0 = 1.4. This small discrepancy in the binding energy is probably due to the neglect of the vibrational energy of the nuclei.

So far we have not taken into account the electron spin. To do so let us take α and β to be the two spin eigen functions corresponding to the eigen values +1/2 (spin up state) and –1/2 (spin down state). With these eigen functions, we can construct two sets of spin eigen functions,

$$\text{I.}\quad \begin{aligned} &\alpha(1)\,\alpha(2)\ \uparrow\uparrow \\ &\beta(1)\,\beta(2)\ \downarrow\downarrow \\ &\frac{1}{\sqrt{2}}[\alpha(1)\,\beta(2)+\alpha(2)\,\beta(1)]\ \uparrow\downarrow \end{aligned} \qquad \ldots (28a)$$

$$\text{II.}\quad \frac{1}{\sqrt{2}}[\alpha(1)\,\beta(2)-\alpha(2)\,\beta(1)]\ \uparrow\downarrow \qquad \ldots (28b)$$

The first set is symmetric in the spin coordinates of the two electrons whereas the second set is antisymmetric. In order to make the complete wave functions antisymmetric in the case of electrons as reqnired by Pauli's exclusion principle, the spatial wave function φ_+^0 should be multiplied by the antisymmetric spin function (28*b*) and similarly φ_-^0 should be multiplied by the symmetric spin functions (28*a*). Thus we get

$$\begin{aligned} \psi_A &= (2+2S^2)^{-1/2}\,\{\varphi_{100}(r_{a_1})\,\varphi_{100}(r_{b_2})+\varphi_{100}(r_{a_2})\,\varphi_{100}(r_{b_1})\} \\ &\quad\times\frac{1}{\sqrt{2}}[\alpha(1)\,\beta(2)-\alpha(2)\,\beta(1)]\ \uparrow\downarrow \end{aligned} \qquad \ldots (29a)$$

corresponding to the energy,

$$E_+ = \frac{-e^2}{a_0}+\frac{e^2}{R}+\frac{(C+J)}{(1+S^2)}$$

and

$$\begin{aligned} \psi_A &= (2-2S^2)^{-1/2}\,\{\varphi_{100}(r_{a_1})\,\varphi_{100}(r_{b_2})-\varphi_{100}(r_{a_2})\,\varphi_{100}(r_{b_1})\} \\ &\quad\times\alpha(1)\,\alpha(2)\ \uparrow\uparrow \\ &\quad\times\beta(1)\,\beta(2)\ \downarrow\downarrow \end{aligned} \qquad \ldots (29b)$$

$$\times \frac{1}{\sqrt{2}}[\alpha(1)\,\beta(2) + \alpha(2)\,\beta(1)] \uparrow\downarrow$$

corresponding to the energy.

$$E_- = \frac{-e^2}{a_0} + \frac{e^2}{R} + \frac{(C-J)}{(1-S^2)}$$

There are thus three repulsive states $(C - J)$ for one attractive state $(C + J)$, that is, the chance is only 1/4 that the two hydrogen atoms interact with one another in the way that a stable molecule is formed. In the $(C + J)$ state the z-component of spin is zero (the two spins being antiparallel, it is spin singlet) and in the $C - J$ state, the z-component has three values, viz. -1, 0 and $+1$ (that is, it is spin triplet), respectively. It is thus seen that the normal state of the hydrogen molecule is a singlet state, the spins of the two electrons being opposed whereas the repulsive state is a triplet state. This also explains the saturation property of the molecule. Since in this molecule the two spins are antiparallel, hence when a third atom combines, two spins will have the same direction (would be parallel) producing repulsion.

15.5 THE HELIUM ATOM

The helium atom consists of a nucleus of charge $Ze (Z = 2)$ surrounded by two electrons as shown in Fig. 15.3. The Hamiltonian for the system would be given by

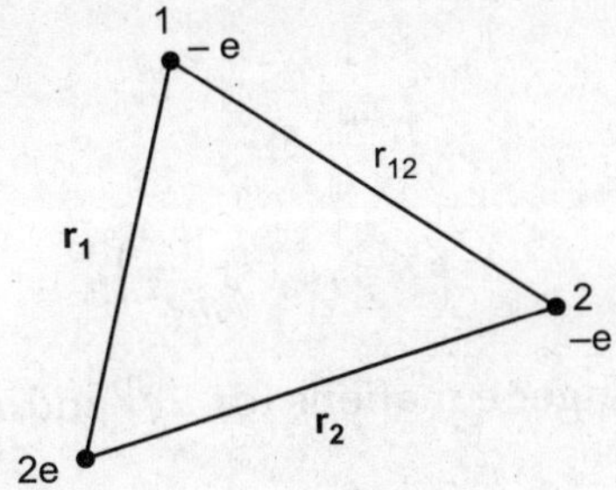

Fig. 15.3 *The coordinate system for helium atom*

$$\hat{H} = -\frac{\hbar^2}{2m}\nabla_1^2 - \frac{\hbar^2}{2m}\nabla_2^2 - \frac{Ze^2}{r_1} - \frac{Ze^2}{r_2} + \frac{e^2}{|\mathbf{r}_1 - \mathbf{r}_2|} \quad \text{... (30)}$$

where $\mathbf{r}_1$ and $\mathbf{r}_2$ represent the position vectors of the two electrons with respect to the nucleus which is assumed to be stationary and $r_{12} = |\mathbf{r}_1 - \mathbf{r}_2|$ is the distance between the two electrons. The first four terms in eqn. (30) represent the kinetic and potential energies of electrons in the field of the nucleus. The last term is the energy associated with the electrostatic repulsion between the electrons.

Now the term $\frac{e^2}{r_{12}}$ is of the same order of magnitude as the other two potential energy terms, however, we shall begin by neglecting it (finally we will incorporate this as perturbation). The resulting approximation may be called independent particle model. In this approximation, the Schrödinger eqnation reads.

$$\left[\left(\frac{-\hbar^2}{2m}\nabla_1^2 - \frac{Ze^2}{r_1}\right) + \left(\frac{-\hbar^2}{2m}\nabla_2^2 - \frac{Ze^2}{r_2}\right)\right]\varphi_1^{(0)}(\mathbf{r}_1, \mathbf{r}_2)$$

$$= E^{(0)}\,\varphi_1^{(0)}(\mathbf{r}_1, \mathbf{r}_2) \qquad \text{... (31)}$$

where the superscript '0' corresponds to the unperturbed values. The above eqnation simply means,

$$(H_1^{(0)} + H_2^{(0)})\,\varphi_1^{(0)} = E^{(0)}\,\varphi_1^{(0)}$$

and may be separated by writing

$$\varphi_1^{(0)}(\mathbf{r}_1, \mathbf{r}_2) = u_b^{(0)}(\mathbf{r}_1)\,u_d^{(0)}(\mathbf{r}_2)$$

or simply

$$\varphi_1^{(0)}(1, 2) = u_b^{(0)}(1)\,u_d^{(0)}(2) \qquad \text{... (32)}$$

where $u_b^{(0)}(1)$ and $u_d^{(0)}(2)$ are hydrogen atom wave functions (b corresponds to quantum numbers n_1, l_1, m_1 and similarly d corresponds to quantum numbers n_2, l_2, m_2) modified to take account of the increased nuclear charge. (In particular, the ground state wave function

$$\varphi^{(0)}(1, 2) = (\pi a^3)^{-1/2}\,e^{-r_1/a}\,(\pi a^3)^{-1/2}\,e^{-r_2/a}$$

$$= \frac{1}{\pi a^3}\,e^{-(r_1 + r_2)/a}$$

$$a = \frac{a_0}{Z},\ a_0 = \frac{\hbar^2}{me^2}\Bigg).$$

Accordingly the Schrödinger eqnations for $u_b^{(0)}$ and $u_d^{(0)}$ are given by

$$\left(\frac{-\hbar^2}{2m}\nabla_1^2 - \frac{Ze^2}{r_1}\right)u_b^{(0)}(1) = E_{n_1}^{(0)}\,u_b^{(0)}(1) \qquad \text{... (33a)}$$

and

$$\left(\frac{-\hbar^2}{2m}\nabla_2^2 - \frac{Ze^2}{r_2}\right)u_d^{(0)}(2) = E_{n_2}^{(0)}\,u_d^{(0)}(2) \qquad \text{... (33b)}$$

with $E^{(0)} = E_{n_1}^{(0)} + E_{n_2}^{(0)}$

Since $\hat{H}(1, 2) = \hat{H}(2, 1)$, that is, the Hamiltonian is symmetric with respect to the exchange of two electrons, the wave function

$$\varphi_2^{(0)} = u_b^{(0)}(2)\,u_d^{(0)}(1) \qquad \text{... (34)}$$

corresponding to electron 1 in the state d and 2 in the state b is also the solution of the Schrödinger eqnation corresponding to the same energy eigen value, $E^{(0)} = E_{n_1}^{(0)} + E_{n_2}^{(0)}$.

From the above discussion it is clear that the system is two fold degenerate. Since this degeneracy arises due to the exchange of two electrons being identical, is called exchange degeneracy. This degeneracy exists as long as we neglect the electron-electron interaction. As soon as we take this interaction into account, the levels would split up as we would see shortly. The linear combination of the solutions (32) and (34), that is,

$$\varphi_{\pm}^{(0)} = \frac{1}{\sqrt{2}}(\varphi_1^{(0)} \pm \varphi_2^{(0)})$$

$$= \frac{1}{\sqrt{2}}[(u_b^{(0)}(1)\, u_d^{(0)}(2) \pm u_b^{(0)}(2)\, u_d^{(0)}(1)] \quad \text{... (35)}$$

would also be the solutions of eqn. (31) belonging to the same energy eigen value, $E^{(0)} = E_{n_1}^{(0)} + E_{n_2}^{(0)}$, the factor $1/\sqrt{2}$ has been introduced so that the wave functions are normalized. The solutions $\varphi_+^{(0)}$ and $\varphi_-^{(0)}$ are symmetric and antisymmetric wave functions with respect to the interchange of spatial coordinates of the two electrons . It has already been pointed out that it is essential that we use either the symmetric or the antisymmetric solutions as the particles are indistinguishable in quantum mechanics.

Treating $\frac{e^2}{r_{12}}$ as the perturbation which arises due to the interaction of the electrons themselves, the first order correction to the ground state energy can be found out by solving the secular eqnation,

$$\begin{vmatrix} \lambda H'_{11} - \lambda E^{(1)} & \lambda H'_{12} \\ \lambda H'_{21} & \lambda H'_{22} - \lambda E^{(1)} \end{vmatrix} = 0 \quad \text{... (36)}$$

The off diagonal term,

$$\lambda H'_{12} = \iint \varphi_+^{(0)} \frac{e^2}{r_{12}} \varphi_-^{(0)}\, d\tau_1\, d\tau_2$$

$$= -\iint \varphi_+^{(0)} \frac{e^2}{r_{12}} \varphi_-^{(0)}\, d\tau_1\, d\tau_2$$

$$= -\lambda H'_{12}$$

as (–)ve sign is introduced in $\varphi_-^{(0)}$ after interchanging the particles. $\varphi_+^{(0)}$ is of course unaffected. Thus $\lambda H'_{12} = 0$ and similarly $\lambda H'_{21} = 0$. The secular eqnation (36), therefore takes the form,

$$\begin{vmatrix} \lambda H'_{11} - \lambda E^{(1)} & 0 \\ 0 & \lambda H'_{22} - \lambda E^{(1)} \end{vmatrix} = 0 \quad \text{... (36a)}$$

Thus the first order correction in the ground state energy due to the perturbation $\frac{e^2}{r_{12}}$ is given by

$$\lambda E^{(1)} = \lambda H'_{11} \text{ and } \lambda H'_{22}$$

or in general,

$$\lambda E_{\pm}^{(1)} = \iint \varphi_{\pm}^{(0)} \frac{e^2}{r_{12}} \varphi_{\pm}^{(0)} \, d\tau_1 \, d\tau_2$$

$$= \frac{1}{2}\iint (\varphi_1^{(0)} \pm \varphi_2^{(0)}) \frac{e^2}{r_{12}} (\varphi_1^{(0)} \pm \varphi_2^{(0)}) \, d\tau_1 \, d\tau_2$$

$$= \frac{1}{2}\iint \varphi_1^{(0)} \frac{e^2}{r_{12}} \varphi_1^{(0)} \, d\tau_1 \, d\tau_2 + \frac{1}{2}\iint \varphi_2^{(0)} \frac{e^2}{r_{12}} \varphi_2^{(0)} \, d\tau_1 \, d\tau_2$$

$$\pm \iint \varphi_1^{(0)} \frac{e^2}{r_{12}} \varphi_2^{(0)} \, d\tau_1 \, d\tau_2 \pm \frac{1}{2}\iint \varphi_2^{(0)} \frac{e^2}{r_{12}} \varphi_1^{(0)} \, d\tau_1 d\tau_2$$

After interchanging 1 and 2, we see that the first and second integrals are the same and similarly third and forth terms are same. Hence

$$\lambda E_{\pm}^{(1)} = \iint \varphi_1^{(0)} \frac{e^2}{r_{12}} \varphi_1^{(0)} \, d\tau_1 \, d\tau_2 \pm \iint \varphi_1^{(0)} \frac{e^2}{r_{12}} \varphi_2^{(0)} \, d\tau_1 \, d\tau_2$$

$$= \iint \left\{u_b^{(0)}(1)\right\}^2 \frac{e^2}{r_{12}} \left\{u_d^{(0)}(2)\right\}^2 \, d\tau_1 \, d\tau_2$$

$$\pm \iint u_b^{(0)}(1)\, u_d^{(0)}(2) \frac{e^2}{r_{12}} u_b^{(0)}(2)\, u_d^{(0)}(1) \, d\tau_1 \, d\tau_2$$

$$= C \pm J \qquad \text{... (37)}$$

where C is called the Coulomb integral and is always positive. J is called the exchange integral. When the electrons are very close to each other, r_{12} becomes small and hence contribution to J from e^2/r_{12} will be large and is found to be positive. When the electrons are far apart, that is, when r_{12} is large, e^2/r_{12} becomes small and hence the contribution of the exchange integral is negligible and as a result the sign of J is not of much significance. From the above argument, we conclude that the sign of J is in general positive. Hence the total energy (after including first order correction) is given by

$$E_{\pm} = E_{n_1}^{(0)} + E_{n_2}^{(0)} + C \pm J \qquad \text{... (38)}$$

which shows that the degeneracy has been lifted (or the level splits) because of the presence of e^2/r_{12}. The eigen function in the upper state corresponding to the energy,

$$E_{+} = E_{n_1}^{(0)} + E_{n_2}^{(0)} + C + J \qquad \text{... (38a)}$$

would be given by

$$\varphi_+^{(0)} = \frac{1}{\sqrt{2}}[(u_b^{(0)}(1)\,u_d^{(0)}(2) + u_b^{(0)}(2)\,u_d^{(0)}(1)] \quad \text{... (35a)}$$

which is symmetric. The eigen function in the lower state corresponding to the energy,

$$E_- = E_{n_1}^{(0)} + E_{n_2}^{(0)} + C - J \quad \text{... (38b)}$$

would be given by

$$\varphi_-^{(0)} = \frac{1}{\sqrt{2}}[(u_b^{(0)}(1)\,u_d^{(0)}(2) - u_b^{(0)}(2)\,u_d^{(0)}(1)] \quad \text{... (35b)}$$

which is antisymmetric.

The above discussion shows that due to the interaction, the energy level is displaced by C and splits into two parts as shown in Fig. 15.4; the magnitude of splitting is 2J.

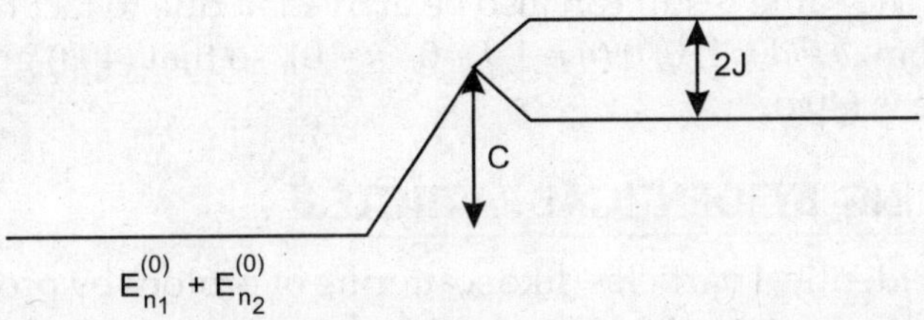

Fig. 15.4 *The energy perturbation due to the interaction between the two electrons*

Let us now take the spin of the electron into account. Let α and β be two spin eigen functions corresponding to the eigen values + 1/2 (spin up) and – 1/2 (spin down). With α and β, we can construct two sets of spin eigen functions,

I. $\alpha(1)\,\alpha(2)\ \uparrow\uparrow$

$\beta(1)\,\beta(2)\ \downarrow\downarrow$ ... (39a)

$\frac{1}{\sqrt{2}}[\alpha(1)\,\beta(2) + \alpha(2)\,\beta(1)]\ \uparrow\downarrow$

II. $\frac{1}{\sqrt{2}}[\alpha(1)\,\beta(2) - \alpha(2)\,\beta(1)]\ \uparrow\downarrow$ ... (39b)

According to the Pauli exclusion principle, the complete wave function of the electron should be antisymmetric. Since the eigen functions given by eqn. (39*a*) are symmetric in spin coordinates of the two electrons and the eigen function given by eqn. (39*b*) is antisymmetric, we should multiply (39*a*) by (35*b*) and (39*b*) by (35*a*) to make the complete wave function antisymmetric. Thus the complete wave function of the helium atom would be given by

$$\psi_A = \frac{1}{\sqrt{2}}[u_b^{(0)}(1)\,u_d^{(0)}(2) + u_b^{(0)}(2)\,u_d^{(0)}(1)]$$

$$\times \frac{1}{\sqrt{2}}[\alpha(1)\,\beta(2) - \alpha(2)\,\beta(1)]\ \uparrow\downarrow \quad \text{... (40a)}$$

and
$$\psi_A = \frac{1}{\sqrt{2}}[u_b^{(0)}(1)\,u_d^{(0)}(2) - u_b^{(0)}(2)\,u_d^{(0)}(1)]$$
$$\times\,\alpha(1)\,\alpha(2)\uparrow\uparrow$$
$$\times\,\beta(1)\,\beta(2)\downarrow\downarrow \qquad \text{... (40}b\text{)}$$
$$\times\,\frac{1}{\sqrt{2}}[\alpha(1)\,\beta(2) + \alpha(2)\,\beta(1)]\uparrow\downarrow$$

In the first case, the z-component of spin is zero and is thus a spin singlet (called para helium). In the second case, the z-component has values $-1, 0, +1$ and is thus a spin triplet (called ortho helium). It is the experimental fact that the ground state of the helium does not split in the magnetic field (or the analysis of the Zeeman effect of helium shows that the ground state is always a singlet), conseqnently it must be singlet rather than a triplet, that is, (40a) represents the ground state wave function of the helium atom. The same result can also be arrived at due to fact that for the ground state of helium atom, $b = d \equiv 1, 0, 0$ ($n = 1, l = 0, m = 0$), so that (40b) gives zero. Thus the only allowed state is (40a).

15.6 SCATTERING BY IDENTICAL PARTICLES

In the scattering of identical particles (like scattering of protons by protons or α-particles by α-particles), there are two situations which are indistinguishable from each other as shown in Fig. 15.5. In the COM system, the incident particle and the target have eqnal and opposite velocities. In Fig. 15.5(a), the particle observed at the detector is the incident particle (particle number 1) while in Fig. 15. 5(b), it is the recoiled target (particle number 2). Due to the indistinguishability of the two situations (or particles), we must construct the appropriate wave functions which are either symmetric or antisymmetric with respect to the interchange of particles for the calculation of the scattering cross-section.

For process shown in Fig. 15.5(a), the asymptotic form of the scattered wave function is

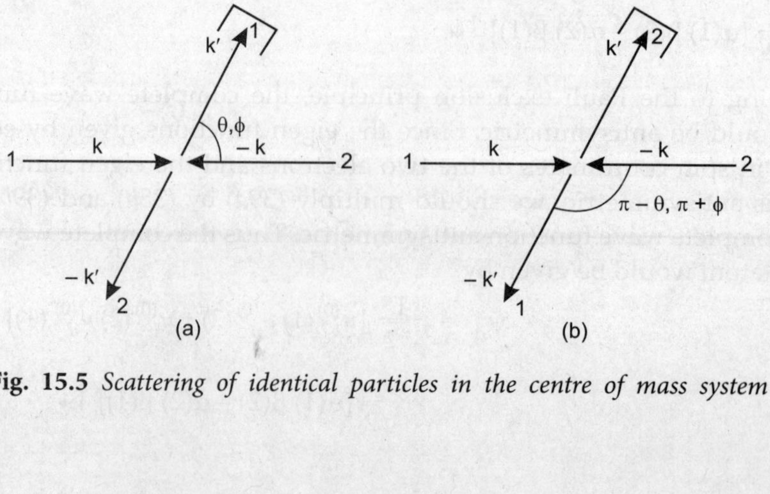

Fig. 15.5 *Scattering of identical particles in the centre of mass system*

$$\varphi_a^{(s)}(\mathbf{k}, \mathbf{k}', \mathbf{r}) = f_a(\mathbf{k}, \mathbf{k}')\frac{e^{ikr}}{r} \qquad \text{... (41)}$$

where $\mathbf{r} = \mathbf{r}_1 - \mathbf{r}_2$ represents the relative coordinate of the two particles, a in the subscript corresponds to Fig. 15.5(*a*) and (*s*) in the superscript corresponds to scattered. The process shown in Fig. 15.5(*b*) differs from that in Fig. 15.5(*a*) in the sense that in the former the particles 1 and 2 are interchanged at the time of scattering. Thus

$$\varphi_b^{(s)}(\mathbf{k}, \mathbf{k}', \mathbf{r}) = P\,\varphi_a^{(s)}(\mathbf{k}, \mathbf{k}', \mathbf{r}) \qquad \text{... (42)}$$

where P corresponds to the exchange operator which interchanges particles 1 and 2. But interchanging the two particles is eqnivalent to reversing the relative coordinate $\mathbf{r} = \mathbf{r}_1 - \mathbf{r}_2$ of the two particles as $\mathbf{r}_1$ becomes $\mathbf{r}_2$ and $\mathbf{r}_2$ becomes $\mathbf{r}_1$ and since $\mathbf{k}'$ has the same direction as $\mathbf{r}$ (the scattered wave being spherical), the said interchange is eqnivalent to reversing the direction of $\mathbf{k}'$. Thus

$$P\varphi_a^{(s)}(\mathbf{k}, \mathbf{k}', \mathbf{r}) = \varphi_a^{(s)}(\mathbf{k}, -\mathbf{k}', \mathbf{r}) \qquad \text{... (43)}$$

The necessary symmetric or antisymmetric combinations of $\varphi_a^{(s)}$ and $\varphi_b^{(s)}$ in the present case are given by

$$\varphi_S^{(s)}(\mathbf{k}, \mathbf{k}', \mathbf{r}) = \varphi_a^{(s)}(\mathbf{k}, \mathbf{k}', \mathbf{r}) + \varphi_b^{(s)}(\mathbf{k}, \mathbf{k}', \mathbf{r}) \qquad \text{... (44)}$$

$$\varphi_A^{(s)}(\mathbf{k}, \mathbf{k}', \mathbf{r}) = \varphi_a^{(s)}(\mathbf{k}, \mathbf{k}', \mathbf{r}) - \varphi_b^{(s)}(\mathbf{k}, \mathbf{k}', \mathbf{r})$$

where A and S (capital) correspond to the symmetric and antisymmetric wave functions, respectively. Using eqns. (42) and (43), one gets

$$\varphi_S^{(s)}(\mathbf{k}, \mathbf{k}', \mathbf{r}) = \varphi_a^{(s)}(\mathbf{k}, \mathbf{k}', \mathbf{r}) + \varphi_b^{(s)}(\mathbf{k}, -\mathbf{k}', \mathbf{r}),$$

$$\varphi_A^{(s)}(\mathbf{k}, \mathbf{k}', \mathbf{r}) = \varphi_a^{(s)}(\mathbf{k}, \mathbf{k}', \mathbf{r}) - \varphi_b^{(s)}(\mathbf{k}, -\mathbf{k}', \mathbf{r})$$

Further using eqn. (41) and after deleting the subscript a, one gets,

$$\varphi_S^{(s)} = f(\mathbf{k}, \mathbf{k}')\frac{e^{ikr}}{r} + f(\mathbf{k}, -\mathbf{k}')\frac{e^{ikr}}{r}$$

$$\varphi_A^{(s)} = f(\mathbf{k}, \mathbf{k}')\frac{e^{ikr}}{r} - f(\mathbf{k}, -\mathbf{k}')\frac{e^{ikr}}{r}$$

Putting $\quad f_S(\mathbf{k}, \mathbf{k}') = f(\mathbf{k}, \mathbf{k}') + f(\mathbf{k}, -\mathbf{k}')$

and $\quad f_A(\mathbf{k}, \mathbf{k}') = f(\mathbf{k}, \mathbf{k}') - f(\mathbf{k}, -\mathbf{k}')$

one gets

$$\varphi_S^{(s)} = f_S(\mathbf{k}, \mathbf{k}')\frac{e^{ikr}}{r} \qquad \text{... (45a)}$$

$$\varphi_A^{(s)} = f_A(\mathbf{k}, \mathbf{k}')\frac{e^{ikr}}{r} \qquad \text{... (45b)}$$

The differential cross-sections for the scattering of identical particles are, then given by

$$\sigma_S(\mathbf{k}, \mathbf{k}') = |f_S(\mathbf{k}, \mathbf{k}')|^2$$

$$= |f(\mathbf{k}, \mathbf{k}')|^2 + |f(\mathbf{k}, \mathbf{k}')|^2 + 2Re\,\{f^*(\mathbf{k}, \mathbf{k}')f(\mathbf{k}, -\mathbf{k}')\} \quad \text{... (46a)}$$

and
$$\sigma_A(\mathbf{k}, \mathbf{k}') = |f_A(\mathbf{k}, \mathbf{k}')|^2$$
$$= |f(\mathbf{k}, \mathbf{k}')|^2 + |f(\mathbf{k}, -\mathbf{k}')|^2 - 2Re\,\{f^*(\mathbf{k}, \mathbf{k}')f(\mathbf{k}, -\mathbf{k}')\} \quad \text{... (46b)}$$

where $Re\,\{\ \}$ represents the real part of the quantity with in the curly brackets.

The above cross-sections (quantum mechanical) are to be compared with the corresponding classical expression,

$$\sigma_{cl}(\mathbf{k}, \mathbf{k}') = \sigma(\mathbf{k}, \mathbf{k}') + \sigma(\mathbf{k}, -\mathbf{k}')$$
$$= |f(\mathbf{k}, \mathbf{k}')|^2 + |f(\mathbf{k}, -\mathbf{k}')|^2 \quad \text{... (47)}$$

as the two situations corresponding to Fig. 15.5 (*a*) and (*b*) are distinguishable in classical mechanics and one has only to count the total number of particles (incident or target) reaching the detector. Note that the difference between quantum mechanical and classical cross-sections is due to the presence of the interference term (between scattering amplitudes of incident and target particles) arising due to the fact that in the former the two situations depicted in Figs. 15.5(*a*) and (*b*) cannot be distinguished.

In terms of the usual angles θ, ϕ which specify the direction of scattering with respect to incident direction,

$$f(\mathbf{k}, \mathbf{k}') \equiv f_k(\theta, \phi) \quad \text{... (48a)}$$

and
$$f(\mathbf{k}, -\mathbf{k}') \equiv f_k(\pi - \theta, \pi + \phi) \quad \text{... (48b)}$$

Further, in the case of central forces, the scattering amplitude being independent of ϕ,

$$f(\mathbf{k}, \mathbf{k}') \equiv f_k(\theta);\ f(\mathbf{k}, -\mathbf{k}') \equiv f_k(\pi - \theta) \quad \text{... (49)}$$

Thus we have,

$$\sigma_S(\theta) = |f_k(\theta) + f_k(\pi - \theta)|^2, \quad \text{... (49a)}$$
$$\sigma_A(\theta) = |f_k(\theta) + f_k(\pi - \theta)|^2, \quad \text{... (49b)}$$
$$\sigma_{cl}(\theta) = |f_k(\theta)|^2 + |f_k(\pi - \theta)|^2, \quad \text{... (49c)}$$

Let us apply the above results to the spin – 1/2 particles. We have seen in the cases of hydrogen molecule and helium atom that the spin part of the wave function could be a triplet (symmetric spin functions) or a singlet state (antisymmetric spin function). The singlet state should be associated with the symmetric space function (as the complete wave function in the case of Fermions must be antisymmetric) so that the corresponding cross-section is given by

$${}^1\sigma_{1/2'} = \sigma_S(\theta) \quad \text{... (50)}$$

where the superscript 1 corresponds to the singlet state and 1/2 in the subscript corresponds to the spin value. Similarly, the triplet spin state is to be associated with the antisymmetric space function. Therefore,

$${}^3\sigma_{1/2}(\theta) = \sigma_A(\theta) \quad \text{... (51)}$$

where the superscript 3 of course corresponds to a triplet spin state.

The observed cross-section would be the weighted average of these two cases. That is (as the singlet has statistical weight one and the triplet has three),

$$\begin{aligned}\sigma_{1/2}(\theta) &= \frac{1}{4}\sigma_S(\theta) + \frac{3}{4}\sigma_A(\theta) \\ &= \frac{1}{4}|f_k(\theta) + f_k(\pi-\theta)|^2 + \frac{3}{4}|f_k(\theta) - f_k(\pi-\theta)|^2 \\ &= |f_k(\theta)|^2 + |f_k(\pi-\theta)|^2 + \frac{1}{2}Re\{f_k^*(\theta)\, f_k(\pi-\theta)\} \\ &\quad - \frac{3}{2}Re\{f_k^*(\theta)\, f_k(\pi-\theta)\} \\ &= |f_k(\theta)|^2 + |f_k(\pi-\theta)|^2 - Re\{f_k^*(\theta)\, f_k(\pi-\theta)\} \qquad \text{... (52)}\end{aligned}$$

At $\theta = \frac{\pi}{2}$, the above eqnation gives,

$$\sigma_{1/2}\left(\frac{\pi}{2}\right) = \left|f_k\left(\frac{\pi}{2}\right)\right|^2 \qquad \text{... (53)}$$

From eqn. (49*c*),

$$\sigma_{cl}\left(\frac{\pi}{2}\right) = 2\left|f_k\left(\frac{\pi}{2}\right)\right|^2 \qquad \text{... (54)}$$

The results (53) and (54) can be utilized to find experimentally the spin of a particle like proton. The fact that the observed cross-section at $\theta = \frac{\pi}{2}$ is approximately half that of the value given by eqn. (54), indicates that the proton is a spin –1/2 particle.

Further let us apply the results obtained above to the case of spin zero particles. In this case the complete wave function is symmetric (hence one has to take into account the symmetric space function) so that

$$\begin{aligned}\sigma_0(\theta) &= \sigma_S(\theta) \\ &= |f_k(\theta) + f_k(\pi-\theta)|^2\end{aligned}$$

At $\theta = \pi/2$,

$$\sigma_0\left(\frac{\pi}{2}\right) = 4\left|f_k\left(\frac{\pi}{2}\right)\right|^2 \qquad \text{... (55)}$$

Thus the quantum theory predicts twice as much scattering as we would expect from classical theory (see eqn. (54)), the experimental data confirms the quantum result. From eqns. (53), (54) and (55) it follows that

$$\sigma_{1/2}\left(\frac{\pi}{2}\right) : \sigma_{cl}\left(\frac{\pi}{2}\right) : \sigma_0\left(\frac{\pi}{2}\right) = 1:2:4 \qquad \text{... (56)}$$

Chapter 16

RELATIVISTIC QUANTUM MECHANICS

16.1 KLEIN–GORDON RELATIVISTIC EQUATION

The non-relativistic quantum theory which we have developed so far fails in some major features, viz.

(*i*) It fails to explain the spin of an elementary particle such as an electron and conseqnences there of (that is, the corresponding magnetic moment).

(*ii*) It cannot give a satisfactory account of high energy phenomenon such as pair production, bremmstrahlung, etc.

(*iii*) It even fails to explain the spectrum of hydrogen atom correctly.

To overcome these difficulties a need was, therefore felt to develop relativistic quantum mechanics (as the phenomenon occurring at high energies must be based upon relativistic wave equation). According to Einstein's theory of relativity, a relationship between the total energy E, momentum p and the rest mass m_0 of a free particle, is expressed by the equation.

$$E^2 = p^2 c^2 + m_0^2 c^4 \qquad \text{... (1)}$$

where c is speed of light.

Schrödinger[1] proposed the relativistic form of the wave equation basing it on eqn. (1) by replacing the classical variables E and p by operators, $E \to i\hbar \frac{\partial}{\partial t}$ and $\mathbf{p} \to -i\hbar \vec{\nabla}$ as in the non-relativistic case. By doing so, we find

$$(i\hbar)^2 \frac{\partial}{\partial t} \frac{\partial}{\partial t} = c^2 (-i\hbar)^2 \vec{\nabla} \cdot \vec{\nabla} + m_0^2 c^4$$

Dividing the above equation by $\hbar^2 c^2$ throughout and rearranging, we get

$$\nabla^2 - 1/c^2\, \partial^2/\partial t^2 = m_0^2 c^2/\hbar^2$$

[1]E. Schrödinger, Ann. Physik 81, 109 (1926)

Upon introducing the wave function $\psi(\mathbf{r}, t)$, we have the following relativistic equation,

$$(\nabla^2 - 1/c^2\, \partial^2/\partial t^2)\, \psi(\mathbf{r}, t) = m_0^2\, c^2/\hbar^2\, \psi(\mathbf{r}, t) \qquad \text{... (2)}$$

The above equation was independently discovered at about the same time by Fock, Klein and Gordon; it is sometimes called the Schrödinger-Fock or more freqnently the Klein-Gordon[2] equation as the implications of the said equation were studied by the later scientists.

Let us write down the Klein-Gordon equation in the covariant form. For obtaining the same, we introduce four dimensional coordinates x_1, x_2, x_3, x_4, where $x_1 = x$, $x_2 = y$, $x_3 = z$ and $x_4 = ict$

Eqn (2) then takes the form,

$$\left(\frac{\partial^2}{\partial x_1^2} + \frac{\partial^2}{\partial x_2^2} + \frac{\partial^2}{\partial x_3^2} + \frac{\partial^2}{\partial x_4^2}\right)\psi = m_0^2\, \frac{c^2}{\hbar^2}\, \psi \qquad \text{... (3)}$$

Note that there is complete symmetry in all four second derivatives with respect to space and time (unlike the non-relativistic Schrödinger equation which is first order in time and second order in space). The eqn. (3) can be written as

$$\psi\,(x_\upsilon) = m_0^2\, c^2/\hbar^2\, \psi(x_\upsilon) \qquad \text{... (4)}$$

where υ, the Greek index ranges over four values, $\upsilon = 1, 2, 3, 4$ (as compared to Roman index $i = 1, 2, 3$ appearing in x_i) and

$$= \frac{\partial^2}{\partial x_1^2} + \frac{\partial^2}{\partial x_2^2} + \frac{\partial^2}{\partial x_3^2} + \frac{\partial^2}{\partial x_4^2} = \frac{\partial}{\partial x_\upsilon}\frac{\partial}{\partial x_\upsilon} = \partial_\upsilon \partial_\upsilon,\ \partial_\upsilon = \frac{\partial}{\partial x_\upsilon} \qquad \text{... (4a)}$$

called the *D'* Alembertian operator can be considered as the four dimensional form of the Laplacian operator ∇^2. Note that eqns. (4) use the Einstein summation convention.

The Klein-Gordon equation, it turns out can describe the behaviour of particles with zero spin such as π and K – mesons, It has nothing to say about the spin of a particle. At the time, the Klein-Gordon equation was introduced no particles with zero spin were known and hence the resulting theory fell into disfavour.

16.2 THE PROBABILITY DENSITY AND PROBABILITY CURRENT DENSITY

We know that in the non-relativistic Schrödinger equation, it is possible to define the position probability density and probability current density by using the continuity equation. to follow this, we start with the Klein–Gordon equation for a free particle, that is,

$$\left(\nabla^2 - \frac{1}{c^2}\frac{\partial^2}{\partial t^2}\right)\psi = m_0^2\, \frac{c^2}{\hbar^2}\, \psi \qquad \text{... (2)}$$

and its complex conjugate equation,

[2] O. Klein, Z. Physik 37, 895 (1926); W. Gordon, Z. Physik 40, 121 (1926)

$$\left(\nabla^2 - \frac{1}{c^2}\frac{\partial^2}{\partial t^2}\right)\psi^* = m_0^2\,\frac{c^2}{\hbar^2}\,\psi^* \qquad \text{... (5)}$$

Multiplying (2) from the left by ψ^* and (5) from the right by ψ, then substracting (5) from (2), one gets

$$(\psi^*\nabla^2\psi - \psi\nabla^2\psi^*) - \frac{1}{c^2}\left(\psi^*\frac{\partial\psi}{\partial t^2} - \psi\frac{\partial\psi^*}{\partial t^2}\right) = 0$$

or

$$\vec{\nabla}\cdot(\psi^*\vec{\nabla}\psi - \psi\vec{\nabla}\psi^*) - \frac{1}{c^2}\frac{\partial}{\partial t}\left(\psi^*\frac{\partial\psi}{\partial t} - \psi\frac{\partial\psi^*}{\partial t}\right) = 0 \qquad \text{... (6)}$$

where we have made use of the vector identity,

$$\vec{\nabla}\cdot\phi\mathbf{a} = \vec{\nabla}\phi\cdot\mathbf{a} + \phi\vec{\nabla}\cdot\mathbf{a} = \mathbf{a}\cdot\vec{\nabla}\phi + \phi\vec{\nabla}\cdot\mathbf{a} \qquad \text{... (7)}$$

After multiplying by $\hbar/2im_0$, eqn. (6) can written as

$$\vec{\nabla}\cdot\mathbf{j} + \frac{\partial\rho}{\partial t} = 0 \qquad \text{... (8)}$$

which is essentially equation of continuity with the probability current density,

$$\mathbf{j} = \frac{\hbar}{2\,im_0}(\psi^*\vec{\nabla}\psi - \psi\vec{\nabla}\psi^*)$$

$$= \frac{1}{2}\left[\psi^*\left(-\frac{i\hbar}{m_0}\vec{\nabla}\right)\psi + \psi\left(\frac{i\hbar}{m_0}\vec{\nabla}\right)\psi^*\right]$$

$$= \frac{1}{2}\left(\psi^*\frac{\mathbf{p}}{m_0}\psi + \psi\frac{\mathbf{p}^*}{m_0}\psi^*\right)$$

$$= Re\left(\psi^*\frac{\mathbf{p}\psi}{m_0}\right)$$

$$= Re\,(\psi^*\,\mathbf{V}\,\psi) \qquad \text{... (9)}$$

and probability density,

$$\rho = \frac{i\hbar}{2m_0c^2}\left(\psi^*\frac{\partial\psi}{\partial t} - \psi\frac{\partial\psi^*}{\partial t}\right)$$

$$= \frac{1}{2}\left[\psi^*\left(\frac{i\hbar}{m_0c^2}\frac{\partial}{\partial t}\right)\psi - \psi\left(-\frac{i\hbar}{m_0c^2}\frac{\partial}{\partial t}\right)\psi^*\right]$$

$$= \frac{1}{2}\left(\psi^* \frac{E}{m_0c^2}\psi + \psi \frac{E^*}{m_0c^2}\psi^*\right)$$

$$= Re\left(\psi^* \frac{E}{m_0c^2}\psi\right) \quad \text{... (10)}$$

In the non-relativistic limit the density given by eqn. (10) reduces to that obtained in the non-relativistic theory (the well known expression of the non-relativistic quantum theory for the probability density).

$$\rho = Re\left(\psi^* \frac{E}{m_0c^2}\psi\right) \to \psi^*\psi \quad \text{... (11)}$$

as $E \to m_0c^2$ (since $E = T + m_0c^2$ and $T << m_0c^2$)

It should, however be noted that the ρ defined in eqn. (10) depends upon the time derivative of $\psi\left(E \equiv \frac{i\hbar\partial}{\partial t}\right)$, hence it may not necessarily be positive (or in other words E from eqn. (1) could be positive or negative, hence from eqn. (10), ρ could be positive or negative). Therefore, ρ cannot be interpretted as position probability density as probability is never negative. However, if eqn. (10) is multiplied by the charge e then it can be interpretted as charge density can have either sign (ρ = 0 for neutral particles). Even this way ρ can still have different signs at different points which is hardly satisfactory in the description of a single particle of given charge. The expression (9) for **j**, the probability current density is of course identical with the non-relativistic form.

16.3 KLEIN–GORDON EQUATION AND HYDROGEN ATOM

The Klein-Gordon equation in the presence of electromagnetic field is given by

$$\left(\frac{i\hbar\partial}{\partial t} - e\phi\right)^2 \psi = c^2\left(-i\hbar\vec{\nabla} - \frac{e}{c}\mathbf{A}\right)\psi + m_0^2c^2\psi \quad \text{... (12)}$$

where E and **p** have been replaced by $E - e\phi$ and $\vec{p} - \frac{e}{c}\vec{A}$, respectively. $\vec{A}$ and ϕ are called vector and scalar potentials and e being the charge on the particle. Finally E and **p** have been replaced by the corresponding operators.

Assuming that $\mathbf{A} = 0$ and ϕ is spherically symmetric as in the case of hydrogen like atoms, eqn. (12) becomes

$$\left(i\hbar\frac{\partial}{\partial t} - e\phi\right)^2\psi = (-c^2\hbar^2\nabla^2 + m_0^2c^4)\psi \quad \text{... (13)}$$

The above equation can be simplified by assuming that ψ has the form (ϕ being independent of t)

$$\psi(\mathbf{r}, t) = u(\mathbf{r})\, e^{-iEt/\hbar} \quad \text{... (14)}$$

On inserting (14) into (13), we find that u (**r**) must satisfy

$$(E - e\phi)^2 \mathbf{u}(\mathbf{r}) = (-c^2\hbar^2\nabla^2 + m_0^2 c^4)u(\mathbf{r})$$

which can be written as

$$\nabla^2 u(\mathbf{r}) = \left[\frac{-(E - e\phi)^2 + m_0^2 c^4}{\hbar^2 c^2}\right] u(\mathbf{r}) \quad \text{... (15)}$$

In the case of hydrogen – like atoms,

$$e\phi = -Ze^2/r \quad \text{... (16)}$$

that is, the potential is spherically symmetric. Hence, we can separate eqn. (15) into radial and angular parts by assuming

$$u(\mathbf{r}) = u(r, \theta, \phi) = R_l(r)\, Y_l^{m_l}(\theta, \phi) \quad \text{... (17)}$$

Substituting eqns. (16) and (17) in (15), the equation for R_l (r) or the radial equation becomes (the equation for $Y_l^{m_l}(\theta, \phi)$ is well known and yields the terms involving l).

$$\left[\frac{1}{r^2}\frac{d}{dr}\left(r^2\frac{d}{dr}\right) - \frac{l(l+1)}{r^2}\right]R_l(r)$$

$$= \left[\frac{m_0^2 c^2}{\hbar^2} - \frac{E^2}{\hbar^2 c^2} - \frac{2EZe^2}{\hbar^2 c^2 r} - \frac{Z^2 e^4}{\hbar^2 c^2 r^2}\right]R_l(r)$$

Dividing above equation by α_1^2, we get

$$\left[\frac{1}{\alpha_1^2 r^2}\frac{d}{d(\alpha_1 r)}\left\{\alpha_1^2 r^2 \frac{d}{d(\alpha_1 r)}\right\} - \frac{l(l+1)}{\alpha_1^2 r^2}\right]R_l(r)$$

$$= \left[\frac{m_0^2 c^4 - E^2}{\hbar^2 c^2 \alpha_1^2} - \frac{2ZEe^2}{\hbar^2 c^2 \alpha_1(\alpha_1 r)} - \frac{Z^2 e^4}{\hbar^2 c^2 \alpha_1^2 r^2}\right]R_l(r)$$

Now introducing the variable,

$$\rho = \alpha_1 r \quad \text{... (18)}$$

and upon transposing, the radial equation takes the form

$$\left[\frac{d^2}{d\rho^2} + \frac{2}{\rho}\frac{d}{d\rho} - \frac{m_0^2 c^4 - E^2}{\hbar^2 c^2 \alpha_1^2} + \frac{2}{\rho}\left(\frac{EZe^2}{\hbar^2 c^2 \alpha_1}\right) - \frac{l(l+1) - Z^2 e^4/\hbar^2 c^2}{\rho^2}\right]R_l(\rho) = 0 \quad \text{... (19)}$$

This is identical with the non-relativistic Schrödinger equation (eqn. (50) of Sec. 5.3) in the case of hydrogen atom except for the values of constants. In this case

$$\alpha_1^2 = \frac{4}{\hbar^2 c^2}(m_0^2 c^4 - E^2),\ \lambda = \frac{2EZe^2}{\hbar^2 c^2 \alpha_1} \quad \text{... (20)}$$

and

$$l(l+1) \rightarrow l(l+1) - \frac{Z^2 e^4}{\hbar^2 c^2} \quad \text{... (21)}$$

and of course $m_0^2 c^4 - E^2$ is positive as for the bound states, E must be less than the rest mass energy m_0c^2.

Following the same procedure as that for solving the non-relativistic Schrödinger equation for hydrogen atom, that is finding the solutions for $\rho \to \infty$, $\rho \to 0$ etc. and then applying the admissibility conditions on the wave function, it can be shown that

$$\lambda = n \to k + l' + 1, k = 0, 1, 2, \ldots \qquad \ldots (22)$$

(in the case of non-relativistic Schrödinger equation for the said problem, the principal quantum number $n = k + l + 1$). in which l' is positive solution of

$$l'(l' + 1) = l(l + 1) - \frac{Z^2 e^4}{\hbar^2 c^2}$$

or

$$l'^2 + l' - \left\{l(l+1) - \frac{Z^2 e^4}{\hbar^2 c^2}\right\} = 0$$

or

$$l' = -\frac{1}{2} \pm \frac{1}{2}\left[(2l+1)^2 - \frac{4Z^2 e^4}{\hbar^2 c^2}\right]^{1/2} \qquad \ldots (23)$$

(of which one is positive and the other is negative for $l > 0$. For $l = 0$, both the l' values given by (23) are negative and hence $R_i\,(r)$ which behaves like $\rho^{l'}$ near ρ (or r) = 0, is singular at the origin. However, the second term under the radical sign is quite small, hence the use of upper sign in (23) gives a value of l' which is close to zero for physically interesting values of Z whereas the lower sign makes l' close to –1) Using eqns. (22) and (20), one gets

$$\lambda^2 = n^2 = \left(\frac{2\,EZe^2}{\hbar^2 c^2}\right)^2 \frac{1}{\alpha_1^2}$$

$$= \left(\frac{2\,EZe^2}{\hbar^2 c^2}\right)^2 \frac{1}{\left[\frac{4}{\hbar^2 c^2}(m_0^2 c^4 - E^2)\right]}$$

or $$E^2\left(1 + \frac{Z^2 e^4}{n^2 \hbar^2 c^2}\right) = m_0^2 c_4$$

or $$E = m_0 c^2 \left(1 + \frac{Z^2 \alpha^2}{n^2}\right)^{-1/2}, \quad \alpha = \frac{e^2}{\hbar c} \simeq \frac{1}{137} \qquad \ldots (24)$$

where α is the Sommerfeld fine structure constant. Expanding above by using the binomial expansion

$$E = m_0 c^2 \left(1 - \frac{Z^2 \alpha^2}{n^2} + \frac{3\,Z^4 \alpha^4}{8n^4} \ldots\right) \qquad \ldots (25)$$

Substituting the value of l', (the positive value) into eqn. (22), we obtain

$$n = k + \left[-\frac{1}{2} + \frac{1}{2}\left\{4\left(l+\frac{1}{2}\right)^2 - 4Z^2\alpha^2\right\}^{1/2}\right] + 1$$

$$= k + \frac{1}{2} + \left(l+\frac{1}{2}\right)\left[1 - \frac{Z^2\alpha^2}{\left(l+\frac{1}{2}\right)^2}\right]^{1/2}$$

where for $\frac{e^2}{\hbar c}$, we have put α. Note that n in this case is not an integer. Using the binomial expansion

$$n = k + \frac{1}{2} + \left(l+\frac{1}{2}\right)\left[1 - \frac{Z^2\alpha^2}{2\left(l+\frac{1}{2}\right)^2}\right]$$

$$= k + \frac{1}{2} + l + \frac{1}{2} - \frac{Z^2\alpha^2}{2\left(l+\frac{1}{2}\right)}$$

$$= (k+l+1) - \frac{Z^2\alpha^2}{2\left(l+\frac{1}{2}\right)}$$

$$= n'\left[1 - \frac{Z^2\alpha^2}{2n'\left(l+\frac{1}{2}\right)}\right]$$

where $n' = k + l + 1$ is an integer. From the above equation

$$\frac{1}{n^2} = \frac{1}{n'^2}\left[1 - \frac{Z^2\alpha^2}{2n'\left(l+\frac{1}{2}\right)}\right]^{-2}$$

Using the binomial expansion

$$\frac{1}{n^2} = \frac{1}{n'^2}\left[1 + \frac{Z^2\alpha^2}{n'\left(l+\frac{1}{2}\right)}\right]$$

$$= \frac{1}{n'^2} + \frac{Z^2\alpha^2}{n'^3\left(l+\frac{1}{2}\right)} \quad \text{... (26)}$$

Putting the value of $1/n^2$ from the above equation into (25) and neglecting the second term in (26) while putting for $1/n^4$, one gets

$$E = m_0c^2\left[1 - \frac{1}{2}Z^2\alpha^2\left\{\frac{1}{n'^2} + \frac{Z^2\alpha^2}{n'^3\left(l+\frac{1}{2}\right)}\right\} + \frac{3}{8}\frac{Z^4\alpha^4}{n'^4}\right]$$

which is correct to the fourth power in α. This can be written as

$$E = m_0c^2\left[1 - \frac{Z^2\alpha^2}{2n'^2} - \frac{Z^4\alpha^4}{2n'^3\left(l+\frac{1}{2}\right)} + \frac{3}{8}\frac{Z^4\alpha^4}{n'^4}\right]$$

$$= m_0c^2\left[1-\frac{Z^2\alpha^2}{2n'^2}-\frac{Z^4\alpha^4}{2n'^4}\left\{\frac{n'}{\left(l+\frac{1}{2}\right)}-\frac{3}{4}\right\}\right]$$

which can further be written as

$$E = m_0c^2\left[1-\frac{Z^2\alpha^2}{2n^2}-\frac{Z^4\alpha^4}{2n^4}\left\{\frac{n}{\left(l+\frac{1}{2}\right)}-\frac{3}{4}\right\}\right] \quad \text{... (27)}$$

where we have removed the prime from integer n because there is no chance for confusion any more.

The first term in (27) or in $E - m_0c^2$, namely

$$-\frac{m_0c^2}{2n^2}\frac{Z^2e^4}{\hbar^2c^2} = -\frac{m_0Z^2e^4}{2\hbar^2 n^2}$$

gives just the Balmer levels of the hydrogen atom (in the non-relativistic case,

$$E_n = -\frac{\mu Z^2e^4}{2\hbar^2 n^2}, \quad n = 1, 2, 3, \ldots)$$

but eqn. (27) has an additional term depending on l which gives fine structure to each of these levels. The term is (we would soon notice that the term containing 3/4 would vanish)

$$\varepsilon = -m_0c^2Z^4\alpha^4/2n^3\left(l+\frac{1}{2}\right) \quad \text{... (28)}$$

The total spread due to fine structure (that is, change in E as l goes from $0-(n-1)$ for given n) can be calculated as follows:

$$\begin{aligned}\varepsilon_{max} &= -m_0c^2Z^4\alpha/n^3\,(2l+1)_{max}\\ &= -m_0c^2Z^4\alpha/n^3\,[2(n-1)+1]\\ &= -m_0c^2Z^4\alpha/n^3\,(2n-1) \quad \text{... (29)}\\ \varepsilon_{min} &= -m_0c^2Z^4\alpha/n^3\,(2l+1)_{min}\\ &= -m_0c^2Z^4\alpha/n^3\,(2\times 0+1)\\ &= -m_0c^2Z^4\alpha/n^3 \quad \text{... (30)}\end{aligned}$$

Thus the spread is

$$\begin{aligned}\Delta E &= \varepsilon_{max} - \varepsilon_{min}\\ &= m_0c^2\,\frac{Z^4\,\alpha^4}{n^3}\left\{1-\frac{1}{(2n-1)}\right\}\\ &= m_0c^2\,\frac{Z^4\,\alpha^4}{n^3}\left\{\frac{(n-1)}{\left(n-\frac{1}{2}\right)}\right\} \quad \text{... (31)}\end{aligned}$$

which is more than twice what the Dirac theory predicts for spin $\frac{1}{2}$ particles in a Coulomb field and it is the later which is in agreement with the experimental data on hydrogen like atoms. This shows that the theory is incomplete in some detail. The Klein-Gordon equation is really known to describe spin-zero particles like the π-meson (whereas electrons do have spin $-\frac{1}{2}$).

16.4 DIRAC RELATIVISTIC EQUATION[3]

No doubt the Klein-Gordon equation satisfied the need for relativistic invariance but there are, at least two drawbacks in the theory:

(*i*) The probability density is not positive definite due to the fact that the wave equation is second order differential equation in time and hence ρ contains first derivative of time. Also ρ $(\mathbf{r}, t)$ not being of the form $\psi^*\psi$, one cannot define expectation values in the usual fashion.

(*ii*) It has nothing to say about the spin of particle.

Dirac reasoned that in order to overcome difficulty (*i*), the equation should be a first order differential equation in time and for relativistic invariance then should be first order differential equation in space coordinates also. The spin of course comes as a by-product of his work and in turn overcomes difficulty (*ii*). Hence the proper equation should have the form,

$$i\hbar\, \partial\, \psi\, (\mathbf{r}, t)/\partial t \;=\; H\psi\, (\mathbf{r}, t) \qquad \text{... (32)}$$

in which

$$H = (p^2c^2 + m_0^2\, c^4)^{1/2} \;=\; (-\, c^2\hbar^2\nabla^2 + m_0^2\, c^4)^{1/2} \qquad \text{... (32a)}$$

where $\mathbf{p}$ has been replaced by the operator $-\, i\hbar\, \vec{\nabla}$.

Within the realm of real number system, it is impossible to express $-\, c^2\hbar^2\nabla^2 + m_0^2 c^4$ as a perfect square of a linear differential operator (that is, linear in space derivative or momentum) hence Dirac took the essential step in the right direction by writing H as

$$H \;=\; c\, \vec{\alpha} \cdot \mathbf{p} + \beta m_0 c^2 \qquad \text{... (33)}$$

reqniring simply that when (33) is squared, agreement is obtained; that is,

$$c^2p^2 + m_0^2\, c^4 \;=\; (c\, \vec{\alpha} \cdot \mathbf{p} + \beta m_0 c^2)^2 \qquad \text{... (34)}$$

where β and the coefficients α_x, α_y, α_z of the momentum components are as yet undetermined.

Let us discover properties of the quantities $\vec{\alpha}$ and β by straight forward expansion, observing the order of multiplication as being possibly important if these quantities do not commute. From the above equation,

---

[3]P.A.M. Dirac, Proc. Roy. Soc. London A117, 610 (1928)

$$c^2\left(p_x^2+p_y^2+p_z^2\right)+m_0^2c^4$$
$$= [c(\alpha_x p_x + \alpha_y p_y + \alpha_z p_z) + \beta m_0 c^2] \times c(\alpha_x p_x + \alpha_y p_y + \alpha_z p_z) + \beta m_0 c^2]$$
$$= c^2\alpha_x^2 p_x^2 + c^2\alpha_y^2 p_y^2 + c^2\alpha_z^2 p_z^2 + c^2(\alpha_x\alpha_y + \alpha_y\alpha_x) p_x p_y$$
$$+ c^2(\alpha_x\alpha_z + \alpha_z\alpha_x) p_x p_z + c^2(\alpha_y\alpha_z + \alpha_z\alpha_y) p_y p_z + m_0 c^3(\alpha_x\beta + \beta\alpha_x) p_x$$
$$+ m_0 c^3(\alpha_y\beta + \beta\alpha_y) p_y + m_0 c^3(\alpha_z\beta + \beta\alpha_z) p_z + \beta^2 m_0^2 c^4$$

Comparing the coefficients of $p_x^2, p_y^2, p_z^2, m_0^2 c^4, p_x p_y, p_y p_z, p_z p_x, p_x, p_y$ and p_z on both sides of the above equation, we get

$$\alpha_x^2 = \alpha_y^2 = \alpha_z^2 = \beta^2 = 1 \qquad \text{... (35a)}$$

$$\alpha_x\alpha_y + \alpha_y\alpha_x = \alpha_y\alpha_z + \alpha_z\alpha_y = \alpha_z\alpha_x + \alpha_x\alpha_z - 0 \qquad \text{... (35b)}$$

and

$$\alpha_x\beta + \beta\alpha_x = \alpha_y\beta + \beta\alpha_y = \alpha_z\beta + \beta\alpha_z = 0 \qquad \text{... (35c)}$$

where 1 and 0 are quantities of the proper types.

The relations (35) determine the essential algebraic properties of the quantities $\vec{\alpha}$ and β. Since $\vec{\alpha}$ and β anticommute rather than commuting with each other, they cannot be numbers. Hence they can be taken as matrices. Accordingly 1 and 0 would mean unit and null matrices, respectively.

In order that H be Hermitian, each matrix $\vec{\alpha}$ $(\alpha_x, \alpha_y, \alpha_z)$ and β has to be Hermitian (since $\vec{\alpha}$ and β are Hermitian matrices, they must, therefore be square ones). The Dirac equation, therefore, reads as

$$i\hbar\,\partial\psi/\partial t = (-i\hbar\, c\,\vec{\alpha}\cdot\vec{\nabla} + \beta m_0 c^2)\,\psi \qquad \text{... (36)}$$

where $\alpha_x, \alpha_y, \alpha_z$ and β are Hermitian matrices and satisfy eqn. (35). Their explicit forms are as yet undetermined.

16.5 DIRAC'S $\vec{\alpha}$ AND β MATRICES

To get a clear idea of the nature of Dirac eqn. (36), one needs to determine the explicit forms of the matrices obeying eqns. (35). From eqn. (35 *c*), we have

$$\alpha_x\beta = -\beta\alpha_x$$

Assuming that the Dirac matrices have inverses, multiplying above by α_x^{-1} from the left, we get

$$\alpha_x^{-1}\alpha_x\beta = -\alpha_x^{-1}\beta\alpha_x$$

Using the relation, $\alpha_x^{-1}\alpha_x = 1$, we get

$$\beta = -\alpha_x^{-1}\beta\alpha_x.$$

On taking trace of the matrices on both sides and further using the relation, trace (ABC) = trace (CAB), we get

$$\text{trace}\,\beta = -\,\text{trace}\,(\alpha_x\alpha_x^{-1}\beta) \qquad \text{... (37)}$$

which using the relation, $\alpha_x \alpha_x^{-1} = 1$ again, leads to trace $\beta = -$ trace β

The only way to satisfy the above condition is that trace $\beta = 0$

That is, whatever be the form of β, the sum of diagonal elements must be zero. Let us consider β to be the diagonalized matrix, that is,

$$\beta = \begin{pmatrix} b_1 & 0 & - & - & - & - & - \\ 0 & b_2 & 0 & - & - & - & - \\ 0 & & & & & & \\ 0 & & & & & & \\ \cdots & & & & & & b_N \end{pmatrix} \quad \text{... (38)}$$

Therefore,

$$\beta^2 = \begin{pmatrix} b_1^2 & 0 & 0 & - & - & - \\ 0 & b_2^2 & 0 & - & - & - \\ 0 & & & & & \\ 0 & & & & & \\ \cdots & & & & & b_N^2 \end{pmatrix} \quad \text{... (39)}$$

Since $\beta^2 = 1$ (that is, the unit matrix, I), the diagonal terms in (39) must be unity, that is,

$$b_i^2 = 1 \quad \text{... (39a)}$$

or

$$b_i = \pm 1, i = 1, 2, \ldots, N$$

Let m be the number of +1 elements and n be the number of –1 elements. Then

$$m + n = N \quad \text{... (40)}$$

and from the condition (37), one gets

$$\sum_i b_i = m \times (+1) + n \times (-1) = m - n = 0$$

or

$$m = n \quad \text{... (41)}$$

From eqns. (40) and (41), one obtain

$$N = 2m \quad \text{... (42)}$$

which amounts to the conclusion that the matrix is even dimensional (or is not 3 × 3, 5 × 5 and so on). Though the simplest type of even dimensional matrix can be 2 × 2 but it can be shown that the Dirac matrices cannot be 2 × 2. This is because of the fact that there are just three anticommuting matrices with dimension 2, while we need four. The three Pauli matrices

$$\sigma_x = \begin{pmatrix} 0 & 1 \\ 1 & 0 \end{pmatrix}, \sigma_y = \begin{pmatrix} 0 & -i \\ i & 0 \end{pmatrix}, \sigma_z = \begin{pmatrix} 1 & 0 \\ 0 & -1 \end{pmatrix}$$

do anticommute ($\sigma_x \sigma_y + \alpha_y \alpha_x = 0$ and so on) and the square of each one is the unit matrix $(\sigma_x^2 = \sigma_y^2 = \sigma_z^2 = 1)$. That is, they satisfy the same relations (35*a*, *b*) as satisfied by α_x, α_y and α_z. Hence one may represent α_x, α_y, α_z by σ_x, σ_y, σ_z. But if one takes any other 2 × 2 matrix with arbitrary elements and reqnires that it should anticommute with $\alpha_x(\sigma_x)$, $\alpha_y(\sigma_y)$ and $\alpha_z(\sigma_z)$, one immediately finds that all its elements must vanish. One can show that with dimension 4 (or greater), one can find four anticommuting matrices.

Let us take,

$$\beta = \begin{pmatrix} 1 & 0 & 0 & 0 \\ 0 & 1 & 0 & 0 \\ 0 & 0 & -1 & 0 \\ 0 & 0 & 0 & -1 \end{pmatrix} = \begin{pmatrix} I & 0 \\ 0 & -I \end{pmatrix} \quad \text{... (42a)}$$

The matrix for α_x may be written as (since $\vec{\alpha}$ and β anticommute, hence if β is diagonal matrix, $\vec{\alpha}$ cannot be diagonal (cannot be measured simultaneously and accurately),

$$\alpha_x = \begin{pmatrix} \alpha_{x_{11}} & \alpha_{x_{12}} \\ \alpha_{x_{21}} & \alpha_{x_{22}} \end{pmatrix} \quad \text{... (43)}$$

where $\alpha_{x_{11}}, \alpha_{x_{12}},$ —— are the matrices of rank two (2 × 2 matrices)

Using the relation, $\alpha_x \beta + \beta \alpha_x = 0$ and substituting α_x and β from eqns. (43) and (42) into it, one gets

$$\begin{pmatrix} \alpha_{x_{11}} & 0 \\ 0 & -\alpha_{x_{22}} \end{pmatrix} = 0 \text{ (Null matrix)}$$

which implies that

$$\alpha_{x_{11}} = \alpha_{x_{22}} = 0 \quad \text{... (44a)}$$

Similarly had we started with the relations, $\alpha_y \beta + \beta \alpha_y = 0$ and $\alpha_z \beta + \beta \alpha_z = 0$, we could have obtained

$$\alpha_{y_{11}} = \alpha_{y_{22}} = 0 \quad \text{... (44b)}$$

and

$$\alpha_{z_{11}} = \alpha_{z_{22}} = 0 \quad \text{... (44c)}$$

Further using the relation, $\alpha_x^2 = 1$ and substituting the values of $\alpha_{x_{11}}$ and $\alpha_{x_{22}}$ from eqn. (44*a*) into eqn. (43), we obtain

$$\begin{pmatrix} \alpha_{x_{12}} \alpha_{x_{21}} & 0 \\ 0 & \alpha_{x_{21}} \alpha_{x_{12}} \end{pmatrix} = 1,$$

hence

$$\alpha_{x_{12}} \alpha_{x_{21}} = \alpha_{x_{21}} \alpha_{x_{12}} = 1 \quad \text{... (45a)}$$

Similarly had we started with the relations $\alpha_y^2 = 1$ and $\alpha_z^2 = 1$, we would have got

$$\alpha_{y_{12}} \alpha_{y_{21}} = \alpha_{y_{21}} \alpha_{y_{12}} = 1 \qquad \text{... (45b)}$$

and

$$\alpha_{z_{12}} \alpha_{z_{21}} = \alpha_{z_{21}} \alpha_{z_{12}} = 1 \qquad \text{... (45c)}$$

For the sake of simplicity, let us take, $\alpha_{x_{12}} = \alpha_{x_{21}}$ and similarly $\alpha_{y_{12}} = \alpha_{y_{21}}$ and $\alpha_{z_{12}} = \alpha_{z_{21}}$. With this simplification, one gets

$$\alpha_{x_{12}}^2 = \alpha_{y_{12}}^2 = \alpha_{z_{12}}^2 = 1 \qquad \text{... (46)}$$

The use of relation, $\alpha_x \alpha_y + \alpha_y \alpha_x = 0$ along with the relations (44*a*, *b*), would give

$$\begin{pmatrix} \alpha_{x_{12}} \alpha_{y_{12}} + \alpha_{y_{12}} \alpha_{x_{12}} & 0 \\ 0 & \alpha_{x_{12}} \alpha_{y_{12}} + \alpha_{y_{12}} \alpha_{x_{12}} \end{pmatrix} = 0$$

which emplies that

$$\alpha_{x_{12}} \alpha_{y_{12}} + \alpha_{y_{12}} \alpha_{x_{12}} = 0 \qquad \text{... (47a)}$$

Similarly had we started with the relations, $\alpha_y \alpha_z + \alpha_z \alpha_y = 0$ and $\alpha_z \alpha_x + \alpha_x \alpha_z = 0$ along with the relations (44), we would have got

$$\alpha_{y_{12}} \alpha_{z_{12}} + \alpha_{z_{12}} \alpha_{y_{12}} = 0 \qquad \text{... (47b)}$$

and

$$\alpha_{z_{12}} \alpha_{x_{12}} + \alpha_{x_{12}} \alpha_{z_{12}} = 0 \qquad \text{... (47c)}$$

Note that the relations (46) and (47) are same as that for σ_x, σ_y and σ_z. Therefore, we may identify $\alpha_{x_{12}}$ with σ_x, $\alpha_{y_{12}}$ with σ_y and $\alpha_{z_{12}}$ with σ_z. That is,

$$\alpha_{x_{12}} \equiv \sigma_x,\ \alpha_{y_{12}} \equiv \sigma_y\ \alpha_{z_{12}} \equiv \sigma_z \qquad \text{... (48)}$$

Accordingly the four Dirac matrices would be given by

$$\beta = \begin{pmatrix} I & 0 \\ 0 & -I \end{pmatrix} = \begin{pmatrix} 1 & 0 & 0 & 0 \\ 0 & 1 & 0 & 0 \\ 0 & 0 & -1 & 0 \\ 0 & 0 & 0 & -1 \end{pmatrix} \qquad \text{... (42)}$$

$$\alpha_x = \begin{pmatrix} 0 & \sigma_x \\ \sigma_x & 0 \end{pmatrix} = \begin{pmatrix} 0 & 0 & 0 & 1 \\ 0 & 0 & 1 & 0 \\ 0 & 1 & 0 & 0 \\ 1 & 0 & 0 & 0 \end{pmatrix} \qquad \text{... (49a)}$$

$$\alpha_y = \begin{pmatrix} 0 & \sigma_y \\ \sigma_y & 0 \end{pmatrix} = \begin{pmatrix} 0 & 0 & 0 & -i \\ 0 & 0 & i & 0 \\ 0 & -i & 0 & 0 \\ i & 0 & 0 & 0 \end{pmatrix} \qquad \text{... (49b)}$$

$$\alpha_z = \begin{pmatrix} 0 & \sigma_z \\ \sigma_z & 0 \end{pmatrix} = \begin{pmatrix} 0 & 0 & 1 & 0 \\ 0 & 0 & 0 & -1 \\ 1 & 0 & 0 & 0 \\ 0 & -1 & 0 & 0 \end{pmatrix} \quad \text{... (49c)}$$

The three α matrices can be written as

$$\vec{\alpha} = \begin{pmatrix} 0 & \vec{\sigma} \\ \vec{\sigma} & 0 \end{pmatrix} \quad \text{... (50)}$$

16.6 POSITION PROBABILITY DENSITY AND CURRENT DENSITY

The Dirac relativistic equation is given by

$$i\hbar \frac{\partial \psi(\vec{r}, t)}{\partial t} = -ic\hbar\, \vec{\alpha} \cdot \vec{\nabla} \psi(\mathbf{r}, t) + \beta m_0 c^2\, \psi(\mathbf{r}, t) \quad \text{... (36)}$$

Now since $\vec{\alpha}$ and β are represented by 4 × 4 matrices, the above equation has no meaning unless the wave function ψ itself is a matrix with four rows and one column, that is,

$$\psi(\mathbf{r}, t) = \begin{pmatrix} \psi_1(\mathbf{r}, t) \\ \psi_2(\mathbf{r}, t) \\ \psi_3(\mathbf{r}, t) \\ \psi_4(\mathbf{r}, t) \end{pmatrix} \quad \text{... (51)}$$

Proceeding as with the schrödinger equation, let us write down the Hermitian adjoint (or conjugate) of eqn. (36),

$$-i\hbar \frac{\partial}{\partial t} \psi^{\dagger} = ic\hbar\, \vec{\nabla}\, \psi^{\dagger} \cdot \vec{\alpha} + m_0 c^2\, \psi^{\dagger}\, \beta \quad \text{... (52)}$$

where we have used the relation, $(AB)^{\dagger} = B^{\dagger} A^{\dagger}$ and also $\alpha^{\dagger} = \alpha$, $\beta^{\dagger} = \beta$ as α and β are Hermitian. The matrix which is Hermitian conjugate (or transpose complex conjugate) of (51) would be one row matrix:

$$\psi^{\dagger} = (\psi_1^* \, \psi_2^* \, \psi_3^* \, \psi_4^*) \quad \text{... (51a)}$$

Multiplying (36) by ψ^{+} from the left and (52) by ψ from the right and then taking the difference, we obtain

$$i\hbar \frac{\partial}{\partial t} (\psi^{\dagger} \psi) = -ic\hbar\, (\psi^{\dagger}\, \vec{\alpha} \cdot \vec{\nabla} \psi + \vec{\nabla}\, \psi^{\dagger} \cdot \vec{\alpha}\, \psi)$$

where we have cancelled the term, $m_0 c^2\, \psi^{+}\, \beta\, \psi$. Further cancelling $i\hbar$ on both the sides of the above equation, we get

$$\frac{\partial}{\partial t} (\psi^{\dagger} \psi) + [\psi^{\dagger}\, c\, \vec{\alpha} \cdot (\vec{\nabla} \psi) + (\vec{\nabla}\, \psi^{\dagger}) \cdot c\, \vec{\alpha}\, \psi] = 0$$

or

$$\frac{\partial}{\partial t} (\psi^{\dagger} \psi) + \vec{\nabla} \cdot (\psi^{\dagger}\, c\, \vec{\alpha}\, \psi) = 0$$

Comparing above with the continuity equation,

$$\frac{\partial \rho}{\partial t} + \vec{\nabla}.\mathbf{j} = 0$$

we get,

$$\begin{aligned} \rho &= \psi^{\dagger}\psi \\ &= (\psi_1^* \psi_2^* \psi_3^* \psi_4^*) \begin{pmatrix} \psi_1 \\ \psi_2 \\ \psi_3 \\ \psi_4 \end{pmatrix} \\ &= \sum_{\mu} \psi_\mu^* \psi_u \text{ (actually } \psi_\mu^* \psi_u) \end{aligned} \qquad \text{... (54a)}$$

where μ, the Greek index goes from 1 – 4. ρ given by above equations is evidently positive definite and can be interpretted as position probability density. That is, the interpretational difficulty experienced with the Klein-Gordon equation does not arise here. And

$$\mathbf{j} = \psi^{\dagger}\, c\,\vec{\alpha}\,\psi = \rho\mathbf{V},\ \mathbf{V} = c\,\vec{\alpha} \qquad \text{... (54b)}$$

Since, $\alpha_x^2 = \alpha_y^2 = \alpha_z^2 = 1$, the eigen values of α are ± 1. Hence the eigen values (or expectation values) of **V** would be ± *c*, that is, <**V**> = ± *c* or in other words, the average velocity of Dirac particle is ± *c*. Hence the operator **V** can be interpretted as particle velocity. Accordingly **j** can be interpretted as the probability current density.

16.7 THE DIRAC EQUATION IN COVARIANT FORM

The Dirac relativistic equation can be written as

$$i\hbar\frac{\partial \psi}{\partial t} = \left(c\,\alpha_x \frac{\hbar}{i}\frac{\partial}{\partial x} + c\,\alpha_y \frac{\hbar}{i}\frac{\partial}{\partial y} + c\,\alpha_z \frac{\hbar}{i}\frac{\partial}{\partial z} + \beta m_0 c^2\right)\psi \qquad \text{... (55)}$$

In the above equation, the time derivative is multiplied by a constant whereas the space derivative is multiplied by a matrix. Therefore, the said equation is not symmetrical w.r.t. space and time derivatives. To make it symmetrical, let us multiply the above equation by β from the left. Doing so,

we get

$$i\hbar\beta\frac{\partial \psi}{\partial t} = \left(c\,\beta\alpha_x \frac{\hbar}{i}\frac{\partial}{\partial x} + c\,\beta\alpha_y \frac{\hbar}{i}\frac{\partial}{\partial y} + c\beta\alpha_z \frac{\hbar}{i}\frac{\partial}{\partial z} + m_0 c^2\right)\psi$$

where we have used the relation, $\beta^2 = 1$.

Dividing the above equation by *ic*, using $x_1 = x, x_2 = y, x_3 = z, x_4 = i\,c\,t$ language and then rearranging the terms, we get

$$\left[-i\left(\beta\alpha_x \frac{\hbar}{i}\frac{\partial}{\partial x_1} + \beta\alpha_y \frac{\hbar}{i}\frac{\partial}{\partial x_2} + \beta\alpha_z \frac{\hbar}{i}\frac{\partial}{\partial x_3}\right) + \beta\frac{\hbar}{i}\frac{\partial}{\partial x_4} - im_0 c\right]\psi = 0 \qquad \text{... (56)}$$

Let us now define four new matrices by

$$\gamma_1 = -i\beta\alpha_x = \begin{pmatrix} 0 & -i\sigma_x \\ i\sigma_x & 0 \end{pmatrix} = \begin{pmatrix} 0 & 0 & 0 & -i \\ 0 & 0 & -i & 0 \\ 0 & i & 0 & 0 \\ i & 0 & 0 & 0 \end{pmatrix} \quad \text{... (57}a\text{)}$$

$$\gamma_2 = -i\beta\alpha_y = \begin{pmatrix} 0 & -i\sigma_y \\ i\sigma_y & 0 \end{pmatrix} = \begin{pmatrix} 0 & 0 & 0 & -1 \\ 0 & 0 & 1 & 0 \\ 0 & 1 & 0 & 0 \\ -1 & 0 & 0 & 0 \end{pmatrix} \quad \text{... (57}b\text{)}$$

$$\gamma_3 = -i\beta\alpha_z = \begin{pmatrix} 0 & -\sigma_z \\ i\sigma_z & 0 \end{pmatrix} = \begin{pmatrix} 0 & 0 & -i & 0 \\ 0 & 0 & 0 & i \\ i & 0 & 0 & 0 \\ 0 & -i & 0 & 0 \end{pmatrix} \quad \text{... (57}c\text{)}$$

and finally,

$$\gamma_4 = \beta = \begin{pmatrix} 1 & 0 & 0 & 0 \\ 0 & 1 & 0 & 0 \\ 0 & 0 & -1 & 0 \\ 0 & 0 & 0 & -1 \end{pmatrix} \quad \text{... (57}d\text{)}$$

Using (57), we may write eqn. (56) as

$$\left(r_1 \frac{\hbar}{i}\frac{\partial}{\partial x_1} + \gamma_2 \frac{\partial}{\partial x_2} + \gamma_3 \frac{\hbar}{i}\frac{\partial}{\partial x_3} + \gamma_4 \frac{\hbar}{i}\frac{\partial}{\partial x_4} - im_0 c\right)\psi = 0$$

There being symmetry in the four space time derivatives, we may write the above equation (using Einstein summation convention–the Greek index μ goes from 1 – 4) as

$$\left(\gamma_\mu \frac{\hbar}{i}\frac{\partial}{\partial x_\mu} - im_0 c\right)\psi = 0$$

Further multiplying the above equation by i/h, we get

$$\left(\gamma_\mu \frac{\partial}{\partial x_\mu} + m_0 \frac{c}{\hbar}\right)\psi = 0 \quad \text{... (58)}$$

Eqn. (58) is called Dirac relativistic equation in γ-matrix form (or covariant form). One may, however, note that the γ-matrices are Hermitian.

16.8 SPIN ANGULAR MOMENTUM – EXISTENCE (DIRAC EQUATION IN CENTRAL FORCE FIELD)

Let us consider motion of a particle in a central force field described by the potential energy $V(r)$. The Dirac Hamiltonian in this case would be given by

$$H = c\vec{\alpha} \cdot \mathbf{p} + \beta m_0 c^2 + V(r) \quad \text{... (59)}$$

When particle is moving in such a force field an orbital angular momentum **L** would be associated with it. Let the components of **L** be L_x, L_y and L_z. The equation of motion of any of the components (say z) of $\vec{L}$ in the Heisenberg representation would be given by

$$
\begin{aligned}
i\hbar\, \dot{L}_z &= [L_z, H] \\
&= c\alpha_x\,[L_z, p_x] + c\alpha_y\,[L_z, p_y] + c\alpha_z\,[L_z, p_z] + \\
&\quad m_0c^2\,[L_z, \beta] + [L_z, V(r)]
\end{aligned}
$$

Since $L_z \left(= \dfrac{-i\hbar\partial}{\partial\phi}\right)$ is independent of r, $[L_z, V(r)] = 0$. Also using the relations, $[L_z, \beta] = [L_z, p_z] = 0$, that is, using the fact that L_z commutes with $V(r)$, β and p_z, one gets

$$
\begin{aligned}
i\hbar\, \dot{L}_z &= c\,\alpha_x\,[L_z, p_x] + c\,\alpha_y\,[L_z, p_y] \\
&= c\,\alpha_x\,(i\hbar\, p_y) + c\,\alpha_y\,(-\,i\hbar\, p_x) \\
&= i\hbar\, c\,(\alpha_x p_y - p_x\,\alpha_y) \\
&= i\hbar\, c\,(\vec{\alpha} \times \mathbf{p})_z
\end{aligned}
$$

or

$$\dot{L}_z = c\,(\vec{\alpha} \times \vec{p})_z$$

Similarly we can work with x and y components of the orbital angular momentum and in general, we can show that the time rate of change of angular momentum would be given by

$$\frac{d\mathbf{L}}{dt} = c\,(\vec{\alpha} \times \mathbf{p}) \quad \text{... (60)}$$

which implies that **L** is not a constant of motion (or is not conserved), a fact which is contrary to the non-relativistic quantum theory. However, we expect on physical grounds that it should be possible to define an angular momentum (to be finally called total angular momentum) which is constant in a central force field. This simply means that we must find out another operator such that its time rate of change is negative of the R.H.S. of eqn. (60). The sum of this and **L** would then be a constant of motion and can be interpretted as the total angular momentum.

To achieve the desired result, let us define a new matrix by

$$\vec{\sigma}' = \begin{pmatrix} \vec{\sigma} & 0 \\ 0 & \vec{\sigma} \end{pmatrix} \quad \text{... (61)}$$

The equation of motion of any of the components (say z) of $\vec{\sigma}'$ in the Heisenberg representation would be given by

$$
\begin{aligned}
i\hbar\, \dot{\sigma}'_z &= [\sigma'_z, H] \\
&= cp_x\,[\sigma'_z, \alpha_x] + cp_y\,[\sigma'_z, \alpha_y] + c\,\alpha_z\,[\sigma'_z, \alpha_z] + m_0c^2\,[\sigma'_z, \beta] + [\sigma'_z, V(r)]
\end{aligned}
$$

Using the fact that σ'_z commutes with every quantity in the above equation, except α_x, α_y, that is,

$$[\sigma'_z, \alpha_z] = [\sigma'_z, \beta] = [\sigma'_z, V(r)] = 0,$$

one gets

$$\begin{aligned} i\hbar\, \dot{\sigma}'_z &= cp_x\, [\sigma'_z, \alpha_x] + cp_y\, [\sigma'_z, \alpha_y] \\ &= cp_x\, (2i\, \alpha_y) + cp_x\, (-2i\alpha_x) \\ &= -2\, ic\, (\vec{\alpha} \times \mathbf{p})_z \end{aligned}$$

Similarly, we can work with x and y components of $\vec{\sigma}'$ and hence, in general, we can show that the time rate of change of $\vec{\sigma}'$ would be given by

$$\frac{d\vec{\sigma}'}{dt} = \frac{-2c}{\hbar}(\vec{\alpha} \times \mathbf{p})$$

or

$$\frac{d}{dt}\left(\frac{1}{2}\hbar\vec{\sigma}'\right) = -c\,(\vec{\alpha} \times \mathbf{p}) \qquad \text{... (62)}$$

From eqns. (60) and (62) after addition, we obtain

$$\frac{d}{dt}\left(\mathbf{L} + \frac{1}{2}\hbar\vec{\sigma}'\right) = 0 \qquad \text{... (63)}$$

which implies that, $\mathbf{L} + \frac{1}{2} + \hbar\vec{\sigma}'$ is a constant of motion and can be interpretted as the total angular momentum,

$$\mathbf{J} = \mathbf{L} + \frac{1}{2}\hbar\vec{\sigma}' \qquad \text{... (64)}$$

out of which the operator,

$$\mathbf{S} = \frac{1}{2}\hbar\vec{\sigma}' \qquad \text{... (65)}$$

is known as the spin angular momentum of the electron (since **S** does not arise from the orbital motion, it is called intrinsic angular momentum or spin of the particle).

Further since $\sigma'^2_z = 1$, the eigen values of σ'_z are + 1 and – 1 and hence those of S_z are $+\frac{1}{2}\hbar$ and $-\frac{1}{2}\hbar$. From these facts, it is evident that the Dirac equation describes particles of spin $\frac{1}{2}$ such as the electron. The theory based on this equation is commonly referred to as the Dirac theory of electron.

16.9 DIRAC PARTICLE IN AN ELECTROMAGNETIC FIELD (SPIN MAGNETIC MOMENT)

We know that Dirac relativistic equation for a free particle is given by

$$E\psi = (c\vec{\alpha}\cdot\mathbf{p} + \beta m_0 c^2)\psi \qquad \text{... (36)}$$

where the operators, $E \equiv i\hbar\dfrac{\partial}{\partial t}$ and $\mathbf{p} \equiv -i\hbar\vec{\nabla}$.

In the presence of an electromagnetic field, one has to replace the energy operator E by $E - e\phi$ and the momentum operator **p** by $(\mathbf{p} - e/c\ \mathbf{A})$ where **A** and ϕ are called vector and scalar potentials and e is the charge on the particle. Making the reqnired replacements, eqn. (36) becomes

$$[E - e\phi - \vec{\alpha}\cdot(c\mathbf{p} - e\mathbf{A}) - \beta m_0 c^2]\psi = 0 \qquad \text{... (66)}$$

On multiplying the above equation from the left by $[E - e\phi - \vec{\alpha}\cdot(c\mathbf{p} - e\mathbf{A}) + \beta m_0 c^2]$ and then simplifying, we get

$$\begin{aligned}&[(E - e\phi)^2 - \{\vec{\alpha}\cdot(c\mathbf{p} - e\mathbf{A})\}^2 - \beta^2 m_0^2 c^4\\ &+ \vec{\alpha}\cdot(c\mathbf{p} - e\mathbf{A})(E - e\phi) - (E - e\phi)\vec{\alpha}\cdot(c\mathbf{p} - e\mathbf{A})\\ &- (c\mathbf{p} - e\mathbf{A})\{\alpha\beta + \beta\alpha)m_0c^2]\psi = 0\end{aligned}$$

Using the relations, $\alpha_x\beta + \beta\alpha_x = 0$ etc. and also $\beta^2 = 1$, we obtain

$$\begin{aligned}&[(E - e\phi)^2 - \{\vec{\alpha}\cdot(c\mathbf{p} - e\mathbf{A})\}^2 - m_0^2 c^4\\ &+ \vec{\alpha}\cdot(c\mathbf{p} - e\mathbf{A})(E - e\phi) - (E - e\phi)\vec{\alpha}\cdot(c\mathbf{p} - e\mathbf{A})]\psi = 0\end{aligned} \qquad \text{... (67)}$$

Making use of the identity,

$$(\vec{\alpha}\cdot\mathbf{B})(\vec{\alpha}\cdot\mathbf{C}) = \mathbf{B}\cdot\mathbf{C} + i\vec{\sigma}'\cdot(\mathbf{B}\times\mathbf{C})$$

where the operators **B** and **C** commute with $\vec{\alpha}$ but not necessarily with each other and finally identifying **B** and **C** with $(c\mathbf{p} - e\mathbf{A})$, the second term in the above expression can be written as

$$\begin{aligned}[\vec{\alpha}\cdot(c\mathbf{p} - e\mathbf{A})]^2 &= (c\mathbf{p} - e\mathbf{A})^2 + i\vec{\sigma}'\cdot\{(c\mathbf{p} - e\mathbf{A})\times(c\mathbf{p} - e\mathbf{A})\}\\ &= (c\mathbf{p} - c\mathbf{A})^2 + i\vec{\sigma}'\cdot\{-ce(\mathbf{A}\times\mathbf{p} + \mathbf{p}\times\mathbf{A})\}\end{aligned}$$

out of which the second term does not vanish as **p** is an operator.

Now

$$\begin{aligned}(\mathbf{A}\times\mathbf{p} + \mathbf{p}\times\mathbf{A})\varphi &= \mathbf{A}\times\mathbf{p}\varphi + \mathbf{p}\times\mathbf{A}\varphi\\ &= \mathbf{A}\times - i\hbar\vec{\nabla}\varphi - i\hbar\vec{\nabla}\times\mathbf{A}\varphi\\ &= -i\hbar\,\mathbf{A}\times(\vec{\nabla}\varphi) - i\hbar\,[(\vec{\nabla}\times\mathbf{A})\varphi - \mathbf{A}\times(\vec{\nabla}\varphi)]\\ &= -i\hbar\,(\vec{\nabla}\times\mathbf{A})\varphi\end{aligned}$$

and hence

$$\mathbf{A}\times\mathbf{p} + \mathbf{p}\times\mathbf{A} = -i\hbar\,\vec{\nabla}\times\mathbf{A}$$

Thus the second term in eqn. (67) simplifies to

$$[\vec{\alpha}\cdot(c\mathbf{p} - c\mathbf{A})]^2 = (c\mathbf{p} - c\mathbf{A})^2 - e\hbar c\,\vec{\sigma}'\cdot\mathbf{H} \qquad \text{... (68a)}$$

where $\mathbf{H} = \vec{\nabla} \times \mathbf{A}$ is the magnetic field

The last two terms in eqn. (67) can be written as

$$\vec{\alpha} \cdot (c\mathbf{p} - e\mathbf{A})(E - e\phi) - (E - e\phi)\,\vec{\alpha} \cdot (c\mathbf{p} - e\mathbf{A})$$

$$= -e\vec{\alpha} \cdot (-E\mathbf{A} + \mathbf{A}E) - ce\vec{\alpha}\,(-\phi\mathbf{p} + \mathbf{p}\phi) \quad \text{... (69)}$$

Now

$$\begin{aligned}(E\mathbf{A} - \mathbf{A}E)\varphi &= E(\mathbf{A}\varphi) - \mathbf{A}(E\varphi)\\ &= i\hbar\frac{\partial}{\partial t}(\mathbf{A}\varphi) - i\hbar\,\mathbf{A}\left(\frac{\partial\varphi}{\partial t}\right)\\ &= i\hbar\left[\left(\frac{\partial\mathbf{A}}{\partial t}\right)\varphi + \mathbf{A}\left(\frac{\partial\varphi}{\partial t}\right)\right] - i\hbar\,\mathbf{A}\left(\frac{\partial\varphi}{\partial t}\right)\\ &= i\hbar\left(\frac{\partial\mathbf{A}}{\partial t}\right)\varphi\end{aligned}$$

hence

$$(E\mathbf{A} - \mathbf{A}E) = \frac{i\hbar\partial\mathbf{A}}{\partial t}$$

Similarly,

$$\begin{aligned}(\mathbf{p}\phi - \phi\mathbf{p})\varphi &= \mathbf{p}(\phi\varphi) - \phi(\mathbf{p}\varphi)\\ &= -i\hbar\,\vec{\nabla}(\phi\varphi) + i\hbar\,\phi(\vec{\nabla}\varphi)\\ &= -i\hbar\,[(\vec{\nabla}\phi)\varphi + \phi(\vec{\nabla}\varphi)] + i\hbar\,\phi(\vec{\nabla}\varphi)\\ &= -i\hbar\,(\vec{\nabla}\phi)\varphi\end{aligned}$$

Hence

$$\mathbf{p}\phi - \phi\mathbf{p} = -i\hbar\,(\vec{\nabla}\phi)$$

Thus the last two terms in eqn. (67) simplify to

$$\begin{aligned}&\vec{\alpha} \cdot (c\mathbf{p} - e\mathbf{A})(E - e\phi) - (E - e\phi)\,\vec{\alpha} \cdot (c\mathbf{p} - e\mathbf{A})\\ &= ie\hbar\,\vec{\alpha} \cdot \frac{\partial\mathbf{A}}{\partial t} + i\hbar c\,\vec{\alpha} \cdot \vec{\nabla}\phi\\ &= -i\hbar ec\,\vec{\alpha} \cdot \mathbf{E}_f \qquad \text{... (68b)}\end{aligned}$$

where $\mathbf{E}_f = -\vec{\nabla}\phi - \dfrac{1}{c}\dfrac{\partial\mathbf{A}}{\partial t}$, is called the electric field.

Substituting eqns. (68*a*, *b*) into eqn. (67), one gets Dirac equation in electromagnetic field as

$$[(E - e\phi)^2 - (c\mathbf{p} - e\mathbf{A})^2 - m_0^2c^4 + e\hbar\,c\vec{\sigma}'\,.\,\mathbf{H} - ie\hbar\,c\,\vec{\alpha} \cdot \mathbf{E}_f]\psi = 0 \quad \text{... (70)}$$

It can be seen that the first three terms are exactly the same as in eqn. (12), that is, Klein-Gordon equation in the presence of electromagnetic field. The physical significance of the last two terms would be inferred from consideration of the non-relativistic limit of eqn. (70).

The energy of the particle is given by

$$E = T + m_0c^2 \qquad \text{... (71)}$$

where T denotes the kinetic energy and m_0c^2, the rest energy of the particle.

After subtracting $e\phi$ from both the sides of the above equation and then squaring, we get

$$(E - e\phi)^2 = (T - e\phi + m_0c^2)^2$$

$$= (T - e\phi)^2 + m_0^2 c^4 + 2m_0c^2 (T - e\phi)$$

In the non-relativistic limit $(T - e\phi)^2$ can be neglected in comparison to $2m_0c^2 (T - e\phi)$ as $2m_0c^2$ is much bigger than $(T - e\phi)$. Therefore,

$$(E - e\phi)^2 \simeq m_0^2 c^4 + 2m_0c^2 (T - e\phi)$$

or

$$(E - e\phi)^2 - m_0^2 c^4 \simeq 2m_0c^2 (T - e\phi) \qquad \text{... (72)}$$

Substituting above into eqn. (70) and then dividing the resulting equation by $2m_0c^2$, we get

$$T\psi = \left[\frac{(\mathbf{p} - e/c\,\mathbf{A})^2}{2m_0} + e\phi - \frac{e\hbar}{2m_0c}\vec{\sigma}' \cdot \mathbf{H} + \frac{ie\hbar}{2m_0c}\vec{\alpha} \cdot \mathbf{E}_f\right]\psi \qquad \text{... (73)}$$

which is called Schrödinger-Pauli equation and is very similar to the non-relavistic Schrödinger equation with two extra terms that involve **H** and $\mathbf{E}_f$. As $\vec{\alpha} = \frac{\mathbf{V}}{c}$. the fourth term containing $\mathbf{E}_f$ would be proportional to $\frac{1}{c^2}$ and can, therefore be ignored in the non-relativistic limit. The resulting equation is termed as the Pauli equation[4].

The third term on the R.H.S. of eqn. (73) is the potential energy term. $-\vec{\mu}_s \cdot \mathbf{H}$ (or interaction energy of the magnetic moment μ_S with the magnetic field **H**) associated with the magnetic dipole of moment

$$\vec{\mu_S} = \frac{e\hbar}{2m_0c}\vec{\sigma}' = \frac{e}{m_0c} \times \frac{1}{2}\hbar\vec{\sigma}'$$

$$= \frac{e}{m_0c}\mathbf{S} \qquad \text{... (74)}$$

where **S** represents the spin angular momentum.

Since component of **S** along any direction (say z, that is, S_z) has values $\pm\frac{1}{2}$ (that is, eigen values of S_z, are $\pm\frac{1}{2}\hbar$), we have

[4]W. Pauli, Handbuch der Physik 2nd ed. Vol. 24, J. Springer, Berlin, 1 (1993).

$$\vec{\mu}_S \text{ (observed)} = \langle(\mu_S)_z\rangle = \pm\frac{e\hbar}{2m_0c} = \pm\mu_B \qquad \text{... (75)}$$

where μ_B is known as the Bohr magneton.

Thus the Dirac theory predicts that for $s = \frac{1}{2}\left(\text{spin} - \frac{1}{2}\right)$ particles the magnetic moment is always given by eqn. (75). This prediction is in very good agreement with the observed magnetic moment of the electron. Thus eqn. (73) not only establish a meaningful non-relativistic limit of the Dirac equation but shows that the Dirac particles (at least positive energy ones) are electrons.

16.10 ZITTERBEWEGUNG

Let us now study the motion of an electron according to the Dirac's relativistic equation. The component of the velocity of the said particle along x-axis in the Heisenberg representation is given by

$$i\hbar\dot{x} = [x, H]$$

$$= [x, c\alpha_x p_x + c\alpha_y p_y + c\alpha_z p_z + \beta m_0c^2]$$

As x commutes with p_y, p_z and β, we get

$$i\hbar\dot{x} = c\alpha_x [x, p_x]$$

$$= c\alpha_x (i\hbar)$$

or
$$\dot{x} = c\alpha_x \qquad \text{... (76a)}$$

Similarly, we can show that

$$\dot{y} = c\alpha_y \qquad \text{... (76b)}$$

and
$$\dot{z} = c\alpha_z \qquad \text{... (76c)}$$

The relations (76), in general amount to $\mathbf{V} = c\vec{\alpha}$ and imply that in the Dirac's theory the components of velocity do not commute with each other but are operators because of the fact that the α's are matrices.

The equation of motion of α_x in the Heisenberg representation would be given by

$$i\hbar\dot{\alpha}_x = [\alpha_x, H]$$

$$= 2\alpha_x H - (\alpha_x H + H\alpha_x) \qquad \text{... (77)}$$

Now

$$\alpha_x H + H\alpha_x = \alpha_x(c\alpha_x p_x + c\alpha_y p_y + c\alpha_z p_z + \beta m_0c^2)$$

$$+ (c\alpha_x p_x + c\alpha_y p_y + c\alpha_z p_z + \beta m_0c^2)\alpha_x$$

and as $\alpha_x^2 = 1$, $\alpha_x\alpha_y + \alpha_y\alpha_x = \alpha_x\alpha_z + \alpha_z\alpha_x = \alpha_x\beta + \beta\alpha_x = 0$, one gets

$$\alpha_x H + H\alpha_x = 2cp_x \qquad \text{... (78)}$$

Substituting (78) into (77), we obtain

$$i\hbar\dot{\alpha}_x = 2\alpha_x H - 2cp_x \qquad \text{... (79)}$$

Differentiating above w.r.t. time, we obtain

$$i\hbar\ddot{\alpha}_x = 2\dot{\alpha}_x H + 2\alpha_x \dot{H} - 2c\,\dot{p}_x$$

Since

$$i\hbar\,\dot{H} = [H, H] = 0$$

$$i\hbar\,\dot{p}_x = [p_x, H] = 0,$$

that is, H and p_x are time independent (because energy and momentum are conserved), one obtains

$$\ddot{\alpha}_x = -2i\dot{\alpha}_x H/\hbar \qquad \text{... (80)}$$

The solution of the above equation is given by

$$\dot{\alpha}_x = (\dot{\alpha}_x)_{t=0}\, e^{-2iHt/\hbar}$$

which on substituting into eqn. (79) leads to

$$i\hbar\,(\dot{\alpha}_x)_{t=0}\, e^{-2iHt/\hbar} = 2\alpha_x H - 2cp_x$$

or

$$\alpha_x H = i\frac{\hbar}{2}(\dot{\alpha}_x)_{t=0}\, e^{-2iHt/\hbar} + cp_x$$

Operating the above eqn. by H^{-1}, the reciprocal of H from the right, we get

$$\alpha_x = cp_x H^{-1} + i\frac{\hbar}{2}(\dot{\alpha}_x)_{t=0}\, e^{-2iHt/\hbar} H^{-1} \qquad \text{... (81)}$$

Putting above into eqn. (76*a*), one gets

$$\dot{x} = c^2 p_x H^{-1} + i\hbar\,\frac{c}{2}(\dot{\alpha}_x)_{t=0}\, e^{-2iHt/\hbar} H^{-1} \qquad \text{... (82)}$$

The expectation value of the velocity would, therefore be given by

$$\langle \dot{x} \rangle = \langle v_x \rangle = c^2\frac{p_x}{E} + i\hbar\,\frac{c}{2E}(\dot{\alpha}_x)_{t=0}\, e^{-2iEt/\hbar} \qquad \text{... (83)}$$

in which the first term gives the classical expression for the velocity of the relativistic particle $\left(\mathbf{v} = \frac{\partial E}{\partial \boldsymbol{p}} \text{ hence } v_x = \frac{\partial E}{\partial p_x} = \frac{\partial}{\partial p_x}(c^2p_x^2 + c^2p_y^2 + c^2p_z^2 + m_0^2c^4)^{1/2}\right)$. The actual velocity operator, $v_x = c\alpha_x$ differs from this through the second term (exponential term) which oscillates extremely rapidly, the freqnency of oscillations being $\frac{2E}{\hbar} \geq 2m_0c^2/\hbar$ which is too fast to be observed and for all practical purposes, the velocity operator effectively is $c^2p_x H^{-1}$. On integrating eqn. (82), one gets

$$x = c^2p_x\, t\, H^{-1} - c\,\frac{\hbar^2}{4}(\dot{\alpha}_x)_{t=0}\, e^{-2iHt/\hbar} H^{-2}$$

Thus, expectation value of the displacement would be given by

$$\langle x \rangle = \frac{c^2 p_x t}{E} - \frac{c\hbar^2}{4E^2}(\dot{\alpha}_x)_{t=0}\, e^{-2iEt/\hbar} \quad \text{... (84)}$$

The last term corresponds to what is known as "Zitterbewegung" (a German word having meaning—vibratory motion). As indicated earlier also, the freqnency of this motion being very large, the average value of the term would be zero. Hence the displacement effectively (or in the actual measurement) is $\frac{c^2 p_x t}{E}$ $\left(\frac{c^2 p_x}{E}\right.$ is nothing but $\left.\langle v_x\rangle\right)$ and the particle moves forward also.

16.11 DIRAC FREE PARTICLE SOLUTIONS

The Dirac's relativistic equation for a free particle is given by

$$(c\vec{\alpha}\cdot\mathbf{p} + \beta m_0 c^2)\,\psi(\mathbf{r}, t) = E\psi(\mathbf{r}, t) \quad \text{... (36)}$$

where E and $\mathbf{p}$ are energy and momentum operators and ψ is a column matrix with four rows.

Let us assume plane wave (or free particle) solutions of the form.

$$\psi_\upsilon(\mathbf{r}, t) = u_\upsilon\, e^{i(\mathbf{k}\cdot\mathbf{r} - wt)}$$

$$= u_\upsilon e^{i(\mathbf{p}\cdot\mathbf{r} - Et)/\hbar}, \quad \upsilon = 1, 2, 3, 4 \quad \text{... (85)}$$

where u_υ (which are independent of space-time coordinates) are numbers (called spinors or very precisely four-component spinor) and are truly speaking eigen functions of the energy and momentum operators E and $\vec{p}$ having eigen values $\hbar w$ and $\hbar\vec{k}$, respectively. Substituting eqn. (85) into (36) leads to

$$(c\alpha\cdot\mathbf{p} + \beta m_0 c^2)u_\upsilon = E\, u_\upsilon \quad \text{... (86)}$$

which is just a short hand notation for four homogeneous linear differential equations.

Without any loss of generality, let us assume for convenience that the electron is moving in the z-direction. In that case, $p_x = p_y = 0$ and $p_z = p$ (say). Accordingly

$$c\,\vec{\alpha}\cdot\mathbf{p} + \beta m_0 c^2 = cp\,\alpha_z + \beta\, m_0 c^2$$

$$= cp\begin{pmatrix} 0 & 0 & 1 & 0 \\ 0 & 0 & 0 & -1 \\ 1 & 0 & 0 & 0 \\ 0 & -1 & 0 & 0 \end{pmatrix} + m_0 c^2 \begin{pmatrix} 1 & 0 & 0 & 0 \\ 0 & 1 & 0 & 0 \\ 0 & 0 & -1 & 0 \\ 0 & 0 & 0 & -1 \end{pmatrix}$$

$$= \begin{pmatrix} m_0 c^2 & 0 & cp & 0 \\ 0 & m_0 c^2 & 0 & -cp \\ cp & 0 & -m_0 c^2 & 0 \\ 0 & -cp & 0 & -m_0 c^2 \end{pmatrix}$$

Using above result, eqn. (86) becomes

$$\begin{pmatrix} m_0c^2 & 0 & cp & 0 \\ 0 & m_0c^2 & 0 & -cp \\ cp & 0 & -m_0c^2 & 0 \\ 0 & -cp & 0 & -m_0c^2 \end{pmatrix}\begin{pmatrix} u_1 \\ u_2 \\ u_3 \\ u_4 \end{pmatrix} = E\begin{pmatrix} u_1 \\ u_2 \\ u_3 \\ u_4 \end{pmatrix}$$

If we expand the above equation, we get four coupled equations as follows:

$$(m_0c^2 - E)u_1 + 0u_2 + cp\, u_3 + 0u_4 = 0 \qquad \text{... (87a)}$$

$$0\, u_1 + (m_0c^2 - E)u_2 + 0u_3 - cp\, u_4 = 0 \qquad \text{... (87b)}$$

$$cp\, u_1 + 0\, u_2 + (-m_0c^2 - E)u_3 + 0u_4 = 0 \qquad \text{... (87c)}$$

$$0\, u_1 - cp\, u_2 + 0\, u_3 + (-m_0c^2 - E)u_4 = 0 \qquad \text{... (87d)}$$

These four simultaneous linear equations would have a non-trivial solution only if the determinant of coefficients of u_1, u_2, u_3 and u_4 vanishes, that is,

$$\begin{vmatrix} m_0c^2 - E & 0 & cp & 0 \\ 0 & m_0c^2 - E & 0 & -cp \\ cp & 0 & -m_0c^2 - E & 0 \\ 0 & -cp & 0 & -m_0c^2 - E \end{vmatrix} = 0 \qquad \text{... (88)}$$

which on simplification gives,

$$(E^2 - c^2p^2 - m_0^2c^4)^2 = 0 \qquad \text{... (89)}$$

which is a biquadratic equation in E. Hence would give four solutions for E. Two solutions for $E = E_+$ and two for $E = E_-$.

$$E_+ = +(c^2p^2 + m_0^2c^4)^{1/2} \quad \text{(two solutions)} \qquad \text{... (90a)}$$

$$E_- = -(c^2p^2 + m_0^2c^4)^{1/2} \quad \text{(two solutions)} \qquad \text{... (90b)}$$

Thus there are four states, in two the energy is positive and in the other two the energy is negative for a given momentum p.

Let us solve for u's as follows. From eqns. (87*a*, *b*)

$$u_3 = \frac{(-E + m_0c^2)\,u_1}{-cp} = \frac{cp\,(-E + m_0c^2)\,u_1}{-c^2p^2}$$

and

$$u_4 = \frac{(-E + m_0c^2)\,u_2}{cp} = \frac{-cp\,(-E + m_0c^2)\,u_2}{-c^2p^2}$$

Using the relation, $E^2 = c^2p^2 + m_0^2c^4$ or $-c^2p^2 = -E^2 + m_0^2c^4 = (E + m_0c^2)(-E + m_0^2c^2)$, we get

$$u_3 = \frac{cpu_1}{E + m_0c^2}, \quad u_4 = \frac{-cpu_2}{E + m_0c^2} \qquad \text{... (91a)}$$

Thus we have solved eqns. (87*a*, *b*) for two of the unknowns in terms of the other two. Let us now focus our attention on (87*c*, *d*). From these equations,

$$u_1 = \frac{(E+m_0c^2)\,u_3}{cp} = \frac{cp\,(E+m_0c^2)\,u_3}{c^2p^2}$$

and
$$u_2 = \frac{(E+m_0c^2)\,u_4}{-\,cp} = \frac{-\,cp\,(E+m_0c^2)\,u_4}{c^2p^2}$$

Using the relations, $c^2p^2 = E^2 - m_0^2c^4 = (E - m_0c^2) \times (E + m_0c^2)$, we get

$$u_1 = \frac{cpu_3}{E - m_0c^2}, \qquad u_2 = \frac{-\,cpu_4}{E - m_0c^2} \qquad \text{... (91}b)$$

Writing all the above solutions together:

$$u_1 = \frac{cpu_3}{(E - m_0c^2)}, \qquad u_2 = \frac{-\,cpu_4}{(E - m_0c^2)} \qquad \text{... (91}b)$$

$$u_3 = \frac{cpu_1}{(E + m_0c^2)}, \qquad u_4 = \frac{-\,cpu_2}{(E + m_0c^2)} \qquad \text{... (91}a)$$

We may choose two *u*'s arbitrarily and obtain the following for $E = E_+ = +\,(c^2p^2 + m_0^2c^4)^{1.2}$. Note that for the particle at rest, that is, at $p = p_z = 0$ only (87*c*, *d*) would not vanish and hence one considers the said equations. For choice 1, take

$$u_1 = 1, \qquad u_2 = 0$$

Then
$$u_3 = \frac{cp}{(E + m_0c^2)} = \frac{cp}{(E_+ + m_0c^2)}$$

and
$$u_4 = 0$$

and we have

$$u^1 = \begin{pmatrix} 1 \\ 0 \\ \dfrac{cp}{(E_+ + m_0c^2)} \\ 0 \end{pmatrix} \qquad \text{... (92}a)$$

For choice 2, take

$$u_1 = 0, \qquad u_2 = 1$$

Then
$$u_3 = 0$$

and
$$u_4 = \frac{-\,cp}{(E + m_0c^2)} = \frac{-\,cp}{(E_+ + m_0c^2)}$$

and we have
$$u^2 = \begin{pmatrix} 0 \\ 1 \\ 0 \\ \dfrac{-cp}{(E_+ + m_0c^2)} \end{pmatrix} \qquad \text{... (92b)}$$

Similarly for $E = E_- = -(c^2p^2 + m_0^2c^4)^{1/2}$, we obtain the following. Note that for $E = E_-$ eqns. (87*a*,*b*) at $p = p_z = 0$ do not vanish and hence one considers the said equations. For choice 3, take

$$u_3 = 1, \quad u_4 = 0$$

Then
$$u_1 = \frac{cp}{(E - m_0c^2)} = \frac{cp}{(E_- - m_0c^2)}$$

and
$$u_2 = 0$$

Thus, we have
$$u^3 = \begin{pmatrix} \dfrac{-cp}{(E_- - m_0c^2)} \\ 0 \\ 1 \\ 0 \end{pmatrix} \qquad \text{... (92c)}$$

For the fourth choice, take

$$u_3 = 0, \quad u_4 = 1$$

Then
$$u_1 = 0$$

and
$$u_2 = \frac{-cp}{(E - m_0\,c^2)} = \frac{-cp}{(E_- - m_0c^2)}$$

Thus we have
$$u^4 = \begin{pmatrix} 0 \\ \dfrac{-cp}{(E_- - m_0c^2)} \\ 0 \\ 1 \end{pmatrix} \qquad \text{... (92d)}$$

If interested, one may find the normalization constant. The normalization of Dirac wave functions per unit volume would be given by

$$\psi^\dagger \psi = u^\dagger u = u_1^* u_1 + u_2^* u_2 + u_3^* u_3 + u_4^* u_4 = 1,$$

or, in particular

$$u^{1\dagger} u^1 = 1$$

gives

$$N^2\begin{pmatrix}1 & 0 & \frac{cp}{(E_+ + m_0c^2)} & 0\end{pmatrix}\begin{pmatrix}1\\0\\ \frac{cp}{(E_+ + m_0c^2)}\\0\end{pmatrix} = 1$$

or

$$N^2\left(1+0+\frac{c^2p^2}{(E_+ + m_0c^2)^2}+0\right) = 1$$

or

$$N = \left[1+\frac{c^2p^2}{(E_+ + m_0c^2)^2}\right]^{-1/2}$$

$$= \left[\frac{\{(E_+ + m_0c^2)^2 + c^2p^2\}}{(E_+ + m_0c^2)^2}\right]^{-1/2}$$

Using the relation $E_+^2 = c^2p^2 + m_0^2c^4$,

$$N = \left[\frac{2E_+^2 + 2E_+ m_0c^2)}{(E_+ + m_0c^2)^2}\right]^{-1/2}$$

$$= \left[\frac{(E_+ m_0c^2)}{2E_+}\right]^{1/2} \qquad \text{... (93)}$$

One may now go back to eqns. (92*a*, *b*) and include *N* if one wants but details of the problem will not be changed. In general, one can write,

$$N = \left[1+\frac{c^2p^2}{(|E| + m_0c^2)^2}\right]^{-1/2} \qquad \text{... (93a)}$$

Similarly, one can study the orthogonality of u^1, u^2, u^3 and u^4. Now

$$u^{1\dagger}u^2 = \begin{pmatrix}1 & 0 & \frac{cp}{(E_+ + m_0c^2)^2} & 0\end{pmatrix}\begin{pmatrix}0\\1\\0\\ \frac{-cp}{(E_+ + m_0c^2)}\end{pmatrix} = 0 \qquad \text{... (94)}$$

Thus u^1 and u^2 are orthogonal. Same is the case for any two Dirac functions. Thus all the four distinct Dirac wave functions (or spinors) are mutually orthogonal.

Let us now take the non-relativistic limit. In that case,

$$\frac{cp}{(E_+ + m_0c^2)} \simeq \frac{cm_0v}{2m_0c^2} = \frac{v}{2c} \to 0 \text{ as } \frac{v}{c} \to 0$$

where E_+ has been taken as m_0c^2 (the kinetic energy part has been neglected). Thus eqn. (92) take the forms,

$$u^1 = \begin{pmatrix} 1 \\ 0 \\ 0 \\ 0 \end{pmatrix}, \quad u^2 = \begin{pmatrix} 0 \\ 1 \\ 0 \\ 0 \end{pmatrix} \qquad \text{... (95a)}$$

$$u^3 = \begin{pmatrix} 0 \\ 0 \\ 1 \\ 0 \end{pmatrix}, \quad u^4 = \begin{pmatrix} 0 \\ 0 \\ 0 \\ 1 \end{pmatrix}$$

The physical distinction between the two solutions for each sign of energy can be seen by utilizing the spin matrix σ'_z which is given as

$$\sigma'_z = \begin{pmatrix} \sigma_z & 0 \\ 0 & \sigma_z \end{pmatrix} = \begin{pmatrix} 1 & 0 & 0 & 0 \\ 0 & -1 & 0 & 0 \\ 0 & 0 & 1 & 0 \\ 0 & 0 & 0 & -1 \end{pmatrix} \qquad \text{... (96)}$$

Now
$$\sigma'_z u^1 = \begin{pmatrix} 1 & 0 & 0 & 0 \\ 0 & -1 & 0 & 0 \\ 0 & 0 & 1 & 0 \\ 0 & 0 & 0 & -1 \end{pmatrix}\begin{pmatrix} 1 \\ 0 \\ 0 \\ 0 \end{pmatrix} = u^1 \qquad \text{... (97a)}$$

and
$$\sigma'_z u^2 = -u^2$$

Similarly
$$\sigma'_z u^3 = u^3 \qquad \text{... (97b)}$$

and
$$\sigma'_z u^4 = -u^4$$

Eqns. (97) show that u is an eigen function of σ'_z with eigen values + 1 for two solutions and eigen values – 1 for the other two solutions. The spin angular momentum being $\mathbf{S} = \frac{1}{2}\hbar\vec{\sigma}'$, from eqns. (97*a*), we conclude that for $E = E_+$ in one case the spin is up and in the other case the spin is down. Similarly, from eqns. (97*b*) it is evident that for $E = E_-$ in one case the spin is up and in the other case the spin is down. Therefore, all the four states, two for $E = E_+$ and two for $E = E_-$ are distinct. Truly speaking the first two cases represent electron with spin up and down states, that is, for given energy and momentum there are two spin orientations and similarly the other two cases correspond to positron (would be clear soon) with spin up and down states.

16.12 DISCUSSION OF NEGATIVE ENERGY STATES

We have seen that energy spectrum of a free Dirac particle consists of two branches corresponding to the two signs in energy (see Fig. 16.1), that is,

$$E = E_{\pm} = \pm (c^2p^2 + m_0^2c^4)^{1/2}$$

The positive branch starts at $E = m_0c^2$ (when $p = 0$, that is, corresponding to the Dirac particles at rest) and extends to $+\infty$ (for $p \to \infty$) while the negative energies begin at $E = -m_0c^2$ and go down to $-\infty$ (for $|p| \to \infty$). There is, thus a forbidden gap of width $2m_0c^2$ between the two branches within which no energy levels exist.

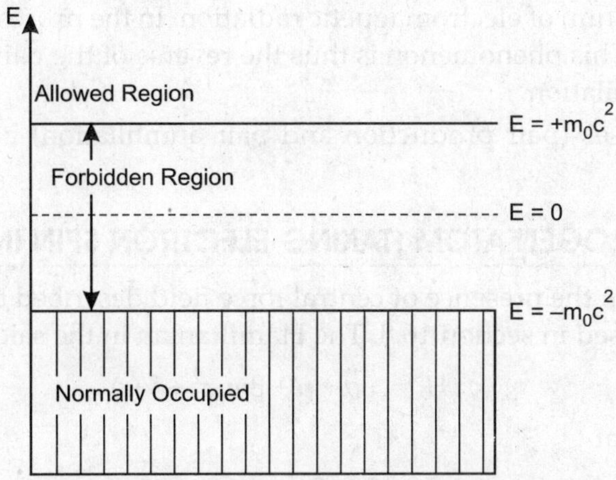

Fig. 16.1 *Dirac Picture of Positive and Negative Energy States of an Electron*

Now a Dirac particle can make a radiative transition (by spontaneously emitting a photon) from a state of positive energy to a state of negative energy. Transition to $E = -\infty$ would result in the release of an infinite amount of energy. This type of situation is not found in nature.

To overcome this difficulty Dirac postulated (assumed) that all negative energy states are normally occupied (the normal state is called vacuum) in which case the Pauli exclusion principle prevents transitions into such occupied states. Thus transitions to negative energy states are prevented. Dirac also assumed that deviations from vacuum and not the vacuum itself, can be observed.

One could make the occupant of the one of the negative energy states to jump to a (vacant) positive energy state by absorbing a photon and thus create a 'hole' in the negative energy sea. The amount of energy (or energy of photon) reqnired for this purpose must necessarily exceed the width $2m_0c^2$ of the gap between the negative and positive energy branches. The 'hole' represents a deviation from normal; its creation means the removal of certain amount of charge from the sea which is eqnivalent to the creation of an eqnal amount of positive charge. Thus the 'hole' clearly represents itself as a particle of charge $+e$ and mass m_0 (hole is denoted by e^+ where as the electron is denoted by e^-). The electron whose expulsion created the 'hole' becomes observable too since it goes to a positive energy state. The whole process may, therefore be described as the disappearance of the quantum of electromagnetic radiation (energy supplied) with creation of a pair of observable particles, a positive energy electron and another particle differing from the electron only in the sign of its charge. No such particle was known at the time Dirac's theory was formulated but it was discovered soon afterwards in cosmic rays and named positron (antiparticle). The discovery of the conversion of high energy gamma rays into electron-position pair was a real triumph of the Dirac's theory. The technique or phenomenon is called pair production.

However, the newly created 'hole' in negative sea will not last long as an electron from the positive energy state will soon fall into it, giving up the excess energy ($\geq 2m_0c^2$) in the form of a quantum of electromagnetic radiation. In the process both electron and the hole disappear. This phenomenon is thus the reverse of the pair production and is known as pair annihilation.

Both the processes (pair production and pair annihilation) have been observed experimentally.

16.13 THE HYDROGEN ATOM (TAKING ELECTRON SPIN INTO ACCOUNT)

The Dirac equation in the presence of central force field described by potential energy $V(r)$ has been discussed in section 16.8. The Hamiltanian in the said case was taken as

$$H = c\vec{\alpha} \cdot \mathbf{p} + \beta m_0 c^2 + V(r) \qquad \text{... (59)}$$

and it was shown that

$$\mathbf{J} = \mathbf{L} + \mathbf{S} \qquad \text{... (98}a\text{)}$$

is a constant of motion or $\mathbf{J}$ does commute with $\mathbf{H}$ and where the spin angular momentum

$$\mathbf{S} = \frac{1}{2}\hbar\vec{\sigma}' = \frac{1}{2}\hbar\begin{pmatrix}\vec{\sigma} & 0\\ 0 & \vec{\sigma}\end{pmatrix} \qquad \text{... (98}b\text{)}$$

Now
$$J^2 = (L+S)\cdot(L+S)$$
$$= L^2 + 2L\cdot S + S^2$$

Substituting the value of $\mathbf{S}$ from eqn. (98b) and the eigen value of S^2, we get

$$J^2 = L^2 + \hbar\,\mathbf{L}\cdot\vec{\sigma}' + S^2$$
$$= L^2 + \hbar\,\mathbf{L}\cdot\vec{\sigma}' + s(s+1)\hbar^2$$
$$= L^2 + \hbar\,\mathbf{L}\cdot\vec{\sigma}' + \frac{3}{4}\hbar^2 \qquad \text{... (99)}$$

where the value of spin, $s = \frac{1}{2}$ has been put. Using the operator identity

$$(\vec{\sigma}'\cdot\mathbf{A})(\vec{\sigma}'\cdot\mathbf{B}) = \mathbf{A}\cdot\mathbf{B} + i\vec{\sigma}'\cdot(\mathbf{A}\times\mathbf{B})$$

and then taking , $\mathbf{A} = \mathbf{B} = \mathbf{L}$, we get

$$(\vec{\sigma}'\cdot\mathbf{L})^2 = L^2 + i\vec{\sigma}'\cdot(\mathbf{L}\times\mathbf{L})$$

Further Using the operator equation $\mathbf{L}\times\mathbf{L} = i\hbar\mathbf{L}$, the above equation results in

$$(\vec{\sigma}' \cdot \mathbf{L})^2 = L^2 - \hbar(\vec{\sigma}' \cdot \mathbf{L})$$

Substituting the value of L^2 from the above equation into eqn. (99), we get

$$J^2 = (\vec{\sigma}' \cdot \mathbf{L} + \hbar)^2 - \frac{1}{4}\hbar^2$$

Thus $J^2 + \frac{1}{4}\hbar^2$ can be expressed as the square of the operator $(\vec{\sigma}' \cdot \mathbf{L} + \hbar)$, that is,

$$J^2 + \frac{1}{4}\hbar^2 = (\vec{\sigma}' \cdot \mathbf{L} + \hbar)^2 = K^2 \text{ (say)} \qquad \text{... (100)}$$

or the operator,

$$K = \beta(\vec{\sigma}' \cdot \mathbf{L} + \hbar) \qquad \text{... (101)}$$

The inclusion of the factor β (which is same as that used in eqn. (59)) has the valuable conseqnence that the operator K commutes with the hamiltonian H (and hence they will have simultaneous eigen functions). Note that the square of $\beta(\vec{\sigma}' \cdot \mathbf{L} + \hbar)$ is given by eqn. (100)

Now since, $J^2 = j(j+1)\hbar^2$, from eqn. (100), K^2 would be given by

$$K^2 = j(j+1)\hbar^2 + \frac{1}{4}\hbar^2$$

$$= \left\{\left(j + \frac{1}{2}\right)\hbar\right\}^2$$

The eigen values of the operator K are, therefore eqnal to $k\hbar$ with

$$k = \pm\left(j + \frac{1}{2}\right)$$

$$= \pm 1, \pm 2, \ldots \qquad \text{... (102)}$$

where we have used the fact that j is a half-odd integer, that is, $j = \frac{1}{2}, \frac{3}{2}, \frac{5}{2} \ldots$ (as we know from the vector addition of the orbital angular momentum (which can take the values $l\hbar$, $l = 0, 1, 2, \ldots$) with the spin angular momentum $\left(\frac{1}{2}\hbar\right)$ that the possible values of the total angular momentum are given by $j\hbar$ with $j = \frac{1}{2}, \frac{3}{2}, \frac{5}{2} \ldots$)

Note that we have two values of k for every value of j. This corresponds to the two values of $\vec{\sigma}' \cdot \mathbf{L}$. We have from eqn. (99)

$$\hbar\vec{\sigma}' \cdot \mathbf{L} = J^2 - L^2 - \frac{3}{4}\hbar^2$$

Hence

$$\left\langle \frac{1}{\hbar}(\vec{\sigma}' \cdot \mathbf{L}) \right\rangle = j(j+1) - l(l+1) - \frac{3}{4}$$

$$= -\left(j+\frac{3}{2}\right), \quad \text{for } l = j+\frac{1}{2} \qquad \text{... (103a)}$$

$$= \left(j-\frac{1}{2}\right), \quad \text{for } l = j-\frac{1}{2}$$

Therefore, from eqn. (100),

$$k = -\left(j+\frac{1}{2}\right), \quad \text{for } l = j+\frac{1}{2} \qquad \text{... (104a)}$$

and

$$k = \left(j+\frac{1}{2}\right), \quad \text{for } l = j-\frac{1}{2} \qquad \text{... (104b)}$$

To obtain the radial equation (or to obtain the energy eigen values in the present case), let us express H in terms of K and certain other operators that involve only radial variables. To achieve this, we define two Hermitian operators, called radial momentum operator (radial component of **p**).

$$p_r = (-i\hbar\vec{\nabla})_r = -i\hbar\left(\frac{1}{r}+\frac{\partial}{\partial r}\right)$$

$$= \frac{1}{r}\left(-i\hbar r\frac{\partial}{\partial r} - i\hbar\right)$$

$$= \frac{1}{r}(\mathbf{r}\cdot\mathbf{p} - i\hbar) \qquad \text{... (105a)}$$

and radial velocity operator (radial component of $\vec{\alpha}$),

$$\alpha_r = \vec{\alpha}\cdot\hat{r} = (\vec{\alpha}\cdot\mathbf{r})/r \qquad \text{... (105b)}$$

From problem 16.1,

$$(\vec{\alpha}\cdot\mathbf{A})(\vec{\alpha}\cdot\mathbf{B}) = \mathbf{A}\cdot\mathbf{B} + i\vec{\sigma}'\cdot(\mathbf{A}\times\mathbf{B}) \qquad \text{... (106)}$$

where $\vec{\alpha}$, represents three Dirac matrices α_x, α_y, α_z and $\vec{A}$ and $\vec{B}$ are usual three dimensional vectors. Putting $\mathbf{A} = \mathbf{r}$ and $\mathbf{B} = \mathbf{p}$ in the above operator identity, we obtain

$$(\vec{\alpha}\cdot\mathbf{r})(\vec{\alpha}\cdot\mathbf{p}) = \mathbf{r}\cdot\mathbf{p} + i\vec{\sigma}'\cdot(\mathbf{r}\times\mathbf{p})$$

$$= r\,p_r + i\hbar + i\vec{\sigma}'\cdot\mathbf{L}$$

where use has been made of eqn. (105a) and the fact that $\mathbf{L} = \mathbf{r}\times\mathbf{p}$

Making use of eqn. (105*b*) and multiplying eqn. (101) by β(using the relation $\beta^2 = 1$), the above equation reduces to

$$r\alpha_r(\vec{\alpha}\cdot\mathbf{p}) = rp_r + i\beta K$$

Further multiplying both sides of the above equation by α_r and using the relation $\alpha_r^2 = 1$ (which can be shown by using the relation (106)) and then dividing the resulting expression by r, we get

$$\vec{\alpha} \cdot \mathbf{p} = \alpha_r p_r + i\alpha_r \beta \frac{K}{r} \qquad \text{... (107)}$$

Now substituting $\vec{\alpha} \cdot \mathbf{p}$ from above into eqn. (59) and also $V(r) = -\frac{Ze^2}{r}$ for the case of hydrogen like atoms, we get

$$H = c\,\alpha_r p_r + icr^{-1}\alpha_r \beta K + \beta m_0 c^2 - \frac{Ze^2}{r} \qquad \text{... (108)}$$

Note that the operator K commutes with the operators β, α_r and p_r and thus also with the Hamiltonian H. Further noting that the quantities appearing above are related as

$$\alpha_r^2 = \beta^2 = 1, \alpha_r\beta + \beta\alpha_r = 0, \qquad \text{... (109)}$$

we can choose representations of α_r and β which satisfy the above relations. A simple one is

$$\beta = \begin{pmatrix} I & 0 \\ 0 & -I \end{pmatrix} \quad \text{and} \quad \alpha_r = \begin{pmatrix} 0 & -iI \\ iI & 0 \end{pmatrix} \qquad \text{... (110)}$$

Taking

$$\varphi(\mathbf{r}) = \begin{pmatrix} \varphi_a\ (r) \\ \varphi_b\ (r) \end{pmatrix} \qquad \text{... (111)}$$

where we have written the four component wave function φ in terms of the two component parts φ_a and φ_b, our eigen value equation as $V(r)$ is time independent, using eqn. (108) becomes,

$$H\,\varphi(\mathbf{r}) = E\varphi(\mathbf{r})$$

or

$$\left(c\,\alpha_r p_r + icr^{-1}\alpha_r\beta K + \beta m_0 c^2 - \frac{Ze^2}{r}\right)\begin{pmatrix} \varphi_a \\ \varphi_b \end{pmatrix} = E\begin{pmatrix} \varphi_a \\ \varphi_b \end{pmatrix} \qquad \text{... (112)}$$

Since K has eigen values $k\hbar$ (K and H, of course have simultaneous eigen functions as they commute),

$$\left(c\,\alpha_r p_r + ic\hbar\, k\, r^{-1}\alpha_r\beta + \beta m_0 c^2 - \frac{Ze^2}{r}\right)\begin{pmatrix} \varphi_a \\ \varphi_b \end{pmatrix} = E\begin{pmatrix} \varphi_a \\ \varphi_b \end{pmatrix}$$

or

$$\left[cp_r\begin{pmatrix} 0 & -iI \\ iI & 0 \end{pmatrix} + \frac{ic\hbar k}{r}\begin{pmatrix} 0 & iI \\ iI & 0 \end{pmatrix} + \begin{pmatrix} \left(m_0c^2 - \frac{Ze^2}{r}\right)I & 0 \\ 0 & \left(-m_0c^2 - \frac{Ze^2}{r}\right)I \end{pmatrix}\right]$$

$$\begin{pmatrix} \varphi_a \\ \varphi_b \end{pmatrix} = E\begin{pmatrix} \varphi_a \\ \varphi_b \end{pmatrix} \qquad \text{... (113)}$$

where α_r and β have been replaced by the corresponding matrices and also we have used the relation,

$$\alpha_r\beta = \begin{pmatrix} 0 & iI \\ iI & 0 \end{pmatrix} \qquad \text{... (110)}$$

On carrying out the matrix multiplications, we get the following two coupled equations for φ_a and φ_b,

$$\left(E - m_0c^2 + \frac{Ze^2}{r}\right)\varphi_a + \left(icp_r + \frac{\hbar ck}{r}\right)\varphi_b = 0 \qquad \text{... (114a)}$$

$$\left(E + m_0c^2 + \frac{Ze^2}{r}\right)\varphi_b - \left(icp_r - \frac{\hbar ck}{r}\right)\varphi_a = 0 \qquad \text{... (114b)}$$

Since the only operator (p_r) involved in the above equations is derivative with respect to r, we need only concern with the radial parts of φ_a and φ_b. and hence setting

$$\varphi_a(r) = \frac{F(r)}{r}, \varphi_b(r) = \frac{G(r)}{r} \qquad \text{... (115)}$$

and further operating p_r on φ_a and φ_b, that is

$$p_r\varphi_a = p_r\left(\frac{F}{r}\right) = -\frac{i\hbar}{r}\left(\frac{r\partial}{\partial r} + 1\right)\left(\frac{F}{r}\right) = -\frac{i\hbar}{r}\frac{dF}{dr}$$

and
$$p_r\varphi_b = p_r\left(\frac{G}{r}\right) = -\frac{i\hbar}{r}\frac{dG}{dr}$$

we obtain

$$\left(E - m_0c^2 + \frac{Ze^2}{r}\right)F + \hbar c\left(\frac{dG}{dr} + \frac{kG}{r}\right) = 0 \qquad \text{... (116a)}$$

and
$$\left(E + m_0c^2 + \frac{Ze^2}{r}\right)G - \hbar c\left(\frac{dF}{dr} - \frac{kF}{r}\right) = 0 \qquad \text{... (116b)}$$

The above equations form the counter part of the radial wave equation of the non-relativistic theory as they are functions of r only. In other words, the above equations are the radial equations for an electron moving in a central force field.

As in the non-relativistic case, let us look at the asymptotic behaviour of F and G. As $r \to \infty$, eqns. (116) become

$$\frac{dG}{dr} \approx \frac{1}{\hbar c}(m_0c^2 - E)F, \qquad \text{... (117a)}$$

$$\frac{dF}{dr} \approx \frac{1}{\hbar c}(m_0c^2 + E)G$$

or

$$\frac{dG}{dr} \approx \frac{\hbar c}{(m_0c^2 + E)} \frac{d^2 F}{dr^2},$$

Substituting $\frac{dG}{dr}$ from above into eqn. (117a), we get

$$\frac{d^2 F}{dr^2} \approx \frac{(m_0^2 c^4 - E^2)}{\hbar^2 c^2} F$$

or

$$\frac{d^2 F}{dr^2} - \frac{(m_0^2 c^4 - E^2)}{\hbar^2 c^2} F = 0 \quad \text{... (118}a\text{)}$$

Similarly it can be shown that

$$\frac{d^2 G}{dr^2} - \frac{(m_0^2 c^4 - E^2)}{\hbar^2 c^2} G = 0 \quad \text{... (118}b\text{)}$$

Introducing a dimensionless variable

$$\rho = \frac{(m_0^2 c^4 - E^2)^{1/2}}{\hbar c} r \quad \text{... (119)}$$

and in the present case of bound states, that is, $|E| < m_0c^2$, eqn. (118) reduce to

$$\frac{d^2 F}{d\rho^2} - F = 0 \quad \text{... (120}a\text{)}$$

and

$$\frac{d^2 G}{d\rho^2} - G = 0 \quad \text{... (120}b\text{)}$$

The asymptotic solutions, therefore would be given by

$$F(\rho), G(\rho) \sim e^{-\rho} \quad \text{... (121)}$$

Note that we have rejected the solution $e^{+\rho}$ which would make $F, G \to \infty$ for $\rho \to \infty$ and hence would violate the condition that the solution should be finite everywhere. This suggests that the solutions for all values of ρ can be taken as

$$F(\rho) = f(\rho)e^{-\rho}, G(\rho) = g(\rho)\, e^{-\rho} \quad \text{... (122)}$$

where $f(\rho)$ and $g(\rho)$ are some functions of ρ and are yet to be determined. Substituting above into eqns. (116) give

$$\left(-\upsilon + \frac{Z\alpha}{\rho}\right) f(\rho) + \left(\frac{d}{d\rho} + \frac{k}{\rho} - 1\right) g(\rho) = 0 \quad \text{... (123}a\text{)}$$

$$\left(\frac{1}{\upsilon} + \frac{Z\alpha}{\rho}\right) g(\rho) - \left(\frac{d}{d\rho} - \frac{k}{\rho} - 1\right) f(\rho) = 0 \quad \text{... (123}b\text{)}$$

where $$\upsilon = \left[\frac{m_0 c^2 - E}{m_0 c^2 + E}\right]^{1/2}, \alpha = \frac{e^2}{\hbar c} \approx \frac{1}{137} \quad \text{... (123c)}$$

is the fine structure constant and is not to be confused with the matrix $\vec{\alpha}$ and where we have used the relations,

$$\frac{dG}{dr} = \frac{dG}{d\rho} \times \frac{d\rho}{dr} \quad \text{and} \quad \frac{dF}{dr} = \frac{dF}{d\rho} \times \frac{d\rho}{dr}$$

As in the non-relativistic treatment of hydrogen atom, we seek series solutions to eqns. (123*a*, *b*) of the form.

$$f(\rho) = \rho^q (f_0 + f_1\rho + f_2\rho^2 + \ldots) = \rho^q \sum_{i=0}^{\infty} f_i \rho^i, f_0 \neq 0 \quad \text{... (124a)}$$

$$g(\rho) = \rho^q (g_0 + g_1\rho + g_2\rho^2 + \ldots) = \rho^q \sum_{i=0}^{\infty} g_i \rho^i, g_0 \neq 0 \quad \text{... (124b)}$$

where $f_0, f_1, \ldots, g_0, g_1, \ldots$ are constants and q is to be found out. Substituting the above expansions into eqns. (123*a*, *b*) and eqnating the coefficients of ρ^{q+i-1}, we obtain for $\upsilon > 0$ (v of course is positive for the bound state problems)

$$(q + i + k)\, g_i - g_{i-1} - \upsilon f_{i-1} + Z\alpha f_i = 0 \quad \text{... (125a)}$$

$$-(q + i - k) f_i + f_{i-1} + \frac{1}{\upsilon} g_{i-1} + Z\alpha g_i = 0 \quad \text{... (125b)}$$

Considering the coefficients of the lowest term ρ^{q-1} (when $i = 0$), we get

$$Z\,\alpha f_0 + (q + k) g_0 = 0 \quad \text{... (126a)}$$

$$Z\,\alpha g_0 - (q - k) f_0 = 0 \quad \text{... (126b)}$$

These equations have the reqnired non-vanishing solutions for f_0 and g_0 only if the determinant of their coefficients vanishes. This gives

$$q^2 = k^2 - Z^2\alpha^2$$

or $$q = (k^2 - Z^2\alpha^2)^{1/2} \quad \text{... (127)}$$

Note that the solution, $q = -(k^2 - Z^2\alpha^2)^{1/2}$ has been rejected as it is not admissible since that would make $f(\rho)$ and $g(\rho)$ diverge at the origin (see eqns. (124)).

Multiplying (125*b*) by v and then adding it to (125*a*), we obtain a relation between the coefficients f_i and g_i:

$$f_i[(q + i - k)\,\upsilon - Z\alpha] = g_i\,[(q + i + k) + \upsilon Z\alpha] \quad \text{... (128)}$$

Substituting above into eqn. (125*a*), we get

$$(q + i + k) f_i \frac{(q + i - k)\upsilon - Z\alpha}{(q + i + k) + \upsilon Z\alpha} - f_{i-1} \frac{(q + i - 1 - k)\upsilon - Z\alpha}{(q + i - 1 + k) + \upsilon Z\alpha} - \upsilon f_{i-1} + Z\alpha f_i = 0$$

Collecting the coefficients of f_i and f_{i-1} and then simplifying, we get

$$f_i \frac{\{(q+i)^2 - k^2\}\,\upsilon + \upsilon Z^2\alpha^2}{(q+i+k)+\upsilon Z\alpha} = f_{i-1}\frac{2\upsilon\,(q+i-1) - Z\alpha\,(1-\upsilon^2)}{(q+i-1+k)+\upsilon Z\alpha}$$

For $i \to \infty$,

$$f_i\, i^2\, \frac{\upsilon}{i} = f_{i-1}\frac{2\upsilon i}{i}$$

or

$$\frac{f_i}{f_{i-1}} = \frac{2}{i} \qquad \text{... (129a)}$$

Similarly substituting (128) into eqn. (125*b*) and taking $i \to \infty$, one finally gets

$$\frac{g_i}{g_{i-1}} = \frac{2}{i} \qquad \text{... (129b)}$$

This ratio is same as the ratio of the consecutive terms in the expansion of $e^{2\rho}$ for $i \to \infty$,

$$e^{2\rho} = 1 + 2\rho + \frac{2^2\rho^2}{2!} + \ldots \frac{2^{i-1}\rho^{i-1}}{(i-1)!} + \frac{2^i\rho^i}{i!}$$

Hence

$$\frac{\text{coefficient of } \rho^i}{\text{coefficient of } \rho^{i-1}} = \frac{2^i\,(i-1)!}{i!\,2^{i-1}} = \frac{2}{i} \qquad \text{... (130)}$$

This simply means that if we let the series (124*a*) and (124*b*) go to an infinite number of terms, the functions $f(\rho)$ and $g(\rho)$ would behave like $e^{2\rho}$ and hence the solutions $F(\rho)$ and $G(\rho)$ would diverge like e^{ρ} (see eqn. (121) at infinity ($\rho \to \infty$). We, therefore conclude that if F and G are to be admissible wave functions, both the series (124*a*, *b*) must terminate at some i, say $i = n'$ where n' is a fixed integer.

Setting $i = n' + 1$ and then putting $f_{n'+1} = 0$ (and hence $g_{n'+1} = 0$ according to eqn. (128)) in eqn. (125*a*), we obtain.

$$g_{n'} = -\upsilon f_{n'} \qquad \text{... (131)}$$

Using eqn. (128) with $i = n'$ we obtain

$$f_{n'}\,[q + n' - k)\,\upsilon - Z\alpha] = g_{n'}\,[(q+n'+k) + \upsilon Z\alpha]$$

Putting the value of $g_{n'}$ from (131) into above, we get

$$(q+n'-k)\upsilon - Z\alpha = -\upsilon[(q+n'+k) + \upsilon Z\alpha]$$

On cancelling the term $k\upsilon$ on both the sides and then substituting value of υ from eqn. (123*c*), we get

$$2(q+n')\left\{\frac{(m_0c^2 - E)}{(m_0c^2 + E)}\right\}^{1/2} = Z\alpha\left\{1 - \frac{m_0c^2 - E}{m_0c^2 + E}\right\}$$

$$= \frac{2Z\alpha E}{(m_0c^2 + E)}$$

Squaring both the sides, we have

$$(q + n')^2 \frac{(m_0c^2 - E)}{(m_0c^2 + E)} = \frac{Z^2 \alpha^2 E^2}{(m_0c^2 + E)^2}$$

or
$$E = m_0c^2 \left[1 + \frac{Z^2 \alpha^2}{(q + n')^2}\right]^{-1/2} \quad \text{... (132)}$$

which amounts to $E < m_0c^2$.

Substituting the value of q from eqn. (127) and then k from eqn. (102), we get

$$E = m_0c^2 \left[1 + Z^2 \alpha^2 \Big/ \left\{\sqrt{\left(j + \frac{1}{2}\right)^2 - Z^2 \alpha^2} + n'\right\}^2\right]^{-1/2} \quad \text{... (133)}$$

In order to compare the above result with the corresponding expression in the Schrödinger theory (non-relativistic case), we define the principal quantum number

$$n = n' + |k| = n' + \left(j + \frac{1}{2}\right) \quad \text{... (134)}$$

Note that the denominator in eqn. (133) deviates only slightly from (134). Since the minimum value of n' (or i) is zero, we have

$$n \geq j + \frac{1}{2} \quad \text{... (135)}$$

and thus the minimum value of n $\left(j\right.$ being total angular momemtum quantum number can take the values $j = \frac{1}{2}, \frac{3}{2} \ldots\left.\right)$ is unity (it can have any positive value excluding zero and has, therefore, been identified as principal quantum number).

Now eqn. (133) can be written as

$$E = m_0c^2 \left[1 + Z^2 \alpha^2 \Big/ \left\{n' + \left(j + \frac{1}{2}\right)\sqrt{1 - \frac{Z^2 \alpha^2}{\left(j + \frac{1}{2}\right)^2}}\right\}^2\right]^{-1/2}$$

Expanding the term under the radical sign and then substituting $n = n' + \left(j + \frac{1}{2}\right)$,

$$E = m_0c^2 \left[1 + Z^2 \alpha^2 \Big/ \left\{n - \frac{Z^2 \alpha^2}{2\left(j + \frac{1}{2}\right)}\right\}^2\right]^{-1/2}$$

Further using binomial expansion,

$$E = m_0c^2\left[1 - \frac{Z^2\alpha^2}{2\left\{n - \frac{Z^2\alpha^2}{2\left(j+\frac{1}{2}\right)}\right\}^2} + \frac{3}{4\times 2}\frac{Z^4\alpha^4}{\left\{n - \frac{Z^2\alpha^2}{2\left(j+\frac{1}{2}\right)}\right\}^4}\right]$$

$$= m_0c^2\left[1 - \frac{Z^2\alpha^2}{2n^2}\left\{1 - \frac{Z^2\alpha^2}{2n\left(j+\frac{1}{2}\right)}\right\}^{-2} + \frac{3}{4\times 2}\frac{Z^4\alpha^4}{n^4}\left\{1 - \frac{Z^2\alpha^2}{2n\left(j+\frac{1}{2}\right)}\right\}^{-4}\right]$$

$$= m_0c^2\left[1 - \frac{Z^2\alpha^2}{2n^2}\left\{1 + \frac{2Z^2\alpha^2}{2n\left(j+\frac{1}{2}\right)}\right\} + \frac{3Z^4\alpha^4}{4\times 2\,n^4}\left\{1 + \frac{4Z^2\alpha^2}{2n\left(j+\frac{1}{2}\right)}\right\}\right]$$

where we have neglected terms of the order higher than $Z^4\alpha^4$. Further neglecting terms of the order higher than $Z^4\alpha^4$ in the last term, we get

$$E = m_0c^2\left[1 - \frac{Z^2\alpha^2}{2n^2} - \frac{Z^4\alpha^4}{2n^4}\left\{\frac{n}{\left(j+\frac{1}{2}\right)} - \frac{3}{4}\right\}\right] \qquad \text{... (136)}$$

The leading term in $E - m_0c^2$, namely $\frac{-m_0c^2Z^2\alpha^2}{2n^2}$ $(= -m_0e^4/2n^2\hbar^2$ in the case of hydrogen atom) gives just the non-relativistic term (which gives Balmer levels of the hydrogen atom). The term which is of the order of $Z^4\alpha^4$ gives the fine structure (relativistic correction) to each of these levels caused by the dependence on $|k|\left(=j+\frac{1}{2}\right)$. The total spread in energy of the fine structure levels for a given value of n can be calculated from eqn. (136), using the fact that the minimum value of $|k| = j + \frac{1}{2} = 1$ and the maximum value of $|k| = j + \frac{1}{2} = n$ (with $n' = 0$ in eqn. (134)). After putting minimum and maximum values of $|k|$ in the relevant term and subtracting one from the other, the fine structure spreading

$$\Delta E = \frac{m_0c^2 Z^4\alpha^2}{n^3}\frac{(n-1)}{2n} \qquad \text{... (137)}$$

which is substantially less than that obtained from the Klein-Gordon equation for a spin less particle in a Coulomb field which contains $n - \frac{1}{2}$ in the denominator instead of $2n$ here. The result (137) is in agreement with the experimental value.

Labelling the energy levels as

$$2S_{1/2}\left(\text{for } n=2,\ j=\frac{1}{2} \text{ and } l=0\right)$$

$$2P_{1/2}\left(\text{for } n=2,\ j=\frac{1}{2} \text{ and } l=1\right)$$

$$2P_{3/2}\ \left(\text{for } n = 2,\ j = \frac{3}{2} \text{ and } l = 1\right)$$

since j can take all the values from $l + s$ to $|l - s|$.

We note that, while non-relativistic theory predicts that all the above three states are degenerate: (n being same), eqn. (136) predicts that only $2S_{1/2}$ and $2P_{1/2}$ are degenerate (n and j being same – E is same) while $2P_{3/2}\ \left(j = \frac{3}{2}\right)$ lies lower by about 10^{-4} eV.

In 1947, Lamb and Rutherford[5] observed that $2S_{1/2}$ lies about 4.4×10^{-9} eV below the $2P_{1/2}$ state of the hydrogen atom which is, of course not given by eqn. (136). This small difference called 'Lamb shift' was explained with the help of quantum electrodynamics (where one considers the interaction of electron with the quantized radiation field).

PROBLEMS

16.1 If $\vec{\alpha}$ represents three Dirac matrices $\alpha_x, \alpha_y, \alpha_z$ and **B** and **C** are usual three dimensional vectors, show that

$$(\vec{\alpha} \cdot \mathbf{B})\,(\vec{\alpha} \cdot \mathbf{C}) = \mathbf{B} \cdot \mathbf{C} + i\vec{\sigma}' \,.\, (\mathbf{B} \times \mathbf{C})$$

where $$\vec{\sigma}' = \begin{pmatrix} \vec{\sigma} & 0 \\ 0 & \vec{\sigma} \end{pmatrix}$$

is a 4×4 matrix whereas σ represents three 2×2 Pauli spin matrices $\sigma_x, \sigma_y, \sigma_z$.

16.2 Prove the operator identity,

$$(\vec{\sigma}' \cdot \mathbf{A})\,(\vec{\sigma}' \cdot \mathbf{B}) = \mathbf{A} \cdot \mathbf{B} + i\vec{\sigma}' \cdot (\mathbf{A} \times \mathbf{B})$$

where **A** and **B** are any two operators which commute with $\vec{\sigma}'$ (and $\vec{\sigma}$) but not necessarily with each other.

SOLUTIONS

16.1 $$(\vec{\alpha} \cdot \mathbf{B})(\vec{\alpha} \cdot \mathbf{C}) = (\alpha_x B_x + \alpha_y B_y + \alpha_z B_z)\,(\alpha_x C_x + \alpha_y C_y + \alpha_z C_z)$$
$$= \alpha_x^2 B_x C_x + \alpha_y^2 B_y C_y + \alpha_z^2 B_z C_z$$
$$+ \alpha_x \alpha_y\,(B_x C_y - B_y C_x) + \alpha_y \alpha_z\,(B_y C_z - B_z C_y)$$
$$+ \alpha_z \alpha_x\,(B_z C_x - B_x C_z)$$

where use has been made of the relations, $\alpha_x\,\alpha_y = -\,\alpha_y\,\alpha_x$, etc. Further using the relation, $\alpha_x^2 = \alpha_y^2 = \alpha_z^2 = 1$, we get

[5]W.E Lamb, Jr and R.C. Rutherford, Phys. Rev 72, 241 (1947); W.E. Lamb, Jr, Rep. Prog. Phys. 14, 19 (1951).

$$(\vec{\alpha} \cdot \mathbf{B})(\vec{\alpha} \cdot \mathbf{C}) = B_xC_x + B_yC_y + B_zC_z + \alpha_x\alpha_y(B_xC_y - B_yC_x)$$
$$+ \alpha_y\alpha_z(B_yC_z - B_zC_y) + \alpha_z\alpha_x(B_zC_x - B_xC_z)$$

Now

$$\alpha_x\alpha_y = \begin{pmatrix} 0 & \sigma_x \\ \sigma_x & 0 \end{pmatrix}\begin{pmatrix} 0 & \sigma_y \\ \sigma_y & 0 \end{pmatrix}$$
$$= \begin{pmatrix} \sigma_x\sigma_y & 0 \\ 0 & \sigma_x\sigma_y \end{pmatrix}$$
$$= \begin{pmatrix} i\sigma_z & 0 \\ 0 & i\sigma_z \end{pmatrix}$$
$$= i\sigma'_z$$

Similarly,

$$\alpha_y\alpha_z = i\sigma'_x,\ \alpha_z\alpha_x = i\sigma'_y$$

Therefore,

$$(\vec{\alpha} \cdot \mathbf{B})(\vec{\alpha} \cdot \mathbf{C}) = (\mathbf{B} \cdot \mathbf{C}) + i\sigma'_z(B_xC_y - B_yC_x) + i\sigma'_x(B_yC_z - B_zC_y)$$
$$+ i\sigma'_y(B_zC_x - B_xC_z)$$
$$= (\mathbf{B} \cdot \mathbf{C}) + i\begin{vmatrix} \sigma'_x & \sigma'_y & \sigma'_z \\ B_x & B_y & B_z \\ C_x & C_y & C_z \end{vmatrix}$$
$$= \mathbf{B} \cdot \mathbf{C} + i\vec{\sigma'} \cdot (\mathbf{B} \times \mathbf{C})$$

16.2 $(\vec{\sigma'} \cdot \mathbf{A})(\vec{\sigma'} \cdot \mathbf{B}) = (\sigma'_x A_x + \sigma'_y A_y + \sigma'_z A_z)(\sigma'_x B_x + \sigma'_y B_y + \sigma'_z B_z)$

$$= \sigma'^2_x A_xB_x + \sigma'^2_y A_yB_y + \sigma'^2_z A_zB_z$$
$$+ \sigma'_x\sigma'_y A_xB_y + \sigma'_y\sigma'_x A_yB_x + \sigma'_y\sigma'_z A_yB_z$$
$$+ \sigma'_z\sigma'_y A_zB_y + \sigma'_z\sigma'_x A_zB_x + \sigma'_x\sigma'_z A_xB_z$$

Using the relation, $\sigma'^2_x = \sigma'^2_y = \sigma'^2_z = 1$ and further that

$$\sigma'_x\sigma'_y + \sigma'_y\sigma'_x = \begin{pmatrix} \sigma_x & 0 \\ 0 & \sigma_x \end{pmatrix}\begin{pmatrix} \sigma_y & 0 \\ 0 & \sigma_y \end{pmatrix} + \begin{pmatrix} \sigma_y & 0 \\ 0 & \sigma_y \end{pmatrix}\begin{pmatrix} \sigma_x & 0 \\ 0 & \sigma_x \end{pmatrix}$$
$$= \begin{pmatrix} \sigma_x\sigma_y + \sigma_y\sigma_x & 0 \\ 0 & \sigma_x\sigma_y + \sigma_y\sigma_x \end{pmatrix} = 0$$

or $\sigma'_x\sigma'_y = -\sigma'_y\sigma'_x$ and similarly $\sigma'_y\sigma'_z = -\sigma'_z\sigma'_y$, $\sigma'_z\sigma'_x = -\sigma'_x\sigma'_z$, we get

$$(\vec{\sigma}' \cdot \mathbf{A})(\vec{\sigma}' \cdot \mathbf{B}) = A_xB_x + A_yB_y + A_zB_z + \sigma'_x\sigma'_y(A_xB_y - A_yB_x) + \sigma'_y\sigma'_z(A_yB_z - A_zB_y) + \sigma'_z\sigma'_x(A_zB_x - A_xB_z)$$

Further

$$\sigma'_x\sigma'_y = \begin{pmatrix} \sigma_x\sigma_y & 0 \\ 0 & \sigma_x\sigma_y \end{pmatrix} = \begin{pmatrix} i\sigma_z & 0 \\ 0 & i\sigma_z \end{pmatrix} = i\sigma'_z$$

and similarly $\sigma'_y\sigma'_z = i\sigma'_x$, $\sigma'_z\sigma'_x = i\sigma'_y$. Therefore,

$$(\vec{\sigma}' \cdot \mathbf{A})(\vec{\sigma}' \cdot \mathbf{B}) = \mathbf{A} \cdot \mathbf{B} + i\vec{\sigma}' \cdot (\mathbf{A} \times \mathbf{B})$$

❑❑❑

Chapter 17

Second Quantization

17.1 LAGRANGIAN FORMULATION OF FIELD

The quantum theory presented so far has dealt with systems which consist of material particles. We wish now to extend the theory so that it can be applied to quantize ψ(to be called ψ field or in general a wave field), such as that described by the non-relativistic Schrödinger equation (or by Klein-Gordon and Dirac relativistic equations). We shall see in the forthcoming section (in the non-relativistic case) that this field quantization technique (if we quantize the electromagnetic field, photons result) converts a one particle theory into a many particle theory.

Before we proceed further, let us first describe the general procedure for the quantization of the equations of motion of a classical system (material particles). We start with the Lagrangian function $L(q, \dot{q}, t)$ for the system where q stands tor the set of generalized coordinates $\{q_i\}$, $\dot{q}_i = \frac{\partial q_i}{\partial t}$ are the generalized velocities and t denotes the time. p_i, the momenta canonical conjugate to the coordinates of the system are found from the Lagrangian L using the relation, $p_i = \frac{\partial L}{\partial \dot{q}_i}$ and then a Hamiltonion function, $H = \sum_i p_i \dot{q}_i - L$ is set up. There after the classical Hamilton's canonical equations, $\dot{q}_i = \frac{\partial H}{\partial p_i}$ and $-\dot{p}_i = \frac{\partial H}{\partial q_i}$ are obtained and these are then written in terms of the Poisson bracket notation, that is, $\dot{q}_i = \{q_i, H\}$, $\dot{p}_i = \{p_i, H\}$

where

$$\{u, H\} = \sum_i \left(\frac{\partial u}{\partial q_i} \frac{\partial H}{\partial p_i} - \frac{\partial u}{p_i} \frac{\partial H}{\partial q_i} \right)$$

is known as the Poisson bracket of u and H. They are, then converted into quantum equations by the substitution of commutator brackets divided by $i\hbar$ for the Poisson brackets.

We would now show that how this procedure can be applied in its totality to $\psi(\mathbf{r}, t)$, a wave field which is specified by its amplitude $\psi(\mathbf{r}, t)$ at all points $\mathbf{r}$ of space and the dependence of these amplitudes on time. This is analogous to the specification of a system of particles by the generalized coordinates q_i and their dependence on time. The field evidently has an infinite number of degrees of freedom since there are infinite number of space points and is thus analogous to a system that consists of an infinite number of particles. It is, therefore natural to use the amplitude $\psi(\mathbf{r}, t)$ at all points $\mathbf{r}$ as coordinates (or field coordinates) in analogy with the particle coordinates $q_i(t)$. For the present, we shall assume that the wave field $\psi(\mathbf{r}. t)$ is real. If complex fields occur, they may be decomposed into a pair of real fields.

Except for some modifications which are needed due to the fact that the coordinates of the field (ψ) are themselves functions of the space coordinates $\mathbf{r}$, the procedure outlined above for the case of a classical system (a system of particles) can be adopted both for obtaining classical field equations and then for their quantization. The dependence of ψ on $\mathbf{r}$, which is a continuous variable, necessitates two types of modifications,

(*i*) To introduce a Lagrangian density $\mathcal{L}$. The Lograngian L of the field would be an integral of $\mathcal{L}$ over entire space.

(*ii*) $\mathcal{L}$ would have to be function not only of ψ, $\dot{\psi}$ and t but also of grad ψ which must be treated on the same footing as $\dot{\psi}$.

Thus,

$$\mathcal{L} \equiv \mathcal{L}(\psi, grad\ \psi, \dot{\psi}, t),\ \dot{\psi} = \frac{\partial \psi}{\partial t} \quad \text{... (1)}$$

$$L = \int \mathcal{L}\, d\tau \quad \text{... (2)}$$

The Hamilton's variational principle then gives,

$$\delta \int_{t_1}^{t_2} L\, dt = \delta \int_{t_1}^{t_2} \mathcal{L}\, dt\, d\tau = \int_{t_1}^{t_2}\int \delta\mathcal{L}\, dt\, d\tau = 0 \quad \text{... (3)}$$

where the infinitesimal variation $\delta\psi$ in ψ is subject to the restrictions,

$$\delta\psi(\mathbf{r}, t_1) = \delta\psi(\mathbf{r}, t_2) = 0, \quad \text{... (4)}$$

that is, there are no variations at the ends.

Now we can write

$$\delta\mathcal{L}(\psi, \text{grad}\ \psi, \dot{\psi}, t) = \frac{\partial \mathcal{L}}{\partial \psi}\delta\psi + \sum_{k=1}^{3} \frac{\partial \mathcal{L}}{\partial\left(\frac{\partial \psi}{\partial x_k}\right)} \delta\left(\frac{\partial \psi}{\partial x_k}\right) + \frac{\partial \mathcal{L}}{\partial \dot{\psi}}\delta\dot{\psi} \quad \text{... (5)}$$

where for the sake of convenience x, y, z have been replaced by x_1, x_2, x_3 and where $\frac{\partial \psi}{\partial x_k}$ represents $(\text{grad}\ \psi)_k$. Also

$$\delta\dot{\psi} = \delta\left(\frac{\partial \psi}{\partial t}\right) = \frac{\partial}{\partial t}(\delta\psi),\ \delta\left(\frac{\partial \psi}{\partial x_k}\right) = \frac{\partial}{\partial x_k}(\delta\psi) \quad \text{... (5a)}$$

as in the δ variations one goes from one path to the other at the same time (the said variation is independent of the differentiation process).

Using eqns. (5), the Hamilton's variational principle reads,

$$\int_{t_1}^{t_2}\int \left[\frac{\partial \mathcal{L}}{\partial \psi}\delta\psi + \sum_{k=1}^{3} \frac{\partial \mathcal{L}}{\partial\left(\frac{\partial \psi}{\partial x_k}\right)} \frac{\partial}{\partial x_k}(\delta\psi) + \frac{\partial \mathcal{L}}{\partial \dot{\psi}}\frac{\partial}{\partial t}(\delta\psi)\right] dt\, dx_1\, dx_2\, dx_3 = 0 \quad \text{... (6)}$$

Integrating a term in the above equation in the summation by parts w.r.t. the space coordinate, we get

$$\int \frac{\partial \mathcal{L}}{\partial\left(\frac{\partial \psi}{\partial x_1}\right)} \frac{\partial}{\partial x_1}(\delta\psi)\, dx_1\, dx_2\, dx_3 = \int \frac{\partial \mathcal{L}}{\partial\left(\frac{\partial \psi}{\partial x_1}\right)} \delta\psi\, dx_2\, dx_3 - \int \frac{\partial}{\partial x_1}\left\{\frac{\partial \mathcal{L}}{\partial\left(\frac{\partial \psi}{\partial x_1}\right)}\right\}\delta\psi\, d\tau$$

$$= -\int \frac{\partial}{\partial x_1}\left\{\frac{\partial \mathcal{L}}{\partial\left(\frac{\partial \psi}{\partial x_1}\right)}\right\}\delta\psi\, d\tau \quad \text{... (7a)}$$

as the surface term (or integral) on the R.H.S. of the above equation vanishes because ψ vanishes at infinity (as ψ represents a wave packet, a wave confined to a small region of the space).

Now integrating the last term of eqn. (6) by parts w.r.t. time and then using condition (4), we get

$$\int_{t_1}^{t_2}\left(\frac{\partial \mathcal{L}}{\partial \dot{\psi}}\right)\frac{\partial}{\partial t}(\delta\psi)\, dt = \left.\frac{\partial \mathcal{L}}{\partial \dot{\psi}}\delta\psi\right|_{t_1}^{t_2} - \int_{t_1}^{t_2}\frac{\partial}{\partial t}\left(\frac{\partial \mathcal{L}}{\partial \dot{\psi}}\right)\delta\psi\, dt$$

$$= -\int_{t_1}^{t_2}\frac{\partial}{\partial t}\left(\frac{\partial \mathcal{L}}{\partial \dot{\psi}}\right)\delta\psi\, dt \quad \text{... (7b)}$$

Substitution of (7) into eqn. (6) leads to

$$\int_{t_1}^{t_2}\int\left[\frac{\partial \mathcal{L}}{\partial \psi} - \sum_{k=1}^{3}\frac{\partial}{\partial x_k}\left\{\frac{\partial \mathcal{L}}{\partial\left(\frac{\partial \psi}{\partial x_k}\right)}\right\} - \frac{\partial}{\partial t}\left(\frac{\partial \mathcal{L}}{\partial \dot{\psi}}\right)\right]\delta\psi\, dt\, d\tau = 0 \quad \text{... (8)}$$

As eqn. (3) is valid for any arbitrary variation δψ at each point in space, this requires the vanishing of, the integrand in the square bracket in the above equation.

Therefore,

$$\frac{\partial \mathcal{L}}{\partial \psi} - \sum_{k=1}^{3}\frac{\partial}{\partial x_k}\left\{\frac{\partial \mathcal{L}}{\partial\left(\frac{\partial \psi}{\partial x_k}\right)}\right\} - \frac{\partial}{\partial t}\left(\frac{\partial \mathcal{L}}{\partial \dot{\psi}}\right) = 0 \quad \text{... (9)}$$

or $$\frac{\partial \mathcal{L}}{\partial \psi} - \nabla \cdot \frac{\partial \mathcal{L}}{\partial (\nabla \psi)} - \frac{\partial}{\partial t}\left(\frac{\partial \mathcal{L}}{\partial \dot{\psi}}\right) = 0 \qquad \text{... (10)}$$

This is the classical field equation in terms of the Lagrangian density $\mathcal{L}$ and is called Euler's equation and is analogue of the Lagrange's equation in classical mechanics.

Before we proceed further, let us see whether eqn. (9) can be written in the covariant form. From the said equation,

$$\frac{\partial \mathcal{L}}{\partial \psi} - \sum_{123} \frac{\partial}{\partial x_1}\left(\frac{\partial \mathcal{L}}{\partial \psi,_1}\right) - \frac{\partial}{\partial x_4}\left(\frac{\partial \mathcal{L}}{\partial \psi,_4}\right) = 0, \; x_4 = ict$$

where we have used the notation,

$$\psi,_\mu = \frac{\partial \psi}{\partial x_\mu}$$

Thus, $$\frac{\partial \mathcal{L}}{\partial \psi} - \frac{\partial}{\partial x_\mu}\left(\frac{\partial \mathcal{L}}{\partial \psi,_\mu}\right) = 0 \qquad \text{... (9a)}$$

It may be noted that it is more convenient to use the above covariant form for the relativistic fields.

In order to persue further the analogy with the classical or particle mechanics, it is desirable to rewrite eqn. (10) in terms of the Lagrangian L rather than the Lagrangian density $\mathcal{L}$. The functional derivative of L w.r.t. ψ is defined as

$$\frac{\not\partial L}{\not\partial \psi} = \frac{\partial \mathcal{L}}{\partial \psi} - \nabla \cdot \frac{\partial \mathcal{L}}{\partial (\nabla \psi)} \qquad \text{... (11a)}$$

Similarly, the functional derivative of L w.r.t. $\dot{\psi}$ is defined as

$$\frac{\not\partial L}{\not\partial \dot{\psi}} = \frac{\partial \mathcal{L}}{\partial \dot{\psi}} - \nabla \cdot \frac{\partial \mathcal{L}}{\partial (\nabla \dot{\psi})}$$

$$= \frac{\partial \mathcal{L}}{\partial \dot{\psi}} \qquad \text{... (11b)}$$

as the second term vanishes because of the fact that $\mathcal{L}$ is not a function of $\nabla \dot{\psi}$

Substituting eqn. (11) into eqn. (10) leads to

$$\frac{\not\partial L}{\not\partial \psi} - \frac{\partial}{\partial t}\left(\frac{\not\partial L}{\not\partial \dot{\psi}}\right) = 0 \qquad \text{... (12)}$$

which closely resembles with the Lagrange's equation for a system of particles. Note that the only difference is that the functional derivatives instead of ordinary derivatives have appeared in the case of fields.

17.2 HAMILTONIAN FORMULATION OF FIELD

In order to quantize the classical field equation, that is, eqn. (10) by the method indicated in the preceding section, we must first go over to the Hamiltonian formulation. The Hamiltonian of the field is defined as

$$H = \int x \, d\tau \qquad \text{... (13)}$$

where X, the Hamiltonian density in analogous to $H = \sum_i p_i \dot{q}_i - L$ in the case of particle mechanics (or material particles), is given by

$$X = \pi \dot{\psi} - \mathcal{L} \qquad \text{... (14)}$$

where

$$\pi = \frac{\partial \mathcal{L}}{\partial \dot{\psi}} \qquad \text{... (14a)}$$

is called momentum density conjugate to ψ or is called conjugate field.

Using Hamilton's variational principle and eqn. (14), we get

$$\int_{t_1}^{t_2}\int \delta(\pi\dot{\psi} - X)\, dt \, d\tau = 0$$

or

$$\int_{t_1}^{t_2}\int (\pi \, \delta\dot{\psi} + \dot{\psi} \, \delta \, \pi - \delta X)\, dt \, d\tau = 0 \qquad \text{... (15)}$$

Now the Hamiltonian density in analogous to $H = H(q, p, t) = \sum_i p_i \dot{q}_i - L$ in the case of particle mechanics, is given by

$$X = X(\psi, \pi, t) \qquad \text{... (16)}$$

Hence

$$\delta X = \frac{\partial X}{\partial \psi}\delta\psi + \frac{\partial X}{\partial \pi}\delta\pi \qquad \text{... (16a)}$$

Substituting above and $\delta\dot{\psi} = \delta\left(\frac{\partial \psi}{\partial t}\right) = \frac{\partial}{\partial t}(\delta\psi)$, eqn. (15) takes the form

$$\int_{t_1}^{t_2}\int \left\{\pi \frac{\partial}{\partial t}(\delta\psi) + \dot{\psi}\,\delta\pi - \frac{\partial X}{\partial \psi}\delta\psi - \frac{\partial X}{\partial \pi}\delta\pi\right\} dt \, d\tau = 0 \qquad \text{... (17)}$$

Integrating the first term of the above equation by parts w.r.t. time, we get

$$\int_{t_1}^{t_2}\int \pi \frac{\partial}{\partial t}(\delta\psi)\, dt \, d\tau = \int \pi \, \delta\psi \Big|_{t_1}^{t_2} d\tau - \int_{t_1}^{t_2}\int \dot{\pi}\,\delta\psi \, dt \, d\tau$$

$$= -\int_{t_1}^{t_2}\int \dot{\pi}\,\delta\psi\,dt\,d\tau \qquad \text{... (18)}$$

where the condition (4) has been used.

Substituting above into eqn. (17), we get

$$\int_{t_1}^{t_2}\int\left[-\dot{\pi}\,\delta\psi + \dot{\psi}\,\delta\pi - \frac{\partial X}{\partial \psi}\,\delta\psi - \frac{\partial X}{\partial \pi}\,\delta\pi\right] dt\,d\tau = 0$$

Rearranging the terms, we get

$$\int_{t_1}^{t_2}\int\left[\delta\psi\left(-\dot{\pi} - \frac{\partial X}{\partial \psi}\right) + \delta\pi\left(\dot{\psi} - \frac{\partial X}{\partial \pi}\right)\right] dt\,d\tau = 0$$

As ψ and π are independent variables and $\delta\psi$ and $\delta\pi$ being arbitrary variations, one gets

$$\dot{\psi} = \frac{\partial X}{\partial \pi} = \frac{\partial X}{\partial \pi} - \nabla\cdot\frac{\partial X}{\partial(\nabla\pi)} = \frac{\partial\!\!\!/ H}{\partial\!\!\!/ \psi}, \qquad \text{... (19a)}$$

$$-\dot{\pi} = \frac{\partial X}{\partial \psi} = \frac{\partial X}{\partial \psi} - \nabla\cdot\frac{\partial X}{\partial(\nabla\psi)} = \frac{\partial\!\!\!/ H}{\partial\!\!\!/ \psi}, \qquad \text{... (19b)}$$

where the definitions for the functional derivatives of H w.r.t. π and ψ and also that X is not a function of $\nabla\pi$ and $\nabla\psi$ have been used. Eqns. (19) are called classical field equations in Hamiltonian form.

17.3 POISSON BRACKET FORMULATION OF FIELD VARIABLES

Let F be any dynamical variable of the field, then

$$\text{F} = \int f\,d\tau \qquad \text{... (20)}$$

where f is the corresponding density of the variable and for the sake of simplicity is assumed not to be function of grad ψ or grad π, that is,

$$f = f(\psi, \pi, t) \qquad \text{... (21)}$$

The time rate of change of F would, therefore be given by

$$\dot{F} = \int\left(\frac{\partial f}{\partial t} + \frac{\partial f}{\partial \psi}\,\dot{\psi} + \frac{\partial f}{\partial \pi}\,\dot{\pi}\right) d\tau$$

$$= \frac{\partial F}{\partial t} + \int\left(\frac{\partial f}{\partial \psi}\,\dot{\psi} + \frac{\partial f}{\partial \pi}\,\dot{\pi}\right) d\tau$$

$$= \frac{\partial F}{\partial t} + \int\left(\frac{\partial\!\!\!/ F}{\partial\!\!\!/ \psi}\,\dot{\psi} + \frac{\partial\!\!\!/ F}{\partial\!\!\!/ \pi}\,\dot{\pi}\right) d\tau \qquad \text{... (22)}$$

where the definitions of the functional derivatives of F with ψ and π have been utilized.

Using eqns. (19), the above equation becomes

$$\dot{F} = \frac{\partial F}{\partial t} + \int \left(\frac{\delta F}{\delta \psi} + \frac{\delta H}{\delta \pi} - \frac{\delta F}{\delta \pi} \frac{\delta H}{\delta \psi} \right) d\tau$$

$$= \frac{\partial F}{\partial t} + \{F, H\} \qquad \text{... (23)}$$

where $\{F, H\}$ is called Poisson bracket for two field variables and is defined by

$$\{F, H\} = \int \left(\frac{\delta F}{\delta \psi} \frac{\delta H}{\delta \pi} - \frac{\delta F}{\delta \pi} \frac{\delta H}{\delta \psi} \right) d\tau \qquad \text{... (24)}$$

If f does not depend explicitly on time $\left(\frac{\partial f}{\partial t} = 0 \right)$, in that case eqn. (23) reads as

$$\dot{F} = \{F, H\} \qquad \text{... (25)}$$

which simply means that H is a constant of motion if it does not depend explicity on time, in this case H is the total energy of the field.

17.4 QUANTUM EQUATIONS FOR THE FIELD

The equation of motion for any quantum dynamical variable F is obtained if the Poisson bracket in eqn. (25) is replaced by the commutator bracket divided by $i\hbar$. Thus

$$i\hbar \dot{F} = [F, H] \qquad \text{... (26)}$$

Let us divide the whole three dimensional space into large number of small enough cells. Let ψ_i and ψ_k represent the average values of ψ at the ith and kth cells having volumes $\delta\tau_i$ and $\delta\tau_k$, respectively. Let π_k be the momentum density at the kth cell. Hence the momentum canonical conjugate to ψ_k will be given by

$$P_k = \pi_k \delta\tau_k \qquad \text{... (27)}$$

In analogy with the commutation relation, $[q_i, p_k] = i\hbar\, \delta_{ik}$ in the particle theory which has been obtained by replacing the conjugate coordinates by the operators satisfying the usual commutation relations of quantum mechanics (or by replacing the Poisson bracket $\{q_i, p_k\}$ by the commutator bracket divided by $i\hbar$), in the present case using eqn. (27), we obtain

$$[\psi_i, \pi_k\, \delta\tau_k] = i\hbar\, \delta_{ik}$$

or

$$[\psi_i, \pi_k]\, \delta\tau_k = i\hbar\, \delta_{ik}$$

Summing over all values of k (that is, taking sum over all the cells) on both sides of the above equation, we get

$$\sum_k [\psi_i, \pi_k]\, \delta\tau_k = \sum_k i\hbar\, \delta_{ik}$$

$$= i\hbar$$

where we have used the property of Kronecker delta function. For $\delta\tau_k \to 0$, that is, in the continuum limit, the above cell summation can be replaced by the volume integral. Therefore,

$$\int [\psi(\mathbf{r}, t), \pi(\mathbf{r}', t)]\, d\mathbf{r}' = i\hbar \qquad \text{... (28)}$$

where **r** and **r′** are the position vectors of *i*th and *k*th cells, respectively. From the above equation, it follows that

$$[\psi(\mathbf{r}, t), \pi(\mathbf{r}', t)] = i\hbar\, \delta(\mathbf{r} - \mathbf{r}') \qquad \text{... (29)}$$

where $\delta(\mathbf{r} - \mathbf{r}')$ is the Dirac delta function. One may integrate both sides of the above equation and using the definition of the Dirac delta function, that is,

$$\int \delta(\mathbf{r} - \mathbf{r}')\, d\mathbf{r}' = 1,$$

would get eqn. (28).

Similarly in analogous to the commutation relation, $[q_i, q_k] = 0$ in the particle theory, in the present case, one gets

$$[\psi_i, \psi_k] = 0$$

Thus if we now let the cell volumes approach zero, that is, $\delta\tau_i, \delta\tau_k \to 0$, we get

$$[\psi(\mathbf{r}, t), \psi(\mathbf{r}', t)] = 0 \qquad \text{... (30)}$$

Similarly, in analogous to the commutation relation, $[p_i, p_k] = 0$ in the particle theory, in the present case, we have

$$[\pi_i\, \delta\tau_i, \pi_k\, \delta\tau_k] = 0$$

Again for $\delta\tau_i, \delta\tau_k \to 0$, we get

$$[\pi(\mathbf{r}, t), \pi(\mathbf{r}', t)] = 0 \qquad \text{... (31)}$$

Eqns. (29 – 31) together can be written as

$$[\psi(\mathbf{r}, t), \psi(\mathbf{r}', t)] = [\pi(\mathbf{r}, t), \pi(\mathbf{r}', t)] = 0,$$

$$[\psi(\mathbf{r}, t), \pi(\mathbf{r}', t) = i\hbar\, \delta(\mathbf{r} - \mathbf{r}') \qquad \text{... (32)}$$

which are called quantum conditions for the canonical field variables at two points **r** and **r′** at the same time *t*.

The commutator in eqn. (26) can be calculated with the help of eqn. (32) when explicit expressions for *F* and *H* in terms of ψ and π are given. Thus eqns. (26) and (32) completely describe the behaviour of quantized field which is specified by the Hamiltonian *H*.

Eqns. (32) simply mean that we have converted the wave field from a numerical function to an operator that can be represented by a matrix.

17.5 FIELD WITH MORE THAN ONE COMPONENT

So far we have dealt with a field that can be described by a single real amplitude. If the field has more than one component (or a case of several independent fields) $\psi_1, \psi_2, \ldots$, the Lagrangian density would have the form

$$\mathcal{L} = \mathcal{L}(\psi_1, \nabla\psi_1, \dot{\psi}_1, \psi_2, \nabla\psi_2, \dot{\psi}_2, \ldots, \psi_N, \nabla\psi_N, \dot{\psi}_N, t) \quad \ldots (33)$$

Then if each of the field components is varied independently, the variational principle, that is, eqn. (3) leads to an equation of the form (9) or (12) for each of $\psi_1, \psi_2 \ldots$ that is,

$$\frac{\partial \mathcal{L}}{\partial \psi_1} - \sum_{k=1}^{3} \frac{\partial}{\partial x_k} \left\{ \frac{\partial \mathcal{L}}{\partial\left(\frac{\partial \psi_1}{\partial x_k}\right)} \right\} - \frac{\partial}{\partial t}\left(\frac{\partial \mathcal{L}}{\partial \dot{\psi}_1}\right) = 0$$

or $$\frac{\delta L}{\delta \psi_1} - \frac{\partial}{\partial t}\left(\frac{\delta L}{\delta \dot{\psi}_1}\right) = 0$$

The momentum density π_s conjugate to each ψ_s and the Hamiltonian density x_s would be given by

$$\pi_s = \frac{\partial \mathcal{L}}{\partial \dot{\psi}_s}, x_s = \sum_s \pi_s \dot{\psi}_s - \mathcal{L} \quad \ldots (34)$$

and the Hamilton's equations consist of a pair like eqns. (19) for each s, that is.

$$\dot{\psi}_s = \frac{\delta H}{\delta \pi_s}, -\dot{\pi}_s = \frac{\delta H}{\delta \psi_s}$$

Similarly

$$\{F, H\} = \sum_s \int \left(\frac{\delta F}{\delta \psi_s}\frac{\delta H}{\delta \pi_s} - \frac{\delta F}{\delta \pi_s}\frac{\delta H}{\delta \psi_s}\right) d\tau \quad \ldots (35)$$

and the eqn. (26) is unchanged.

The quantum conditions (or commutation relations) (32) are replaced by

$$[\psi_s(\mathbf{r}, t), \psi_{s'}(\mathbf{r}', t)] = [\pi_s(\mathbf{r}, t), \pi_{s'}(\mathbf{r}', t)] = 0$$

$$[\psi_s(\mathbf{r}, t), \pi_{s'}(\mathbf{r}', t)] = i\hbar\, \delta_{ss'}\, \delta(\mathbf{r} - \mathbf{r}')$$

where $s, s' = 1, 2, \ldots, N$.

Example of field with more than one component is the electromagnetic field described by four potentials A_1, A_2, A_3 (three components of the vector potential $A(\mathbf{r}, t)$) and ϕ (the scalar potential) or a single complex field ψ_c. expressed in terms of the real fields ψ_1 and ψ_2 as

$$\psi_c = \frac{1}{\sqrt{2}}(\psi_1 + i\psi_2), \psi_c^* = \frac{1}{\sqrt{2}}(\psi_1 - i\psi_2) \quad \ldots (36)$$

17.6 QUANTIZATION OF THE NON-RELATIVISTIC SCHRÖDINGER EQUATION

As an example of the application of the field quantization technique, we choose a non-relativistic field, namely the Schrödinger field, the classical field equation of which is given by

$$i\hbar \frac{\partial \psi(\vec{r}, t)}{\partial t} = -\frac{\hbar^2}{2m} \nabla^2 \psi(\mathbf{r}, t) + V(\mathbf{r})\, \psi(\mathbf{r}, t) \qquad \text{... (37)}$$

From the point of view of the quantum mechanics the above equation is the quantized equation of motion of a particle of mass m moving in an external field represented by the potential V. The said equation has been obtained by a process of quantization (could be called first quantization) where by the classical variables E and $\mathbf{p}$ have been replaced by the corresponding operators. But here we look upon it as a classical field equation describing the motion of some kind of material fluid. It can then be quantized according to the procedure described in the earlier sections. Since it is the second time the equation is being quantized, the field quantization of this equation is referred to as the second quantization. We shall see that the second quantization of eqn. (37) leads to the appearance of the field as an assembly of non-interacting indistinguishable (identical) particles.

We start by guessing the Lagrangian density which when substituted in the Euler's equation of motion, that is in eqn. (10) and its complex conjugate gives Schrödinger eqn. (37) and its complex conjugate. The Lagrangian density $\mathcal{L}$ may be taken to be of the form,

$$\mathcal{L} = i\hbar\psi^* \dot{\psi} - \frac{\hbar^2}{2m} \nabla\psi^* \cdot \nabla\psi - V(\mathbf{r})\, \psi^* \psi \qquad \text{... (38)}$$

The Euler's equations of motion due to the variations of ψ and ψ^* are given by

$$\frac{\partial \mathcal{L}}{\partial \psi} - \nabla \cdot \frac{\partial \mathcal{L}}{\partial (\nabla\psi)} - \frac{\partial}{\partial t}\left(\frac{\partial \mathcal{L}}{\partial \dot{\psi}}\right) = 0 \qquad \text{... (39)}$$

and

$$\frac{\partial \mathcal{L}}{\partial \psi^*} - \nabla \cdot \frac{\partial \mathcal{L}}{\partial (\nabla\psi^*)} - \frac{\partial}{\partial t}\left(\frac{\partial \mathcal{L}}{\partial \dot{\psi}^*}\right) = 0 \qquad \text{... (40)}$$

From eqn. (38), substituting the values of

$$\frac{\partial \mathcal{L}}{\partial \psi^*} = i\hbar\dot{\psi} - V\psi, \ \frac{\partial \mathcal{L}}{\partial (\nabla\psi^*)} = -\frac{\hbar^2}{2m} \nabla\psi \ \textit{and} \ \frac{\partial \mathcal{L}}{\partial \dot{\psi}^*} = 0$$

into eqn. (40), we get

$$\frac{i\hbar\partial\psi}{\partial t} = \frac{-\hbar^2}{2m} \nabla^2 \psi + V\psi$$

which is same as eqn. (37).

Similarly from eqn. (38), substituting the values of

$$\frac{\partial \mathcal{L}}{\partial \psi} = -V\psi, \ \frac{\partial \mathcal{L}}{\partial (\nabla\psi)} = \frac{-\hbar^2}{2m} \nabla\psi^* \text{ and } \frac{\partial \mathcal{L}}{\partial \dot{\psi}} = i\hbar\, \psi^*$$

into eqn. (39), we get

$$\frac{-i\hbar\partial\psi^*}{\partial t} = \frac{-\hbar^2}{2m}\nabla^2\psi^* + V\psi^* \qquad \text{... (37a)}$$

which is complex conjugate of eqn. (37).

This shows that the variation w.r.t. ψ^* gives eqn. (37) and the variation w.r.t. ψ gives its complex conjugate. Hence the form of the Lagrangian density which we have chosen is correct.

The momentum density canonical conjugate to ψ or conjugate field $\pi(\mathbf{r}, t)$ using eqn. (38), is given by

$$\pi(\mathbf{r}, t) = \frac{\partial \mathcal{L}}{\partial \dot{\psi}} = i\hbar\,\psi^* \qquad \text{... (41)}$$

Since $\dot{\psi}^*$ does not appear in the Lagrangian density, the momentum density canonical conjugate to ψ^* will be identically zero, that is, $\pi^* = \frac{\partial \mathcal{L}}{\partial \dot{\psi}^*} = 0$. It is, therefore impossible to satisfy the commutation relation,

$$[\psi^*(\mathbf{r}, t), \pi^*(\mathbf{r}', t)] = -i\hbar\,\delta(\mathbf{r} - \mathbf{r}') \qquad \text{... (42)}$$

or the corresponding classical Poisson bracket relation. Accordingly ψ^*, π^* cannot be regarded as a pair of canonical conjugate variables. They can easily be eliminated from the Hamiltonian as π^* never appears and eqn. (41) gives ψ^* in terms of π (it would be seen at a slightly later stage).

The Hamiltonian density by substituting the values of $\mathcal{L}$ and π from eqns. (38) and (41), can be expressed as

$$x = \pi\dot{\psi} - \mathcal{L}$$

$$= i\hbar\,\psi^* - i\hbar\,\psi^*\,\dot{\psi} + \frac{\hbar^2}{2m}\nabla\psi^*\cdot\nabla\psi + V\psi^*\,\psi$$

$$= \frac{\hbar^2}{2m}\nabla\psi^*\cdot\nabla\psi + V\psi^*\,\psi \qquad \text{... (43)}$$

where we have cancelled the first two terms.

Substituting the value of ψ^* from eqn. (41) into the above equation, we get

$$x = -\frac{i\hbar}{2m}\nabla\pi\cdot\nabla\psi - \frac{i}{\hbar}V\pi\psi \qquad \text{... (44)}$$

Using above, the classical field equations in the Hamiltonian formulation are given by

$$\dot{\psi} = \frac{\partial X}{\partial \pi} - \nabla\cdot\frac{\partial X}{\partial(\nabla\pi)}$$

$$= \frac{-i}{\hbar}V\psi + \frac{i\hbar}{2m}\nabla^2\psi$$

or
$$\frac{i\hbar\partial\psi}{\partial t} = \frac{-\hbar^2}{2m}\nabla^2\psi + V\psi \qquad \text{... (37)}$$

and
$$\dot{\pi} = -\left\{\frac{\partial X}{\partial\psi} - \nabla\cdot\frac{\partial X}{\partial(\nabla\psi)}\right\}$$
$$= \frac{i}{\hbar}V\pi - \frac{i\hbar}{2m}\nabla^2\pi$$

which on substituting the values of π and $\dot{\pi}$ from eqn. (41) leads to

$$\frac{-i\hbar\partial\psi^*}{\partial t} = \frac{-\hbar^2}{2m}\nabla^2\psi^* + V\psi^* \qquad \text{... (37a)}$$

Note that the eqns. (37) are nothing but the classical equation and its complex conjugate for the Schrödinger field. We have thus shown from the above analysis that the Lagrangian density expressed by eqn. (38), the canonical variables and the Hamiltonian density derived from it, are in complete agreement with the Schrödinger eqn. (37).

Since ψ is now an operator rather than a numerical function, ψ^* is to be interpretted as the Hermitian adjoint of ψ rather than its complex conjugate. Let us denote the Hermitian adjoint of ψ by $\psi^\dagger$. In that case the Hamiltonian density expressed by eqn. (43) would take the form,

$$X = \frac{\hbar^2}{2m}\nabla\psi^\dagger\cdot\nabla\psi + V\psi^\dagger\psi \qquad \text{... (43a)}$$

Therefore, the total Hamiltonian would be given by

$$H = \int\left(\frac{\hbar^2}{2m}\nabla\psi^\dagger\,.\,\nabla\psi + V\psi^\dagger\psi\right)d\tau \qquad \text{... (45)}$$

Now H is Hermitian as $H^\dagger = H$. The quantized Hamiltonian given by the above equation is the operator that represents total energy of the field; it should not be confused with the usual operator H which is energy operator for a single particle that is described by the Schrödinger equation.

The quantum condition,

$$[\psi(\mathbf{r}, t), \pi(\mathbf{r}', t)] = i\hbar\,\delta(\mathbf{r}-\mathbf{r}')$$

in the present case would take the form,

$$[\psi(\mathbf{r}, t), \psi^\dagger(\mathbf{r}', t)] = \delta(\mathbf{r}-\mathbf{r}')$$

where use has been made of eqn. (41). The quantum condition,

$$[\psi(\mathbf{r}, t), \psi(\mathbf{r}', t)] = 0$$

remains unchanged and the quantum condition,

$$[\pi(\mathbf{r}, t), \pi(\mathbf{r}', t)] = 0$$

takes the form

$$[\psi^\dagger(\mathbf{r}, t), \psi^\dagger(\mathbf{r}', t)] = 0$$

where once again use has been made of eqn. (41). The qantum conditions thus together can be written in the present case as

$$[\psi(\mathbf{r}, t), \psi(\mathbf{r}', t)] = [\psi^\dagger(\mathbf{r}, t), \psi^\dagger(\mathbf{r}', t)] = 0,$$

$$[\psi(\mathbf{r}, t), \psi^\dagger(\mathbf{r}', t)] = \delta(\mathbf{r} - \mathbf{r}') \qquad \text{... (46)}$$

17.7 THE N-REPRESENTATION

So far we have not given any explicit representation for the new operators ψ and H. Let us take this aspect now and expand the field operator $\psi(\mathbf{r}, t)$ in terms of a complete orthonormal set of functions $u_k(\mathbf{r})$ which we assume for definiteness to be discrete. Thus

$$\psi(\mathbf{r}, t) = \sum_k a_k(t)\, u_k(\mathbf{r}) \qquad \text{... (47a)}$$

and

$$\psi^+(\mathbf{r}, t) = \frac{-i}{\hbar} \pi(\mathbf{r}, t)$$

$$= \sum_k a_k^\dagger(t)\, u_k^*(\mathbf{r}) \qquad \text{... (47b)}$$

where u_k and u_k^* are numerical functions of space coordinates. a_k and $a_k^\dagger$ are operators which are functions of time.

Multiplying (47*a*) by $u_k^*(r)$ and integrating over the whole range of the variable and finally using the orthonormality relation,

$$\int u_i^*(\mathbf{r})\, u_k(\mathbf{r})\, d\tau = \delta_{ik'}$$

one gets $a_i(t)$. Then replacing i by k on both sides of the resulting equation, we get

$$a_k(t) = \int \psi(\mathbf{r}, t)\, u_k^*(\mathbf{r})\, d\tau \qquad \text{... (48a)}$$

Similarly, one can show that

$$a_k^\dagger(t) = \int \psi^\dagger(\mathbf{r}, t)\, u_k(\mathbf{r})\, d\tau$$

Let us now evaluate the commutation relation,

$$[a_{k'}, a_l^\dagger] = \int\int u_k^*(\mathbf{r})\, u_l(\mathbf{r}')\, d\tau\, d\tau' \,[\psi(\mathbf{r}, t), \psi^\dagger(\mathbf{r}', t)]$$

which on using the quantum condition,

$$[\psi(\mathbf{r}, t), \psi^\dagger(\mathbf{r}', t)] = \delta(\mathbf{r} - \mathbf{r}')$$

gives

$$\begin{aligned}[a_k, a_l^\dagger] &= \iint u_k^*(\mathbf{r})\, u_l(\mathbf{r}')\, d\tau\, d\tau'\, \delta(\mathbf{r}-\mathbf{r}') \\ &= \int u_k^*(\mathbf{r})\, u_l(\mathbf{r})\, d\tau \\ &= \delta_{kl} \qquad \text{... (49a)}\end{aligned}$$

where we have used the relation,

$$\int \delta(\mathbf{r}-\mathbf{r}')\, u_l(\mathbf{r}')\, d\tau' = u_l(\mathbf{r})$$

and the orthonormality of u_k.

In the similar fashion, it can be shown that

$$[a_k, a_l] = [a_k^\dagger, a_l^\dagger] = 0 \qquad \text{... (49b)}$$

that is, a_k and a_l, and similarly $a_k^\dagger$ and $a_l^\dagger$ commute for all k and l.

Note that the operator character of $\psi(\mathbf{r}, t)$ and $\pi(\mathbf{r}, t)$ is being reflected in the fact that the amplitudes a_k and $a_k^\dagger$ are behaving as operators as is evident from the commutators (49).

Let us now define a new operator (to be called number operator)

$$N = \int \psi^\dagger \psi\, d\tau \qquad \text{... (50)}$$

which we assume represents the number of particles in the field. Substitution of eqns. (47) in the above equation gives

$$\begin{aligned}N &= \sum_{kk'} a_k^\dagger a_{k'} \int u_k^*(\mathbf{r})\, u_{k'}(\mathbf{r})\, d\tau \\ &= \sum_{kk'} a_k^\dagger a_{k'}\, \delta_{kk'} \\ &= \Sigma\, a_k^\dagger\, a_k \\ &= \Sigma N_k \qquad \text{... (51)}\end{aligned}$$

where
$$N_k = a_k^\dagger\, a_k \qquad \text{... (52)}$$

By virtue of the commutation rules (49) for a_k's, the N_k's for different k's commute,

$$\begin{aligned}[N_k, N_{k'}] &= [a_k^\dagger a_k, a_{k'}^\dagger a_{k'}] \\ &= [a_k^\dagger a_k, a_{k'}^\dagger]\, a_{k'} + a_{k'}^\dagger [a_k^\dagger a_k, a_{k'}] \\ &= a_k^\dagger [a_k, a_{k'}^\dagger]\, a_{k'} + [a_k^\dagger, a_{k'}^\dagger]\, a_k \\ &\quad + a_{k'}^\dagger a_k^\dagger [a_k, a_{k'}] + a_{k'}^\dagger [a_k^\dagger, a_{k'}]\, a_k \\ &= a_k^\dagger a_{k'}\, \delta_{kk'} + 0 + 0 - a_{k'}^\dagger a_k\, \delta_{kk'} \\ &= 0 \qquad \text{... (53)}\end{aligned}$$

Since each N_k commutes with all others, they can be diagnolized (or measured simultaneously or would have simultaneous eigen kets). We label such a state (or eigen ket) by the eigen values $n_1, n_2, \ldots\ldots\ldots, n_{k'}, \ldots\ldots\ldots$, so that we put

$$\psi(n_1, n_2, \ldots\ldots\ldots, n_k, \ldots\ldots\ldots) \equiv |n_1, n_2, \ldots\ldots\ldots, n_k, \ldots\ldots\ldots> \quad \ldots (54)$$

Let us find the eigen values of the operator N_k by solving the eigen value equation,

$$N_k \, \psi(n_k) = n_k \, \psi(n_k) \quad \ldots (55)$$

where $\psi(n_k)$ is the normalized eigen function and n_k is the eigen value of N_k. Multiplying the above equation by $\psi^*(n_k)$ from the left and then integrating over the entire space, we get

$$\begin{aligned} n_k &= \int \psi(n_k)^* \, N_k \, \psi(n_k) \, d\tau \\ &= \int \psi(n_k)^* \, a_k^\dagger a_k \, \psi(n_k) \, d\tau \\ &= \int |a_k \, \psi(n_k)|^2 \, d\tau \geq 0 \end{aligned} \quad \ldots (56)$$

which follows from the definition of the scalar product in the Hilbert space. That is, the eigen values of N_k are all non-negative. Now using the commutation relations (49), we get

$$\begin{aligned} [a_k, N_k] &= [a_k, a_k^\dagger a_k] \\ &= [a_k, a_k^\dagger] \, a_k + a_k^\dagger \, [a_k, a_k] \\ &= a_k \end{aligned}$$

Hence

$$a_k \, N_k - N_k \, a_k = a_k$$

or

$$N_k a_k = a_k (N_k - 1) \quad \ldots (57)$$

Let us operate N_k on $a_k \, \psi(n_k)$. In that case using above equation,

$$N_k (a_k \, \psi(n_k)) = a_k (N_k - 1) \, \psi(n_k)$$

Further using the eigen value eqn. (55), we get

$$\begin{aligned} N_k (a_k \, \psi(n_k)) &= a_k (n_k - 1) \, \psi(n_k) \\ &= (n_k - 1) \, (a_k \, \psi(n_k)) \end{aligned} \quad \ldots (58)$$

as $(n_k - 1)$ is a number.

Thus, if $\psi(n_k)$ is an eigen function of N_k, then $a_k \, \psi(n_k)$ is also an eigen function but it belongs to the eigen value $n_k - 1$. In other words, it simply means that when N_k operates on $a_k \, \psi(n_k)$ the eigen value of N_k is reduced by 1 (or the number of particles in the *kth* state is decreased by 1). Hence, we can regard a_k as the destruction or annihilation operator.

Similarly, one can show that

$$[a_k^\dagger, N_k] = -a_k^\dagger$$

or

$$a_k^\dagger \, N_k - N_k \, a_k^\dagger = -a_k^\dagger$$

or

$$N_k a_k^\dagger = a_k^\dagger \, (N_k + 1) \quad \ldots (59)$$

Let us operate N_k on $a_k^+ \psi(n_k)$. In that case using above equation, we get

$$N_k(a_k^\dagger \psi(n_k)) = a_k^\dagger(N_k + 1)\psi(n_k)$$

Further using eqn. (54), we get

$$\begin{aligned} N_k(a_k^\dagger \psi(n_k)) &= a_k^\dagger(n_k + 1)\psi(n_k) \\ &= (n_k + 1)(a_k^\dagger \psi(n_k)) \qquad \text{... (60)} \end{aligned}$$

as $(n_k + 1)$ is a number.

The above equation implies that $a_k^\dagger \psi(n_k)$ is also an eigen function of N_k but belongs to the eigen value $n_k + 1$. In other words, it means that when N_k operates on $a_k^\dagger \psi(n_k)$, the eigen value of N_k is increased by 1 (or the number of particles in the *kth* state increases by 1). Hence we can regard a_k^+ as the creation operator.

Since according to eqn. (56), the eigen value spectrum is bounded from the below, there must exist an eigen function $\psi(0)$ such that $a_k \psi(0) = 0$. That is, it is a lowest normalized state with no particle which can be destroyed. Then, of course.

$$N_k \psi(0) = a_k^\dagger a_k \psi(0) = 0$$

that is, $\psi(0)$ is an eigen function and belongs to the eigen value $n_k = 0$. State $\psi(0) = |0\rangle$ is called no particle or vacuum state. Applying on to $\psi(0)$ the operator $a_k^\dagger$, we obtain a function $\psi(1) = a_k^\dagger \psi(0)$ such that because of eqn. (60), $N_k \psi(1) = N_k (a_k^\dagger \psi(0))$ $= 1\, a_k^\dagger \psi(0) = 1\, \psi(1)$. Thus $\psi(1)$ is an eigen function and belongs to the eigen value $n_k =$ 1. Proceeding in the similar fashion, we can show that the eigen value spectrum of the operator N_k is the set of all non-negative integral numbers:

$$n_k \ : \ 0, 1, 2, \ldots, \qquad \text{... (61)}$$

The field Hamiltonian H given by eqn. (45) takes the following form

$$H = \sum_{kk'} a_k^\dagger a_{k'} \int \left(\frac{\hbar^2}{2m} \nabla u_k^* \cdot \nabla u_{k'} + V u_k^* u_{k'} \right) d\tau \qquad \text{... (62)}$$

where we have substituted for ψ and ψ^+ from eqn. (47).

Integrating the first term in the integrand of the above equation, we get

$$\begin{aligned} \int \nabla u_k^* \cdot \nabla u_k^* \, d\tau &= \int u_k^* \nabla u_{k'} \, ds - \int u_k^* \nabla^2 u_{k'} \, d\tau \\ &= -\int u_k^* \nabla^2 u_{k'} \, d\tau \end{aligned}$$

as u_k vanishes at infinite bounding surface.

Substituting above into eqn. (62), we obtain

$$H = \sum_{kk'} a_k^\dagger a_{k'} \int u_k^* \left(\frac{-\hbar^2}{2m} \nabla^2 u_{k'} + V u_{k'} \right) d\tau$$

Further making use of the time independent Schrödinger equation for the k' state (or assuming that $u_{k'}$ are eigen functions of $\left(\frac{-\hbar^2}{2m}\nabla^2 + V\right)$ with eigen value $E_{k'}$), that is

$$\frac{-\hbar^2}{2m}\nabla^2 u_{k'} + V u_{k'} = E_{k'} u_{k'},$$

we get

$$\begin{aligned} H &= \sum_{kk'} a_k^\dagger a_{k'} \int u_k^* E_{k'} u_{k'} \, d\tau \\ &= \sum_{kk'} a_k^\dagger a_{k'} E_{k'} \delta_{kk'} \\ &= \sum_k a_k^\dagger a_k E_k \\ &= \sum_k N_k E_k \end{aligned} \qquad \text{... (63)}$$

Thus the total energy (the eigen value of the Hamiltonian H) of the field in the state (54), that is, $|n_1, n_2, \ldots\ldots\ldots, n_{k'} \ldots\ldots\ldots>$ is given by

$$E = \langle H \rangle = \sum_k n_k E_k \qquad \text{... (64)}$$

(Note that since N and H commute (see eqn. (74) of the section 17.8) they would have simultaneous eigen kets). Eqn. (64) shows that n_k is the number of particles in the state u_k with energy E_k (that is, a field state (54) is one in which there are n_1 particles of energy E_1, n_2 particles of energy E_2 and so on). Thus N_k can be regarded as particle number operator in kth state (see section 6.8). This effectively means that $n = \Sigma n_k$ is the total number of particles in the field or the eigen value of the operator N is n. Since a given particle state u_k can be occupied by an arbitrary number of particles with the same energy, the field represents an assembly of bosons.

17.8 SYSTEM OF FERMIONS

We have seen in the preceding section that the commutator relations

$$[a_k, a_l^\dagger] = \delta_{kl}, \qquad \text{... (49a)}$$

$$[a_k, a_l] = [a_k^\dagger, a_l^\dagger] = 0 \qquad \text{... (49b)}$$

lead to a system of bosons. For a system of fermions (particles that obey Pauli exclusion principle), the occupation number (say n_k), that is, number of particles in any state should be restricted to 0 and 1. It has been shown by Jordan and Wigner[1] that this condition could be achieved by replacing the above commutator relations by the following anticommutation relations,

$$[a_k, a_l^\dagger]_+ = \delta_{kl}, \qquad \text{... (65a)}$$

$$[a_k, a_l]_+ = [a_{k'}^\dagger, a_l^\dagger]_+ = 0 \qquad \text{... (65b)}$$

[1]P. Jordan and E. Wigner, Z. Physik 47, 631 (1928)

Here also all the operators refer to the same time. Of course, $[A, B]_+ = AB + BA$ is the anticommutator of two operators A and B. When $l = k$, from (65a)

$$a_k a_k^\dagger + a_k^\dagger a_k = 1 \quad \text{... (66a)}$$

and from (65b)

$$a_k a_k + a_k a_k = a_k^\dagger a_k^\dagger + a_k^\dagger a_k^\dagger = 0$$

or

$$a_k a_k = a_k^\dagger a_k^\dagger = 0 \quad \text{... (66b)}$$

We interpret and define $N_k = a_k^+ a_k$ as before. Each N_k commutes with all others so that they can be diagonolized (or measured simultaneously). Eigen value of N_k can be obtained by evaluating the square of the number operator,

$$\begin{aligned} N_k^2 &= N_k N_k = a_k^\dagger a_k a_k^\dagger a_k \\ &= a_k^+ (1 - a_k^\dagger a_k) a_k \\ &= a_k^\dagger a_k \\ &= N_k \end{aligned} \quad \text{... (67)}$$

where use has been made of eqns. (66).

If N_k is diagonal and has the eigen value n_k, it is clear that N_k^2 would also be diagonal with eigen value n_k^2. Thus the above equation is equivalent to the algebraic equation

$$n_k^2 = n_k$$

or

$$n_k (n_k - 1) = 0$$

which is quadratic and hence has two roots

$$n_k = 0, 1 \text{ for any } k. \quad \text{... (68)}$$

Thus the eigen values of each N_k are only 0 and 1 and the particles obey Pauli exclusion principle. Hence the number operator N_k can be represented by a diagonal matrix

$$N_k = a_k^\dagger a_k = \begin{pmatrix} 0 & 0 \\ 0 & 1 \end{pmatrix} \quad \text{... (69)}$$

Also a_k and $a^\dagger$, that is, the annihilation and creation operators should be such that their squares are zero (see eqn. (66(b)), and they satisfy eqns. (66a) and eqn. (69),

$$a_k = \begin{pmatrix} 0 & 1 \\ 0 & 0 \end{pmatrix}, \quad a_k^\dagger = \begin{pmatrix} 0 & 0 \\ 1 & 0 \end{pmatrix} \quad \text{... (70)}$$

The two eigen functions of the operator (69) referring to the eigen values 0 and 1, respectively (that is, one state containing no particle and the other containing one particle) are

$$\psi(0) = \begin{pmatrix} 1 \\ 0 \end{pmatrix} \quad \psi(1) = \begin{pmatrix} 0 \\ 1 \end{pmatrix} \quad \text{... (71)}$$

as

$$N_k \psi(0) = 0\, \psi(0) \quad \text{and} \quad N_k \psi(1) = 1\, \psi(1)$$

Now $$a_k^\dagger \psi(0) = \begin{pmatrix} 0 & 0 \\ 1 & 0 \end{pmatrix}\begin{pmatrix} 1 \\ 0 \end{pmatrix} = \begin{pmatrix} 0 \\ 1 \end{pmatrix} = \psi(1) \quad \text{... (72}a\text{)}$$

and $$a_k \psi(1) = \begin{pmatrix} 0 & 1 \\ 0 & 0 \end{pmatrix}\begin{pmatrix} 0 \\ 1 \end{pmatrix} = \begin{pmatrix} 1 \\ 0 \end{pmatrix} = \psi(0) \quad \text{... (72 }b\text{)}$$

Also $$a_k \psi(0) = a_k^\dagger \psi(1) = 0 \text{ for all } k.$$

From the above relations, it follows that

$$a_k \psi(n_k) = n_k\, \psi(1 - n_k), \quad \text{... (73)}$$

$$a_k^\dagger \psi(n_k) = (1 - n_k)\, \psi(1 - n_k), \text{ with } n_k = 0, 1.$$

Thus a_k and $a_k^\dagger$ again play the role of destruction and creation operators and that an empty state cannot be further emptied and a filled state cannot be further filled.

One must note that in the actual problem, the number of states of the system is infinite and it is not convenient to write the explicit matrices like those in (70). One can, however, find the effect of operation with a_k and $a_k^\dagger$ on a *ket* $|n_1, n_2, \ldots\ldots\ldots, n_k, \ldots\ldots\ldots\rangle$ that has eigen value 0 or 1 for the operator N_k.

It is necessary to mention here that the Hamiltonian H and the total energy E of the field even in this case are given by the equations,

$$H = \int \left(\frac{\hbar^2}{2m} \nabla\psi^\dagger \cdot \nabla\psi + \nabla\psi^\dagger\, \psi \right) d\tau$$

and $$E = \sum_k n_k\, E_k$$

where the same method as that used for obtaining (64) in the case of bosons has been used here. The only restriction in this case is that $n_k = 0$ or 1. Also the anticommutation relations (65) have been obtained from the anticommutators

$$[\psi(\mathbf{r}, t), \psi^+(\mathbf{r}', t)]_+ = i\hbar\, \delta(\mathbf{r} - \mathbf{r}'),$$

$$[\psi(\mathbf{r}, t), \psi(\mathbf{r}', t)]_+ = [\pi(\mathbf{r}, t), \pi(\mathbf{r}', t)]_+ = 0$$

by using once again the same method as used in the case of bosons.

Like in the case of bosons, here also (that means in both the cases bosons as well as fermions), we can define an operator N representing the total number of particles by

$$N = \Sigma N_k$$

From above and from eqn. (63)

$$[N, H] = \sum_{k,l} [N_k, N_l]\, E_l = 0 \quad \text{... (74)}$$

where use has been made of eqn. (53). further the use of equation,

$$i\hbar\, \dot{N} = [N, H]$$

which has been obtained by substituting N for F in eqn. (26), implies (see eqn. (74)) that the total number of particles in the field is conserved (or remains constant in time).

❑❑❑

APPENDIX 1

BLACK BODY RADIATION

All of us are familiar with the glow of a hot piece of metal (say iron bar), which emits visible light whose color varies with the temperature of the metal going from red to yellow to white as it becomes hotter and hotter. In fact, other frequencies to which our eyes do not respond are also present. Truly speaking objects (substances) radiate also electromagnetic radiation, continuously at all temperatures though which frequencies predominate depends on the temperature. At room temperature most of the radiation is in the Infrared region of the spectrum and hence is invisible.

The ability of a body to radiate is closely related to its ability to absorb radiation. This is to be expected, since a body at a constant temperature is in thermal equilibrium with its surroundings and must absorb energy from surroundings at the same rate as it emits energy. It is convenient to consider an ideal body as one that absorbs all radiation incident upon it, regardless of frequency. Such a body is called a black body.

The point of introducing the idealized black body in a discussion of thermal radiation is that we can now disregard the precise nature of the radiating body, since all black bodies behave identically. A black body can be approximated by a, hollow object with a very small hole leading to its interior (Fig. A 1.1). Any radiation striking

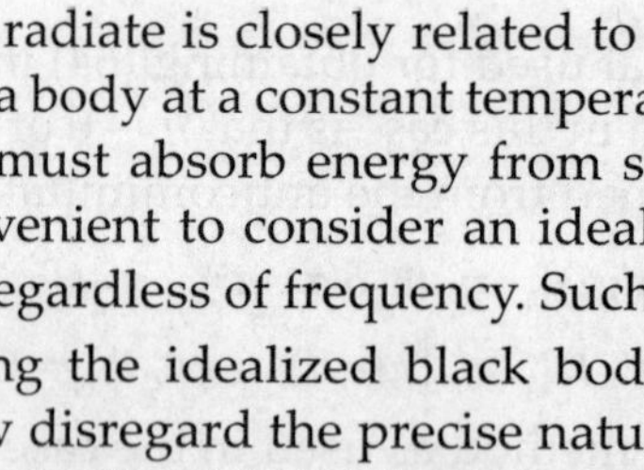

Fig. A 1.1 *A hole in the wall of a hollow object is an excellent approximation of a black body*

the hole enters the cavity where it is trapped by reflection back and forth until it is absorbed. The cavity walls are constantly emitting and absorbing radiation, and it is this property of this radiation (called black body radiation) that we are interested.

Experimentally we can sample black body radiation simply by inspecting what emerges from the hole in the cavity and the same, that is, the spectral distribution of energy in the black body radiation, is shown in Fig. A 1.2 for two temperatures (All solids exhibit nearly the same spectrum of continuous radiation when at the same temperature). The principles of classical physics (which treats the radiation inside the cavity as if it were the assembly of oscillators, each at a different frequency) are unable to account for the observed black body spectrum. In fact, it was this particular failure of classical physics that led the German physicist Max Planck, in 1901 to suggest that light emission is a quantum or photon phenomenon. That is, each oscillator could have only one or another of a set of discrete energy values (thai is the energies of the oscillators are quantized) differing in energy by $h\upsilon$. We shall use Maxwell-Boltzmann statistical mechanics to derive the Planck radiation formula, which predicts the same spectrum as that found by experiment.

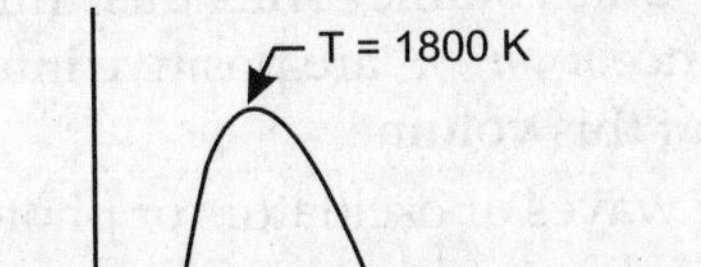

Fig. A 1.2 *The spectral distribution of energy in the black body radiation for two temperatures*

To explain the black body spectrum, we begin by considering a black body, as a radiation, filled cavity (say in the form of a cube having each side a) at temperature T (Fig. 1.1). Because the cavity walls are assumed to be perfect reflectors, the radiation must consist of standing electromagnetic waves (the reflected and incident waves interfere and stationary or standing waves are formed. This simply means that the enclosure would, be filled with large number of photons of various wavelengths or frequencies). In order for a node to occur at each wall (the situation is analogous to a string fixed at both the ends), the path, length from wall to wall in any direction, must be an integral multiple of half-wavelengths. As assumed earlier if the cavity is a cube a long on each edge, this condition means that for standing waves in the x, y and z directions respectively, the possible wavelengths are such that $\frac{\lambda}{2} n_x = a$.

$$n_x = \frac{2a}{\lambda} = 1,2,3,... = \text{number of half wavelengths in } x\text{-direction}$$

$$n_y = \frac{2a}{\lambda} = 1,2,3,... = \text{number of half wavelengths in } y\text{-direction.}$$

$$n_z = \frac{2a}{\lambda} = 1,2,3,... = \text{number of half wavelengths in } z\text{-direction.}$$

For a standing wave in any arbitrary direction, it must be true that

$$n_x^2 + n_y^2 + n_z^2 = (2a/\lambda)^2 \text{ where } \left. \begin{array}{l} n_x = 0, 1, 2, ... \\ n_y = 0, 1, 2, ... \\ n_z = 0, 1, 2... \end{array} \right\} \quad ... (A\ 1.1)$$

in order that the wave terminates in a node at its ends, (Of course, if $n_x = n_y = n_z = 0$, there is no wave, though it is possible for any one or two of the n's to equal zero).

The above equation defines a sphere of radius $R = 2a/\lambda = 2a\upsilon/c$. Coordinates of each point instead of (x, y, z) would be (n_x, n_y, n_z) in (n_x, n_y, n_z)*–space* in this case. Each permissible set (n_x, n_y, n_z) defines one stationary wave (or one mode of vibration or state of the system) inside the enclosure. Since n_x, n_y, n_z are integers, hence there would be one such point per unit volume, Thus their number would be given by the total volume of the sphere. Since n_x, n_y, n_z are positive integers, the required number is contained only in the octant of this volume.

The number of stationary waves or oscillators or photons between R and $R + dR$ (or frequencies between υ and $\upsilon + d\upsilon$) would, therefore be given by

$$\begin{aligned} g(\upsilon)d\upsilon &= \frac{1}{8}(4\pi R^2\, dR) \\ &= \frac{1}{8}4\pi(2a\upsilon/c)^2\, 2a/c\, d\upsilon \\ &= 4\pi V\upsilon^2\, d\upsilon/c^3 \end{aligned} \quad ... (A\ 1.2)$$

where we have made use of the relation $R = 2a\upsilon/c$ or $dR = 2a/c\, d\upsilon$ and the fact that $V = a^3$ is the volume of the cavity.

This result would have been the final one had the waves were longitudinal. But we are concerned with em waves which are transverse in nature. Thus for each value of k (or λ or υ) there are infact two independent transverse vibrations. Therefore, the actual number of stationary waves (or oscillators) with frequencies between υ and $\upsilon + d\upsilon$ would be given by

$$g(\upsilon)d\upsilon = \frac{8\pi V\upsilon^2 d\upsilon}{c^3} \quad ... (A\ 1.3)$$

which is nothing but number of states in the frequency range between υ and $\upsilon + d\upsilon$ (if we divide the above expression by V, what we get is nothing but the density of standing waves in cavity).

The next step is to find the average energy per standing wave (or per oscillator). To achieve this as pointed out earlier Planck had to assume that the oscillators in the cavity walls were limited to energies of $\in_n = nh\upsilon$ where $n = 0, 1, 2,....$ That is, an oscillator emits radiation of frequency υ when it drops from one energy state to the next lower one and it jumps to next higher state when it absorbs radiation of frequency υ. Further he used Maxwell-Boltzmann distribution law to find that the number of oscillators with the energy $\in_n$ proportional to $e^{-\in_n/k_BT}$ temperature T. Thus the average energy per oscillator (and hence per standing wave) in the cavity would be given by

$$\overline{\in} = \frac{\in}{N} = \frac{\sum_{n=0}^{\infty} \in_n e^{-\in_n/k_BT}}{\sum_{n=0}^{\infty} e^{-\in_n/k_BT}} \qquad ... (A\ 1.4)$$

where N is the total number of oscillators and $\in$ their total energy

Taking $x = e^{-h\upsilon/k_BT}$, the above equation becomes

$$\in = \frac{\sum_{n-0}^{\infty} \in_n x^n}{\sum_{n=0}^{\infty} x^n} = \frac{0 + h\upsilon x + 2h\upsilon x^2 + 3h\upsilon x^3 + ...}{1 + x + x^2 + ...}$$

$$= \frac{h\upsilon x(1 + 2x + 3x^2 + ...)}{1 + x + x^2 + ...}$$

$$= \frac{h\upsilon x(1-x)^{-2}}{(1-x)^{-1}}$$

$$= \frac{h\upsilon x}{(1-x)} = \frac{h\upsilon}{(e^{h\upsilon/k_BT} - 1)} \qquad ... (A\ 1.5)$$

If we multiply the above expression by (A 1.3) and divide by the volume V of the cavity, we get the corresponding spectral energy density $u(\upsilon)d\upsilon$, which is the energy per unit volume in the radiation between frequency υ and $\upsilon + d\upsilon$, given by

$$u(\upsilon)d\upsilon = 8\pi h\upsilon^3/c^3 \frac{d\upsilon}{e^{h\upsilon/k_BT} - 1} \qquad ... (A\ 1.6)$$

This is the Planck radiation formula or Planck's law of radiation which agrees with the experiment. Note that the above result does not depend on the shape of cavity and hence the shape is immaterial

WIEN'S DISPLACEMENT LAW

An interesting feature of the black body spectrum at a given temperature is the wavelength λ_{max} for which the energy density is greatest. To find λ_{max}, let us express (A 1.6) in terms of wavelength. We obtain

$$u(\lambda)d\lambda = \frac{8\pi h}{c^3}(c/\lambda)^3(c/\lambda^2)d\lambda/(e^{hc/\lambda k_BT} - 1)$$

$$= \frac{8\pi hc}{\lambda^5} d\lambda / (e^{hc/\lambda k_B T} - 1) \quad \text{... (A 1.7)}$$

where we have used the relation, $\upsilon = c/k$ or $d\upsilon = -c/\lambda^2\, d\lambda$ in which minus sign is because of the fact that when υ increases λ decreases

Setting $du(\lambda)/d\lambda = 0$ for $\lambda = \lambda_{max}$, we obtain

$$\lambda^{-5}(-1)(e^{hc/k_B T} - 1)^{-2} \times hc/k_B T \times (-1/\lambda^2)\, e^{hc/\lambda k_B T}$$

$$+ (e^{hc/\lambda k_B T} - 1)^{-1} (-5)\lambda^{-6} = 0$$

or
$$\lambda^{-6} (e^{hc/\lambda k_B T} - 1)^{-1} \left[\frac{hc}{\lambda k_B T} (e^{hc/\lambda k_B T} - 1)^{-1} e^{hc/\lambda k_B T} - 5\right] = 0$$

or
$$\frac{hc}{\lambda_{max} k_B T} = 5(1 - e^{-hc/\lambda_{max} k_B T})$$

which is a transcendental equation. After solving we get

$$\frac{hc}{\lambda_{max} k_B T} = 4.965$$

which is more conveniently expressed as

$$\lambda_{max} T = \frac{hc}{4.965 k_B} = 2.898 \times 10^{-3} \text{ m.K} \quad \text{... (A 1.8)}$$

This is known as Wien's displacement law. It qualitatively expresses the empirical fact that the peak in the black body spectrum shifts to progressively shorter wavelengths (higher frequencies) as the temperature is increased in Fig. *A* 1.2. Since the value of the r.h.s of eqn. (*A* 1.8) is accurately known, a measurement of the value of λ_{max} can be used to determine the temperature of any black body. This method is used to determine the surface temperature of stars and other inaccessible hot objects.

STEFAN-BOLTZMANN LAW

Another result we can obtain from (*A* 1.6) is the total energy per unit volume or energy density *a* of radiation in cavity. This is integral of energy density over all frequencies. Thus

$$u = \int_0^\infty u(\upsilon)\, d\upsilon = \frac{8\pi h}{c^3} \int_0^\infty \frac{\upsilon^3\, d\upsilon}{e^{h\upsilon/k_B T} - 1}$$

Putting $x = h\upsilon/k_B T$ or $dx = hd\upsilon/k_B\, T$. we get

$$u = 8\pi k_B{}^4 T^4 / h^3 c^3 \int_0^\infty \frac{x^3\, dx}{e^x - 1}$$

$$= 8\pi k_B^4 T^4 / h^3 c^3 \times \pi^4/15$$

$$= 8\pi^5 k_B^4 / (15 h^3 e^3) T^4$$

$$= aT^4 \quad \text{... (A 1.9)}$$

where $a = 8\pi^5 k_B^4/(15c^3h^3) = 7.56 \times 10^{-16}\ J/(m^3\ K^4)$

The total energy density is proportional to the fourth power of the absolute temperature of the black body. This leads to the Stefan-Boltzmann law, namely the power e radiated per unit area by a black body is proportional to T^4, that is

$$e = \sigma T^4 \qquad \text{... (A 1.10)}$$

where σ is Stefan's constant and is equal to $5.67 \times 10^{-8}\ W(m^2.\ K^4)$

Both Wien's displacement law and the stefan-Boltzmann law are evident in qualitative fashion in Fig. A1.2, the maxima in the various curves shift to higher frequencies and the total area underneath them increase rapidly with rising temperature.

As pointed out earlier, the classical approach treats the radiation inside an enclosure (cavity) as if it were an assembly of oscillators (or standing waves) each at different frequency. If we use the law of equipartition of energy (or classical physics) that is, the average energy of an oscillator (or energy associated with each degree of vibration) is k_BT, then the energy per unit volume within, the frequency range $d\upsilon$ (using eqn. A 1.2) comes out to be

$$u(\upsilon)d\upsilon = 8\pi\ \upsilon^2/c^3\ k_BT\ d\upsilon \qquad \text{... (A 1. 11)}$$

This is the Rayleigh-Jeans formula which shows that as the frequency increases toward the ultraviolet end of the spectrum, the energy density increases as υ^2. In the limit of infinitely high frequencies, $u(\upsilon)$ goes to infinity (see Fig. A 1.3). This discrepancy of classical physics is known as the ultraviolet catastrophe. The observed data of course, shows that the energy density falls to zero as $\upsilon \rightarrow \infty$ (see Fig. A 1.3). Also from the Planck radiation formula (A 1.6) at high frequencies, $h\upsilon >> k_BT$ and hence $e^{h\upsilon/k_BT} \rightarrow \infty$, which means that $u(v) \rightarrow 0$ as observed. Thus no more ultraviolet catastrophe. Further at low frequencies where the Rayleigh-Jeans formula (A 1.11) is a good approximation to the observed results (Fig. A 1.3), $hv << k_BT$ or $hv/k_BT << 1$. Thus

$$e^{h\upsilon/k_BT} \simeq 1 + h\upsilon/k_BT$$

and hence

$$1(e^{h\upsilon/k_BT} - 1) \simeq 1/(1 + \frac{h\upsilon}{k_BT} - 1) = k_BT/h\upsilon$$

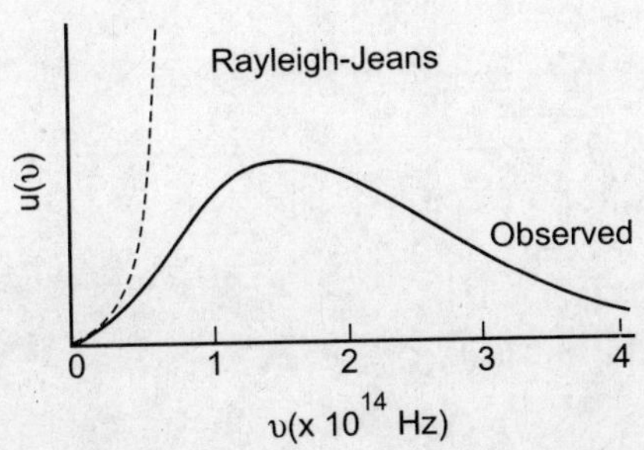

Fig. A 1.3 *Comparison of the Rayleigh-Jeans results for the spectrum of black body radiation at 1500K with the observed results.*

Therefore at low frequencies, Planck's radiation formula (*A* 1.6) becomes

$$u(\upsilon)d\upsilon = 8\pi h\upsilon^3/c^3 (k_B T/h\upsilon)\, d\upsilon = \frac{8\pi\upsilon^2}{c^3} k_B T d\upsilon$$

which is nothing but the Rayleigh-Jeans formula.

It is necessary to point out here that though Planck got the right formula but his derivation from today's understanding is seriously flawed. The energies of the harmonic oscillators in the cavity are to be taken as $\in_n = (n + 1/2)h\upsilon$, not $nh\upsilon$. Including the zero point energy of ½ $h\upsilon$ (corresponding to $n = 0$) does not lead to the average energy, $\in = h\upsilon/(e^{h\upsilon/k_BT} - 1)$ when Maxwell-Boltzmann statistics is used. The correct procedure is to consider the em waves in the cavity as a photon gas subject to Bose-Einstein statistics, since the spin of the photon is 1. The end result is of course, the same.

Example A1.1

How many independent stationary waves with wavelengths between 9.5 and 10.5 mm can occur in a cubical caviiy 1 m on each side?

Solution

From eqn. (*A* 1.3), the number of stationary waves between frequencies υ and $\upsilon + d\upsilon$ in a cubical cavity *L m* on each side, is given by

$$g(\upsilon)\, d\upsilon = 8\pi L^3\upsilon^2\, d\upsilon/c^3$$

Therefore, the given number between wavelengths λ and $\lambda + d\lambda$

$$g(\lambda)d\lambda = 8\pi L^3 (c/\lambda)^2/c^3 \times (c/\lambda^2)d\lambda$$

$$= 8\pi L^3\, d\lambda/\lambda^4$$

Thus the number in the present case would be

$$8 \times 3.14 \times 1^3 \times 1 \times 10^{-3}/(10 \times 10^{-3})^4 = 2.5 \times 10^6$$

where $d\lambda = (10.5 - 9.5)$ mm = 1 mm and $\lambda = \frac{(9.5 + 10.5)}{2}$ mm = 10 mm.

APPENDIX 2

PHOTOELECTRIC EFFECT

Electrons are emitted from a metal surface (first observed by Hertz, when ultraviolet light was made to incident on the zinc plate. Afterwards Milliken investigated the effect with a number of alkali metals over a wide range of light frequencies and won the Nobel prize in 1923) when light of sufficiently high frequencies is incident on it. This phenomenon is known as the photoelectric effect and the emitted electrons are called photoelectrons.

The photoelectric effect can be demonstrated by shining light on a surface of plate (called emitter-*E*) in an evaluated chamber (see Fig. *A* 2.1) which also has another plate (called collector, *C*). These two electrodes are connected to a source of variable voltage. The flow of current in the circuit can be measured with the help of the ammeter *A*. The current (called photoelectric current) flow here implies the flow of the charged particles from *E* to *C*.

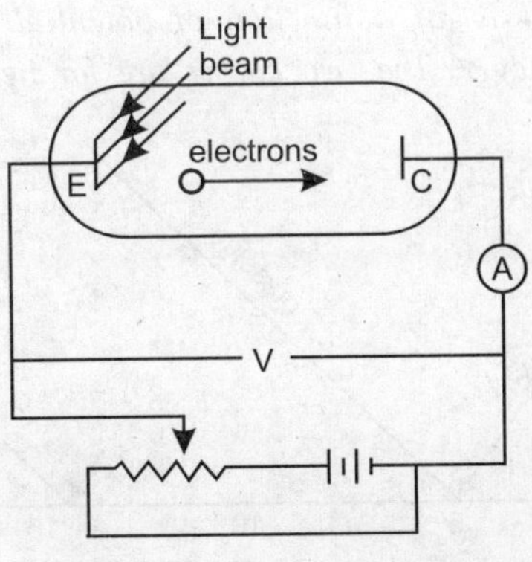

Fig. A 2.1 *Apparatus for the study of photoelectric effect E, emitter, C, collector V, potential difference; A, ammeter*

While conducting the experiment, a potential difference is applied between the two plates through a source of variable voltage and the photoelectric current i_p is measured-for various values of the potential difference, both positive (that is, the collector being at a potential higher than that of the emitter on which the light is incident) and negative. The typical curves between the applied voltage V and the resulting photoelectric current i_p, are shown in Fig. A 2.2 for the two values of the light intensities (W/m^2), I_1 and I_2 ($I_1 < I_2$) for a given value of the frequency of light, It may be noted that the current flows even when the polarity of the collector is negative. In such a situation photoelectrons are repelled by the collector and many of these return to the emitter (some of the photoelectrons that emerge from the emitter have enough energy to reach the collector despite its negative polarity, and they constitute the measured current). If this negative potential difference is increased further in magnitude, most of the photoelectrons emitted go back to the emitter and in turn the flow of the current is reduced. If we continue to increase the negative potential difference then at some stage, none of the photoelectrons not even the fastest ones, would reach the collector. As a result, the current slops flowing. The magnitude of the potential difference V_0 at which the current just becomes zero is called the stopping (or extinction) voltage (for most of the substances, it is of the order of several volts). This extinction voltage corresponds to the maximum photoelectron kinetic energy.

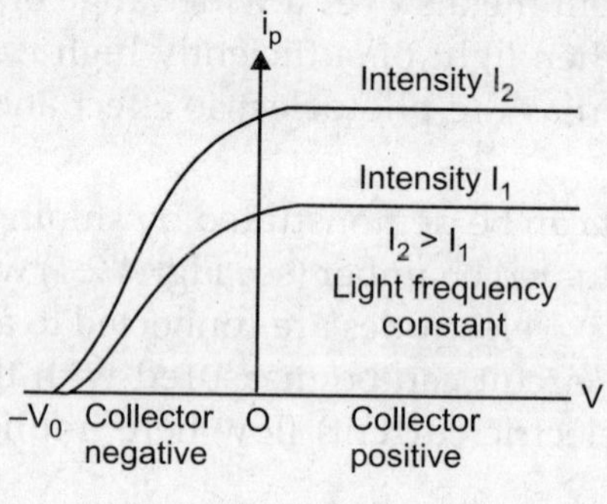

Fig. A 2.2 *Photoelectric current is an a function of potential difference between collector and emitter for a fixed light frequency v. The two curves are for light intensities I_1 and I_2 ($I_1 < I_2$)*

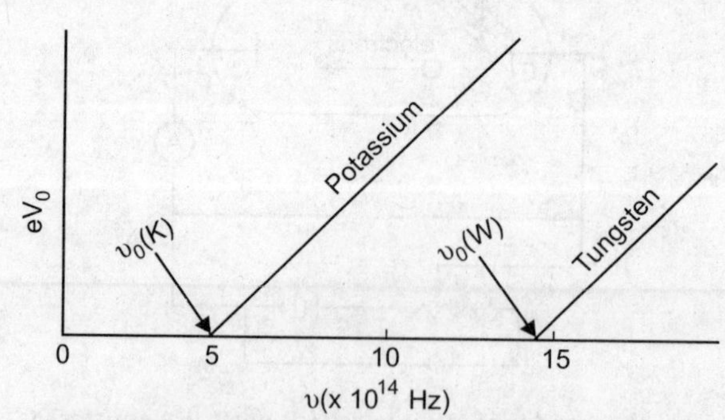

Fig. A 2.3 eV_0 as a function of v for potassium and tungsten emitters.

Now let us vary the frequency υ of the incident light and keep the intensity of light constant. The result of this experiment is illustrated in Fig. A 2.3. As the frequency of the light increases the stopping potential V_0 also increases. Furthermore, V_0 is a linear functions of υ. The data for two materials of the emitter is shown in the same Figure.

Let us now summarize the above experimental observations:

(*i*) One finds that within the limits of experimental accuracy (~ 10^9 s). there is no time interval between the arrival of the light at a metal surface (emitter) and the emission of photoelectrons. The observation that electrons are emitted is of course not surprising as all substances are made of atoms and atom contains electrons. A metal being electrically neutral has both positive and negative charges. Therefore, when an electron is to be removed from a metal, it needs to be pulled away from the positive charge and this requires an amount of work to be done (or energy). In other words there is a binding energy of electrons to a substance, that is, that energy must be given to an electron in order to remove it from the substance (solid). We denote this binding energy of an electron to solid by $e\phi$ where ϕ is measured in volts, and $e\phi$ is called the work function. More specifically. $e\phi$ is defined as lite minimum energy that must be given to an electron in the solid in order to remove it from the solid. If there are electrons with various energies inside the solid (as in meal, where energies from zero to several electron volts occur in the same metal), $e\phi$ represents the additional energy that must be given to the most energetic of these electron in order to remove it. In the case of sodium (Na) metal its value is 2.3 eV, Therefore unless this amount of energy is supplied by the light beam, the electron will not leave the metal.

According to classical physics, the transfer of energy from the light wave to the electron should take a long time, of the order of seconds†. This means that the electron would take a sizeable time before it is emitted. Thus classical theory fails to explain the experimental observation.

(*ii*) Figure A 2.2 shows that a bright light (light of higher intensity) yields more photoelectrons (higher photocurrent) than a dim one (light of lesser intensity) of the same frequency, but the maximum energy of an emitted electron (eV_0) remains the same. This means that the maximum energy of the emitted electron can not be increased by just increasing the intensity of light. This is contrary to what one would expect from the Maxwell's electromagnetic theory (classical physics) according to which if the intensity of light increases, the energy of the emitted electrons should increase as the force which incident light exerts on the electrons in metal is proportional to the magnitude of the electric field vector $\vec{E}$ and the magnitude of this vector increases when the intensity of the light is increased (intensity being proportional to the square of the amplitude of $\vec{E}$). Thus classical theory fails to explain the experimental observation.

(*iii*) Figure A 2.3 shows that higher the frequency of the incident light, the more energy (eV_0) the photoelectrons have. Blue light results in faster electrons than

---

†Consider a 100 W bulb emitting light in all directions. Suppose that a metallic sodium (Na) surface is kept at a distance of 1 m from the bulb. Find the time needed by an individual electron in a Na atom to receive an energy 2.3 eV in order to leave the metal.

Assuming that the light from the bulb is spreading uniformly in all directions, the intensity of light at a spherical surface of radius r = 1 m would be

$$I = \frac{100}{4\pi(1)^2} \sim 8J/(\text{sec}-\text{m}^2)$$

Let us assume that all the energy is absorbed by the top layer of the surface. The number of atoms per unit area of the surface would be given by

$$N = \frac{1}{\pi \times (10^{-10})^2} \sim 3 \times 10^{19} \text{ atoms/m}^2$$

where radius of the atom is taken to be 10^{-10} m

The power absorbed by the atom is then

$$\frac{8J/(\text{sec}-\text{m}^2)}{3 \times 10^{19} \text{ atoms/m}^2} \sim 2.7 \times 10^{-19} \text{ J/sec} \sim 1.7 \text{ eV/sec}$$

Therefore the time needed by an electron to gain an energy of 2.3 eV would be

2.3 eV/(1.7 eV/sec) ~ 1.3 sec

Note that the assumption we have made is not realistic as the energy absorbed by an atom can not all get transferred to a single electron. As such, the actual energy required by an electron would be much less. This simply means that the actual time required would be much higher than that we have estimated.

red light. Further at frequencies below a certain critical frequency υ_0, which is characteristic of each particular metal, no electrons are emitted. Above υ_0 the photoelectrons range in energy from zero to a maximum value that increases with increasing frequency. Further V_0 is found to be a linear function of v. Figure A 2.3 depicts the data for two emitter materials. Note that the slope of the lines is independent of the emitter material. This observation also cannot be explained based on the classical physics because it predicts that the maximum energy of the emitted electron (or stopping potential V_0) depends only on the intensity of light, in other words, the frequency of the light wave finds no significance in the classical explanation.

The above observations were explained by Einstein in 1905 by assuming that a light beam should not be considered as a wave supplying energy to a substance in continuous manner instead, it should be thought as consisting of particles (corpuscles) called photons, each of energy $h\upsilon$ and velocity equal to that of light, that is c. υ being the frequency of the light, beam.

Let us now apply this photon theory to explain the experimental observation (*i*). it is well understood as the wave energy is now assumed to be concentrated in photons. An electron in an atom of metal is emitted when a photon collides with it. The electron absorbs an energy hv from the photon and can leave the metal. The collision is instantaneous so that there is no time lag between the emission of the electron and its collision with the photon.

Result (*ii*), that is the photoelectric current is proportional to the intensity of light, can be understood by noting that an increase in intensity leads to an increase in the number of photons in direct proportion. With more photons, the number of collisions rises and hence the number of electrons emitted. The current being directly proportional to the number of electrons emitted, is hence proportional to the intensity. Further all photons of frequency υ have the same energy, so changing the intensity of a monochromatic light beam will change the number of photoelectrons but not their energies. Thus the maximum photoelectron kinetic energy eV_0 is independent of the intensity of the light beam.

Not all the electrons hit by photons actually leave the metal. Those that are not emitted may just move further inside the metal. So far. we have-only mentioned how an electron acquires energy but have not discussed the dynamics of electron emission. Let us now see how the photon theory leads to the experimental observation (*iii*).

When the radiation falls on a metal surface, the photon in the beam is completely absorbed by an electron in the emitter. The part of the energy of the absorbed photon is used by the electron lo become free itself from the metal, the balance energy will appear as kinetic energy of the outgoing electron. The kinetic energy of the photoelectron would be maximum for the least tightly bound electron. The energy required to make free the least tightly bound electron from the metal or the minimum energy required to remove an electron from the metal or the energy that must be given to the most energetic electron to release it from the metal, as mentioned earlier, is called work function of the metal and is denoted by $e\phi$. Thus

$$h\upsilon = T_{max} + e\phi$$

or $$T_{max} = h\upsilon - e\phi \quad \text{... (A 2.1)}$$

where $h\upsilon$ is the photon energy and T_{max} is the maximum photoelectron energy.

The above equation is the Einstein photoelectric equation for which he received (not for his theory of relativity), the 1921 Nobel Prize in Physics.

As explained above the maximum photoelectron energy

$$T_{max} = eV_0 \quad \text{... (A 2.2)}$$

$$e\phi = h\upsilon_0 \quad \text{... (A 2.3)}$$

where υ_0 is called the critical (cut off or threshold frequency) below which no photoelectrons are emitted. The greater the work function of a metal, the more energy is needed to leave its surface and higher the critical frequency for photoelectron emission to occur

Substitution of eqns. (*A* 2.2) and (*A* 2.3) into eqn. (*A* 2.1) yields

$$T_{max} = eV_0 = h\upsilon - e\phi \quad \text{... (A 2.4a)}$$

and $$T_{max} = eV_0 = h\upsilon\,(v - v_0) \quad \text{... (A 2.4b)}$$

The above eqns. (*A* 2.4*a.b*) clearly agree well with the data of the Fig, A 2.3 since they predict that V_0 should be a linear function of υ and that the slope $dV_0/d\upsilon = h$ should be independent of the material of emitter.

The work function $e\phi$ is the intercept of the lines on eV_0 (*y*) axis in Fig. A 2.3 and is different for different emitter materials.

Example A 2.1

Find the energy of a 7000 Å photon.

Solution

The energy of the given photon

$$E - h\upsilon = hc/\lambda$$

$$\frac{6.626 \times 10^{-34}\ J.s}{1.6 \times 10^{-19}\ J/eV} \times \frac{3 \times 10^8\ m/s}{7000 \times 10^{-10} m} = 1.77 \text{ eV}$$

Example A 2.2

A 1000 W radio transmitter operates at a frequency of 880 kHz. How many photons per second does it emit?

Solution

The energy of photon

$$E = h\upsilon = (6.626 \times 10^{-34}\ J.s) \times (880 \times 10^{3}\ s^{-1})$$

$$= 5.83 \times 10^{-28}\ J$$

Therefore, the number of photons emitted per second

$$n_p = \frac{P}{E}$$

where P is the power of the transmitter. Thus

$$n_p = \frac{1000\ J/s}{5.83 \times 10^{-28}\ J/\text{photon}} = 1.7 \times 10^{30}\ \text{photons/s}$$

Example A 2.3

The threshold wavelength for photoelectric emission in tungsten is 230 nm. What wavelength of light must be used in order for electrons with a maximum energy of 1.5 eV to be ejected?

Solution

$$\upsilon_0 = c/\lambda_0$$

where λ_0 is the threshold wavelength. Therefore,

$$\upsilon_0 = \frac{3 \times 10^{8}\ m/s}{230 \times 10^{-9}\ m} = 1.3 \times 10^{15}\ \text{Hz}$$

Using eqn. (*A* 2.4*b*), one gets

$$h\upsilon = \text{T}_{\text{max}} + h\upsilon_0$$

and hence

$$hc/\lambda = \frac{(6.626 \times 10^{-34}\ J.\ s) \times (3 \times 10^{8}\ m/s)}{(1.6 \times 10^{19}\ J/eV)\ \lambda}$$

$$= 1.5\ eV + \frac{(6.626 \times 10^{34}\ J.s) \times (1.3 \times 10^{15}\ s^{-1})}{(1.6 \times 10^{-19}\ J/eV)}$$

or $$(1.24 \times 10^{-6}\ eV.m)/\lambda = 1.5\ eV + 5.38\ eV$$

or $$\lambda = 180\ \text{nm}$$

Example A 2.4

An ultraviolet light of wavelength 2000 Å is incident on a Nickel emitter having work function 5 eV. Find the potential difference to be applied to stop the fastest photoelectrons emitted by the emitter.

Solution

The energy of the photon.

$$E = h\upsilon = hc/\lambda = \frac{(6.626 \times 10^{-34}\ J.s) \times (3 \times 10^8\ m/s)}{(1.6 \times 10^{-19}\ J/eV) \times (2000 \times 10^{-10}\ m)}$$

$$= 6.21\ eV$$

From eqn. (*A* 2.4*a*)

$$eV_0 = h\upsilon - e\phi,$$

we get

$$eV_0 = 6.21\ eV - 5\ eV$$

Therefore, the stopping potential

$$V_0 = 1.21 \text{ Volts}$$

Example A 2.5

1.3 mW of 400 nm light is directed at a photoelectric cell. If 0.1 per cent of the incident photons produce photoelectron, find the current in the cell.

Solution

The energy of the photon,

$$E = h\upsilon = hc/\lambda = \frac{(6.626 \times 10^{-34}\ J.s) \times (3 \times 10^8\ m/s)}{400 \times 10^{-9}\ m}$$

$$= 4.97 \times 10^{-19}\ \text{J}$$

Therefore, the number of photons that reach the cell per second is

$$n_p = P/E$$

where P is the power of the light source thus

$$n_p = \frac{1.5 \times 10^{-3}\ J/s}{4.97 \times 10^{-19}\ J/\text{photon}} = 3.02 \times 10^{15} \text{ photons/s}$$

The rate at which the photoelectrons are emitted is

$$n_e = (0.001)\ n_p$$

$$= 0.001 \times 3.02 \times 10^{15} \text{ photoelectrons/s}$$

Therefore the current in the photocell is

$$i_p = en_e$$

where e is the charge on the electron. Thus

$$i_p = 1.6 \times 10^{-19} \text{ coulomb} \times 0.001 \times 3.02 \times 10^{15} \text{ photoelectrons/s}$$

$$= 0.48\ \mu\text{A}$$

Appendix 2

APPENDIX 3

COMPTON EFFECT

Compton in 1923 observed that when X-rays are scattered by loosely bound electrons in atom the wavelength of the scattered radiation exceeds that of the incident radiation. This scattering with shift in wavelength is called the Compton effect.

The experimental set up for studying Compton effect is illustrated in Fig. A 3.1. A beam of X-rays of single known wavelength is allowed to fall on a target through collimators. The wavelengths of the scattered X-rays are determined at various values of the scattering angle θ. The results shown in Fig. A 3.2 exhibit the wavelength shift. At each angle the scattered X-rays also include many that have the initial wavelength. This fact would lie explained at the end.

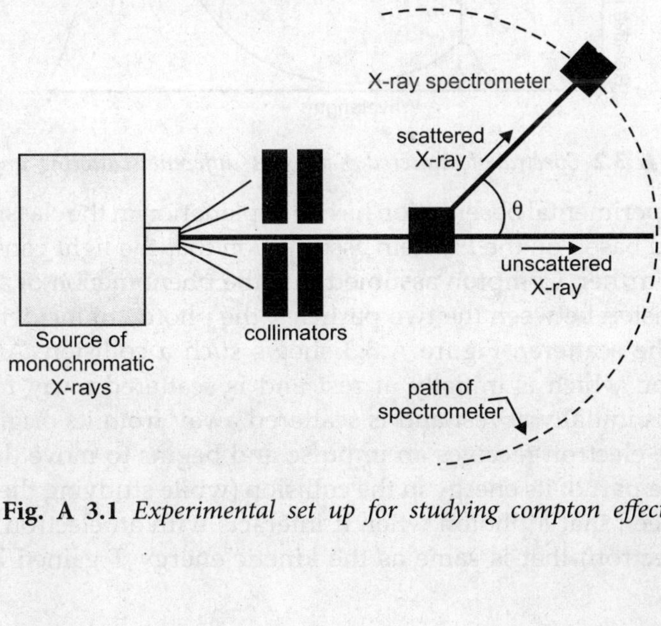

Fig. A 3.1 *Experimental set up for studying compton effect*

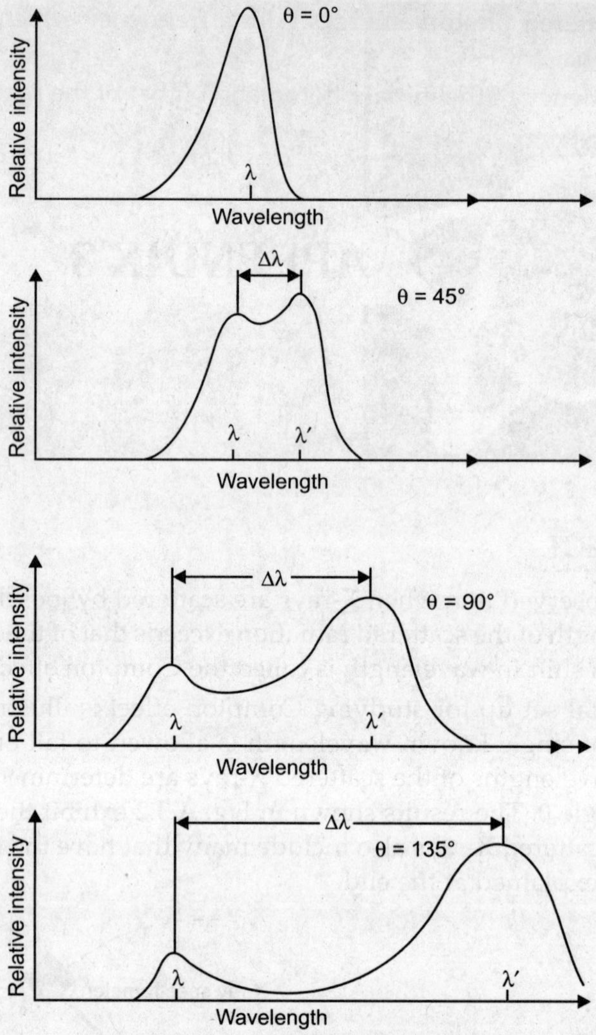

Fig. A 3.2 *Spectra of scattered X-rays at different scattering angles*

The above experimental observation has no explanation in the classical physics. The same is explained based on the Einstein assumption that the light consists of particles called photons. Further, Compton assumed that the phenomenon of scattering is due to the elastic collision between the two particles, the photon of incident radiation and the electron of the scatterer. Figure A 3.3 shows such a collision. An *X*-ray photon strikes an electron which is initially at rest and is scattered away from its original direction which is initially at rest and is scattered away from its original direction of motion while the electron receives an impulse and begins to move. In this situation, the photon loses a part of its energy in the collision (while studying the photoelectrons effect, we have seen that a photon when it interacts with an electron given its entire energy to the electron) that is same as the kinetic energy T gained by the electron.

Therefore, the scattered photon emerges with a frequency which is less than that of the incident photon.

If υ is the frequency of the initial photon and υ' that of the scattered photon, then

Loss in photon energy = gain in kinetic energy of the electron

or $$h\upsilon - h\upsilon' = T \qquad \text{... (A 3.1)}$$

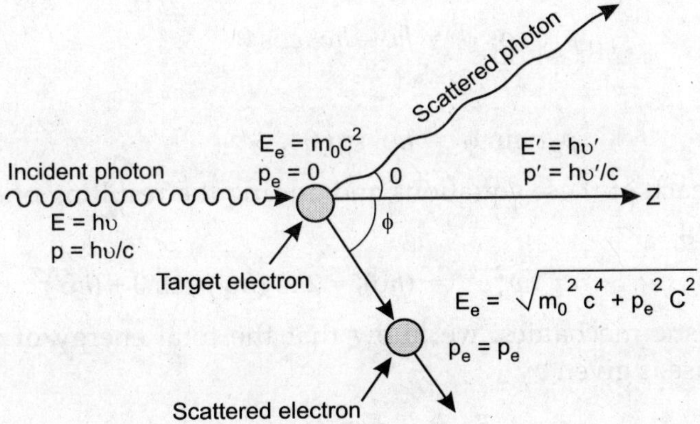

Fig. A 3.3 *The scattering of a photon by an electron, called the Compton scattering*

Before we proceed further, let us first find an expression for the momentum of a photon. From relativistic mechanics we know that the energy of a particle is given by

$$E = \frac{m_0 c^2}{\left(1 - \frac{\upsilon^2}{c^2}\right)^2} \qquad \text{... (A 3.2)}$$

where m_0 is the rest mass of the particle.

From the above expression, if a particle has to have a finite energy and if it moves with the velocity of light, its rest mass must be zero. Thus m_0 must be zero for photon as it moves with the velocity of light. Further from the relativistic energy expression

$$E = (p^2c^2 + m_0^2c^4)^{\frac{1}{2}} \qquad \text{... (A 3.3)}$$

the photon momentum with $m_0 = 0$ and $E = h\upsilon$ would be given by

$$p = E/c = h\upsilon/c \qquad \text{... (A 3.4)}$$

From the law of conservation of momentum along the z axis, that is in the original photon direction.

Initial momentum = Final momentum

$$\frac{h\upsilon}{c} + 0 = \frac{h\upsilon'}{c}\cos\theta + p_e\cos\phi \qquad \text{(A 3.5)}$$

and along the direction perpendicular to the z-axis

Initial momentum = final momentum

$$0 = \frac{h\upsilon'}{c}\sin\theta - p_e \sin\phi \qquad (A\ 3.6)$$

where angle θ is that between the directions of the initial and scattered photons, called angle of scattering and ϕ is the between the directions of the initial photon and the recoil electron.

Multiplying eqns. (*A* 3.5) and (*A* 3.6) by c, we can rewrite them as

$$p_e c \cos\phi = h\upsilon - h\upsilon' \cos\theta$$

and

$$p_e c \sin\phi = h\upsilon' \sin\theta$$

By squaring each of these equations and adding the resulting ones, the angle ϕ is eliminated leaving

$$p_e^2 c^2 = (h\upsilon)^2 = 2h\upsilon\,(h\upsilon')\cos\theta + (h\upsilon')^2 \qquad \text{... } (A\ 3.7)$$

From relativistic mechanics, we know that the total energy of the particle, the electron in this case is given by

$$E_e = T + m_0 c^2 \qquad \text{... } (A\ 3.8)$$

and also by

$$E_e = (p_e^2 c^2 + m_0^2 c^4)^{1/2} \qquad \text{... } (A\ 3.9)$$

Equating the two expressions for the total energy of the particle, one gets

$$(T + m_0 c^2)^2 = p_e^2 + m_0^2 c^4$$

or

$$p_e^2 c^2 = T^2 + 2m_0\, c^2 T$$

Substituting the value of T from eqn. (*A* 3.1) into above, one gets

$$p_e^2 c^2 = (h\upsilon)^2 - 2(h\upsilon)\,(h\upsilon') + (h\upsilon')^2 + 2m_0 c^2\,(h\upsilon - h\upsilon') \qquad \text{... } (A\ 3.10)$$

Equating the two expressions (*A* 3.7) and (*A* 3.10) for $p_e^2 c^2$, one obtains

$$2m_0 c^2\,(h\upsilon - h\upsilon') = 2\,(h\upsilon)\,(h\upsilon')\,(1 - \cos\theta)$$

Dividing the above equation by $2h^2c^2$, one gets

$$\frac{m_0 c}{h}\left(\frac{\upsilon}{c} - \frac{\upsilon'}{c}\right) = \frac{\upsilon}{c} \times \frac{\upsilon'}{c} \times (1 - \cos\theta)$$

Since $\upsilon/c = 1/\lambda$ and $\upsilon'/c = 1/\lambda'$, therefore

$$\frac{m_0 c}{h}\left(\frac{1}{\lambda} - \frac{1}{\lambda'}\right) = \frac{1 - \cos\theta}{\lambda\lambda'}$$

or

$$\lambda' - \lambda = \Delta\lambda = \frac{h}{m_0 c}(1 - \cos\theta) \qquad \text{... } (A\ 3.11)$$

which gives the change or shift in wavelength, called Compton shift expected for a photon that is scattered through an angle θ by a particle of rest mass m_0. Note that this change is independent of the wavelength λ of the incident photon. Equation (*A* 3.11) gives excellent agreement with the experimental observations and therefore provides another evidence supporting the particle nature of the radiation, The quantity

$$\lambda_C = \frac{h}{m_0 c} \quad \text{... (A 3.12)}$$

is called the Compton wavelength of the scattering particle. For an electron $\lambda_C = 0.024$ Å. In terms of λ_c eqn. (*A* 3.11) becomes

$$\lambda' - \lambda = \Delta\lambda = \lambda_C (1 - \cos\theta) \quad \text{... (A 3.13)}$$

From above, we infer that the greatest wavelength change corresponds to the scattering angle θ = 180° and is equal to $2\lambda_C$ = 0.048 Å. That is it is the situation in which the collision between photon and electron is head on and the incident photon is reversed in direction. Change of this magnitude or less wavelength shift are easily observable only in X-rays. In the case of visible light where λ is very large compared to Δλ, λ' = λ + Δλ = λ. Therefore the shift is insignificant. Whereas in the case of X-rays, Δλ is comparable to λ. and hence the shift is significant and measurable.

As pointed out earlier, the results expressed in Fig. A 3.2 show that at each angle of scattering the scattered X-rays also include many that have the initial wavelength. If can be understood easily, in deriving eqn. (*A* 3.11) it was assumed that the scattering particle is able to move freely, which, is reasonable since many of the electrons in matter are only loosely bound to their parent atoms. Other electrons, however, are very tightly bound and when stuck by photon, the entire atom recoils instead of the single electron. In this situation, the value of m_0 to be used in eqn. (*A* 3.11) is that of the entire atom which is several thousands of times greater than that of an electron and the resulting Compton shift is accordingly too small to be detected. Further we know that the atoms with large atomic number Z have more tightly bound electrons, and therefore the Compton effect is easily observed with atoms of small Z. A carbon (graphite) block was, therefore used in the experiment described here.

Example A 3.1

A beam of X-rays is scattered by a target. The scattered X-rays at 45° from the beam direction have a wavelength 0.022 Å. What is the wavelength of the X-rays in the direct beam?

Solution

From eqn. (*A* 3.13)

$$\lambda = \lambda' - \lambda_C (1 - \cos\theta),$$

and hence

$$\lambda = 0.022 \text{ Å} - 0.293\,\lambda_C$$

where $\lambda_C = 0.024$ Å

Therefore $\lambda = 0.022 \text{ Å} - 0.007 \text{ Å} = 0.015 \text{ Å}$

Example A 3.2

An X-ray photon of initial frequency 3×10^{19} Hz collides with an electron at rest and is scattered through 90°. Find its new frequency.

Solution

Initial wavelength of the X-ray photon

$$\lambda = c/\upsilon = \frac{3 \times 10^8 \text{ ms}^{-1}}{3 \times 10^{10} \text{ s}^{-1}} = 0.1 \text{ Å}$$

From eqn. (*A* 3.13), the new wavelength of the scattered photon at 90°.

$$\lambda' = \lambda + \lambda_C (1 - \cos\theta)$$

$$= 0.10 \text{ Å} + 0.024 \text{ Å}$$

$$= 0.124 \text{ Å}$$

Therefore, the new frequency of the scattered photon.

$$\upsilon' = c/\lambda' = \frac{3 \times 10^8 \text{ m/s}}{0.124 \times 10^{10} \text{ m}} = 2.42 \times 10^{19} \text{ Hz}$$

Example A 3.3

At what scattering angle will incident 100-keV X-rays leave a target with an energy of 90 keV?

Solution

Wavelength of the incident X-rays.

$$\lambda = c/\upsilon = \frac{hc}{h\upsilon} = \frac{hc}{E} = \frac{(6.626 \times 10^{-34} \textit{ J.s})(3 \times 10^8 \textit{ ms}^{-1})}{100 \times 10^3 \textit{ eV} \times 1.6 \times 10^{-19} \textit{ J/eV}}$$

$$= 0.124 \text{ Å}$$

Similarly the wavelength of the scattered photon

$$\lambda' = c/\upsilon' = \frac{hc}{h\upsilon'} = \frac{hc}{E'} = \frac{(6.626 \times 10^{-34} \textit{ J.s})(3 \times 10^8 \textit{ m/s})}{900 \times 10^3 \textit{ eV} \times 1.6 \times 10^{-19} \textit{ J/eV}}$$

$$= 0.138 \text{ Å}$$

From eqn. (*A* 3.13) the scattering angle θ would be given by

$$\lambda' - \lambda = \lambda_C (1 - \cos\theta)$$

or $$1 - \cos\theta = 0.014/0.024 = 0.583$$

or $$\cos\theta = 0.417$$

Thus, $$\theta = 65.5°$$

Example A 3.4

X-rays of wavelength 0.1 Å are scattered from a target. Find maximum wavelength present in the scattered X-rays. Also find maximum kinetic energy of the recoil electrons.

Solution

From eqn. (A 3.13), the wavelength of the scattered X-rays is given by

$$\lambda' = \lambda + \lambda_C (1 - \cos\theta)$$

It would be maximum where $(1 - \cos\theta) = (1 - \cos 180°) = 2$ and hence

$$\lambda' = \lambda + 2\lambda_C = 0.1 \text{ Å} + 2 \times 0.024 \text{ Å}$$

$$= 0.148 \text{ Å}$$

The maximum recoil kinetic energy of the electron, in this case would be

$$T_{max} = h(\upsilon - \upsilon')$$

as v' corresponds to the minimum frequency, Therefore.

$$T_{max} = hc\left(\frac{1}{\lambda} - \frac{1}{\lambda'}\right)$$

$$= \frac{(6.626 \times 10^{-34} \ J.s) \times (3 \times 10^{8} \text{ ms}^{-1})}{(1.6 \times 10^{-19} \ J/eV)}$$

$$\left(\frac{1}{0.1 \times 10^{-10} \ m} - \frac{1}{0.148 \times 10^{-10} \ m}\right)$$

$$= 40.3 \text{ keV}$$

APPENDIX 4

SPECIFIC HEAT OF SOLIDS

The specific heat per kmol of a solid is defined as

$$C = \frac{dQ}{dT}$$

where dQ is the heat required to raise the temperature of 1 kmol of a solid by an amount equal to dT. If the process is carried out at constant volume, then the term PdV would vanish and as a result from the Ist law of thermodynamics.

$$dQ = dE$$

where dE is the increase in the internal energy of the system. Therefore, the specific heat per kmol called the molar specific heat of a solid at constant volume, C_V, is given by

$$C_V = \left(\frac{\partial E}{\partial T}\right)_V \qquad \text{... (A 4.1)}$$

This is the energy that must be added to 1 kmol of the solid whose volume is held fixed to raise its temperature by 1 K. Note that the specific heat at constant pressure C_P would be little higher than C_V in solids as it includes the work associated with a volume change as well as the change in internal energy.

The internal energy of a solid resides in the vibrations of its constituent particles, which may be atoms, ions or molecules. We shall refer to them as atoms here for the sake of convenience Further, the specific heat is composed of two parts, namely, me specific heat of vibration and the electronic specific beat (This contribution comes in the ease of metals because of the presence of free electrons in them). We would, however, restrict our discussion here to the specific heat of vibration.

Figure A 4.1 shows the variation of the molar specific heat at constant volume C_V (actually specific heat of vibration here) of solids as function of temperature. At temperatures about 200 K and above (which if course includes room temperature), the value of C_V is close to $3R$ for all solids. Where R is the universal gas constant. Since $R = N_A\, k_B = 8.31 \times 10^3$ J/(kmol.K) = 1.99 kcal/ (kmol.K), therefore $C_V = 3R = 5.97$ kcal/ (kmal.K) at high temperatures. N_A is Avogadro's number and k_B is Boltzmann constant. The fact that C_V is nearly equal to 3R for most of the solids at room temperature and above, is known as the Dulong-Petit law in their honor.

The deviation from this law in low temperature regions is evident from the figure. As T decreases, C_V, at a temperature lower than 200 K drops sharply and vanishes entirely at absolute zero. Further, near absolute zero (~ 20K) C_V is proportional to T^3.

Let us now evaluate C_V theoretically and compare the value so obtained with the experimental result. First let us do so using the classical physics: The model used to describe the solid is one in which each atom is bound to its site by a harmonic force. When the solid is heated, the atoms vibrate around their sites. These vibrations may be resolved into components along three perpendicular axes, so that we may represent each atom by three harmonic oscillators. From classical physics, we know that average energy associated with a one-dimensional oscillator in a system of them in thermal equilibrium at temperature T is equal to k_BT (1/2 k_BT corresponding to kinetic energy and 1/2 k_BT corresponding to the potential energy). That is,

$$\overline{\epsilon} = k_BT \quad \text{... } (A\ 4.2)$$

Therefore, the average energy per atom in a solid on the basis of the above discussion would be $3k_BT$. A kmol of solid contains Avogadro's number N_A of atoms, and its total internal energy E at temperature T, accordingly, ought to be

$$E = 3\, N_A\, k_BT = 3RT \quad \text{... } (A\ 4.3)$$

Substituting eqn. (A 4.3) into eqn. (A 4.1), one gets

$$C_V = 3R \quad \text{... } (A\ 4.4)$$

This result is certainly in agreement with experiment at high temperatures (200 K and above) but it fails completely at low temperatures. This clearly shows that there is something wrong with the analysis leading to eqn. (A 4.4).

EINSTEIN MODEL

Einstein in 1907 realized that the basic flaw in the derivation of eqn. (A 4.4) lies in the use of the value k_BT for the average energy per oscillator in solid. This flaw is the same as that responsible for the incorrect Rayleigh-Jeans formula for black body radiation. Einstein in his model treated the atoms as independent oscillators and assumed that the energy of the oscillator is given by quantum mechanics rather than by the classical result (A 4.2). According to quantum mechanics, the energy †† of an isolated oscillator is restricted to the values

††Here we are using the exact quantum mechanical expression for energy of an oscillator instead of expression $\epsilon_n = nh\nu$ used by Einstein. This however, would not make any difference as we would see so far the computation of the specific heat is concerned.

$$\in_n = (n + 1/2)h\upsilon, n = 0, 1, 2, ... \qquad ... (A\ 4.5)$$

As we know, the atomic oscillators in a solid are not isolated, instead they are continually exchanging energy with the thermal bath surrounding the solid. As a result, the energy of the oscillator is continually changing and its average value in thermal equilibrium at temperature T is given by

$$\overline{\in_n} = \in/N = \sum_{n=0}^{\infty} \in_n e^{-\in_n/k_B T} / \sum_{n=0}^{\infty} e^{-\in_n/k_B T}$$

where N is the total number of oscillators and $\in$ their total energy. The exponential $e^{\in_n/k_B T}$ is the well known Boltzmann factor which gives the probability that the energy state $\in_n$ is occupied. Substituting the value of $\in_n$ from eqn. (A 4.5) into the above equation, one gets

$$\in = \sum_{n=0}^{\infty}\left(n+\frac{1}{2}\right)h\upsilon e^{-\left(n+\frac{1}{2}\right)h\upsilon/k_B T} \sum_{n=0}^{\infty} e^{-\left(n+\frac{1}{2}\right)h\upsilon/k_B T}$$

$$= \frac{1}{2}h\upsilon + \frac{\sum_{n=0}^{\infty} nh\upsilon\, e^{-nh\upsilon/k_B T}}{\sum_{n=0}^{\infty} e^{-nh\upsilon/k_B T}}$$

$$= \frac{1}{2}h\upsilon + k_B T^2 \frac{\partial}{\partial T}\left[In \sum_{n=0}^{\infty} nh\upsilon\, e^{-nh\upsilon/k_B T}\right]$$

But

$$\sum_{n=0}^{\infty} e^{-nh\upsilon/k_B T} = 1 + e^{-h\upsilon/k_B T} + e^{-2h\upsilon/k_B T} + ...$$

$$= \frac{1}{1-e^{-h\upsilon/k_B T}}$$ as this is an infinite geometric

series in which the ratio between the successive terms is $e^{-h\upsilon/k_B T}$ so that

$$\overline{\in} = \frac{1}{2}h\upsilon + k_B T^2 \frac{\partial}{\partial T} \text{In} \frac{1}{1-e^{-h\upsilon/k_B T}}$$

$$= \frac{1}{2}h\upsilon + k_B T^2 \partial/\partial T \text{ In} (1-e^{-h\upsilon/k_B T})^{-1}$$

$$= \frac{1}{2}h\upsilon - k_B T^2 \partial/\partial T \text{ In} (1-e^{-h\upsilon/k_B T})$$

$$= \frac{1}{2}h\upsilon + \frac{h\upsilon}{e^{h\upsilon/k_B T}-1} \qquad ... (A\ 4.6)$$

From now onwards we would drop out the term 1/2 $h\upsilon$ called the zero point energy since this after differentiation with temperature for finding specific heat gives zero. Therefore.

$$\overline{\in} = \frac{h\upsilon}{e^{h\upsilon/k_B T}-1} \qquad ... (A\ 4.7)$$

The total energy of a kmol of solid, therefore becomes

$$E = 3N_A \in = 3N_A \frac{h\upsilon_E}{e^{h\upsilon_E/k_BT}-1} \quad ...(A\ 4.8)$$

where we have used the term υ_E for υ which denotes the common frequency of all the oscillators, called Einstein frequency and hence its molar specific heat

$$C_V = \left(\frac{\partial E}{\partial T}\right)_V = 3N_A h\upsilon_E (-1)(e^{h\upsilon_E/k_BT}-1)^2$$

$$\times \frac{h\upsilon_E}{k_B}\left(-\frac{1}{T^2}\right)e^{h\upsilon_E/k_BT}$$

$$= 3R\left(\frac{h\upsilon_E}{k_BT}\right)^2 e^{h\upsilon_E/k_BT}/(e^{h\upsilon_E/k_BT}-1)^2 \quad ...(A\ 4.9)$$

This is Einstein specific heat formula. Introducing the Einstein temperature θ_E, where $k_B\theta_E = h\upsilon_E$ (θ_E would be different for different substances as υ_E is) the above equation can be written as

$$C_V = 3R(\theta_E/T)^2 e^{\theta E/T}(e^{\theta E/T}-1)^2$$

$$= 3RE(\theta_E/T)$$

$$= 3RE(x) \quad ...(A\ 4.10a)$$

where

$$E(x) = x^2e^x/(e^x-1)^2 \quad ...(A\ 4.10b)$$

which is known as the Einstein function.

It we plot C_V as a function of temperature using the above equation, we obtain a curve of the same general shape as depicted by Fig. A 4.1. This simply indicates flat the theory is in agreement with experiment, at least qualitatively over the entire temperature range. Further let us examine the behaviour of C_V as given by eqn. (A 4.10) in extreme temperature limits.

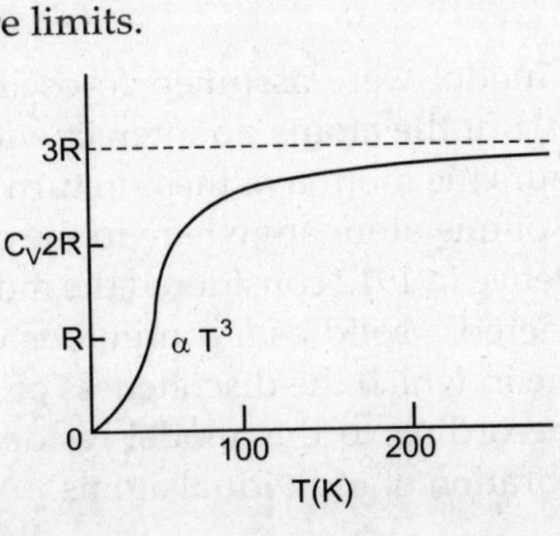

Fig. A 4.1 *Molar specific heat at constant volume C_V (actually specific heat of vibration here) of solids as function of temperature*

In the high temperature lim, where $T >> \theta_E$ or $x << 1$, one may expand the exponential e^x in the power series of x and retain only the largest terms. Doing so one finds that

$$E(x) = x^2 \frac{1+x}{(1+x-1)^2}$$

Further neglecting x in comparison to 1 in the numerator and on simplification, one gets.

$$E(x) - 1$$

which results in

$$C_V = 3R, \text{ which matches with the classical result.}$$

In the low temperature limit, in which $T << \theta_E$ or $x >> 1$, the exponent e^x in eqn. (*A* 4.10*b*) is much greater than unity. Therefore, neglecting 1 in comparison to e^x in the denominator, one gets

$$E(x) - x^2 e^{-x}$$

which result in

$$C_V - 3R\,(\theta_E/T)^2\, e^{-\theta_E/T} \qquad \text{... (A 4.11)}$$

Because of the exponential $e^{-\theta_E/T}$, the specific heat approaches zero very rapidly, exponentially, in fact and vanishes at $T = 0K$. Although the fact that $C_V \to 0$ as $T \to 0K$ agrees with the experiment but the manner in which C_V approaches zero does not. Equation (*A* 4.11) indicates that C_V approaches zero exponentially, while experiments show that C_V approaches zero as T^3, in other words at very low temperatures, Einstein model predicts a specific heat that is much smaller than the observed value This is the basic weakness of the Einstein model (Though it is in good agreement with experiment: over most of the temperature range). This disagreement is removed by the Debye Model, to which we turn now.

DEBYE MODEL

The atoms in the Einstein model were assumed to oscillate independently of each other at a fixed frequency v. But the atoms do interact with each other, the motion of one atom affects its neighbours the motion of these in turn affect their neighbours, and so forth, so that the motion of one atom anywhere in the solid in fact, affects all other atoms present. Therefore, Debye in 1912 considered the motion of the lattice as a whole or in other words he considered a solid as a continuous elastic body. That is, Debye went to the opposite extreme in which the discreteness of the lattice was ignored. The internal energy of a solid according to this model resides in elastic standing waves instead of residing in the vibration of individual atoms.

In analogous to the standing waves in the frequency interval between υ and $\upsilon + d\upsilon$ in the cavity in the case of black body radiation (see eqn. *A* 1.2). the number of standing waves or possible modes of vibration here would be given by

$$g(\upsilon)d\upsilon = 4\pi V \upsilon^2 \, d\upsilon/v^3 \quad \text{... (A 4.12)}$$

where V is the volume of the solid and v is the isotropic velocity of propagation of the elastic waves.

The above result would have been the final one had the waves were longitudinal but here we are concerned with elastic waves in solid. These are of two kinds, longitudinal and transverse, and range in frequency from zero to a maximum υ_m (the inter atomic spacing in a solid sets a lower limit to the possible wavelengths and hence an upper limit to the frequencies). Further, we should keep in mind that for each frequency, there are two transverse modes and one longitudinal mode. Therefore, instead of (*A* 4.12), one has

$$g(\upsilon)\, d\upsilon = 4\pi V\left(\frac{1}{v_l^3} + \frac{2}{v_t^3}\right)\upsilon^2 d\upsilon \quad \text{... (A 4.13)}$$

where v_l and v_t denote the velocities of propagation of the longitudinal and transverse waves, respectively.

Debye assumed that these waves, like em waves have energies quantized in units of hv. A quantum of acoustic energy in solid is called a phonon and it travels with the speed of sound since sound waves are clastic in nature. He finally asserted that a phonon gas has the same statistical behaviour as a photon (a quantum of em energy) gas or a system of harmonic oscillators in thermal equilibrium so that average energy $\overline{\in}$ per standing wave is (see eqn. A 1.5) given by

$$\overline{\in} = \frac{h\upsilon}{e^{h\upsilon/k_B T} - 1} \quad \text{... (A 4.14)}$$

The total energy of vibration for the lattice (or the total internal energy of the solid) would, therefore be given by

$$E = \int_0^{\upsilon_m} \overline{\in}(\upsilon) g(\upsilon) d\upsilon \quad \text{... (A 4.15)}$$

Substituting the values of $g(\upsilon)\, d\upsilon$ and $\overline{\in}$ from eqns. (*A* 4.13) and (*A* 4.14) into above, one gets

$$E = 4\pi V\left(\frac{1}{v_l^3} + \frac{2}{v_t^3}\right)\int_0^{\upsilon_m} \frac{h\upsilon^3\, d\upsilon}{e^{h\upsilon/k_B T} - 1} \quad \text{... (A 4.16)}$$

Before we can evaluate the integral in the above expression, we need to know the value of υ_m, the maximum or upper cut-off frequency. This was determined by Debye, by requiring that the total number of modes (or oscillators) included must be equal to the number of degrees of freedom for the entire solid which is equal to $3N_A$ because each atom has three degrees of freedom (each atom is regarded as a three-dimensional oscillator). This leads to

$$\int_0^{\upsilon_D} g(\upsilon) d\upsilon = 3\, N_A$$

which on using eqn. (*A* 4.13) gives

$$4\pi V\left(\frac{1}{v_l^3}+\frac{2}{v_t^3}\right)\int_0^{\upsilon_D}\upsilon^2 d\upsilon = 3\,N_A$$

$$4\pi V\left(\frac{1}{v_l^3}+\frac{2}{v_t^3}\right)\upsilon_D^3/3 = 3\,N_A$$

Therefore,

$$\upsilon_D^3 = \frac{9N_A}{4\pi V}\left(\frac{1}{v_l^3}+\frac{1}{v_t^3}\right) \quad \text{... } (A\ 4.17)$$

where the cut off frequency υ_m from now onwards would be denoted by υ_D and is called Debye frequency. It may however, be pointed out that if the upper limit to frequency were allowed to go to infinity, the number of modes would become infinite and also the energy (see eqn. *A* 4.16) which would not make any sense physically.

Returning now to eqn. (*A* 4.16) and using eqn. (*A* 4.17), the total internal energy of the solid becomes.

$$E = \frac{9N_A}{\upsilon_D^3}\int_0^{\upsilon_D}\frac{h\upsilon^3\,d\upsilon}{e^{h\upsilon/k_BT}-1} \quad \text{... } (A\ 4.18)$$

and therefore, the molar specific heat C_V, which is found by differentiating the above equation with respect to temperature *T*, is

$$C_V = \frac{9N_A}{\upsilon_D^3}\int_0^{\upsilon_D} h\upsilon^3(-1)\frac{1}{(e^{h\upsilon/k_BT}-1)^2}\,e^{h\upsilon/k_BT}\,(h\upsilon/k_B)(-1/T^2)\,d\upsilon$$

$$= \frac{9N_A}{\upsilon_D^3}\times\frac{h^2}{k_BT^2}\int_0^{\upsilon_D}\frac{\upsilon^4\,e^{h\upsilon/k_BT}}{(e^{-h\upsilon/k_BT}-1)}\,d\upsilon \quad \text{... } (A\ 4.19)$$

One can simplify the appearance of the above equation by changing to dimensionless variable

$$x = h\upsilon/k_BT \text{ or } dx = h d\upsilon/k_BT$$

and by defining the Debye temperature θ_D as

$$k_B\theta_D = h\upsilon_D$$

or

$$\theta_D = h\upsilon_D/k_B \quad \text{... } (A\ 4.20)$$

which is different for different solids.

The upper limit of integration is then equal to $x_m = h\upsilon_D/k_BT = \theta_D/T$. As a result eqn. (*A* 4.19) takes the form

$$C_V = 9R\,(T/\theta_D)^3\int_0^{\theta_D/T}\frac{x^4\,e^x}{(e^x-1)^2}\,dx$$

$$= 9RD\,(\theta_D/T) \quad \text{... } (A\ 4.21a)$$

where

$$D(\theta_D/T) = (T/\theta_D)^3 \int_0^{\theta_D/T} \frac{x^4 e^x}{(e^x - 1)^2} dx \qquad \text{... (A 4.21b)}$$

is known as the Debye function.

The expression (*A* 4.21) is the Debye formula for the specific heat of solids. The integral appearing in the expression can be evaluated numerically and then one can plot C_x as a function of the reduced temperature T/θ_D. The general agreement between theory and experimental results is found to be quite satisfactory at all temperatures.

Let us now consider the limiting cases

(*i*) At high temperatures, $T >> \theta_D$, therefore the upper limit of the integral in eqn. (*A* 4.21) is very small. Thus x in the integrand is small over the entire range of integration and hence one can make the approximation, $e^x - 1 + x$ and also neglect x in comparison to 1 in the numerator. Equation (*A* 4.21) then results in

$$C_V \simeq 9\,R\,(T/\theta_D)^3 \int_0^{\theta_D/T} \frac{x^4 \times 1}{(1 + x - 1)^2} dx$$

$$\simeq 9\,R\,(T/\theta_D)^3 \int_0^{\theta_D/T} x^2\,dx$$

$$\simeq 3\,R \qquad \text{... (A 4.22)}$$

which is in agreement with the Dulong-Petit law.

(*ii*) At low temperatures, $T << \theta_D$ and hence the upper limit of the integral in equation (*A* 4.21) approaches infinity. The integral has the standard value $4\pi^4/15$ which leads to

$$C_V \simeq 9\,R\,(T/(\theta)_D)^3 \times 4\pi^4/15$$

$$\simeq \frac{12}{5}\pi^4 R\,(T/(\theta)_D)^3 \qquad \text{... (A 4.23)}$$

which is the famous Debye T^3 law and is the result we have sought.

Despite its impressive success, the Debye model also remains only an approximation. The nature of approximation lies in assuming the solid as an elastic continuum or in assuming the continuum dispersion relation (range of frequencies from zero corresponding to the infinite wavelength to maximum allowed frequency) to hold true for all possible modes of excitation, in order to improve upon the Debye model, one needs to consider the correct dispersion relation and the corresponding $g(\upsilon)$. The agreement with the experimental results, obtained in this manner is decidedly better than that given by the Debye model, particularly in the intermediate temperature region.

No doubt the result: (*A* 4.21) is not the final one as pointed out above but it is sufficient for the purpose of displaying the fact that the classical physics was inadequate For explaining the experimental observations and there was need of the quantum ideas.

APPENDIX 5

THE DOUBLE WELL POTENTIAL (MODEL OF AMMONIA MOLECULE NH_3)

The Ammonia molecule NH_3 has three atoms at the corners of a pyramid with N atom at its apex (see Fig. A 5.1). The heavier N atom is regarded as fixed and the lighter H atoms form a rigid equilateral triangle with its axis passing through N. The potential energy, therefore depends on x only, the distance between the N atom and the plane containing three H atoms. The general shape of the potential energy function $V(x)$ versus x is as shown in Fig. A 5.2. The two minima in $V(x)$ represent the two symmetrical configurations (1st with the N atom above the plane of H atoms and the 2nd with the N atom below the plane of H atoms – see Fig. A 5.1) of the molecule in which it is stable. The finite potential barrier at $x = 0$ corresponds to the repulsion of N by H atoms when it is in their plane and the large value of $V(x)$ corresponds to the chemical binding forces.

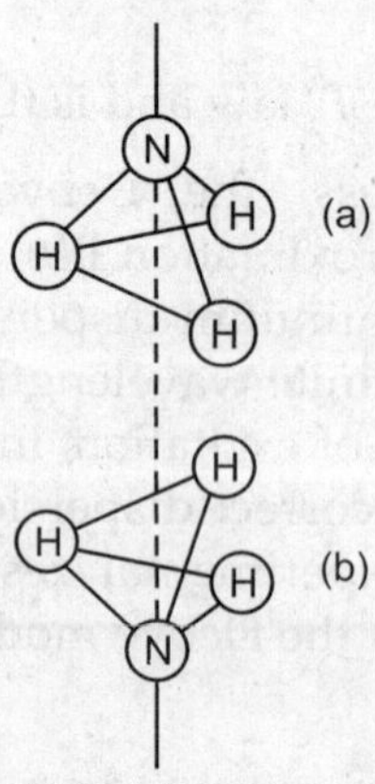

Fig. A 5.1 *The two configurations (states) of Ammonia molecule NH_3 (a) with N atom above the plane of H atom (b) with N atom below the plane of H atoms*

From Fig. A 5.2, one infers that the problem can be simplified or $V(x)$ may be approximated as equivalent to double well potential (see Fig. A 5.3) characterized by

$$V(x) = \begin{cases} V_0 \text{ for } 0 < |x| < b \\ 0 \text{ for } b < |x| < b + a \\ \infty \text{ for } |x| > b + a \end{cases} \quad \text{... (A 5.1)}$$

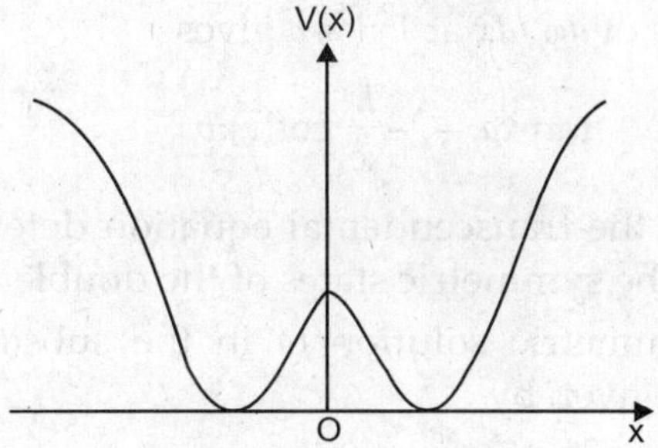

Fig. A 5.2 *$V(x)$ versus x for the Ammonia molecule NH_3. The two minima in $V(x)$ represent the two symmetrical configurations of the molecule in which it is stable*

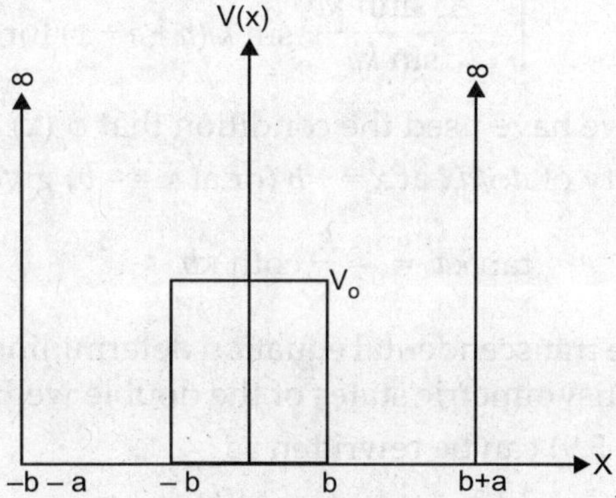

Fig. A 5.3 *The double well potential*

Since $V(-x) = V(x)$, the solutions in the present case would be either symmetric or antisymmetric functions of x. The Schrödinger equation for $E < V_0$ has forms

$$d^2\varphi/dx^2 - \kappa^2 \varphi(x) = 0 \text{ for } 0 < |x| < b \quad \text{... (A 5.2)}$$

and

$$d^2\varphi/dx^2 + k^2 \varphi(x) = 0 \text{ for } 0 < |x| < b + a \quad \text{... (A 5.3)}$$

where

$$\kappa^2 = 2m/\hbar^2 (V_0 - E) \quad \text{... (A 5.4)}$$

and

$$k^2 = 2m\, E/\hbar^2 \quad \text{... (A 5.5)}$$

The solution symmetric in x or symmetric solution (s in the subscript of φ corresponds to symmetric) would be of the form.

$$\varphi_s(x) = \begin{cases} A_s \sinh \kappa x \text{ for } 0 < |x| < b \\ \dfrac{A_s \sinh \kappa b}{\sin ka} \sin k(b + a - |x|) \text{ for } b < |x| < b + a \end{cases} \quad \text{... (A 5.6)}$$

where we have used the condition that $\varphi_s(x)$ is continuous at $|x| = b$. Also the fact that since $V(x) \to \infty$ at $|x| = b + a$, the wave function has to vanish at $|x| = b + a$.

Further the continuity of $d\varphi/dx$ at $|x| = b$ gives

$$\tan ka = -\frac{k}{\kappa} \coth \kappa b \quad \text{... (A 5.7)}$$

which is nothing but the transcendental equation determining the energy eigen values corresponding to the symmetric states of the double well.

Similarly, the antisymmetric solution (a in the subscript of φ corresponds to antisymmetric) would be given by

$$\varphi_a(x) = \begin{cases} A_a \sinh \kappa x \text{ for } 0 < |x| < b \\ \dfrac{A_a \sinh \kappa b}{\sin ka} \sin k(b + a - x) \text{ for } b < x < b + a \\ -\dfrac{A_a \sinh \kappa b}{\sin ka} \sin k(b + a + x) \text{ for } b < a < -b \end{cases} \quad \text{... (A 5.8)}$$

where, once again, we have used the condition that $\varphi_a(x)$ is continuous at $x = \pm b$.

Further the continuity of $d\varphi/dx$ at $x = +b$ (or at $x = -b$) gives

$$\tan ka = -\frac{k}{\kappa} \coth \kappa b \quad \text{... (A 5.9)}$$

which represents the transcendental equation determining the energy eigen values corresponding to the antisymmetric states of the double well.

Eqns. (A 5.7) and (A 5.9) can be rewritten as

$$-\xi \cot \xi = (\alpha^2 - \xi^2)^{1/2} \tanh (\alpha^2 - \xi^2)^{1/2}\, b/a \quad \text{(symmetric states)} \quad \text{... (A 5.10)}$$

and

$$-\xi \operatorname{cor} \xi = (\alpha^2 - \xi^2)^{1/2} \coth (\alpha^2 - \xi^2)^{1/2}\, b/a \quad \text{(antisymmetric states)} \quad \text{... (A 5.11)}$$

where

$$\xi = ka = (2m E a^2/\hbar^2)^{1/2}, \quad \alpha = (2m V_0 a^2/\hbar^2)^{1/2} \quad \text{... (A 5.12)}$$

For $b/a \to \infty$, that is, when the two wells are at large separation, eqns. (A 5.10) and (A 5.11) reduce to

$$-\xi \cot \xi = (\alpha^2 - \xi^2)^{1/2} \quad \text{... (A 5.13)}$$

that is, the energy levels are the same or we have degenerate states. When the two wells are brought together, the tunnel effect across the potential barrier (or coupling between the wells) removes the degeneracy and each level splits into two levels.

The above conclusion can be verified by the following example. Consider a case when $\alpha = 3\pi$ and $b/a = 0.2$. Solving eqns. (A 5.10) and (A 5.11) with these values leads to

$$\xi \approx \begin{cases} 2.82\,(s), 5.60\,(s), 8.24\,(s) \\ 2.85\,(a), 5.68\,(a), 8.48\,(a) \end{cases} \quad \text{... (A 5.14)}$$

where *s* and *a* correspond to symmetric and antisymmetric states.

As a result, the energy eigen values would be given by

$$\varepsilon \approx (\hbar^2/2m\,a^2)/E \times \begin{cases} 7.96\,(s), 31.36\,(s), 67.84\,(s) \\ 8.12\,(a), 32.29\,(a), 71.85\,(a) \end{cases} \quad \text{... (A 5.15)}$$

From the above result, one finds the splitting is given by

$$\Delta\varepsilon = (\hbar^2/2m\,a^2)/E \times \{0.16, 0.93, 4.0\} \quad \text{... (A 5.16)}$$

which indicated that because of the coupling between the two wells (or *b*/*a* being finite), each level splits into two; the splitting becoming more for higher energy states.

Figs. (A5.4) and (A5.5) represent the spatial variation of the first four wave functions. In Fig. (A 5.4), the solid and dashed curves correspond to the first energy eigen state ε_1^s (symmetric function of *x*) and the second energy eigen state ε_2^a (antisymmetric function of *x*), respectively. Similarly, in Fig. (A 5.5), the solid and dashed curves correspond to the third energy eigen state ε_3^s (symmetric function of *x*), and the fourth energy eigen state ε_4^a (antisymmetric function of *x*), respectively.

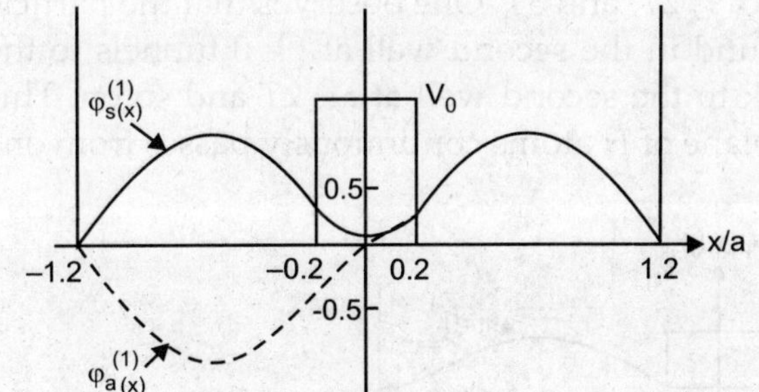

Fig. A 5.4 *The solid curve corresponds to ε_1^s and the dashed curve to ε_2^a. V_0 here amounts to $V_0/(\hbar^2/2ma^2)$*

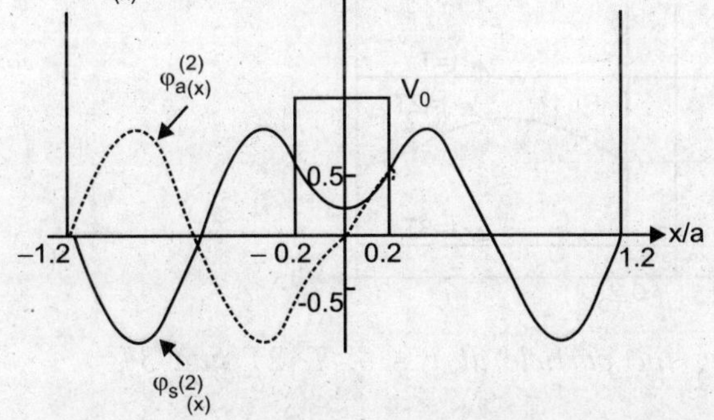

Fig. A 5.5 *The solid curve corresponds to ε_3^s and the dashed curve to ε_4^a. V_0 here amounts to $V_0/(\hbar^2/2ma^2)$*

Suppose at $t = 0$, the molecule is in the state

$$\Psi\,(x, t = 0) = \frac{1}{\sqrt{2}}\,[\varphi_s^{(1)}\,(x) + \varphi_a^{(1)}\,(x)] \quad \text{... (A 5.17)}$$

where $\varphi_s^{(1)}(x)$ and $\varphi_a^{(1)}(x)$ represent the lowest symmetric and the lowest antisymmetric wave functions, respectively (see Fig. A 5.4). The factor $1/\sqrt{2}$ makes it normalized.

After a time t, the wave function

$$\Psi(x, t) = 1/\sqrt{2}\left[\varphi_s^{(1)}(x)\, e^{-i E_s^{(1)} t/\hbar} + \varphi_a^{(1)}\, e^{-i E_a^{(1)} t/\hbar}\right] \quad \text{... (A 5.18)}$$

which truly speaking represents the solution of the time dependent Schrödinger equation corresponding to the double-well potential.

Eqn. (A 5.18) can be written in the form

$$\Psi(x, t) = 1/\sqrt{2}\, e^{-E_s^{(1)} t/\hbar}\, [\varphi_s^{(1)}(x) + \varphi_a^{(1)}\, e^{-i\pi t/T}] \quad \text{... (A 5.19a)}$$

in which $$T = \pi\hbar\,(E_a^{(1)} - E_s^{(1)}) \quad \text{... (A 5.19b)}$$

From eqn. (*A* 5.19*a*), one finds that at $t = 0, 2T, 4T, ...$

$$|\psi(x, t)|^2 = 1/2\,[\varphi_s^{(1)}(x) + \varphi_a^{(1)}(x)]^2 \quad \text{... (A 5.20)}$$

and the particle is most likely to be found in the second well. On the other hand the probability of finding the particle at $t = T, 3T, 5T, ...$ is given by

$$|\psi(x, t)|^2 = 1/2\,[\varphi_s^{(1)}(x) - \varphi_a^{(1)}(x)]^2 \quad \text{... (A 5.21)}$$

and the particle is most likely to be found in the first well.

Thus the particle tunnels through the barrier in a time T. Fig. A 5.6 gives the probability of finding the particle at $t = 0, T, 2T$, and $3T$. One observes that the particle which has high probability of being found in the second well at $t = 0$ tunnels to the first well at $t = T$ and then tunnels back to the second well at $t = 2T$ and so on. This effectively amounts to the fact that the plane of H atoms continuously passes from one side of the N atom to the other.

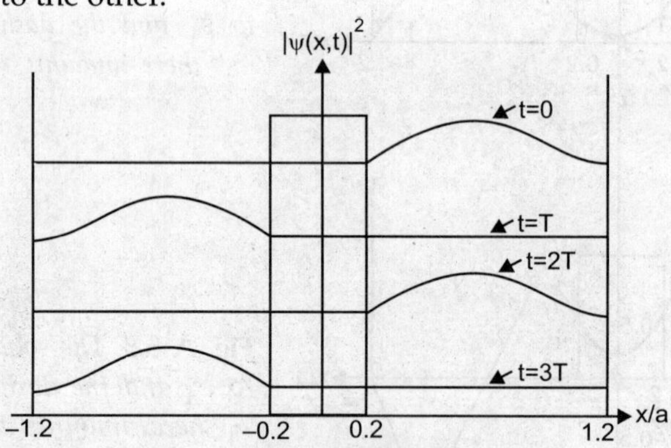

Fig. A 5.6 *Probability of finding the particle at* $t = 0, T, 2T$ *and* $3T$

APPENDIX 6

MOTION OF ELECTRON IN A PERIODIC POTENTIAL (THE KRÖNIG – PENNEY MODEL)

A 6.1 Bloch's Theorem

Each conduction (outermost or valence) electron in a metallic solid finds itself in the field of ion cores, that is, in a periodic potential, periodicity being that of the lattice. In the case of sodium Na, by ion core we mean Na nucleus + 10 electrons ($1s^2\ 2s^2\ 2p^6$), that is, Na^+ and all other outer electrons. It turns out that the effect of electron-electron interaction can be well approximated by a periodic potential which represents the average effect on one outer electron because of all other outer electrons. This periodic potential is assumed to be the same for each outer electron. We thus have a model where each outer electron say of Na atom moves independently in the same periodic potential consisting of due to the ion cores and due to the average effect of the other outer electrons.

Let us now study the motion of an electron under such a periodic potential. We would follow *F*. Bloch in doing so and would confine our discussion to one dimension only for the sake of simplicity. Consider a metallic solid consisting of N ion cores located at the lattice sites a, $2a$, ..., Na where N is large and an electron moving in a periodic potential $V(x)$ produced by the ion cores and as well as due to its interaction with other outer electrons. Such a potential is shown in Fig. A 6.1. The potential is periodic (except when we are close to the boundary ion cores) with period a, that is

$$V(x+a) = V(x) \qquad \text{... (A 6.1)}$$

where x is measured from some arbitrary origin.

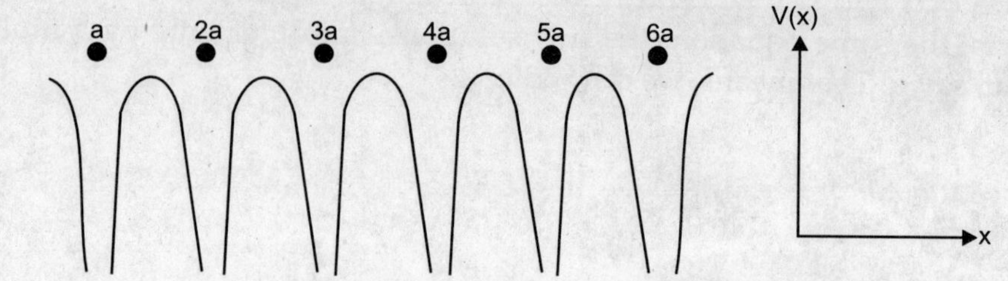

Fig. A 6.1 *Periodic potential secn by the electron in a metallic solid*

The solution of the Schrödinger equation becomes simpler if we make the periodicity exact. This can be done by assuming that the lattice sites in a one-dimensional metallic crystal are arranged in the form of a ring (see Fig. A 6.2) with the (N + l)th lattice site coinciding with the lattice site *a*. In this situation, the crystal does not have a boundary and all the lattice sites on the ring are similar. Thus the periodicity condition (A 6.1) is strictly followed.

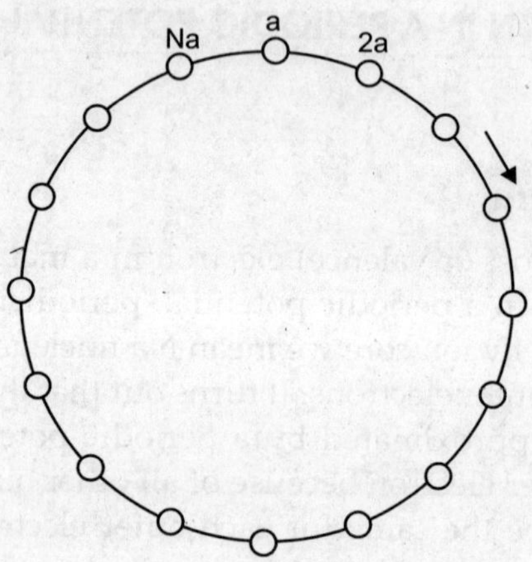

Fig. A 6.2 *Lattice sites in one-dimensional crystal arranged in the form of ring*

The Schrödinger equation for an electron in a periodic potential $V(x)$ is

$$\frac{d^2\varphi}{dx^2} + \frac{2m}{\hbar^2}(E - V(x))\,\varphi(x) = 0 \qquad \text{... (A 6.2)}$$

where $\varphi(x)$ represents the wave function of electron with energy E.

Replacing x in the above equation by $x + a$, we get

$$\frac{d^2\varphi\,(x+a)}{d\,(x+a)^2} + \frac{2m}{\hbar^2}(E - V\,(x+a))\,\varphi\,(x+a) = 0$$

Noting that $d/d\,(x + a) = d/dx$ and $V(x + a) = V(x)$, the above equation reads

$$\frac{d^2\varphi\,(x+a)}{dx^2} + \frac{2m}{\hbar^2}(E - V(x))\,\varphi\,(x+a) = 0 \qquad \text{... (A 6.3)}$$

Since φ(x) and φ(x + a) satisfy the Schrödinger equation with the same energy E, that is, the same equation, the two solutions can differ only by a multiplicative constant, say α. That means, we can write.

$$\varphi(x+a) = \alpha\varphi(x),\ \varphi(x+Na) = \alpha^N \varphi(x) \quad \text{... (A 6.4)}$$

Since the lattice is in the form of a ring, the wave function has the same value at both the ends. That is,

$$\psi(x+Na) = \alpha^N \psi(x) = \psi(x) \quad \text{... (A 6.5)}$$

Hence $\alpha^N = 1$ and using $e^{2\pi in} = 1$, where n is an integer, one obtains

$$\alpha^N = 2^{\pi in} \text{ or } \alpha = e^{2\pi in/N},\ n = 0, \pm 1, \pm 2, \ldots \quad \text{... (A 6.6)}$$

Therefore,

$$\varphi(x+a) = e^{2\pi in/N}\varphi(x) \quad \text{... (A 6.7)}$$

It is more convenient to write φ(x) in terms of another function, $u_k(x)$ defined as

$$\varphi(x) = e^{ikx} u_k(x) \quad \text{... (A 6.8)}$$

where

$$k = \frac{2\pi n}{Na} \quad \text{... (A 6.9)}$$

If we rewrite condition (A 6.7) using eqn. (A 6.8), we get

$$e^{ik(x+a)} u_k(x+a) = e^{2\pi in/N} e^{ikr} u_k(x)$$

or

$$e^{ika} u_k(x+a) = e^{2\pi in/N} u_k(x)$$

Noting that $ka = 2\pi n/N$ from eqn. (A 6.9), we get

$$u_k(x+a) = u_k(x)$$

The wave function (A 6.8) can be thought as equivalent to a plane wave modulated by a function $u_k(x)$ having periodicity of the lattice. Such a wave function φ(x) is referred to as the Bloch function and the result that it can be written in terms of e^{ikx} into a periodic function $u_k(x)$ is known as the Bloch theorem.

A 6.2 Krönig – Penney Square – Well Periodic Potential

The potential under which the conduction electron is supposed to move in a metallic solid is assumed to consist of a periodic array of square wells and barriers. The period of the potential is a. This approximation of the periodic potential shown in Fig. A 6.1 is known as the Krönig-Penney potential and is depicted in Fig. A 6.3. The potential energy function $V(x)$ in the said model is given by

$$\begin{aligned} V(x) &= 0,\ a \le x \le a-b \\ &= V_0,\ -b \le x \le 0 \end{aligned} \quad \text{... (A 6.10)}$$

with the periodicity condition

$$V(x+a) = V(x) \quad \text{... (A 6.11)}$$

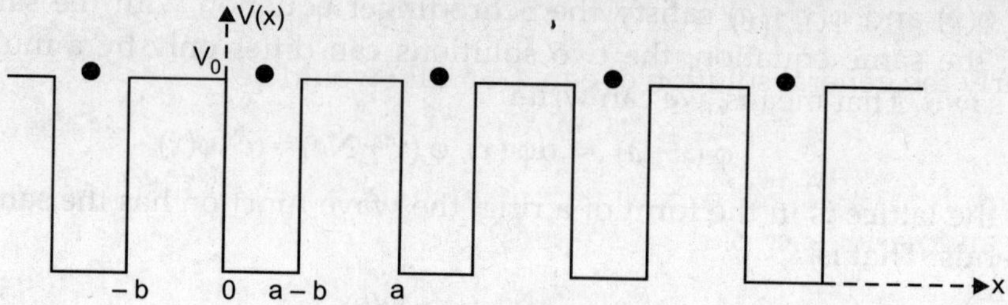

Fig. A 6.3 *The Krönig-Penney periodic potential*

The Schrödinger equations corresponding to the bound state (electron bound to the lattice), that is, $E < V_0$ in the two regions are

$$\frac{d^2\varphi}{dx^2} + \alpha^2\varphi = 0, \quad 0 < x \le a - b \qquad \text{... (A 6.12a)}$$

and

$$\frac{d^2\varphi}{dx^2} - \beta^2\varphi = 0, \quad -b < x < 0 \qquad \text{... (A 6.12b)}$$

Where $\alpha^2 = 2\,mE/\hbar^2$ and $\beta^2 = 2m\,(V_0 - E)/\hbar^2$

According to Bloch's theorem for a periodic potential given by eqn. (*A* 6.11), the solution of eqns. (*A* 6.12*a*) and (*A* 6.12*b*) must be of the form

$$\varphi_1 = e^{ikx}\, u_{k_1}(x), \quad 0 \le x \le a - b \qquad \text{... (A 6.13a)}$$

$$\varphi_2 = e^{ikx}\, u_{k_2}(x), \quad -b \le x \le 0 \qquad \text{... (A 6.13b)}$$

Substitution of eqn. (A 6.13*a*) into eqn. (*A* 6.12*a*) and eqn. (*A* 6.13*b*) into eqn. (*A* 6.12*b*) gives

$$\frac{d^2u_{k_1}}{dx^2} + 2ik\,\frac{du_{k_1}}{dx} + (\alpha^2 - k^2)\,u_{k_1} = 0, \quad 0 \le x \le a - b \qquad \text{... (A 6.14a)}$$

$$\frac{d^2u_{k_2}}{dx^2} + 2ik\,\frac{du_{k_2}}{dx} - (\beta^2 + k^2)\,u_{k_2} = 0, \quad -b \le x \le 0 \qquad \text{... (A 6.14b)}$$

To solve eqn. (*A* 6.14*a*), let us assume a solution of the form

$$u_{k_1}(x) = e^{mx} \qquad \text{... (A 6.15)}$$

Substituting above into eqn. (*A* 6.14*a*), one gets

$$m^2 + 2\,ikm + (\alpha^2 - k^2) = 0$$

Therefore,

$$m = -ik + i\alpha = i\,(\alpha - k),\ -i(\alpha + k)$$

Hence, the general solution of eqn. (*A* 6.14*a*) is (if $e^{-i(\alpha - k)x}$ and $e^{-i(\alpha + k)x}$ are the solutions, then the linear combination of these would also be the solution)

$$u_{k_1}(x) = A\,e^{i(\alpha - k)x} + B\,e^{-i(\alpha + k)x}, \quad 0 \le x \le a - b \qquad \text{... (A 6.16a)}$$

and similarly the general solution of eqn. (*A* 6.14*b*) would be

$$u_{k_2}(x) = C\,e^{(\beta - ik)x} + D\,e^{-(\beta + ik)x}, \quad -b \le x \le 0 \qquad \text{... (A 6.16b)}$$

where *A*, *B*, *C* and *D* are constants and are to be chosen such that the condition of continuity and periodicity of the wave function $u_k(k)$ is satisfied.

Applying boundary conditions that the wave functions and their derivatives must be continuous at $x = 0$, that is

$$u_{k_1}(x)\,|_{x=0} = u_{k_2}(x)\,|_{x=0} \text{ and } \left.\frac{du_{k_1}(x)}{dx}\right|_{x=0} = \left.\frac{du_{k_2}(x)}{dx}\right|_{x=0}$$

one gets

$$A + B = C + D \qquad \text{... (A 6.17a)}$$

and

$$i(\alpha - k)A - i(\alpha + k)B = (\beta - ik)\,C - (\beta + ik)D \qquad \text{... (A 6.17b)}$$

Because of the periodicity of the wave functions, that is

$$u_{k_1}(x)\,\big|_{x=a-b} = u_{k_2}(x)\,\big|_{x=-b} \text{ and } \left.\frac{du_{k_1}(x)}{dx}\right|_{x=a-b} = \left.\frac{du_{k_2}(x)}{dx}\right|_{x=-b}$$

one gets

$$A\,e^{i(\alpha - k)(a-b)} + B\,e^{-(\alpha + k)(a-b)} = C\,e^{(\beta - ik)(-b)} + D\,e^{(\beta + ik)(-b)} \qquad \text{... (A 6.17c)}$$

and

$$i(\alpha - k)A\,e^{i(\alpha - k)(a-b)} - i(\alpha + k)B\,e^{-i(\alpha + k)(a-b)} = (\beta - ik)C\,e^{(\beta - ik)(-b)} - (\beta + ik)D\,e^{-(\beta + ik)(-b)} \qquad \text{... (A 6.17d)}$$

For a nontrivial solution of eqns. (*A* 6.17), the determinant of the coefficients of *A*, *B*, *C* and *D* should vanish. On simpliciation, one gets

$$\frac{\beta^2 - \alpha^2}{2\alpha\beta}\sinh\beta b\,\sin\alpha\,(a-b) + \cosh\beta b\,\cos\alpha\,(a-b) = \cos ka \qquad \text{... (A 6.18)}$$

From the above equation, one can find αa or the allowed values of energy. Since it is a difficult task to solve the above equation, therefore Krönig and Penney assumed that $V_0 \to \infty$ and $b \to 0$ (that is, the potential barriers become delta functions) while $V_0 b$ (area of the potential barrier) is finite. Hence α^2 can be neglected compared to β^2 and also for $b \to 0$, Sinh $\beta b \to \beta b$ and Cosh $\beta b \to 1$ (taking $\beta^2 b \propto V_0 b$ finite but $\beta b \to 0$). As a result the condition (*A* 6.18) takes the form

$$\beta^2/2\alpha\beta\;\beta b\sin\alpha a + \cos\alpha a = \cos ka$$

Cancelling β in the 1st part and then multiplying and dividing by *a* and finally rearranging the terms, we get

$$\beta^2 \frac{ab}{2} \frac{\sin \alpha a}{\alpha a} + \cos \alpha a = \cos ka$$

or
$$P \frac{\sin \alpha a}{\alpha a} + \cos \alpha a = \cos ka \quad \text{... (A 6.19)}$$

where $P = \text{Lim}_{\substack{b \to 0 \\ V_0 \to \infty}} \frac{2m}{\hbar^2} V_0 \frac{ab}{2}$ is measure of the quantity $V_0 b$, which is area of the potential barrier

The physical significant of the quantity P is that if it is increased, the area of the potential barrier is increased and hence the given electron is bound more tightly to a particular potential well. In the extreme case for which $P \to \infty$, eqn. (*A* 6.19) has solutions only if Sin $\alpha\, a = 0$, that is, if

$$\alpha a = \pm n\pi, \quad n = 1, 2, 3, ...$$

The energy spectrum is then given by

$$\alpha^2 a^2 = n^2 \pi^2$$

or
$$E_n = n^2 h^2 / 8\text{m}\, a^2, \quad n = 1, 2, 3, ...$$

The energy levels in this case are discrete (the energy spectrum becomes a line spectrum) and the result is same as that in the case of a particle in a constant potential box of dimension *a*. This is an expected result because for large *P*, tunnelling through the barriers become improbable.

When $P \to 0$, the potential barrier becomes very weak. This simply amounts to the fact that we have a free electron model. In this case, from eqn. (*A* 6.19), we get

$$\cos \alpha a = \cos ka$$

or
$$\alpha = k$$

Therefore,

$$\alpha^2 = 2\text{m}\, E/\hbar^2 = k^2$$

or
$$E = \hbar^2 k^2 / 2\text{m}$$

which implies that *k* defined in the present context is nothing but the wave number (or wave vector in the three dimensional case) and the energy spectrum is continuous.

To study the allowed values of αa, which in turn gives the allowed values of energy, we plot l.h.s. of eqn. (*A* 6.19) as a function of αa for an arbitrary value of $P = 3\pi/2$ (see Fig. A 6.4). Since cos *ka* (for real values of *k*) must lie between –1 and +1, the condition (A 6.19) is satisfied whenever the value of αa is such that the l.h.s. lies between –1 and +1 (as indicated by horizontal lines in Fig. A 6.4). There are regions of αa where the l.h.s. of eqn. (*A* 6.19) does not lie between –1 and +1. For these values of αa and hence for the corresponding values of energy *E*, no solutions exist. In other words, there are regions of energy which are prohibited. Thus the energy spectrum of electron consists of regions of allowed energy levels followed by forbidden regions of energy. That is, we have alternate energy bands and energy gaps as shown in Fig. A 6.4.

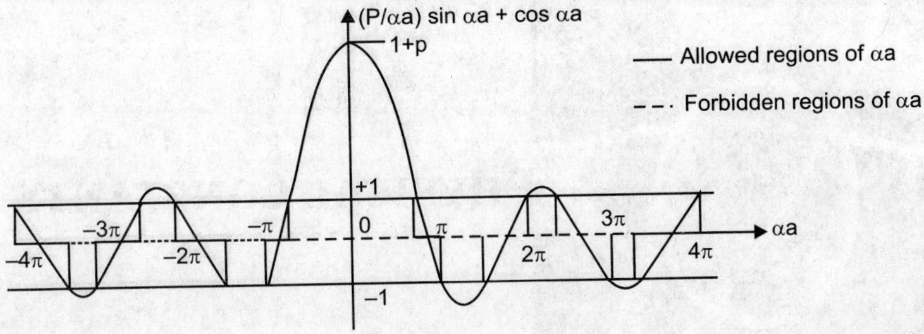

Fig. A 6.4 *Plot of function* $P\dfrac{\sin \alpha a}{\alpha a} + \cos \alpha a$ *versus* αa *for arbitrary value of* $P = 3\pi/2$

One can also plot $E(k)$ versus k from eqn. (*A* 6.19) and the result is as shown in Fig. A 6.5. Boundaries of allowed values of αa are given by,

$$\cos ka = \pm 1$$

or

$$ka = \pm n\pi, \quad n = 1, 2, 3, \ldots.$$

or

$$k = \pm n\pi/a, \quad n = 1, 2, 3, \ldots$$

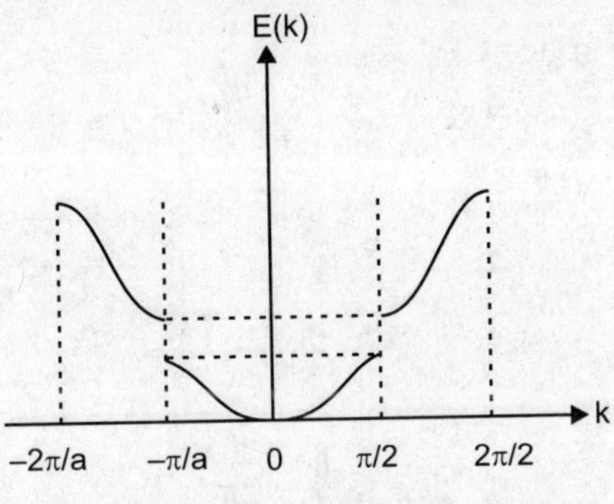

Fig. A 6.5 *Plot of the energy E(k) of the electron versus wave number k*

The region $-\pi/a$ to $+\pi/a$ is referred to as the 1st Brillouin zone. The second zone consists of two parts, one extending from $-\pi/a$ to $-2\pi/a$ and the other part from $+\pi/a$ to $+2/a$ and so on.

PHYSICAL CONSTANTS

Quantity	Symbol	Value
Electron rest mass	m_e	9.1×10^{-31} kg
Proton rest mass	m_p	1.673×10^{-27} kg
Neutron rest mass	m_n	1.675×10^{-27} kg
Speed of light in free space	c	2.998×10^{8} m/s
Electron volt	eV	1.6×10^{-19} J
Fermi	F	10^{-15} m
Electronic charge	e	1.6×10^{-19} C
Planck's constant	h	6.626×10^{-34} J.s
	$\hbar = \dfrac{h}{2\pi}$	1.054×10^{-34} J.s
Boltzmann constant	k_B	1.38×10^{-23} J.K^{-1}
Angstrom	Å	10^{-10} m
Avogadro number	N_A	6.023×10^{26} (kg – mol)$^{-1}$
Atomic mass unit	amu	1.66×10^{-27} kg

USEFUL INTEGRALS

$$\int_0^\infty e^{-ax} x^n \, dx = \frac{n!}{a^{n+1}}$$

$$\int_0^\infty x^{2n} e^{-px^2}\, dx = \frac{(2n-1)!!}{2(2p)^n}\left(\frac{\pi}{p}\right)^{1/2}, p > 0$$

$$\int_0^\infty x^{2n+1} e^{-px^2}\, dx = \frac{n!}{2p^{n+1}}, p > 0$$

USEFUL EXPANSIONS

$$e^x = 1 + x + \frac{x^2}{2!} + \frac{x^3}{3!} + ..., -\infty < x < +\infty$$

$$\sin x = x - \frac{x^3}{3!} + \frac{x^5}{5!} - \frac{x^7}{7!} + ...,$$

$$\cos x = 1 - \frac{x^2}{2!} + \frac{x^4}{4!} - \frac{x^6}{6!} + ...,$$

$$\sinh x = x + \frac{x^3}{3!} + \frac{x^5}{5!} + \frac{x^7}{7!} + ...,$$

$$\cosh x = 1 + \frac{x^2}{2!} + \frac{x^4}{4!} + \frac{x^6}{6!} + ...,$$

$$\sin\theta = \frac{(e^{i\theta} + e^{-i\theta})}{2i},$$

$$\cos\theta = \frac{(e^{i\theta} + e^{-i\theta})}{2},$$

Taylor Series

$$f(x) = f(a + x - a) = f(a) + \frac{(x-a)}{1!} f'(x)|_{x=a} + \frac{(x-a)^2}{2!} f''(x)|_{x=a} + ...,$$

Taylor Series (Another form)

$$f(x+a) = f(a) + \frac{x}{1!} f'(x)|_{x=a} + \frac{x^2}{2!} f''(x)|_{x=a} + ...$$

VALUABLE REFERENCES

1. A. Beiser, Concepts of Modern Physics 2nd Ed., McGraw-Hill, Kogakusha, Tokyo, Japan 1973 (6th ed) 2003.
2. A.J. Dekker, Solid State Physics, Macmillan & Co. Ltd. London 1958.
3. A.K. Ghatak and S. Lokanathan, Quantum Machanics: Theory and Applications (5th ed) Macmillan India Ltd., New Delhi 2004.
4. A.S. Davydov, Quantum Mechanics, Translated, edited and with additions by D. Ter Haar. Pergamon Press, Oxford 1965.
5. B.K. Agarwal and Hari Prakash, Quantum Mechanics, Prentice-Hall of India, New Delhi 1997.
6. C. Cohen-Tannoudgi, B. Diu and F. Laloe, Quantum Mechanics Vols. I & II 2nd ed., translated by S.R. Hemley, N. Ostrowsky and D. Ostrowsky, John Wiley & Sons, New York 1977.
7. E. Mertzbacher, Quantum Mechanics 2nd Ed., John Wiley & Sons. New York 1970.
8. Gurbachan S. Chaddha, University Physics, Narosa Publishing House (P) Ltd. New Delhi 2015.
9. H.F. Hameka, Introduction to Quantum Theory. Harper & Row, Publishers, Inc., New York 1967.
10. H.L. Strauss, Quantum Mechanics: An Introduction, Prentice Hall, Inc., Englewood Cliffs, N.J., U.S.A.1968.
11. H.S. Mani and G.K. Mehta, Introduction to Modern Physics, Affiliated East-West Press, New Delhi 1988.
12. J.A. Eisele, Modern Quantum Mechanics with Applications to Elementary Particle Physics, Wiley Interscience, New York 1969.
13. J.D. McGervey. Introduction to Modern Physics, Academic Press, Inc., New York 1971.

14. J.L. Powell and B. Craseman, Quantum Mechanics, Addison Wesley Publishing Co., Inc., Reading Mass 1961.
15. L.D. Landau and E.M. Lifshitz, Quantum Mechanics (Non-relativistic Theory), Pergamon Press, Oxford 1965.
16. L.I. Schiff, Quantum Mechanics 3rd Ed., McGraw-Hill, Kogakusha, Tokyo 1981.
17. L. Pauling and E.B. Wilson Jr., Introduction to Quantum Mechanics. McGraw-Hill, New York 1935.
18. M.A. Omar, Elementary Solid State Physics-Principles and Applications, Addison-Wesley Publishing Co, Inc. Philippines 1975.
19. M.E. Rose, Elementary Theory of Angular Momentum, John Wiley & Sons. New York 1957.
20. N. Tralli and F.R. Pomilla, Atomic Theory, McGraw-Hill Book Co., New York 1969.
21. P. Gombas and D. Kisdi, Wave Mechanics and Applications, translated by J. Schanda, Pergamon Press. Oxford 1973.
22. P.M. Mathews and K. Venkatesan. A Testbook of Quantum Mechanics. Tata McGraw-Hill Publ. Co. Ltd. New Delhi 1977.
23. P. Roman, Advanced Quantum Theory. Addison-Wesley, Reading, Mass, 1965.
24. P. Stehle, Quantum Mechanics, Holden Day, Inc., California 1956.
25. R.L. Sproul, Modern Physics 2nd Ed., John-Wiley & Sons, Inc., New York 1963.
26. V.A. Fock, Fundamentals of Quantum Mechanics, Mir Publishers, Moscow 1978.
27. V.K. Thankappan, Quantum Mechanics 2nd Ed., New Age International (P) Ltd., New Delhi 1993.

INDEX

Index

P

Q

R

S